ADVANCES IN EXPERIMENTAL MEDICINE AND BIOLOGY

Recent Volumes in this Series

Volume 454
OXYGEN TRANSPORT TO TISSUE XX
Edited by Antal G. Hudetz and Duane F. Bruley

Volume 455
RHEUMADERM: Current Issues in Rheumatology and Dermatology
Edited by Carmel Mallia and Jouni Uitto

Volume 456
RESOLVING THE ANTIBIOTIC PARADOX:
Progress in Understanding Drug Resistance and Development of New Antibiotics
Edited by Barry P. Rosen and Shahriar Mobashery

Volume 457
DRUG RESISTANCE IN LEUKEMIA AND LYMPHOMA III
Edited by G. J. L. Kaspers, R. Pieters, and A. J. P. Veerman

Volume 458
ANTIVIRAL CHEMOTHERAPY 5: New Directions for Clinical Application and Research
Edited by John Mills, Paul A. Volberding, and Lawrence Corey

Volume 459
IMPACT OF PROCESSING ON FOOD SAFETY
Edited by Lauren S. Jackson, Mark G. Knize, and Jeffrey N. Morgan

Volume 460
MELATONIN AFTER FOUR DECADES: An Assessment of Its Potential
Edited by James Olcese

Volume 461
CYTOKINES, STRESS, AND DEPRESSION
Edited by Robert Dantzer, Emmanuele Wollman, and Raz Yirmiya

Volume 462
ADVANCES IN BLADDER RESEARCH
Edited by Laurence S. Baskin and Simon W. Hayward

Volume 463
ENZYMOLOGY AND MOLECULAR BIOLOGY OF CARBONYL METABOLISM 7
Edited by Henry Weiner, Edmund Maser, David W. Crabb, and Ronald Lindahl

Volume 464
CHEMICALS VIA HIGHER PLANT BIOENGINEERING
Edited by Fereidoon Shahidi, Paul Kolodziejczk, John R. Whitaker, Agustin Lopez Munguia, and Glenn Fuller

A Continuation Order Plan is available for this series. A continuation order will bring delivery of each new volume immediately upon publication. Volumes are billed only upon actual shipment. For further information please contact the publisher.

ENZYMOLOGY AND MOLECULAR BIOLOGY OF CARBONYL METABOLISM 7

ENZYMOLOGY AND MOLECULAR BIOLOGY OF CARBONYL METABOLISM 7

Edited by

Henry Weiner
Purdue University
West Lafayette, Indiana

Edmund Maser
Philipps University of Marburg
Marburg, Germany

David W. Crabb
Indiana University School of Medicine
Indianapolis, Indiana

and

Ronald Lindahl
University of South Dakota School of Medicine
Vermillion, South Dakota

Kluwer Academic / Plenum Publishers
New York, Boston, Dordrecht, London, Moscow

Proceedings of the Ninth International Symposium on Enzymology and Molecular Biology of Carbonyl Metabolism, held June 20 – 24, 1998, in Varallo Sesia, Vercelli, Italy

ISBN 0-306-46113-7

233 Spring Street, New York, N.Y. 10013

10 9 8 7 6 5 4 3 2 1

A C.I.P. record for this book is available from the Library of Congress.

Printed in the United States of America

PREFACE

Prior to the start of the eighth meeting, I had the good sense to ask Professor Rosa Angela Canuto of Turin, Italy if she would help me organize the ninth meeting. She quickly suggested that both she and Dr. Guiliana Muzio, also of Turin, help plan the meeting. Each of our previous eight meetings was a unique experience for the participants. The science was always outstanding and the presentations and discussions were excellent. By moving each meeting to a different part of the world we were able to experience exciting foods and cultural aspects of the world in addition to the science. The ninth meeting was no exception. We met from June 18 to 22 in the small mountain city of Varallo, Italy, the birth place of Dr. Canuto. Holding the scientific sessions in a several-hundred-year-old converted mansion and having an afternoon trip to either Lago Maggiore or Monte Rosa made some aspects of this meeting extremely memorable. An additional unique aspect of the social portion of the meeting was our ability to invite the townspeople to share with us a concert performed in an old church.

Though the social and cultural aspects of the meeting were outstanding, the purpose of the meeting was to exchange scientific information about the status of the three enzyme systems. Over 50 talks and 40 posters provided the participants with a wide variety of presentations dealing with enzymology, molecular biology, and metabolic aspects of these carbonyl metabolizing oxido-reductases. Much new information was presented, including three-dimensional structures of enzymes not previously reported, drug interactions with the enzymes, and new aspects of gene regulation along with sequence alignments, metabolism and enzyme mechanisms, to mention a few. In addition, there were serious discussions about developing standardized nomenclatures for the different enzyme families.

On behalf of all the participants I want to thank Drs. Canuto and Muzio for organizing an outstanding meeting and providing us with a gastronomic and culturally exciting experience. We are not a society, hence we have no financial base to support these meetings. Dr. Canuto found some local sponsors and their generosity helped make this an outstanding affair. We especially wish to acknowledge the Sindaco del Comune di Varallo and Fondazione Cassa di Risparmio di Torino and sincerely thank them for helping to make our ninth meeting a memorable experience.

I want to thank all the participants for making this another outstanding meeting. I also want to thank my three co-editors for helping me review all the manuscripts.

Our tenth meeting will be held in Taos, New Mexico from July 1–5, 2000. We invite our fellow scientists interested in the topic of the meeting to join us. Please write me to receive additional information about the tenth meeting.

Henry Weiner
Scientific Organizer
Biochemistry Department
Purdue University
West Lafayette, Indiana 47907-1153

CONTENTS

Alcohol Dehydrogenase

Short and Medium Chain Dehydrogenase

Reductases

ENZYMOLOGY AND MOLECULAR BIOLOGY OF CARBONYL METABOLISM 7

1

THE BIG BOOK OF ALDEHYDE DEHYDROGENASE SEQUENCES

An Overview of the Extended Family

John Perozich,[1] Hugh Nicholas,[2] Ronald Lindahl,[3] and John Hempel[4]

[1]Department of Molecular Genetics and Biochemistry
University of Pittsburgh School of Medicine
Pittsburgh, Pennsylvania 15261
[2]Pittsburgh Supercomputing Center
Carnegie-Mellon University
Pittsburgh, Pennsylvania 15213
[3]Department of Biochemistry and Molecular Biology
University of South Dakota
Vermillion, South Dakota 57069
[4]Department of Biological Sciences
University of Pittsburgh
Pittsburgh, Pennsylvania 15260

1. INTRODUCTION

Traditionally, research on aldehyde dehydrogenases (ALDHs, EC 1.2.1) has focused on three groups: class 1, 2, and 3 ALDHs. Class 1 and 2 ALDHs are very closely related homotetramers and both participate in ethanol metabolism, though class 1 ALDHs can oxidize a wide range of metabolites, including retinal (reviewed Lindahl, 1982; Yoshida et al., 1998). Class 3 ALDHs appear at first glance to be highly divergent from the other 2 classes. Class 3 ALDHs share only about 25% sequence identity with class 1 and 2 ALDHs and are homodimers. However, in main-chain folding, the structures of the class 2 and 3 monomers are nearly identical (Hempel et al., 1999).

Though these three classes have received most of the attention, a number of other ALDH families with various metabolic roles exist. Now, four ALDH tertiary structures have been solved in close succession (Liu et al., 1997; Steinmetz et al., 1997; Moore et al., 1999; Cod betaine ALDH - PDB accession 1A4S). To aid in the determination of ALDH function and to determine the relationships between the various ALDH families, we

Enzymology and Molecular Biology of Carbonyl Metabolism 7
edited by Weiner *et al.* Kluwer Academic / Plenum Publishers, New York, 1999.

aligned all available ALDH protein sequences. At the time the initial study was performed (Perozich et al., 1998), 145 ALDHs were aligned (though this number is now at about 185 and still growing). This alignment project can help not only in defining relationships within the entire ALDH extended family, but can also aid in identifying residues and sequence motifs which evolution has deemed necessary for the function of all ALDHs.

2. DISCUSSION

2.1. Structural/Functional Conservations

Sixteen residues are at least 95% conserved in the 145 ALDHs examined (Perozich et al., 1998). More than half of this number (9 of 16) are conserved glycines and prolines essential for maintaining critical turns and loops in the tertiary structure of ALDHs. Among these are two invariant glycines: Gly-187/245 (class 3/class 1 & 2 numbering) which is integral to the coenzyme-binding Rossmann fold and Gly-240/299 which is required to allow the ALDH main chain to "twist back on itself" and position the catalytic cysteine, Cys-243/302, for catalysis.

As might be expected, the residues which play a direct role in catalysis are highly conserved. Cys-243/302 and Asn-114/168 are conserved in all ALDHs with demonstrated enzymatic activity. Asn-114/169 is the closest residue to the thiol, and may serve to coordinate the carbonyl oxygen atom of the substrate aldehyde (Steinmetz et al., 1997; Hempel et al., 1999). Phe-335/401, another invariant residue, interacts with the nicotinamide portion of NAD(P). Though the conservation of Glu-209/268 currently falls just shy of 95% in the current sample, mutagenesis studies have supported its possible role as a general base for the catalytic mechanism (Wang & Weiner, 1995). Methylmalonyl semialdehyde dehydrogenases (MMSALDHs) lack Glu-209/268 and are the only ALDH family that does not produce a free acid—instead yielding a CoA ester (Kedishvili et al., 1992). This glutamic acid has been proposed to deprotonate the catalytic thiol (Steinmetz et al., 1997) or it may serve to aid in the expulsion of the free acid product (Hempel et al., 1993). The remaining invariant residue is Glu-333/399. This residue is indicated to be involved in cofactor binding in the bovine class 2 structure (Steinmetz et al., 1997) and by mutagenesis of the human class 2 enzyme, though the authors also feel that it may play another role (Ni et al., 1997). Based on its invariant conservation and close position to the catalytic thiol, it has also been proposed at this meeting that Glu-333/399 (instead of Glu-209/268) may activate the catalytic thiol through a water molecule (Hempel et al., 1999).

Nearly all of these highly conserved residues fall within seven of the ten most conserved sequence motifs in ALDHs. Though they are distributed across almost the entire length of the ALDH sequences (Figure 1), the motifs cluster around the active site of the molecule and play essential roles in ALDHs, including coenzyme-binding (Motifs 2, 4 and 8) and catalysis (Motifs 1, 5, 6 and 8). These motifs appear to reflect essential ALDH structure/function elements, as most contain a conserved turn or loop (Perozich et al., 1998).

2.2. ALDH Families

We have chosen to use the term "extended family" when referring to all ALDHs, as the term "superfamily" is better used with a group of proteins with related structure, but sometimes very diverse activity or function. As all ALDHs, with the noted exception of MMSALDHs, likely use the same chemical mechanism, we feel "extended family" is more appropriate. The alignment of 145 ALDH sequences supported the existence of at

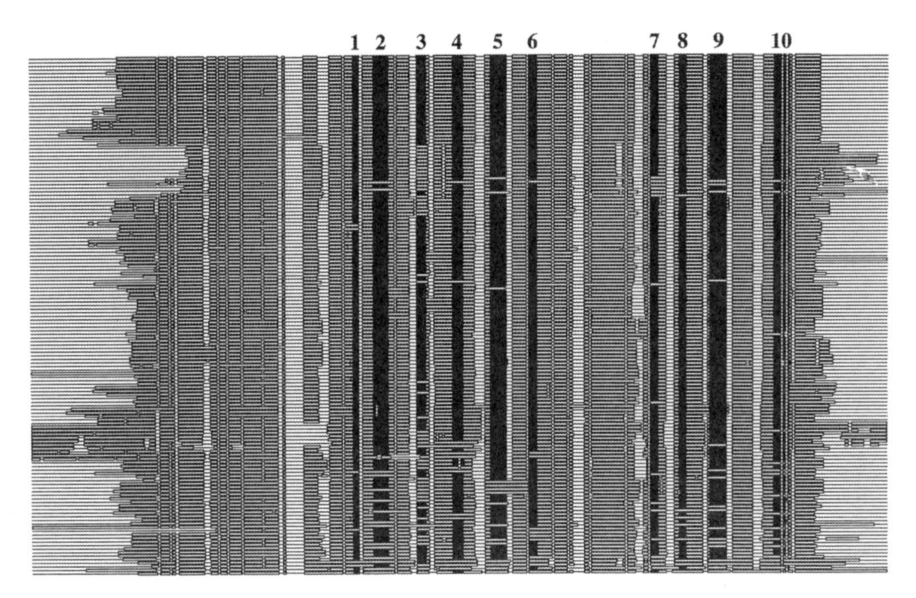

Figure 1. Summary view of all 145 ALDH sequences highlighting the positions of the 10 conserved motifs (black). Each motif is numbered above the top line. Regions of protein sequence are represented by boxes, while gaps are lines.

least 13 ALDH families within the entire extended family (Figure 2). These thirteen families are split into two main trunks of the phylogenetic tree, the "Class 3" and "Class 1/2" trunks. Divergence into these two trunks does not correlate with subcellular localization, quaternary structure, or coenzyme preference. In fact, some ALDH families, such as MMSALDHs and γ-glutamyl semialdehyde dehydrogenases (GGSALDHs), have orthologs from bacteria through humans. Other families appear specific to a certain kingdom. For example, class 1 ALDHs have only been found in animals, Fungal ALDHs in fungi, and Aromatic ALDHs in bacteria (Perozich et al., 1998).

There is some, but not perfect, correlation to substrate specificity. All ALDH families in the "Class 3" trunk, except for class 3 ALDHs themselves, are substrate-specific enzymes. The "Class 1/2" trunk contains mostly variable substrate enzymes, with the exceptions of 2-hydroxymuconic semialdehyde dehydrogenases (HMSALDHs) and formyltetrahydrofolate dehydrogenases (FTDHs). Substrate-specific ALDH families do appear to have diverged earlier in evolution than the variable substrate families (Perozich et al., 1998). Thus, it would appear that the driving force in ALDH evolution has been to selectively metabolize different aldehydes in both the intracellular and extracellular environments. The substrate-specific ALDHs, such as betaine ALDHs, MMSALDHs and GGSALDHs, would have been necessary for basic cellular metabolic pathways, necessitating their earlier emergence. As the extracellular environment of the cells continued to change through the millennia, the variable substrate ALDHs, such as class 1, 2, and 3 ALDHs, which are involved primarily in detoxification of xenobiotics, needed to continue to adapt to the presence of all other aldehydes. Hence, their divergence to yield the ALDHs we see today occurred much later in evolution.

2.3. Physiological Importance of ALDHs

The ALDH family consists of many diverse enzyme families, as summarized in Figure 2. These ALDHs appear to function broadly in four different enzymatic capacities: de-

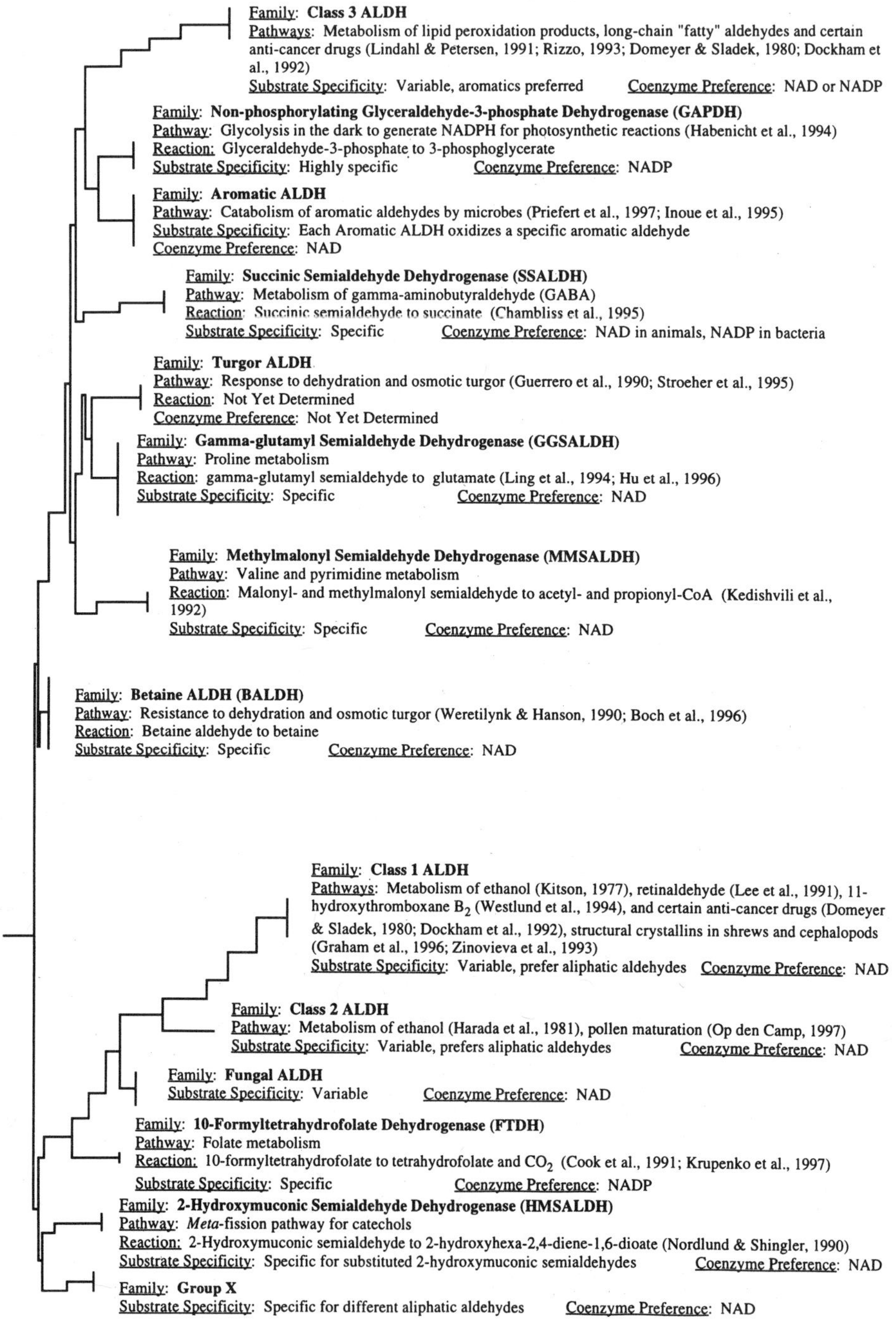

Figure 2. Summary tree of ALDH families. A short list of summary information for each ALDH family is presented at that family's relative position on the phylogenetic tree of the ALDH extended family (Perozich et al., 1998).

toxification, intermediary metabolism, osmotic protection and NADPH generation, as well as in a structural capacity in certain eye lenses. The variable substrate ALDHs, class 1, 2, and 3 isozymes, are involved in detoxification, including the metabolism of acetaldehyde, other dietary aldehydes, xenobiotics, lipid peroxidation products and certain anti-cancer drugs. Also, HMSALDHs and Aromatic ALDHs are involved in pathways that remove aromatic xenobiotics.

In basic metabolic pathways, GGSALDHs and MMSALDHs are both involved in amino acid metabolism. GAPDHs function in glycolysis. SSALDHs and γ-aminobutyraldehyde dehydrogenase (Pietruszko et al., 1997) participate in the catabolism of inhibitory neurotransmitter GABA. FTDHs are involved in folate metabolism. Class 1 ALDHs can also synthesize retinoic acid, relevant both in development and vision, and metabolize 11-hydroxythromboxane B2.

In adaptation to osmotic stress, including dehydration and high salinity, organisms accumulate small compounds intracellularly, including proline and betaine. Proline, which is catabolized by GGSALDHs, may serve as a carbon source in these times of stress (Straub et al., 1996). Betaine, synthesized by BALDHs, serves an osmoprotectant role in all organisms. In addition, Turgor ALDHs, named for their induction in response to changes in osmotic turgor, also serve a yet unknown role in this osmotic adaptation. Next, several different enzymes have been recruited by various organisms to serve as structural proteins in the eye. Cephalopods and shrews have both utilized class 1 ALDHs as structural lens proteins, Ω and η crystallins, respectively. Also, class 1 and 3 ALDHs are major soluble proteins in lens and cornea, respectively. These ALDHs may aid in the detoxification of peroxidation products generated by UV light absorption (Cooper & Baptist, 1991; King & Holmes, 1997). Finally, non-phosphorylating GAPDHs generate NADPH necessary for photosynthetic reactions.

The importance of ALDHs is manifest in several autosomally-inherited diseases caused by ALDH deficiencies. Sjögren-Larsson syndrome (SLS) is caused by deleterious mutations in the human microsomal class 3 (fatty) ALDH. This autosomal recessive disorder is characterized by ichthyosis at early stages, followed by mental retardation and spasticity (De Laurenzi et al., 1996). Currently, this effect is thought to result from accumulation of long-chain aldehydes which then react with membrane-embedded amines such as phosphatidyl ethanolamine (James & Zoeller, 1997). Another autosomal recessive disorder, type II hyperprolinemia, is caused by the loss of GGSALDH function. Patients with this disorder exhibit 10–40 fold higher plasma levels of proline and Δ^1-pyrroline-5-carboxylate and may suffer from mental retardation and seizures (Phang et al., 1995). SSALDH-deficiency leads to an intracellular accumulation of succinic semialdehyde and an increase in 4-hydroxybutyrate in physiological fluids. These changes affect the central nervous system, causing altered motor activity and speech delay (Jakobs et al., 1993). Clearly ALDHs are essential components of cells for a variety of different reasons.

A website, **www.psc.edu/biomed/pages/research/Col_HBN_ALDH.html**, will contain figures and tables associated with several aspects of our alignment project. With the ALDH families separated and the motifs defined, our next effort will analyze characteristic conservations specific/unique to the different ALDH families.

ACKNOWLEDGMENTS

Sequence analysis and evolutionary studies were done through the resources of the Pittsburgh Supercomputing Center and the NIH Division of Research Resources (grant RR06009). Additional support was provided by the National Institute on Alcohol Abuse and Alcoholism (NIAAA) grant AA06985.

REFERENCES

Boch, J., Kempf, B., Schmid, R. and Bremer, E. (1996) Synthesis of the osmoprotectant glycine betaine in *Bacillus subtilis*: Characterization of the *gbsAB* genes. *J Bacteriol* **178,** 5121–5129.

Chambliss, K.L., Caudle, D.L., Hinson, D.D., Moomaw, C.R., Slaughter, C.A., Jakobs, C. and Gibson, K.M. (1995) Molecular cloning of the mature NAD(+)-dependent succinic semialdehyde dehydrogenase from rat and human. cDNA isolation, evolutionary homology, and tissue expression. *J Biol Chem* **270,** 461–467.

Cook, R.J., Lloyd, R.S. and Wagner, C. (1991) Isolation and characterization of cDNA clones for rat liver 10-formyltetrahydrofolate dehydrogenase. *J Biol Chem* **266,** 4965–4973.

Cooper, D.L. and Baptist, E.W. (1991) Degenerate oligonucleotide-directed PCR cloning of the BCP54/AlDH 3 cDNA. *PCR Meth Appl* **1,** 57–62.

De Laurenzi, V., Rogers, G.R., Hamrock, D.J., Marekov, L.N., Steinert, P.M., Compton, J.G., Markova, N. and Rizzo, W.B. (1996) Sjögren-Larsson Syndrome is caused by mutations in the fatty aldehyde dehydrogenase gene. *Nature Genet* **12,** 52–57.

Dockham, P.A., Lee, M.-O. and Sladek, N.E. (1992) Identification of human AlDHs that catalyze the oxidation of aldophosphamide and retinaldehyde. *Biochem Pharmacol* **43,** 2453–2469.

Domeyer, B.E. and Sladek, N.E. (1980) Metabolism of 4-OH-cyclophosphamide and aldophosphamide *in vitro*. *Biochem Pharmacol* **29,** 2903–2912.

Graham, C., Hodin, J. and Wistow, G. (1996) A retinaldehyde dehydrogenase as a structural protein in a mammalian eye lens. Gene recruitment of eta-crystallin. *J Biol Chem* **271,** 15623–15628.

Guerrero, F.D., Jones, J.T. and Mullet, J.E. (1990) Turgor-responsive gene transcription and RNA levels increase rapidly when pea shoots are wilted. Sequence and expression of three inducible genes. *Plant Mol Biol* **15,** 11–26.

Habenicht, A., Hellman, U. and Cerff, R. (1994) Non-phosphorylating GAPDH of higher plants is a member of the aldehyde dehydrogenase family with no sequence homology to phosphorylating GAPDH. *J Mol Biol* **237,** 165–171.

Harada, S., Agarwal, D.P. and Goedde, H.W. (1981) AlDH deficiency as cause of facial flushing reaction to alcohol in Japanese. *Lancet* **ii,** 982.

Hempel, J., Nicholas, H. and Lindahl R. (1993) Aldehyde dehydrogenases: Widespread structural and functional diversity within a shared framework. *Prot Sci* **2**; 1890–1900.

Hempel, J., Perozich, J., Chapman, T., Rose, J., Liu, Z.-J., Boesch, J.S., Wang, B.-C. and Lindahl, R. (1999) Aldehyde dehydrogenase catalytic mechanism: A proposal. *Adv Exp Med Biol* (this meeting).

Hu, C.A., Lin, W.W. and Valle, D. (1996) Cloning, characterization, and expression of cDNAs encoding human delta 1-pyrroline-5-carboxylate dehydrogenase. *J Biol Chem* **271,** 9795–9800.

Inoue, J., Shaw, J.P., Rekik, M. and Harayama, S. (1995) Overlapping substrate specificities of benzaldehyde dehydrogenase (the *xylC* gene product) and 2-hydroxymuconic SADH (the *xylG* gene product) encoded by TOL plasmid pWW0 of *Pseudomonas putida*. *J Bacteriol* **177,** 1196–1201.

Jakobs, C., Jaeken, J. and Gibson, K.M. (1993) Inherited disorders of GABA metabolism. *J Inher Metab* **16,** 704–715.

James, P.F. and Zoeller, R.A. (1997) Isolation of animal cell mutants defective in long-chain fatty acid aldehyde dehydrogenase. Sensitivity fatty aldehydes and Schiff's base modification of phospholipids: Implications for Sjögren-Larsson Syndrome. *J Biol Chem* **272,** 23532–23539.

Kedishvili, N.Y., Popov, K.M., Rougraff, P.M., Zhao, Y., Crabb, D.W. and Harris, R.A. (1992) CoA-dependent methylmalonate semialdehyde dehydrogenase, a unique member of the aldehyde dehydrogenase superfamily. *J Biol Chem* **267,** 19724–19729.

King, C.T. and Holmes, R.S. (1997) Human corneal and lens aldehyde dehydrogenases: Purification and properties of human lens ALDH1 and differential expression as major soluble proteins in human lens (ALDH1) and cornea (ALDH3). *Adv Exp Med Biol* **414,** 19–27.

Kitson, T.M. (1977) The disulfiram-ethanol reaction: A review. *J Stud Alcohol* **38,** 96–113.

Krupenko, S.A., Wagner, C. and Cook, R.J. (1997) Expression, purification, and properties of the aldehyde dehydrogenase homologous carboxyl-terminal domain of rat 10-formyltetrahydrofolate dehydrogenase. *J Biol Chem* **272,** 10266–10272.

Lee, M.-O., Manthey, C.L. and Sladek, N.E. (1991) Identification of mouse liver AlDHs that catalyze the oxidation of retinaldehyde to retinoic acid. *Biochem Pharmacol* **42,** 1279–1285.

Lindahl, R. (1992) Aldehyde dehydrogenases and their role in carcinogenesis. *Crit Rev Biochem Mol Biol* **27,** 283–335.

Lindahl, R. and Petersen, D.R. (1991) Lipid aldehyde oxidation as a physiological role for class 3 AlDH. *Biochem*

Ling, M., Allen, S.W. and Wood, J.M. (1994) Sequence analysis identifies the proline dehydrogenase and delta-1-pyrroline-5-carboxylate dehydrogenase domains of the multifunctional *Escherichia coli* PutA protein. *J Mol Biol* **243,** 950–956.

Liu, Z.-J., Sun, Y.-J., Rose, J., Chung, Y.-J., Hsiao, C.-D., Chang, W.-R., Kuo, I., Perozich, J., Lindahl, R., Hempel, J. and Wang, B.-C. (1997) The first structure of an aldehyde dehydrogenase reveals novel interactions between NAD and the Rossmann fold. *Nature Struct Biol* **4,** 317–326.

Moore, S.A., Baker, H.M., Blythe, T., Kitson, K.E., Kitson, T.M. and Baker, E.N. (1999) Structure of a class 1 mammalian aldehdye dehydrogenase at 2.35Å resolution. *Adv Exp Med Biol* (this meeting).

Ni, L., Sheikh, S. and Weiner, H. (1997) Involvement of glutamate 399 and lysine 192 in the mechanism of human liver mitochondrial aldehyde dehydrogenase. *J Biol Chem* **272,** 18823–18826.

Nordlund, I. and Shingler, V. (1990) Nucleotide sequences of the meta-cleavage pathway enzymes 2-hydroxymuconic semialdehyde dehydrogenase and 2-hydroxymuconic semialdehyde hydrolase from *Pseudomonas* CF600. *Biochim Biophys Acta* **1049,** 227–230.

Op den Camp, R.G. and Kuhlemeier, C. (1997) Aldehyde dehydrogenase in tobacco pollen. *Plant Mol Biol* **35,** 355–365.

Perozich, J., Nicholas, H., Wang, B.-C., Lindahl, R. and Hempel, J. (1998) Relationships within the aldehyde dehydrogenase extended family. *Prot Sci* (submitted).

Phang, J.M., Yeh, G.C. and Scriver, C.R. (1995) Disorders of proline and hydroxyproline metabolism. In: Scriver, C.R., Beaudet, A.L., Sly, W.S. and Valle, D., eds. *The Metabolic and Molecular Bases of Inherited Disease,* 7th edition. New York: McGraw-Hill. pp 1107–1124.

Pietruszko, R., Kikonyogo, A., Chern, M.-K. and Izaguirre, G. (1997) Human aldehyde dehydrogenase E3: Further characterization. *Adv Exp Med Biol* **414,** 243–252.

Priefert, H., Rabenhorst, J., Steinbuchel, A. (1997) Molecular characterization of genes of *Pseudomonas* sp. strain HR199 involved in bioconversion of vanillin to protocatechuate. *J Bacteriol* **179,** 2595–2607.

Rizzo, W.B. (1993) Sjögren-Larsson syndrome. *Sem Dermatol* **12,** 210–218.

Steinmetz, C.G., Xie, P., Weiner, H. and Hurley, T.D. (1997) Structure of mitochondrial aldehyde dehydrogenase: the genetic component of ethanol aversion. *Structure* **15,** 701–711.

Straub, P.F., Reynolds, P.H., Althomsons, S., Mett, V., Zhu, Y., Shearer, G. and Kohl, D.H. (1996) Isolation, DNA sequence analysis, and mutagenesis of a proline dehydrogenase gene (*putA*) from *Bradyrhizobium japonicum. Appl Environ Microbiol* **62,** 221–229.

Stroeher, V.L., Boothe, J.G. and Good, A.G. (1995) Molecular cloning and expression of a turgor-responsive gene in *Brassica napus. Plant Mol Biol* **27,** 541–551.

Wang, X.-P. and Weiner, H. (1995) Involvement of glutamate 268 in the active site of human liver mitochondrial (class 2) aldehyde dehydrogenase as probed by site-directed mutagenesis. *Biochemistry* **34,** 237–243.

Weretilnyk, E.A. and Hanson, A.D. (1990) Molecular cloning of a plant betaine-aldehyde dehydrogenase, an enzyme implicated in the adaptation to salinity and drought. *Proc Natl Acad Sci USA* **87,** 2745–2749.

Westlund, P., Fylling, A.C., Cederlund, E. and Jörnvall, H. (1994) 11-hydroxy-thromboxane B_2 dehydrogenase is identical to cytosolic aldehyde dehydrogenase. *FEBS Lett* **345,** 99–103.

Zinovieva, R.D., Tomarev, S.I. and Piatigorsky, J. (1993) Aldehyde dehydrogenase-derived omega-crystallins of squid and octopus. Specialization for lens expression. *J Biol Chem* **268,** 11449–11455.

2

REACTION-CHEMISTRY-DIRECTED SEQUENCE ALIGNMENT OF ALDEHYDE DEHYDROGENASES

Pritesh Shah and Regina Pietruszko

Center of Alcohol Studies
and Department of Molecular Biology and Biochemistry
Rutgers University
607 Allison Road
Piscataway, New Jersey 08854-8001

1. INTRODUCTION

Aldehyde dehydrogenases (ALDH) occur in multiple forms throughout a large variety of organisms. In recent years many have been sequenced and compared to arrive at the consensus of invariable residues that are essential for catalysis; a conserved feature during evolution. Thus, 16 different aldehyde dehydrogenases were compared by Hempel et al., (1993) who found 23 invariable amino acid residues by alignment of sequences available at that time.

Classification of members of enzyme families such as alcohol dehydrogenases, glutathione transferases, and cytochrome P450 monoxygenases have been made with relatively few complications. Similar attempts to classify aldehyde dehydrogenases were more difficult. The first classification attempt (Workshop on Aldehyde Dehydrogenase, 1988) excluded semialdehyde, steroid aldehyde, vitamin aldehyde, and formaldehyde dehydrogenases. A more recent attempt (Vasiliou et al., 1995), based on primary structure, excludes enzymes such as aspartic semialdehyde dehydrogenase, phosphorylating glyceraldehyde-3-phosphate dehydrogenase and formaldehyde dehydrogenase.

Unlike, other enzymes on which attempts to classify aldehyde dehydrogenase were modelled, reactions catalyzed by aldehyde dehydrogenases are not identical but fall within five distinct groups (Table 1 and Enzyme Nomenclature, 1992). NAD or NADP or both, participate in all five reactions. Only group 1 enzymes (Table 1) catalyze unidirectional irreversible reaction with NAD(P) as coenzyme and with participation of water. Water is required for hydrolysis of enzyme-acyl intermediate. All other enzymes (Table 1) catalyze a reversible reaction for which additional coenzymes (e.g., coenzyme A or glutathione or

Enzymology and Molecular Biology of Carbonyl Metabolism 7
edited by Weiner *et al.* Kluwer Academic / Plenum Publishers, New York, 1999.

Table 1. Reactions catalyzed by aldehyde dehydrogenases

EC Number 1.2.1:	Reaction catalyzed	
3-5, 7, 8, 15, 16, 19-24, 26, 28, 29, 31-33, 36, 39, 40, 45-49, 53, 54[1]	Aldehyde + NAD(P) + H_2O ⟶ Acid + NAD(P)H + H^+	1.
10, 17, 18, 25, 27 42, 44, 57[2]	Aldehyde + NAD(P) + CoA ⟷ Acyl CoA + NAD(P)H + H^+	2.
9, 11, 12, 13, 38, 41[3]	Aldehyde + orthophosphate + NAD (P) ⟷ Acyl phosphate + NAD(P)H + H^+	3.
1.	Formaldehyde + Glutathione + NAD ⟷ S-formylglutathione + NADH + H^+	4.
30.	Aryl aldehyde + NADP + AMP + pyrophosphate + H_2O ⟷ Aromatic Acid + NADPH + ATP + H^+	5.

Numbers in [ ] show last digit of the EC number. [1] Substrates include: aliphatic, aromatic aldehydes [3, 4, 5], benzaldehyde [7,28], betaine aldehyde [8], hydroxyaldehydes [3,22], ketoaldehydes [23], malonate semialdehyde [15], succinate semialdehyde [16,24], glutamic-5-semialdehyde [*], 4-aminobutyraldehyde [19], glutarate [20], glycolaldehyde [3,21] lactaldehyde [3,22], 2-oxaldehyde [23], 2,5-dioxovalerate [25], aryl aldehyde [29], aminoadipic semialdehyde [31], aminomuconic semialdehyde [32], 4-dehydropentoate [33], retinaldehyde [36], phenylacetaldehyde [39], 3α,7α, 12α -trihydroxycholestan-26-al [40], 4-carboxy-2-hydroxymuconate-6-semialdehyde [45], formaldehyde [46], 4-trimethylammoniobutyraldehyde [47], 4-hydroxyphenyl-acetaldehyde [53] , γ-guanidinobutyraldehyde [54], and glyceraldehyde-3-phosphate (7).

[2]Substrates include: acetaldehyde [10], glyoxylate [17], malonate semialdehyde [18], 2-oxoisovalerate [25], methylmalonate semialdehyde [27], hexadecanal [42], cinnamaldehyde [44], butanal [57].

[3]Substrates include: glyceraldehyde-3-phosphate [9, 12, 13], aspartic semialdehyde [11], N-acetyl-L-glutamate-5-semialdehyde [38], glutamic-5-semialdehyde [41].

*Enzyme not listed in 1992 Enzyme Nomenclature, see Kurys et al., (1989) to confirm the reaction catalyzed; see Gross (1972) and Strittmatter and Ball (1955) to confirm reactions 4 and 5.

adenosine monophosphate) are needed. When a reaction is reversible, (except for reaction 5) the product is usually not a free carboxylic acid, but an acyl derivative. It has been postulated that in reaction 2, semimercaptal derived from coenzyme A sulfhydryl group and the aldehyde is a true substrate (Burton and Stadtman, 1953). Water does not participate in reactions 2–5. Reaction 5 generates ATP as one of the products (Gross, 1972). There are some aldehydes (Enzyme Nomenclature, 1992), like acetaldehyde, propionaldehyde, butyraldehyde, hexadecanal and cinnamaldehyde that are substrates for enzymes catalyzing reactions 1, 2 and 3; glutamic-γ-semialdehyde is a substrate for enzymes catalyzing reaction 1 as well as enzymes catalyzing reaction 3. Recently glyceraldehyde-3-phosphate dehydrogenase from *Thermoproteus tenax* (Brunner et al., 1998) which catalyzes reaction 1 has been described; glyceraldehyde-3-phosphate is usually associated with reaction 3 in Table 1.

In the previously published alignment (Hempel et al., 1993) only structural features were considered. Because of similarity of its primary structure to that of aldehyde dehydrogenases, catalyzing reaction 1, methylmalonate semialdehyde dehydrogenase, which catalyzes a reversible reaction 2 (Table 1) with participation of Coenzyme A, instead of water was included in the alignment. In this paper we show how attention to the reaction catalyzed alters the pattern of totally conserved residues obtained by computer alignment of aldehyde dehydrogenase sequences.

2. MATERIALS AND METHODS

Amino acid sequences of enzymes from different sources catalyzing reaction of group 1 (Table 1) were selected on the basis of the reaction catalyzed and downloaded from GenBANK/PIR databases. The multiple sequence alignment was performed using

Macaw Program (Schuler et al., 1991) with a cut off score of 30. The numbers used for identification of amino acid residues are those of Hempel et al., (1993). The enzymes used for sequence comparison were: U24266, human glutamic-γ-semialdehyde dehydrogenase; RDBYC, yeast glutamic-γ-semialdehyde dehydrogenase; P17445, *E. coli* betaine aldehyde dehydrogenase; S14629, chicken cytosolic ALDH; U07235, mouse mitochondrial ALDH; P11884, rat mitochondrial ALDH; S00364, horse mitochondrial ALDH; DEHUE1, human cytosolic ALDH; DEHUE2, human mitochondral ALDH; U34252, human amino aldehyde dehydrogenase; L34821, rat succinic semialdehyde dehydrogenase; M23995, rat phenobarbital inducible ALDH; L42009, retinaldehyde dehydrogenase; M74542, human ALDH3; M73714, rat microsomal ALDH; J03637, rat ALDH3.

3. RESULTS AND DISCUSSION

Our alignment (Table 2) results in 25 totally conserved residues of which 13 are glycines, one is an alanine and two are prolines. Hempel et al. (1993) alignment consisted of 23 residues of which 11 were glycines and three were prolines; there were no conserved alanines. There are 17 amino acid residues common to both alignments. Those, other than glycines include: K192, C302, P383, E399, F401, P403, N454 and S471. In the present alignment 8 residues are gained, four of which are glycines and one is alanine; other gained residues include T244, E268 and Q300. The following 6 residues from the Hempel et al., (1993) alignment are lost: R84, P158, G223, G292, T384, N421. Please note that most of the gained residues in the present alignment, are in the active site area of aldehyde dehydrogenase (Abriola et al., 1987; Steinmetz et al., 1997).

Both cysteine 302 and glutamate 268 were previously identified (Hempel and Pietruszko, 1981; Hempel et al., 1982; Mackerell et al., 1986; Abriola et al., 1987; Blatter et al., 1990 and 1992) as essential for catalysis in our laboratory by chemical modification. Since then, the essentiality of both of these residues has been confirmed by site directed mutagenesis (Farres et al., 1995; Wang and Weiner, 1995; Vedadi and Meighen, 1997). No other amino acid residue, essential to catalysis, has been identified by site directed mutagenesis. Recent report on X ray crystallography of bovine mitochondrial aldehyde dehydrogenase (Steinmetz et al., 1997) clearly shows that E268 along with C302 and T244 are located within the active site pocket of the enzyme. Of the residues identified by Hempel et al., (1993) alignment, only C302 was found to be located at the bottom of the active site pocket. Both E268 and T244 were eliminated as totally conserved residues from Hempel et al., (1993) alignment only because methylmalonate semialdehyde dehydrogenase catalyzing reaction 2 was also included in the comparison. T244 equivalent is replaced by valine while E268 equivalent is replaced by asparagine in methylmalonyl semialdehyde dehydrogenase. On the basis of our present alignment, it appears possible that T244 may be the third residue (in addition to C302 and E268) which is directly involved in catalysis. Although chemical modification experiments never demonstrated any enhanced reactivity of this residue, crystallography of bovine mitochondrial aldehyde dehydrogenase showed that T244 was located at the bottom of the active site pocket (Steinmetz et al., 1997). Glutamate 476 was also visualized in the active site pocket, however, on the basis of our alignment, we envisage no role for this residue in catalysis. Although Q300 is present in all enzymes used for our alignment, it is absent from *Vibrio cholerae* aldehyde dehydrogenase whose catalysis of reaction 1 is well established (Parsot and Mekalanos, 1991). This suggests that Q300 may not be essential for catalysis. Site-directed mutagenesis of totally conserved (Hempel et al., 1993) residues R84, K192, T384,

Table 2. Comparison of totally conserved residues from our alignment with those of Hempel et al., (1993)

Hempel et al., (1993)	Present alignment	New residues	New not G or A
R (84)	–		
P (158)	–		
G (160)..........................	**G (160)**		
–	A (182)	A (182)	
G (186)..........................	**G (186)**		
K (192)..........................	**K (192)**		
G (223)	–		
–	T (244)	T (244)	T (244)
G (245)..........................	**G (245)**		
–	E (268)	E (268)	E (268)
G (271)..........................	**G (271)**		
–	G (272)	G (272)	
G (292)	–		
–	G (299)	G (299)	
–	Q (300)	Q (300)	Q (300)
C (302)..........................	**C (302)**		
G (370)..........................	**G (370)**		
–	G (371)	G (371)	
P (383)..........................	**P (383)**		
T (384)	–		
E (399)..........................	**E (399)**		
F (401)..........................	**F (401)**		
G (402)..........................	**G (402)**		
P (403)..........................	**P (403)**		
N (421)	–		
G (449)..........................	**G (449)**		
N (454)..........................	**N (454)**		
G (466)..........................	G (466)		
–	G (467)	G (467)	
S (471)..........................	**S (471)**		
G (472)..........................	**G (472)**		
23	25	8	3

Residues present in both alignments are connected by a dotted line and are shown in bold. Residues gained in the present alignment are shown as – in the first column; residues lost from the previous alignment are shown as – in the present alignment column. Total residues in each alignment are shown in the last line of the Table. The numbering system used is that of Hempel et al., (1993).

E399 and S471 residues (Sheikh, et al., 1997; Ni et al., 1997) found that K192 and E399 were important for hydride transfer; please note that these two residues along with S471 (but not R84 or T384) are also conserved residues resulting from our alignment (Table 2).

Enzyme evolution is most probably directed by function. The groups that are necessary for catalysis of the irreversible reaction with participation of water in the process, would be different from those necessary to catalyze a reversible reaction. Also, it is well known that divergent as well as convergent evolution is possible. Substrate specificity cannot be used as a classification criterion because the same substrates are utilized by different enzymes (Table 1); the most recent example of which is glyceraldehyde-3-phosphate dehydrogenase from *Thermoproteus tenax* (Brunner et al., 1998). The primary structure of *Thermoproteus* enzyme fully reflects its similarity to enzymes catalyzing reaction 1 in the subunit size and the conserved residues. However, publication (Brunner et al., 1998) appeared too late to appear in our primary structure alignment. If primarily an attention is paid to catalyzed reaction rather than to substrate specificity or structure, classification of aldehyde dehydrogenases would be an easy process comparable to that of alcohol dehydrogenases and other similar enzymes. Such classification would be also more informative to those who are interested in the mechanisms of enzyme action.

ACKNOWLEDGMENTS

Financial support of USPHS NIAAA Grant 1R01 AA00186 is gratefully acknowledged.

REFERENCES

Abriola, D.P., Fields, R., Stein, S., McKerell, A.D.Jr., and Pietruszko, R. (1987) Active site of human aldehyde dehydrogenase. Biochemistry 26, 5679–5684.

Blatter, E.E., Abriola, D.P., and Pietruszko, R. (1992) Aldehyde dehydrogenase: Covalent intermediate in aldehyde dehydrogenation and ester hydrolysis. Biochem. J. 282, 353–360.

Blatter, E.E., Tasayco, M.L., Prestwich, G., and Pietruszko, R. (1990) Chemical modification of aldehyde dehydrogenase by a vinyl ketone analog of an insect pheromone. Biochem. J. 272, 351–358.

Brunner, N.A., Brinkmann, H., Siebers, B., and Hensel, R. (1998) NAD^+-dependent glyceraldehyde-3-phosphate dehydrogenase from *Thermoproteus tenax*. J. Biol. Chem. 273, 6149–6156.

Burton, R.M., and Stadtman, E. R. (1953) The oxidation of acetaldehyde to acetyl coenzyme A. J. Biol. Chem. 202, 873–890.

Enzyme Nomenclature. (1992) International Union of Biochemistry, Academic Press Inc.

Farres, J., Wang, T.Y., Cunningham, S.J., and Weiner, H. (1995) Investigation of the active site cysteine residue of rat liver mitochondrial aldehyde dehydrogenase by site-directed mutagenesis. Biochemistry 34, 2592–2598.

Gross, G.G. (1972) Formation and reduction of intermediate acyladenylate by aryl-aldehyde NADP oxidoreductase from *Neurospora crassa*. Eur. J. Biochem. 31, 585–592.

Hempel, J., Nicholas, H., and Lindahl, R. (1993) Aldehyde dehydrogenases: Widespread structural and functional diversity within a shared framework. Protein Science 2, 1890–1900.

Hempel, J., and Pietruszko, R. (1981) Selective chemical modification of human liver aldehyde dehydrogenases E1 and E2 by iodoacetamide. J. Biol. Chem. 256, 10889–10896.

Hempel, J.D., Pietruszko, R. Fietzek, P., and Jornvall, H. (1982) Identification of a segment containing a reactive cysteine residue in human liver cytoplasmic aldehyde dehydrogenase (isoenzyme E1). Biochemistry 21, 6834–6838.

Kurys, G., Ambroziak, W., and Pietruszko, R. (1989) Human aldehyde dehydrogenase. Purification and characterization of of a third isozyme with low Km for γ-aminobutyraldehyde. J. Biol. Chem. 264, 4715–4721.

Liu, Z-J., Sun, Y-J., Rose, J., Chung, Y-J., Hsiao, C-D., Chang, W-R., Kuo, I., Perozich, J., Lindahl, R., Hempel, J., and Wang, B-C. (1997) The first structure of and aldehyde dehydrogenase reveals novel interactions between NAD and the Rossmann fold. Nature Structural Biology 4, 317–326.

MacKerell, A.D.Jr., MacWright, R.S., and Pietruszko, R. (1986) Bromoacetophenone as an affinity reagent for human liver aldehyde dehydrogenase. Biochemistry 25, 5182–5189.

Ni, L., Sheikh, S., and Weiner, H. (1997) Involvement of glutamate 399 and lysine 192 in the mechanism of human liver mitochondrial aldehyde dehydrogenase. J. Biol. Chem. 272, 18823–18826.

Parsot, C., and Mekalanos, J.J., (1991) Expression of the *Vibrio cholerae* gene encoding aldehyde dehydrogenase is under control of ToxR, the cholera toxin transcriptional activator. J. Bact. 173, 2842–2851.

Schuler, G.D., Altschul, S.F., and Lipman, D.J. (1991) A workbench for multiple alignment construction and analysis. Proteins: Struc. Func. Genet. 9, 180–190.

Sheikh, S., Ni, L., Hurley, T.D., and Weiner, H. (1997) The potential roles of conserved amino acids in human mitochondrial aldehyde dehydrogenase. J. Biol. Chem. 272, 18817–18822.

Steinmetz, C.G., Xie, P., Weiner, H., and Hurley, T.D. (1997) Structure of mitochondrial aldehyde dehydrogenase: the genetic component of ethanol aversion. Structure 5, 701–711.

Strittmatter, P., and Ball, E.G. (1955) Formaldehyde dehydrogenase, a glutathione-dependent enzyme system. J. Biol. Chem. 213, 445–461.

Vasiliou, V., Weiner, H., Marselos, M., and Nebert, D. (1995) Mammalian aldehyde dehydrogenase genes: classification based on evolution, structure and regulation. Eur. J. Drug Met. Pharmacokinet. Special Issue pp 53–64.

Vedadi, M., and Meighen, E. (1997) Critical glutamic acid residues affecting the mechanism and nucleotide specificity of *Vibrio harvei* aldehyde dehydrogenase. Eur. J. Biochem. 246,698–704.

Wang, X., and Weiner, H. (1995) Involvement of glutamate 268 in the active site of human mitochondrial (class 2) aldehyde dehydrogenase as probed by site-directed mutagenesis. Biochemistry 34, 237–243.

Workshop on Aldehyde Dehydrogenase (1988) Progress in Clinical and Biological Research 290:pp xix-xxi, 1989.

3

THREE-DIMENSIONAL STRUCTURE OF MITOCHONDRIAL ALDEHYDE DEHYDROGENASE

Mechanistic Implications

Thomas D. Hurley,[1] Curtis G. Steinmetz,[1] and Henry Weiner[2]

[1]Department of Biochemistry and Molecular Biology
Indiana University School of Medicine
Indianapolis, Indiana 46202
[2]Department of Biochemistry
Purdue University
West Lafayette, Indiana 47907

1. INTRODUCTION

Aldehyde dehydrogenases (ALDH) comprise a diverse set of enzymes which catalyze the $NAD(P)^+$ dependent oxidation of aldehydes. Crystal structures for representatives of three of the basic classes of ALDH isoenzymes now exist. The dimeric Class 3 structure being first presented at the 1996 meeting (Liu, *et al.* 1997a) and the tetrameric Class 1 and 2 structures are presented at this meeting. However, each of these structures represent enzymes which catalyze essentially identical chemical reactions. The basic mechanism involves the nucleophilic attack of an active site thiolate toward the aldehydic carbonyl carbon, resulting in a covalently bound thio-hemiacetal. Following hydride transfer to the coenzyme, the thio-hemiacetal collapses to yield an acylated enzyme intermediate. For most of the ALDH family members, and especially for those enzymes whose 3-D structures are known, the acyl-enzyme intermediate is hydrolyzed by an activated water molecule to generate the corresponding acid product. However, several family members transfer the acyl-enzyme intermediate to another acceptor, such as to coenzyme-A in methyl-malonyl semialdehyde dehydrogenase. Thus, only the initial chemical event, namely the attack of the aldehyde by the active site cysteine residue, is a common reaction in all members of the aldehyde dehydrogenase gene family and that the subsequent deacylation reaction may utilize uniquely different catalytic residues. Therefore, lack of residue

Enzymology and Molecular Biology of Carbonyl Metabolism 7
edited by Weiner *et al.* Kluwer Academic / Plenum Publishers, New York, 1999.

conservation in all members of the gene family may not necessarily eliminate that residue from catalytic involvement within certain subsets of enzymes that catalyze the same reaction chemistry.

We present here the X-ray crystal structure of the bovine mitochondrial form of aldehyde dehydrogenase (ALDH2) in the presence and absence of its cofactor NAD^+. We then propose a plausible chemical mechanism based on our crystallographic structures and by comparison to the other existing structures. We believe that this mechanism is generally applicable only to those members of the ALDH gene family which activate a water molecule to hydrolyze the acyl-enzyme intermediate. Although, the initial nucleophilic attack by the active site cysteine residue to form the thio-hemiacetal would be generally applicable to all members of the ALDH gene family.

2. RESULTS AND DISCUSSION

2.1. Overall Description of the Structure

The subunit positioning of the Class 2 tetrameric bovine enzyme resembles a dimer of dimers (Figure 1). Each dimer shares overall structural similarity to the Class 3 structure. Three distinct domains can be identified in each subunit; a coenzyme-binding domain, a catalytic domain, and a oligomerization domain (Figure 2). As their names imply, each domain would appear to contribute unique functional properties to the overall structure (Steinmetz, *et al.* 1997). The coenzyme-binding domain comprises residues 1–135 and residues 161–270. This domain contributes almost all of the important contacts between the enzyme and the bound coenzyme molecule. The catalytic domain comprises residues 273–470 and includes the important active-site loop containing Cys302. The core of both domains is comprised of a five-stranded variant of the classic "Rossmann-fold". Lastly, the oligomerization domain is comprised of residues 136–157 and residues 486–495. This three-stranded

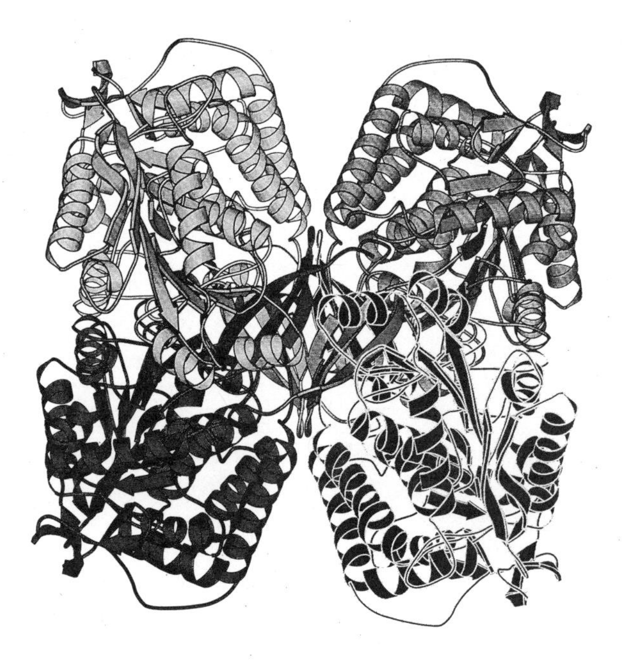

Figure 1. A ribbon diagram of the tetrameric Class 2 aldehyde dehydrogenase looking down the dimer-dimer interface. Individual subunits are identified by different gray-scales.

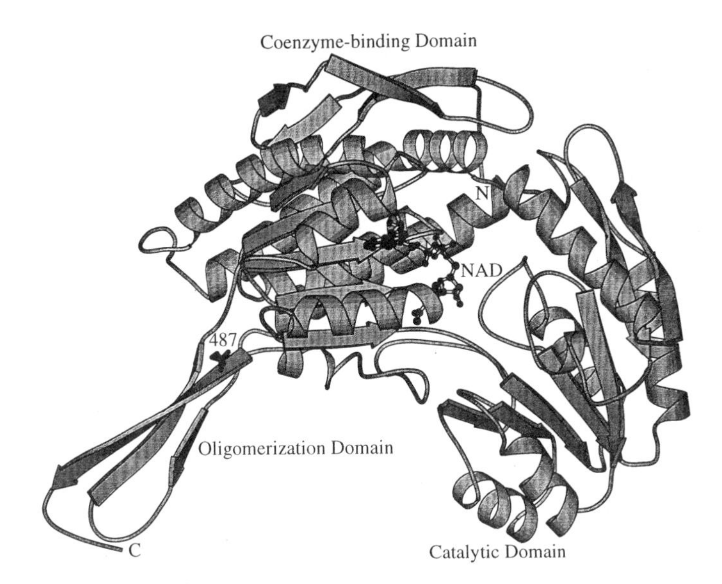

Figure 2. A ribbon diagram of an individual subunit of the Class 2 aldehyde dehydrogenase. Individual domains within the structure are labeled as well as the position of the bound cofactor (NAD) and the position of Glu487 (487).

antiparallel β-sheet domain forms much of the intersubunit contacts between monomers in both the dimer and the tetramer. The structural similarity between the Class 2 and 3 enzymes breaks down near the N- and C-terminal portions of the structures. The Class 2 enzyme possesses a 56 amino acid extension at the N-terminus relative to the Class 3 enzyme and the Class 3 enzyme possesses a 17 amino acid extension at its C-terminus relative to the Class 2 enzyme. The N-terminal extension in the Class 2 structure caps the top of the coenzyme-binding domain and clearly folds as a integral part of this domain, rather than a separate isolated structure. The C-terminal extension in the Class 3 structure loops across to the other subunit in the structure and may participate in substrate binding (Liu, *et al.* 1997b). It is not clear whether significant functions can be ascribed to these sequence extensions, but the positioning of the C-terminal extension in the Class 3 enzyme would prevent tetramer formation if present in the Class 2 enzyme.

2.2. Coenzyme Binding

The structure of the bovine ALDH2 isoenzyme was determined in the presence and absence of its cofactor, NAD^+ (Steinmetz, *et al.* 1997). Since the enzyme was initially crystallized only in the presence of NAD^+, this was first accomplished by removing NAD^+ from the crystals and then later, by crystallizing the enzyme in the absence of NAD^+. The resulting coenzyme-free structures are identical at 2.65 Å, which surprisingly, were also identical in conformation to the coenzyme-bound form of the enzyme. This was clearly a surprising result for a dehydrogenase and remains so. To date, we have seen no evidence of any difference in protein conformation in any of our structures, including the recombinant human ALDH2 enzyme we are now working with. However, this constitutes only negative evidence and we cannot rule out that a conformational change does occur or that important subtle changes in the structure of the enzyme are not visible at 2.65 Å.

The unusual binding mode of NAD(H) to the ALDH "Rossmann-fold" has been well documented in the Class 3 structure (Liu, et al. 1997b) and the overall features of coenzyme binding across the "Rossmann-fold" are similar in the Class 2 enzyme and will not be discussed further here. However, there are some important differences at the level of the atomic contacts in the two enzyme structures that have created controversy over the involvement of particular amino acids in the catalytic mechanism. Clearly, the controversy over the catalytic mechanism is primarily due to a limited amount of information, both from the structural and the functional viewpoint, and is not likely to be resolved until important catalytic complexes can by trapped by structural or spectroscopic techniques. In this chapter we will discuss what is observed in the bovine Class 2 structure, (now confirmed in the human Class 2 structure), and how it influences our view of the catalytic mechanism.

The coenzyme molecule was only completely ordered in the bovine ALDH2 structure in the presence of the metal cation, samarium. In the absence of the metal ion, only the AMP-portion of the cofactor was visible in the electron density maps (Figure 3). This influence of metal cations has now been confirmed in the recombinant human structure with the more physiological metal ion, manganese (T. Hurley, unpublished observations). In the Class 2 enzyme, the AMP-portion of the cofactor is deeply buried in the enzyme cleft between Pro226 and Val249 and adenosine ribose oxygens form direct hydrogen bonding contacts with the side chains of Lys192, Glu195 and the backbone carbonyl oxygen atom of Ile166 (Figure 4). One of the adenosine-phosphate oxygens interacts with the

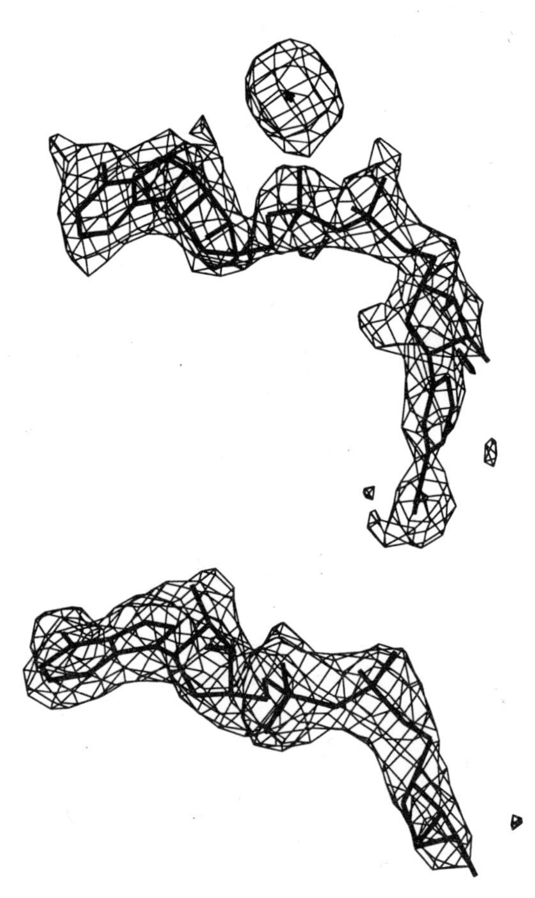

Figure 3. "Omit-map" electron density (contoured at 1 standard deviation of the map) for NAD^+ bound to the Sm-NAD^+ structure (top) and the NAD^+ structure in the absence of metal ions (bottom).

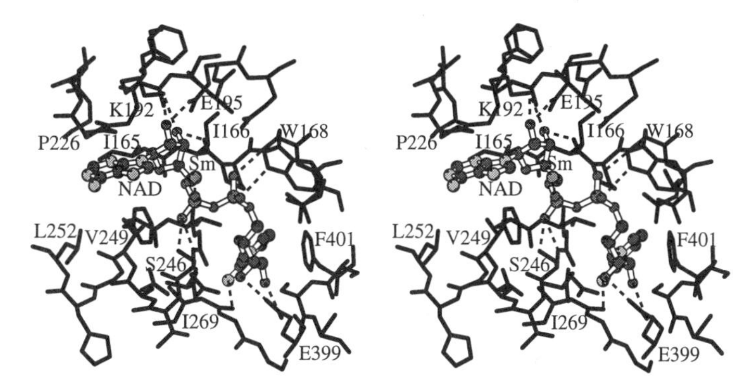

Figure 4. A stereo diagram of the NAD$^+$ binding site in the Class 2 ALDH structure. Individual amino acids are labeled using the single amino acid code and their corresponding sequence number in the bovine enzyme.

metal cation, samarium, and the other phosphate oxygen accepts two hydrogen bonds from Ser246, one donated by its peptide nitrogen and the other donated by its side-chain hydroxyl group. The nicotinamide phosphate oxygens are directly hydrogen bonded only to the indole nitrogen of Trp168. The peptide nitrogen of Trp168 is within 3.8 Å of one phosphate oxygen, but this is too far for a direct interaction. In our higher resolution human structure, this latter interaction appears to be water mediated. The nicotinamide ribose is held in position by donating hydrogen bonds to Glu399. The top and bottom faces of the ribose ring are held in position by van der Waals contacts with Gly245 and the side chain of Phe401. Interestingly, Gly245 is the first residue in the GxxxxG sequence suggested to be the coenzyme-binding fingerprint in ALDH2 (Hempel, et al. 1993). It is, perhaps, not coincidental that this Gly residue is actually performing the same function as the first Gly residue in the more famaliar GxGxxG coenzyme binding motif, namely close Van der Waals contact with a ribose ring. The unique feature of ALDH is that this Gly is interacting with the *nicotinamide* ribose ring, rather than the *adenosine* ribose ring in all other known "Rossmann-folds". The nicotinamide ring is found sandwiched between the side chains of Cys302 and Thr244 and the carboxamide group of the cofactor forms a hydrogen bond between its amide nitrogen and the backbone carbonyl oxygen of Ile269. Also nearby the nicotinamide ring are the side chains of Asn169 and Glu268.

2.3. Active-Site Structure and Mechanistic Implications

Access to the active site residues can occur from either face of the subunit in the absence of cofactor. However, once cofactor is bound to ALDH2, access to the active site from the cofactor binding side is severely restricted as the nicotinamide ring forms the base of the binding site. Thus, it is likely that substrate enters the active site from the face opposite to where coenzyme is observed to bind in both the Class 3 and Class 2 enzymes. In the Class 2 enzyme, this substrate-binding site is lined by hydrophobic residues; including Val120, Met124, Phe170, Leu173, Met174, Trp177, Phe296, Phe459, and Phe465. The only exception being Asp457, whose side chain points inward toward the side chains of Cys301 and Cys303 and would not appear to directly contact many of the preferred substrates for the Class 2 isoenzyme. At the base of the substrate-binding pocket, in the vicinity of the nicotinamide ring, are several polar residues that seem properly poised to participate in catalysis. These include the side chains of Asn169, Thr244, Glu268, Cys302 and Glu476. Among these amino acids, only Asn169 and Cys302 are completely conserved in all catalytically active members of the ALDH gene family (J. Perozich *et al.* 1999 and J. Hempel *et al.* 1999). The potential roles of these two amino acids will be discussed first, as there is little controversy over their proposed roles.

It has long been hypothesized that the active site of ALDH contains an Cys residue that acts as a nucleophile to covalently attack the incoming aldehydic carbonyl group (Racker, 1955). This residue was subsequently identified as Cys302 in the Class 1 and 2 isoenzymes by a number of methods (Hempel, *et al.* 1981, Hempel, *et al.*, 1982, and Farrés, *et al.* 1995). Cys302 sits in the middle of a loop of residues formed by residues 298 to 305. The side chain points out toward the solvent accessible entrance to the substrate-binding site. In the Class 2 isoenzyme, positions 301 and 303 are also Cys residues and the main chain conformation of this loop results in a structure where all the side chain sulfhydryls and the main chain amide nitrogens point in the same relative direction as Cys302. The side chain of Cys302 is in van der Waals contact with the nicotinamide ring in the Class 2 structure. Asn169 lies directly above Cys302 and the nicotinamide ring such that if a covalently bound thio-hemiacetal transition-state for propionaldehyde is model-built into the active site, the oxyanion of the transition-state is held in position by a hydrogen bond donated from the side chain amide nitrogen of Asn169 (Figure 5). An important consequence of the resulting model structure is that the oxyanion of the transition-state can also be stabilized by an additional hydrogen bond donated from the amide nitrogen of Cys302 (Figure 5). It is our belief, but by no means proven, that this special positioning of Cys302 within this loop structure—with all the peptide nitrogens oriented toward the Cys302 thiol—creates a structural and electrostatic environment that would stablize the negatively charged oxyanion of the transition-state prior to hydride transfer. This same structure could also stabilize the thiolate form of Cys302 enough to permit its spontaneous deprotonation under physiological conditions. A similar mechanism has been proposed to explain the low pKa observed for the active site thiol present in tyrosine phosphatases (Stuckey, *et al.* 1994).

The roles of Thr244, Glu268, and Glu476 are less obvious for two important reasons. First, they are not strictly conserved residues in all members of the ALDH gene fam-

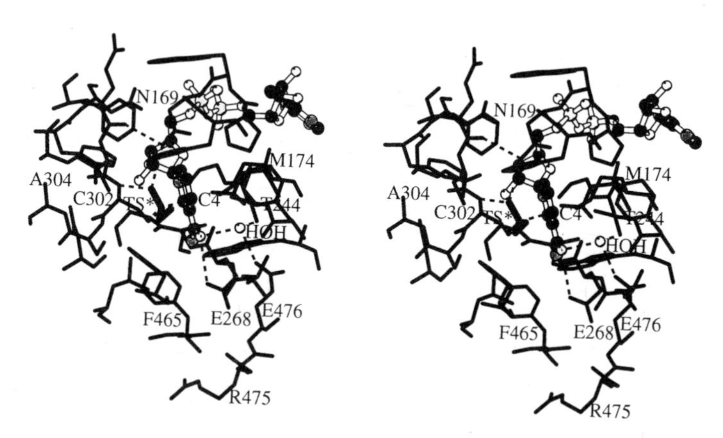

Figure 5. A stereo diagram of the active site in the Class 2 ALDH structure. Individual amino acids are labeled using the single amino acid code and their corresponding sequence number in the bovine enzyme. A model of a transition-state structure for the substrate propionaldehyde (TS*) is represented in the bold bonds and the "helper water" molecule is labelled as HOH.

ily. Secondly, the nature of the differences observed in the binding of NAD^+ to the Class 2 and Class 3 enzymes create two divergent views as to where the activation of the water molecule necessary for hydrolysis of the acyl-enzyme intermediate occurs. However, this reaction is not a conserved feature of the ALDH gene family. Thus, maybe it is not surprising that the residues that participate in the activation of a water molecule are not completely conserved. It is my belief (T.H.) that a careful alignment of ALDH family members based on the chemistry catalyzed by the enzymes would be useful in delineating the roles of these "non-conserved" residues.

We hypothesize that the activation of a water molecule occurs in the vicinity of Glu268, and Glu268 functions as the general base responsible for the activation. The role of Glu268 as a general base (and not Glu399) in the ALDH reaction is supported by mutagensis studies on the Class 2 enzyme (Wang *et al.*, 1995). The average distance between the carboxylate oxygen of Glu268 and the thiol of Cys302 in our tetrameric binary complex is 6.7 Å. We believe that the side chains of residues 244 and 476 aid in this reaction by helping to stablize the position of the hydrolytic water molecule through a second water molecule hydrogen bonded to their side chains (Figure 5). A brief, but by no means complete, survey of ALDH sequences suggests that a polar residue at both these positions is observed in those ALDH isoenzymes which hydrolyze the acyl-enzyme intermediate with a water molecule. This putative "helper water" molecule is similarly positioned between Thr186 and Tyr412 (Class 3 numbering) in the active site of the Class 3 enzyme structure. We propose that the hydrolytic water molecule would occupy a position roughly equivalent to the position occupied by the carbonyl oxygen atom of the carboxamide group on the nicotinamide ring of NAD^+ in our structure. In this position, the hydrolytic water would be 3.4 Å from the thiol of Cys302, 3.3 Å from the carboxylate oxygen of Glu268 (Figure 6) and 3.5 Å from the "helper" water. There is, in fact, a water molecule bound near Glu209 (equivalent to Glu268) in the "A" subunit of the Class 3 isoenzyme structure. An alignment of the Class 2 and Class 3 structures shows that the water molecule observed near Glu209 in the Class 3 structure aligns very close the the position observed for the carboxyamide oxygen of NAD^+ in the Class 2 structure.

We must stress, however, that all is not perfect with our proposal since this mechanism requires the movement of the nicotinamide ring, following hydride transfer, from its

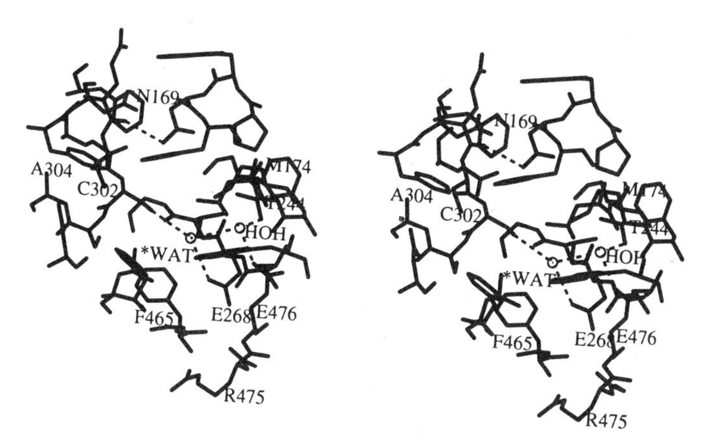

Figure 6. A stereo diagram of the hypothetical position for the "activated water" molecule that hydrolyzes the acyl-enzyme intermediate. This water is observed in three of the four subunits of the apo-enzyme structure, but was not included in the final model because only water molecules observed in all four subunits were included in the final coordinate set.

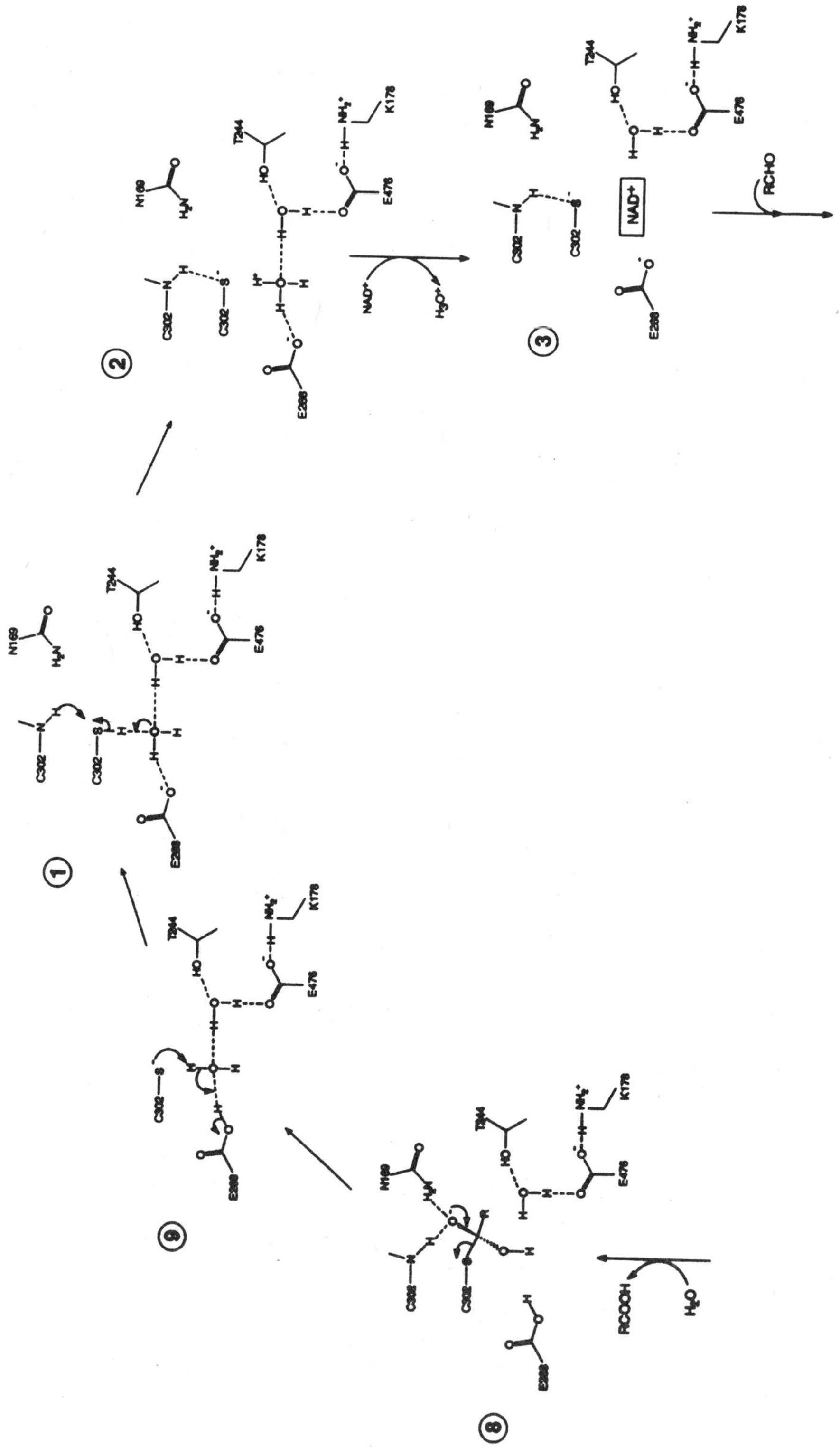
1
2
3
8
9
N169
T244
K178
E476
E268
C302
NAD+
NAD^+
H_3O^+
RCHO
RCOOH
H_2O

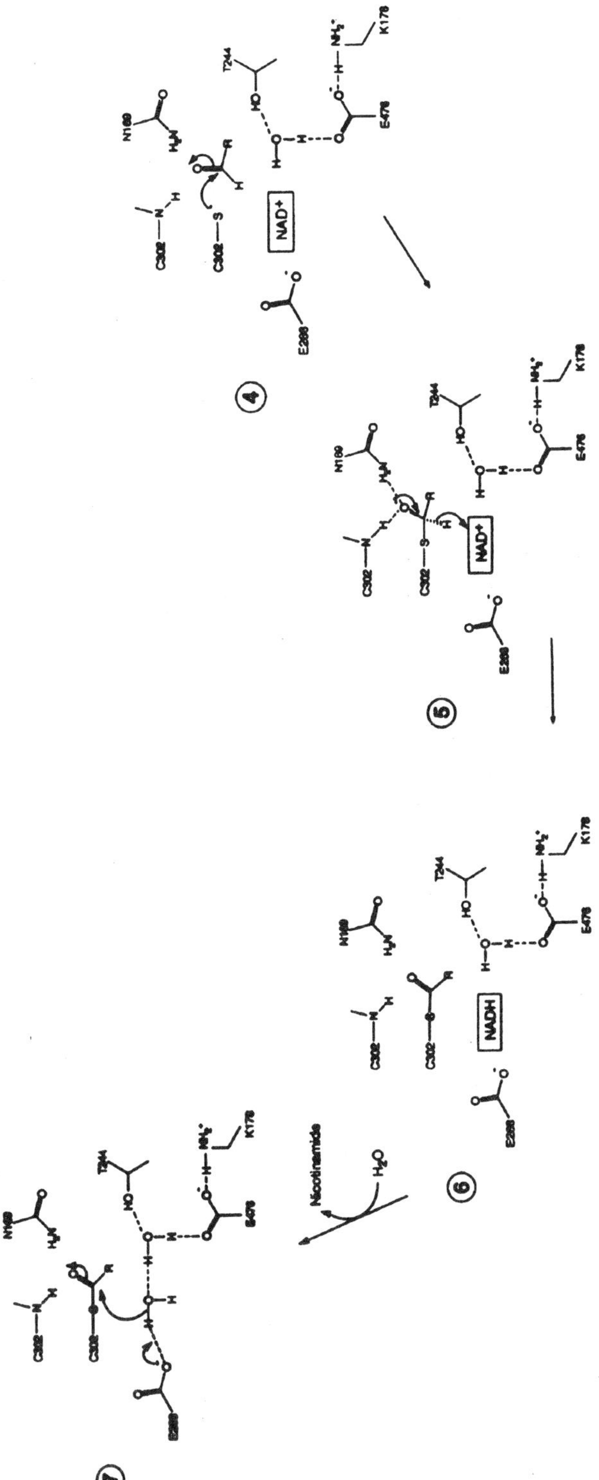

4
5
6
7
C302
N169
T244
E476
K178
E268
NAD+
NADH
Nicotinamide

observed position in our current structure to an alternate position. This movement is conjecture based on the fact that our structure clearly would not permit the activation of a water molecule due to steric considerations. There is ample precedent for alternate conformations of the cofactor as seen not only in the differences between the Class 2 and Class 3 structures, but also in the two distinct conformations observed for the cofactor in the Class 1 structure (Moore, *et al.* 1999). In addition to this structural evidence, the nature of the chemical changes that occur upon hydride transfer would also lend support to this idea. First, the nicotinamide ring becomes uncharged and significantly less polar upon its reduction to NADH. The immediate surroundings of the nicotinamide ring in the Class 2 structure are polar and somewhat negatively charged due to the proximity of Glu268 and Glu476. Thus, a change from a positively charged species to a neutral species and a changc in its polarity might force the reduced ring to seek more favorable surroundings. Secondly, the enzyme-bound substrate species collapses from a tetrahedral thio-hemiacetal transition state to a planar acyl-enzyme intermediate. This change in spatial arrangement of the enzyme-bound species may also help to drive the nicotinamide ring from its position near Cys302. Lastly, two different positions for oxidized and reduced cofactor would appear to explain why many mutations to the ALDH2 sequence change NAD^+ binding, but not NADH binding (Farrés, *et al.*, 1994 and Sheikh, *et al.*, 1997).

A schematic representation of our interpretation of the ALDH2 catalytic mechanism is shown on pages 22 and 23. This mechanism would be applicable to those ALDH enzymes that hydrolyze the acyl-enzyme intermediate with an activated water molecule.

ACKNOWLEDGMENTS

The authors wish to acknowledge support from the following NIH grants: R29-AA10399, P50-AA07611, R37-AA02342.

REFERENCES

Farres, J., Wang, X.-P., Takahashi, K., Cunningham, S.J., Wang, T.T.Y., and Weiner, H. (1994) Effect of changing glutamate to lysine in rat and human liver mitochondrial aldehyde dehydrogenase: a model to study human (Oriental type) class 2 aldehyde dehydrogenase. *J. Biol. Chem.*, 269:13854–13860.

Farres, J., Wang, T.T.Y., Cunningham, S.J., and Weiner, H. (1995) Investigation of the active site cysteine residue of rat liver aldehyde dehydrogenase as probed by site-directed mutagenesis. *Biochemistry*, 34:2592–2598.

Hempel, J. and Pietruszko, R. (1981) Selective chemical modification of human liver aldehyde dehydrogenases E1 and E2 with iodoacetamide. J. Biol. Chem., 256:10889–10896.

Hempel, J., Peitruszko, R., Fietek, P., and Jornvall, H. (1982) Identification of a segment containing a reactive cysteine residue in human liver cytoplasmic aldehyde dehydrogenase (isoenzyme E1). *Biochemistry*, 21:6834–6838.

Hempel, J., Nicholas, H., and Lindahl, R. (1993) Aldehyde dehydrogenases: a widespread structural and functional diversity within a shared framework. *Protein Sci.*, 2:1890–1900.

Hempel, J., Perozich, J., Chapman, T., Rose, J., Boesch, J.S., Liu, Z.J., Lindahl, R., and Wang, B.C. (1999) Aldehyde dehydrogenase catalytic mechanism: a proposal. Adv. Exp. Med. Biol. (this volume).

Liu, Z.J., Hempel, J., Sun, J., Rose, J., Hsiao, C.D., Chang, W.R., Chung, Y.J., Kuo, I., Lindahl, R., and Wang, B.C. (1997a) Crystal Structure of a class 3 aldehyde dehydrogenase at 2.6 Å resolution. *Adv. Exp. Med. Biol.*, 414:1–7.

Liu, Z.J., Sun, Y.J., Rose, J., Chung, Y.J., Hsiao, C.D., Chang, W.R., Kuo, I., Perozich, J., Lindahl, R., Hempel, J., and Wang, B.C. (1997b) The first structure of an aldehyde dehydrogenase reveals novel interactions between NAD and the Rossmann fold. *Nature Str. Biol.*, 4:317–326.

Moore, S.A., Baker, H.M., Blythe, T.J., Kitson, K.E., Kitson, T.E., and Baker, E.N. (1999) The structure of sheep liver cytosolic aldehyde dehydrogenase reveals the basis for the retinaldehyde specificity of class 1 ALDH enzymes. Adv. Exp. Med. Biol., (this volume).

Perozich, J., Nicholas, H., Wang, B.C., Lindahl, R., and Hempel, J. (1999) The aldehyde dehydrogenase extended family: an overview. Adv. Exp. Med. Biol. (this volume).

Racker, E. (1955) Actions and properties of pyridine-nucleotide linked enzymes. *Phys. Rev.* 35:1–56.

Sheikh, S., Ni., L., Hurley, T.D., and Weiner, H. (1997) The potential roles of the conserved amino acids in human liver aldehyde dehydrogenase. *J. Biol. Chem.*, 272:18817–18822.

Steinmetz, C.S., Weiner, H., and Hurley, T.D. (1997) Structure of mitochondrial aldehyde dehydrogenase: the genetic component of ethanol aversion. *Structure*, 5:701–711.

Stuckey, J.A., Schubert, H.L., Fauman, E.B., Zhang, Z.Y., Dixon, J.E., and Saper, M.A. (1994) Crystal Structure of Yersinia protein tyrosine phosphatase at 2.5 Å and the complex with tungstate. *Nature*, 370:571–575.

Wang, X.-P., and Weiner, H. (1995) Involvement of glutamate 268 in the active site of human liver mitochondrial aldehyde dehydrogenase as probed by site-directed mutagenesis. *Biochemistry*, 34:237–243.

4

A STRUCTURAL EXPLANATION FOR THE RETINAL SPECIFICITY OF CLASS 1 ALDH ENZYMES*

Stanley A. Moore,[1] Heather M. Baker,[1†] Treena J. Blythe,[1]
Kathryn E. Kitson,[3] Trevor M. Kitson,[2] and Edward N. Baker[1†]

[1]Institute of Molecular Biosciences
[2]Institute of Fundamental Sciences
[3]Institute of Food, Nutrition and Human Health
Massey University
Private Bag 11-222
Palmerston North, New Zealand

1. INTRODUCTION

The retinoic acid signalling pathway in vertebrates utilises the RXR/RAR family of ligand dependent transcription factors which bind 9-*cis* (RXR,RAR) or all-*trans* (RAR) retinoic acid via a ligand binding domain, and direct the transcription of target genes via a DNA binding domain (Kastner *et al.*, 1995; Manglsdorf *et al.*, 1995). Retinoic acid is derived from vitamin A (retinol) and its pleiotropic effects include spinal chord and retina development during embryogenesis, neuronal cell differentiation, and maintenance of epithelial cell type in adult tissues (Hofmann and Eichele, 1994). Although the retinoic acid signalling pathway is reasonably well understood, it is less clear how spatio-temporal gradients of retinoic acid are maintained during developmental processes, and what are the major enzymes responsible for retinoic acid synthesis in various tissues (Duester, 1996). Recently, the cloning and characterisation of several aldehyde dehydrogenases (ALDH's) showing a high specificity for retinaldehyde as substrate has been reported. Each of these enzymes has been shown to belong to the broad family of cytosolic or Class 1 ALDH enzymes (Duester, 1996; Blythe, 1997). Studies on mouse embryos colocalise retinaldehyde-

---

* A more detailed account of the sheep ALDH1 strucuture and its implications for ALDH enzyme mechanism and substrate specificity has been submitted for publication.

† Present Address: School of Biological Sciences, University of Auckland, Auckland, New Zealand.

Enzymology and Molecular Biology of Carbonyl Metabolism 7
edited by Weiner *et al.* Kluwer Academic / Plenum Publishers, New York, 1999.

specific ALDH activity with high concentations of retinoic acid in the developing spinal chord (McCaffery and Drager, 1994), and retinaldehyde dehydrogenase is a positional marker in the mouse embryonic retina (McCaffery *et al.*, 1991). Retinal-specific ALDH's can be divided into three groups based on their amino acid sequences and pI values (Blythe, 1997). The archetypal Class I ALDH is found predominantly in the liver of higher vertebrates including horse, beef, sheep, and man, and has a pI of 5.2. Two retinal-specific variants of ALDH1 have been characterised in mouse and rat. One (RalDH-1) is about 90% identical at the amino acid sequence level to the archetypal Class 1 enzymes just mentioned, but has a pI of about 8.3 (McCaffery *et al.*, 1991; Rongnoparut and Weaver, 1991; Bhat *et al.*, 1995; Penzes *et al.*, 1997). The other (RalDH-2) is only about 70% identical to the ALDH1 enzymes from other mammals, and has a pI of 5.1 (Zhao *et al.*, 1997; Wang *et al.*, 1996).

ALDHs are found in most mammalian tissues. In humans alone, eight distinct ALDH alleles have been characterised (Yoshida *et al.*, 1991). The liver has high activity, and in this tissue, there are two principal isozymes, ALDH1 and ALDH2. ALDH1 is cytosolic and ALDH2 is produced with a leader peptide sequence and targeted to the mitochondrial matrix. Human ALDH1 has been shown to have a sub-micromolar K_m for both all-*trans* retinal (Yoshida *et al.*, 1992) and 9-*cis* retinal (Blythe, 1997). Kinetic studies on sheep ALDH1 indicate that it has high specific activity for both all-*trans* and 9-*cis* retinal with K_m value of 0.14 μM for both substrates (Blythe, 1987). Acetaldehyde is a much poorer substrate for human ALDH1, with a K_m of 180 μM (Klyosov *et al.*, 1996). In contrast, kinetic data suggests that acetadehyde is the preferred substrate of human ALDH2, the K_m value for that substrate being 0.2 μM (Klyosov *et al.*, 1996). The presence of an alcohol-sensitivity phenotype mapping to the human ALDH2 locus (Yoshida *et al.*, 1984) strongly indicates that the major biological function of ALDH2 is to metabolise ethanol-derived acetaldehyde. Human ALDH2 also has essentially no activity for all-*trans* retinaldehyde.(Yoshida *et al.*, 1992), underscoring the unique substrate specificities of these two enzymes.

Both ALDH1 and ALDH2 are homotetramers composed of 500 or 501 amino acid subunits, and they require NAD^+ as a cofactor (Yoshida *et al.*, 1991). The aligned ALDH1 and ALDH2 amino acid sequences are roughly 69% identical, suggesting a high similarity of their three dimensional structures. The inducible ALDH3 family of enzymes is much more distantly related to ALDH1 and ALDH2, being only 30% identical at the amino acid level, and lacking an N-terminal segment present in ALDH1 and ALDH2. Whereas the latter are homotetramers, ALDH3 is dimeric (Liu *et al.*, 1997).

2. MATERIALS AND METHODS

2.1. Isolation, Purification, and Crystallisation of Sheep Liver ALDH1

Sheep liver cytosolic aldehyde dehydrogenase was isolated and purified as described previously (Kitson and Kitson, 1994). Active fractions were identified by enzyme assay using acetaldehyde as substrate and monitoring the appearance of NADH spectrophotometrically at 340 nm. The final fractions were assayed for activity and checked for purity by SDS gel electrophoresis and isoelectric focusing. The purest fractions were then pooled, dialysed against 50 mM Na/K phosphate, pH 7.4, 1 mM DTT, and concentrated to 10 mg/mL protein using microconcentrators. Crystals were grown using the hanging drop vapour diffusion technique as described previously (Baker *et al.*, 1994). The protein concentration was 10 mg/mL with 1 mM NAD^+, and the reservoir solution contained 170 mM $MgCl_2$, 100 mM bis-tris-propane pH 6–7, and 6–7% w/v monomethylether PEG-5000. Drops were typically 4 to 6 μL and contained equal volumes of the protein solution and the precipitant. The best crystals were triclinic (a=80.69, b=87.62, c=90.84 Å, α=110.5, β=105.6, γ=103.6°) and contained a tetramer in the P1 unit cell. Crystals of the triclinic cell were flash frozen at 113 K, using 30% v/v glycerol as a cryoprotectant, and diffraction data were recorded using a Rigaku RaxisIIC imaging plate detector, with 0.1mm collimated Cu-K_α radiation (λ=1.5418 Å) from a Rigaku RU-200B rotating anode generator operating at 50kV, 150mA. All data were indexed, integrated, and postrefined with DENZO/SCALEPACK (Otwinowski and Minor, 1997). Profile-fitted intensities were converted to amplitudes using the CCP4 program Truncate (CCP4, 1994; French and Wilson, 1978).

2.2. Structure Determination and Refinement

We solved the structure of ALDH1 initially in an orthorhombic cell (results not shown), using the bovine mitochondrial ALDH2 atomic coordinates as a search model (kindly provided by Dr. Thomas Hurley). Structure solution and refinement using the orthorhombic data were straightforward and we switched to the better-quality triclinic data when the Free-R was about 31% in the orthorhombic cell. The rotation function solution using the partially refined orthorhombic tetramer was straightforward in AMoRe (Navaza and Saludjian, 1997) with a peak height of 63 (rms=2.0). The next highest peak was 10.0. Least squares refinement in TNT (Tronrud *et al.*, 1987; Tronrud, 1992) using a maximum likelihood target (Pannu and Read, 1996), followed by model building, and eventually refinement in CNS (Adams *et al.*, 1997) resulted in a Free R-factor of 25.9% for a randomly chosen 10% subset of the data. The quality of the electron density maps was excellent and a model for the complete polypeptide, using the published sheep cytosolic ALDH amino acid sequence (Staynor and Tweedie, 1995) excluding residues -1 to 7 was built. A total of 140 water molecules, corresponding to 35 unique and identical waters per ALDH1 monomer, were added to the structure. During positional refinement of the model, strict NCS restraints were maintained, giving a final rms difference between monomers of 0.009 Å. Although at least two conformations of the nicotinamide half of the NAD^+ were observable in difference and omit electron density maps, only the highest occupancy NAD^+ conformer was refined. Analysis of the final coordinates with PROCHECK (Laskowski *et al.*, 1993) revealed excellent stereochemistry, with 89% of residues in the most favoured regions, and 10.2% in additional allowed regions.

3. RESULTS AND DISCUSSION

3.1. General Features of the Sheep ALDH1 Structure

A superposition of the refined sheep liver ALDH1 model with the bovine ALDH2 atomic coordinates yields a root mean square (RMS) difference in main chain atoms (N, CA, C, O) of 0.74 Å for the entire polypeptide chain. Omitting sixteen outlying residues from the superposition increases the structural agreement markedly, the RMS pairwise difference drops to 0.58 Å. For ease of describing the sheep ALDH1 structure, we will use the standard residue numbering for the tetrameric aldehyde dehydrogenases (e.g. Asn169, Glu268, Cys302) and adopt the secondary structure assignments of Hurley and colleagues (Steinmetz *et al.*, 1997). This results in the N-terminal residue of sheep ALDH1 starting at position -1.

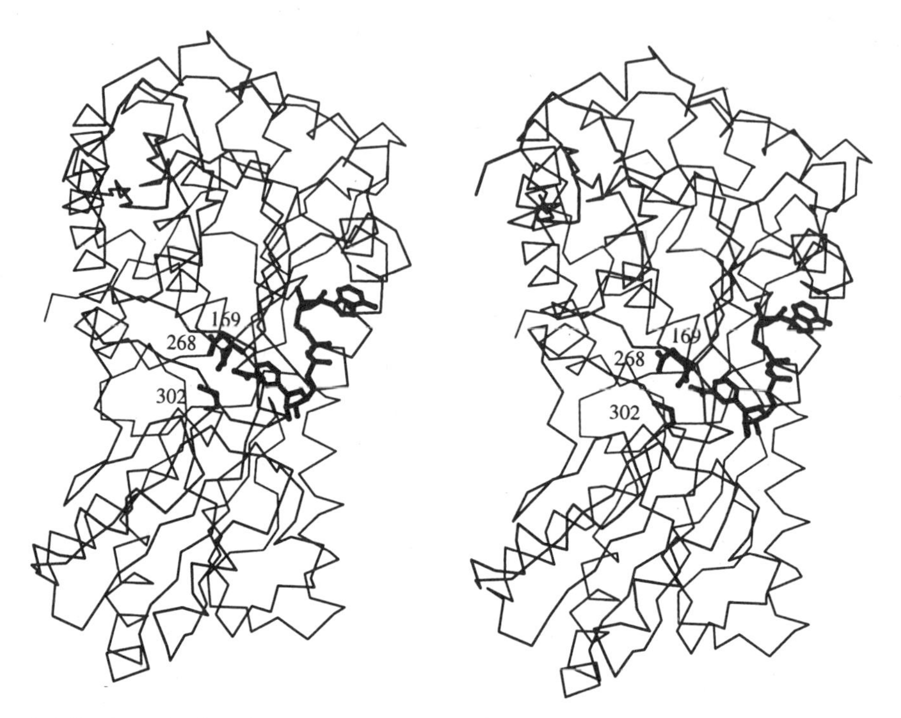

Figure 1. A Cα plot of the sheep ALDH1 monomer and NAD^+ (major conformation). The side chains for Asn169, Glu268, and Cys302 are drawn in bold along with the NAD cofactor.

As in the bovine ALDH2 structure, ALDH1 is made up of three domains: an N-terminal NAD-binding domain (residues 8–135 and 159–270) that is novel to the ALDH family and contains a parallel five-stranded β sheet (Liu *et al.*, 1997), a catalytic domain spanning residues 271–470 that contains a parallel six-stranded β sheet, and an oligomerisation domain made up of a three-stranded antiparallel β sheet (residues 140–158 and 486–495) (Figure 1).

The N-terminal NAD-binding domains of ALDH enzymes exhibit a unique mode of cofactor binding that appears to be distantly related to the canonical NAD binding domain found in most NAD(P) utilising dehydrogenases, and has been discussed in detail (Liu *et al.*, 1997; Steinmetz *et al.*, 1997). Briefly, the Rossmann fold of the N-terminal domain lacks the last αβ unit, and hence contains only five strands in the β sheet. In addition the first α helix of the fold (αD) does not participate in nucleotide binding, and is covered over by the N-terminal three-helix bundle (helices αA to αC). The pyrophosphate moiety of the NAD cofactor is instead bound by residues just preceding helix αG. The sheep ALDH1 structure expands the generality of the unique ALDH NAD^+ binding motif, as the general mode of NAD^+ binding observed first in ALDH3 (Liu *et al.*, 1997) and then ALDH2 (Steinmetz *et al.*, 1997) is conserved for the cofactor in our structure.

3.2. The Substrate Access Tunnel

The most striking feature of the sheep ALDH1 three-dimensional structure is the large scale change in the entrance tunnel for the substrate relative to ALDH2. The entrance to this tunnel is made up of two helices and a surface loop (Figure 2). A portion of

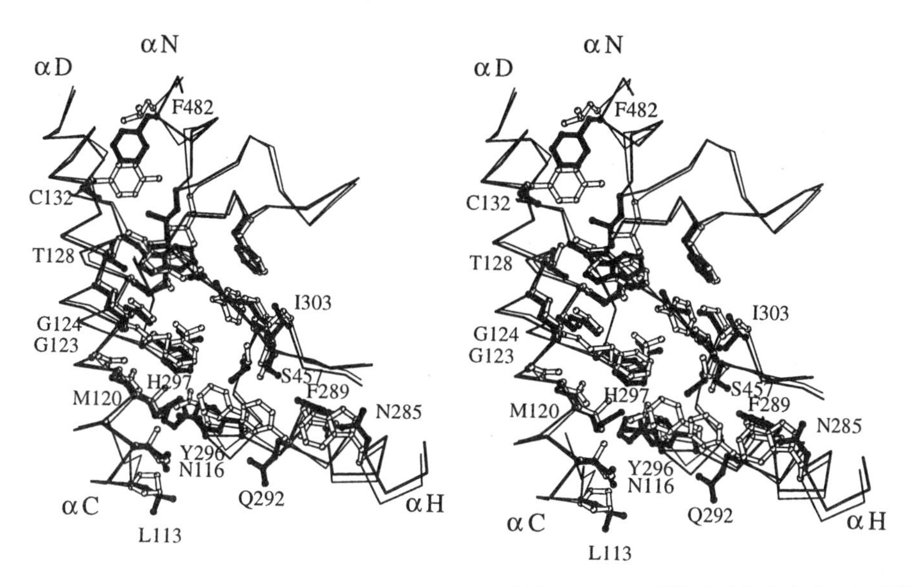

Figure 2. A stereo overlay of the active site tunnel in ALDH1 (thick Cα trace and black side chains) and ALDH2 (thin Cα trace and white side chains) showing the major differences in side chains between the two enzymes. Residues that differ in the two structures are labeled for ALDH1.

the oligomerisation domain of another monomer is also nearby to the surface loop containing residues 455–461. Helix αC (residues 114–135) makes up the left hand side of the opening and is part of a three-helix bundle near the beginning of the N-terminal domain. The back of the tunnel is made up by helix αD (residues 170–185), which incidentally packs against helix αC. Helix αD is also the first α helix of the canonical Rossman fold of the N-terminal domain. Helix αH (residues 282 to 296) comes from the catalytic domain, immediately preceding the active site nucleophile, Cys302, and makes up the bottom of the substrate entrance tunnel as viewed in Figure 2. The right hand side of the tunnel is made up of a surface loop (residues 455–461) that precedes helix αN near the oligomerisation domain (Figure 2), and this loop rests against helix αH. The base of the tunnel is composed of two β strands containing Thr244 and Glu268 respectively.

There are multiple amino acid substitutions between the ALDH1 and ALDH2 enzyme families that map to these portions of the enzyme structure, facing into the substrate access tunnel (Figure 2). These amino acid substitutions are found in almost all of the ALDH1 (retinal metabolising) and ALDH2 (acetaldehyde metabolising) amino acid sequences. For helix αC, these substitutions, going from bovine ALDH2 to sheep ALDH1 respectively, are Pro113 to Leu, Ile116 to Asn, Val120 to Met, Asp123 to Gly, Met124 to Gly, Cys128 to Thr, and Tyr132 to Cys. Similarly, for helix αH, the observed substitutions are Trp285 to Asn, Gln289 to Phe, Phe292 to Gln, Phe296 to Tyr, and Asn297 to His. The substitutions on the surface loop preceeding the oligomerisation interface are, Asp457 to Ser, Phe459 to Val, Gly460 to Ser, and Leu482 to Phe. In nine of these sixteen substitutions, the ALDH2 side chains are more bulkier than their ALDH1 counterparts, and they almost exclusively adopt conformations that result in a more "constricted" entrance tunnel than in ALDH1. Several amino acid side chains that are identical in the ALDH1 and ALDH2 sequences again adopt conformations in ALDH2 that make the entrance tunnel smaller than in ALDH1 (Figure 2). Included in this group are Lys112, Met174, Trp177,

Cys301, and Leu477. In addition to amino acid substitutions, small rigid body movements of secondary structure elements contribute to the opening up of the substrate tunnel in ALDH1 relative to ALDH2. These factors combine such that for ALDH1, the entrance tunnel is much larger than for bovine ALDH2, thereby allowing access to larger, more bulky substrates (Figures 2 and 3). These observations are in accord with all-*trans* retinal being the a very good substrate for ALDH1 and being a very poor substrate for ALDH2 (Yoshida *et al.*, 1992). Modeling all-*trans*-retinal into both active site tunnels illustrates that the ALDH1 tunnel can easily accommodate all-*trans*-retinal, but the tunnel in ALDH2 is too small to accommodate such a bulky substrate, and energetically-unfavourable steric clashes result (Figure 3).

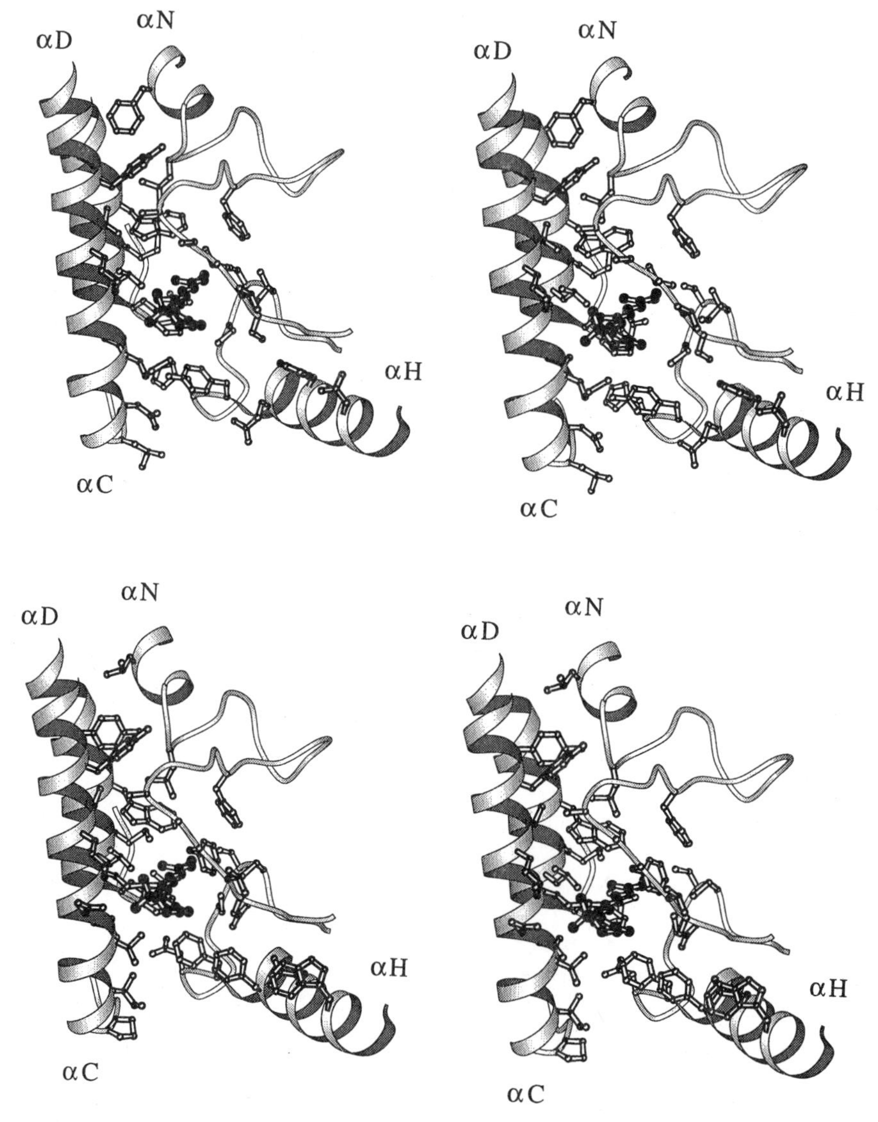

Figure 3. A) Docking of all-*trans* retinal in the ALDH1 active site entrance. ALDH1 is shown similar to Figure 2 while retinal is depicted as thicker lines. B) As in A, but with ALDH2. Not all side chains are shown for clarity.

The size of the substrate entrance tunnel also readily explains the differential reactivities of ALDH1 and ALDH2 towards disulfiram. Disulfiram and its *in vivo* metabolites are bulky, hydrophobic compounds, that react quickly with ALDH1 and more slowly with ALDH2 (Vallari and Pietruszko, 1982; Kitson, 1983; Lam *et al.*, 1997). This fits with earlier suggestions that it is their inherent bulkiness and size that predisposes all-*trans*-retinal, disulfiram, and other compounds to preferential reaction with ALDH1 versus ALDH2 (Kitson and Kitson, 1996).

In conclusion, we propose that helices αC and αH, and the loop preceding helix αN are major determinants of substrate specificity in Class 1 and Class 2 ALDH enzymes, as nearly all of the amino acid substitutions listed above are generally obeyed in the ALDH1 and ALDH2 families, consistent with their being determinants for retinal or acetaldehyde specificity respectively.

3.3. Evidence for Two Distinct Bound Conformations of NAD^+

During the course of our refinement of the NAD^+ cofactor, it became apparent that there was evidence of discrete disorder in the nicotinamide half of the NAD^+. After modeling of the major NAD^+ conformer (seen in Figure 4), a difference electron density peak was large enough at the 5 σ level in $|F|_{obs} - |F|_{calc}$ difference electron density maps to accommodate a phosphate anion, but modeling a phosphate anion at this position made bad steric clashes with the position of our modeled nicotinamide ring (Figure 4). The large difference peak was persistent, and once we had superimposed the ALDH2 NAD^+ coordinates (kindly provided by Dr. Thomas Hurley) onto our sheep ALDH1 model, the ALDH2 nicotinamide phosphate was coincident with our large difference electron density peak. Hence we concluded that the large difference peak must be due to a second, more weakly occupied conformation of the nicotinamide phosphate. Indeed, there is weaker, discontinuous $|F|_{obs} - |F|_{calc}$ difference electron density at the 3 σ level corresponding to the ribose and the nicotinamide of the "second" conformation that almost exactly overlaps with the atomic coordinates for the nicotinamide half of the ALDH2 cofactor (Figure 4). We conclude that the sheep ALDH1 structure contains two major conformations of NAD^+ differing only in the conformation of the nicotinamide half of the cofactor (including the phosphate, ribose and nicotinamide), the adenine portion being superimposable. The less occupied of the two conformations corresponds to the NAD^+ conformation observed in the bovine ALDH2 structure (Steinmetz *et al.*, 1997). From our electron density maps, we conclude that the major conformer is likely at 75–80% occupancy, and the minor conformer is at roughly 20–25% occupancy.

The reorientation of the major NAD^+ conformer observed in our structure results in a movement of over 5 Å for the nicotinamide ribose relative to bovine ALDH2. Interactions between the nicotinamide ribose and the protein include the side chains of residues Glu399, Gln349, and Lys352, as in ALDH2, but are strikingly different in details of hydrogen bonds and van der Waals contacts (Figure 4). In addition, the side chain of Phe401 makes different contacts with the nicotinamide ribose ring in the ALDH1 structure. Hence the same amino acids participate in nicotinamide binding in both ALDH1 and ALDH2, but with distinctly different details of interaction. The end result is that in ALDH1, the C4 atom of the nicotinamide is too far from Cys302 SG (d= 7.1 Å in ALDH1 vs d=3.7 Å in ALDH2) for direct hydride transfer from a thiohemiacetal intermediate to occur. Hence we are likely looking at a catalytically non-productive mode of NAD^+ binding for the major NAD^+ conformation observed in sheep ALDH1.

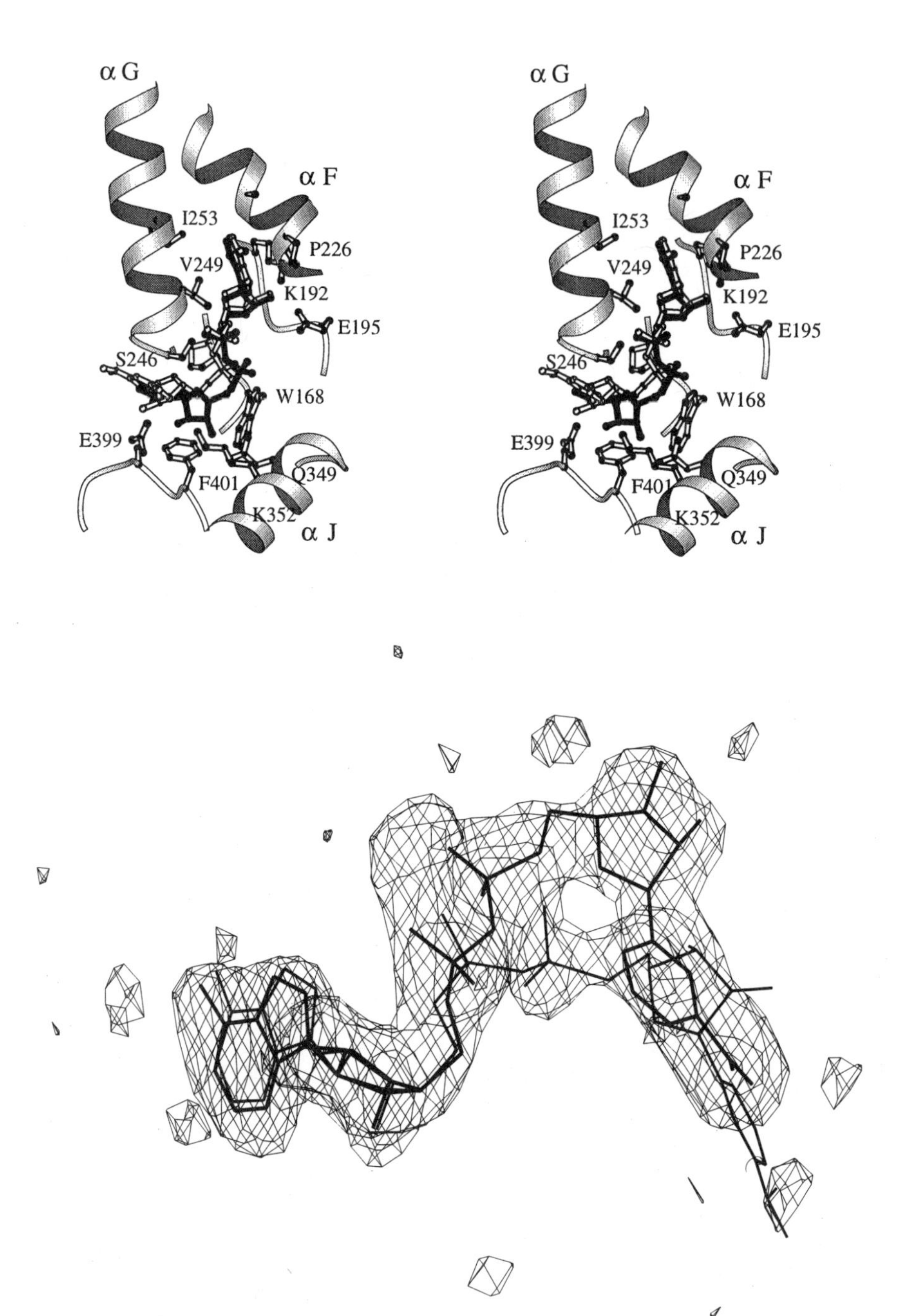

Figure 4. A) A stereo view of the major NAD$^+$ binding mode in ALDH1 (with black bonds) along with the overlaid ALDH2 NAD$^+$ coordinates (white bonds). Key residues of the protein that interact with the cofactor are labeled. B) An omit map of the NAD$^+$ binding region in ALDH1. The major NAD$^+$ conformer is in thick, the minor conformer in thin lines. The electron density is contoured at the 3 σ level.

In general, NAD^+ binding appears to be both flexible and highly variable for the nicotinamide half of the cofactor in the aldehyde dehydrogenases. ALDH1, ALDH2 and ALDH3 all exhibit different binding modes for the nicotinamide half of the cofactor, although the adenine moiety interacts with the protein almost identically in all three enzyme structures determined to date. This suggests that movement of the nicotinamide ring in and out of the active site is a conserved and important component of the functional enzyme.

3.4. The Side Chain of the General Base Glu268 Is Disordered

Mutagenesis experiments with human liver ALDH2 have shown that Glu268 acts as a general base in both thiohemiacetal formation and thioester hydrolysis (Wang and Weiner, 1995; Abriola *et al.*, 1987). In bovine ALDH2, the active site general base, Glu268, is buried from solvent in the active site, and too far from Cys302 to directly act as a base (Steinmetz *et al.*, 1997). Although our Glu268 side chain is modeled as the same rotamer as is observed in bovine ALDH2, it exhibits unexpectedly weak electron density and high B-factors for a residue buried from solvent (Figure 5) (average B-factor 38 $Å^2$ compared to neigbouring amino acids with average B-factors about 25 $Å^2$). There is no 2Fo-Fc or

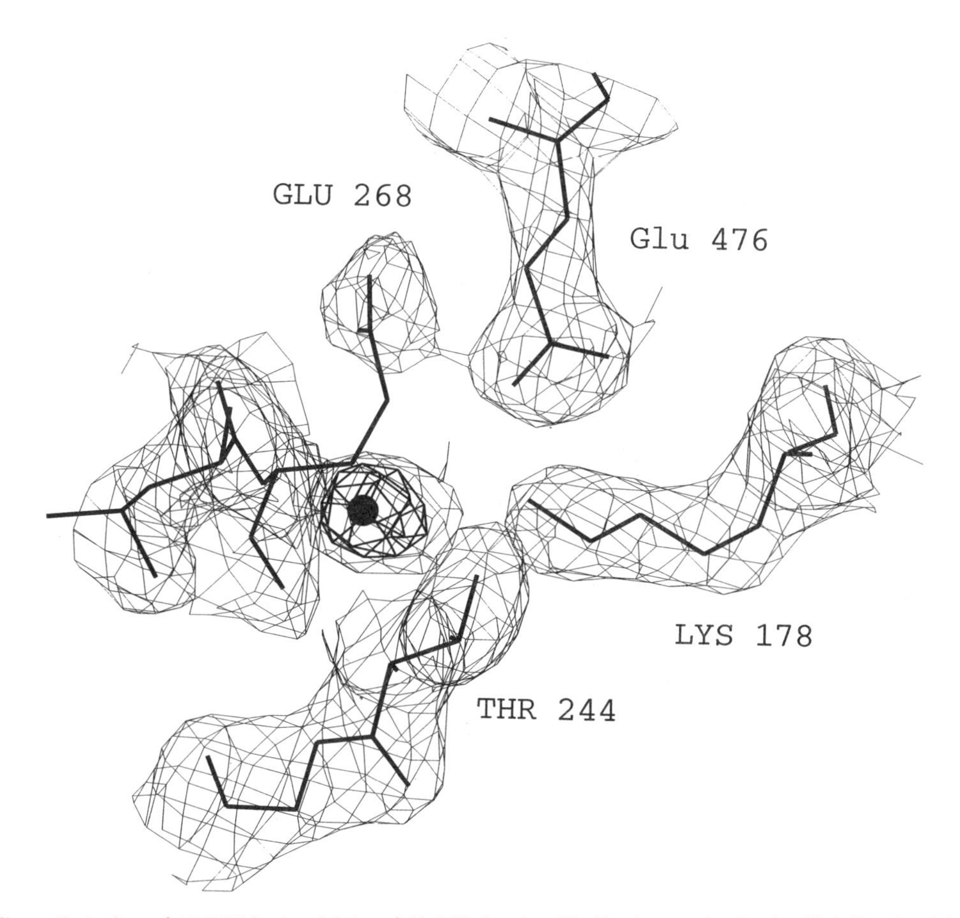

Figure 5. A view of ALDH1 in the vicinity of Glu268 showing 2Fo-Fc electron density for Glu268 and neighbouring residues. Density attributable to a candidate deacylating water molecule is also visible.

difference electron density for the CA-CB bond if the atomic coordinates of the Glu268 side chain are included or omitted during least squares refinement. Glu268 is likely hydrogen bonded to the nicotinamide ring amide nitrogen of our low occupancy nicotinamide conformer, as observed in the ALDH2 structure. However, there are no chemically reasonable hydrogen bonds observed between the Glu268 side chain and the rest of the protein molecule, suggesting at least one reason for the observed structural disorder.

Our observation that Glu268 exhibits disorder in ALDH1, and taking into consideration chemical requirements for both the dehydrogenase and deacylation steps, leads us to suggest that movement of the NAD cofactor and Glu268 are coupled in the ALDH reaction mechanism. Specifically, during hydride transfer from the Cys302 thiohemiacetal to the NAD^+ nicotinamide, the side chain of Glu268 must be "tucked away" from the nicotinamide ring of the cofactor, as it is primarily observed to be in all ALDH crystal structures. Before deacylation of the thioester can take place, the reduced nicotinamide ring must at least partially exit the active site pocket to give room for the general base, Glu268, to abstract a proton from a water molecule that will then deacylate the thioester. Hence we believe that the side chain of Glu268 has two conformations, a passive conformer that "stays out of the way" to allow hydride transfer to the nicotinamide, and an active conformer that participates in proton abstraction from a thioester-deacylating water molecule.

ACKNOWLEDGMENTS

This research was supported by the Health Research Council of New Zealand, and by the Howard Hughes Medical Institute through the award of an International Research Scholarship to ENB. We also gratefully acknowledge Dr. Thomas Hurley and colleagues for providing the atomic coordinates of bovine ALDH2 prior to their release from the PDB, and Arijit Dasgupt who grew several of the triclinic ALDH1 crystals used in this study.

REFERENCES

Abriola, D. P., Fields, R., Stein, S., MacKerell, A. D. and Pietruszko, R. (1987). Active site of human liver aldehyde dehydrogenase. Biochemistry **26**, 5679–5684.

Adams, P. D., Pannu, N. S., Read, R. J. and Brünger, A. T. (1997). Cross-validated maximum likelihood enhances crystallographic simulated annealing refinement. Proc. Natl. Acad. Sci. **94**, 5018–5023.

Baker, H. M., *et al.* and Baker, E. N. (1994). Crystallization and preliminary X-ray diffraction studies on cytosolic (Class 1) aldehyde dehydrogenase from sheep liver. J. Mol. Biol. **241**, 263–264.

Bhat, P. V., Labrecque, J., Boutin, J.-M., Lacroix, A. and Yoshida, A. (1995). Cloning of a cDNA encoding rat aldehyde dehydrogenase with high activity for retinal oxidation. Gene **166**, 303–306.

Blythe, T. J. (1997). Studies on proteins involved in retinoid and alcohol metabolism. PhD Thesis, Massey University, Palmerston North, New Zealand.

Collaborative Computational Project Number 4 (1994) The CCP4 suite: Programs for protein crystallography. *Acta Cryst.* **D50**, 760–763.

Duester, G. (1996). Involvement of alcohol dehydrogenase, short-chain dehydrogenase/reductase, aldehyde dehydrogenase, and cytochrome P450 in the control of retinoid signaling by activation of retinoic acid synthesis. Biochemistry **35**, 12221–12227.

Farres, J., Wang, T. T. Y., Cunningham, S. J. and Weiner, H. (1995). Investigation of the active site cysteine residue of rat liver mitochondrial aldehyde dehydrogenase by site-directed mutagenesis. Biochemistry **34**, 2592–2598.

French, S. and Wilson, K. (1978). On the treatment of negative intensity observations. *Acta Cryst.* **A34**, 517–525.

Hofmann, C. and Eichele, G. (1994). Retinoids in Development. In *The Retinoids:Biology, Chemistry and Medicine, 2nd Ed.* pp 387–441. Eds, Sporn, M. B., Roberts, A. B. & Goodman, D. S. . Raven Press, New York.

Kastner, P., Mark, M. and Chambon, P. (1995). Nonsteroid nuclear receptors: what are genetic studies telling us about their role in real life? Cell **83**, 859–869.

Kitson, T. M. (1983). Mechanism of inactivation of sheep liver cytoplasmic aldehyde dehydrogenase by disulfiram. Biochem. J. **213**, 551–554.

Kitson, T. M. and Kitson, K. E. (1994). Probing the active site of cytoplasmic aldehyde dehydrogenase with a chromophoric reporter group. Biochem. J. **300,** 25–30.

Kitson, T. M. and Kitson, K. E. (1996). The action of cytosolic aldehyde dehydrogenase on resorufin acetate. In Enzymology and Molecular Biology of Carbonyl Metabolism, **6**, 201–208, Weiner, H. *et al.*, Ed., Plenum Press, New York.

Klyosov, A. A., Rashkovetsky, L. G., Tahir, M. K. and Keung, W.-M. (1996). Possible role of liver cytosolic and mitochondrial aldehyde dehydrogenases in acetaldehyde metabolism. Biochemistry **35**, 4445–4456.

Lam, J. P., Mays, D. C. and Lipsky, J. J. (1997). Inhibition of recombinant human mitochondrial and cytosolic aldehyde dehydrogenases by two candidates for the active metabolites of disulfiram. Biochemistry **36**, 13748–13754.

Laskowski, R. A., MacArthur, M. W., Moss, D. S. and Thornton, J. M. (1993). PROCHECK: a program to check the stereochemical quality of protein structures. J. Appl. Cryst. **26**, 283–291.

Liu, Z.-J., *et al.*, and Wang, B.-C. (1997). The first structure of an aldehyde dehydrogenase reveals novel interactions between NAD and the Rossmann fold. Nature Struct. Biol. **4**, 317–326.

Mangelsdorf, D. J. *et al.* and Evans, R. D. (1995). The nuclear receptor superfamily: the second decade. Cell **83**, 835–839.

McCaffery, P., Tempst, P., Lara, G. and Drager, U. C. (1991). Aldehyde dehydrogenase is a positional marker in the retina. Development **112**, 693–702.

McCaffery, P. and Drager, U. C. (1994). Hot spots of retinoic acid synthesis in the developing spinal chord. Proc. Natl. Acad. Sci. **91**,7194–7197.

Navaza, J. and Saludjian, P. (1997). AMoRe: An automated molecular replacement program package. Meth. Enzymol. **276**, 581–594.

Otwinowski, Z. and Minor, W. (1997). Processing of X-ray diffraction data collected in oscillation mode.*Meth. Enzymol.* **276**, 307–326.

Pannu, N. S. and Read, R. J. (1996). Improved structure refinement through maximum liklihood. *Acta. Cryst.* **A52**, 659–668.

Penzes, P., Wang, X., Sperkova, Z. and Napoli, J. L. (1997). Cloning of a rat cDNA encoding retinal dehydrogenase isozyme type I and its expression in *E. Coli*. Gene **191**, 167–172.

Read, R. J. (1986). Improved Fourier coefficients for maps using phases from partial structures with errors. *Acta Cryst.* **A42**, 140–149.

Rongnoparut, P. and Weaver, S. (1991). Isolation and characterization of a cytosolic aldehyde dehydrogenase-encoding cDNA from the mouse liver. Gene **101**, 261–265.

Sheikh, S., Ni, L., Hurley, T. D. and Weiner, H. (1997). The potential roles of the conserved amino acids in human liver mitochondrial aldehyde dehydrogenase. J. Biol. Chem. **272**, 18817–18822.

Staynor, C. K and Tweedie, J. W. (1995). Cloning and characterization of the cDNA for sheep liver cytosolic aldehyde dehydrogenase. In Enzymology and Molecular Biology of Carbonyl Metabolism, **5**, 61–66, Weiner, H. *et al.*, Ed., Plenum Press, New York.

Steinmetz, C. G., Xie, P., Weiner, H. and Hurley, T. D. (1997). Structure of mitochondrial aldehyde dehydrogenase: the genetic component of ethanol aversion. Structure **5**, 701–711.

Tronrud, D. E. (1992). Conjugate-direction minimization: an improved method for refinement of macromolecules. *Acta Cryst.* **A48**, 912–916.

Tronrud, D. E. & Ten Eyck, L. F. and Matthews, B. W. (1987). An efficient general-purpose least squares refinement program for macromolecular structures. Acta Cryst. **D52**, 743 -748.

Vallari, R. C. and Pietruszko, R. (1982). Human aldehyde dehydrogenase: mechanism of inhibition by disulfiram. Science **216**, 637–639.

Wang, X., Penzes, P. and Napoli, J. L. (1996). Cloning of a cDNA encoding an aldehyde dehydrogenase and its expression *in Escherichia coli*. J. Biol. Chem. **271**, 16288–16293.

Wang, X. and Weiner, H. (1995). Involvement of glutamate 268 in the active site of human liver mitochondrial (Class 2) aldehyde dehydrogenase as probed by site-directed mutagenesis. Biochemistry **34**, 237–243.

Yoshida, A., Hsu, L. C. and Yasunami, M. (1991). Genetics of human alcohol-metabolizing enzymes. Prog. Nucl. Ac. Res. Mol. Biol. **40**, 255–285.

Yoshida, A., Huang, I.-Y. and Ikawa, M. (1984). Molecular abnormality of an inactive aldehyde dehydrogenase variant commonly found in orientals. Proc. Natl. Acad. Sci. **81**, 258–261.

Yoshida, A., Hsu, L. C. and Dave, V. (1992). Retinal oxidation activity and biological role of human cytosolic aldehyde dehydrogenase. Enzyme **46**, 239–244.

Yoshida, A., Dave, V., Ward, R. J. and Peters, T. J. (1989). Cytosolic aldehyde dehydrogenase (ALDH1) variants found in alcohol flushers. Ann. Hum. Genet. **53**, 1–7.

Zhao, D., McCaffery, P., Ivins, K. J., Neve, R. L., Hogan, P., Chin, W. W. and Drager, U. C. (1996). Molecular identification of a major retinoic-acid-synthesizing enzyme, a retinaldehyde-specific dehydrogenase. Eur. J. Biochem. **240**, 15–22.

5

STRUCTURE AND FUNCTION OF BETAINE ALDEHYDE DEHYDROGENASE

An Enzyme within the Multienzyme Aldehyde Dehydrogenase System

Lars Hjelmqvist,[1] Mustafa El-Ahmad,[1] Kenth Johansson,[2] Annika Norin,[1] Ramaswamy S,[2] and Hans Jörnvall[1]

[1]Department of Medical Biochemistry and Biophysics
Karolinska Institutet
S-171 77 Stockholm, Sweden
[2]Department of Molecular Biology
Swedish University of Agricultural Sciences
S-751 24 Uppsala, Sweden

1. INTRODUCTION

Aldehyde and alcohol dehydrogenases (ALDHs and ADHs) are both highly multiple enzymes with many different forms in metabolically linked functional pathways. However, they are derived from separate protein families with distinct properties. Those of the vertebrate ADHs have been well characterized since long, and their different classes belong to the MDR protein family of known structures from several sources. However, the properties and multiplicity of the ALDHs have only recently been partly characterized. Presently, just three of the at least twelve different ALDH classes (Yoshida et al., 1998) are known in three-dimensional structure (Liu et al., 1997; Steinmetz et al., 1997; Johansson et al., 1998), several have not been purified and characterized enzymatically, and little is known about their species or class variabilities.

We have therefore studied the cod liver betaine aldehyde dehydrogenase, constituting both a distinct species variant (the first non-mammalian ALDH characterized in detail) and a distinct class (the first betaine aldehyde dehydrogenase three-dimensional structure). The characterization of this cod liver enzyme allows conclusions on the structural variability and the class relationships of the aldehyde dehydrogenase system. The results show that the betaine aldehyde dehydrogenase constitutes a distinct enzyme of old origin within the multienzyme system.

Enzymology and Molecular Biology of Carbonyl Metabolism 7
edited by Weiner *et al.* Kluwer Academic / Plenum Publishers, New York, 1999.

2. MATERIALS AND METHODS

2.1. Protein Purification

Liver from Baltic cod (*Gadus morhua*) was the starting material for purification of betaine aldehyde dehydrogenase in a four-step chromatography scheme after initial homogenization and centrifugation steps. The chromatographies utilized DEAE-Sepharose, AMP-Sepharose, High-load Q-Sepharose and Mono Q HR 5/5 in buffers and systems as described elsewhere (Hjelmqvist et al., in preparation). Purity was analyzed by SDS/polyacrylamide gel electrophoresis.

2.2. Enzymatic Characterization

Enzymatic activities were determined at 25 °C monitoring NAD^+ reduction in 50 mM sodium phosphate, pH 7.5, by measurement of the 340 nm absorbance using Beckman DU 64 and 68 spectrophotometers. During purification, the activity was followed by formaldehyde oxidation. Kinetic parameters of the pure enzyme with aldehydes were measured at 2.4 mM NAD^+ and calculated with the program ENZYME (Lutz et al., 1986).

2.3. Primary Structure Analysis

The protein was carboxymethylated (Hjelmqvist et al., in preparation) and separate samples were used for digestions with CNBr in 70% formic acid and with five different proteases in 0.1 M ammonium bicarbonate. The proteases utilized were Lys-C, Glu-C, and Asp-N proteases, chymotrypsin and trypsin. All digests were separated by reverse phase HPLC on C4 and C18 columns. Pure peptides obtained were analyzed for composition after hydrolysis, for N-terminal sequence by degradations in gas-phase and solid-phase sequencers, for C-terminal sequence by degradation in the recently available Applied Biosystems Procise 494 C instrument, and for molecular masses by MALDI-TOF mass spectrometry, all as described in further detail elsewhere (Hjelmqvist et al., in preparation).

2.4. Crystallization and Three-Dimensional Structure Analysis

Crystals of betaine ALDH were grown by the hanging drop vapor diffusion method at 14 °C from solutions containing 0.1 M HEPES, pH 7.5, 9.5% isopropanol, 20% PEG 4000 at 4 mg/ml protein (and 1 mM NAD^+ for the NAD^+ co-crystals).

Data from x-ray diffraction analysis of the native crystals were collected at 100K on BM14 at the European Synchrotron Radiation Facility (Grenoble) using a Princeton CCD camera (Johansson et al., 1998). Data of the NAD^+ co-crystals were collected at room temperature on a RAXIS IIC area detector with a Rigaku rotating anode. The crystal structure was solved by molecular replacement, using the bovine class 2 enzyme (Steinmetz et al., 1997) as search model with the native data set.

2.5. Structural Comparisons

The primary structure was aligned with those of other ALDHs from the SWISS-PROT database using the program CLUSTAL W (Thompson et al., 1994), which was also used to express phylogenetic tree relationships. Their confidence limits were evaluated by bootstrap sampling (Felsenstein, 1985) and plotting was performed with the program NJPLOT (Perriere and Gouy, 1996).

3. RESULTS AND DISCUSSION

3.1. Purification and Enzyme Identification

A major ALDH from cod liver was purified by monitoring its dehydrogenase activity with formaldehyde. After four steps of ion-exchange and affinity chromatography, the preparation was obtained in a yield of about 4 mg from 400 g liver after a 260-fold purification and was pure as judged by SDS/polyacrylamide gel electrophoresis.

The enzyme was identified as a betaine ALDH by the highest k_{cat} value against betaine (49 min^{-1}) versus lower values for all other aldehydes tested (2.9 to 11 min^{-1}). These properties resemble those of the human class 9 ALDH (Kurys et al., 1989; Yoshida et al., 1998) also identified as a betaine ALDH (Chern and Pietruszko, 1995; Pietruszko et al., 1997). The enzyme purified had a broad substrate specificity, compatible with a partly hydrophobic substrate binding pocket (below).

3.2. Primary Structure

Sequence analysis of peptides obtained from six different proteolytic digests identified the amino acid sequence as a 503-residue subunit with an acetylated N-terminus. The C-terminus was positively identified by C-terminal sequencer degradation (Boyd et al., 1992). The primary structure obtained was used for interpretation of the crystallographic data (below) and is deposited in the SWISS-PROT data bank (accession number P56533). Comparison of the sequence with that of other ALDHs revealed that the greatest identity is with the human class 9 enzyme (70% residue identity), confirming the identity of the cod protein as a betaine ALDH. Towards the most similar *E. coli* enzyme, the residue identity is 51%. These values are similar to those that have been found for the constant type of ADH (class III) from a similar species spread, and prove that the betaine ALDH also belongs to the constant type of dehydrogenases (cf. Yin et al., 1991). The presence of betaine ALDH in distant vertebrate lines (fish and human) and the low divergence values also show that this ALDH is of old origin. In this respect it partly resembles not only class III ADH but also the mitochondrial form of ALDH (class 2), and glycolytic enzymes (aldolase class I and lactate dehydrogenase type B) but not the cytosolic class 1 ALDH form (Yin et al., 1991), and not the well-known class I ADHs or CYP1A1 cytochrome P450s. Combined, all these properties suggest that betaine ALDH has a defined metabolic role common to the vertebrate system in general. Phylogenetic relationships for relevant ALDH forms are shown in Figure 1.

3.3. Tertiary Structure

Crystals suitable for data collection from x-ray diffraction analysis were obtained and analyzed both as native crystals (0.6 × 0.2 × 0.1 mm) and NAD^+ co-crystals (0.8 × 0.4 × 0.1 mm). The crystals belong to the triclinic space group P1, with cell dimensions a = 84.2 Å, b = 86.2 Å, c = 88.3 Å, α = 105.2°, β = 115.1°, and γ = 100.0° for the native enzyme which gave diffraction data to a resolution of 2.1 Å. There are four molecules per asymmetric unit related by a proper 222 non-crystallographic symmetry. The NAD^+ co-

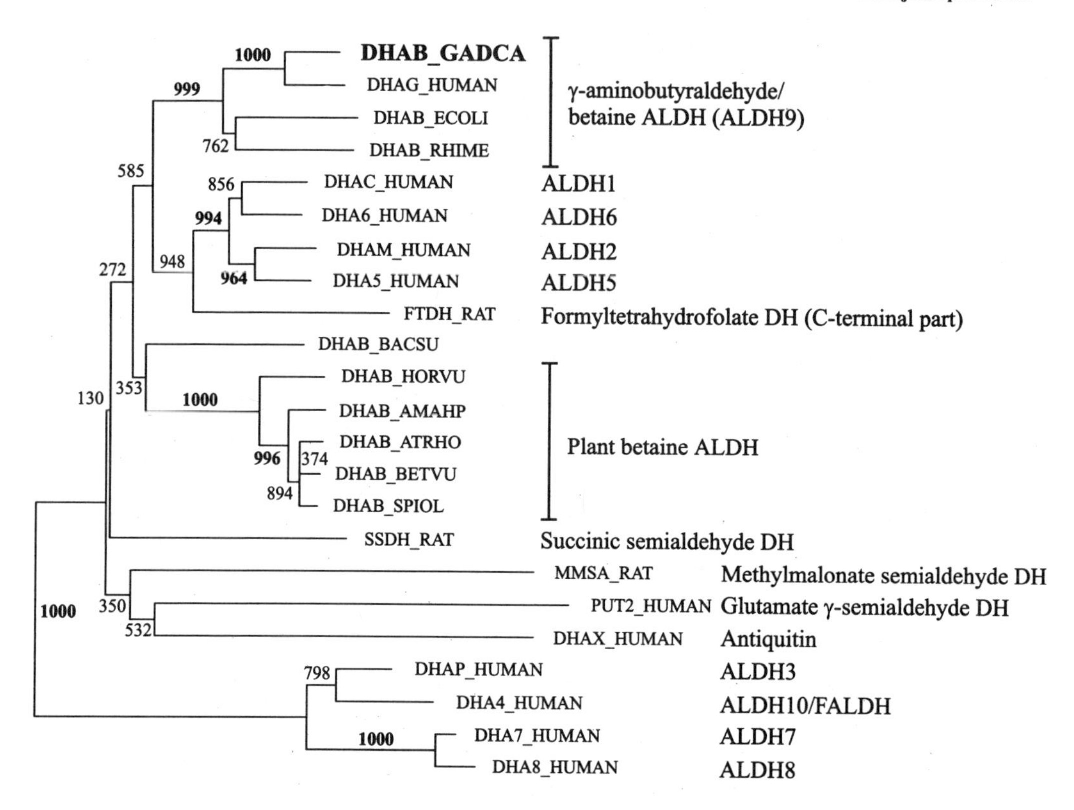

Figure 1. Phylogenetic tree showing the relationships between betaine ALDHs and human (or rat) forms of the other known ALDH classes. Numbers indicate results from bootstrap analysis.

crystals were similarly investigated and gave a resolution of 2.8 Å. The structure was solved by molecular replacement using the native data set and the bovine class 2 ALDH conformation. Combined the analyses gave both the native structure and that of the NAD^+ complex (Figure 2).

The results show the subunit to have the three-domain fold typical of the other two ALDHs known in tertiary structure, with a coenzyme-binding domain, a catalytic domain and an oligomerization domain (Liu et al., 1997; Steinmetz et al., 1997; Johansson et al., 1998). Consequently, ALDH of three different classes (classes 2, 3, 9), with widely different functions and with different quaternary structures (dimer for class 3, tetramer for the other two classes) all show highly similar and largely superimposable structures, establishing this overall fold as that typical for ALDHs in general.

Further data on all relationships are given separately (Johansson et al., 1998) but two properties may be of particular interest to note now. One is the mode of coenzyme binding, the other the residues at the substrate-binding site. In regard to the coenzyme binding, the previous two structures have differed by showing the coenzyme bound to class 3 further out in the cleft than to class 2 and with the nicotinamide ring at different sites and in different conformations (*syn* in class 3 and *anti* in class 2). The present data on betaine ALDH with bound NAD^+ resemble those of the class 2 enzyme (Steinmetz et al., 1997) suggesting this to be the most common mode of coenzyme binding. In regard to the substrate binding, the residues at the active site differ between the classes in a manner

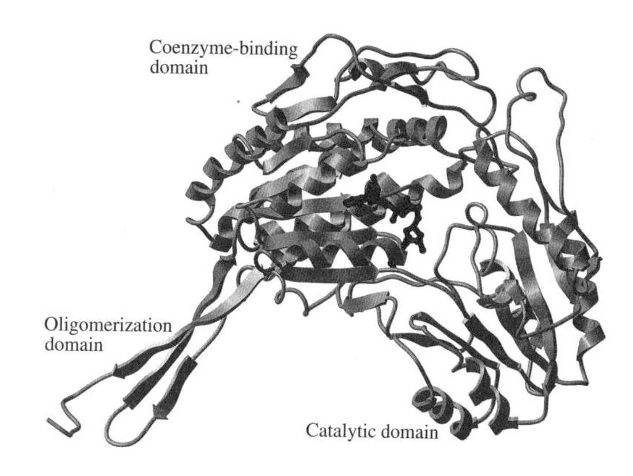

Figure 2. Tertiary structure of the betaine ALDH subunit. The bound NAD^+ is shown in a ball and stick model. The drawing was made using ICMLite (Molsoft, L.L.C., Metuchen, NJ).

compatible with the substrate differences, again confirming the overall structures and explaining the functional differences. Thus, when a betaine aldehyde molecule is modeled to the active site, it is surrounded by a large, essentially hydrophobic pocket. Although this pocket has several similarities to that of class 2 and class 3, it is smaller in those structures. Accordingly, Met174 in class 2 at the position corresponding to that of Ile171 in the cod enzyme prevents access of betaine aldehyde to this site in the class 2 enzyme.

3.4. Conclusion

We have purified and analyzed betaine ALDH from cod liver. The results identify this protein as an enzyme with the typical ALDH fold, but of low species variability compatible with important functions of betaine turnover in vertebrates in general.

ACKNOWLEDGMENTS

This work was supported by grants from the Swedish Medical Research Council (Project 13X-3532), the Swedish Alcohol Research Fund (Project 96/18:1), the Swedish Society of Medicine, the foundations of Magn. Bergvall, Clas Groschinsky, Lars Hierta, Fredrik och Ingrid Thuring, Åke Wiberg, and Berth von Kantzow (fellowship to L.H.).

REFERENCES

Boyd, V. L., Bozzini, M.-L., Zon, G., Noble, R. L. and Mattaliano, R. J. (1992) Sequencing of peptides and proteins from the carboxy terminus. *Anal. Biochem.* **206**, 344–352.

Chern, M.-K. and Pietruszko, R. (1995) Human aldehyde dehydrogenase E3 isozyme is a betaine aldehyde dehydrogenase, *Biochem. Biophys. Res. Commun.* **213**, 561–568.

Felsenstein, J. (1985) Confidence limits on phylogenies: an approach using the bootstrap, *Evolution* **39**, 783–791.

Johansson, K., El-Ahmad, M., Ramaswamy, S., Hjelmqvist, L., Jörnvall, H. and Eklund, H. (1998) Structure of betaine aldehyde dehydrogenase at 2.1 Å resolution, *Prot Sci., in press*.

Kurys, G., Ambroziak, W. and Pietruszko, R. (1989) Human aldehyde dehydrogenase. Purification and characterization of a third isozyme with low K_m for γ-aminobutyraldehyde, J. *Biol. Chem.* **264**, 4715–4721.

Liu, Z.-J., Sun, Y.-J., Rose, J., Chung, Y.-J., Hsiao, C.-D., Chang, W.-R., Kuo, I., Perozich, J., Lindahl, R., Hempel, J and Wang, B.-C. (1997) The first structure of an aldehyde dehydrogenase reveals novel interactions between NAD and the Rossmann fold, *Nature Struct. Biol.* **4**, 317–326.

Lutz, R. A., Bull, C. and Rodbard, D. (1986) Computer analysis of enzyme-substrate-inhibitor kinetic data with automatic model selection using IBM-PC compatible microcomputers, *Enzyme* **36,** 197–206.

Perriere, G. and Gouy, M. (1996) WWW-Query: an on-line retrieval system for biological sequence banks, *Biochimie* **78**, 364–369.

Pietruszko, R., Kikonyogo, A., Chern, M.-K. and Izaguirre, G. (1997) Human aldehyde dehydrogenase E3. Further characterization, *in Enzymology and Molecular Biology of Carbonyl Metabolism* **6** (Weiner, H., Lindahl, R., Crabb, D. W., and Flynn, T. G., Eds.) pp. 243–252, Plenum Press, New York

Steinmetz, C. G., Xie, P., Weiner, H. and Hurley, T. D. (1997) Structure of mitochondrial aldehyde dehydrogenase: the genetic component of ethanol aversion, *Structure* **5**, 71–711.

Thompson, J. D., Higgins, D. G. and Gibson, T. J. (1994) CLUSTAL W: improving the sensitivity of progressive multiple sequence alignment through sequence weighting, position-specific gap penalties and weight matrix choice. *Nucleic Acids Res.* **22**, 4673–4680.

Yin, S.-J., Vagelopoulos, N., Wang, S.-L. and Jörnvall, H. (1991) Structural features of stomach aldehyde dehydrogenase distinguish dimeric aldehyde dehydrogenase as a 'variable' enzyme. 'Variable' and 'constant' enzymes within the alcohol and aldehyde dehydrogenase families, *FEBS Lett.*, **283**, 85–88.

Yoshida, A., Rzhetsky, A., Hsu, L. C. and Chang, C. (1998) Human aldehyde dehydrogenase gene family, *Eur. J. Biochem.* **251** 549–552.

EVALUATION OF THE ROLES OF THE CONSERVED RESIDUES OF ALDEHYDE DEHYDROGENASE

Thomas D. Hurley[1] and Henry Weiner[2]

[1]Department of Biochemistry and Molecular Biology
Indiana University School of Medicine
Indianapolis, Indiana 46202
[2]Department of Biochemistry
Purdue University
West Lafayette, Indiana 47907

1. INTRODUCTION

Aspects of the mechanism of action of mammalian liver aldehyde dehydrogenase have been studied since the first mammalian liver enzyme was purified in our laboratory. The initial studies suggested that the enzyme functioned with covalent catalysis and that a cysteine might be the nucleophile (Feldman and Weiner, 1972). It was later shown by chemical modifications that cysteine 302 was at the active site and it most likely functioned as the nucleophile (Hempel and Pietruszko, 1981). Later, it was shown that glutamate 268 appeared also to be involved in the catalytic process (Abriola et al., 1990). Once a number of sequences of diverse ALDHs were aligned it was found that all possessed cysteine 302 and most contained the glutamate residue (Vasiliou et al., 1995). The few that did not possess a glutamate were enzymes that did not produce a free acid as the final product.

The sequence alignment of many ALDHs revealed that few residues were complete conserved among all the known enzyme forms. The alignment included representatives from many different classes of ALDH. Of the three best studied classes of enzyme are the tetrameric class 1 and 2 enzymes from liver cytosol and mitochondria, respectively. Each has 500 amino acids per subunit. The class 3 forms of the enzyme, also well studied, are dimers possessing around 460 residues. In the past two years the structure of all three enzyme forms have been solved and presented at these meetings. The structure of the class 3 rat enzyme was solved first and presented at the 1996 meeting (Liu et al., 1997a) and published in 1997 (Liu, et al., 1997b). The structure of the class 2 beef enzyme was also pub-

Enzymology and Molecular Biology of Carbonyl Metabolism 7
edited by Weiner *et al.* Kluwer Academic / Plenum Publishers, New York, 1999.

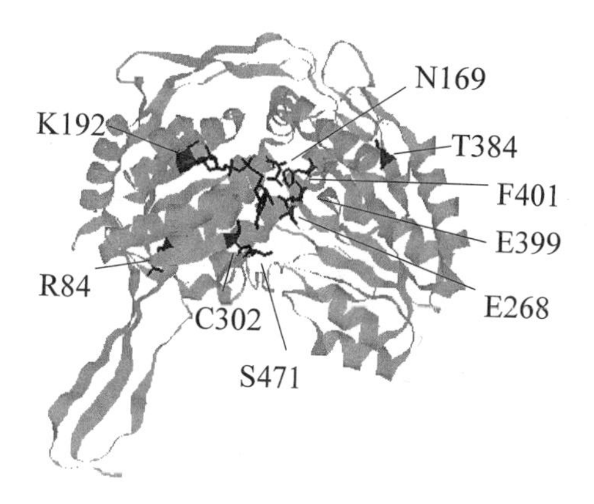

Figure 1. A single subunit of tetrameric mitochondrial aldehyde dehydrogenase illustrating the position of NAD and the two active site components, cysteine 302 and glutamate 268. The other residues are those conserved in all known aldehyde dehydrogenases.

lished in 1997 (Steinmetz et al., 1997) and was presented for the first time at this meeting (Hurley et al., 1999). The structure of the class 1 sheep cytosolic form was also presented at this meeting (Moore et al., 1999). The basic structure of an individual subunit from each enzyme form is very similar. The position of the residues that are common to all ALDHs, excluding glycine and prolines are shown for a subunit of the class 2 enzyme along with NAD in Figure 1.

The class 2 enzyme is essentially composed of a pair of dimers, with each dimer pair being somewhat similar in structure to the dimer found in the class 3 form. A major difference between the structures is the extended C-terminal region of the class 3 enzyme that presumably prevents tetramer formation.

In as much as there were only a few residues that were conserved in addition to the 14 glycines and prolines, mutational studies were undertaken to try to assess the roles of some of the various residues in the catalytic process. The primary data in Table 1 has been published. Most of the mutational work was performed with the human enzyme that shares 95% sequence homology with the beef liver enzyme. Some of the conclusions reached in our studies differ from what is concluded by the investigators who solved the structure of the dimeric class 3 rat liver enzyme (Liu, et al., 1997). These differences are primarily related to the general base necessary for the reaction and the NAD-ribose binding residues.

Table 1. Kinetic properties for mutants of human liver mitochondrial aldehyde dehydrogenase[a]

Enzyme	Km (NAD) μM	k_{cat} (min^{-1})	Kia (uM)[b]	Kiq (uM)[c]
Native	28	180	11	3
R84Q	32	57	ND[d]	ND
K192Q	3600	35	590	9
T384A	160	35	ND	ND
E399Q	120	21	ND	ND
S471A	1470	28	280	17
E487K[e]	7400	19	548	16

[a]From Sheikh et al., 1997.
[b]Kia is the dissociation constant for NAD found from analysis of kinetic data.
[c]Kiq is the dissociation constant for NADH found by fluorescence binding.
[d]ND means not determined. The values were only measured for the mutants with very large Km values for NAD.
[e]From Farrés et al., 1994.

2. METHODS

2.1. Recombinantly Expressed ALDHs

The cDNAs coding for native and mutant forms of human class 2 enzyme were cloned into *E. coli* BL21 cells and the enzymes were expressed and purified as described in the publications from our laboratory mentioned in the text.

2.2. Assays

The enzyme was assayed for both its ability to oxidize aldehydes in the presence of NAD and to hydrolyze p-nitrophenyl acetate. The assays were performed at pH 7.4. To determine the rate limiting step in the dehydrogenase reaction the enzyme was assayed with chloroacetaldehyde, an aldehyde that is oxidized more rapidly if deacylation is the rate limiting step. Deuterated substrates were also employedto determine if they were oxidized more slowly than the corresponding hydrogen-containing substrate. If they were, it would be indicative of hydride transfer being the rate limiting step.

2.3. Structure

The figures showing the location of the various residues were constructed using the program Rasmol and were based on the co-ordinants of the beef liver enzyme which are on file in the Brookhaven Data Base.

3. RESULTS

3.1. Active Site Nucleophile and General Base

Chemical modification studies followed by sequencing suggested that cysteine 302 and glutamate 268 could function as the active site nucleophile and general base, respectively. The role of the general base is to active the cysteine residue so it can attack the aldehyde to form a thiohemiacetal intermediate. After hydride transfer the base would be involved in the activation of water to hydrolyze the acyl intermediate. It has been shown that deacylation, k_7 in Figure 2, is rate limiting for the class 2 enzyme. In Figure 3 is shown the relation between the location of NAD and these two residues. As discussed in a preceding chapter (Hurley, et al., 1999) we have to postulate that there is movement of the

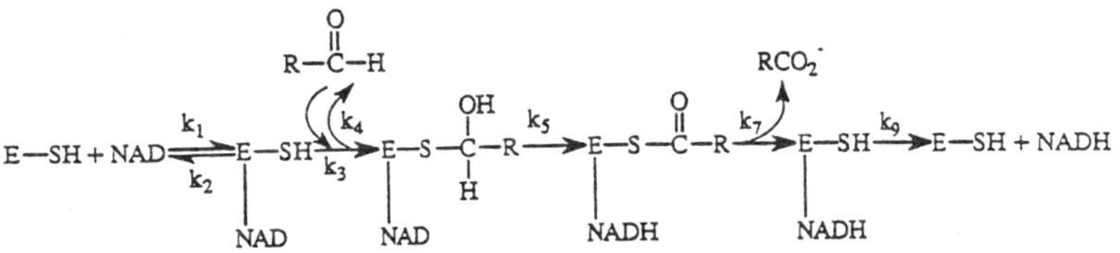

$$\mathrm{E{-}SH + NAD} \underset{k_2}{\overset{k_1}{\rightleftharpoons}} \underset{\mathrm{NAD}}{\mathrm{E{-}SH}} \xrightarrow[k_3]{\mathrm{R{-}C(=O){-}H},\ k_4} \underset{\mathrm{NAD}}{\mathrm{E{-}S{-}C(OH)(H){-}R}} \xrightarrow{k_5} \underset{\mathrm{NADH}}{\mathrm{E{-}S{-}C(=O){-}R}} \xrightarrow[\mathrm{RCO_2^-}]{k_7} \underset{\mathrm{NADH}}{\mathrm{E{-}SH}} \xrightarrow{k_9} \mathrm{E{-}SH + NADH}$$

Figure 2. A scheme showing the ordered binding of substrate to the enzyme. The rate limiting step is k_7, the deacylation of covalently bound product.

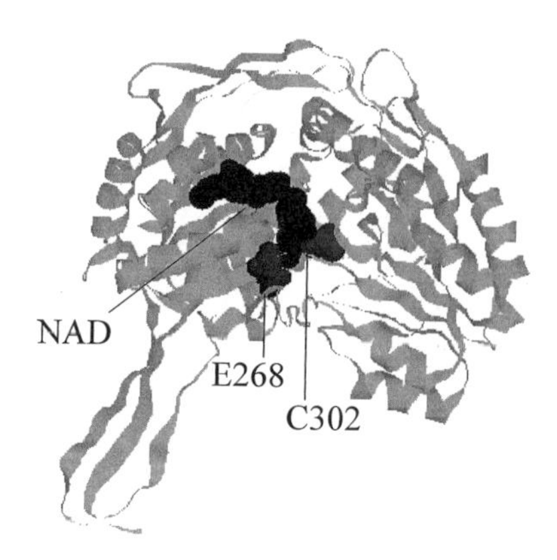

Figure 3. Space filling model showing the spacial relation between NAD and the two active site components.

nicotinamide ring after the hydride transfer step occurs in order to invoke these two residues being at the active site.

The evidence that led us to conclude that cysteine 302 was at the active site included the fact that the C302A mutant had no measurable activity and that some activity was recovered when serine replaced the alanine (Farrés, et al., 1995). The C302S mutant possessed about 2% the activity of the native enzyme and the Km for aldehyde increased. This is consistent with serine being a poorer nucleophile so that the hemiacetal formation actually became rate limiting for the enzyme.

The evidence for glutamate 268 being a general base included the fact that any mutation to the residue produced an enzyme with less that 2% the specific activity of the native enzyme (Wang and Weiner, 1995). These mutants had the same Km for substrate and NAD. Further, there was no partial reaction in the mutants. That is, no accumulation of an Enzyme-NADH complex was found. Similarly, there was no accumulation of an acyl enzyme when p-nitrophenyl acetate was employed as the substrate for the esterase reaction. These findings show that the residue was needed for both the acylation and deacylation portion of the reaction. Lastly, glutamate 268 is found in every ALDH that produces a free acid, supporting a role of the residue in water activation. The involvement of the residue in the action of class 2 ALDH is discussed in more detail elsewhere in this volume (Hurley et al., 1999).

3.2. Kinetic Properties of the Enzyme with Mutations to Conserved Residues

The Km for NAD and propionaldehyde as well as the specific activity for a number of mutant forms of the enzyme is presented in Table 1. It can be noted that the specific activity of all the mutant forms of the enzyme decreased while the Km for propionaldehyde did not change. Only mutations at position 192 and 471 caused the Km for NAD to dramatically increase. Position 487 is not a conserved residue, but is the position that is altered in the Asian variant of the enzyme. The recombinantly expressed Asian variant of ALDH (E487K) had a very large Km for NAD (Farrés et al., 1994).

At the time we performed the measurements we did not know the structure of the enzyme. While analyzing the data the structure was solved. Hence we could locate the residues in the subunit as was presented in Figure 1.

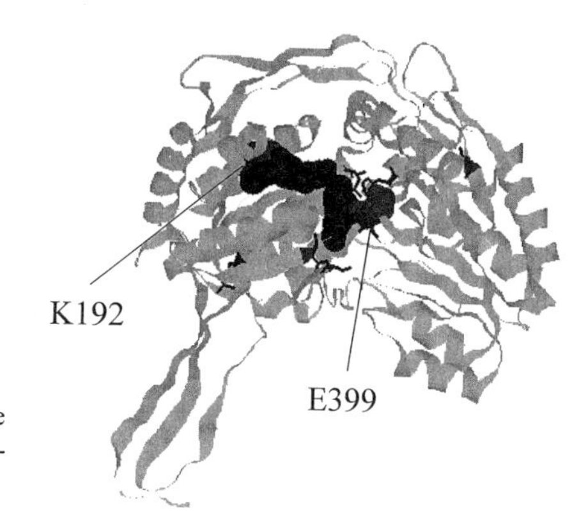

Figure 4. Two conserved residues were found to bind to the ribose rings of NAD. These were K192 binding to adenine ribose and E399 binding to the nicotinamide ribose.

3.3. NAD Binding Residues

Two of the conserved residues were found to be in contact with the ribose rings of NAD. These were K192, binding to the adenine ribose and E399, binding to the nicotinamide ribose (Figure 4). Mutation of the lysine caused the Km as well as the Kia for NAD to increase dramatically (Sheikh et al., 1997; Li et al., 1997). Less of an affect was found when E399 was changed. It is possible that the nicotinamide half (NMN) of the coenzyme is held less rigidly.

It was unexpected to have found that the rate limiting step of the enzyme changed when either of the two NAD binding residues was mutated. The step became hydride transfer (k_5) when acetaldehyde was the substrate while it is deacylation (k_7) for the native enzyme. Still even more surprising was to find that the rate limiting step for the mutant enzyme remained deacylation when benzaldehyde was the substrate. No explanation for either of these two observations can be made. Since hydride transfer becomes rate limiting when either residue was mutated, one can assume that the proper orientation of the NAD moiety is necessary to stabilize the transition state necessary for the hydrogen to be transferred from the carbonyl to the nicotinamide ring. Perhaps with the aromatic substrate, which also happen to be large in size and more hydrophobic than the methyl group in acetaldehyde, the structure is more rigid and hydride transfer is accommodated whether or not the ribose binding residues are native or mutated.

From the structure of the rat class 3 enzyme the investigator concluded that two different residues interact with the NAD riboses. These are two non-conserved residues, E140 and R292, that correspond to E195 and K352 in the class 2 enzyme. These investigators feel that the residue corresponding to E399 in class 2 ALDH is not a ribose binding residue, but is the general base. We proposed that the general base was E268. Arguments for the class 3 enzyme using the equivalent of residue 399 as a general base can be found in this volume (Hempel et. al., 1999).

3.4. Threonine 384

This residue is located far from the active site and not in a region of the subunit that interacts with other subunits. Converting the threonine into either a serine or an alanine caused changes to occur in Km for either substrate and caused kcat to decrease. The rate limiting step remained k_7, deacylation. This residue appears to play a structural role as it is bonded to the backbone of proline 383, also a conserved residue. Residue 384 is located approximately 20 Å from cysteine 302, yet a mutation to it caused the specific activity to decrease by 2–5 fold and the Km for NAD to increase 7–16 fold depending upon whether the mutation was to a serine or to an alanine, respectively. We can not offer a structural explanation for these findings.

3.5. Arginine 84

The residue is located distal to the active site and appears to interact through a water molecule to the C-terminal serine 500. This serine is located not in the same subunit but in one which does not make up the dimer pair. A mutation to Arg84 produced an enzyme with just a 3-fold decrease in activity and essentially no change in the Km for the substrates. It is of potential interest to note that the interaction between this arginine and the serine residue would not exist in the class 3 enzyme for this would occur only in a tetrameric enzyme. The residue, though, is conserved and hence must serve some other structural role in the dimeric forms of the enzyme, such as capping a helix. as it also does in the class 2 enzyme.

3.6. Serine 471

This residue is located 9 Å from the active site and interacts with residues 269 and 270. Mutations to the serine caused a drastic increase to the Km for NAD and a dramatic decrease in the specific activity of the enzyme. The Km for NAD increased 50 fold, similar to what was found when the ribose binding lysine 192 was mutated. The activity decreased by a factor of 40 and 6 with the threonine and alanine mutants, respectively. It is not possible to explain the kinetic effects observed, but the change in kcat could be related to the need of glutamate 268 in the active site. Altering the interaction between the serine and residues at 269 and 270 could cause a movement of the essential general base. How this causes a change in the Km for NAD is not known, for a mutation to glutamate 268 did not cause a change in the Km value. The effect is not just on Km for NAD but also on Kd. We found by analysis of the two substrate kinetic data that Kia (the dissociation constant for NAD from the enzyme) increased from 11 μM to 280 μM for the S471A mutant. Similarly, with the K192Q mutant the Kia term increased to 590 μM, showing that the change in Km really reflected a change in the binding ability of NAD.

3.7. Glutamate 487

This residue is not conserved, but has been studied extensively in our laboratory (Farrés et al., 1994; Wang et al., 1996). It is the residue that has been changed to a lysine in the Asian variant of the enzyme. The variant form of the enzyme was thought to be inactive, but when it was recombinantly expressed it was found to have a low specific activity (10%) and a very high Km for NAD (7400 μM). This mutation also resulted in an increase in Kia (40 fold), again showing that the binding of coenzyme was affected. We showed that a mutant with a glutamine residue at position 487 possessed properties that were more native like. Thus, it was not the loss of a glutamate residue but the presence of the lysine that caused the alterations to occur.

From the structure we were able to suggest a possible cause of the change in properties of the Asian variant. In the active form of the enzyme glutamate 487 is salt bonded to two arginine residues. Arginine 264 is located in the same subunit while arginine 475 is located in the other subunit that comprises the dimer pair. Neither of these arginine residues is conserved but is present in all the liver tetrameric class 1 and 2 forms of the enzyme. We have yet to solve the structure of the Asian variant so we do not know with certainty what alterations have occurred as a result of having the lysine at position 487. However, we can suggest that disrupting the salt bond to arginine 475 could affect catalysis since the residue at position 476 bonds to a water that also binds to the general base, glutamate 268. A movement of the guanidino portion of arginine 475 could bring it into van der Waal distance to the nitrogen of the nicotinamide ring of the NAD moiety. This could cause a repulsion that could explain the increased in Kia. Consistent with this suggestion is the fact the Kd for NADH did not increase significantly in the Asian variant compared to the glutamate-active form of the enzyme. In NADH the nicotinamide nitrogen does not carry a positive charge.

We propose that the reason for the lower activity found in the Asian variant is due to the interaction between the lysine now at position 487 and the arginine in a different subunit located at position 475 (Figure 5). The intersubunit interaction would explain why in a heterotetramers composed of subunits of each type does the lysine-subunit dominate. That is, the activity of the heterotetramer is not the average of the activity of the individual subunits. The interactions between the subunits (E487-R475) and within a subunit (E487-R264) could be involved. Structural data is needed to verify our hypothesis.

4. CONCLUSIONS

From the mutational analysis of the residues that are conserved among all the known ALDHs we concluded that the nucleophile is cysteine 302 and the essential general base is glutamate 268. The structure of the class 2 mitochondrial beef ALDH shows that these residues most likely perform the role proposed. Further, the structure revealed that asparagine 169, a conserved residue we did not study, could function as a general acid acting to polarize the carbonyl bond of the substrate. The structure revealed that two of the conserved residues, lysine 192 and glutamate 399, interacted with the ribose residues of NAD. The structure also revealed that phenylalanine 401, another conserved residue we

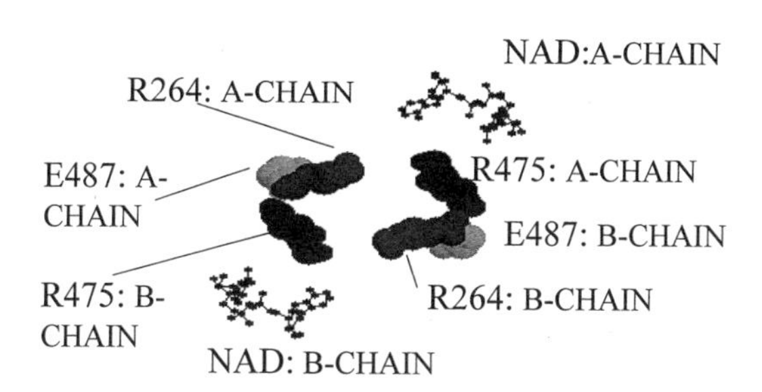

Figure 5. In the Asian variant of aldehyde dehydrogenase the glutamate at position 487 is a lysine. The glutamate salt bonds to two arginine residues. One is in the same subunit, R264, and the other (R475) is in the subunit that makes up the dimer-pair. In the left panel is shown a pair of dimers with the glutamate and two arginines presented as space filling models and each subunit shown in different tone of gray. In the right panel are shown the residues along with NAD. For spacial clarity, the latter is presented as a stick model. It is proposed, but not yet proven, that when the residue is a lysine, the arginine residues might move and the positive charge of the guanidino group of 475 could be within van der Waals distance to the positive nicotinamide ring. A and B refer to the different subunits in the dimer pair.

did not investigate, is located near the nicotinamide ribose ring, perhaps serving as an anchor for it. Of perhaps greater interest was finding that the other conserved residues with potentially reactive side chances were not in contact with the active site. Some were as far a 20 Å from cysteine 302. Though no structure has yet been determined for the various mutants, it appears that these other conserved residues perform a structural role and are not directly part of the catalytic site.

REFERENCES

Abriola, D. P., MacKerell, A. D. Jr., and Pietruszko, R. , 1990, Correaltion of loss of activity of human aldehyde dehydrogenase with reaction of bromoacetophenone with glutamatic acid-268 and cysteine-302 residues, *Biochem J.*, 266:179–187.

Farrés, J., Wang, T. Y., Cunningham, S. J., and Weiner, H., 1995, Investigation of active site cysteine residues of rat mitochondrial aldehyde dehydrogenase by site-directed mutagenesis, *Biochemistry*, 34:2592–2598.

Farrés, J., Wang, X., Takahashi, K., Cunningham, S. J., Wang, T. T., and Weiner, H., 1994, Effects of changing glutamate 487 to lysine in rat and human liver mitochondrial aldehyde dehydrogenase, a model to study human (Oriental type) class 2 aldehyde dehydrogenase, *J. Biol. Chem.*, 269:13854–13860.

Feldman, R. I., and Weiner, H., 1972, Horse liver aldehyde dehydrogenase, II: kinetics and mechanistic implications of the dehydrogenase and esterase activity, *J. Biol. Chem.*, 247:267–272.

Hempel, J., and Pietruszko, R., 1981, Selective chemical modification of human liver aldehyde dehydrogenases E1 and E2 by iodoacetamide, *J. Biol. Chem.*, 256:10889–10896.

Hempel, J., Perozich, J., Chapman, T., Rose, J., Boesch, J. S., Liu, Z. J., Lindahl, R., Wang, B. C., Aldehyde dehydrogenase catalytic mechanism: a proposal, 1998, These Proceedings.

Hurley, T. D., Steinmetz, C. G., and Weiner, H., Three-dimensional structure of mitochondrial aldehyde dehydrogenase: mechanistic implications, 1998, These Proceedings.

Li, N., Sheikh, S., and Weiner, H., 1997, Involvement of glutamate 399 and lysine 192 in the mechanism of human liver mitochondrial aldehyde dehydrogenase, *J. Biol. Chem.*, 272:18823–18826.

Liu, Z. J., Hempel, J., Sun, J., Rose, J., Hsiao, D., Chang, W. R., Chung, Y. J., Kuo, I., Lindahl, R., and Wang, B. C., 1997, Crystal structure of a class 3 aldehyde dehydrogenase at 2.6Å resolution, *Enzymology and Molecular Biology of Carbonyl Metabolism 6*, (Weiner H, Lindahl R, Crabb, DW, and Flynn, TG Eds) pp 1–7, Plenum Press, New York.

Liu, Z. J., Sun, Y. J., Rose, J., Chung, Y. J., Hsiao, C. D., Chang, W. R., Kuo, I., Perozich, J., Lindahl, R., Hempel, J., and Wang, B. C., 1997, The first structure of an aldehyde dehydrogenase reveals novel interactions between NAD and the Rossmann fold, *Nature Structural Biology*, 4:317–326.

Moore, S. A., Baker, H. M., Blythe, T. J., Kitson, K. E., Kitson, T. M., and Baker, E. N., A structural explanation for the retinal specificity of class 1 ALDH enzymes, 1998, These Proceedings.

Sheikh, S., Ni, L., Hurley, T. D., and Weiner, H., 1997, The potential roles of the conserved amino acids in human liver mitochondrial aldehyde dehydrogenase, *J. Biol. Chem.*, 272:18817–18822.

Steinmetz, C. G., Xie, P. G., Weiner, H., and Hurley, T. D., 1997, Structure of mitochondrial aldehyde dehydrogenase: the genetic component of ethanol aversion, *Structure*, 5:701–711.

Vasiliou, V., Weiner, H., Marselos, and Nebert, D. W., 1995, Mammalian aldehyde dehydrogenase genes: classification based on evolution, structure and regulation, *Eur. J. Drug*, 26:53–64.

Wang, X., Sheikh, S., Saigal, D., Robinson, L., and Weiner, H., 1996, Heterotetramers of human liver mitochondrial (class 2) aldehyde dehydrogenase expressed in *E. coli* , a model to study the heterotetramers expected to be found in Oriental people, *J. Biol. Chem.*, 271: 31172–31178.

Wang, X., and Weiner, H., 1995, Involvement of glutamate 268 in the active site of human liver mitochondrial (class 2) aldehyde dehydrogenase as probed by site-directed mutagenesis, *Biochemistry*, 34:237–243.

7

ALDEHYDE DEHYDROGENASE CATALYTIC MECHANISM

A Proposal

John Hempel,[1] John Perozich,[1] Toby Chapman,[2] John Rose,[3]
Josette S. Boesch,[4] Zhi-Jie Liu,[3] Ronald Lindahl,[4] and Bi-Cheng Wang[3]

[1]Department of Biological Sciences
[2]Department of Chemistry
University of Pittsburgh
Pittsburgh, Pennsylvania 15260
[3]Department of Biochemistry and Molecular Biology
University of Georgia
Athens, Georgia 30602
[4]Department of Biochemistry and Molecular Biology
University of South Dakota
Vermillion, South Dakota 57069

1. INTRODUCTION

Elsewhere in this volume we detail findings from an alignment of 145 ALDH sequences (Perozich et al., 1999), and previously at these meetings we reported that the crystal structure of a class 3 ALDH (E-NAD binary complex) revealed a non-traditional mode of NAD-binding within an open β/α domain otherwise familiar in the NAD-binding "Rossmann folds" of other dehydrogenases (Liu et al., 1997a). The variability of residues in the substrate-binding site clearly indicates evolutionary tailoring of the substrate specificities of individual ALDHs. However, farther to the interior of the active site—between the catalytic thiol and NAD molecule where hydride transfer from aldehyde to NAD occurs—strict conservations are compatible with a common chemical mechanism (Liu et al., 1997b). The position of NAD in an isomorphous class 3 ALDH derivative and the emergence of Asn-114/169 as a strictly conserved residue prompted us to consider the catalytic mechanism we present here.

Enzymology and Molecular Biology of Carbonyl Metabolism 7
edited by Weiner *et al.* Kluwer Academic / Plenum Publishers, New York, 1999.

Table 1. RMSD values calculated between five residues, Asn 114/169, Glu 209/268, Cys 243/302, Glu 333/399, Phe 335/401 (258 atoms total) from subunits of tetrameric class 2 and dimeric class 3 aldehyde dehydrogenases (PDB accessions *1ag8* and *1ad3*, respectively)

Abbreviations:		
2a = *1ag8* chain A		3a = *1ad3* chain A
2b = *1ag8* chain B		3b = *1ad3* chain B
2c = *1ag8* chain C		
2d = *1ag8* chain D		
RMSD values:		
3a–3b = 0.27Å	3a–2a = 0.64Å	
		3a–2b = 0.80Å
2a–2b = 0.57Å		3a–2c = 0.71Å
2a–2c = 0.31Å		3a–2d = 0.82Å
2a–2d = 0.60Å		

2. RESULTS

2.1. Similarity of Class 2 and 3 Subunits

Class 1 and 2 ALDHs may be broadly distinguished from class 3 forms in having an extra 56 residues at the N-terminus (and lacking a shorter segment at the C-terminus), and by their tetrameric, instead of dimeric, quaternary structure. Overall, class 3 sequences display ~28% positional identity vs. either class 1 or 2 sequences, excluding N and C-terminal overhanging segments. Nevertheless, structure-based comparison of the class 2 (PDB accession 1ag8) and class 3 (PDB accession 1ad3) tertiary structures reveals an RMSD of just 1.85Å. Using five highly conserved residues in the active site, RMSD values calculated for 258 atoms are even smaller (Table 1). On this basis, some subunits of the class 3 and class 2 structures are nearly as similar to each other as different subunits of the class 2 structure are to each other. Relevant distances in the class 3 structure are shown in Figure 1.

2.2. Conservation and Importance of Asn-114

When the tertiary structure of the class 3 ALDH was solved, we never anticipated that the residue closest to Cys-243 would be Asn-114 (Asn-169 in the class 2 structure), or that it would even lie in the active site, since that residue had not been listed with those strictly conserved (Hempel et al., 1993). However, two database revisions now place this residue among those strictly conserved in all active ALDHs (summarized, Perozich et al., 1998).

An N114D class 3 mutant retains no more than 0.6% activity vs. the native expressed enzyme, using our standard assay with benzaldehyde and NADP (Rose et al., 1990). Measurements in greater detail have not yet been made due to the scant activity, but the effect appears to directly affect k_{cat}, since a 5-fold increase in either aldehyde or coenzyme failed to increase the rate.

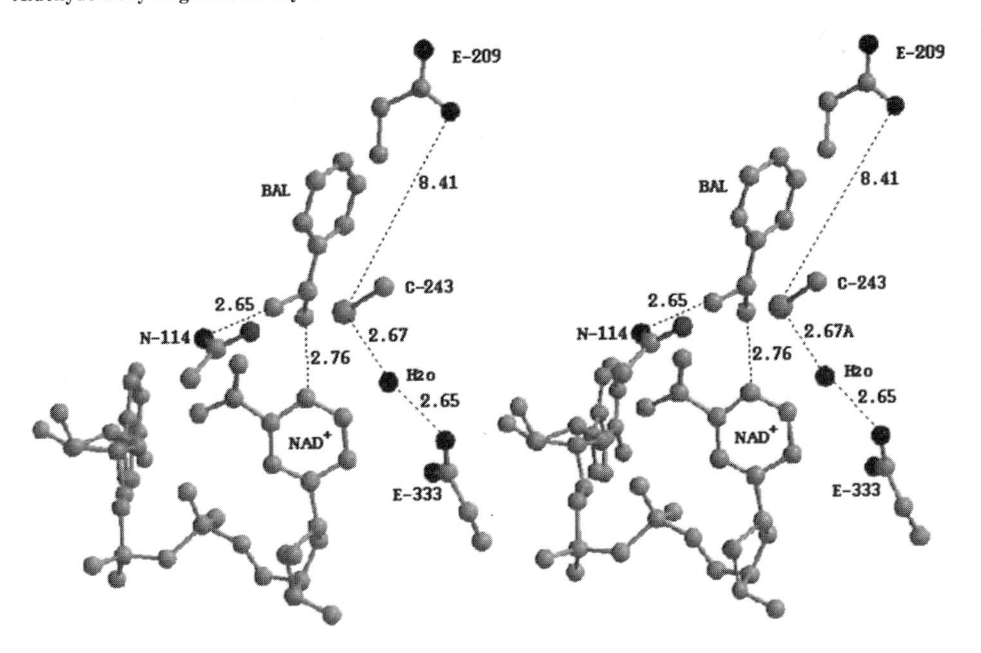

Figure 1. Stereo view of the active site residues of class 3 ALDH. Substrate (benzaldehyde, BAL) was modelled into the site; NAD is positioned from the crystallographic coordinates of the E-NAD binary complex (mercury derivative).

2.3. NAD Occupancy

E-NAD binary complex structures are available for a class 3, a class 2 and a class 1 ALDH (Liu et al., 1997b; Steinmetz et al., 1997; Moore et al., 1999) while the cod betaine ALDH apoenzyme tertiary structure is electronically-available (PDB entry 1A4S). No crystal structures are available of the ternary complex. With class 1, 2 and 3 ALDHs, the adenine ribose portion of the NAD molecule is easily seen. However, the density reflecting the nicotinamide portion of the cofactor is poor in in the case of the native E-NAD class 2 and 3 ALDH binary complexes. In the very recently reported class 1 binary structure (Moore et al., 1999) the nicotinamide ring is reported in two different orientations within each tetramer. In contrast, the density of the nicotinamide ring and its ribose is quite clear in one of the isomorphous (Hg) derivatives of the class 3 ALDH. In this derivative, a mercury atom is bound to the catalytic Cys-243, between the sulfur atom and the nicotinamide ring, where the substrate carbonyl carbon would be bound during catalysis. A mercury atom at this position presents no steric problems vs. carbon; relevant van der Waals radii are: Hg=1.10, O =1.40, N =1.5, C =1.8Å. The nicotinamide ring stacks offset with the strictly-conserved Phe-335, which may play a significant role as a weak cation sink, considered further below.

3. DISCUSSION

3.1. Mechanism

These observations together with the distances shown in Figure 1 allow us to propose a detailed mechanism for ALDHs (Figure 2). Benzaldehyde was earlier modelled in

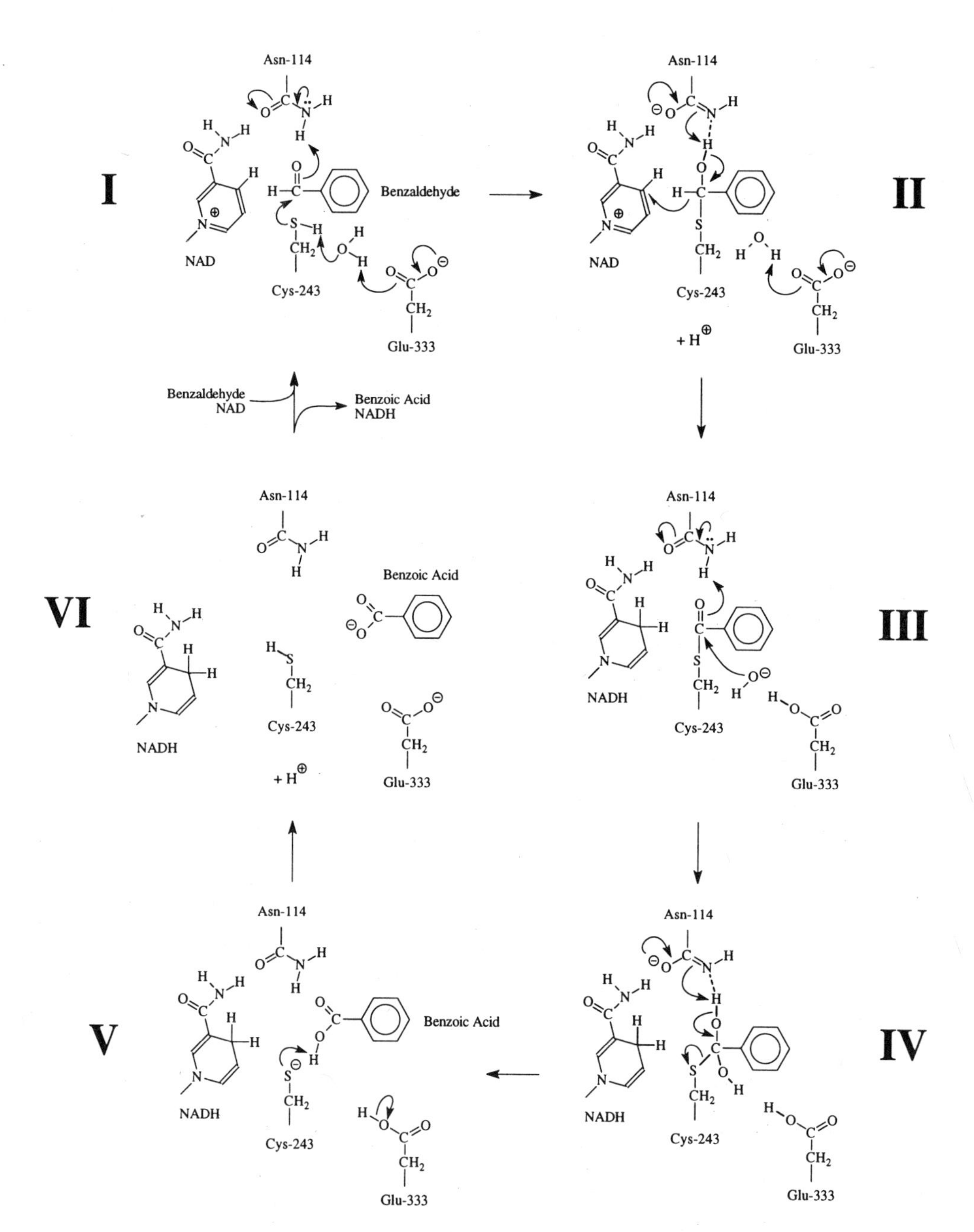

Figure 2. Proposed mechanism of aldehyde oxidation based on the class 3 ALDH tertiary structure, modelling of benzaldehyde into the catalytic site, and the newly-appreciated strict conservation of Asn-114.

the catalytic site (Liu et al., 1997b). The substrate carbonyl O projects directly between the side-chain O and N of Asn-114, prompting deduction of further mechanistic details. The thiolate of the catalytic Cys-243 can be generated in concert with Glu-333, strictly conserved and 4.5Å away (5.3Å on modelled binding of aldehyde), through an intervening water molecule (**I**). Furthermore, on aldehyde binding, the resultant decreased solvation of these residues should increase the nucleophilicity of the thiol (Kitson, 1983) while also increasing the pKa of the glutamate carboxyl. Thus, the thiolate attacks the substrate carbonyl carbon (**I**), to form a transition-state tetrahedral intermediate. Although elements are available for the intermediate to protonate as shown (**II**), it is more likely that this event does not go to completion and that transient stabilization occurs via H bonding to the substrate carbonyl O of the intermediate from the side-chain amide NH of the strictly conserved Asn-114, with a partial positive charge on the amide nitrogen balanced by a partial negative charge on the side-chain carbonyl O, which is within H-bonding distance of the NAD amide nitrogen (NN7) (Liu et al., 1997b). Either way, on hydride transfer (**II**) the thiolester intermediate forms (**III**). We note that our crystal data show the hydride virtually coplanar with the nicotinamide ring, which may be taken to be inconsistent with the A-side specificity of the enzyme (Jones et al., 1987). However, only slight movement of the ring or the aldehyde group gives directionality to the hydride. Such movement may well occur in concert with transfer, since, as noted above, phenylalanine in hydrophobic environments may act as a cation sink or "hydrophobic anion" (Dougherty, 1996). This aspect seems relevant since, with hydride transfer, the nicotinamide ring loses its cationic character, which should facilitate NADH displacement. After hydride transfer, the thiolester (**III**) is poised for base atalysis initiated by abstraction of a proton from a water molecule between the Cys-243 sulfur and Glu-333 carboxylate (**II** and **III**), creating a second tetrahedral intermediate (**IV**). (Alternatively, it has been proposed that Glu-268 activates a water for this purpose.) Collapse of this intermediate yields the product acid which may reprotonate the thiolate (**V**). On dissociation of products (**VI**), the enzyme can bind reactants again.

At **V**, an alternative can be entertained: if the water molecule between the thiolate and Glu-333 is replaced at this point, the water may serve to reprotonate the thiol instead of the product acid, leaving the Cys and Glu at the same state as shown in (**I**). In free solution the product carboxyl will ionize instantly. However, in the substrate-binding cavity, its carboxyl is likely sufficiently buried to elevate its pKa. If this is the case, without other assistance the product could remain bound in its hydrophobic environment (or at least exit at a diminished rate). An appealing candidate to facilitate deprotonation of the product acid is Glu-209, one of the residues lining the substrate-binding funnel (Liu et al., 1997b), but is believed by others to act as the general base (see below). This residue was detected by its chemical reactivity over 10 years ago (Abriola et al., 1987), and is strictly conserved in all ALDHs which release a free acid. The notable exceptions are the methyl malonyl semialdehyde dehydrogenases (Kedishvilli et al., 1992) which release product as the CoA ester, which in this scenario would not require a proton-abstracting residue. While this point is speculative, only modification of **V** in Figure 2 would be required in that case. (Protonation of Cys-243 in concert with deprotonation of Glu-333 via an intervening water, and deprotonation of the product acid by Glu 209).

The overall proposal (Figure 2) was initially based on the relationship of the nicotinamide ring in the class 3 structure with Asn-114, Glu-209, Glu-333, Cys-243, and Phe-335 and model building. The mechanism as detailed here differs from that outlined in connection with the class 2 ALDH structure (Steinmetz et al., 1997). It follows that outline in many respects, but differs regarding roles of Glu-209/268 and 333/399 (this notation denotes position numbers in class 3 and 2 ALDHs, respectively) and in recognition of the significance of residue conservation. Instead of Glu333/399, as proposed here, Glu 209/268 was proposed as the general base, while a role in hydrogen-bonding the nicotinamide ribose was ascribed to Glu333/399. However, we note: **1)** the five residues are oriented very similarly in the two structures, with RMSD values based on these residues between some pairs of class 2 ALDH subunits effectively equal to those between some pairs of class 2/class 3 subunits (Table 1, Figure 2), **2)** in the class 3 crystallographic structure, the Glu-209 carboxylate is 8.3Å from Cys-243, and it points *away from* the thiol; Glu-209 can be rotated to within 2.8Å of the Cys-243 sulfur, but the presence of aldehyde in our modelled ternary complex blocks this potential interaction, **3)** the position of the nicotinamide half of NAD appears to be the largest difference between these two regions (Figure 2), with the Hg-derivative of the class 3 enzyme providing the best data available at this time for approximation of a ternary complex. An E399Q mutant of class 2 ALDH elevated Km, NAD only about fourfold, while decreasing k_{cat} tenfold, and those authors also conceded that Glu-399 may *"play some role in catalysis other than binding to the coenzyme"* (Ni et al., 1997). The equivalent mutation in *V harveyi* ALDH has similarly mild effects on Km for coenzyme while affecting k_{cat} even more (Vedadi and Meighen, 1997). **4)** despite the long-anticipated critical role of Glu-209/268 in ALDH catalysis, it is *not* strictly conserved (Hempel et al., 1993); negatively-charged residues are not present at this position in methylmalonyl semialdehyde dehydrogenases, which release product as the CoA ester. Assuming that the catalytic thiol requires activation as in **I**, this should also apply to the methylmalonyl semialdehyde dehydrogenase family, thus a different residue must fill this role.

For these reasons we conclude that our mechanism indicating a role of Glu333/399 as the general base in the mechanism of all ALDHs warrants consideration. This is a fundamental difference vs. the mechanism outlined for the class 2 structure (which shows only small RMS differences vs. the class 3 structure) and conclusions based on site-directed mutagenesis alone (Ni et al., 1997; Wang & Weiner, 1995). Therefore, crystallographic data on a ternary complex would be invaluable in resolving this issue.

ACKNOWLEDGMENTS

Supported by AA06985 and the Georgia Research Foundation. We thank Dr. Pangala Bhat for sharing unpublished data and David Deerfield for helpful discussions regarding van der Waals radii.

REFERENCES

Abriola, D.P., Fields, R., Stein, S., MacKerell, A.D. and Pietruszko, R. (1987) Active site of human aldehyde dehydrogenase. Biochemistry **26**, 5679–5686.

Dougherty, D.A. (1996) Cation-π interactions in chemistry and biology: a new view of Benzene, Phe, Tyr and Trp. Science **271**, 163–168.

Hempel, J., Nicholas, H. and Lindahl, R. (1993) Aldehyde dehydrogenases: widespread structural and functional diversity within a shared framework. Protein Science **2**, 1890–1900.

Jones, K.H., Lindahl, R., Baker, D.C. and Timkovich, R. (1987) Hydride transfer stereospecificity of rat liver aldehyde dehydrogenases. *J. Biol. Chem.* **262**, 10911–10913.

Kedishvilli, N.Y., Popov, K.M., Rougraff, P.M., Zhao, Y., Crabb, D.W. and Harris, R.A. (1992) CoA- dependent methylmalonate-semialdehyde dehydrogenase, a unique meber of the aldehyde dehydrogenase superfamily. J. Biol. Chem. **267**, 19724–19729.

Kitson, T. (1983) The inactivation of aldehyde dehydrogenase by disulfiram in the presence of glutathione. Biochem. J **199**, 255–258.

Liu, Z., Hempel, J., Sun, J., Rose, J., Hsiao, D., Chang, W-R., Chung, Y-J., Kuo, I., Lindahl, R., and Wang, B-C. (1997a) Crystal structure of a class 3 aldehyde dehydrogenase at 2.6Å resolution. Adv. Exp. Med. Biol **414**, 1–7.

Liu, Z., Sun, Y-J., Rose, J., Chung, Y-J., Hsiao, C-D., Chang, W-R., Kuo, I., Perozich, J., Lindahl, R., Hempel, J. and Wang, B-C. (1997b) The first structure of an aldehyde dehydrogenase reveals novel interactions between NAD and the Rossmann fold. Nature Structural Biology **4**, 317–326.

Moore, S.A., Baker H.M., Blythe T., Kitson K.E., Kitson T.M. and Baker E.N. (1999) Structure of a class 1 mammalian aldehyde dehydrogenase at 2.35Å resolution. Adv. Exp. Med. Biol (this volume).

Ni, L, Sheikh, S. and Weiner, H. (1997) Involvement of Glutamate 399 and Lysine 192 in the mechanism of human liver mitochondrial aldehyde dehydrogenase. J. Biol. Chem. **272**, 18823–18826.

Perozich, J., Nicholas, H., Wang, B-C., Lindahl, R., and Hempel, J. (1999) The aldehyde dehydrogenase extended family: an overview. Adv. Exp. Med. Biol. (this volume).

Perozich, J., Nicholas, H., Wang, B-C., Lindahl, R., and Hempel, J. (1998) Relationships within the aldehyde dehydrogenase extended family. Protein Science (in press).

Rose J.P., Hempel, J., Kuo, I., Lindahl, R., and Wang, B-C. (1990) Preliminary crystallographic analysis of class 3 rat liver aldehyde dehydrogenase. Proteins **8**, 305–308.

Steinmetz ,C.G., Xie, P., Weiner, H. and Hurley, T.D. (1997) Structure of mitochondrial aldehyde dehydrogenase: the genetic component of ethanol aversion. Structure **5**, 701–711.

Vedadi, M. and Meighen, E. (1997) Critical glutamic acid residiues affecting the mechanism and nucleotide specificity of *Vibrio harveyi* aldehyde dehydrogenase. Eur. J. Biochem. **246**. 698–704.

Wang, X. and Weiner, H. (1995) Involvement of Glutamate 268 in the acitive site of human liver mitochondrial aldehyde dehydrogenase as probed by site-directed mutagenesis. Biochemistry **34** , 237–243.

8

INHIBITION OF HUMAN MITOCHONDRIAL ALDEHYDE DEHYDROGENASE BY METABOLITES OF DISULFIRAM AND STRUCTURAL CHARACTERIZATION OF THE ENZYME ADDUCT BY HPLC-TANDEM MASS SPECTROMETRY

Dennis C. Mays,[1] Andy J. Tomlinson,[2] Kenneth L. Johnson,[2] Jennifer Lam,[1] James J. Lipsky,[1] and Stephen Naylor[1,2]

[1]Clinical Pharmacology Unit
Department of Pharmacology
[2]Biomedical Mass Spectrometry Facility
Department of Biochemistry and Molecular Biology
Mayo Clinic and Foundation
Rochester, Minnesota 55905

1. INTRODUCTION

Ingestion of disulfiram blocks the metabolism of acetaldehyde, the product of ethanol metabolism, by inhibiting hepatic mitochondrial aldehyde dehydrogenase (ALDH2), a key enzyme by virtue of its low K_m for acetaldehyde (Mascher & Kikuta, 1992; Greenfield & Pietruszko, 1977). Disulfiram is rapidly reduced *in vivo* to *N,N*-diethyldithiocarbamate (DDC) (Cobby *et al.*, 1977) which is further metabolized as shown in Scheme 1. The general consensus is that disulfiram is too short-lived *in vivo* to account for the inhibition of ALDH2 and that one its metabolites is the ultimate inhibitor (Yourick & Faiman, 1991; Hart & Faiman, 1992).

In this chapter, we describe the effects of disulfiram and its metabolites on recombinant human ALDH2 (rhALDH2). In addition, we determined the nature of the irreversible inactivation of rhALDH2 by MeDTC sulfoxide (MeDTC-SO), the leading candidate for the active metabolite of disulfiram *in vivo*. Finally, we identified the location and structure of the covalent modification of rhALDH2 by MeDTC-SO using HPLC-tandem mass spectrometry (HPLC-MS/MS).

Enzymology and Molecular Biology of Carbonyl Metabolism 7
edited by Weiner *et al.* Kluwer Academic / Plenum Publishers, New York, 1999.

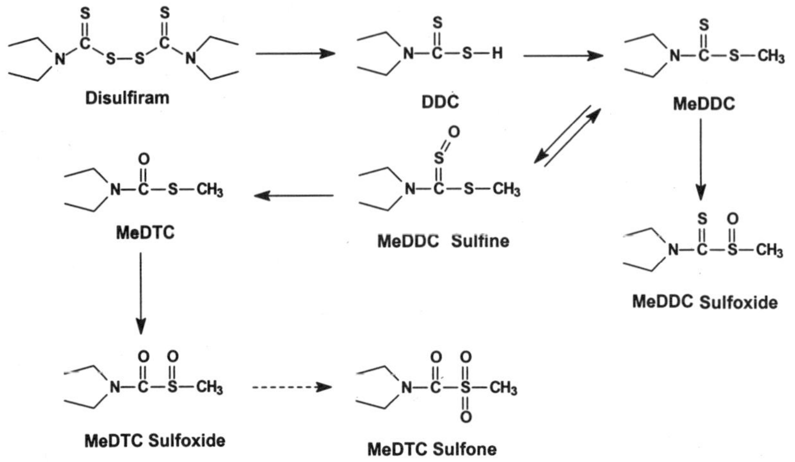

Disulfiram
DDC
MeDDC
MeDTC
MeDDC Sulfine
MeDDC Sulfoxide
MeDTC Sulfoxide
MeDTC Sulfone

Scheme 1. Pathways of activation of disulfiram. The confirmed and proposed pathways of metabolism are shown in the solid and dashed arrows, respectively.

2. EXPERIMENTAL PROCEDURES

2.1. Materials

The human ALDH2 cDNA in pT7–7 (Zheng *et al.*, 1993) was a generous gift from Dr. Henry Weiner (Department of Biochemistry, Purdue University, West Lafayette, IN). The enzyme was expressed in E. coli and purified as previously described (Lam *et al.*, 1997). MeDDC (Faiman *et al.*, 1983), MeDDC-SO (Mays *et al.*, 1998), MeDDC sulfine (Mays *et al.*, 1998), MeDTC (Hart *et al.*, 1990), MeDTC-SO (Mays *et al.*, 1996), and MeDTC-SO_2 (Mays *et al.*, 1995) were synthesized as previously described and determined to be >99% pure by HPLC-UV. Disulfiram and DDC were obtained from Sigma (St. Louis, MO). The BCA Protein Assay kit and Slide-A-Lyzer cassettes were obtained from Pierce (Rockford, IL). Staphylococcus aureus V8 (Glu C) endopeptidase and pepsin were obtained from Boehringer Mannheim.

2.2. ALDH Activity

The ALDH activity was measured using a microtiter-based assay as described previously (Nelson & Lipsky, 1995) with modifications. Briefly, purified recombinant human ALDH was added to Buffer G (0.05 M sodium pyrophosphate, pH 8.8) typically in a final volume of 700 μL. Typically, triplicate aliquots (200 μL) were added to the wells of a 96-well microtiter plate. Acetaldehyde (160 μM final concentration) and NAD (500 μM) were added together in 25 μL of Buffer G to start the dehydrogenase reaction. The protein concentration in the final incubation mixture for rhALDH2 was, unless indicated otherwise, 0.02–0.06 μM (as the tetramer). NADH generated from the oxidation of acetaldehyde was monitored by measuring the change in absorbance at 340 nm for 3 min.

2.3. Inhibitor Studies

Purified rhALDH2 was preincubated at 22 °C in Buffer G with inhibitor (0.1–1000 μM final concentration) or methanol (vehicle). The inhibitor was added in 7 μL of methanol bringing the total volume to 700 μl. At timed intervals from 0 to 30 min, triplicate aliquots of 200 μL were transferred into wells of a microtiter plate. The ALDH reaction was initiated by adding NAD and acetaldehyde in 25 μl of Buffer G. In the protection studies, glutathione (0.7 mM final concentration) or a mixture of acetaldehyde (2 mM) and NAD (1 mM) was added to rhALDH2 prior to the inhibitor.

2.4. Dialysis of Inhibited ALDH

rhALDH2 was preincubated at 22 °C for 15–30 min with inhibitor at a concentration that produced >70% inhibition, or methanol (vehicle control). A 1-ml portion of each mixture was transferred to a dialysis cassette and placed in a beaker with 150 ml of Buffer G at 22 °C. Another portion of the preincubation mixture was not dialyzed and maintained at 22 °C for 30 min. After 15 min, the dialysis buffer was replaced with 150 ml of fresh Buffer G and the sample was dialyzed for another 15 min. ALDH activity was measured in the control sample and inhibitor-treated samples before and after dialysis.

2.5. Mass Spectrometry

2.5.1. HPLC-MS and HPLC-MS/MS Analysis of Proteolytic Digests rhALDH2. The analysis of the rhALDH2 by mass spectrometry has been described in detail (Tomlinson *et al.*, 1997). Peptides were analyzed by HPLC-MS and HPLC-MS/MS with a Finnigan MAT 95Q hybrid mass spectrometer equipped with a Finnigan MAT electrospray interface. MS_1 was a magnetic sector instrument for selection of precursor ions and MS_2 was a combination of an octapole collision cell for fragmenting the precursor ion and a quadrapole mass filter for analyzing fragments ions produced in the collision cell.

2.5.2. ALDH Digestion Conditions. Twenty μL of either native or MeDTC-SO treated rhALDH2 was diluted with 40 μL of an aqueous solution containing 20 mM ammonium acetate and 1% acetic acid (pH 3.7) and incubated at 37 °C for 2 hr with 10% w/w Staphylococcus aureus V8 (Glu C) endopeptidase or pepsin. The resultant peptide mixtures were concentrated under vacuum to 20 μL and analyzed by MS under the conditions detailed above.

3. RESULTS AND DISCUSSION

3.1. ALDH Inhibition Studies

DDC, MeDDC, and MeDTC were very weak inhibitors of rhALDH2 with IC_{50}'s > 1000 μM under the conditions used in this study (Table 1). MeDDC sulfine was a relatively weak inhibitor with an IC_{50} of 62±14μM after preincubation for 30 min. MeDDC sulfoxide was a potent inhibitor of recombinant human ALDH2 with an IC_{50} of 4.1±0.9 μM after preincubation with the enzyme for 15 min (Table 1). Disulfiram, MeDTC-SO, and MeDTC-SO_2 were also very potent inhibitors of rhALDH2. In our study, the inhibition of rhALDH2 did not require the cofactor NAD, indicating that the dehydrogenase activity was not necessary for inhibition. The inhibition of rhALDH2 by the parent drug and its metabolites was time-dependent (Table 1) and not reversible by extensive dialysis. Furthermore, glutathione substantially reduced or completely blocked the inhibition of rhALDH2 by all compounds tested except MeDTC-SO (Table 1).

Table 1. Inhibition of rhALDH2

Inhibitor	IC_{50} (μM)[1]	Time-dependent	Irreversible[2]	Reactivity with GSH[3]
DDC	>1000	ND	ND	ND
MeDDC	>1000	ND	ND	ND
MeDTC	>1000	ND	ND	ND
MeDDC Sulfine	62±14	yes	yes	+
MeDDC-SO	4.1±0.9	yes	yes	+
Disulfiram	1.4±0.4	yes	yes	+++
MeDTC-SO	1.3±0.6	yes	yes	~0
MeDTC-SO_2	0.40±0.10	yes	yes	+++

[1]All inhibitors were preincubated with rhALDH2 at 22 °C for 15 min except for MeDDC sulfine which was for 30 min. Values are the mean ± S.D. of 3 to 4 assays.
[2]Determined by dialysis.
[3]Relative values with +++ as the highest. ND = not determined.

The inhibition of rhALDH2 by MeDTC-SO and MeDTC-SO_2 was studied in more detail. The kinetics of inactivation of rhALDH2 by both compounds was pseudo-first-order with respect to inhibitor concentration as shown in Figure 1. Treating the inhibitor as an affinity labeling reagent and the enzyme as a reagent, the reaction can be described as follows (Kitz & Wilson, 1962).

Reaction 1 (Plapp, 1982):

$$E + I \underset{k_2}{\overset{k_1}{\rightleftharpoons}} E{\bullet}I \xrightarrow{k_3} E - I, \qquad K_I = k_2/k_1$$

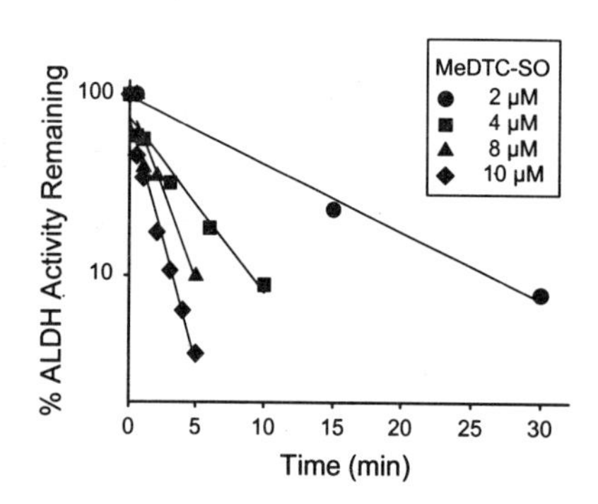

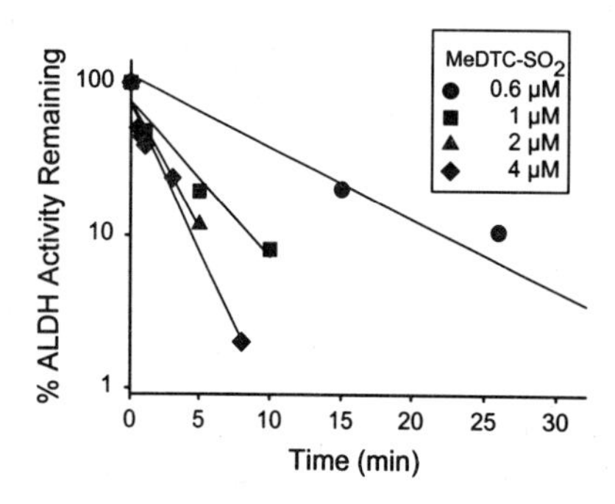

Figure 1. Time-course of inactivation of rhALDH2 by MeDTC-SO and MeDTC-SO_2. Reproduced with permission from (Lam *et al.*, 1997). Copyright 1997 American Chemical Society.

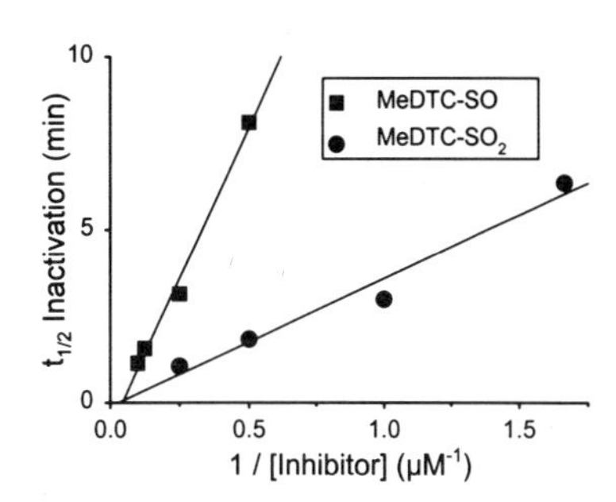

Figure 2. Kitz and Wilson plots of the inactivation of rhALDH2. The half-life rhALDH2 versus 1/[I] for the data in Figure 2 are shown. Reprinted with permission from (Lam *et al.*, 1997). Copyright 1997 American Chemical Society.

The enzyme E and the inhibitor I form a reversible complex (E•I), which proceeds to an irreversibly inactivated enzyme (E-X). The pseudo-first-order rate constant, k_{obs}, obtained from the plots of ALDH activity versus time (Figure 1) is related to k_3 and K_I by equation 1 (Plapp, 1982).

$$k_{obs} = k_3[I]/(K_I + [I]) \tag{1}$$

This equation predicts hyperbolic saturation kinetics. Plots of $t_{1/2}$ versus 1/[I] should be linear with the y-axis intercept equal to the $t_{1/2}$ of inactivation and the x-axis intercept equal to $-1/K_I$. However, the Kitz and Wilson plots of the inactivation of rhALDH2 by MeDTC-SO and MeDTC-SO_2 intercepted the y-axis near zero, an indication that inactivation of rhALDH2 was apparently not saturable (Figure 2)(Kitz & Wilson, 1962). This may have been due, in part, to the difficulty of measuring the rate of inactivation of ALDH at higher concentrations of inhibitor due to the rapidity of the reaction. Because saturation kinetics was not observed, plots of k_{obs} of inactivation versus [I] were made (Figure 3). The slope of this plot is equal to k_3/K_I when $[I] \ll K_I$ (see equation 1) and approximates the pseudo-bimolecular rate for the inhibitor-enzyme pairs (Plapp, 1982). The respective values for k_3/K_I were 2.9×10^3 and 9.8×10^2 $s^{-1}M^{-1}$ for rhALDH2-MeDTC-SO_2 and rhALDH2-MeDTC-SO, indicating that the former reagent pair is about 3 times more reactive. However, the resistance of MeDTC-SO to scavenging by thiols supports the notion that it is the more likely inhibitor *in vivo*. Therefore, we chose to study the interaction of MeDTC-SO with rhALDH2 in the subsequent mass spectrometry experiments.

3.2. Mass Spectrometry

In a previous chapter in this series (Lipsky *et al.*, 1996) and elsewhere (Tomlinson *et al.*, 1997), we have described our initial investigation by HPLC-MS into the mechanism of the inhibition of recombinant human ALDH2 by MeDTC sulfoxide. In those experiments,

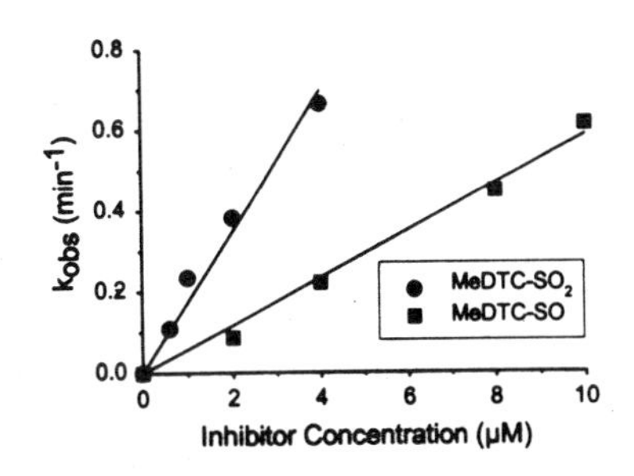

Figure 3. k_{obs} versus [I] for inactivation of rhALDH2. The k_{obs} (= $0.693/t_{1/2}$) versus [I] are plotted for the time-dependent inactivation of rhALDH2 for the corresponding data in Figure 2. Reprinted in modified form with permission from (Lam *et al.*, 1997). Copyright 1997 American Chemical Society.

rhALDH2 (1 μM) was incubated with 40 μM MeDTC-SO or vehicle (methanol) at pH 7.4 at 22 °C for 25 min. This treatment inhibited rhALDH2 activity by 80–90% and resulted in an increase of approximately 100 Da in molecular mass of the intact protein. The molecular mass of the MeDTC-SO-inhibited rhALDH2 was 54,533 Da (± 0.01%) by HPLC-MS compared to 54,432 Da for the uninhibited control. In addition, both native rhALDH2 and MeDTC-SO-modified rhALDH2 were digested with Glu C and analyzed by HPLC-MS. In the digest of the native protein, an incomplete peptide digestion fragment was detected by multiply charged ions at $[MH_3]^{3+}$ = 1609 and $[MH_4]^{4+}$ = 1207 which corresponded to a molecular weight of 4825 Da (Lipsky *et al.*, 1996; Tomlinson *et al.*, 1997). This molecular weight is consistent with a Glu-C generated peptide spanning the active site region of rhALDH2 from Leu^{269} to Glu^{312} LGGKSPNIIMSDADME-WAVEQAHFALFFNQGQCCC-AGSRTFVQE (Hempel *et al.*, 1993). LC-MS analysis of the endopeptidase Glu C digest of rhALDH2 treated with MeDTC-SO revealed multiply charged ions at $[MH_3]^{3+}$ = 1642 and $[MH_4]^{4+}$ = 1232, which were not present in the native protein digest. These latter ions were consistent with the Glu C-derived peptide described above for the native protein but modified by the addition of a mass of 99 Da, resulting in a molecular mass of 4924 Da. The results with the intact protein and the peptide from the Glu C digest can be rationalized by carbamoylation rhALDH2 by MeDTC-SO. It is noteworthy that this peptide contains the essential active site nucleophile Cys^{302} in the highly conserved active site region of ALDH2 (Hempel & Pietruszko, 1981; Hempel *et al.*, 1993; Farrés *et al.*, 1995).

To produce smaller peptides suitable for determining the location and structure of the putative covalent modification, samples of native and MeDTC-SO treated rhALDH2 were treated as described above and digested with pepsin at 37 °C for 2 hr at pH 3.7. In

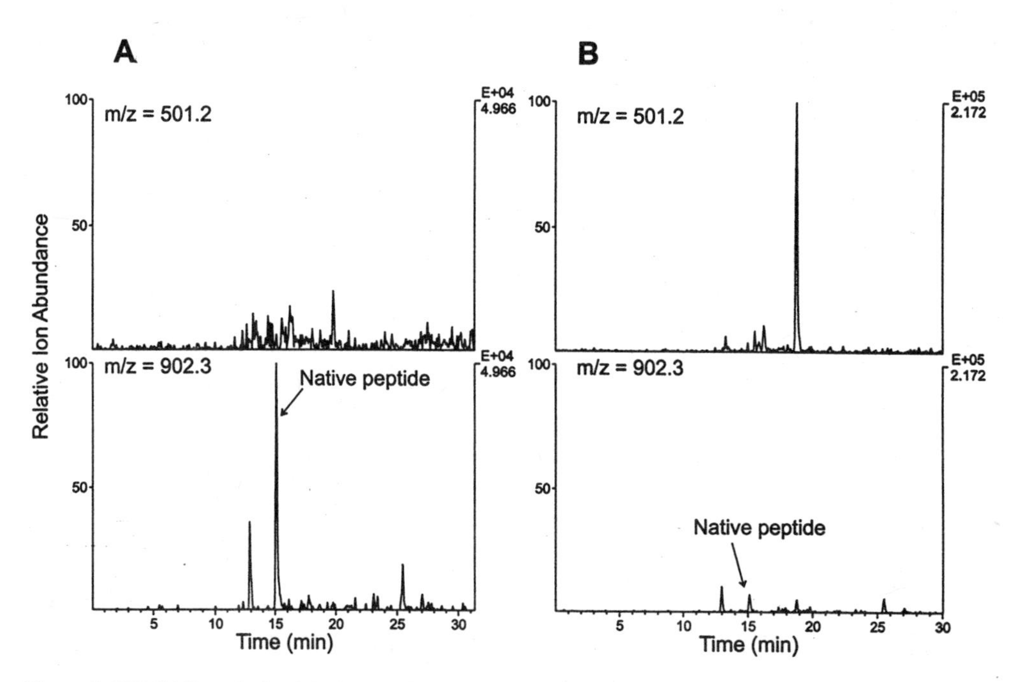

Figure 4. HPLC-MS analysis of rhALDH2 after pepsin digestion. HPLC-MS ion chromatograms of the pepsin digests of native rhALDH2 (A) and MeDTC-SO-treated rhALDH2 (B). The native active site peptide at m/z 902.3 (MH^+) and the modified active site peptide at m/z 501.2 ($[MH_2]^{2+}$) eluted at 15.1 and 18.5 min, respectively.

the HPLC-MS analysis of the pepsin digest of the native rhALDH2, a prominent ion at $MH^+ = 902.3$ eluted at 15.1 min, which corresponded to native FNQGQCCC. (Figure 4A) The HPLC-MS analysis of the pepsin digest of the MeDTC-SO-treated rhALDH2 revealed a doubly charged ion of m/z 501.2 ($[MH_2]^{2+} = 501.2$, equivalent to $MH^+ = 1001.4$) at retention time 18.5 min (Figure 4B). This ion corresponded to the carbamoylated peptide FNQGQCCC spanning the active site region of residues 296 to 303, but with an additional 99 Da. There was a small signal at m/z 902.3 corresponding to the unmodified FNQGQCCC ($MH^+ = 902.3$) in the pepsin digest of the MeDTC-SO treated rhALDH2.

Using conditions similar to those described by us previously (Tomlinson & Naylor, 1995) the doubly charged precursor ion at $[MH_2]^{2+} = 501.2$ from the pepsin digest of MeDTC-SO-treated rhALDH2 was subjected to HPLC-MS/MS. With the low collision energy used in these experiments, fragmentation of a peptide ion occurs primarily along the peptide backbone yielding ions according to Scheme 2. The product ion spectrum for the adducted peptide $[MH_2]^{2+} = 501.2$ is shown in Figure 5A. Even for this relatively simple peptide, the mass spectrum was complex. However, among the important ions were those which were identical to the unmodified sequence FNQGQC of the native peptide: 1) a series of singly charged ions of b_2-b_5 at m/z 262, 390, 447 and 575, respectively; of $b_2{}^*$ - $b_4{}^*$, $b_6{}^*$ at m/z 245, 373, 430, and 661, respectively; and 2) the doubly charged ions of $[b_4{}^*]^{2+}$, $[b_5{}^*]^{2+}$, and $[a_6{}^*]^{2+}$ at m/z 216, 280, and 367, respectively. Several ions indicated the presence of a 99-Da adduct (shown in bold in Figure 5A): 1) singly charged ions of $y_2{}^*$, y_2, y_3 and y_6 at m/z 307, 324, 427 and 740, respectively; and 2) the doubly charged ions of $[y_3]^{2+}$ and $[a_7{}^*]^2$ at m/z 206 and 418, respectively. These results are consistent with the modification occurring at Cys^{302}. Additional important ions in Figure 5A are at m/z 72 and 100 which are characteristic fragments of the *N,N*-diethylcarbamoyl group (Jin *et al.*, 1994). One unexplained ion is at m/z 208.3, which is consistent with an unmodified $y_2{}^*$. It is possible that this ion was formed by the further fragmentation of MH^+ at m/z 902.3 which had previously undergone loss of adduct in the mass analyzer. Another possible explanation is that the adduct was lost directly from $y_2{}^*$.

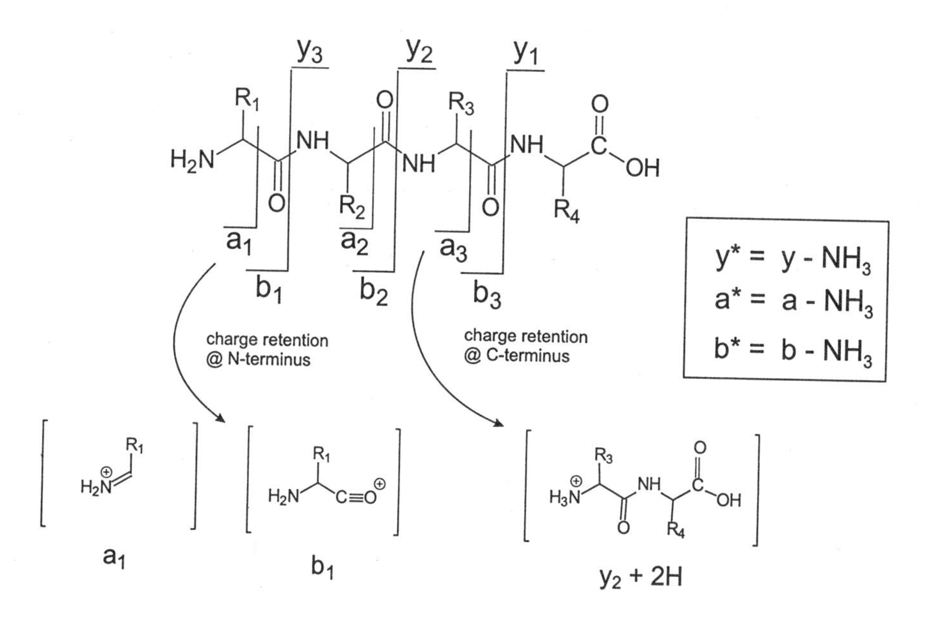

Scheme 2. Notation for the peptide fragment ions in MS/MS (Roepstroff & Fohlman, 1984).

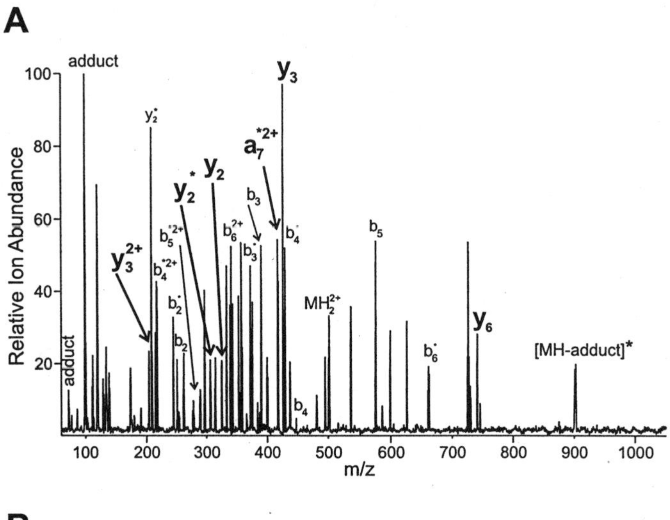

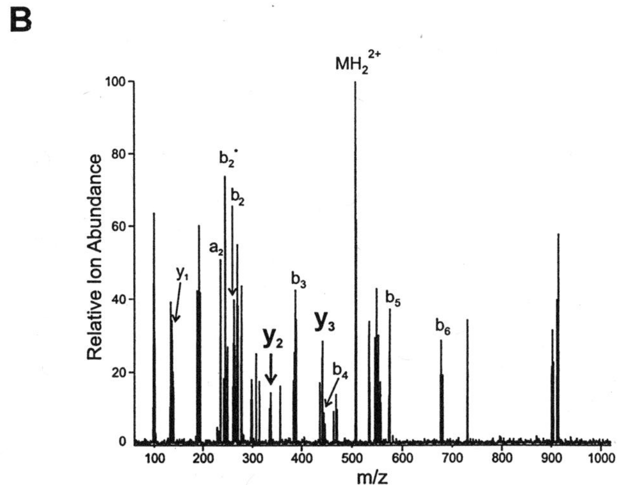

Figure 5. HPLC-MS/MS analysis of the active site peptide FNQGQCCC . The product ion spectra of (A) the adducted peptide ($[MH_2]^{2+}$ = 501.2) and of (B) the methylated adducted peptide ($[MH_2]^{2+}$ = 508.2) derived from the pepsin digest of MeDTC-SO-treated rhALDH2. Ions that are 99 Da larger than those expected from the native peptide are labeled in bold. Reproduced with modifications from Biochemical Pharmacology (Tomlinson *et al.*, 1997) with permission from Elsevier Science.

The location of the modified amino acid was further investigated by treating the pepsin digests of MeDTC-SO-modified and native rhALDH2 with methanol-HCl. This procedure converts C-terminus and side chain carboxylic acids to their methyl ester derivatives and has been used to improve "y" series ions in MS/MS analysis (Tomlinson & Naylor, 1995). A doubly charged ion at m/z 508.2 was detected in the HPLC-MS of the digest of the MeDTC-SO treated rhALDH2 that was absent from the digest of native rhALDH2 (data not shown). The ion at m/z 508.2, presumably the C-terminus methylated adducted peptide FNQGQCCC(=O)OMe, was analyzed by HPLC-MS/MS (Figure 5B). Among the important ions in this mass spectrum are the "y"-series ions y_2 and y_3 (shown in bold) at m/z 338 and 441, respectively, which were both 99 Da higher than expected for

y_2 and y_3 from the methylated native peptide. In addition, an ion at m/z 136 was detected, which corresponded to the unadducted carboxy terminus of the methylated peptide (i.e. residue 303), thus confirming the location of the adduct on Cys^{302}. These data are consistent with carbamoylation of rhALDH2 by MeDTC-SO at Cys^{302} as shown below.

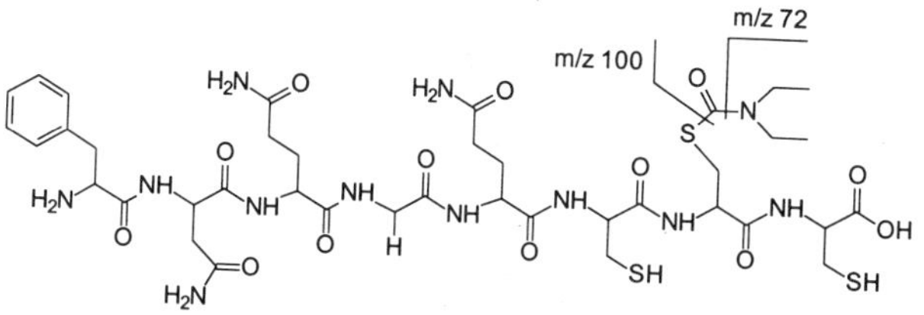

m/z 72
m/z 100

This super-reactive cysteine at residue 302 was first identified in ALDH2 employing the group-specific reagent iodoacetamide (Hempel & Pietruszko, 1981) and later using a vinyl ketone affinity reagent (Blatter *et al.*, 1990). The specificity of the reaction of this particular cysteine with MeDTC-SO is of interest and may have implications for understanding in more detail the mechanism of inhibition of ALDH by disulfiram-related compounds.

In summary, disulfiram and several of its metabolites inactivated human mitochondrial ALDH, the pharmacological target of alcohol aversion therapy, in a time-dependent, irreversible manner. However, glutathione prevented inactivation of ALDH2 by all inhibitors except MeDTC-SO, suggesting that this inhibitor is more likely the active metabolite *in vivo*. HPLC-MS analyses of MeDTC-SO-treated intact rhALDH2 and its proteolytic digests demonstrated that the protein had been covalently modified. Further analyses revealed a doubly charged ion at $[MH_2]^{2+} = 501.2$, corresponding to the carbamoylated peptide FNQGQCCC (residues 296–303). HPLC tandem mass spectrometry analyses of this peptide confirmed the presence of a carbamoyl group at Cys^{302}. Overall, these results are consistent with a mechanism of inactivation of ALDH by disulfiram *in vivo* that involves carbamoylation of the essential active site Cys^{302} by MeDTC sulfoxide.

ACKNOWLEDGMENTS

These investigations were supported by NIH AA09543. We thank Dr. Henry Weiner for the valuable discussions of this research and his gift of the clone of the rhALDH2.

REFERENCES

Blatter, E.E., Tasayco J., M.L., Prestwich, G. & Pietruszko, R. (1990). Chemical modification of aldehyde dehydrogenase by a vinyl ketone analogue of an insect pheromone. *Biochem.J.* **272**, 351–358.

Cobby, J., Mayersohn, M. & Selliah, S. (1977). The rapid reduction of disulfiram in blood and plasma. *J.Pharmacol.Exp.Ther.* **202**, 724–731.

Faiman, M.D., Artman, L. & Maziasz, T. (1983). Diethyldithiocarbamic acid-methyl ester distribution, elimination, and LD50 in the rat after intraperitoneal administration. *Alcohol.Clin.Exp.Res.* **7**, 307–311.

Farrés, J., Wang, T.T.Y., Cunningham, S.J. & Weiner, H. (1995). Investigation of the Active Site Cysteine Residue of Rat Liver Mitochondrial Aldehyde Dehydrogenase by Site-Directed Mutagenesis. *Biochemistry* **34**, 2592–2598.

Greenfield, N.J. & Pietruszko, R. (1977). Two aldehyde dehydrogenases from human liver. Isolation via affinity chromatography and characterization of the isozymes. *Biochim.Biophys.Acta.* **483**, 35–45.

Hart, B.W. & Faiman, M.D. (1992). In vitro and in vivo inhibition of rat liver aldehyde dehydrogenase by S-methyl N,N-diethylthiolcarbamate sulfoxide, a new metabolite of disulfiram. *Biochem.Pharmacol.* **43**, 403–406.

Hart, B.W., Yourick, J.J. & Faiman, M.D. (1990). S-methyl-N,N-diethylthiolcarbamate: a disulfiram metabolite and potent rat liver mitochondrial low Km aldehyde dehydrogenase inhibitor. *Alcohol* **7**, 165–169.

Hempel, J., Nicholas, H. & Lindahl, R. (1993). Aldehyde dehydrogenases: Widespread structural and functional diversity within a shared framework. *Protein Sci.* **2**, 1890–1900.

Hempel, J.D. & Pietruszko, R. (1981). Selective chemical modification of human liver aldehyde dehydrogenases E1 and E2 by iodoacetamide. *J.Biol.Chem.* **256**, 10889–10896.

Jin, L., Davis, M.R., Hu, P. & Baillie, T.A. (1994). Identification of novel glutathione conjugates of disulfiram and diethyldithiocarbamate in rat bile by liquid chromatography-tandem mass spectrometry. Evidence for metabolic activation of disulfiram in vivo. *Chem.Res.Toxicol.* **7**, 526–533.

Kitz, R. & Wilson, I.B. (1962). Esters of methanesulfonic acid as irreversible inhibitors of acetylcholinesterase. *J.Biol.Chem.* **237**, 3245–3249.

Lam, J.P., Mays, D.C. & Lipsky, J.J. (1997). Inhibition of recombinant human mitochondrial and cytosolic aldehyde dehydrogenase by two candidates for the active metabolites of disulfiram. *Biochemistry* **36**, 13748–13754.

Lipsky, J.J., Mays, D.C., Holt, J.L., Tomlinson, A.J., Johnson, K.L., Veverka, K.A. & Naylor, S. (1996). Inhibition of and interaction with human recombinant mitochondrial aldehyde dehydrogenase by methyl diethylthiocarbamate sulfoxide. *Adv.Exp.Med.Biol.* **414**, 209–216.

Mascher, H. & Kikuta, C. (1992). New, high-sensitivity high-performance liquid chromatographic method for the determination of acyclovir in human plasma, using fluorometric detection. *J.Chromatogr.* **583**, 122–127.

Mays, D.C., Nelson, A.N., Fauq, A.H., Shriver, Z.H., Veverka, K.A., Naylor, S. & Lipsky, J.J. (1995). *S*-Methyl *N,N*-diethylthiocarbamate sulfone, a potential metabolite of disulfiram and potent inhibitor of low K_m mitochondrial aldehyde dehydrogenase. *Biochem.Pharmacol.* **49**, 693–700.

Mays, D.C., Nelson, A.N., Lam-Holt, J.P., Fauq, A.H. & Lipsky, J.J. (1996). *S*-Methyl-*N,N*-diethylthiocarbamate sulfoxide and *S*-methyl-*N-N*-diethylthiocarbamate sulfone, two candidates for the active metabolite of disulfiram. *Alcohol.Clin.Exp.Res.* **20**, 595–600.

Mays, D.C., Ortiz-Bermudez, P., Lam, J.P., Tong, I.H., Fauq, A.H. & Lipsky, J.J. (1998). Inhibition of recombinant human mitochondrial aldehyde dehydrogenase by two intermediate metabolites of disulfiram. *Biochem.Pharmacol.* **55**, 1099–1103.

Nelson, A.N. & Lipsky, J.J. (1995). Microtiter plate-based determination of multiple concentration-inhibition relationships. *Anal.Biochem.* **231**, 437–439.

Plapp, B.V. (1982). Application of affinity labeling for studying structure and function of enzymes. *Methods Enzymol.* **87**, 469–499.

Roepstroff, P. & Fohlman, J. (1984). Proposal for a common nomenclature for sequence ions in mass spectra of peptides [letter]. *Biomed.Mass.Spectrom.* **11**, 601–601.

Tomlinson, A.J., Johnson, K.L., Lam-Holt, J., Mays, D.C., Lipsky, J.J. & Naylor, S. (1997). Inhibition of human mitochondrial aldehyde dehydrogenase by the disulfiram metabolite S-methyl-N-N-diethylthiocarbamoyl sulfoxide. Structural characterization of the enzyme adduct by HPLC-tandem mass spectrometry. *Biochem.Pharmacol.* **54**, 1253–1260.

Tomlinson, A.J. & Naylor, S. (1995). A strategy for sequencing peptides from dilute mixtures at the low femtomole level using membrane preconcentration-capillary electrophoresis-tandem mass spectrometry (mPC-CE-MS/MS). *J.Liquid Chromatogr.* **18**, 3591–3615.

Yourick, J.J. & Faiman, M.D. (1991). Disulfiram metabolism as a requirement for the inhibition of rat liver mitochondrial low Km aldehyde dehydrogenase. *Biochem.Pharmacol.* **42**, 1361–1366.

Zheng, C.-F., Wang, T.T.Y. & Weiner, H. (1993). Cloning and expression of the full-length cDNAs encoding human liver class 1 and class 2 aldehyde dehydrogenase. *Alcohol.Clin.Exp.Res.* **17**, 828–831.

9

MECHANISM OF INHIBITION OF RAT LIVER CLASS 2 ALDH BY 4-HYDROXYNONENAL

Stephen W. Luckey, Ronald B. Tjalkens, and Dennis R. Petersen

Molecular Toxicology and Environmental Health Sciences
University of Colorado Health Sciences Center
Denver, Colorado 80262

1. INTRODUCTION

Lipid peroxidation is a pathological process that results in the peroxidation of cellular membrane lipids ultimately giving rise to a number of reactive, cytotoxic, aldehydic products (Esterbauer *et al.,* 1991). 4-hydroxy-2-*trans*-nonenal (4-HNE) and malondialdehyde (MDA), the most abundant aldehydes produced during lipid peroxidation, have a relatively long half-life and are capable of diffusing to distant sites within the cell of origin or into adjacent cells. 4-HNE can produce a variety of adverse cellular effects which have been summarized in detail elsewhere (Schauer *et al.,* 1990) and include the inhibition of various enzymes (Vander Jagt *et al.,* 1997). The ability of certain biogenic aldehydes to produce diverse biological and cytotoxic effects can be attributed to their α, β-unsaturated configuration that gives the compounds strong electrophilic properties (Esterbauer *et al.,* 1991). Thus, investigators attribute these adverse effects to the formation of aldehyde adducts with cellular protein nucleophiles through covalent alkylation of sulfhydryl, primary amino, and histidyl groups of proteins (Hartley *et al.,* 1997; Uchida and Stadtman, 1993).

Based on the cytotoxic nature of 4-HNE, it is important for cells to maintain very low concentrations of this lipid peroxidation endproduct. Our laboratory has shown, using isolated hepatocytes, that 4-HNE is rapidly metabolized to less reactive intermediates by alcohol dehydrogenase, aldehyde dehydrogenase, and glutathione-*S*-transferase pathways (Hartley *et al.,* 1995). These data established that the aldehyde dehydrogenase and glutathione-*S*-transferase pathways are the primary detoxification routes of 4-HNE. Multiple forms of ALDH present in the cytosolic, mitochondrial and microsomal fractions display a wide range of affinities for 4-HNE (Mitchell and Petersen, 1987). However, based on V/K values, the high-affinity form of ALDH present in rat liver mitochondria (Class 2 ALDH) is catalytically the most active in oxidizing 4-HNE. We have also reported that 4-HNE is a potent competitive and, to a lesser extent, a mixed-type inhibitor of acetaldehyde oxidation by the high affinity ALDH with an apparent K_i value of 0.48 μM (Mitchell and Petersen, 1991).

Enzymology and Molecular Biology of Carbonyl Metabolism 7
edited by Weiner *et al.* Kluwer Academic / Plenum Publishers, New York, 1999.

Since 4-HNE inhibition of high-affinity mitochondrial ALDH compromises the ability of the cell to detoxify 4-HNE allowing the aldehyde to further damage cells, it is important to determine the mechanism of ALDH inactivation. The observation that 4-HNE is both a substrate and an inhibitor for the high affinity mitochondrial ALDH, prompted us to investigate whether the observed inhibition by 4-HNE may be mechanism-based or occur as a result of nonspecific, covalent interactions with the enzyme.

2. MATERIALS AND METHODS

2.1. Preparation of 4-Hydroxynonenal

4-hydroxynonenal was synthesized and liberated from its diacetal form as described previously (Mitchell and Petersen, 1991). Pure 4-HNE was diluted in methanol and quantitated spectrophotometrically by measuring absorbance at 224 nm (e = 13,750 $M^{-1}cm^{-1}$).

2.2. Preparation of Rat Liver Mitochondrial and Purification of High-Affinity Mitochondrial Aldehyde Dehydrogenase

Rats were sacrificed by intraperitoneal injection of pentobarbital sodium (75 mg/kg). Livers were quickly removed and placed in ice-cold (4°C) 0.25 M sucrose. Wet weight was determined and a 10% homogenate was prepared and the mitochondrial subcellular fraction was obtained by differential centrifugation as described previously (Mitchell and Petersen, 1987).

High-affinity mitochondrial aldehyde dehydrogenase was purified as previously described (Mitchell and Petersen, 1987) with several modifications. Briefly, the mitochondrial fraction was sonicated and centrifuged at 100,000*g* for 60 minutes to collect mitochondrial membranes. The supernatant was subjected to ammonium sulfate precipitation (30% to 55% saturation) and centrifuged at 20,000 *g* for 20 minutes. The 30% to 55% pellet, enriched with high-affinity ALDH, was resuspended in 3 ml of 1 mM sodium-phosphate buffer (pH 7.4) containing 2 mM mercaptoethanol. Removal of ammonium sulfate was accomplished by desalting column chromotography using Bio-Gel P-6DG. Further purification included application to a DEAE-cellulose anion-exchange chromatography column equilibrated with 1 mM sodium-phosphate buffer (pH 7.4) containing 2 mM mercaptoethanol. The high-affinity mitochondrial was eluted using 50 mM sodium-phosphate buffer (pH 6.2) containing 2 mM mercaptoethanol. Those fractions with activity were then applied to a 5'-AMP sepharose 4B affinity column equilibrated with 100 mM potassium-phosphate buffer. The column was sequentially washed with 100 mM potassium-phosphate buffer, 450 mM potassium-phosphate buffer, and 25 mM potassium-phosphate buffer, whereafter high-affinity enzyme was eluted with 25 mM potassium-phosphate buffer containing 0.5 mg/ml NAD^+. In all instances, appropriate column eluent fractions were assayed for ALDH activity. Protein concentrations were determined by BCA Protein Assay kit as required by the instructions. Denaturing polyacrylamide gel electrophoresis (7.5% polyacrylamide gels) was used to ascertain the purity of ALDH protein.

2.3. Assay for Aldehyde Dehydrogenase Activity

ALDH activity was assayed for the oxidation of propionaldehyde by monitoring the production of NADH spectrophotometrically at 340 nm. The activity reaction mixture contained 1 ml of 50mM sodium-phosphate buffer (pH 7.4), 1 mM NAD^+, protein (10–100 μL), and 1 mM propionaldehyde. The inhibition experiments contained incubations of purified ALDH protein (40 μg), 50 mM sodium-phosphate buffer (pH 7.4), and 4-HNE concentrations ranging from 25 to 250 μM. To determine mechanism-based inactivation, 1 mM NAD^+ was added to the proper incubations. After the appropriate incubation times at 30°C, aliquots (100 μL) were obtained and promptly assayed for ALDH activity. ALDH activity is expressed in nmol of NADH formed/min/mg protein.

2.4. Analysis of Sulfhydryl Groups

Inhibition reaction mixtures were constructed as previously discussed and an aliquot (12.5 μg of protein) was derivatized with 5,5'-dithio-*bis*-2-nitrobenzoic acid (DTNB) in a Tris buffer (0.2 mM Tris, 1 mM EDTA, pH 8.0) for detection of protein sulfhydryls. After 15 minutes, the absorbance at 412 nm was measured. For quantification, glutathione was utilized for production of a standard curve.

3. RESULTS AND DISCUSSION

Conceptually, mechanism-based inactivators are substrates for the enzyme which during the catalytic cycle convert the parent compound into an intermediate which inactivates the enzyme. Our proposition that 4-HNE may be a mechanism-based inhibitor is based on the observation that it is both a substrate for the high affinity mitochondrial ALDH with an apparent K_m of 17.5 μM (Mitchell and Petersen, 1987) as well as an inhibitor with a K_i of 0.48 μM (Mitchell and Petersen, 1991).

Inactivation of ALDH was measured by incubating protein with increasing concentrations of 4-HNE in the presence or in the absence of NAD^+. Samples were withdrawn from the incubations and then measured for ALDH activity by monitoring for NADH production using 1 mM propionaldehyde. The plots of these data are depicted in Figure 1A-D and demonstrate that ALDH inactivation is not mechanism-based. This conclusion is based on the observation that, independent of 4-HNE concentration, ALDH inhibition occurs at a higher rate in the absence of NAD^+. For instance, after 2 hours of pre-incubation with 25 μM 4-HNE (Figure 1A) in the absence of NAD^+, ALDH activity decreased to almost 50% of the initial activity while ALDH activity in pre-incubations containing NAD^+ decreased less than 20%. Similarly, at 250 μM 4-HNE (Figure 1D), ALDH activity completely disappeared by 1 hour when NAD^+ is not present in the pre-incubation. Meanwhile, ALDH preincubated with NAD^+ and 250 μM 4-HNE retained almost 20% activity after 2 hours at this 4-HNE concentration. These results strongly suggest that the inactivation of ALDH by 4-HNE is not mechanism-based but indicates that the 4-HNE molecule itself directly inhibits the enzyme.

Consistent with our previous reports (Mitchell and Petersen, 1991), the data in Figure 2 demonstrate that 4-HNE is a potent mixed-type inhibitor of ALDH. By plotting V_{max} versus enzyme concentration, in the presence of increasing concentrations of inhibitor, a competitive or noncompetitive inhibitor will cause a decrease in the slope of the lines but all lines will intersect at a common origin (Segel, 1976). Irreversible inhibition is evidenced by a pattern of lines with similar slopes but intersections on the *x* axis corresponding to the amount of enzyme irreversibly inactivated. The data in Figure 2 shows that pre-incubation of ALDH with increasing concentrations 4-HNE, in the presence (Figure 2A) or absence (Figure 2B) of NAD^+, for 5 minutes results in a family of lines with differ-

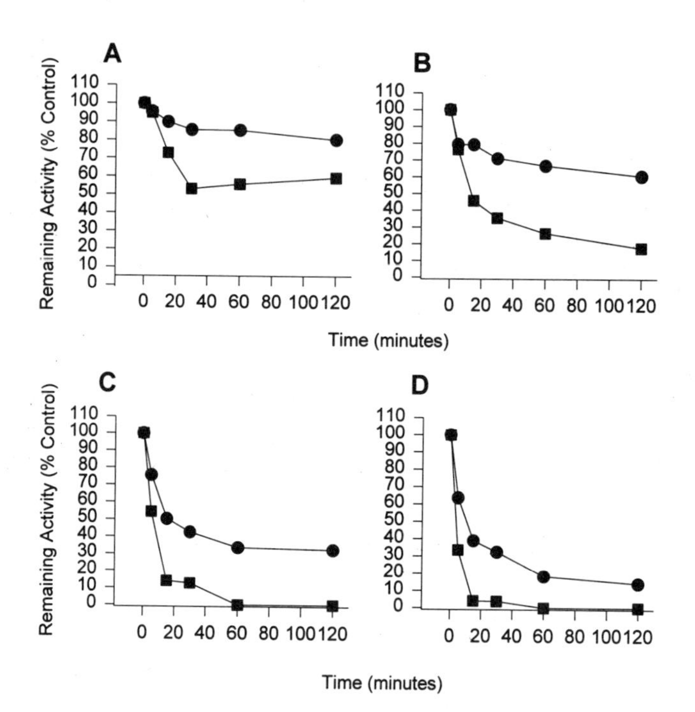

Figure 1. Time course of inactivation of high-affinity mitochondrial ALDH by 4-HNE. Purified ALDH (40μg) was treated with (A) 25 μM, (B) 50 μM, (C) 125 μM, and (D) 250 μM 4-HNE for the appropriate time periods in the presence (●) or absence (■) of NAD^+ (1mM). Activity was determined by production of NADH with 1mM propionaldehyde. The results are corrected for controls in which enzyme was incubated with NAD^+ only under identical experimental conditions.

ent slopes and a common *x* intercept suggesting reversible inhibition that is most likely competitive or noncompetitive in nature. Interestingly and consistent with the data in Figure 1, the decreases in the slope of the lines are markedly less in the incubations containing NAD^+ (Figure 2B and 2D) further demonstrating a protective effect of the coenzyme.

Figure 2 also presents ALDH activities after 2 hours of incubation with increasing concentrations of 4-HNE in the presence (Figure 2C) and absence (Figure 2D) of NAD^+. These graphical data reveal that more prolonged incubation times of ALDH with 4-HNE results in a series of lines with decreasing slopes and significant shifts in the *x* intercept suggestive of increasing amounts of enzyme inactivation through irreversible inhibition. Again, the more marked decrease in the slope of lines in incubations not containing NAD^+ suggests a protective role of this cofactor.

Collectively, the data in Figures 1 and 2 suggest that the inhibition of ALDH by 4-HNE is not mechanism-based. This is based on the observation that the rate and magnitude of ALDH inhibition was much greater in the absence of NAD^+. These results also indicate that NAD^+ may protect the enzyme by either inhibiting covalent adduction or maintaining a catalytically active protein that oxidizes 4-HNE to a lesser toxic metabolite. It is well recognized that the mechanism of action of ALDH involves an ordered addition of cofactor (NAD^+) followed by addition of substrate. Thus, under physiological conditions, ALDH is normally present with NAD^+ bound. Therefore, ALDH is normally protected from an immediate inactivation by electrophilic aldehydes. This notion is consistent with the ability of NAD^+ to protect other enzymes such as glyceraldehyde-3-phosphate de-

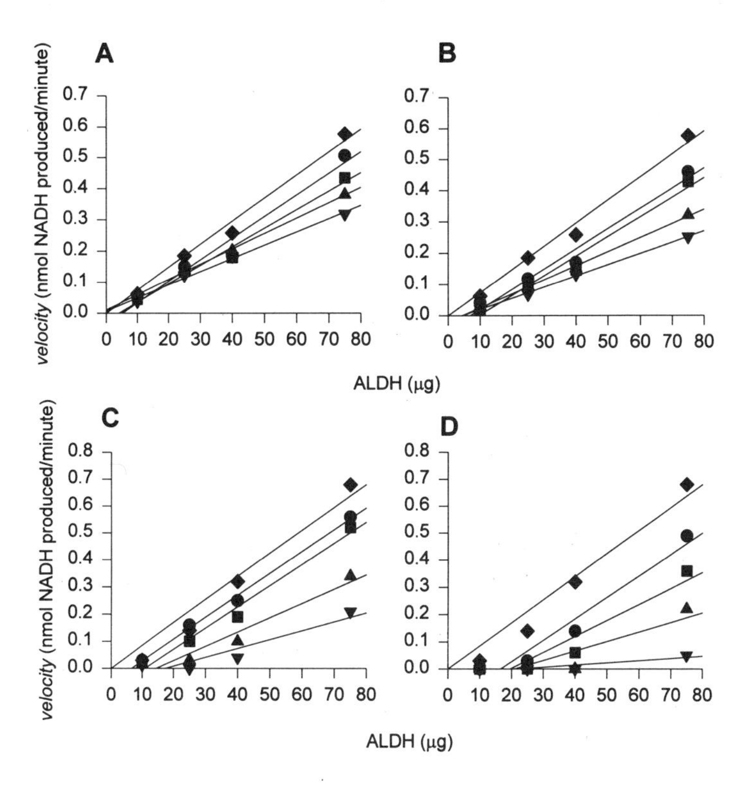

Figure 2. Inhibition of high-affinity mitochondrial ALDH by 4-HNE. Purified ALDH (40 μg) was treated with (◆) 0 μM, (●) 25 μM, (■) 50 μM, (▲) 125 μM, (▼) 250 μM 4-HNE for 5 minutes (A and B) and 2 hours (C and D) in the presence (A and C) or absence (B and D) of NAD^+ (1mM). Activity was determined by production of NADH with 1mM propionaldehyde.

hydrogenase (Uchida and Stadtman, 1993) and most recently a cardiac oxidoreductase (Srivastava *et al.,* 1998) against the inactivation by 4-HNE.

In this context, it has been demonstrated that there are two residues in contact with NAD^+ when it is bound to the coenzyme site in the Class II high affinity mitochondrial ALDH. One residue, lysine 192, is believed to properly anchor NAD^+ for effective hydride transfer (Sheikh *et al.,* 1997). Since 4-HNE has been observed to readily react with the ε-amino group of lysine through Michael addition chemistry, it is possible that lysine 192 may be a critical target for the inactivation of ALDH by 4-HNE.

Other possible targets for ALDH inactivation are the numerous cysteine residues conserved among mammalian class 2 aldehyde dehydrogenases. Of essential importance is cysteine 302 which binds to the aldehyde during the catalytic cycle (Farres *et al.,* 1995). Figure 3 demonstrates that with increasing time and 4-HNE concentration, a significant loss of sulfhydryl groups occurs. This is especially apparent at the 2 hour time point where greater than 50% of the sulfhydryl groups have become adducted by 250 μM of 4-HNE. Certainly, this electrophilic aldehyde could form Michael addition adducts with cysteine 302 and form a covalent bond rendering the nucleophilic group at the active site nonfunctional. While we do not known which sulfhydryl groups in ALDH are adducted, a single or several adducts may produce profound changes in the enzyme causing irreversible inactivation. The susceptibility of ALDH to inactivation by 4-HNE is consistent with another report that glyceraldehyde-3-phosphate dehydrogenase is also inactivated by 4-HNE

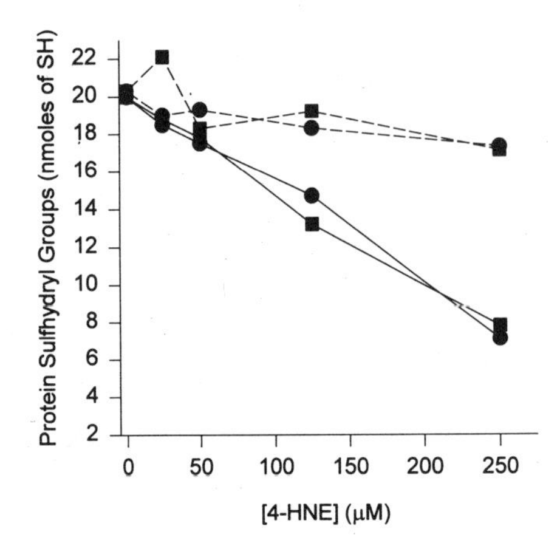

Figure 3. Determination of 4-HNE-mediated changes in protein sulfhydryl groups. Purified ALDH (40 μg) was treated with 0–250 μM 4-HNE for (---) 5 minutes and (—) 2 hours in the (●) presence or (■) absence of NAD^+ (1 mM). Samples were taken ans derivatized with 5, 5'-dithio-*bis*-2-nitrobenzoic acid (DTNB).

(Uchida and Stadtman, 1993). These investigators reported that inactivation of the enzyme was explained by alkylation of cysteine 149 that resides in the active site of the protein. Studies are now underway in our laboratory to identify susceptible amino acid residues in ALDH altered by 4-HNE.

The results presented here support previous reports from our laboratory that the high-affinity mitochondrial ALDH is susceptible to inactivation by 4-HNE produced as a result of prooxidant-initiated lipid peroxidation. The mechanisms of inhibition appear to be complex in that they may be reversible if the exposure to 4-HNE is brief or irreversible following prolonged exposure to 4-HNE. While the physiologic significance of ALDH inhibition by 4-HNE has not been established it is likely that such a process would impair the hepatocytes ability to not only metabolize 4-HNE, but also impedes oxidation of ethanol-derived acetaldehyde.

REFERENCES

Esterbauer, H., Schauer, R.J., and Zollner, H. (1991) Chemistry and biochemistry of 4-hydroxynonenal, malonaldehyde, and related aldehydes. *Free Rad. Biol. Med.* **11**, 81–128.

Farres, J., Wang, T.T., Cunningham, S.J., and Weiner, H. (1995) Investigation of the active site cysteine residue of rat liver mitochondrial aldehyde dehydrogenase by site-directed mutagenesis. *Biochemistry* **34**, 2592–2598.A

Hartley, D.P., Kroll, D.J., and Petersen, D.R. (1997) Prooxidant-initiated lipid peroxidation in isolated rat hepatocytes: detection of 4-hydroxynonenal- and malondialdehyde-protein adducts. *Chemical. Research. in Toxicology* **10**, 895–905.

Hartley, D.P., Ruth, J.A., and Petersen, D.R. (1995) The Hepatocellular Metabolism of 4-Hydroxynonenal by Alcohol Dehydrogenase, Aldehyde Dehydrogenase, and Gluthathione-*S*-Transferase. *Arch. Biochem. Biophysics* **10**, 197–205.

Mitchell, D.Y. and Petersen, D.R. (1987) The oxidation of alpha-beta unsaturated aldehydic products of lipid peroxidation by rat liver aldehyde dehydrogenases. *Toxicology &. Applied. Pharmacology* **87**, 403–410.

Mitchell, D.Y. and Petersen, D.R. (1991) Inhibition of rat hepatic mitochondrial aldehyde dehydrogenase-mediated acetaldehyde oxidation by trans-4-hydroxy-2-nonenal. *Hepatology* **13**, 728–734.

Schauer, R.J., Zollner, H., and Esterbauer, H. (1990) Biological Effects of Aldehydes with Particular Attention to 4-Hydroxynonenal and Malondialdehyde. In: *Membrane Lipid Peroxidation*, 141–163. Edited by Vigo and Pelfrey, Boca Raton, FL, CRC Press.

Segel, I.H. (1976) Enzymes. In: *Biochemical Calculations*, 2nd Ed., 208–319. New York, John Wiley & Sons.

Sheikh, S., Ni, L., Hurley, T.D., and Weiner, H. (1997) The potential roles of the conserved amino acids in human liver mitochondrial aldehyde dehydrogenase. *Journal. of. Biological. Chemistry.* **272**, 18817–18822.

Srivastava, S., Chandra, A., Ansari, N.H., Srivastava, S.K., and Bhatnagar, A. (1998) Identification of cardiac oxidoreductase(s) involved in the metabolism of the lipid peroxidation-derived aldehyde-4-hydroxynonenal. *Biochemical. Journal.* **329**, 469–475.

Uchida, K. and Stadtman, E.R. (1993) Covalent attachment of 4-hydroxynonenal to glyceraldehyde-3-phosphate dehydrogenase. A possible involvement of intra- and intermolecular cross-linking reaction. *Journal. of. Biological. Chemistry.* **268**, 6388–6393.

Vander Jagt, D.L., Hunsaker, L.A., Vander Jagt, T.J., Gomez, M.S., Gonzales, D.M., Deck, L.M., and Royer, R.E. (1997) Inactivation of glutathione reductase by 4-hydroxynonenal and other endogenous aldehydes. *Biochemical. Pharmacology* **53**, 1133–1140.

10

ALDEHYDE INHIBITORS OF ALDEHYDE DEHYDROGENASES

Regina Pietruszko, Darryl P. Abriola, Gonzalo Izaguirre, Alexandra Kikonyogo, Marek Dryjanski, and Wojciech Ambroziak

Center of Alcohol Studies and Department of Molecular Biology and Biochemistry
Rutgers The State University of New Jersey
Piscataway, New Jersey 08854-8001

1. INTRODUCTION

Aldehyde dehydrogenases catalyze aldehyde dehydrogenation as well as ester hydrolysis. Nitrate esters e.g. isosorbide nitrates and nitroglycerin inhibit the enzyme. Mechanism of inhibition which involves aldehyde dehydrogenase catalyzed formation of a reactive species (Mukerjee and Pietruszko, 1994), which inactivates the enzyme via covalent bond formation, was previously discussed (Pietruszko et al., 1995) in this series. In addition, aldehyde dehydrogenases share a common property of being inhibited by aldehyde inhibitors such as chloral, citral, methylglyoxal and 4-dialkylamino benzaldehyde, whose structure resembles that of aldehyde substrates (Figure 1). It was interesting, therefore, to investigate how these compounds inhibit the enzyme. Methylglyoxal is a natural metabolite which is formed enzymatically and non-enzymatically from triose phosphates and from acetone by cytochrome P-450; it also arises from metabolism of threonine. Because of its chemical reactivity toward proteins and nucleic acids, methylglyoxal is considered to be a toxic compound (Kalapos, 1994; Chaplen et al., 1998); modification of proteins and nucleic acids by methylglyoxal has been postulated to be a possible cause of development of diabetic complications (Vander Jagt et al., 1992). Methylglyoxal is metabolized by glyoxalase which catalyzes the conversion of methylglyoxal to D-lactate in the presence of reduced glutathione. Aldose reductase and aldehyde reductase catalyze reduction of methylglyoxal to acetol and D-lactaldehyde. Another metabolic route for methylglyoxal is its oxidation to pyruvate by α-ketoaldehyde dehydrogenase (Monder, 1967). The structure of methylglyoxal is shown in Figure 1.

The monoterpene, citral (3,7-dimethyl-2,6-octadienal), is a reactive and volatile α,β-unsaturated aldehyde that occurs naturally in herbs, plants and citrus fruits. Both natural and

Enzymology and Molecular Biology of Carbonyl Metabolism 7
edited by Weiner *et al.* Kluwer Academic / Plenum Publishers, New York, 1999.

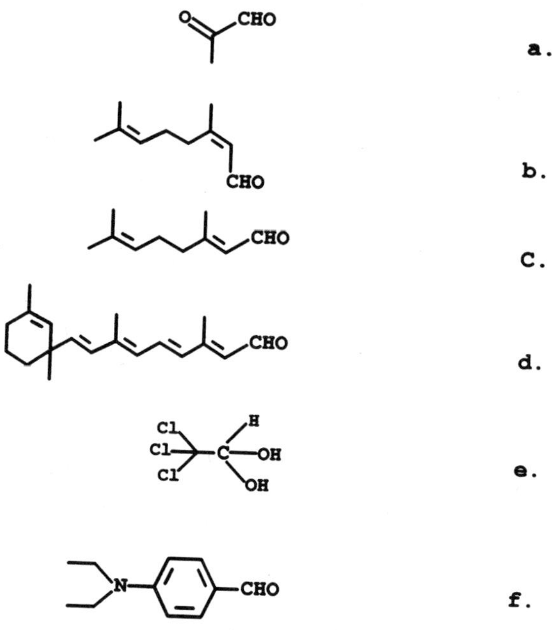

a.

b.

c.

d.

e.

f.

Figure 1. Inhibitors used in this study and their comparison with retinaldehyde. (a) = methylglyoxal; (b) = neral; (c) = geranial; (d) = retinaldehyde; (e) = chloral; (f) = 4-diethylamino benzaldehyde.

synthetic citral occurs as 2:1 mixture of the geometric isomers, geranial (*trans*-3,7-dimethyl-2,6-octadienal) and neral (*cis*-3,7-dimethyl-2,6-octadienal). The *trans*-form of citral has some structural resemblance to all *trans*-retinal. In experiments of mammalian embryogenesis citral is routinely employed as an inhibitor of the NAD-dependent pathway of retinoic acid biosynthesis. At extremely low concentrations (as low as 10μM) citral was found to completely inhibit the NAD-dependent pathway of retinoic acid formation (Chen et al., 1995). The chemical structures of citral isomers and of retinal are also shown in Figure 1.

Both citral (Boyer and Petersen, 1991) and methylglyoxal (Ray and Ray, 1984) were reported to inhibit mitochondrial aldehyde dehydrogenase; inhibition of cytoplasmic aldehyde dehydrogenase by methylglyoxal was also described (Ray and Ray, 1984). The effect of citral on other isozymes of mammalian aldehyde dehydrogenase or the mechanism of its inhibition with either methylglyoxal and citral have never previously been investigated. In this paper we describe the mode of inhibition of aldehyde dehydrogenase by methylglyoxal and citral and discuss the common features of the inhibition. Resolution of the inhibition mechanism has been facilitated by the simultaneous use of three isozymes of human aldehyde dehydrogenase (E1, E2 and E3) which exhibit distinct characteristics with respect to the inhibitors. The results have shown that both compounds constitute a class of inhibitors which are well known in serine protease field (Chase and Shaw, 1969; Bagio et al., 1996) but are novel for dehydrogenases. They are themselves substrates with Km values so low that they are preferentially metabolized but with maximal velocities also low, in some cases requiring more than 1 hour for a single turnover (Kikonyogo et al., 1998).

2. MATERIALS AND METHODS

The techniques employed for investigation of methylglyoxal and citral are published (Izaguirre et al, 1998; Kikonyogo et al., 1998). Experiments at high enzyme concentrations investigating the "burst" during hydride transfer for methylglyoxal (not described in Izaguirre et al., 1998) were monitored at 340 nm in a Gilford spectrophotometer. The reaction mixture was composed of 1mg/ml E2 isozyme of highest specific activity (1.6 μmol NADH/min/mg), NAD (2 mM) and methylglyoxal (1 mM) in 80 mM sodium phosphate buffer, pH 7.2 at 10°C and 25°C. The reaction was initiated by the addition of methylglyoxal.

Evidence for metabolism of 4-diethylamino benzaldehyde was obtained by HPLC analysis of the reaction products following incubation of diethylamino benzaldehyde (400 μM) with E1 (160 μg) and E2 (50 μg) in 0.1M pyrophosphate buffer, pH 9.0; 1mM EDTA; 500μM NAD in 1ml total volume for 16 hr at 25°C. HPLC analysis was carried out on Waters C_{18} reverse phase column employing a gradient from 0.1% trifluoroacetic acid in water to 100% acetonitrile. Enzyme activity was observed from disappearance of 4-diethylamino benzaldehyde peak.

Citral isomers: *trans* isomer, geranial (66%), and *cis* isomer, neral (33%) are not available commercially. For separation of geranial and neral a 5 μ Supelco C_{18} reversed phase column (0.46 cm × 25 cm) was used with a Waters HPLC system (Kikonyogo et al., 1998). The concentrations of geranial and neral were determined by using the E1 isozyme in the presence of excess NAD and calculated from the extinction coefficient of NADH = 6.22 mM^{-1} cm^{-1}.

3. RESULTS AND DISCUSSION

3.1. Inhibition of E1, E2 and E3 Isozymes by Citral, Geranial, Neral, Methylglyoxal, Diethylamino Benzaldehyde, and Chloral

Inhibition of all three isozymes of human aldehyde dehydrogenase by citral and of the E1 and E2 isozymes by methylglyoxal was observed when glycolaldehyde, propionaldehyde or γ-aminobutyraldehyde were used as substrates. When citral was separated into its component isomers (*trans*, Geranial, and *cis,* Neral), both isomers inhibited all three isozymes. Also, employing conditions described by Russo et al., (1988) selective inhibition of the E1 isozyme (vs. the E2 isozyme) with diethylamino benzaldehyde was confirmed in spectrophotometric experiments. Chloral was previously reported to inhibit all three isozymes (Ki E1 = 28 μM (Vallari and Pietruszko, 1981; Ki E2 = 3 μM (Sidhu and Blair, 1975; Ki E3, never determined)

3.2. Preliminary Evidence for Metabolism

There were considerable differences between isozymes in the manner in which inhibition occurred. In the case of E3 isozyme and methylglyoxal it was difficult to observe any significant inhibition because NADH was produced from methylglyoxal and NAD at a considerable rate. Although inhibition of glycolaldehyde activity of the E3 isozyme could be observed with chloral, significant production of NADH also occurred when chloral, E3 isozyme and NAD were incubated together. In fact, kinetic constants could be determined for both methylglyoxal and chloral (see Figure 2 and Table 1). With E1 and E2 isozymes it was much more difficult to observe NADH formation from methylglyoxal; however when enzyme concentrations were increased it could be demonstrated that NADH formation occurred and kinetic constants could be determined. Incubation of the

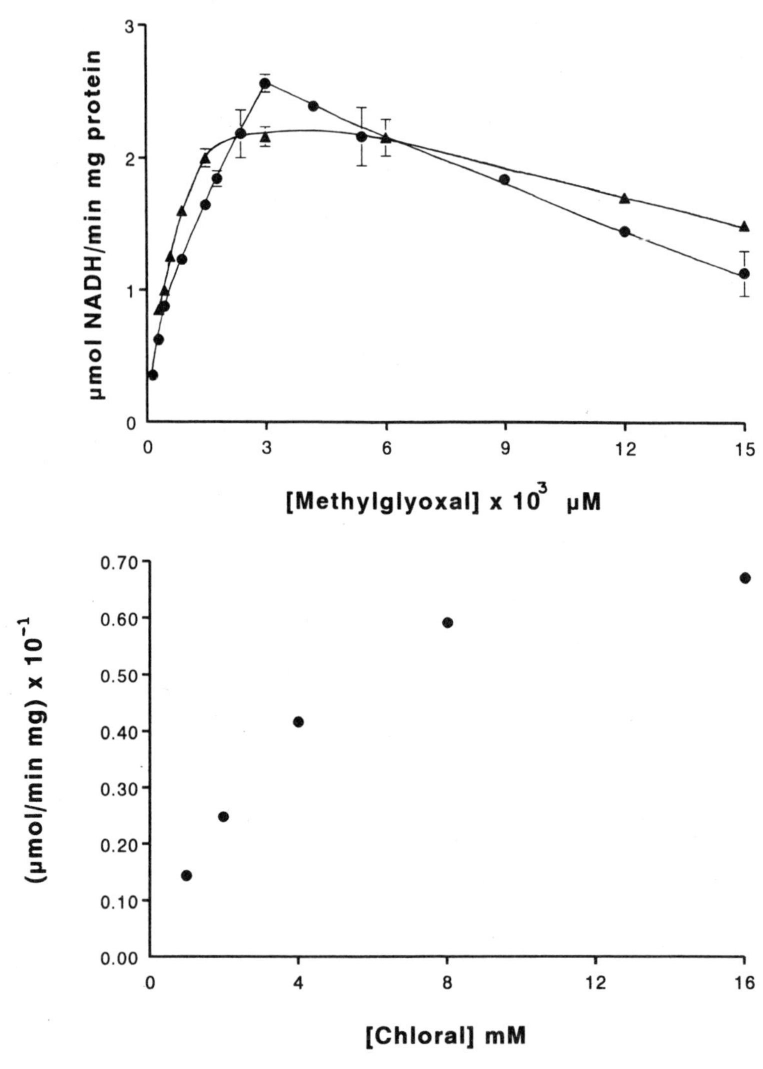

Figure 2. Methylglyoxal and chloral as substrates for the E3 isozyme. A. Activity of the E3 isozyme with methylglyoxal at pH 9.0. (circles = major component of E3; (triangles) = minor component of E3 isozyme. Both components show substrate inhibition at high methylglyoxal concentrations. B. Chloral as substrate for the E3 isozyme. The two components of the E3 isozyme are not separated from each other in this experiment.

E1 and E2 isozymes with NAD and chloral at high enzyme concentrations for long time periods was never systematically done.

When glycolaldehyde at Km concentration (see Table 1) was used with all three isozymes addition of 10 µM citral during the course of reaction completely abolished E3 isozyme activity but only partially abolished the activity of the E1 isozyme. Further experiments showed that E1 isozyme (and to a lesser extent the E2 isozyme) produced NADH from NAD and citral. Production of NADH from NAD and citral by the E3 isozyme could be only demonstrated at very high enzyme concentrations following long incubation time. The isozymes also appeared to differ in the velocity of diethylamino benzaldehyde metabolism; there was faster disappearance of 4-diethylamino benzaldehyde in the presence of the E2 isozyme than in the presence of the E1 isozyme.

Table 1. Comparison of kinetic constants of E1, E2 and E3 Isozymes with methylglyoxal, geranial, neral and chloral with those of known substrates

Enzyme	Substrate	Km (μM)	V (μmol/min/mg)	V/Km
E1	Geranial	0.07	0.063	9.0
	Neral	3.2	0.066	0.02
	Methylglyoxal	48	0.067	0.0014
	Glycolaldehyde	240	0.34	0.0014
	Propionaldehyde	0.8	0.3	0.375
E2	Geranial	0.25*	0.013**	0.052
	Neral	10*	0.016	0.0016
	Methylglyoxal	8.6	0.06	0.007
	Glycolaldehyde	46	2.0	0.04
	Acetaldehyde	1.0	0.3	0.3
E3	Citral	0.1*	0.00007**	0.0007
	Methylglyoxal	590	1.1	0.0019
	Chloral	5750	0.094	0.000016
	Glycolaldehyde	221	0.9	0.004
	γ-Aminobutyraldehyde	10	2.0	0.2

All experiments were done at pH 7.4; * = determined as inhibition constants; ** direct velocity determinations.

3.3. Characterization of Products from Methylglyoxal and Citral

Following incubation of the E1 isozyme with NAD and citral, formation of geranic acid product could be demonstrated via HPLC (3A and 3B). Long incubation of the E2 and E3 isozymes with citral at higher enzyme concentrations also resulted in formation of geranic acid detectable and identifiable via HPLC. Formation of pyruvic acid following incubation of the E3 isozyme with methylglyoxal in the presence of NAD could be easily demonstrated (Figure 3C). The HPLC procedure was also employed to see if E1 and E2 isozymes produced pyruvic acid from methyl glyoxal following incubation with NAD. This was demonstrated by HPLC (Izaguirre et al., 1998). Thus, both citral and methylglyoxal appeared to be alternate substrates, rather than dead end inhibitors. An attempt to identify trichloracetic acid as a product of dehydrogenation of chloral was unsuccessful for technical reasons.

3.4. Inhibition by Alternate Substrates

Inhibition by alternate substrates can be followed successfully, provided a product formation specific to a measured substrate is determined. Inhibition by alternate substrates is always competitive (Segel, 1975) and the Ki value obtained equals the Km value of the alternate substrate, thus allowing determination of Km for substrates with low velocity. In dehydrogenases, NADH is a common product for all substrates. However, on close examination of the approximate velocity of formation of geranic acid by the E2 and E3 isozymes it was found that velocities were less than 1% of glycolaldehyde velocity. Thus, formation of NADH from citral by both E2 and E3 isozymes could be disregarded because it would fall within experimental error of any procedure used for determination of the geranic acid product. This allowed use of a common product (NADH) in kinetic experiments. Also, because the maximal velocity of methylglyoxal dehydrogenation of the E2 isozyme was only 3% of the maximal velocity with glycolaldehyde, the inhibition could be studied following NADH formation. With the E1 isozyme direct determination of kinetic constants was possible.

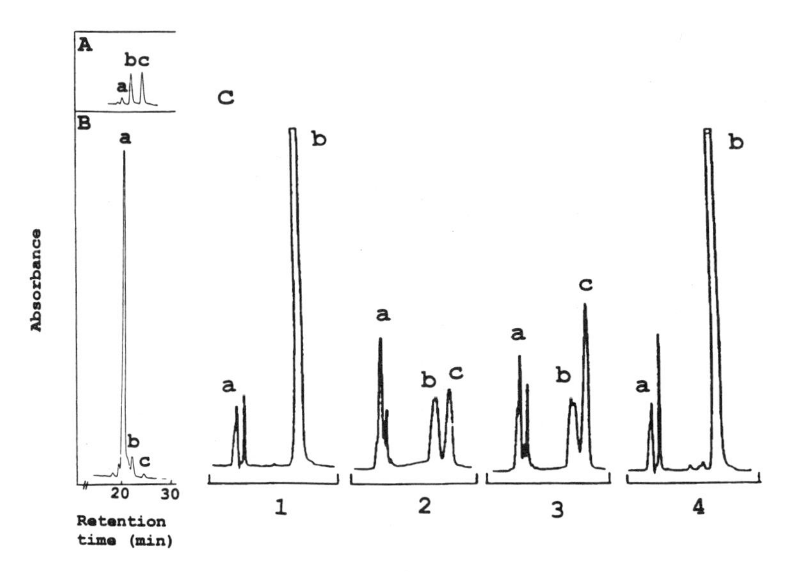

Figure 3. Formation of products from citral and following incubation with enzyme. A and B. Identification of geranic acid as the reaction product following incubation of the E1 isozyme with citral detected at 219 nm absorbance. Incubation was done for 12 h using 100 μg of E1 isozyme per ml at 500 μM citral. (A) = HPLC analysis of control, without enzyme. a = geranic acid; b = neral; c = geranial. (B) = HPLC analysis of experimental incubation with enzyme and NAD. A peak (a) identified as geranic acid of retention time ca. 20.7 min is seen in (B). C. HPLC analysis of pyruvate formation from methylglyoxal following incubation with the E3 isozyme detected at 200 nm absorbance. Incubations were in 1 ml total volume in 0.1M sodium pyrophosphate buffer, pH 9.0 at 25°C. Reaction mixture contained 1mM NAD, 12 mM methylglyoxal, and 80 μg of enzyme. Peak a = methylglyoxal; peak b = NAD and NADH; peak c = pyruvate. Chromatograms shown are: (1) time 0; (2) 12h; (3) 12 h to which one tenth volume of 5 mM pyruvate was added; (4) control in the absence of enzyme at 12h.

3.5. Inhibition by Citral and Methylglyoxal

As shown in Figures 4 (A and B) inhibition by methylglyoxal and citral was noncompetitive. Both slopes and intercepts were produced in all inhibition experiments. With the E2 isozyme and methylglyoxal both Km and V values could be directly determined as well as by inhibition of glycolaldehyde activity of the E2 isozyme by methylglyoxal. It was found that Ki (slope) for methylglyoxal was indeed equal to the directly determined Km value for methylglyoxal. The same appeared to be true for citral and E2 isozyme, where Km and V values could be also directly determined at higher enzyme concentrations. The Ki intercept was usually numerically larger than Ki slope.

3.6. Experiments with Methylglyoxal, Citral, Geranial, and Neral at High Enzyme Concentrations

The dehydrogenation of citral by all three isozymes and of methylglyoxal by E1 and E2 isozymes is slow, making these compounds good inhibitors of reactions that proceed much faster. It was thought possible that a slow dehydrogenation rate might have been the result of slow decomposition of enzyme-acyl intermediate. Because the velocity of the E2 isozyme with citral was low, it was possible to employ large enzyme concentrations and a spectrophotometer in order to determine if there was any "burst" of NADH formation

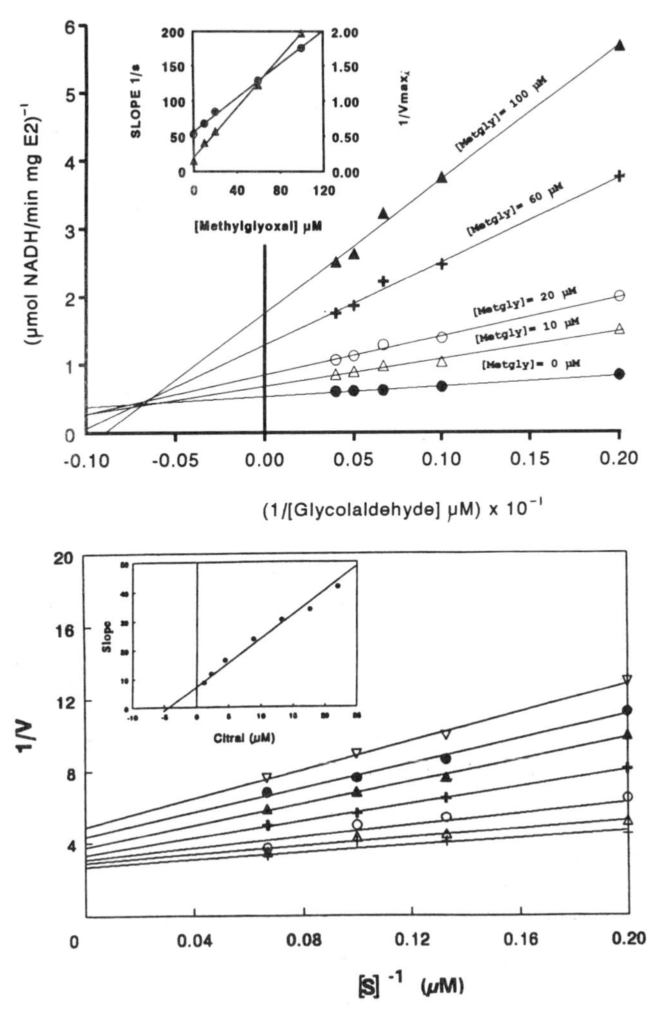

Figure 4. Inhibition of the E2 Isozyme by Methylglyoxal and Citral. A. Inhibition of glycolaldehyde activity of human aldehyde dehydrogenase E2 isozyme by methylglyoxal. Inset: secondary plots of slopes (triangles) and intercepts (circles). B. Inhibition of E2 isozyme by citral. Propionaldehyde was used as the varied substrate in the presence of NAD (2mM). The primary plots show both slope and intercept effects. The slope replot is shown in the inset. Citral concentrations were: 1.1, 2.2, 4.4, 8.8, 13.2, 17.6 and 22 μM from bottom to top.

upon mixing the E2 isozyme in the presence of NAD with citral or its two components, geranial and neral, separated by HPLC. No "burst" was observed with either of these compounds. There was also no "burst" when methylglyoxal was mixed with the E2 isozyme and NAD. These experiments excluded the possibility that the acyl-enzyme intermediate with methylglyoxal or citral would be the enzyme species determining the rate-limiting step of citral or methylglyoxal dehydrogenation. The acyl-enzyme intermediate also could not be the enzyme species that removed the enzyme from the reaction sequence, producing intercept effects in inhibition kinetics. With both methylglyoxal and citral, the rate limiting step occurred before or during the hydride transfer. If the slow formation of the transition state complex was limiting the velocity of reaction with citral and methylglyoxal no intercept effect would be observed, because the varied substrate would be able to compete with the inhibitor at that step. The intercepts indicate that a particularly tight transition state is the rate limiting step in the dehydrogenation of citral and methylglyoxal by aldehyde dehydrogenase.

3.7. Kinetic Constants of Methylglyoxal, Citral, and Chloral with Human Aldehyde Dehydrogenases

Kinetic constants for citral or its component isomers, geranial and neral and for methylglyoxal have been determined and are listed in Table 1. Kinetic constants for glycolaldehyde and also those of a typical well known substrate for each isozyme are listed in Table 1 for comparison. Where velocities were extremely small, e.g. citral with the E3 isozyme, they were determined directly at high substrate concentrations, after establishing absence of the substrate inhibition.

The substrates are compared via V/Km values. For the E3 isozyme, citral, methylglyoxal and chloral are definitely much worse substrates than γ-aminobutyraldehyde, or even glycolaldehyde. With the E2 isozyme geranial, neral and methylglyoxal are much poorer substrates than acetaldehyde but geranial is similar to glycolaldehyde (identical V/Km in Table 1), because of its very low Km, relative to velocity. With the E1 isozyme, neral and methylglyoxal are poorer substrates than propionaldehyde, used here as the representative good substrate, but also comparable to glycolaldehyde. In fact, neral appears to be a better substrate than glycolaldehyde. The results with geranial were a real surprise, the V/Km ratio showing that it is a much better substrate for the E1 isozyme than propionaldehyde. This is especially interesting because of geranial's close structural resemblance to retinaldehyde (Figure 1), which may be its natural substrate. The high V/Km ratio again results from the low Km relative to maximal velocity.

3.8. E1, E2, and E3 Isozymes Differ in Velocities of Methylglyoxal and Citral Dehydrogenation

Methylglyoxal is dehydrogenated by the E3 isozyme at considerable velocity relative to its good substrates. The active site substrate-binding area has to differ considerably in E3 isozyme from those in E1 and E2. This is also demonstrated by the fact that chloral appears to be a substrate, but this still requires a confirmation via identification of the reaction product. The velocity of citral dehydrogenation is fastest with the E1 isozyme where it constitutes about 20% of the velocity with propionaldehyde. The velocity of citral dehydrogenation is ca. 5 times less with the E2 isozyme than with the E1 isozyme and three orders of magnitude less with the E3 isozyme (Kikonyogo et al., 1998).

3.9. Correlation of Substrate Properties with Inhibitory Characteristics

Methylglyoxal, citral and probably chloral and 4-diethylamino benzaldehyde constitute a new class of dehydrogenase inhibitors, which are novel in that they are themselves substrates. They are characterized by low Km values promoting their easy recognition and subsequent preferential utilization by enzymes, and by V values, sometimes so low that in ordinary experimental conditions no indication can be obtained of their substrate-like characteristics. Low Km and low V values make them excellent and specific inhibitors. The inhibitory substrates, well known in serine protease field, usually inhibit by forming a stable enzyme-acyl intermediate with the proteases (Chase and Shaw, 1969; Baggio et al.,

1996). It has been shown during this investigation that slow turnover of citral and methylglyoxal is not due to the stable enzyme-acyl covalent intermediate. No "burst" of NADH formation has been ever detected with methylglyoxal or citral and its separated isomers when experiments were done at high enzyme concentrations. Thus, all the catalytic steps following the hydride transfer, including the enzyme-acyl intermediate, do not contribute to the slow turnover of these substrates. The rate-limiting step in the reaction sequence must occur before the hydride transfer. It is unlikely that in the case of methylglyoxal or citral slow binding to the enzyme.NAD complex constitutes the rate-limiting step in their catalysis. Slow binding alternate substrates could be easily displaced by a competing substrate in inhibition kinetics, eliminating intercept effects. Decomposition of a very tight transition state during the hydride transfer, not displaceable by a competing substrate, is the most likely catalytic step which results in slow turnover of methylglyoxal and citral and their resultant inhibitory characteristics. Thus, inhibition of aldehyde dehydrogenase by methylglyoxal and citral, like that by nitrate esters of isosorbide, is mechanism-based.

REFERENCES

Baggio, R., Shi, Y-Q., Wu, Y., and Abeles, R.H. (1996) From poor substrates to good inhibitors: design of inhibitors for serine and thiol proteases. Biochemistry, USA 35, 3351–3353.

Boyer, C.S., and Petersen, D.R. (1991) The metabolism of 3,7-dimethyl-2,6-octadienal (citral) in rat hepatic mitochondrial and cytosolic fractions. Interactions with aldehyde and alcohol dehydrogenases. Drug Metabolism and Disposition. 19, 81–86.

Chaplen, F.W.R., Fahl, W.E., and Cameron, D.C. (1998) Evidence for high levels of methylglyoxal in cultured Chinese hamster ovary cells. Proc. Ntl. Acad. Sci. USA 95, 5533–5538.

Chase, T., and Shaw, (1969) Comparison of the esterase activities of trypsin, plasmin, and thrombin on guanidinobenzoate esters. Titration of the enzymes. Biochemistry USA 8, 2212–2224.

Chen, H., Namkung, M.J., and Juchau, M.R. (1995) Biotransformation of all-*trans*-retinal to all-*trans*-retinoic acid in rat conceptal homogenates. Biochem. Pharmacol. 50, 1257–1264.

Izaguirre, G., Kikonyogo, A., and Pietruszko, R. (1998) Methylglyoxal as substrate and inhibitor of human aldehyde dehydrogenases: comparison of kinetic properties among the three isozymes. Comp. Biochem. Physiol. in press.

Kalapos, M.P., (1994) Methylglyoxal toxicity in mammals. Toxicology Lett. 73, 3–24.

Kikonyogo, A., Abriola, D.P., Dryjanski, D., and Pietruszko, R. (1998) Mechanism of inhibition of aldehyde dehydrogenase by citral, a retinoid antagonist. Metabolism of citral by aldehyde dehydrogenase. (submitted for publication).

Monder, C., (1967) α-Ketoaldehyde dehydrogenase, an enzyme that catalyzes the enzymic oxidation of methylglyoxal to pyruvate. J. Biol. Chem. 242, 4603–4609.

Mukerjee, N., and Pietruszko, R. (1994) Inactivation of aldehyde dehydrogenase by isosorbide dinitrate. J. Biol. Chem. 269,21664–21669.

Pietruszko, R., Mukerjee, N., Blatter, E.E., and Lehmann, T. (1995) Nitrate esters as inhibitors and substrates of aldehyde dehydrogenase. Advances in Experimental Medicine and Biology 372, 25–34.

Ray, S., and Ray, M. (1984) Oxidation of lactaldehyde by cytosolic aldehyde dehydrogenase and inhibition of cytosolic and mitochondrial aldehyde dehydrogenase by metabolites. Biochim. Biophys. Acta 802, 128–134.

Russo, J.E., Hauquitz, J.F., and Hilton, J. (1988) Inhibition of mouse cytosolic aldehyde dehydrogenase by 4-(diethylamino)benzaldehyde. Biochem. Pharmacol. 37, 1639–1642.

Segel, I.H. (1975) Enzyme Kinetics. Inhibition by alternate substrates. pp 793–813. John Wiley and Sons.

Sidhu, R.S., and Blair, A.H. (1975) human liver aldehyde dehydrogenase: kinetics of aldehyde oxidation. J. Biol. Chem. 250, 7899–7904.

Vallari, R.C., and Pietruszko, R. (1981) Kinetic mechanism of human cytoplasmic aldehyde dehydrogenase. Arch. Biochem. Biophys. 212, 9–19.

Vander Jagt, D.L., Robinson, B., Taylor, K.K., and Hunsaker, L.A., Reduction of trioses by NADPH-dependent aldo-keto reductases. Aldose reductase, methylglyoxal, and diabetic complications. J. Biol. Chem. 267, 4364–4369.

11

COVALENT MODIFICATION OF SHEEP LIVER CYTOSOLIC ALDEHYDE DEHYDROGENASE BY THE OXIDATIVE ADDITION OF COLOURED PHENOXAZINE, PHENOTHIAZINE AND PHENAZINE DERIVATIVES

Trevor M. Kitson,[1] Kathryn E. Kitson,[2] and Gordon J. King

[1]Institute of Fundamental Sciences (Chemistry)
[2]Institute for Food, Health and Human Nutrition
Massey University
Palmerston North, New Zealand

1. INTRODUCTION

Aldehyde dehydrogenase acts as an esterase towards reactive esters such as *p*-nitrophenyl acetate, and although the subject of prolonged debate, it is now generally believed that the dehydrogenase and esterase actions of the enzyme involve the same active site and catalytic groups (see Kitson and Kitson, 1996, and references therein). With *p*-nitrophenyl dimethylcarbamate, the reactivity (compared to the acetate) is so drastically reduced that it takes several hours for this 'substrate' to acylate cytosolic aldehyde dehydrogenase (ALDH-1), and the rate of subsequent deacylation is essentially zero (Kitson et al., 1991). Thus the carbamate acts as an active-site-directed irreversible inactivator of the enzyme. It was thought that resorufin dimethylcarbamate (see Figure 1) would react likewise, but since the molar absorptivity of the resorufin anion (69,700) is so much greater than that of *p*-nitrophenoxide (18,320) (Kitson, 1996), the resorufin carbamate would be a much more sensitive active site titrant than the *p*-nitrophenyl compound. The results reported and discussed below show that this expectation was not borne out. Resorufin dimethylcarbamate does inactivate ALDH-1, but the chemistry of the reaction is more complicated and interesting than the simple acylation process expected from the previous work with the *p*-nitrophenyl equivalent; it may be termed 'oxidative addition'.

Once it became apparent how ALDH-1 interacts with resorufin dimethylcarbamate (and the corresponding methanesulphonate and ethyl ether), we investigated other compounds whose structure suggests they also might undergo oxidative addition with the en-

Enzymology and Molecular Biology of Carbonyl Metabolism 7
edited by Weiner *et al.* Kluwer Academic / Plenum Publishers, New York, 1999.

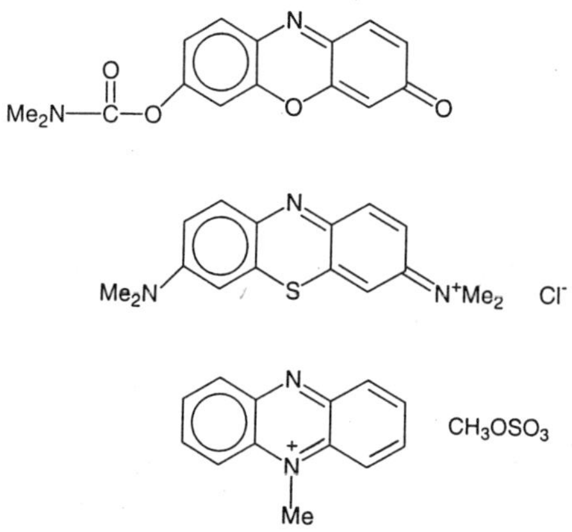

Figure 1. Structure of the compounds referred to in this work.

zyme. One such compound is the redox dye methylene blue, widely used in various metabolic studies both in vitro and in vivo, and which has been reported to be an inhibitor of ALDH (Helander et al., 1993). Another is phenazine methosulphate, frequently used along with nitro blue tetrazolium chloride in activity staining of dehydrogenases on electrophoresis gels. The structures of these compounds are also shown in Figure 1.

2. EXPERIMENTAL

ALDH-1 was isolated from sheep liver as previously described (Kitson and Kitson, 1994). The dimethylcarbamate, methanesulphonate and ethyl ether of resorufin were synthesised as before (Kitson, 1998). All spectra and all spectrophotometric assays of enzyme activity were recorded using a Varian Cary 1 spectrophotometer. Reactions between enzyme and modifiers were carried out in 50 mM phosphate buffer, pH 7.4, at 25 °C. Chemically modified enzyme forms were isolated from excess modifier by gel filtration using Biogel P-6.

3. RESULTS AND DISCUSSION

3.1. Interaction of ALDH-1 with Resorufin Derivatives

Resorufin dimethylcarbamate inactivates ALDH-1 more rapidly than *p*-nitrophenyl dimethylcarbamate (about 1 hour compared to about 18), but there is little or no release of the intensely red resorufin anion. Thus with the former compound, unlike with the latter, the inactivation cannot be due to conversion of the thiol group of the catalytically essential Cys-302 to a -S-CO-NMe_2 derivative (Kitson et al., 1991). Isolation of the enzyme by gel filtration after modification by resorufin dimethylcarbamate shows that it carries a covalently-linked yellow-orange label. The methanesulphonate of resorufin inactivates ALDH-1 more rapidly than the dimethylcarbamate while the ethyl ether of resorufin inactivates the enzyme more slowly (see Figure 2). Since these relative rates parallel the leaving group ability of the various groups attached to the resorufin moiety, we speculated that

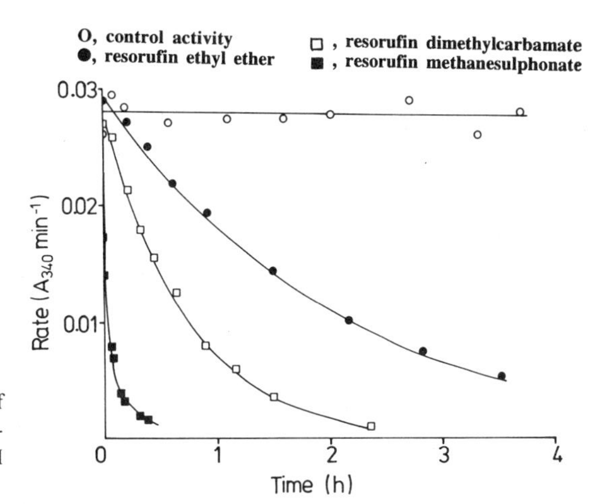

Figure 2. Time course of the inactivation of ALDH-1 (5.2 μM) by various resorufin derivatives (25 μM) in 50 mM phosphate buffer, pH 7.4, 25 °C.

a nucleophilic aromatic substitution mechanism may be involved. In this, a nucleophile (presumably Cys-302) would displace the dimethylcarbamate ion, the methanesulphonate ion, or the ethoxide ion, respectively. Such a mechanism, however, would lead to exactly the same modified form of the enzyme being produced from the three different modifiers; spectrophotometric analysis of the modified forms of the enzyme showed this not to be the case. Reaction with the dimethylcarbamate, the methanesulphonate, and the ethyl ether gives labels with maximal absorptions at 441–3, 437 and 530 nm, respectively,thus eliminating the possibility of the suggested substitution mechanism.

At this point we turned to the reaction between the resorufin compounds and β-mercaptoethanol (as a model for Cys-302) in an attempt to identify the chemical mechanism involved. Investigation of the product of the reaction by NMR spectroscopy (Kitson, 1998) showed unequivocally that the small thiol adds at position 2 of the resorufin ring and that the initial product (which has an aromatic ring and would be colourless) undergoes immediate spontaneous oxidation to the final yellow-orange product shown in Figure 3, presumably with atmospheric oxygen as oxidant; this pathway may be termed 'oxidative addition'. With the dimethylcarbamate and the methanesulphonate, the λ_{max} of the product of reaction with β-mercaptoethanol is essentially the same as that with ALDH-1, showing that the pathway shown in

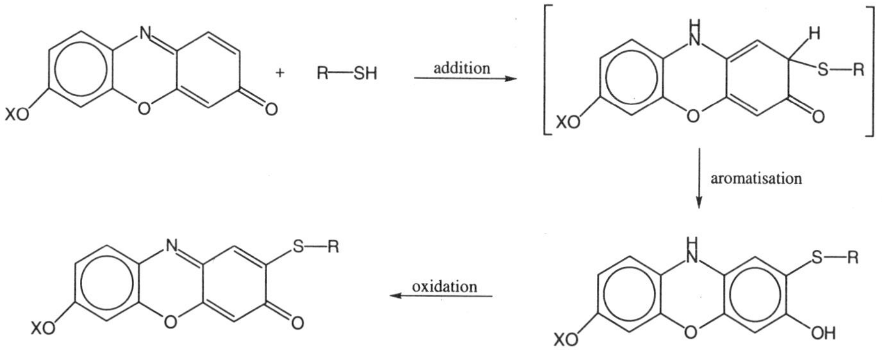

Figure 3. Oxidative addition pathway for the reaction of resorufin derivatives with thiols. X = Me_2NCO-, $MeSO_2$- or Et-. R-SH = β-mercaptoethanol or Cys-302 of ALDH.

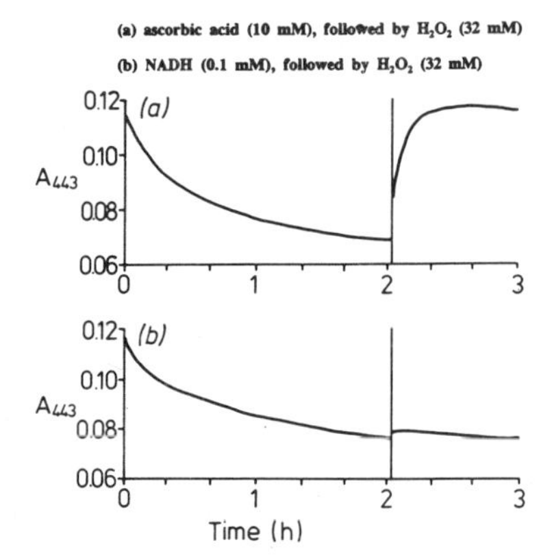

Figure 4. Redox properties of the resorufin label. The upper trace shows the gradual reduction of the coloured label (supplied to ALDH-1 by resorufin dimethylcarbamate) by ascorbic acid, followed by its re-oxidation by hydrogen peroxide. The lower trace shows the reduction of the enzyme-linked label by NADH and the subsequent lack of effect of hydrogen peroxide.

Figure 3 applies equally well to the enzyme. The relative rates of inactivation of the enzyme by the various resorufin compounds studied (referred to above) are now seen to be due to the electronic nature of the substituents; the more electron-releasing the substituent is (e.g. ethoxy), the more the initial nucleophilic attack of cysteine on the resorufin moiety is slowed.

3.2. Redox Properties of the Enzyme-Linked Resorufin Label

As a further confirmation of the veracity of the oxidative addition pathway, we investigated the labelled form of ALDH-1 using reducing and oxidising agents. Figure 4 shows that a high concentration of ascorbic acid slowly reduces the label, but to a position of equilibrium rather than to completion; this presumably reflects the great propensity of the reduction product to become oxidised. Indeed, the subsequent addition of an oxidising agent (hydrogen peroxide) causes the complete reversal of the reduction. These changes are compatible with the chemistry shown in Figure 3. Figure 4 also shows that NADH causes the partial reduction of the enzyme-linked resorufin label, although NADH has no effect on free resorufin dimethylcarbamate. This observation suggests that the resorufin label is positioned very closely to bound NADH in the enzyme complex, thus enabling hydride transfer to take place; this is consistent with the idea that the label is attached to Cys-302. Interestingly, the presence of the nucleotide protects the reduced form of the enzyme-linked resorufin label from reoxidation by hydrogen peroxide (see Figure 4).

3.3. Interaction of ALDH-1 with Resorufin Ethyl Ether

After modification of ALDH-1 by resorufin ethyl ether, the enyme appears pink-violet over the pH range 4–10, with a λ_{max} of 530 nm. This is an unexpected result in view of the fact that reaction of resorufin ethyl ether with β-mercaptoethanol by the oxidative addition pathway leads to a product of λ_{max} 481 nm. However, below pH 4 or after denaturation at pH 10.6, the modified enzyme then displays the expected yellow-orange colour and λ_{max} of 480–1 nm. It is evident that ALDH-1 reacts with the modifier initially according to the scheme in Figure 3, but that at intermediate pH values a further reversible reaction occurs giving the pink-violet form of labelled enzyme. We speculate that the chemistry shown in Figure 5 may account for these changes. In this, a putative enzymic nucleophilic group X opens the central ring of the resorufin system to give a resonance-stabilised phenoxy anion. If the ring-opening is freely reversible, then low pH would move the equilibrium to the left (by the effective removal of group X upon protonation); similarly, high pH

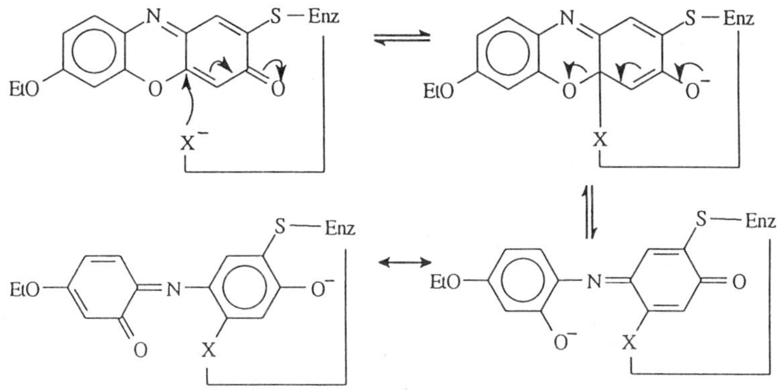

Figure 5. Speculative chemistry to account for the λ_{max} of the ALDH-1/resorufin ethyl ether adduct at intermediate pH values.

causes unfolding of the protein and the physical removal of group X, again shifting the equilibrium to the left. At present we have no information as to the possible identity of group X, but this may conceivably come from future X-ray crystallography studies as the pink-violet form of the enzyme crystallises very nicely (Baker, 1998). A further presently unexplained and intriguing point is that only the ethyl ether and not the dimethylcarbamate or methanesulphonate of resorufin reacts in the way described.

3.4. Interaction of ALDH-1 with Methylene Blue

Methylene blue has previously been reported to be an inhibitor of ALDH in vitro and to lower the activity of the enzyme in vivo (Cronholm, 1993; Helander et al., 1993). We find that with sheep liver ALDH-1, methylene blue is quite a strong inhibitor (against *p*-nitrophenyl pivalate as substrate for the esterase activity of the enzyme); the inhibition approximates to competitive and the K_i is of the order of 4 μM. The results obtained in such a study, however, are of limited precision as it is evident that a slow inactivation of the enzyme is also occurring, as shown in Figure 6. Following modification by methylene blue and gel filtration to isolate the protein, it is found that a covalently linked blue label is associated with the enzyme, as shown in Figure 7, with a λ_{max} slightly longer than that of the free dye. Unlike the case with resorufin dimethylcarbamate, the enzyme/methylene blue adduct seems to be unaffected by NADH, which is surprising in view of the non-enzymatic reaction between the free dye and NADH (Helander et al., 1993). However, the coloured label can be completely bleached by reduction with ascorbic acid and then substantially reoxidised by hydrogen peroxide treatment (Figure 7). (It is evident that the λ_{max} after re-oxidation is not quite the same as originally, and the colour subsequently fades away again under these conditions; it appears that the label is chemically unstable in the presence of a high concentration of hydrogen peroxide.) The observed redox changes, coupled with the structural similarity between methylene blue and derivatives of resorufin, make it very likely that the oxidative addition pathway (Figure 3) established for the resorufin compounds applies also to methylene blue. The present work is the first demonstration that methylene blue covalently inactivates ALDH rather than merely inhibiting it; the possibility of this reaction should be borne in mind in any metabolic studies involving the effect of the dye on alcohol metabolism.

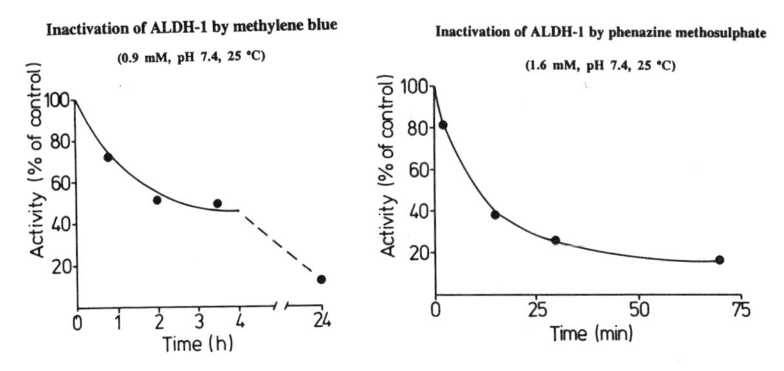

Figure 6. Time course of the inactivation of ALDH-1 by methylene blue and by phenazine methosulphate.

3.5. Interaction of ALDH-1 with Phenazine Methosulphate

Many years ago, one of us (K.E.K.) observed that when polacrylamide gels containing ALDH were soaked in solutions of phenazine methosulphate followed by acetic acid, a pink band developed where the enzyme was located (Crow, 1975). This occurred without the presence of nitro blue tetrazolium chloride (the other usual ingredient of activity stains) or the enzyme's substrates (NAD^+ and acetaldehyde). We decided to examine this phenomenon further, since phenazine methosulphate has some structural similarity to resorufin derivatives and methylene blue (Figure 1), and thus has the potential to undergo oxidative addition with ALDH. Figure 6 shows that the compound inactivates the enzyme faster than does methylene blue; after isolation of the labelled enzyme the spectrum shown in Figure 8 is observed. To the eye, the modified enzyme solution appears brown, but after addition of acetic acid, the same pink colour as seen on gels is produced. Once again, the enzyme-linked label is reducible by ascorbic acid, but this time only a small degree of re-oxidation by hydrogen peroxide appears possible (see Figure 8). Like the case of resorufin dimethylcarbamate, but unlike that of methylene blue, the enzyme-phenazine adduct is susceptible to reduction by NADH (Figure 8), arguing as before that the dye and the cofactor must be in intimately close contact within the enzyme complex.

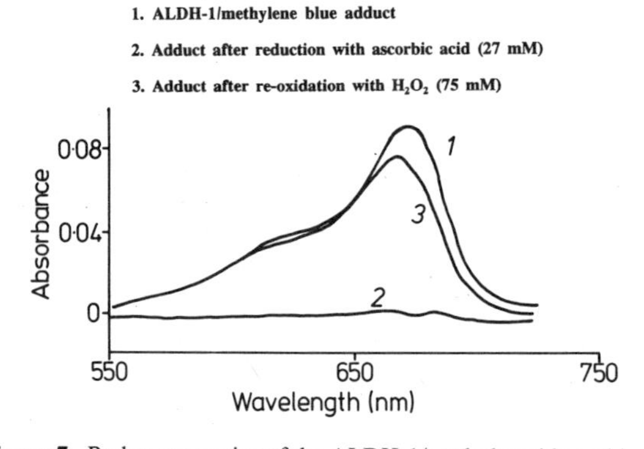

Figure 7. Redox properties of the ALDH-1/methylene blue adduct.

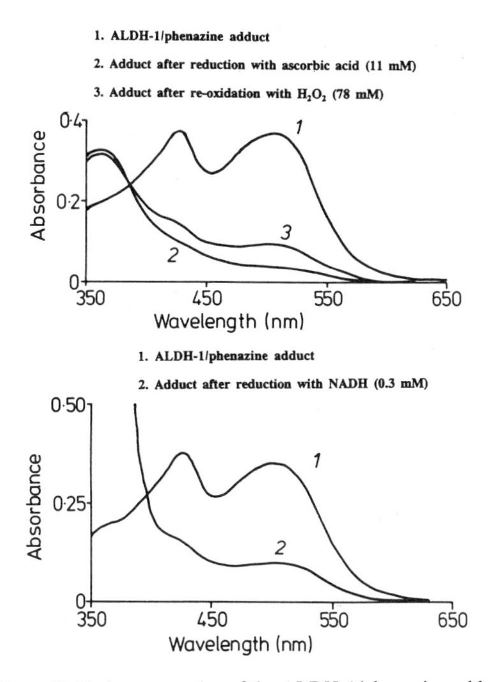

Figure 8. Redox properties of the ALDH-1/phenazine adduct.

3.6. Conclusion

There are several α,β-unsaturated carbonyl compounds that have been reported to be inactivators of ALDH; these include acrolein (propenal) (Mitchell and Petersen, 1988), propiolaldehyde (propynal) (DeMaster and Nagasawa, 1978), and a vinyl ketone (Blatter et al., 1990). The common thiol-specific reagent *N*-methylmaleimide is another such compound that has been investigated in relation to ALDH (Tu and Weiner, 1988). In these cases the enzymic nucleophile is presumed to add to the reagent in a 1,4-manner. The present work shows that with a variety of α,β-unsaturated imine-containing molecules, the initial 1,4-addition is followed immediately by spontaneous oxidation. Each of the compounds studied here has a relatively large tricyclic planar structure which may have a particular affinity for the capacious substrate binding site of ALDH (Moore et al., 1999). It will be of interest to see if the chemistry investigated here with ALDH-1 is also exhibited by ALDH-2, and whether the reactions occur in vivo as well as in vitro (although the latter seems to be already evident in the case of methylene blue; Cronholm, 1993). There may well be naturally occurring α,β-unsaturated compounds that could be endogenous modifiers of ALDH; acrolein (produced during lipid peroxidation and discussed further in the accompanying paper; King et al., 1999) and quinones such as vitamin K and ubiquinone spring to mind. Menadione (vitamin K_3) has been shown to inactivate a phosphatase with an active site thiol group, and exactly the same oxidative addition mechanism as discussed here has been proposed to explain the loss of activity (Ham et al., 1997). Finally, it is conceivable that compounds of similar structure to those investigated here could be developed as 'anti-alcohol drugs' and used therapeutically as alternatives to disulfiram.

REFERENCES

Baker, H.M., 1998, unpublished observations.

Blatter, E.E., Tasayco, M.L., Prestwich, G and Pietruszko, R., 1990, Chemical modification of aldehyde dehydrogenase by a vinyl ketone analogue of an insect pheromone, *Biochem. J.* **272**, 351–358.

Cronholm, T., 1993, Ethanol metabolism in isolated hepatocytes; effects of methylene blue, cyanamide and penicillamine on the redox state of the bound coenzyme and on the substrate exchange at alcohol dehydrogenase, *Biochem. Pharmacol..* **45**, 553–558.

Crow, K.E., 1975, Acetaldehyde metabolism in mammals, Ph. D. thesis, Massey University, Palmerston North, New Zealand.

DeMaster, E.G. and Nagasawa, H.T., 1978, Inhibition of aldehyde dehydrogenase by propiolaldehyde, a possible metabolite of pargyline, *Res. Comm. Chem. Path. Pharmacol,* **21**, 497–505.

Ham, S.W., Park, H.J. and Lim, D.H., 1997, Studies on menadione as an inhibitor of the cdc25 phosphatase, *Bioorg. Chem.* **25**, 33–36.

Helander, A., Cronholm, T. and Tottmat, O., 1993, Inhibition of aldehyde dehydrogenases by methylene blue, *Biochem. Pharmacol.* **46**, 2135–2138.

King, G.J., Norris, G.E., Kitson, K.E., and Kitson, T.M., 1999, Reaction between sheep liver mitochondrial aldehyde dehydrogenase and a chromogenic 'reporter group' reagent. [this volume]

Kitson, T.M., 1996, Comparison of resorufin acetate and *p*-nitrophenyl acetate as substrates for chymotrypsin, *Bioorg. Chem..* **24**, 331–339.

Kitson, T.M., 1998, The oxidative addition reaction between compounds of resorufin (7-hydroxy-3*H*-phenoxazin-3-one) and 2-mercaptoethanol, *Bioorg. Chem..* In the press.

Kitson, T.M., Hill, J.P. and Midwinter, G.G., 1991, Identification of a catalytically essential nucleophilic residue in sheep liver cytoplasmic aldehyde dehydrogenase, *Biochem. J.* **275**, 207–210.

Kitson, T.M. and Kitson, K.E., 1994, Probing the active site of cytoplasmic aldehyde dehydrogenase with a chromophoric reporter group, *Biochem. J.* **300**, 25–30.

Kitson, T.M. and Kitson, K.E., 1996, A comparison of nitrophenyl esters and lactones as substrates of cytosolic aldehyde dehydrogenase, *Biochem. J.* **316**, 225–232.

Mitchell, D.Y. and Petersen, D.R., 1988, Inhibition of rat liver aldehyde dehydrogenase by acrolein, *Drug. Metab. Disp.* **16**, 37–42.

Moore, S.A., Baker, H.M., Blythe, T.J., Kitson, K.E., Kitson, T.M. and Baker, E.N., 1999, The structure of sheep liver cytosolic aldehyde dehydrogenase reveals the basis for the retinal specificity of Class 1 ALDH enzymes. [this volume]

Tu, G.-C. and Weiner, H., 1988, Evidence for two distinct active sites on aldehyde dehydrogenase, *J. Biol. Chem.* **263**, 1218–1222.

12

A THIOESTER ANALOGUE OF AN AMINO ACETYLENIC ALDEHYDE IS A SUICIDE INHIBITOR OF ALDEHYDE DEHYDROGENASE AND AN INDUCER OF APOPTOSIS IN MOUSE LYMPHOID CELLS OVEREXPRESSING THE bcl2 GENE

Gérard Quash,[1] Guy Fournet,[2] Catherine Raffin,[3] Jacqueline Chantepie,[1] Yvonne Michal,[1] Jacques Gore,[2] and Uwe Reichert[3]

[1]Laboratoire d'Immunochimie
INSERM U 329
Faculté de Médecine Lyon-Sud
BP 12, 69921 Oullins Cedex, France
[2]Laboratoire de Chimie Organique 1
UMR 5622 du CNRS Bâtiment 308 (CPE)
43 Bld du 11 novembre 1918
69622 Villeurbanne Cedex, France
[3]CIRD Galderma
Route des Lucioles, BP 87
06902 Sophia-Antipolis, Valbonne, France

1. INTRODUCTION

We have previously shown that methional, $CH_3SCH_2CH_2CHO$, an endogeneous cellular aldehyde derived by the oxidative decarboxylation of 4-methylthio-2-oxobutanoic acid (MTOB), an intermediate in the methionine salvage pathway is a potent inducer of apoptosis when added to cultures of mouse lymphoid cells BAF_3 bo (Quash et al., 1995). MTOB at 100 μM was also capable of alleviating the methionine dependence shown by transformed cells (Ogier et al., 1993) because of its ready transamination to methionine, with glutamine as amine donor (Backlund et al., 1982). Further, the inhibition of this MTOB transaminase with novel transition state inhibitors such as methionine-ethyl ester-pyridoxal induced apoptosis in BAF_3 bo cells but not in BAF_3 bcl_2 which had been trans-

Enzymology and Molecular Biology of Carbonyl Metabolism 7
edited by Weiner *et al.* Kluwer Academic / Plenum Publishers, New York, 1999.

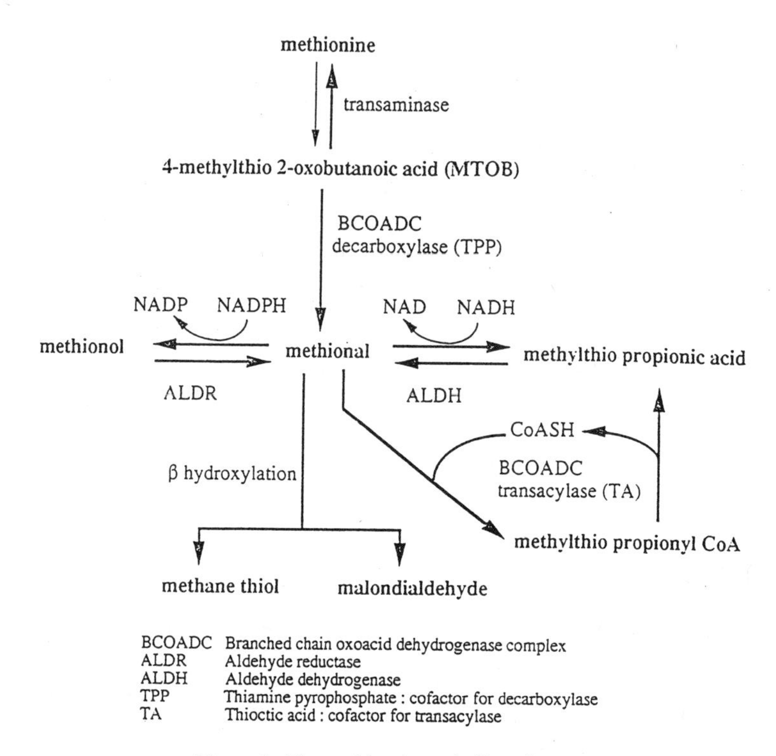

Figure 1. The methional metabolic pathway.

fected with the human bcl_2 gene (Roch et al., 1996). When the reasons for this refractory behaviour of BAF_3 bcl_2 were investigated using[^{14}C] MTOB it was found that the production of [^{14}C] methional from [^{14}C] MTOB was reduced by 50% in BAF_3 bcl_2 and that even the direct addition of 600μM methional could not induce apoptosis in BAF_3 bcl_2 (Roch and al, 1996) Methional is itself further metabolised intracellularly [Figure 1] by 3 pathways: reduction by aldehyde reductase (ALR) to methionol, oxidation by aldehyde dehydrogenase (ALDH1) to methyl thiopropionic acid and β-hydroxylation to malondialdehyde. We therefore tried to see whether the inhibition of the reductase or the dehydrogenase would induce apoptosis in methional treated BAF_3 bcl_2 cells. Using quercitin as an inhibitor of ALR an increase in apotosis was observed in BAF_3 bcl_2, but this result is to be interpreted with caution as quercitin also induces DNA strand breaks directly.With disulfiram as an inhibitor of ALDH a small increase was seen but experimentation could be carried out only with low concentrations (< 100μM) due to the intrinsic toxicity of this compound. It was therefore clear that more specific inhibitors of ALDH were required to try to determine if ALDH played a role in the resistance to the apoptosis-inducing activity of methional.

This need for more specific inhibitors of ALDH had been adressed by us many years ago in the course of a programme undertaken to develop inhibitors of the enzymes involved in metabolizing amino aldehydes arising from polyamine metabolism . This work led us to design and synthesize 4-amino 4-methyl 2-pentyne 1-al (ampal) as an irreversible inhibitor of ALDH. Ampal not only inhibited cellular ALDH but was also an inhibitor of the growth of the human gliobastoma cell line U251 in culture (Quemener et al., 1990). This inhibitory effect of ampal was restricted to the growth of all the transformed cells examined (HeLa,

B16, 293, L1210) because the growth of normal human embryonic fibroblasts (MRC-5) was not inhibited. The inhibitory effect on cell growth was also found *in vivo*. Indeed, ampal increased the life span of Swiss mice grafted with the Krebs ascites tumour and that of B6D2F1 mice grafted with the L1210 lymphoma (Quemener et al., 1989).

The simultaneous presence, however, of a primary amine and an aldehyde group on ampal meant that the compound had to be synthesized as the diethyl acetal derivative and be cleaved extemporaneously. Further, cleavage conditions: 0.2 M HCl at 27°C for 18 h had to be rigorously respected to avoid polymerisation of the free amino aldehyde. This proved to be an important draw-back for undertaking large scale experimentation with ampal. We therefore undertook the synthesis of ampal in the form of a prodrug: thioampal and to facilitate administration, converted it to the formate salt to increase its hydrosolubility. It is the synthesis of thioampal, its action as an enzyme activated irreversible inhibitor of yeast ALDH1, and its apoptosis inducing activity on mouse lymphoid cells BAF_3 transfected with the bcl_2 gene which will be reported here.

2. MATERIALS AND METHODS

2.1. Cell Lines

MRC-5: Normal human embryonic lung fibroblasts were obtained from Institut Mérieux (Lyon, France).

HeLa: Human cervical carcinoma cells were obtained from ATCC (Rockville, MD, USA).

BAF_3 bo: A murine bone marrow-derived cell line (dependent on IL-3 for growth) and BAF_3 bcl_2 transfected with a 1.9 kb human bcl_2 cDNA were a kind gift of Dr. Marvel (E.N.S. Lyon, France).

2.2. Growth of Cells

HeLa & MRC-5 cells were grown as previously described by Ogier et al., (1993). For BAF_3 bo & BAF_3 bcl_2, growth conditions were those used by Roch et al., (1996).

2.3. Chemistry

Compounds 1–9 were obtained as depicted in Figure 2. Full details will be published elsewhere.

2.4. Effect of Compounds on Cell Growth

10^5 cells seeded in 1 ml of culture medium per well of 24 well-plates were maintained at 37°C in a humid atmosphere of air/CO2 (95 : 5). At 4 h. after seeding, compounds were added to the wells at the concentrations indicated. After 4 days at 37°C the cells were washed twice with phosphate buffered saline (PBS) and harvested directly in 0.1 M NaOH for protein estimation (Lowry et al., 1951).

2.5. Effect of Compounds on DNA Fragmentation

BAF_3 bcl_2 cells were seeded at 10^5 cells/ml, labelled with [^{3}H] Thymidine (0.5 µCi /ml culture medium) for 40 h at 37°C. After 2 washes with culture medium, 2.5×10^5 cells

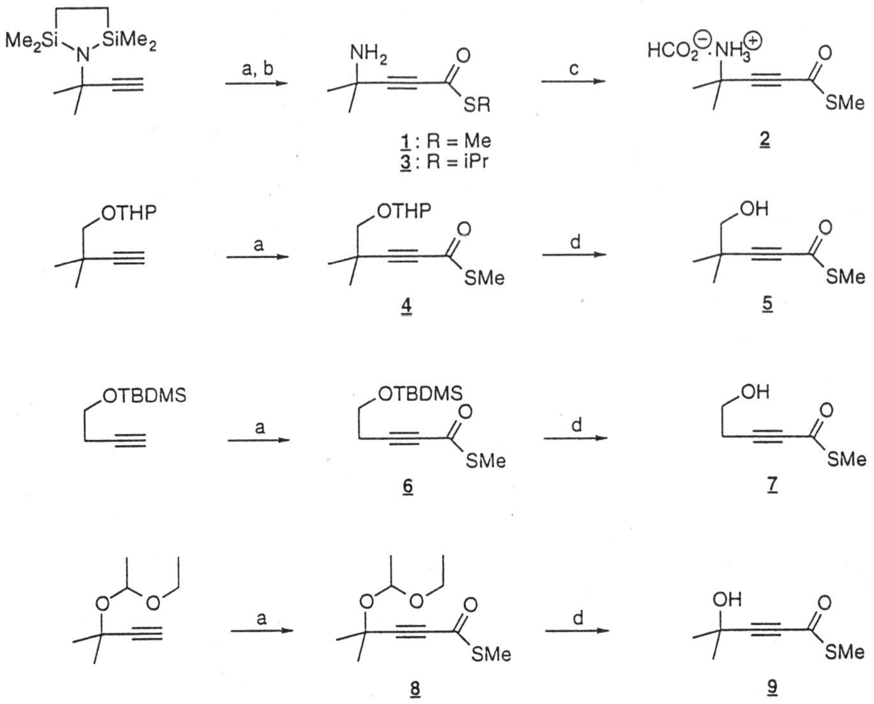

a : nBuLi, THF, -70°C to rt; COS, -70° to 3°C; RI, 3°C
b : SiO_2
c : HCO_2H, Et_2O
d : camphosulfonic acid, aceton/water, reflux

Figure 2. Chemical synthesis of thioampal.

were reseeded in 5 ml culture medium containing thioampal from 100 to 600 μM. in the presence or absence of 200 μM methional. After contact for 6 h, cells were harvested and DNA fragmentation measured as described previously (Roch et al., 1996).

2.6. Enzymatic Studies

To a reaction mixture containing 1 mM EDTA, 100 mM KCl, 2.35 mM NAD in 60 mM sodium phosphate buffer pH 8.5 was added 260 mU baker's yeast ALDH. The reaction was started by the addition of 2 mM propanal and the O.D was read at 340 nm at 0 time (on adding the substrate) and at intervals of 5 and 10 min. Enzyme activity is expressed as Δ O.D/min.

2.7. Enzyme-Activated Inhibiton

For measuring this, the preincubation was carried out at pH 6 instead of pH 8.5, to avoid the spontaneous cleavage of thioampal which occurs at pH 8.5. Accordingly, 260 mU of ALDH (bakers yeast) were incubated at 37°C or at 0°C in 60 mM sodium phosphate buffer pH6 containing 1mM EDTA, 100 mM KCl in a final volume of 200μl. Thioampal at a final concentration of 400μM was added to each of a series of tubes incubated at 0°C and at

37°C. To the control series no thioampal was added. Incubation was carried out at both temperatures for periods of time ranging from 5, 10, 20 and 30 min. At the end of each incubation period 100 μg of BSA was added as a carrier to each tube followed by acetone at –20°C to a final concentration of 80% to precipitate proteins (ALDH & BSA). After standing for 1h at –20°C the tubes were centrifuged at 10.000 rpm for 10 mins, washed once with 80% acetone to eliminate any residual thioampal. The final precipitate was dissolved in 358μl water and immediately added to the complete enzymatic reaction mixture containing NAD, EDTA, KCl, propanal, phosphate buffer pH 8.5 at the concentrations described above.

2.8. Irreversible Inhibition

This was carried out as described for enzyme-activated inhibition but for a fixed period of time: 20 mins, in the presence of thioampal from 100 to 400 μM followed by precipitation and dissolution of the precipitate. The enzyme solution was dialysed against 10 mM phosphate buffer pH6 containing 0.14M NaCl with 2 changes and finally against the phosphate buffer pH8.5 used for the enzymatic reaction.

2.9. Competitive Inhibition

To examine this with excess substrate, glyceraldehyde at 12 and 40 mM was used instead of propanal because of the inhibition of the enzyme with this latter substrate at such high concentrations. To reduce spontaneous hydrolysis of thioampal, the enzymatic reaction was performed at pH7 and not at pH 8.5. Apart from these 2 changes, the enzymatic reaction mixture contained the same concentrations of all the other constituents present in the reaction at pH 8.5.

3. RESULTS AND DISCUSSION

3.1. Screening of Compounds for Their Growth Inhibitory Effect

Compounds were routinely screened for inhibitory efficacy and selectivity using MRC5 fibroblasts as model of normal human cells and HeLa as typical of transformed human cells. As a rapid assay for growth we used the protein content per well after verifying that there was good correlation (r = 0.99) between the number of attached viable cells counted on a haemocytometer, their DNA content measured by the Hoechst assay (West et al., 1985) and their protein content as determined by the Lowry method (1951).

3.2. Growth Inhibitory Activity of Ampal Analogues

Before undertaking any modifications of ampal we first had to determine the functionalities involved in inhibiting growth and ALDH activity

3.3. C-1 Analogues

To determine the functionalities on ampal which were essential for inhibiting growth, we examined the effect of different C1 analogues on cell growth. It is apparent (Table 1) that growth inhibition is dependent on the presence of a bio-cleavable bond because it is lost when the aldehyde group is replaced by an amide, a ketone or is substituted as the diethylacetal, but is retained when the C-1 substituent is an ester.

Table 1. To 1 10^5 cells/ml in 24 well plates were added 4 h after seeding, 10 µl of each of the synthetic compounds dissolved in ethanol to give the final concentrations indicated and a maximum ethanol concentration of 1%. After incubation for 4 days, growth was assessed as described in the Methods section

Compounds	Cells	% inhibition of growth				
		Concentrations tested				
		$0.5.10^{-4}$M	1.10^{-4}M	2.10^{-4}M	4.10^{-4}M	6.10^{-4}M
 H_2N CHO	MRC5 HeLa	0 5	0 11	0 18	4 54	11 82
 H_2N OEt OEt	HeLa	0	0	0	0	0
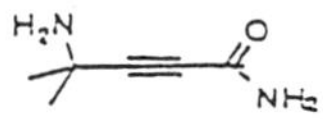 H_2N O NH_2	MRC5 HeLa	0 0	0 0	0 0	0 0	0 8
 H_2N O	MRC5 HeLa	0 0	0 0	0 0	0 11	
 H_2N O OMe	MRC5 HeLa	0 0	0 0	0 0	27 82	46 83

3.4. C-4 Analogues

This was investigated by replacing the amino group on C-4 of ampal with a primary alcohol: CH_2 OH, an ester group: $COOCH_3$, or by this same ester at the end of a methylene side-chain: CH_2-$COOCH_3$. All 3 compounds exhibited growth inhibitory activity but with reduced selectivity as MRC-5 cell growth was also inhibited (Raffin et al., 1995). However, these C-4 analogues were poor inhibitors of ALDH activity requiring mM concentrations for inhibiting enzyme activity by 50%.

These results taken together confirm that the aldehyde on C-1 and the amine on C-4 are important for recognition by ALDH and for growth inhibition, albeit by a mechanism which remains to be elucidated. They also suggest that any substitution on C-1 must be susceptible of being converted to an aldehyde for the compound to retain its activity. Of the possible bio-cleavable bonds which could be introduced, we chose the thio-ester instead of the carboxylic acid ester because on one hand thio esters are high energy bonds, which on cleavage *in vivo* give rise to reactive intermediates and on the other, ALDH has been shown to possess both esterase and dehydrogenase activity in its active site (Duncan, 1985; Kitson & Kitson, 1997).

3.5. Synthesis of Thioampal

Accordingly, thioampal was synthesised as described in the Methods section and the formate salt, because of its hydrosolubility, was chosen for further experimentation on cells in culture.

Table 2. IC_{50} corresponds to the concentration of compound inhibiting cell growth by 50%. Growth was assessed as described for Table 1

	IC 50	
Compounds	MRC-5	HeLa
2	1.15 mM	0.027 mM
5	0.052 mM	0.043 mM
7	0.030 mM	0.022 mM
9	0.020 mM	0.013 mM

3.6. Functionalities on Thioampal Involved in the Selectivity of Growth Inhibition

The selective growth inhibitory activity of *2* would appear to be strictly dependent on the presence of the primary amine group on C-4. Indeed, compounds 5, 7 and 9 exhibited similar growth inhibitory activity for both MRC5 and HeLa cells (see IC50 on Table 2) whereas with thioampal 2 the IC_{50} for HeLa & MRC5 were 0.2mM and 1.15M respectively.

From the results obtained so far it appeared that thioampal also fulfilled the criteria of efficacy and selectivity as a growth inhibitor. The question remained whether ALDH was indeed a target enzyme of thioampal.

3.7. Effect of Thioampal on Baker's Yeast ALDH Activity *in Vitro*: Optimisation of the Assay

Preliminary experiments undertaken to determine the optimum conditions for measuring ALDH activity allowed us to define the assay conditions at 260 mU ALDH/assay, 2.35 mM NAD, 100 mM KCl concentrations of propanal from 0.5 to 2.0 mM, 60 mM sodium phosphate buffer pH 8.5 & an incubation time up to 20 min during which there was a linear increase in enzyme activity.

As regards pH, though the optimum was found to be pH 8,5, preincubation of the enzyme with thioampal was performed at pH6 to prevent the spontaneous hydrolysis of thioampal which occurs at pH 8.5.

3.8. Is Thioampal an Enzyme-Activated Inhibitor of ALDH?

This was investigated as described in the Methods section using 2 mM propanal as substrate and 400μM thioampal as this concentration of inhibitor was found in preliminary experiments to bring about 90% inhibition of enzyme activity. The results (Figure 3) show that the preincubation at 37°C of the enzyme with thioampal at pH 6 for periods varying from 5 to 30 min. brings about a linear decrease in enzyme activity at 5, 10 & 20 min. with practically total inhibition (98%) at 30 min. On the contrary, when ALDH was incubated at 0°C with thioampal at pH6, there was about a 20% decrease in enzyme activity. These results provide evidence that thioampal is converted to an inhibitory form by the action of ALDH.

Further, in order to assess residual activity after preincubation, the enzyme in each reaction mixture was isolated from the preincubation medium by precipitation with ace-

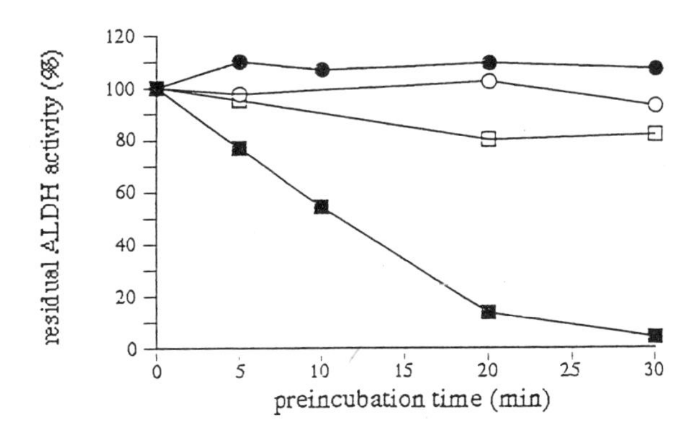

Figure 3. Residual enzyme activity in ALDH preincubated with thioampal. 4 series of tubes with 260 mU of ALDH were incubated at 37°C or at O°C in 200 μl of 60 mM sodium phosphate buffer pH6, containing 1mM EDTA & 100 mM KCl. 400 μM thioampal was added to each of 2 series of tubes, one of which was kept at 37°C ■ and the other at O°C .Tubes without thioampal at 37°C ● at O°C ○. At the end of the incubation period, the enzyme was separated from the reaction mixture and activity measured as described in the optimised assay (Materials and Methods).

tone at –20°C. Yet, it was only the enzyme isolated from the reaction mixture incubated with thioampal at 37°C which lost activity suggesting then that inhibition was irreversible. We therefore undertook experiments to try to obtain direct evidence for this suggestion.

3.9. Is Thioampal an Irreversible Inhibitor of ALDH?

This was carried out as decribed for enzyme-activated inhibition but at the end of the incubation period at 37°C and at 0°C the enzyme was precipated from the incubation media, isolated, redissolved in water and dialysed against 3 changes of 0.01 M phosphate buffer pH6 for 18h. Residual enzyme activity was then measured at pH8.5. It was found that the enzyme activity of samples preincubated with thioampal at 37°C is still inhibited even after dialysis whereas that preincubated at 0°C retained activity. This provides confirmatory evidence for irreversible inhibition of ALDH by thioampal.

3.10. Is Thioampal a Competitive Inhibitor of ALDH?

To examine this at high substrate concentrations, glyceraldehyde was used as substrate and not propanal, because of the inhibition of enzyme activity observed with high concentrations of propanal > 2mM.. Accordingly, glyceraldehyde at 12 & 40 mM was used to examine the effect of thioampal at 200, 300, 400 and 500 μM. The results expressed as a Dixon plot (Figure 4) show that thioampal is a competitive inhibitor of ALDH with an apparent Ki of 120 μM, meaning then that this compound interacts with the active site of ALDH.

Further, as irreversible inhibition was also seen in the presence of 100μg of BSA, (added as a carrier to the preincubation mixture of ALDH & thioampal to aid precipitation by acetone), the active species derived from thioampal must have interacted as it was formed with a nucleophile in the active site of the enzyme even though there were nucleophiles on BSA, present in 200 fold excess.

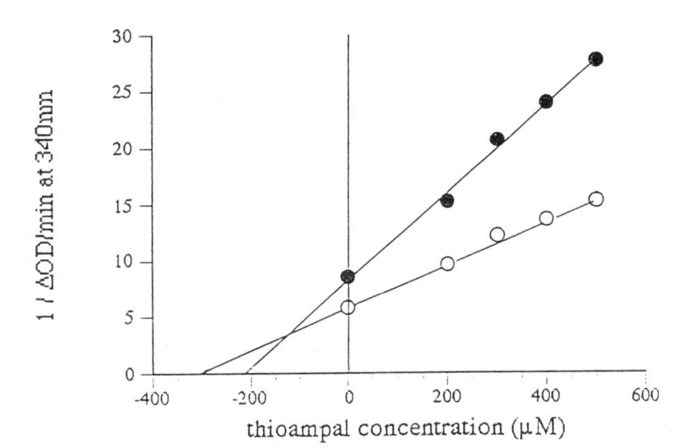

Figure 4. Dixon plot of enzyme activity vs inhibitor concentration. This was measured as described in the optimised assay but with 12 mM ● & 40 mM ○ glyceraldehyde as substrate in the presence of increasing concentrations of thioampal.

3.11. Effect of Methional and Thioampal on DNA Fragmentation in BAF_3 bcl_2 Cells

This was examined as described in the Methods section. It is apparent (Figure 5) that 200μM methional does not induce DNA fragmentation in BAF_3 bcl_2 cells thus confirming our previous results that BAF_3 bcl_2 cells are resistant to the apoptosis inducing activity of methional at concentrations even up to 600 μM (Roch et al.., 1996). On the contrary, in the presence of thioampal alone at increasing concentrations from 100 to 600 μM], there is a regular increase in DNA fragmentation in BAF_3 bcl_2 cells. This increase in DNA fragmentation in BAF_3 bcl_2 further increases in the presence of 200 μM methional (which alone is devoid of activity). Thus a neat synergy is observed.

Since methional is a good substrate for ALDH1 and thioampal has been shown to be an enzyme activated irreversible inhibitor of this enzyme [Figure 3], these results strongly suggest that there may be an impairement in intracellular methional oxidation in bcl_2 transfected cells.

However, we cannot for the moment rule out an effect of thioampal on mitochondrial $ALDH_2$ or even on $ALDH_3$. Indeed, this last enzyme would appear to be the one most intimately involved in apoptosis because it is induced by many different xenobiotics including phenobarbital which has been shown to provoke liver hyperplasia when injected to rats (Deitrich et al., 1977; Huang & Lindhal, 1990). Bursch et al., (1992) have shown

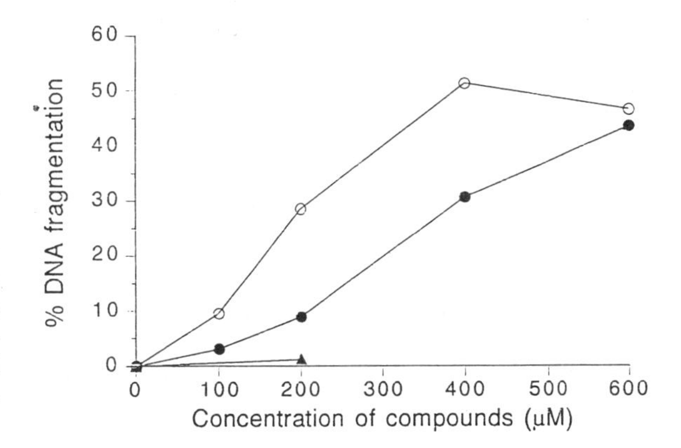

Figure 5. Percent DNA fragmentation in thioampal treated BAF_3 bcl_2 cells. BAF_3 bcl_2 cells with their DNA prelabelled with ^3H thymidine, were treated with thioampal at the concentrations shown, with ○ or without ● 200μM methional. (200 μM methional alone ▲). After 6 h, cells were harvested, washed and lysed. Lysates were spun at 30,000 g for 30 min to separate fragmented from non-fragmented chromatin. Radioactivity in the supernatant and pellet was counted in a liquid scintillation counter.

that when the administration of phenobarbital is stopped, the hyperplastic cells die by apoptosis, but not if the administration of phenobarbital is prolonged. These observations in the literatrure provide additional evidence for unidentified endogeneous aldehydes as mediators of apoptosis, and, taken together with the results obtained so far on methional, support a role for at least this cellular aldehyde in inducing apoptosis in lymphoid cells and more specifically in lymphoid cells in the G2/M phase (Roch et al., 1998).

Nevertheless, direct evidence will be sought using on one hand, cell lines overexpressing the ALDH1, ALDH2, and ALDH3 isozymes and on the other, the crystallographic data presently available on these 3 isozymes.

ACKNOWLEDGMENTS

Financial support for this work was provided by Le Conseil Régional Rhône Alpes, la Ligue Nationale contre le Cancer, Comité de la Drôme et CIRD Galderma, Sophia Antipolis.

REFERENCES

Backlund P.S. Jr., Chang, C.P. and Smith, R.A., (1982), Identification of 2 keto-4-methylthiobutyrate as an intermediate compound in methionine synthesis from 5'-methylthioadenosine, *J. Biol. Chem.* **257** 4196–4202.

Bursch, W., Oberhammer, F. and Schulte-Hermann, R., (1992), Cell death by apoptosis and its protective role against disease, *TIPS* **13** 245–251.

Deitrich, R.A., Bludeau, P., Stock, T. and Roper, M., (1977), Induction of different rat liver supernatant aldehyde dehydrogenases by phenobarbital and tetrachlorodibenzo-p-dioxin, *J. Biol. Chemistry,* **10** 6169–6176.

Duncan, R.J.S, (1985), Aldehyde dehydrogenase: An enzyme with two distinct catalytic activities at a single type of active site, *Biochem. J.* **230** 261–267.

Huang, M. and Lindahl, R., (1990), Aldehyde dehydrogenase heterogeneity in rat hepatic cells, *Arch.of Biochemistry & Biophys.* **277** N°2 296–300.

Kitson, T.M. and Kitson K.E., (1997), Studies of the esterase activity of cytosolic aldehyde dehydrogenase with resorufin acetate as substrate, *Biochem. J,* **322** 701–708.

Lowry, O.H., Rosebrough, N.J., Farr, A.L. and Randall R.J., (1951), Protein measurement with the Folin phenol reagent, *J. Biol. Chem.. **193*** 265–675.

Ogier, G., Chantepie, J., Deshayes, C., Chantegrel, B., Charlot, C., Doutheau, A. and Quash, G., (1993), Contribution of 4-methylthio -2-oxobutanoate and its transaminase to the growth of methionine-dependent cells in culture, *Biochem. Pharmacol.* **45** 1631–1644.

Quash, G., Roch, A.M., Chantepie, J., Michal, Y., Fournet, G. and Dumontet, C., (1995), Methional derived from 4-methylthio-2-oxobutanoate is a cellular mediator of apoptosis in BAF_3 lymphoid cells, *Biochem. J.* **305** 1017–1025.

Quemener, V., Quash G., Moulinoux, J.P., Penlap, V., Ripoll, H., Havouis, R., Doutheau, A. and Gore, J., (1989), *In vivo* antitumor activity of 4-Amino 4-methyl 2-pentyne 1-al, an inhibitor of aldehyde dehydrogenase, *In Vivo* **3** 325–330.

Quemener, V., Moulinoux, J.P., Martin, C., Darcel, F., Guegan, Y., Faivre, J. and Quash, G.A., (1990), ALDH dehydrogenase activity in xenografted human brain tumor in nude mice. Preliminary results in human glioma biopsies, *J. of Neuro-Oncology,* **9** 115–123.

Raffin, C., Bernard, D., Fournet, G., Gore, J., Chantepie, J. and Quash G., (1995), Synthesis of α, α '-difunctional actylenic compounds, selective inhibitors of the growth of transformed cells, *J. Chem. Res.(S.)* 8–9.

Roch, A.M., Quash, G., Michal, Y., Chantepie, J., Chantegrel, B., Deshayes, C., Doutheau, A. and Marvel, J., (1996), Altered methional homoeostasis is associated with decreased apoptosis in BAF_3 bcl_2 murine lymphoid cell, *Biochem. J.***313** 973–981.

Roch, A.M., Panaye, G., Michal, Y. and Quash, G., (1998), Methional, a cellular metabolite, induces apoptosis preferentially in G2/M-synchronized BAF3 murine lymphoid cells, *Cytometry,* **31** 10–19.

West, D.C., Sattar, A. and Kumar, S., (1985,) A simplified *in situ* solubilization procedure for the determination of DNA and cell number in tissue cultured mammalian cells, *Anal. Biochemistry,* **147** 289–295.

13

REACTION BETWEEN SHEEP LIVER MITOCHONDRIAL ALDEHYDE DEHYDROGENASE AND A CHROMOGENIC 'REPORTER GROUP' REAGENT

Gordon J. King,[3] Gillian E. Norris,[1] Kathryn E. Kitson,[2] and Trevor M. Kitson[3]

[1]Institute of Molecular Biosciences
[2]Institute for Food Nutrition and Human Health
[3]Institute of Fundamental Sciences (Chemistry)
Massey University
Palmerston North, New Zealand

1. INTRODUCTION

The cytosolic form of aldehyde dehydrogenase (ALDH-1) has been shown to react slowly with *p*-nitrophenyl dimethylcarbamate, liberating *p*-nitrophenoxide and giving an inactive form of the enzyme in which Cys-302 carries a -CO-NMe$_2$ label (Kitson et al., 1991). This work led to the idea that a cyclic analogue of the carbamate would constitute a 'reporter group' reagent of the type originally envisaged by Burr and Koshland (1964), since the chromophoric *p*-nitrophenoxide moiety would end up covalently bound within the enzyme's active site. The compound in question, namely 3,4-dihydro-3-methyl-6-nitro-2*H*-1,3-benzoxazin-2-one or DMNB (see Figure 1) was synthesised and shown to react in the expected way with esterases such as chymotrypsin (Kitson and Freeman, 1993). Further work with ALDH-1 showed that in this case the pK$_a$ of the *p*-nitrophenol reporter group was perturbed upwards by about 3 pH units (Kitson and Kitson, 1994). This rather dramatic observation was interpreted to mean that the substrate binding site of ALDH-1 is either of very hydrophobic character or contains a negatively charged amino acid sidechain (or conceivably has both characteristics).

The purpose of the work described here was to use DMNB to probe the active site of the mitochondrial form of sheep liver aldehyde dehydrogenase (ALDH-2) and to compare the results obtained with those found with ALDH-1.

Enzymology and Molecular Biology of Carbonyl Metabolism 7
edited by Weiner *et al.* Kluwer Academic / Plenum Publishers, New York, 1999.

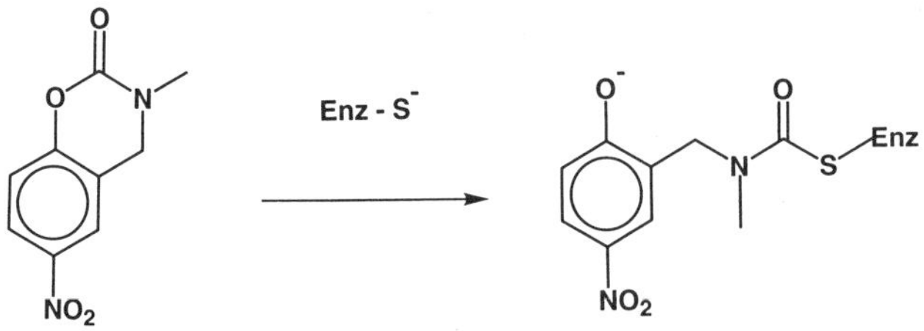

Figure 1. Reaction of DMNB with the active site nucleophile of ALDH.

2. EXPERIMENTAL

DMNB was synthesised as before (Kitson and Freeman, 1993). Mitochondrial ALDH was isolated from frozen sheep liver as follows. (All buffers contained 0.1 mM dithiothreitol.) Liver (1 kg) was homogenised in 2 litres of 2 mM phosphate buffer, pH 7.2, containing 0.3 mM EDTA. The homogenate was centrifuged at 500 x g for 15 min and 20,000 × g for 30 min. Polyethyleneglycol 8000 was added to the supernatant from the second spin to a concentration of 10%. The pellets from the 20,000 x g spin and the PEG cut were re-extracted with 5 mM phosphate buffer, pH 7.2, and PEG added again to a concentration of 10%. The combined supernatants from the PEG precipitations were loaded on to a column of DEAE-cellulose (Iontosorb) equilibrated with 5 mM phosphate buffer, pH 7.2. This buffer was used to wash the column until all haemoglobin was removed. ALDH-2 was eluted with 20 mM acetate buffer, pH 5.6. (If desired, ALDH-1 may then also be eluted from the column by raising the concentration of acetate buffer, pH 5.6, to 40 mM.) The dilute solution of ALDH-2 was adjusted to pH 7 and loaded on to an acetophenone affinity column (Ghenbot and Weiner, 1992). This was washed with 25 mM phosphate buffer, pH 7.3, containing 15 mM NaCl, and subsequently with the same buffer containing 0.75 mM *p*-hydroxyacetophenone (to elute traces of ALDH-1). Increasing the concentration of *p*-hydroxyacetophenone to 10 mM resulted in the elution of ALDH-2. The enzyme was concentrated by ammonium sulphate precipitation, dissolved in 50 mM phosphate buffer, pH 7.4, and stored frozen (above a concentration of 5 mg/ml).

A Varian Cary 1 spectrophotometer was used for spectrophotometric monitoring of the DMNB reporter group and for assay of ALDH activity, using either NAD^+ and acetaldehyde or *p*-nitrophenyl acetate. ALDH-2 was reacted with DMNB (0.8 mM) in 50 mM phosphate buffer at pH 7.4 and 25 °C and subsequently isolated from excess reagent by passage down a Biogel P-6 gel filtration column, eluting with 10 mM phosphate buffer, pH 7.4. Electrospray mass spectrometry was carried out using a Sciex API 300 instrument equipped with a nebulisation-assisted electrospray ion source. Protein samples were in 5 mM ammonium acetate buffer, pH 5.5, containing 25% acetonitrile. The spray needle was held at a potential of 5 kV, and the orifice and ring voltages were held at 30 and 140 volts respectively. Samples were scanned over a m/z range of 1200 to 2200 amu using a step size of 0.1 amu and a dwell time of 1 ms. Data were interpreted with the Sciex software 'BioSpec Reconstruct'.

3. RESULTS AND DISCUSSION

3.1. Characteristics of the Reaction between ALDH-2 and DMNB

Figure 2 shows a typical inactivation profile of ALDH-2 by DMNB; in several repeats of this experiment the inactivation always tailed off with about 50% of the enzymic

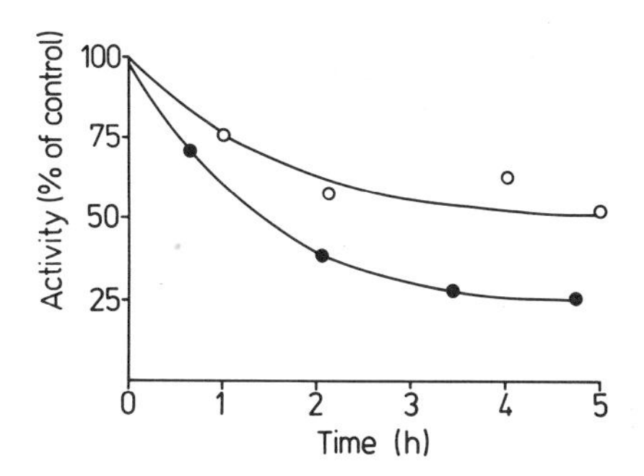

Figure 2. Inactivation of ALDH-2 by DMNB in the absence (open circles) and presence (closed circles) of NAD^+.

activity remaining. Isolation of the modified enzyme by gel filtration showed the presence of a *p*-nitrophenol label with a stoicheiometry of about 0.6 mol per tetramer and a pK_a (immediately after isolation of the protein) of about 7.5, as shown in Figure 3. Upon standing at pH 7.4 for an hour or two, however, the colour of the bound *p*-nitrophenoxide moiety declines, consequent upon a shift in pK_a to about 8, after which there is no further change. By contrast, the remarkable pK_a of about 10 seen with ALDH-1 (Kitson and Kitson, 1994) shows that the *p*-nitrophenol group is at least 100 times less acidic in that case than with ALDH-2. Significant differences in the architecture of the substrate binding sites of ALDH-1 and ALDH-2 have recently become apparent from X-ray crystallography studies (Steinmetz et al., 1997; Moore et al., 1999). That of the cytosolic isozyme is wider and roomier than that of the mitochondrial form, consistent for example with the fact that the bulky molecules retinal and disulfiram are acted on avidly by ALDH-1 (as a substrate and as an inactivator, respectively) but not by ALDH-2. However, there is more hydrophobic character overall to the sidechains surrounding the active site in ALDH-2 than in ALDH-1 which is opposite to what might be predicted based on the relative pK_a values of the DMNB reporter group when bound to the two isozymes. Furthermore, the X-ray structure of the cytosolic isozyme does not reveal the presence of any closely-positioned negatively charged residue that could account for the very low tendency of the bound *p*-nitrophenol group to ionise. It seems that a full explanation of these reporter group observations must await the emergence of X-ray pictures of the DMNB-modified enzymes themselves.

An interesting and unexpected observation that is consistently seen is that parallel with the shift in pK_a referred to above there is a return of enzymic activity to approximately the 100% level. We can completely rule out an explanation for this involving hydrolytic loss of label from the enzyme (thereby freeing the catalytically essential Cys-302) since precipitation of the protein with trichloroacetic acid shows the label is still covalently attached. This is confirmed by the mass spectral results discussed below. One possible explanation is that the label

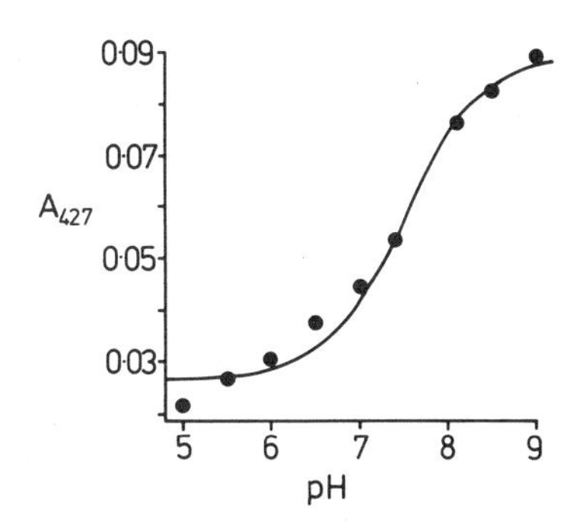

Figure 3. Titration profile of the reporter group supplied to ALDH-2 by DMNB. (Results were obtained as soon as possible after modification.)

migrates from Cys-302 to another amino acid residue. Although a thiocarbamate group is chemically rather inert, it is conceivable that some neighbouring enzymic nucleophilic group could attack it; for example, Cys-301 and Cys-303 spring to mind here, assuming that these residues have the same identity in the sheep ALDH-2 as is known to be the case with the human and horse (Lindahl and Hempel, 1991). It may be relevant here that in sheep liver ALDH-1, which does not display reemergence of activity after DMNB treatment, residue 303 is Ile not Cys (Stayner and Tweedie, 1995). Another explanation, which is currently completely speculative, is as follows. ALDH is known not to display 'full sites reactivity' (Takahashi et al., 1981) in that it appears to have two active sites per tetramer and two sites that are masked or 'dormant' in some way. Presumably this may be because of the precise way in which the quaternary structure is assembled, or it may be that the dormancy is induced only upon binding of substrates. (From the stoicheiometry of the DMNB reaction referred to above, it appears that only one of the active subunits becomes labelled; perhaps this process indirectly renders the other active subunit non-susceptible to the reagent.) It may then be that, following isolation of the enzyme after DMNB treatment, the four monomer units slowly rearrange their positions within the tetramer such that a previously dormant subunit becomes active and the DMNB-modified subunit takes up the position of a hitherto dormant one.

The presence of NADH almost completely prevents reaction of ALDH-2 with DMNB, but as seen in Figure 2, NAD^+ causes somewhat faster and more extensive inactivation. (With ALDH-1, NAD^+ slows the reaction; Kitson and Kitson, 1994). After modification of ALDH-2 in the presence of NAD^+, a stoicheiometry of 1.4 mol of *p*-nitrophenol per tetramer is seen, and now there is no re-emergence of activity upon standing. Thus (with reference to the possible explanations mentioned above) it appears that after DMNB/NAD^+ treatment either the label is permanently attached to Cys-302 and not subject to migration to another residue, or that the arrangement of subunits within the tetramer is fixed and not prone to 'reshuffling'.

3.2. Electrospray Mass Spectrometry of ALDH-2 before and after DMNB Treatment

Figure 4(a) shows the electrospray mass spectrum of native sheep liver ALDH-2 isolated as described above. The largest peak (I) at 54,425 amu is assumed to represent the subunit of the enzyme, but we cannot relate this exactly to the 'expected' molar mass since the sheep liver form of ALDH-2 has not been sequenced. Interestingly, mass spectrometry has previously given an effectively identical molar mass of 54, 432 amu for human recombinant ALDH-2 (Lipsky et al., 1997). Another major peak (II) is observed in Figure 4(a) at 155–158 amu lower than peak I; we propose that this is due to a truncated form of the enzyme that lacks the two N-terminal residues. In the corresponding horse and human isozymes, these are Ser and Ala residues, which together add up to 158 amu. The partial loss of these two residues might be due to proteolysis during enzyme isolation, or it may mean that cleavage of the leader sequence after protein synthesis is not completely specific. The significance of the peaks labelled Ia and IIa will be discussed below.

After DMNB-treatment in the presence of NAD^+, the enzyme gave the mass spectral profile shown in Figure 4(b). In this, peak I is much smaller, and a new large peak (Ib) has

→

Figure 4. Electrospray mass spectrometry of ALDH-2. (a): native enzyme; (b): enzyme after modification by DMNB in the presence of NAD^+. (See text for the identification of the labelled peaks.)

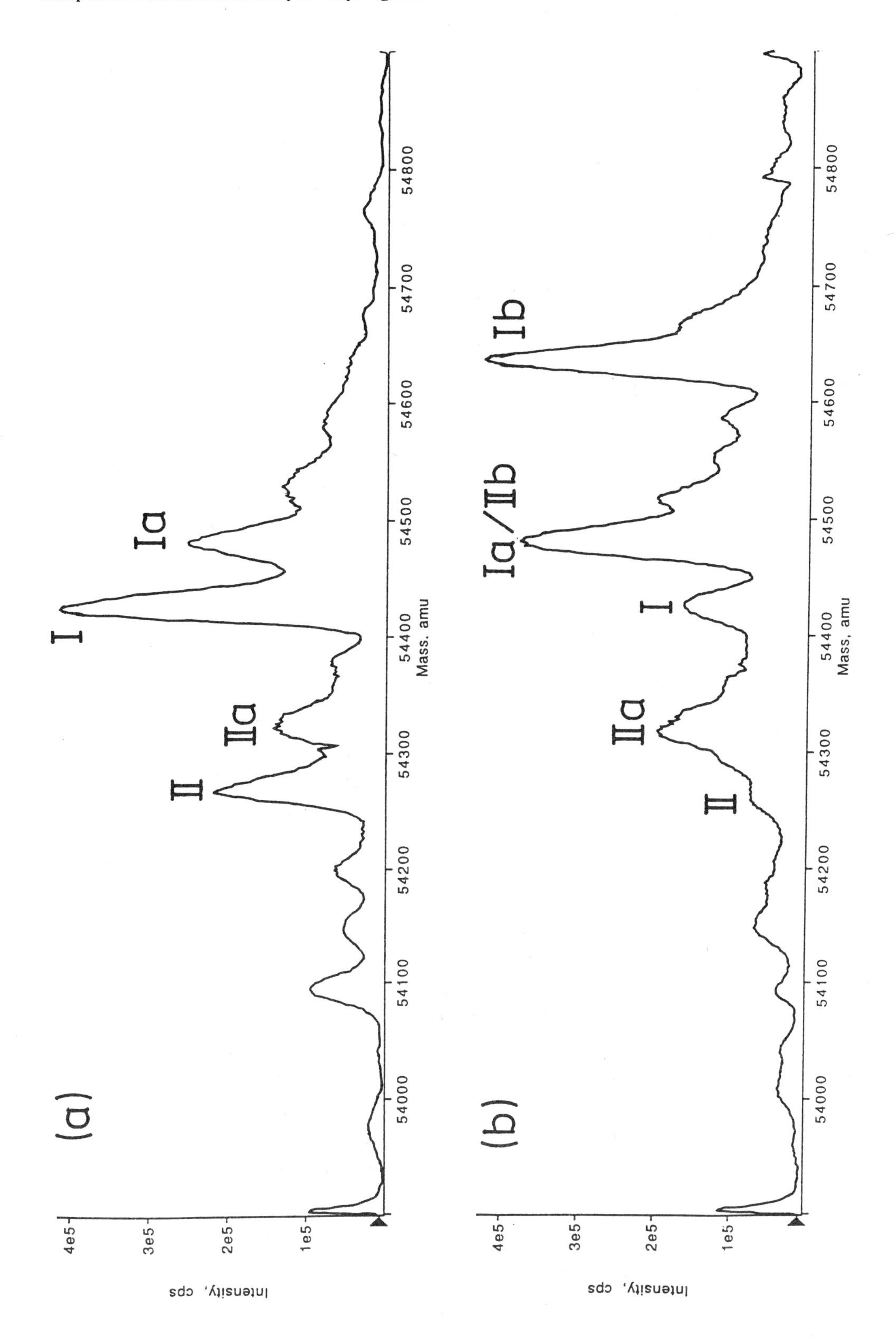
(a)
I
Ia
II
IIa
Intensity, cps
Mass, amu
(b)
Ib
Ia/IIb
I
IIa
II
Intensity, cps
Mass, amu
4e5
3e5
2e5
1e5
54000
54100
54200
54300
54400
54500
54600
54700
54800

appeared at a mass of 211–213 amu greater than I. We interpret this as an enzyme subunit carrying a DMNB moiety (the mass of which is 209 amu), and this observation confirms of course that the interaction between enzyme and modifier is covalent in nature. Similarly, the original peak II is now very small, and we assume that a new peak (labelled IIb) has correspondingly emerged at about 210 amu greater than peak II, but this happens to overlie the original peak Ia. Thus both the complete enzyme and the proposed truncated form are active towards DMNB. In a similar experiment in which modification was carried out in the absence of NAD^+, the mass spectrum showed similar but quantitatively smaller changes. (Peak I did not shrink as much and peak Ib did not grow so large, for example.) This result is consistent with the relative stoicheiometries of labelling observed in the presence and absence of the cofactor, as mentioned above.

Returning now to peaks Ia and IIa , it appcars that thcsc rcprcscnt inactivc forms of ALDH (inactive, at least, in the sense of not reacting with DMNB). Thus IIa is still present in Figure 4(b) and has not diminished in size as I and II have. It is difficult to estimate the size of Ia in Figure 4(b) as it overlies IIb, but it is quite clear that (unlike peak I) it does not give rise to a new peak of mass about 210 amu larger, as this would be at ~54, 690 amu which is not observed. We can estimate from Figure 4(a) that roughly one-third of the total protein is present as these inactive forms. From this we can correct the labelling stoicheiometries referred to above to about 0.9 and 2.1 mol of *p*-nitrophenol per active tetramer (in the absence and presence of NAD^+, respectively).

Peaks Ia and IIa are heavier than I and II, respectively, by 54–57 amu. The simplest explanation for the observed lack of reactivity of Ia and IIa towards DMNB is that the essential nucleophilic group of Cys-302 is modified in some way. We have no direct evidence of what this modification might be, but we note that conversion of a thiol group to the sodium salt of a sulphinic acid group ($-SO_2^- Na^+$) leads to a rise in mass of 54 amu. It is conceivable that (during storage or use in experiments) solutions of ALDH might undergo partial atmospheric oxidation in this way, and it would not be surprising if the particularly reactive thiol group of Cys-302 were to be specifically involved. Cysteinesulphinic acid is a stable compound that is an intermediate in the cellular oxidation of cysteine to sulphate (Patai, 1974; Oae, 1977). It is quite a strong acid (pK_a about 2) and therefore would exist as the anion except at very low pH. Alternatively, it may be that ALDH-2 exists in a partially modified state in vivo (or becomes so during the isolation procedure) due to the attachment of some small reactive endogenous molecule. For example, if the thiol of Cys-302 were to add in a 1,4-manner to the α,β-unsaturated aldehyde acrolein (propenal), this would lead to inactivation and an increase in mass of 56 amu. Acrolein is produced during lipid peroxidation (Poli et al., 1985), and is identifiable in normal human serum (Zlatkis et al., 1980). It has been shown to be a potent inactivator of rat liver ALDHs, particularly the mitochondrial form (Mitchell and Petersen, 1988). If peaks Ia and IIa really do represent ALDH-acrolein adducts, then this would support the proposal that inactivation of ALDH by acrolein may play an important role in enhancing the toxicity of other aldehydes liberated during lipid peroxidation (Mitchell and Petersen, 1988).

REFERENCES

Burr, M. and Koshland, D.E., 1964, Use of "reporter groups" in structure-function studies of proteins, *Proc. Natl. Acad. Sci. (USA),* 52:1017–1024.

Ghenbot, G. and Weiner, H., 1992, Purification of liver aldehyde dehydrogenase by *p*-hydroxyacetophenone-Sepharose affinity matrix and the coelution of chloramphenicol acetyl transferase from the same matrix with recombinantly expressed aldehyde dehydrogenase, *Protein Expression Purif.*, 3:470–478.

Kitson, T.M. and Freeman, G.H., 1993, 3,4-Dihydro-3-methyl-6-nitro-2*H*-1,3-benzoxazin-2-one, a reagent for labeling *p*-nitrophenyl esterases with a chromophoric reporter group - synthesis and reaction with chymotrypsin, *Bioorg. Chem.*, 21:354–365.

Kitson, T.M., Hill, J.P. and Midwinter, G.G., 1991, Identification of a catalytically essential nucleophilic residue in sheep liver cytoplasmic aldehyde dehydrogenase, *Biochem. J.*, 275:207–210.

Kitson, T.M. and Kitson, K.E., 1994, Probing the active site of cytoplasmic aldehyde dehydrogenase with a chromophoric reporter group, *Biochem. J.*, 300:25–30.

Lindahl, R. and Hempel, J., 1991, Aldehyde dehydrogenases: What can be learned from a baker's dozen sequences? *Adv. Exp. Med. Biol.*, 284:1–8.

Lipsky, J.J., Mays, D.C., Holt, J.L., Tomlinson, A.J., Johnson, K.L., Veverka, K.A. and Naylor, S., 1997, Inhibition of and interaction with human recombinant mitochondrial aldehyde dehydrogenase by methyl diethylthiocarbamate sulfoxide, *Adv Exp. Med. Biol..*, 414:209–216.

Mitchell, D.Y. and Petersen, D.R., 1988, Inhibition of rat liver aldehyde dehydrogenase by acrolein, *Drug Metab. Disp.*, 16:37–42.

Moore, S.A., Baker, H.M., Blythe, T.J., Kitson, K.E., Kitson, T.M. and Baker, E.N., 1999, The structure of sheep liver cytosolic aldehyde dehydrogenase reveals the basis for the retinal specificity of Class 1 ALDH enzymes. [this volume]

Oae, S., 1977, Organic Chemistry of Sulfur, Plenum Press, New York and London.

Patai, S., 1974, The Chemistry of the Thiol Group, John Wiley and Sons, London.

Poli, G., Dianzani, M.U., Cheeseman, K.H., Slater, T.F., Lang, J. and Esterbauer, H., 1985, Separation and characterisation of the aldehydic products of lipid peroxidation stimulated by carbon tetrachloride or ADP-iron in isolated hepatocytes and rat liver microsomal suspensions, *Biochem. J.*, 227:629–638.

Stayner, C.K. and Tweedie, J.W., 1995, Cloning and characterisation of the cDNA for sheep liver cytosolic aldehyde dehydrogenase, *Adv. Exp. Med. Biol.*, 372:61–66.

Steinmetz, C.G., Xie, P., Weiner, H. and Hurley, T.D., 1997, Structure of mitochondrial aldehyde dehydrogenase: the genetic component of ethanol aversion, *Structure,* 5:701–711.

Takahashi, K., Weiner, H. and Filmer, D.L., 1981, Effects of pH on horse liver aldehyde dehydrogenase: alterations in metal ion activation, number of functioning active sites, and hydrolysis of the acyl intermediate, *Biochemistry,* 21:6225–6230.

Zlatkis, A., Poole, C.F. and Brazeli, R., 1980, Volatile metabolites in sera of normal and diabetic patients, *J. Chromatogr.*, 182:137–145.

ACTIVITY OF THE HUMAN ALDEHYDE DEHYDROGENASE 2 PROMOTER IS INFLUENCED BY THE BALANCE BETWEEN ACTIVATION BY HEPATOCYTE NUCLEAR FACTOR 4 AND REPRESSION BY PEROSIXOME PROLIFERATOR ACTIVATED RECEPTOR δ, CHICKEN OVALBUMIN UPSTREAM PROMOTER-TRANSCRIPTION FACTOR, AND APOLIPOPROTEIN REGULATORY PROTEIN-1

Jane Pinaire, Wan-Yin Chou, Mark Stewart, Katrina Dipple, and David Crabb

Departments of Medicine and Biochemistry and Molecular Biology
Indiana University School of Medicine
Indianapolis, Indiana

1. INTRODUCTION

We have had a long-standing interest in the regulation of expression of mitochondrial aldehyde dehydrogenase (ALDH2). This enzyme has the lowest K_m for acetaldehyde among the many known aldehyde dehydrogenases, and a syndrome of genetic deficiency in ALDH2 (Crabb et al., 1989) is known to markedly impair oxidation of acetaldehyde after consumption of ethanol (Enomoto et al., 1991). This impairment of acetaldehyde disposition is the cause of the Asian alcohol flush reaction and is very strongly protective against the development of alcoholism (Goedde et al., 1983; Harada et al., 1982; Thomasson et al., 1991). We reason that factors that alter the expression of the enzyme might also influence risk of alcoholism and have therefore studied the control of the *ALDH2* gene, with particular attention to its expression in liver.

This gene is expressed in many tissues, but at highly variable levels. The highest expression is in the liver and kidney, with substantial expression in heart and skeletal muscle, and lower levels of mRNA in most other tissues examined to date (Stewart et al.,

Enzymology and Molecular Biology of Carbonyl Metabolism 7
edited by Weiner *et al.* Kluwer Academic / Plenum Publishers, New York, 1999.

1996b). The *ALDH2* promoter has many attributes of a housekeeping gene. It appears to be imbedded in a CpG island (Dipple and Crabb, 1993), and lacks a TATAA box. However, we have shown that it contains a CCAAT box at about -60 bp from the translation start site that binds the trimeric transcription factor NF-Y (Stewart et al., 1996a). NF-Y has been shown to directly interact with the basal transcriptional apparatus (directly contacting TBP, the TATAA binding protein (Bellorini et al., 1997)). Moreover, the transcription initiation site conforms well to the consensus for an initiator site (Smale and Baltimore, 1989). Thus, the basal promoter for ALDH2 comprises the initiator and the NF-Y binding site. This short promoter (extending about 120 bp upstream from the translation initiation codon) is an active promoter when transfected into cultured cells, whereas shorter promoters extending only to -75 bp are virtually inactive (Stewart et al., 1996a). This region of the promoter may be responsible for the low level, ubiquitous expression of the gene, but does not appear to explain high level liver and kidney expression. We therefore studied more distal regions of the *ALDH2* promoter for their ability to bind proteins and direct tissue-specific expression.

2. METHODS

2.1. Plasmids

pALDH3'- BLCAT (containing four copies of the ALDH2 FP330-3' site from the human *ALDH2* gene ligated to a herpes simplex thymidine kinase promoter upstream of the chloramphenicol acetyltransferase (CAT) gene was previously described (Stewart et al., 1998). pALDH5'- BLCAT was constructed in a similar fashion and contains three copies of the FP330-5' site in the same vector, pBL2-CAT (Pharmacia). The expression plasmids for HNF4, COUP-TFI, ARP-1, and PPARδ were kindly provided by Drs. Frances Sladek, Ming Ter Tsai, S. Karathansis, and Ronald Evans, respectively.

2.2. Isolation of Nuclear Protein Extracts and DNA Binding Assays

Nuclear proteins were isolated from rat liver based on the original protocol of Dignam *et al.* (Dignam et al., 1983) with the modifications as described by Shapiro et al. (Shapiro et al., 1988). The nuclear proteins in the supernatant were precipitated with $(NH_4)_2SO_4$ and then dialyzed overnight against buffer containing 20 mM HEPES, pH 7.9, 20% (v/v) glycerol, 0.1 M KCl, 0.2 mM EDTA, 0.2 mM EGTA, and 2.0 mM DTT. Aliquots were frozen in liquid nitrogen and stored at -70°C. DNase I footprinting was performed as described (Stewart et al., 1998). Electrophoretic mobility shift assays (EMSA) were performed by radiolabeling double-stranded oligonucleotides corresponding to each half of the footprint seen in the 330 bp restriction fragment described above. Nuclear extracts (5 μg) were incubated with 1–2 μg of non-specific competitor DNA (poly (dIC)) in binding buffer on ice for 15 min. For supershift assays, antibodies were added and the mixture was incubated an additional 1–2 hours. Labeled probe (20,000 cpm) was added last and the reaction incubated an additional 15 min on ice. Reaction mixtures were electrophoresed on a non-denaturing 5% acrylamide gel and subjected to autoradiography. Anti-HNF4 antibody was the gift of Dr. F.M. Sladek, anti-COUP-TF antibody was from Dr. Ming Ter Tsai, anti-ARP1 antibody was the gift of Dr. S. Karathanasis, and anti-RXR antibody was from Dr. Pierre Chambon and Santa Cruz Biotechnology.

2.3. Transfection of Tissue-Culture Cells

Cells were grown in MEM supplemented with 10% charcoal-stripped fetal bovine serum, 100 μg/ml streptomycin, and 63 μg/ml penicillin G. The day before transfection, the cells were plated at 10^6 cells/100 mm dish. The cells were transfected with 10 μg of reporter plasmid, 10 μg of receptor expression plasmid, and 5 μg of pSV_2-luc (as an internal control of transfection efficiency) by calcium phosphate precipitation (Gorman et al., 1982). Four hours later the cells were exposed to PBS containing 15% glycerol for 3 min then cultured in fresh MEM with charcoal-stripped serum. Forty eight hours after transfection, the cells were harvested and assayed for luciferase (de Wet et al., 1987) and chloramphenicol acetyl transferase (CAT) activity (Crabb and Dixon, 1987).

3. PROMOTER ELEMENTS INVOLVED IN TISSUE-SPECIFIC EXPRESSION OF ALDH2

We recently reported that a pair of enhancers are found in the upstream region of the ALDH2 promoter at about -300 bp (Stewart et al., 1998). These sites were first discovered by DNase I footprinting, and designated FP330. Because the footprint was over 60 bp in length, oligonucleotides corresponding to the 5' and 3' halves were synthesized for further analysis. The two subregions were named FP330-5' and FP330-3'. Oligonucleotides corresponding to these regions were bound by nuclear proteins present in most tissues, and they cross-competed with each other in electrophoretic mobility shift assays (EMSA), suggesting that similar proteins could bind each oligonucleotide. Methylation intereference experiments demonstrated that each of these sites contained variants of the direct repeat that is part of the binding site for members of nuclear receptor family of transcription factors (Mangelsdorf et al., 1995b). The canonical element is defined as $^{A}/_{G}G^{G}/_{T}TCA$, and the specificity of binding to this element is determined largely, but not exclusively, by the spacing of direct repeats of this motif (Nakshatri and Bhat-Nakshatri, 1998). The nuclear receptors generally bind DNA as dimers (Mangelsdorf et al., 1995b). A number of them bind as homodimers, including hepatocyte nuclear factor 4 (HNF-4), retinoid X receptors (RXR), chicken ovalbumin upstream promoter transcription factor (COUP-TF), and apolipoprotein regulatory protein-1 (ARP-1). Many others bind as heterodimers with RXR including retinoic acid receptors (RARs), thyroid hormone receptor, vitamin D receptor, and the peroxisome proliferator-activated receptor (PPAR) (Kliewer et al., 1992; Mangelsdorf and Evans, 1995a).

The elements present in FP330 are shown in Figure 1. One interpretation of the sequence of the FP330-3' site is that it contains a direct repeat with one spacer nucleotide (DR-1) element. This would be predicted to be bound by homodimers of RXR, HNF-4, COUP-TF, or ARP-1, and heterodimers of RAR/RXR, PPAR/RXR, and possibly COUP/RXR or ARP/RXR (Mangelsdorf et al., 1995b). In fact, we recently demonstrated by EMSA with antibody induced super-shifts that HNF-4, COUP-TF, and RXR bind to the FP330-3' site (Stewart et al., 1998). It seems likely therefore that HNF-4, which is predominantly expressed in liver and kidney (Sladek et al., 1990; Zhong et al., 1994), plays an important role in the high level expression of ALDH2 in those two organs. The roles of COUP-TF and ARP-1, and of the RXR-containing factors have not been characterized to date. The factors binding to FP330-5', a DR-0 and possibly DR-1 element, are under investigation. Based on the published binding preferences for nuclear receptors (Mangelsdorf et al., 1995b), COUP-TF and ARP-1 might be expected to bind a DR-0 site.

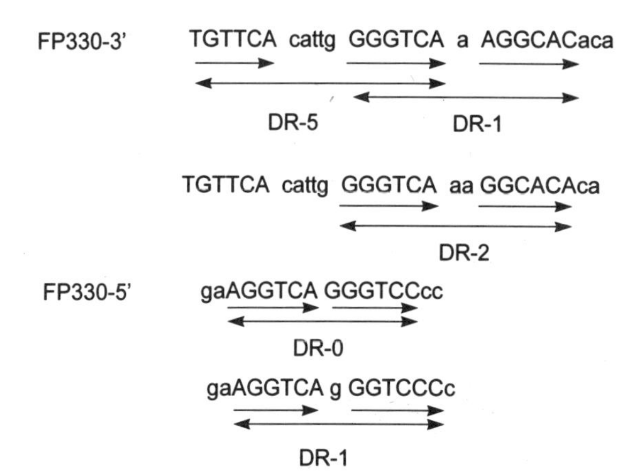

Figure 1. The sequences in the FP330-3' and 5' sites that interact with rat liver nuclear factors are shown. The underlining denotes the conserved motifs. The FP330-5' site appears to consist of a DR-0 or, less likely, a DR-1 element, while the FP330-3' could be interpreted to represent a DR-5, DR-1 or DR-2 element.

4. INTERACTIONS BETWEEN FACTORS BINDING THE FP330-3' SITE

Studies on the promoter activity of the first 600 bp of the ALDH2 promoter cloned upstream of reporter genes were reported earlier. Deletions the 330 bp region containing the FP330-3' and FP330-5' sites reduced expression of reporter constructs in HepG2 and H4IIEC3 hepatoma cells more so than in HeLa cells (Stewart et al., 1996a). This would be consistent with binding of a hepatoma-expressed transcription factor to this region. To better understand the function of the two sites, the FP330 sites were cloned into the vector pBL2-CAT. pALDH3'-BLCAT contained four copies of the FP330-3' element upstream of a thymidine kinase promoter linked to the CAT gene. pALDH5'-BLCAT contained three copies of the FP330-5' element; these constructs were used to test the functional consequences of the presence of these nucleotide sequence in transfection studies with H4IIEC3 hepatoma cells and COS-1 cells (derived from a simian fibroblast cell line).

Co-transfection of the pALDH3'-BLCAT reporter with a HNF-4 expression plasmid resulted in up to 5-fold stimulation of the reporter, and the pALDH5'-BLCAT reporter was inducible up to about 3-fold in the H4IIEC3 cells (Figure 2). The induction of pALDH3'-BLCAT reporter activity was about twice as great in the hepatoma cells compared with the COS-1 cells, and pALDH5'-BLCAT was only marginally induced in the latter cells (Figure 3). HNF-4 did not induce reporter activity of the parent plasmid pBL2-CAT (not shown), indicating that the effect of HNF-4 required the presence of the ALDH promoter elements.

The greater responsiveness of the FP330-3' site to HNF-4 is consistent with the known preference of HNF-4 for DR-1 type binding sites (Mangelsdorf et al., 1995b; Nakshatri and Chambon, 1994). HNF-4 was recently reported to serve as a receptor for acyl-CoA esters (Hertz et al., 1998), with the greatest stimulation of transcriptional activity observed with palmitate supplementation. It is possible that the presence of relatively high levels of fatty acids and triacylglycerol in the H4IIEC3 cells (Galli et al., 1998) accounts for the more marked transactivation of the pALDH3'-BLCAT reporter by HNF-4 in that cell line compared with COS-1 cells. Addition of palmitate to the medium did not further stimulate HNF-4 induced activity of the pALDH3'-BLCAT reporter (not shown). We have

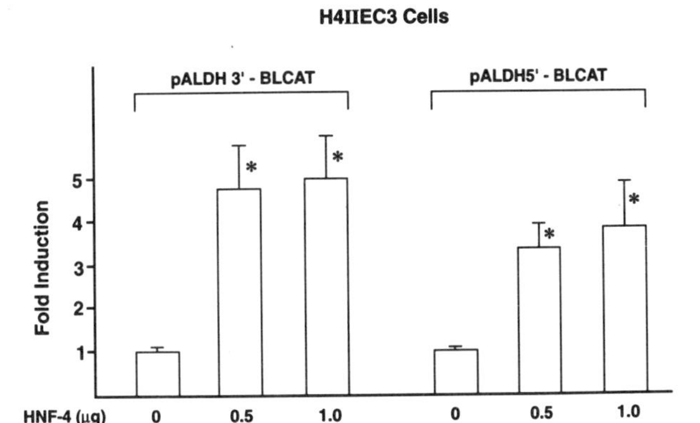

Figure 2. Effect of co-transfection of HNF-4 on the expression of pALDH3'-BLCAT and pALDH5'-BLCAT in H4IIEC3 cells. Fold-induction was calculated from the increase in reporter activity above control cells not co-transfected with the HNF-4 expression plasmid. The amount of HNF-4 expression plasmid used in each transfection is indicated below each bar. These data are the mean and standard error for at least three replicate experiments. An asterisk indicates a statistically significant ($p<0.05$) difference from the control cells by paired t-test.

not yet determined if palmitate stimulates the transcriptional activity of HNF-4 on the ALDH2 promoter in other cell lines.

Consistent with the supershift data, we have found that RXR/RAR, COUP-TF, ARP-1, and PPAR/RXR can bind this FP330-3' (preliminary data not shown). The functional relevance of this is just being determined and studies on interactions between the various factors are underway. We have observed that co-transfection of PPARδ, but not the other PPAR isoforms, abolished HNF-4 stimulation of the reporter. Similarly COUP-TF and ARP-1 reduced the basal expression of the reporter plasmid, and also abolished HNF-4 stimulation. If confirmed, these data suggest that ALDH2 expression may be regulated in part by the relative levels of these transcription factors in different tissues (Figure 4).

In liver and kidney, with high HNF-4 levels, this site is probably occupied by that factor and ALDH2 is highly expressed. In many other tissues in which COUP-TF or ARP-1 is expressed, expression may be repressed. PPARδ is widely expressed (Kliewer et al., 1994). This class of receptor is known to bind a number of lipophilic compounds such as fatty acids and prostaglandin metabolites, as well as a large number of compounds classed as peroxisome proliferators (Forman et al., 1997); whether the suppressive effect of PPARδ on ALDH2 expression can be modulated by exogenous or endogenous ligands remains to be determined.

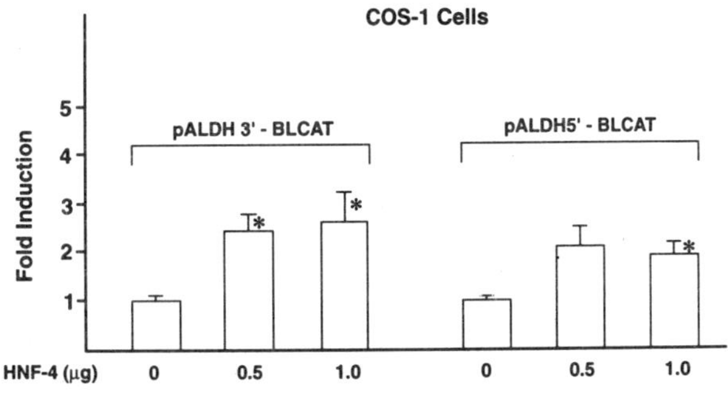

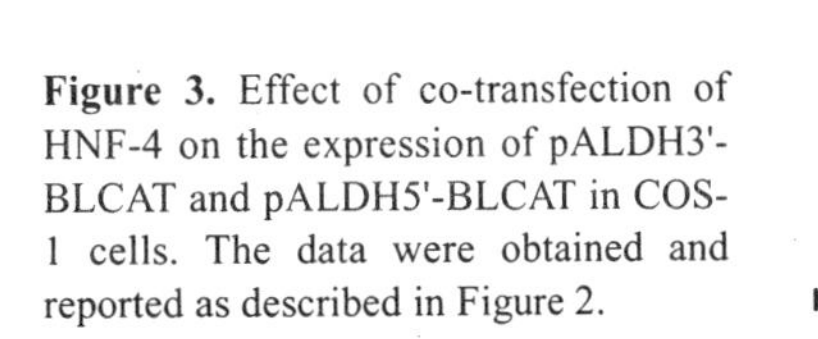

Figure 3. Effect of co-transfection of HNF-4 on the expression of pALDH3'-BLCAT and pALDH5'-BLCAT in COS-1 cells. The data were obtained and reported as described in Figure 2.

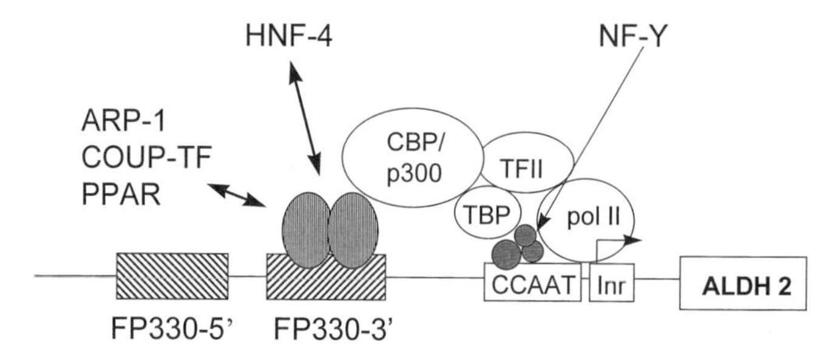

Figure 4. A model for the control of ALDH2 expression through the nuclear receptor response element in FP330-3'. HNF-4 activates expression of the promoter, while ARP-1, COUP-TF, and PPARδ oppose this activation. This figure also shows the interactions between the CCAAT box, the transcription factor NF-Y, and the basal transcriptional machinery (TBP = TATAA binding protein; TFII = transcription factor II complex, and pol II = RNA polymerase II), as well as the proposed role of CBP/p300 in mediating HNF-4 activation of the promoter.

5. CONCLUSIONS

The *ALDH2* promoter contains an important regulatory element in the first 300 bp upstream from the transcription start site. Multiple members of the nuclear receptor family, specifically HNF-4, COUP-TF, and RXR, can bind the FP330-3' site. HNF-4 can strongly activate the promoter via binding to FP330-3', and somewhat less well by way of the FP330-5' element. It appears that the simultaneous presence of COUP-TF, ARP-1, or PPARδ can interfere with the ability of HNF-4 to activate the promoter. The mechanism for this effect remains to be determined. For instance, these factors might simply compete for binding to the DR-1 element in FP330-3'. Alternatively, this site might be able to bind more than one factor at a time, for example at the upstream direct repeat motif as well as the DR-1 element (Figure 1). Because of the ability of multiple factors to bind this promoter sequence, it may be more appropriate to name it a nuclear receptor response element (NRRE), as with a somewhat similar site in the medium chain acyl-CoA dehydrogenase gene (Carter et al., 1994). It seems probable that this site integrates the effects of numerous transcription factors in different tissues to determine the expression of ALDH2. It is noteworthy that several of these factors are ligand-responsive (PPAR for fatty acids, RXR/RAR for retinoic acid, and HNF-4 for acyl-CoAs); this may provide for nutritional and hormonal control of ALDH2 expression.

ACKNOWLEDGMENTS

This work has been supported by grants from the NIAAA: AA06434 (to DWC), T32 AA07462 (to JAP), K05 AA00199 (to Mark Stewart), and the general support of the Indiana Alcohol Research Center (AA 07611). We are indebted to Dr. H. Nakshatri for helpful discussions and shared reagents.

REFERENCES

Bellorini, M., Lee, D.K., Dantonel, J.C., et al., 1997, CCAAT binding NF-Y-TBP interactions: NF-YB and NF-YC require short domains adjacent to their histone fold motifs for association with TBP basic residues. *Nucl Acids Res* 25:2174–2181.

Carter, M.E., Gulick, T., Moore, D. and Kelly, D.P., 1994, A pleiotropic element in the medium-chain acyl coenzyme A dehydrogenase gene promoter mediates transcriptional regulation by multiple nuclear receptor transcription factors and defines novel receptor-DNA binding motifs. *Mol Cell Biol* 14:4360–4372.

Crabb, D.W. and Dixon, J.E., 1987, A method for increasing the sensitivity of chloramphenicol acetyltransferase assays in extracts of transfected cultured cells. *Anal Biochem* 163:88–92.

Crabb, D.W., Edenberg, H.J., Bosron, W.F. and Li, T.-K. , 1989, Genotypes for aldehyde dehydrogenase deficiency and alcohol sensitivity. The inactive *ALDH2*2* allele is dominant. *J Clin Invest* 83:314–316.

de Wet, J.R., Wood, K.V., DeLuca, M., Helinski, D.R. and Subramani, S., 1987, Firefly luciferase gene: structure and expression in mammalian cells. *Mol Cell Biol* 7:725–737.

Dignam, J.D., Lebovitz, R.M. and Roeder, R.G., 1983, Accurate transcription initiation by RNA polymerase II in a soluble extract from isolated mammalian nuclei. *Nucl Acids Res* 11:1475–1489.

Dipple, K.M. and Crabb, D.W., 1993, The mitochondrial aldehyde dehydrogenase gene resides in an HTF island but is expressed in a tissue-specific manner. *Biochem Biophys Res Comm* 193:420–427.

Enomoto, N., Takase, S., Yasuhara, M. and Takada, A., 1991, Acetaldehyde metabolism in different aldehyde dehydrogenase 2 genotypes. *Alcoholism: Clin Exp Res* 15:141–144.

Forman, B.M., Chen, J. and Evans, R.M., 1997, Hypolipidemic drugs, polyunsaturated fatty acids, and eicosanoids are ligands for peroxisome proliferator-activated receptors α and δ. *Proc Nat Acad Sci USA* 94:4312–4317.

Galli, A., Stewart, M.J., Dorris, R. and Crabb, D.W., 1998, High level expression of RXRα and the presence of endogenous ligands contribute to expression of a peroxisome proliferator-activated receptor-responsive gene in hepatoma cells. *Arch Biochem Biophys* 354:288–294.

Goedde, H.W., Agarwal, D.P., Harada, S., et al., 1983, Population genetic studies of aldehyde dehydrogenase isozyme deficiency and alcohol sensitivity. *Am J Hum Genet* 35:769–772.

Gorman, C.M., Merlino, G.T., Willinham, M.C., Pastan, I. and Howard, B.H., 1982, The Rous sarcoma virus long terminal repeat is a strong promoter when introduced into a variety of eukaryotic cells by DNA-mediated transfection. *Proc Nat Acad Sci USA* 79:6777–6781.

Harada, S., Agarwal, D.P., Goedde, H.W., Tagaki, S. and Ishikawa, B., 1982, Possible protective role against alcoholism for aldehyde dehydrogenase isozyme deficiency in Japan. *Lancet* 2:827.

Hertz, R., Magenheim, J., Berman, I. and Bar-Tana, J., 1998 Fatty acyl-CoA thioesters are ligands of hepatic nuclear factor-4 α. *Nature* 392:512–516.

Kliewer, S.A., Forman, B.M., Blumberg, B., et al., 1994, Differential expression and activation of a family of murine peroxisome proliferator-activated receptors. *Proc Nat Acad Sci USA* 91:7355–7359.

Kliewer, S.A., Umesono, K., Mangelsdorf, D.J. and Evans, R.M., 1992, Retinoid X receptor interacts with nuclear receptors in retinoic acid, thyroid hormone and vitamin D3 signaling. *Nature* 355:446–449.

Mangelsdorf, D.J. and Evans, R.M., 1995a, The RXR heterodimers and orphan receptors. *Cell* 83:841–850.

Mangelsdorf, D.J., Thummel, C., Beato, M., et al., 1995b, The nuclear receptor superfamily: the second decade. *Cell* 83:835–839.

Nakshatri, H. and Bhat-Nakshatri P., 1998, Multiple parameters determine the specificity of transcriptional response by nuclear receptors HNF-4, ARP-1, PPAR, RAR, and RXR through common response elements. *Nucl Acids Res.* 26:2491–2499.

Nakshatri, H. and Chambon, P., 1994, The directly repeated RG(G/T)TCA motifs in the rat and mouse cellular retinol binding protein II are promiscuous binding sites for RAR, RXR, HNF-4 and ARP-1 homo- and heterodimers. *J Biol Chem* 269:890–902.

Shapiro, D.J., Sharp, P.A., Wahli, W.W. and Keller, M.J., 1988, A high-efficiency HeLa cell nuclear transcription extract. *DNA* 7:47–55.

Sladek, F.M., Zhong, W.M., Lai, E. and Darnell, J.E., 1990, Liver-enriched transcription factor HNF-4 is a novel member of the steroid hormone receptor superfamily. *Gene Devel* 4:2353–2365.

Smale, S.T. and Baltimore, D., 1989, The initiator as a transcription control element. *Cell* 57:103–113.

Stewart, M.J., Dipple, K.M., Estonius, M., Nakshatri, H., Everett, L.M. and Crabb, D.W., 1998, Binding and activation of the human aldehyde dehydrogenase 2 promoter by hepatocyte nuclear factor 4. *Biochim.Biophys.Acta* (in press).

Stewart, M.J., Dipple, K.M., Stewart, T.R. and Crabb, D.W., 1996a, The role of nuclear factor NF-Y/CP1 in the transcriptional regulation of the human aldehyde dehydrogenase 2-encoding gene. *Gene* 173:155–161.

Stewart, M.J., Malek, K. and Crabb, D.W., 1996b, Distribution of messenger RNAs for aldehyde dehydrogenase 1, aldehyde dehydrogenase 2, and aldehyde dehydrogenase 5 in human tissues. *J.Invest.Med.* 44:42–46.

Thomasson, H.R., Edenberg, H.J., Crabb, D.W., et al., 1991, Alcohol and aldehyde dehydrogenase genotypes and alcoholism in Chinese men. *Am J Hum Genet* 48:677–681.

Zhong, W., Mirkovitch, J. and Darnell, J.E., 1994, Tissue-specific regulation of mouse hepatocyte nuclear factor 4 expression. *Mol Cell Biol* 14:7276–7284.

15

α,β-UNSATURATED ALDEHYDES MEDIATE INDUCIBLE EXPRESSION OF GLUTATHIONE *S*-TRANSFERASE IN HEPATOMA CELLS THROUGH ACTIVATION OF THE ANTIOXIDANT RESPONSE ELEMENT (ARE)

Ronald B. Tjalkens, Stephen W. Luckey, David J. Kroll, and Dennis R. Petersen

Molecular Toxicology and Environmental Health Sciences Program
Department of Pharmaceutical Sciences
University of Colorado Health Sciences Center
Denver, Colorado

1. INTRODUCTION

It is well established that during oxidative stress highly reactive, α,β-unsaturated aldehydes are produced from the β-scission of lipid alkoxy radicals formed from the reaction of carbon-centered lipid radicals with molecular oxygen (Esterbauer *et al.*, 1991; Schauer *et al.*, 1990). Four-hydroxynonenal (4-HNE) and other endogenous and exogenous aldehydes, such as malondialdehyde and acrolein, are also generated in response to a variety of oxidative stimuli including free intracellular iron (Fe^{2+}) (Esterbauer *et al.*, 1991), chronic ethanol (Ingelman-Sundberg and Johansson, 1984), and exposure to halogenated hydrocarbons (Poli *et al.*, 1985). The α,β-unsaturated configuration of these compounds allows them to undergo Michael addition chemistry resulting in the formation of covalent adducts with DNA and protein nucleophiles. The potential of these electrophilic aldehydes to target specific cellular nucelophiles is consistent with the deleterious effects of 4-HNE which includes inhibition of protein synthesis (White and Rees, 1984), disruption of Ca^{2+} homeostasis (Schauer *et al.*, 1990), and neutrophil chemotaxis (Curzio *et al.*, 1987).

Specific enzymatic pathways have been identified that detoxify 4-HNE to metabolites that are less toxic. In freshly isolated hepatocytes (Hartley *et al.*, 1995), 4-HNE is reduced to the respective diol by alcohol dehydrogenase (ADH) oxidized to the corresponding hydroxynonenoic acid by aldehyde dehydrogenase (ALDH). However, the primary detoxification pathway for reactive aldehydes produced during lipid peroxidation is GST-catalyzed conjugation to glutathione (GSH) (Hartley *et al.*, 1995; Siems *et al.*, 1997).

Enzymology and Molecular Biology of Carbonyl Metabolism 7
edited by Weiner *et al.* Kluwer Academic / Plenum Publishers, New York, 1999.

Glutathione *S*-transferases (GST) are a class of phase II metabolic enzymes present in virtually all organs and tissues which protect against toxic effects of endo- and xenobiotic electrophiles. The role of GST in the detoxification of 4-HNE is significant in that this enzyme is inducible by electrophiles interacting with 5'-flanking *cis*-acting regulatory elements termed Antioxidant Response Elements (ARE) (Nguyen *et al.,* 1994; Rushmore and Pickett, 1990). These promoter elements transcriptionally activate gene expression in response to such diverse xenobiotics as β-naphthoflavone (β-NF), *tert*-butylhydroquinone (*t*BHQ), hydrogen peroxide and 12-*O*-tetradecanoylphorbol 13-acetate (TPA) (Rushmore *et al.,* 1991). α,β-Unsaturated aldehydes are strong electrophiles and as such, Michael acceptor compounds that have the potential to act as monofunctional inducers of phase II genes (Prestera and Talalay, 1995; Prochaska and Talalay, 1988). However, these well-characterized products of lipid peroxidation have yet to be systematically evaluated as activators of the ARE.

In the present study we have investigated whether certain α,β-unsaturated aldehydes produced during lipid peroxidation are able to activate the ARE to induce expression of GST A1–1 and A4–4, which are the primary GST isoforms responsible for the conjugation of 4-HNE with glutathione. Data are presented demonstrating that the lipid aldehydes studied are activators of the ARE and transcriptional inducers of GST. These studies, using antibodies to specific isoforms of GST, as well as measurements of GST activity, confirm that the observed transcriptional activation does result in increased functional GST protein.

2. METHODS AND PROCEDURES

2.1. Sub-Cloning of the Antioxidant Response Element into a Luciferase Reporter Vector

Sixty base-pair oligonucleotides corresponding to the 5'-3' and 3'-5' core consensus sequence of the Antioxidant Response Element (ARE) (Paulson *et al.,* 1990; Rushmore and Pickett, 1990) from positions -719 to -680 of the rGST A1 promoter were custom synthesized (BRL Life Sciences, Gaithersburg, MD) and used for sub-cloning into a mammalian luciferase expression system. The addition of 5'-*Bam*H1 and 3'-*Xho*1 sites into the synthetic oligonucleotides enabled directional cloning into pT81Luc, a pA3Luc-derivative vector containing the enhancerless basal promoter (-81 to +52) of Herpes Simplex I thymidine kinase (Nordeen, 1988). Oligonucleotides were annealed, sequentially digested with *Bam*HI and *Xho*I and ligated into pT81Luc which had been previously digested with *Bam*HI and *Xho*I. Annealing of oligonucleotides, restriction reactions and ligations were performed using standard procedures (Sambrook *et al.,* 1989). Restriction endonucleases and T4 DNA ligase were used according to the manufacturer's instructions. The sequence of the oligonucleotide corresponding to the sense strand is presented below. Bold letters denote the added 5'-*Xho*1 and 3'-*Bam*H1 sites. Underlined bold letters correspond to the distal and proximal half-sites of the response element.

5'-GATC**GGATCC**CTTGGAAA**TGGCATTGC**TAATGG**TGACAAAGC**AACTTTCG**CTCGAG**GATC-3'

The construct, designated pARE-T81Luc, was transformed into chemically competent DH5α *E.* coli as previously described (Inoue *et al.,* 1990), grown preparatively in 1200 ml Circle Grow media and purified by double cesium chloride density gradient centrifugation for transfections (Sambrook *et al.,* 1989).

2.2. Tissue Culture and Cytotoxicity Assays

Rat Clone 9 hepatoma cells were maintained in Dulbecco's Modified Eagle's/F12-Ham minimal essential medium plus L-glutamine with nonessential amino acids, sodium pyruvate, 10% heat-inactivated fetal bovine serum, penicillin (10 units/ml), and streptomycin (10 units/ml). Cytotoxicity assays for 4-HNE, acrolein and *trans*-2-hexenal were performed as using 3-[4,5-dimethylthiazol-2-yl]-2,5-diphenyltetrazolium bromide (MTT) as described elsewhere (Mosmann, 1983). Rat Clone 9 cells were plated at 2000 cells/100 μL medium/well in 96-well plates and allowed to attach overnight. Various concentrations of the aldehydes were added in serum-free medium and the cells were treated for 4 hours at 37 ^{0}C and grown for a further 72 hours in complete medium. IC_{50} values determined from cytotoxicity assays were used to construct dosing regimens for each compound in the structure-activity series

2.3. Transfection of Clone 9 Hepatoma Cells

Clone 9 cells were plated the day prior to transfection in 6-well tissue culture plates to a cell density of 3×10^5 cells/well and allowed to recover overnight. Cells were co-transfected with the ARE luciferase reporter construct described above and pCMVβ, an internal control expression vector containing the β-galactosidase gene under control of the cytomegalovirus promoter, using synthetic liposomes. The ratio of DNA to lipid (m/m) was 1:4 and 1μg each of pARE-T81Luc and pCMVβ was used for each transfection. Briefly, medium was removed, plates were washed with serum free medium, and the DNA:lipid mixture in 1 ml serum-free medium was layered onto each well. Transfection was allowed to proceed 4 hours in serum-free medium, followed by 16 hours in complete medium. Cells were then washed with phosphate-buffered saline, treated with the various compounds for 4 hours in serum-free medium, allowed to recover 4 hours in complete medium and then harvested for luciferase and β-galactosidase assays. All transfections and treatments were performed in triplicate for each experiment.

2.4. Luciferase and β-Galactosidase Assays

Following treatment, media was removed and 200 μL aliquot of luciferase lysis buffer was added to each well, incubated for 20 min at room temp and plates were scraped to recover cell lysates. Samples were then flash frozen in liquid nitrogen and stored overnight at -80 ^{0}C. Prior to measurement of luciferase and β-galactosidase activity, samples were thawed quickly at 37 ^{0}C, spun at 14000x*g* for 15 min. Luciferase measurements were performed and recorded as the average of duplicates for each sample. β-Galactosidase activity was measured in 96-well plates using 20 μL of cell lysate. Values for luciferase activity were calculated by first normalizing the assay volume against the total volume of cell lysate for each sample and then by normalizing total luciferase activity for each sample against corresponding β-galactosidase activity as a measure of transfection efficiency.

2.5. Treatment and Harvesting of Hepatoma Cells for Western Blotting and GST Activity Experiments

Clone 9 cells were seeded at $1–2 \times 10^6$/plate in three 100 mm culture dishes for each treatment group 2–3 days prior to the experiment and allowed to reach 70–80% confluency. The aldehydes were diluted from serum-free medium and the cells were treated for 4 hours with doses of the various compounds where maximal induction was observed in luciferase assays as follows: 4-HNE, 5 μM; *t*-2-HE, or 20 μM; 2-PE. Following treatment, medium was removed and replaced with fresh complete medium and cells allowed to recover for 4 hours. Plates were then washed with phosphate-buffered saline, the cells were harvested, pulse-sonicated on ice and centrifuged. Protein content of the resulting supernatants was measured by the BCA method (Pierce Chemical Co., Rockford, IL). Protein samples (50 μg of cellular protein) was loaded onto 12% SDS-PAGE and immobilized on PVDF membranes (Laemmli, 1970). Blots were probed with anti-GST A4–4 (Zimniak *et al.*, 1994). The secondary antibody (Horse Radish Peroxidase-linked goat α rabbit IgG, and enhanced chemiluminescence was used visualize immunopositive staining. Total GST activity towards 1-chloro-2,4-dinitrobenzene (CDNB) was measured in cell homogenates by assaying enzyme activity spectrophotometrically as described previously (Alin *et al.*, 1985; Jakoby, 1978).

2.6. RNA Isolation and Northern Blotting Experiments

Clone 9 cells were treated as described above except that one 100 mm tissue culture dish was used per sample. Isolation of total RNA was performed using the RNeasy Mini and QIAshredder column purification kits (Qiagen, Chatsworth, CA) according to the manufacturer's protocols. RNA samples were eluted from columns in 60 μL RNAse-free water and quantitated by $A_{260/280}$ and prepared for agarose gel electrophoresis. The RNA was subsequently fixed to the nylon membrane by UV crosslinking. The Northern Blots were prehybridized for 4 hours in a hybridization oven at 42 ^{0}C in 10 ml hybridization buffer. Isolated and labeled GST A1 and GST A4 probes were denatured prior to addition to hybridization buffer by heating, placed on ice 2 min and added to the hybridization solution to a concentration of 5×10^6 cpm/ml. Blots were hybridized overnight at 42 ^{0}C, and washed in hybridization bottles and the bands visualized using a phosphor imager.

3. RESULTS AND DISCUSSION

The overall cytotoxic potential of the compounds evaluated in this study is evident from the numerical data presented in Table 1 which presents the concentration (IC_{50} value) of the respective aldehydes that inhibited cell growth by 50% as compared to controls. These data are consistent with previous reports describing the toxicities of various aldehydes (Esterbauer *et al.*, 1991) and document that Clone 9 cells experienced a retardation of cell growth when exposed to micromolar concentrations for all of the compounds evaluated. Acrolein (2-PE) and 4-HNE were demonstrated to be the most cytotoxic, while the toxicity of *t*-2-HE was markedly less. The reactivity of 4-HNE as a Michael reaction acceptor is increased by the presence of the electron-withdrawing 4-hydroxyl group and probably accounts for its relatively high toxicity compared to the unsubstituted aliphatic aldehyde *t*-2-HE. The narrow tolerated dose range and high toxicity of these compounds is consistent with their known ability to covalently alkylate important cellular macromolecules and disrupt of such vital functions as protein synthesis and maintenance of Ca^{2+} homeostasis. These data provided information used to select concentrations of each aldehyde for use in the induction studies such that transfected cells were treated with concentrations of 2-PE, 4-HNE or *t*-2-HE less than the IC_{50}.

Table 1. Cytotoxicity values for α,β-unsaturated carbonyl compounds used in induction studies

	IC_{50} value[a] (μM)		
Cell line	4-HNE	2-PE	*t*-2-HE
Clone 9	6.3 ± 0.7	2.2 ± 0.4	16.0 ± 0.7

[a]Mean±S.E. (*n*=9-12)
Clone 9 cells were treated with each compound as described above. Percent viability relative to control was calculated for the various compounds in each cell line and used to calculate IC_{50} values for growth inhibition.

Experiments involving transient transfections in Clone 9 cells revealed that each aldehyde evaluated activated transcription at the ARE, as indicated by a concentration-dependent increases in luciferase activity (Figure 1). Transfected Clone 9 cells, treated with concentrations of 4-HNE ranging from 0.5–5 μM, displayed a 2.3-fold maximal induction of luciferse activity over untreated controls at 5 μM 4-HNE (Figure 1*A*). *Trans*-2-hexenal (1–20 μM) added to transfected Clone 9 cells elicited a maximal 2.3-fold induction of luciferase activity (Figure 1*B*) and Clone 9 cells treated with a 0.1–3 μM 2-PE displayed a 3.0-fold maximal induction in luciferase activity (Figure 1*C*). In control experiments, the enhancerless, basal construct displayed only background activity levels and showed no inducibility in response to aldehyde treatment, indicating that induction was specifically directed through events mediated by the ARE (data not shown).

Transcripts for GST A1 and A4 were measured to assess whether the induction observed with transfected ARE corresponded to increases in steady state levels of message in response to treatment with the series of aldehydes. Representative northern blots shown in Figure 2 indicate that in Clone 9 cells, message levels for both GST A1 and A4 were elevated above basal levels in response to treatment with all four compounds tested. Densitometric analysis of the bands revealed that, relative to untreated (normalized for RNA loading against 28S and 18S rRNA), message for GST A1 was increased by 1.4-fold by 4-HNE treatment, 1.6-fold by *t*-2-HE, and 1.6-fold by acrolein. Message for GST A4 was similarly increased over control by 1.4-fold in response to 4-HNE treatment, 1.6-fold by *t*-2-HE, 1.3-fold by acrolein.

Immunoblotting for GST A4 in whole cell extracts from Clone 9 cells is shown in Figure 3. Cells were treated with the various compounds as described for northern blots and evaluated for induction of GST A1–1 and A4–4 protein using polyclonal antibodies directed against these enzymes. The resulting immunoblots are presented in Figure 3 and show that basal levels of GST A4–4 is low in Clone 9 cells (lane 2) but markedly inducted in response to all four compounds evaluated (lanes 3–6).

Specificity of the antibody was confirmed by lack of reaction of the anti-GST A4–4 antibody with recombinant GST A1 protein (lane 1). Fold-induction of GST A4–4 elicited by each compound was 4-HNE, 7.1, *t*-2-HE, 7.7, and 2-PE, 8.0. GST A1–1 protein was not detectable in this cell line. Consistent with the observed increases in GST mRNA and GST A1–1 and A4–4 protein, GST activity in Clone 9 cells was significantly increased in response to treatment with each compound evaluated (Figure 4). GST activity in Clone 9 cells was increased over untreated controls by 1.26-fold in response to 4-HNE treatment, 1.43-fold by *t*-2-HE, and 1.18-fold by 2-PE.

The data presented in Figures 1–4 clearly show that α,β-unsaturated aldehydes are activators of the ARE from rGST A1 and efficient inducers of GST A1–1 and A4–4 at the level of transcription and translation. Concentrations of α,β-unsaturated aldehydes are

A.

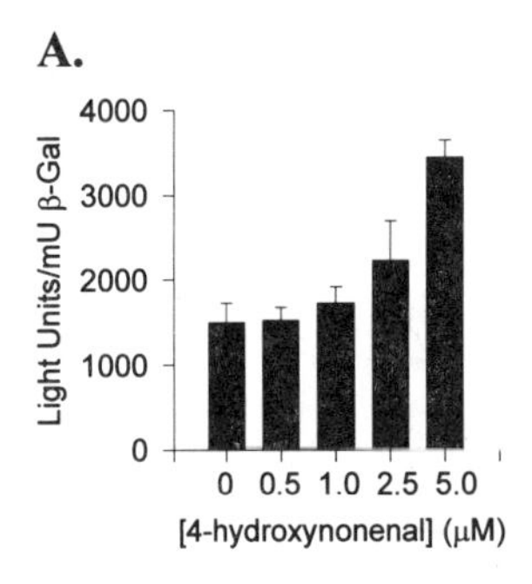

B.

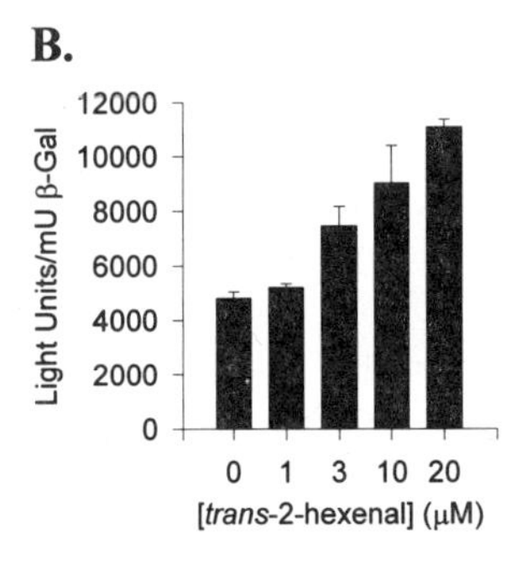

C.

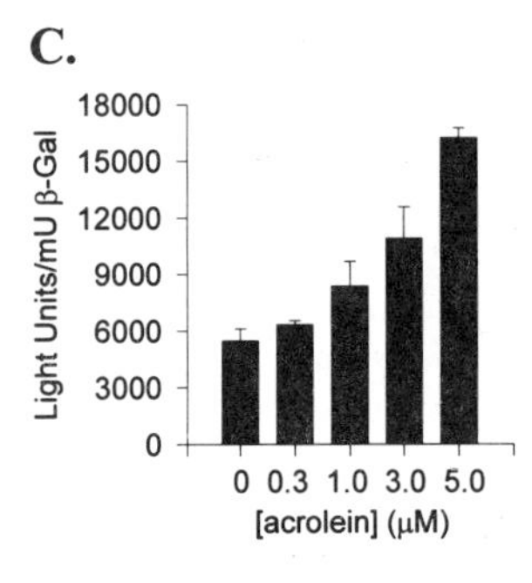

Figure 1. Functional analysis of the antioxidant response element in transiently transfected Clone 9 and HepG2 cells treated with 4-HNE, *t*-2-HE and 2-PE. Clone 9 hepatoma cells were transiently transfected with a chimeric luciferase reporter construct containing the 5'-flanking ARE from the rGST A1 promoter and were treated with increasing doses of the various compounds A. - 4-HNE; B,- *t*-2-HE; and C.- 2-PE. Experiments were performed in triplicate for each compound and the data presented are representative of at least three independent experiments. Data are presented as mean ± S.E. for n=3.

increased in tissues during oxidative stress (Esterbauer *et al.,* 1991) and have been proposed to be responsible for upregulation of GST A4 in animals displaying increased hepatic lipid peroxidation from chronic iron overload (Khan *et al.,* 1995). The data presented here are consistent with our recent observation that GST A1–1 and A4–4 are induced *in vivo* in hepatic tissue from two inbred mouse strains undergoing iron-induced lipid peroxidation (Tjalkens *et al.,* 1998). Further, 4-HNE produced in abundance during lipid peroxidation (Schauer *et al.,* 1990) is the most efficient substrate known for GST A4–4 (Alin *et al*, 1985; Jensson *et al.*, 1986) and, as such, may function to induce expression of this enzyme as part of the general response of phase II genes to "electrophilic stress" (Prestera and Talalay, 1995; Prochaska and Talalay, 1988). The results reported here provide evidence that physiologically relevant concentrations 4-HNE and other products of lipid peroxidation act through the ARE to induce expression of GST. Though responsiveness of this ARE to a wide variety of compounds has been previously shown, evidence for α,β-unsaturated aldehydes as inducers of ARE-mediated GST expression has not been reported.

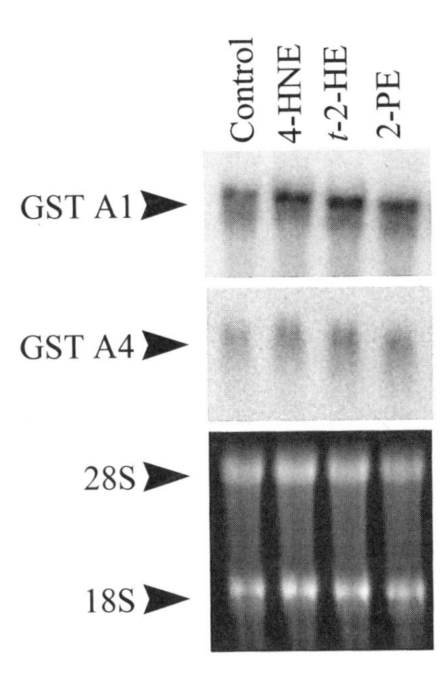

Figure 2. Induction of GST A1 and A4 mRNA in Clone 9 cells treated with 4-HNE, *t*-2-HE and 2-PE. Clone 9 cells were treated with doses of each compound where maximal induction was observed in transfection studies and evaluated for increases in message for GST A1 and A4. Each lane contains 20 μg of RNA. *Lane 1*, Control, or treated with *lane 2*, 4-HNE (5 μM), *lane 3*, *t*-2-HE (20 μM), and *lane 4*, 2-PE (3 μM). Blots were visualized using a Phosphor Imager (Molecular Dynamics, Menlo Park, CA) and analyzed for fold-induction relative to control (normalized for RNA loading against 28S and 18S RNA).

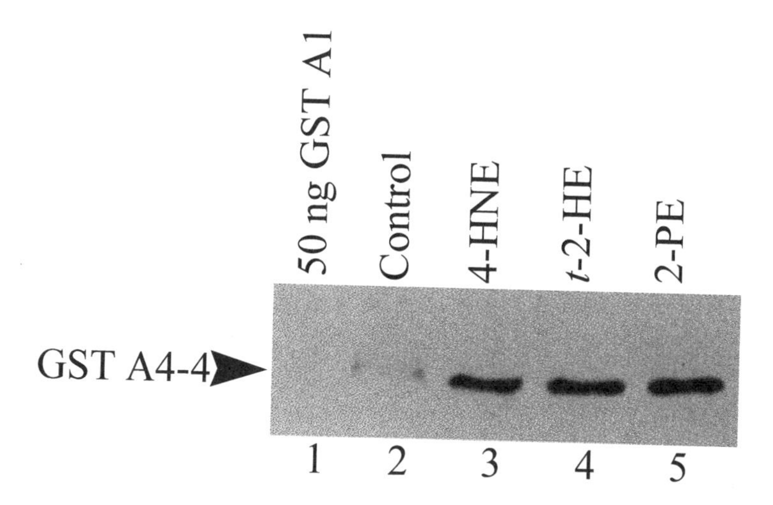

Figure 3. Immunoblot of GSTA4–4 in Clone 9 cells treated with 4-HNE, *t*-2-HE and 2-PE. Clone 9 cells were treated with each compound at the dose which yielded maximal induction in transient transfections and assayed for increases in GSTA4–4 protein by western blotting using a GST A4–4-specific polyclonal antibody. *Lane 1*, 50 ng purified recombinant rGSTA1, *lane 2*, Control, or treated with *lane 3*, 4-HNE (5 μM), *lane 4*, *t*-2-HE (20 μM), and *lane 5*, 2-PE (3 μM). Lanes 2–6 contain 50 μg of protein. Bands were visualized by enhanced chemiluminescence and analyzed for fold induction relative to control by scanning densitometry.

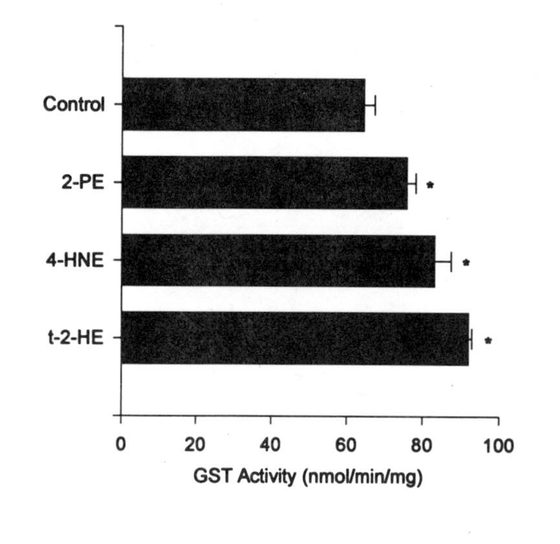

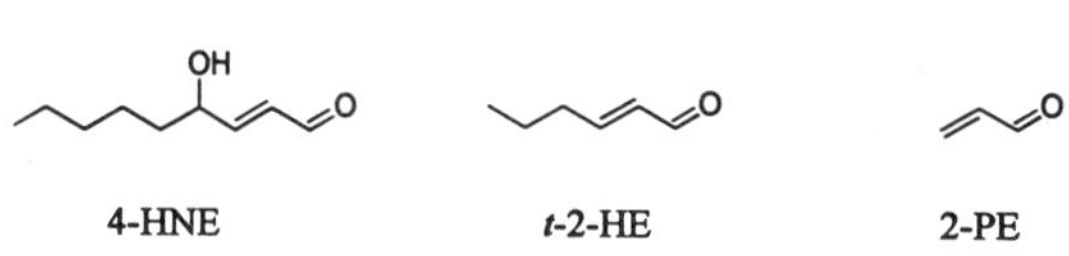

Figure 4. Induction of total GST activity in Clone 9 cells treated with 4-HNE, *t*-2-HE and 2-PE. Clone 9 cells were treated with each compound as in northern and western blot experiments and assayed for GST activity using 1-chloro-2,4-dinitrobenzene (CDNB) as a substrate. Each experiment was performed in triplicate and data from three to four independent experiments were combined to calculate mean and standard error. Statistical significance was determined by Student's T-test. Data and are presented as mean ± S.E. for n=9–12 (*p<0.05).

REFERENCES

Alin, P., Mannervik, B., and Jornvall, H. (1985) Structural evidence for three different types of glutathione transferase in human tissues. *FEBS Letters.* **182**, 319–322.

Curzio, M., Esterbauer, H., Poli, G., Biasi, F., Cecchini, G., Di Mauro, C., Cappello, N., and Dianzani, M.U. (1987) Possible role of aldehydic lipid peroxidation products as chemoattractants. *Int. J. Tissue React.* **9**, 295–306.

Esterbauer, H., Schauer, R.J., and Zollner, H. (1991) Chemistry and biochemistry of 4-hydroxynonenal, malonaldehyde, and related aldehydes. *Free Rad. Biol. Med.* **11**, 81–128.

Hartley, D.P., Ruth, J.A., and Petersen, D.R. (1995) The Hepatocellular Metabolism of 4-Hydroxynonenal by Alcohol Dehydrogenase, Aldehyde Dehydrogenase, and Gluthathione-*S*-Transferase. *Arch. Biochem. Biophysics* **10**, 197–205.

Ingelman-Sundberg, M. and Johansson, I. (1984) Mechanisms of hydroxyl radical formation and ethanol oxidation by ethanol-inducible and other forms of rabbit liver microsomal cytochromes P-450. *J. Biol. Chem.* **259**, 6447–6458.

Inoue, H., Nojima, H., and Okayama, H. (1990) High efficiency transformation of Escherichia coli with plasmids. *Gene* **96**, 23–28.

Jakoby, W.B. (1978) The glutathione S-transferases: a group of multifunctional detoxification proteins. [Review]. *Advances Enzymol. Mol. Biol.* **46**, 383–414.

Khan, F.M., Srivastava, S.K., Singhal, S.S., Chaubey, M., Awasthi, S., Petersen, D.R., Ansari, G.A.S., and Awasthi, Y.C. (1995) Iron-Induced Lipid Peroxidation in Rat Liver Is Accompanied by Preferential Induction of Glutathione-S-Transferase 8–8 Isozyme. *Toxicol. Appl. Pharmacol.* **131**, 63–72.

Laemmli, U.K. (1970) Cleavage of structural proteins during the assembly of the head of bacteriophage T4. *Nature* **227**, 680–685.

Mosmann, T. (1983) Rapid colorimetric assay for cellular growth and survival: application to proliferation and cytotoxicity assays. *J. Immunol. Methods* **65**, 55–63.

Nguyen, T., Rushmore, T.H., and Pickett, C.B. (1994) Transcriptional regulation of a rat liver glutathione S-transferase Ya subunit gene. Analysis of the antioxidant response element and its activation by the phorbol ester 12-O-tetradecanoylphorbol-13-acetate. *J. Biol. Chem.* **269**, 13656–13662.

Nordeen, S.K. (1988) Luciferase Reporter Gene Vectors for Analysis of Promoters and Enhancers. *BioTechniques* **6**, 454–457.

Paulson, K.E., Darnell, J.E., Rushmore, T., and Pickett, C.B. (1990) Analysis of the upstream elements of the xenobiotic compound-inducible and positionally regulated glutathione S-transferase Ya gene. *Mol. Cellular Biol.* **10**, 1841–1852.

Poli, G., Dianzani, M.U., Cheeseman, K.H., Slater, T.F., Lang, J., and Esterbauer, H. (1985) Separation and characterization of the aldehydic products of lipid peroxidation stimulated by carbon tetrachloride or ADP-iron in isolated rat hepatocytes and rat liver microsomal suspensions. *Biochem. J.* **227**, 629–638.

Prestera, T. and Talalay, P. (1995) Electrophile and antioxidant regulation of enzymes that detoxify carcinogens. *Proc. Natl. Acad. Sci. USA* **92**, 8965–8969.

Prochaska, H.J. and Talalay, P. (1988) Regulatory Mechanisms of Monofunctional and Bifunctional Anticarcinogenic Enzyme Inducers in Murine Liver. *Cancer Res.* **48**, 4776–4782.

Rushmore, T.H., Marcia, R.M., and Pickett, C.B. (1991) The Antioxidant Response Element. *J. Biol. Chem.* **266**, 11632–11639.

Rushmore, T.H. and Pickett, C.B. (1990) Transcriptional regulation of the rat glutathione S-transferase Ya subunit gene. Characterization of a xenobiotic-responsive element controlling inducible expression by phenolic antioxidants. *J. Biol. Chem.* **265**, 14648–14653.

Sambrook, J., Fritsch, E.F., and Maniatis, T. (1989) . In: *Molecular Cloning: A Laboratory Manual.*, Second Edition Ed., Cold Spring Harbor Laboratory Press.

Schauer, R.J., Zollner, H., and Esterbauer, H. (1990) Biological Effects of Aldehydes with Particular Attention to 4-Hydroxynonenal and Malondialdehyde. In: *Membrane Lipid Peroxidation*, 141–163. Edited by Vigo-Pelfrey, Boca Raton, FL, CRC Press.

Siems, W.G., Zollner, H., Grune, T., and Esterbauer, H. (1997) Metabolic fate of 4-hydroxynonenal in hepatocytes: 1,4- dihydroxynonene is not the main product. *J. Lipid Res. Mar;38(3):612–22* 612–622.

Tjalkens, R.B., Valerio, Jr., Awasthi, Y.C., and Petersen, D.R. (1998) Association of Glutathione S-Transferase Isozyme-Specific Induction and Lipid Peroxidation in Two Inbred Strains of Mice Subjected to Chronic Dietary Iron Overload. *Toxicol. Appl. Pharmacol.* **150**,

White, J.S. and Rees, K.R. (1984) The Mechanism of Action of 4-Hydroxynonenal in Cell Injury. *Chem. -Biol. Interactions* **52**, 233–241.

Zimniak, P., Singhal, S.S., Srivastava, S.K., Awasthi, S., Sharma, R., Hayden, J.B., and Awasthi, Y.C. (1994) Estimation of Genomic Complexity, Heterologous Expression, and Enzymatic Characterization of Mouse Glutathione *S*-Transferase mGSTA4–4 (GST 5.7). *J. Biol. Chem* **269**, 992–1000.

16

EFFECT OF ARACHIDONIC ACID ALONE OR WITH PROOXIDANT ON ALDEHYDE DEHYDROGENASES IN HEPATOMA CELLS

Rosa A. Canuto,[1] Margherita Ferro,[2] Raffaella A. Salvo,[1] Anna M. Bassi,[2] Mario Terreno,[1] Mario U. Dianzani,[1] Ronald Lindahl,[3] and Giuliana Muzio[1]

[1]Dip. Scienze Cliniche e Biologiche
Università di Torino
Ospedale S. Luigi, Orbassano-Torino, Italy
[2]Dip. Medicina Sperimentale
Università di Genova
Genova, Italy
[3]Department of Biochemistry and Molecular Biology
University of South Dakota
Vermillion, South Dakota

1. INTRODUCTION

Aldehyde dehydrogenase (ALDH) is a family of several isoenzymes (Lindahl, 1992; Vasiliou et al., 1996) important in the cellular defenses against exogenous toxic aldehydes (Lindahl, 1992; Yoshida, 1992) and against endogenous aldehydes, such as those derived from lipid peroxidation (Lindahl and Petersen, 1991; Canuto et al., 1994). The latter appear to influence cell growth and differentiation in some tumour cell lines (Barrera et al., 1994; Canuto et al., 1995).

Compared to normal hepatocytes, in rat hepatoma cells, cytosolic class 3 ALDH appears and mitochondria class 2 ALDH decreases (Huang and Lindahl, 1990; Canuto et al., 1993; Canuto et al., 1994). The activity of class 3 ALDH increases with the degree of deviation in hepatoma cell lines, just as occurs in chemical carcinogenesis induced in the liver (Lindahl, 1979; Wishusen et al., 1983; Canuto et al., 1989; Canuto et al., 1994).

In parallel with the increased appearance of class 3 ALDH in hepatoma cells, there is a decrease of lipid peroxidation due to the reduction of the percentage content of polyunsaturated fatty acids (Canuto et al., 1994). In a previous paper we demonstrated that the restoration of lipid peroxidation in hepatoma cell lines, by enriching them with arachi-

Enzymology and Molecular Biology of Carbonyl Metabolism 7
edited by Weiner *et al.* Kluwer Academic / Plenum Publishers, New York, 1999.

donic acid and treating them with prooxidant, has an inhibitory effect on class 3 ALDH activity (Canuto et al., 1996; Canuto et al., 1998). This is accompanied by cell death or decrease in cell growth. In 7777 hepatoma cells, which already have a low basal content of class 3 ALDH, the products of restored lipid peroxidation almost immediately cause a reduction of cell growth and, when the concentration of these products are such as to determine a complete disappearance of class 3 ALDH, cell death. In JM2 hepatoma cells, which have a high basal level of class 3 ALDH, the products of restored lipid peroxidation inhibit about 50% of class 3 ALDH, and as a consequence, are able to decrease cell growth. The inhibition of class 3 ALDH has been seen to be a decrease of enzyme activity, by using 4-hydroxynonenal or benzaldehyde as substrate, and a decrease of ALDH3 mRNA content (Canuto et al., 1996; Canuto et al., 1998).

In this paper, we examined the effect of enrichment with arachidonic acid in hepatoma cell lines and of the subsequent addition of prooxidant on mitochondrial class 2 ALDH. This enzyme is expressed in a large number of tissues, with the highest levels in liver, kidney, muscle, and heart (Stewart et al., 1996). Based on enzymological (Greenfield and Pietruszko, 1977), metabolic (Cao et al., 1988), and genetics studies (Yoshida et al., 1985), it is believed that class 2 ALDH is mainly responsible for the oxidation of acetaldehyde generated during alcohol oxidation *in vivo*. It is also important in the elimination of 4-hydroxynonenal in normal rat liver (Canuto et al., 1989; Chen and Yu, 1996). Class 2 ALDH levels are lower in hepatoma cells than in normal hepatocytes (Canuto et al., 1989).

2. MATERIALS AND METHODS

2.1. Culture Conditions

Hepatoma cell lines (7777 and JM2) were seeded (day 0) and maintained for 24h in a medium A [DMEM/F12 plus 2 mM glutamine, 1% antibiotic/antimycotic solution] plus 10% of newborn calf serum; 24h later (day 1) they were put into medium B [medium A plus 0.4% of albumin, 1% ITS (insulin, transferrin, sodium selenite), 1% nonessential amino acids, 1% vitamin solution] supplemented or not with arachidonic acid (50 μM); 24h later again (day 2), the cells were put into unsupplemented medium B, given 4 doses of prooxidant (500 μM ascorbate/100 μM iron sulphate) at 12h intervals, and maintained for additional two days in culture (day 3 and 4). The various parameters were determined on cells untreated with arachidonic acid and prooxidant (control cells), on cells treated only with arachidonic acid, and on cells treated with arachidonic acid and prooxidant. Cells were harvested at the following times: after 24h of enrichment with arachidonic acid, 4h after the 1st dose of ascorbate/iron sulphate, 12h after the 2nd dose of ascorbate/iron sulphate, 4h after the 3rd dose of ascorbate/iron sulphate for 7777 hepatoma cells and 12h after the 4th dose of ascorbate/iron sulphate for JM2 hepatoma cells.

2.2. Fatty Acid Content

The percentage content of fatty acids in phospholipids extracted from cells *in toto* and from mitochondria was measured as described in Canuto et al. (1995).

2.3. Preparation of Mitochondrial Fraction

Cell suspensions obtained from hepatoma cell cultures were sedimented by centrifugation at 600 g for 10 min. Homogenates were obtained by disruption with a hand-driven

Tenbroeck glass homogenizer of cell pellets suspended in a volume of hypotonic medium containing: 17.5 mM sucrose, 55 mM mannitol, 5 mM Tris-HCl buffer (pH 7.4), 0.5 mM EGTA, and corresponding to 2.5 times weight of cell pellets. Then the homogenates were diluted to 20% (w/v) with sucrose and mannitol so as to have an isotonic medium, and a mild sonication was carried out; diluted homogenates were centrifuged at 1500 g for 6.5 min (Beckman centrifuge J-6B). From the collected supernatants the mitochondrial fractions were isolated by centrifuging at 28000 g for 2 min (rotor 30, Beckman ultracentrifuge L8–65). The supernatants were discharged and the pellets (mitochondrial fraction) were resuspended in 250 mM sucrose and 20 mM Tris-HCl (pH 7.4) and used for fatty acid and enzyme activity determination.

2.4. Enzyme Activity Determination

ALDH activity was determined as described (Canuto et al., 1994). 0.1 mM 4-hydroxynonenal and 10 mM acetaldehyde were used as substrate.

2.5. Cell Growth

Cell growth was evaluated as number of cells present in the monolayer.

2.6. Western Blot Analysis

The cells were homogenized in a lysing buffer and used for Western blot analysis, which was performed as described in Canuto et al. (1996), except that anti-mitochondrial ALDH-1 (anti-MT-ALDH-1) and anti-mitochondrial ALDH-2 (anti-MT-ALDH-2) antibodies were used.

2.7. Northern Blot Analysis

Northern blot analysis was performed as described in Canuto et al. (1996), except that a DNA probe for mitochondria class 2 ALDH was used.

2.8. Statistical Analysis

All data are expressed as means ± S.D. The significance of differences between group means was assessed by variance analysis, followed by the Newman-Keuls test.

3. RESULTS AND DISCUSSION

Class 3 ALDH is inhibited in both mRNA content and enzymatic activity by aldehydes produced by lipid peroxidation, induced in hepatoma cells by enriching them with arachidonic acid and exposing them to ascorbate/iron sulphate (Canuto et al., 1996). The decrease of class 3 ALDH is parallel not only to the increase of lipid peroxidation, but also to the decrease of cell growth: Aldehydes produced in the cells during the lipid peroxidation process appear to inhibit cell growth and cause cell death in correlation with the class 3 ALDH activity.

This paper examines class 2 mitochondria ALDH and its behaviour in 7777 and JM2 hepatoma cells after enrichment with arachidonic acid and exposure to ascorbate/iron sul-

Table 1. Aldehyde dehydrogenase activity in mitochondria isolated from normal hepatocytes and hepatoma cell lines

Cells	ACA	4-HNE
Hepatocytes	15.20 ± 3.01a	4.31 ± 0.81a
7777 hepatoma	0.59 ± 0.51b	0.69 ± 0.31b
JM2 hepatoma	1.21 ± 0.34b	1.51 ± 0.42b

Data are expressed as nmoles of NADH produced/min/mg of protein and represent mean ± S.D. of 3 experiments. Acetaldehyde (ACA) and 4-hydroxynonenal (4-HNE) were added at the final concentration of 10 mM and 0.1 mM respectively.
For each substrate, means with different letters are statistically different ($p < 0.001$) from one another as determined by variance analysis followed by the Newman-Keuls test.

phate. The activity of class 2 ALDH is decreased using either acetaldehyde or 4-hydroxynonenal as the substrate in both hepatoma cells in comparison with normal hepatocytes (Table 1). This is unlike with class 3 ALDH, whose activity is increased in JM2 hepatoma cells (Canuto et al., 1994). The reason for this different behaviour between the mitochondrial and the cytosolic isoenzyme is not clear.

JM2 and 7777 hepatoma cells were cultured in presence of arachidonic acid for 24h; arachidonic acid enrichment is reported in Table 2. This table shows the increase of arachidonic acid in phospholipids extracted from cells *in toto*, and in phospholipids extracted from mitochondria of both cell lines in comparison with cells untreated with arachidonic acid. The percentage value of hepatocytes has been included to demonstrate that arachidonic acid returns to normal values for phospholipids extracted from cells *in toto*, and reaches about 65 and 53 % of normal value, for mitochondrial phospholipids of 7777 and JM2 hepatoma cells, respectively.

Arachidonic acid was removed from the culture medium at this point, and some flasks of cells received 4 doses of ascorbate/iron sulphate at 12h intervals. Figures 1 and 2 show cell growth. Arachidonic acid alone is able to slow the growth of 7777 cells, as is clear by the end of culture time (Figure 1), whereas it has no effect on growth of JM2 cells

Table 2. Percentage of arachidonic acid in phospholipids extracted from cells *in toto* or from mitochondria

Cells	Control	Enriched
Cells in toto		
Hepatocytes	23.49 ± 1.91a	
7777 hepatoma	7.43 ± 2.01b	22.07 ± 1.98a
JM2 hepatoma	6.82 ± 0.93b	24.50 ± 2.67a
Mitochondria		
Hepatocytes	28.30 ± 3.72a	
7777 hepatoma	1.09 ± 0.15b	18.63 ± 1.98c
JM2 hepatoma	1.23 ± 0.12b	15.08 ± 2.01c

Data are expressed as percentages of total fatty acids and are mean ± S.D. of 5 experiments. The rat hepatoma cells were enriched with arachidonic acid as in Materials and Methods.
Means with different letters are statistically different ($p < 0.001$) from one another as determined by variance analysis followed by the Newman-Keuls test.

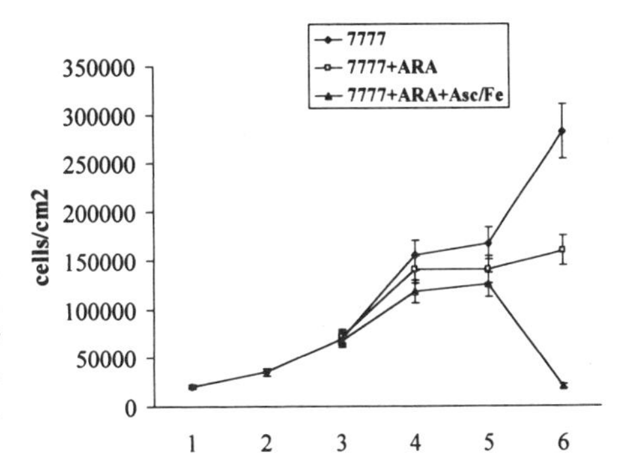

Figure 1. Growth of 7777 hepatoma cells untreated, treated with arachidonic acid (ARA), and treated with arachidonic acid and ascorbate/iron sulphate (ARA+Asc/Fe^{2+}). 1. 24h after seeding; 2. 24h after arachidonic acid enrichment; 3. 4h after the first dose of ascorbate/iron sulphate; 4. 12h after the second dose of ascorbate/iron sulphate; 5. 4h after the third dose of ascorbate/iron sulphate; 6. 12h after the fourth dose of ascorbate/iron sulphate.

(Figure 2). The decrease of 7777 cell numbers by arachidonic acid alone is much less than that caused by the prooxidant induced aldehydes. Enrichment with arachidonic acid does not increase cell growth: i.e., at this concentration, it is not a stimulus to cell growth. The effects of arachidonic acid on cells differ, probably in relation to the concentrations used and the compounds derived from its metabolism. Some researchers support the notion that arachidonic acid is a stimulus to cell growth through the production of prostaglandins (Reich and Martin, 1996); others have found arachidonic acid to inhibit cell growth through lipid peroxidation (Takeda et al., 1993).

In Figure 3, the effect of enrichment with arachidonic acid on mRNA content of class 2 mitochondria ALDH is shown in 7777 cells. Arachidonic acid increases the content of mRNA, and the increase remains almost steady during culture time, even after removal of the arachidonic acid from culture medium. The addition of the first two doses of prooxidant causes a further increase of mRNA content (Figure 4); at this point, cell growth has already decreased, but there are no dead cells (Canuto et al.,1996). With 4 doses of prooxidant by which time there are many dead cells, the cells that were still alive showed a slightly decreased mRNA content compared to dose 2, though still above control values.

In JM2 cells, arachidonic acid alone increases the content of mRNA of class 2 ALDH just as in the 7777 cells, but the increase is not steady; by the end of culture time, the mRNA content of JM2 cells treated with arachidonic acid is decreased and it is almost the same of that of control cells (Figure 5). Treatment with prooxidant after arachidonic acid causes a rapid decrease of mRNA, even if the expression remains slightly above that of untreated cells (Figure 6). The less deviated 7777 hepatoma cells appear to be more susceptible to arachidonic acid than the more deviated JM2, with regard to class 2 ALDH.

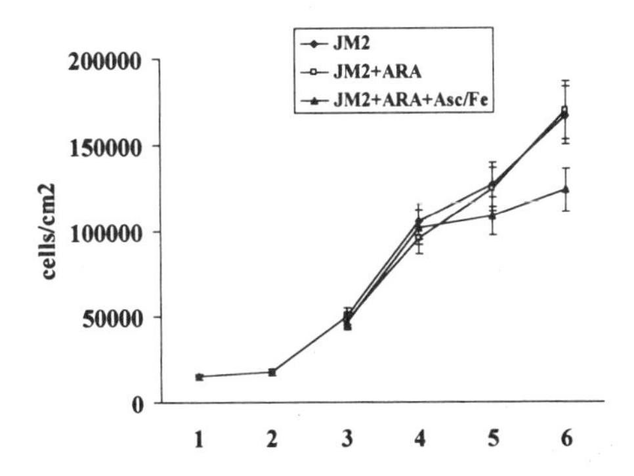

Figure 2. Growth of JM2 hepatoma cells untreated, treated with arachidonic acid (ARA), and treated with arachidonic acid and ascorbate/iron sulphate (ARA+Asc/Fe^{2+}). 1. 24h after seeding; 2. 24h after arachidonic acid enrichment; 3. 4h after the first dose of ascorbate/iron sulphate; 4. 12h after the second dose of ascorbate/iron sulphate; 5. 4h after the third dose of ascorbate/iron sulphate; 6. 12h after the fourth dose of ascorbate/iron sulphate.

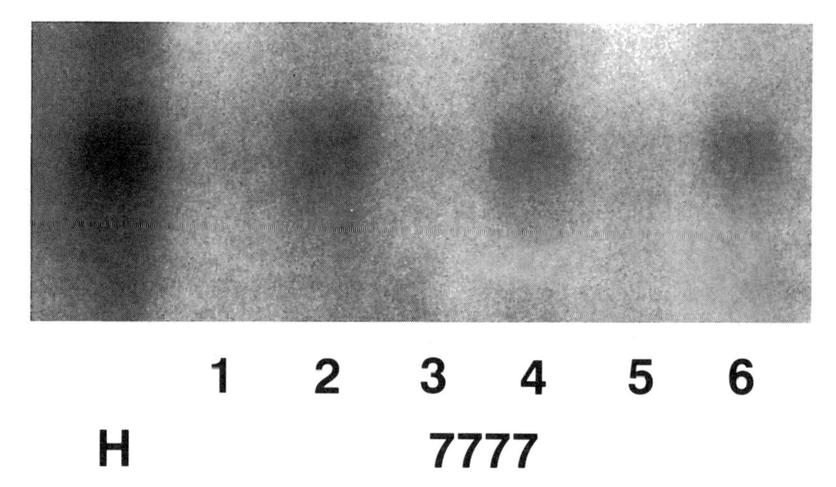

Figure 3. Effect of arachidonic acid on mRNA content of class 2 ALDH in 7777 hepatoma cells. H. Normal hepatocytes; 1. Control cells for lane 2; 2. 7777 hepatoma cells harvested 24h after enrichment with arachidonic acid; 3. Control cells for lane 4; 4. 7777 hepatoma cells harvested 48h after enrichment with arachidonic acid (time corresponding to 12h after the 2nd dose of ascorbate/iron sulphate; 5. Control cells for lane 6; 6. 7777 hepatoma cells harvested 52h after enrichment with arachidonic acid (time corresponding to 4h after the 3rd dose of ascorbate/iron sulphate).

The protein content of class 2 ALDH was also examined using Western blot analysis. Two antibodies versus class 2 mitochondria ALDH were used: one versus class 2 isoenzyme with long half-life, 87h (ALDH-MTI), the other versus class 2 isoenzyme with short half-life, 45 min (ALDH-MTII). In 7777 cells, no variations are evident with either mitochondria isoenzyme (Figure 7); protein content appears not to behave similarly to mRNA content. On the contrary, in JM2 hepatoma cells (Figure 8), some variations in class2 ALDH protein are more evident than in 7777 cells and more in agreement with the mRNA pattern. The protein content is increased by arachidonic acid in both mitochondria

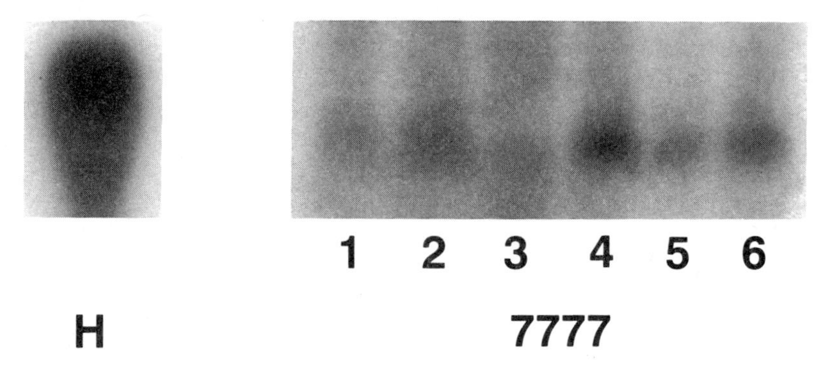

Figure 4. Effect of lipid peroxidation products on mRNA content of class 2 ALDH in 7777 hepatoma cells. H. Normal hepatocytes; 1. Control cells for lane 2; 2. 7777 hepatoma cells harvested 24h after enrichment with arachidonic acid; 3. Control cells for lane 4; 4. 7777 hepatoma cells harvested 12h after the 2nd dose of ascorbate/iron sulphate; 5. Control cells for lane 6; 6. 7777 hepatoma cells harvested 4h after the 3rd dose of ascorbate/iron sulphate.

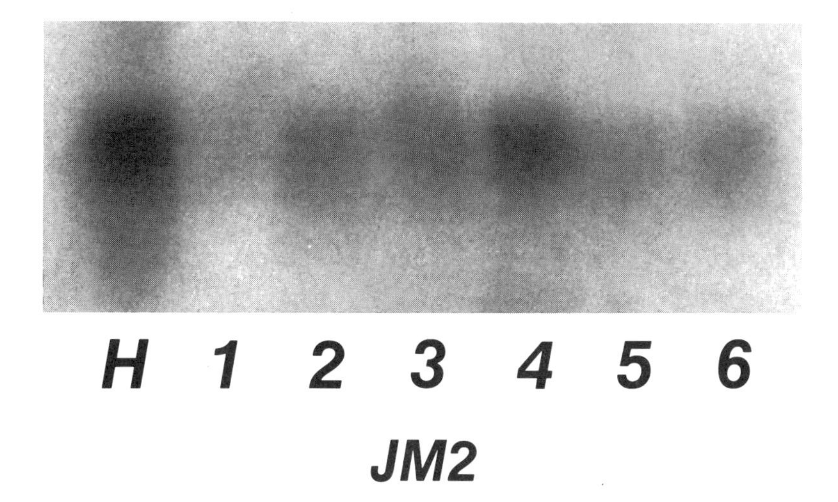

Figure 5. Effect of arachidonic acid on mRNA content of class 2 ALDH in JM2 hepatoma cells. H. Normal hepatocytes; 1. Control cells for lane 2; 2. JM2 hepatoma cells harvested 24h after enrichment with arachidonic acid; 3. Control cells for lane 4; 4. JM2 hepatoma cells harvested 48h after enrichment with arachidonic acid (time corresponding to 12h after the 2nd dose of ascorbate/iron sulphate); 5. Control cells for lane 6; 6. JM2 hepatoma cells harvested 52h after enrichment with arachidonic acid (time corresponding to 12h after the 4th dose of ascorbate/iron sulphate.

isoenzymes in comparison with untreated JM2 cells, and tends to reach the content of hepatocytes. Moreover, there is a decrease of protein content after the 4th dose of ascorbate/iron sulphate, both with control JM2 cells and with JM2 cells treated with arachidonic acid alone.

In conclusion, in hepatoma cells, arachidonic acid is able to induce gene expression of class 2 ALDH, which is a typical isoenzyme of hepatocytes. We may put forward the hy-

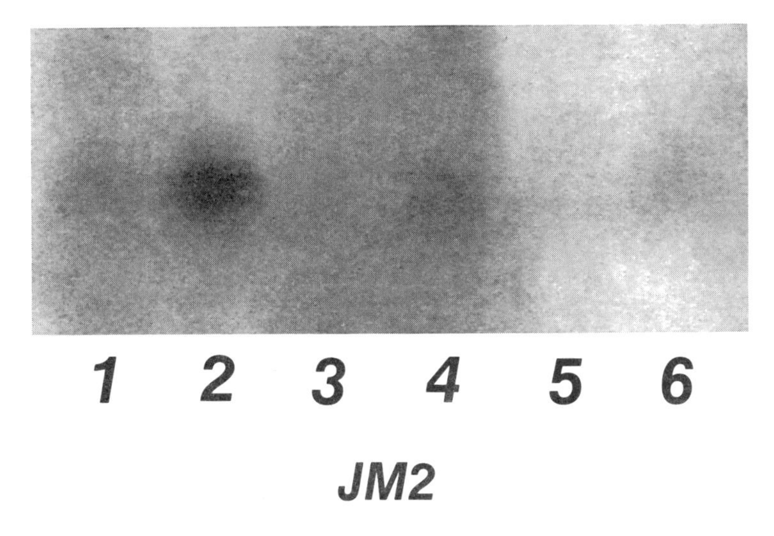

Figure 6. Effect of lipid peroxidation products on mRNA content of class 2 ALDH in JM2 hepatoma cells. 1. Control cells for lane 2; 2. Cells harvested 24h after enrichment with arachidonic acid; 3. Control.cells for lane 4; 4. Cells harvested 12h after the 2nd dose of ascorbate/iron sulphate; 5. Control cells for lane 6; 6. Cells harvested 12h after the 4th dose of ascorbate/iron sulphate.

ALDH MT I

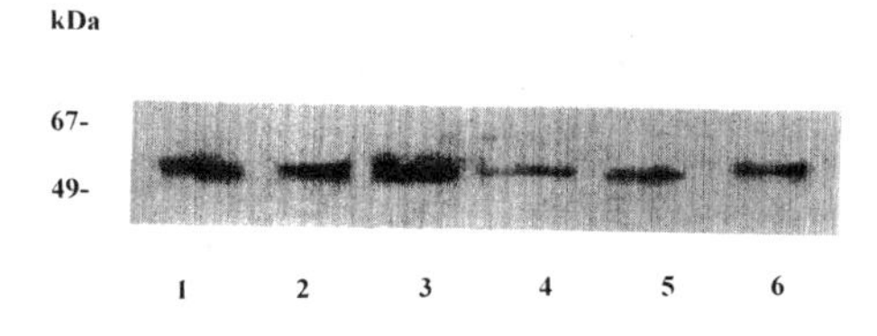

ALDH MT II

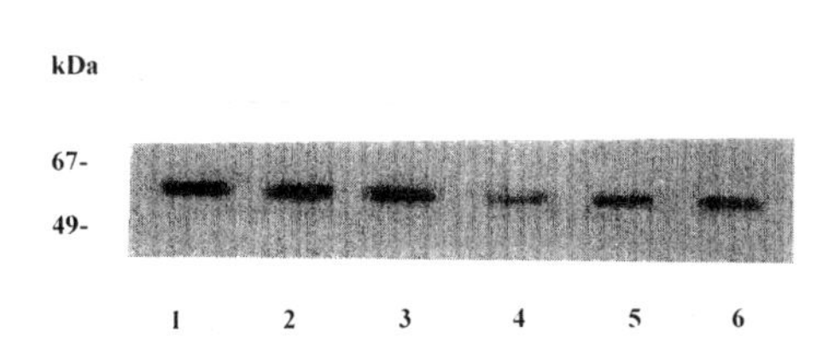

Figure 7. Western-blot analysis of class 2 ALDH (MT-I and MT-II) in 7777 hepatoma cells enriched with arachidonic acid and exposed to ascorbate/iron sulphate. 1. Control cells for lane 2; 2. Cells harvested 24h after enrichment with arachidonic acid; 3. Control cells for lane 4; 4. Cells harvested 12h after the 2nd dose of ascorbate/iron sulphate; 5. Control cells for lane 6; 6. Cells harvested 4h after the 3rd dose of ascorbate/iron sulphate.

ALDH MT I

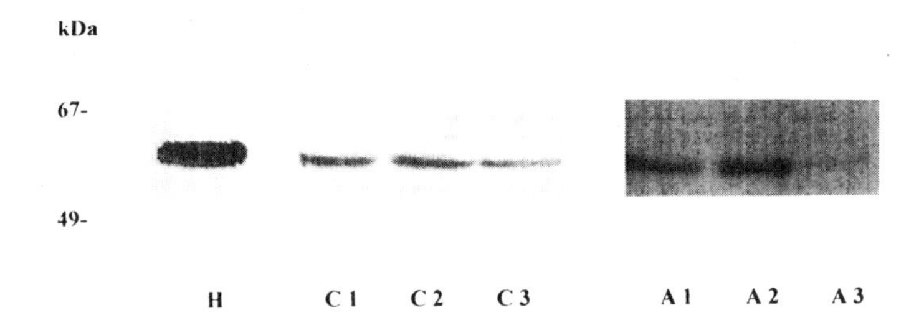

ALDH MT II

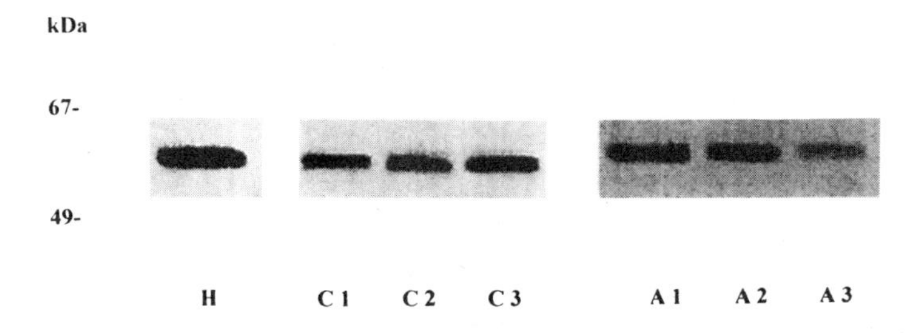

Figure 8. Western-blot analysis of class 2 ALDH (MT-I and MT-II) in JM2 hepatoma cells enriched with arachidonic acid and exposed to ascorbate/iron sulphate. H. Normal hepatocytes; C1. Control cells for lane 2; C2. Control cells for lane 4; C3. Control cells for lane 6; A1. JM2 hepatoma cells harvested 24h after enrichment with arachidonic acid; A2. JM2 hepatoma cells harvested 12h after the 2nd dose of ascorbate/iron sulphate; A3. JM2 hepatoma cells harvested 12h after the 4th dose of ascorbate/iron sulphate.

pothesis that arachidonic acid tends to induce normal phenotype in hepatoma cells, when it is added to the medium of cultured cells and is consequently included in cell phospholipids.

At this time it is not clear how arachidonic acid may increase gene expression of class 2 ALDH, but some possible mechanisms may be: first, the increase of gene expression might be mediated by activation of peroxisome proliferators receptors. Peroxisome proliferator receptors are, indeed, activated by fatty acids (Issemann et al, 1993). The presence of these receptors in 7777 and JM2 hepatoma cells must be determined; they are present in rat hepatocytes, but not in all hepatoma cells. In one highly deviated hepatoma, they were not evidenced by determination of mRNA content by Northern blot analysis (Canuto et al., 1998); second, the very small quantity of lipid peroxidation products, due to the increase of arachidonic acid, might stimulate gene expression. Some researchers have found that 4-hydroxynonenal is able to activate an aldehyde reductase gene (Spycher et al., 1996), to elicit induction of mRNA for *c-jun* protooncogene (Parola et al., 1996) and to stimulate the TGFbeta (Leonarduzzi et al., 1997). These alternatives must be further investigated. Furthermore, enzyme activity of ALDH in mitochondria will also have to be determined to see whether the activation effect of arachidonic acid on gene expression has any repercussions on activity.

ACKNOWLEDGMENTS

This study was supported by grants from the Italian Ministry for University, Scientific and Technological Research (60% and Cofinanziamento 1997), and from the Italian Association for Cancer Research. R. A. Salvo was recipient of a fellowship from the Cavalieri Ottolenghi Foundation, Turin, Italy. We wish to thank Dr. Dennis Petersen for kindly giving us 4-hydroxynonenal.

REFERENCES

Barrera, G., Di Mauro, C., Muraca, R., Ferrero, D., Cavalli, G., Fazio, V.M., Paradisi, L., and Dianzani, M.U. (1991) Induction of differentiation in human HL-60 cells by 4-hydroxynonenal, a product of lipid peroxidation, Exper. Cell Res. **197**,148–152.

Canuto, R.A., Muzio, G., Biocca, M.E., and Dianzani, M.U. (1989) Oxidative metabolism of 4-hydroxy-2,3-nonenal during diethyl-nitrosamine-induced carcinogenesis in rat liver, Cancer Lett. **46**, 7–13.

Canuto, R.A., Muzio, G., Maggiora, M., Poli, G., Biasi, F., Dianzani, M.U., Ferro, M., Bassi, A.M., Penco, S., and Marinari, U.M. (1993) Ability of different hepatoma cells to metabolize 4-hydroxynonenal, Cell Biochem. Funct. **11**, 79–86.

Canuto, R.A., Ferro, M., Muzio, G., Bassi, A.M., Leonarduzzi,G., Maggiora, M., Adamo, D., Poli, G., and Lindahl, R. (1994) Role of aldehyde metabolizing enzymes in mediating effects of aldehyde products of lipid peroxidation in liver cells, Carcinogenesis **15**, 1359–1364.

Canuto, R.A., Muzio, G., Bassi, A.M., Maggiora, M., Leonarduzzi, G., Lindahl, R., Dianzani, M.U., and Ferro, M. (1995) Enrichment of arachidonic acid increases the sensitivity of hepatoma cells to the cytotoxic effects of oxidative stress, Free Rad. Biol. Med. **18**, 287–293.

Canuto, R.A., Ferro, M., Maggiora, M., Federa, R., Brossa, O., Bassi, A.M., Lindahl, R., and Muzio, G. (1996) In hepatoma cell lines restored lipid peroxidation affects cell viability inversely to aldehyde metabolizing enzyme activity, Adv. Exp. Med. Biol. **414**, 113–122.

Canuto, R.A., Muzio, G., Ferro, M., Maggiora, M., Federa, R., Bassi, A.M., Lindahl, R., and Dianzani, M.U. (1998) Inhibition of class-3 aldehyde dehydrogenase and cell growth by restored lipid peroxidation in heaptoma cell lines, Free Rad. Biol. Med. in press.

Canuto, R.A., Muzio, G., Bonelli, G., Maggiora, M., Autelli, R., Barbiero, G., Costelli, P., Brossa, O., and Baccino, F.M. (1998) Peroxisome proliferators induce apoptosis in hepatoma cells, Cancer Detect. Prevent. **22**, 357–366.

Cao, Q.N., Tu, G.C., and Weiner, H. (1988) Mitochondria as the primary site of acetaldehyde metabolism in beef and pig liver slices, Alcohol. Clin. Exp. Res. **12**, 720–724.

Chen, J.J., and Yu, B.P. (1996) Detoxification of reactive aldehydes in mitochondria: effects of age and dietary restriction, Aging **8**, 334–340.

Greenfield, N.J., and Pietruszko, R. (1977) Two aldehyde dehydrogenases from human liver. Isolation via affinity chromatography and characterization of the isoenzymes, Biochim. Biophys. Acta **483**, 35–45.

Huang, M., and Lindahl, R. (1990) Aldehyde dehydrogenase heterogeneity in rat hepatic cells, Arch. Biochem. Biophys. **277**, 296–300.

Issemann,·I., Prince, R.A., Tugwood, J.D., and Green, S. (1993) The peroxisome proliferator activated receptor: retinoid X receptor heterodimer is activated by fatty acids and fibrate hypolipidaemic drugs, J. Mol. Endocrinol. **11**, 37–47.

Leonarduzzi, G., Scavazza, A., Biasi, F., Chiarpotto, E:, Camandola, S., Vogl S., Dargel, R., and Poli, G. (1997) The lipid peroxidation end product 4-hydroxy-2,3-nonenal up-regulates transforming growth factor â1 expression in the macrofage lineage: a link between oxidative injury and fibrosclerosis, FASEB J. **11**, 851–857.

Lindahl, R. (1979) Subcellular distribution and properties of aldehydedehydrogenase from 2-acetylaminofluorene-induced rat hepatomas, Biochem. J. **183**, 55–64.

Lindahl, R., and Petersen, D.R. (1991) Lipid aldehyde oxidation as a physiological role for class 3 aldehyde dehydrogenases, Biochem. Pharmacol. **41**, 1583–1587.

Lindahl, R. (1992) Aldehyde dehydrogenases and their role in carcinogenesis, CRC Crit. Rev. Biochem. Mol. Biol. **27**, 283–335.

Parola, M., Pinzani, M., Marra, F., Casini, A:, Leonarduzzi, G., Robino, G., Albano, E., Bellomo, G., Camandola, S., Poli, G., and Dianzani, M.U. (1996) The pro-fibrogenic action of 4-hydroxy-2,3-alkenals in human hepatic stellate cells (hHSC) involves cytosolic and nuclear mechanisms, Hepatology **24**, 457A.

Reich, R., and Martin, G.R. (1996) Identification of arachidonic acid patway required for the invasive and metastatic activity of malignant tumor cells, Prostaglandins **51**, 1–17.

Spycher, S., Tabataba-Vakili, S., O'Donnell, V.B., Palomba, L., and Azzi, A. (1996) 4-hydroxy-2,3-trans-nonenal induces transcription and expression of aldose reductase, Biochem Biophys. Res. Commun. **226**, 512–516.

Stewart, M.J., Malek, K., and Crabb, D.W. (1996) Distribution of messenger RNAs for aldehyde dehydrogenase 1, aldehyde dehydrogenase 2, and aldehyde dedhydrogenase 5 in human tissues, J. Investig. Med. **44**, 42–46.

Takeda, S., Sim, P.G., Horrobin, D.F., Sanford, T., Chisholm, K., Simmons, V. (1993) Mechanism of lipid peroxidation in cancer cells in response to gamma-linolenic acid analysed by GC-MS: conjugated dienes with peroxyl (or hydroperoxyl) groups and cell-killing effects, Anticancer Res. **13**, 193–200.

Vasiliou, V., Kozak, C., Lindahl, R., and Nebert, D.W. (1996) Mouse microsomal class 3 aldehyde dehydrogenase: AHD3 cDNA sequence, inducibility by dioxin and clofibrate, and genetic mapping, DNA Cell Biol. **15**, 235–245.

Wischusen, S.M., Evces, S., and Lindahl, R. (1983) Changes in aldehyde dehydrogenase activity during diethylnitrosamine- or 2-acetylaminofluorene-initiated rat hepatocarcinogenesis, Cancer Res. **43**, 1710–1715.

Yoshida, A., Ikawa, M., Hsu, L.C., and Tani, K. (1985) Molecular abnormality and cDNA cloning of human aldehyde dehydrogenases, Alcohol **2**, 103–106.

Yoshida, A. (1992) Molecular genetics of human aldehyde dehydrogenase, Pharmacogenetics **2**, 139–147.

17

PREPUBERTAL REGULATION OF THE RAT DIOXIN-INDUCIBLE ALDEHYDE DEHYDROGENASE (ALDH3)

Panayiotis Stephanou,[1] Periklis Pappas,[1] Vasilis Vasiliou,[2] and Marios Marselos[1]

[1]Department of Pharmacology
Medical School
University of Ioannina
GR-451 10 Ioannina, Greece
[2]Department of Pharmaceutical Sciences
School of Pharmacy
University of Colorado Health Sciences Center
4200 East Ninth Avenue
Denver, Colorado 80262

1. INTRODUCTION

Polycyclic aromatic hydrocarbons (PAHs) are ubiquitous environmental pollutants formed primarily from the incomplete combustion of organic material. They are present in tobacco smoke, exhaust fumes and urban air. PAHs are also found in water and soil as a result of the rain and gravitational influence providing other possible routes for human exposure (Williams and Weisburger, 1986). Benzo[a]pyrene (BP) is a potent inducer of drug-metabolizing enzymes. BP up-regulates the expression of [Ah] battery genes, by binding to the aryl-hydrocarbon receptor (AHR) (Nebert et al.,1993). The [Ah] battery consists of at least six genes: two Phase I (CYP1A1 and CYP1A2) and four Phase II genes (ALDH3, NQO-1, UGT1*A6 and GSTA-1). Induction of CYP1A1 and CYP1A2 enzymes provokes increased metabolic activation of PAHs to mutagenic derivatives that may activate protooncogenes or inactivate tumor supressor genes (Nebert, 1989; Hankinson, 1995). On the other hand, induction of the Phase II enzymes of the [Ah] gene battery may be beneficial, since these enzymes are involved in detoxification pathways.

In most mammalian species, hepatic drug metabolism and its induction by exogenous stimuli varies with age, sex and hormonal status. Sex differences in the metabolism of drugs and xenobiotics have been well documented in various species of experimental animals. For

Enzymology and Molecular Biology of Carbonyl Metabolism 7
edited by Weiner *et al.* Kluwer Academic / Plenum Publishers, New York, 1999.

example, in adult rats metabolism of several drugs is faster in males than in females (Kato, 1974). In addition, expression of steroid- and drug-metabolizing enzymes is sex-dependent (Skett, 1988). Such sex differences are induced during puberty by the presence or absence of androgens or estrogens, and the phenomenon has been defined as "imprinting". Presumably, imprinting involves various neuroendocrinological mechanisms, where the balance of gonadal sex hormones appears to play an important role (Skett and Paterson, 1985). The pituitary gland may also play an important role by determining sex differences in the response of hepatic drug-metabolizing enzymes to inducing agents.

In the present study, steroids and BP were given in male and female rats during puberty, in order to determine whether imprinting is a regulatory mechanism involved in the induction of ALDH3 later in adult life. Subsequently, in vitro studies in primary pituitary cell cultures exposed to different concentrations of BP were used to elucidate the effects of BP on the release of gonadotrophins (luteinizing hormone-LH and follicle-stimulating hormone-FSH).

2. MATERIALS AND METHODS

2.1. *In Vivo* Studies

2.1.1. Animals and Treatment. Albino rats of the Wistar/Mol/Io/rr substrain (Marselos, 1976) were used. Animals were housed under constant temperature (20°C) and a light-dark cycle of 12 h. Food and water were available ad libidum. Different groups (10 animals in each group) of male and female rats at puberty (between 20 and 30 days of age) were treated with BP or steroids. Substances were diluted in olive oil and given intraperitoneally (i.p.), every second day for 10 days at the following concentrations: BP (20 mg/kg b.w.), Progesterone, Estradiol, Hydrocortisone, Testosterone, Tamoxifen and Cyproterone (5 mg/kg b.w.), LH (2.1 IU) and Danazole (200 mg/kg b.w.). Controls were treated with olive oil. At 90 days of age all animals were given a single dose of BP (50 mg/kg b.w.). Animals were killed 5 days later and livers were taken out for enzyme activity assay and estrogen and progesterone receptor level measurement.

In another series of experiments female adult rats were treated once with BP (50 mg/kg b.w.), i.p. Five days later trunk blood was collected to measure the serum levels of LH and FSH by ELISA immunoassay.

2.1.2. ALDH3 Enzyme Assay. ALDH3 activity was measured at 37°C, by following the reduction of $NADP^+$ to NADPH at 340 nm in a Shimadzu UV1601 spectrophotometer. The assay mixture contained sodium pyrophosphate buffer (75 mM, pH 8.0), pyrazole (1 mM), $NADP^+$ (2.5 mM) as coenzyme and benzaldehyde (5 mM) as substrate.

2.1.3. Estrogen (ER) and Progesterone (PR) Receptors Levels. Both ER and PR levels were determined by using the activated charcoal method as described by Clark et al., (1982). Multiple point saturation assays used as labeled steroid range (3.5–8.0 nM for [^{3}H]-estradiol and 1×10^{-8}–2.5×10^{-8} M for [^{3}H]-progesterone). Non specific binding was assessed by the addition of unlabelled diethylstilbestrol or progesterone. After a 30 min incubation period of samples at 4°C, 1 ml of dextran-coated charcoal suspension was added and left for 10 min at 4°C. The mixture was then centrifuged for 5 min at 5000xg and 400 μl from the supernatant were counted for radioactivity. The specific binding was calculated by substracting nonspecific binding from total binding.

2.2. *In Vitro* Studies

2.2.1. Primary Pituitary Cell Cultures. The method was adapted from Byrne et al. (1995). Anterior pituitaries were collected from 2-months old female rats, cut into 1 to 2 mm fragments and rinsed several times with Dulbecco's phosphate-buffered saline (DPBS) containing 0.1% bovine serum albumin (BSA) and 7.5 mM glucose. The tissue fragments were dispersed into single cells by using trypsin (4.1 mg/ml in DPBS) and pancreatin (2.5 mg/ml in DPBS), and when dispersion was complete, cells were centrifuged at 400×g for 8 min. Cell pellet was resuspended in 15 ml Dulbecco's modified Eagle's medium nutrient mixture F-12 Ham serum free culture medium (SFDM) containing 0.1% BSA, 10 μg/ml insulin, 5 μg/ml transferrin, 4 nM dexamethasone, 50 pM 3,3',5-triiodothyronine, 1% penicillin, 1% streptomycin, 500 μl/l gentamycin and 10% fetal calf serum and recentrifuged two times. The cell concentration was 3×10^{-6} cells/ml and the viability about 90%.

The cell suspension was then pipetted into a Corning 24-well culture plate at 100 μl (3×10^{5} cells/well) and preincubated at 37(in a water-saturated atmosphere of 95% air and 5% CO_2. After 48 h preincubation period, the medium was gently removed from wells and the cells attached to the bottom were rinsed 3 times with calcium DMEM. The cells were then incubated in DMEM containing different concentrations of BP (0.1–20 μM), GnRH or vehicle (DMSO). The cells were then incubated at 37°C in a water saturated atmosphere of 95% air and 5% CO_2 for 24 h. The medium was then removed carefully from the cells with a Pasteur pipette and stored at -20°C until measurement of the LH and FSH levels by ELISA assay.

2.2.2. ELISA Assay. Concentrations of LH and FSH in sera and in the media were determined by a competitive ELISA assay. Samples of various concentrations of standard LH or FSH were preincubated with an equal volume of LH or FSH antiserum for 1 h at 37°C then added to individual wells of a Nunc 96-well culture plate that had been precoated with competing ligand rat antigen LH or FSH in coating buffer and blocked with PBS containing 1% BSA. The samples and antisera were incubated overnight at 4°C and then after being washed they were incubated with donkey antirabbit IgG conjugated to horseradish peroxidase for 1 h at 37°C. The optical density was read at 492 nm in Micro-Elisa reader. Results are expressed in terms of the appropriate standards rLH-RP-3 and rFSH-RP-2.

2.3. Statistics

All results were expressed as the mean ± SEM. Statistical analysis was performed by Student's two-tailed t-test.

3. RESULTS

3.1. *In Vivo* Studies

The induction of ALDH3 by a single dose of BP, without any pretreatment at puberty, was more pronounced in male than in female adult rats (Figure 1).

Pretreatment with BP at puberty affected the inducibility of ALDH3 by BP in adult life. At 90 days of age, 2 months after the first treatment with BP, a statistically significant

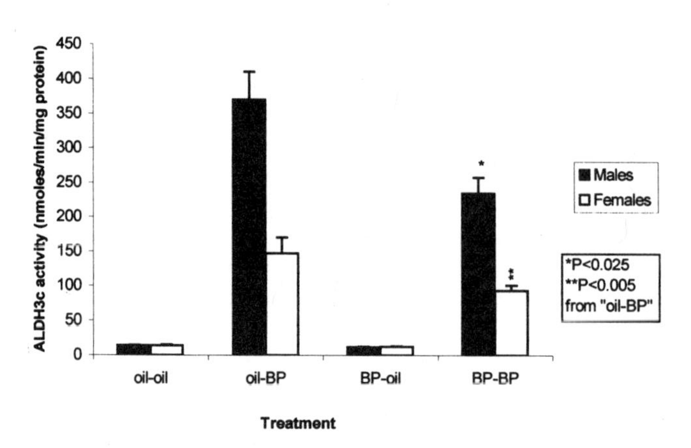

Figure 1. Effects of BP administration on the inducibility of ALDH3. "oil-oil": control group; "oil-BP": group treated with BP only at 90 days of age; "BP-oil": group treated with BP only at puberty; "BP-BP": group treated with BP at puberty and at 90 days of age.

decrease of approximately 40% was found in the ALDH3 inducibility by BP in both sexes (Figure 1). The latter was associated with a 50% and 70% decrease in FSH and LH serum levels respectively (Table 1).

BP was found to decrease the liver cytosolic levels of both estrogen (ER) and progesterone (PR) receptors in both male and female adult rats. ER and PR levels were decreased by 60% and 20%, respectively in both sexes (Figures 2A and 2B). This effect was more prominent after BP pretreatment during puberty. In male rats, ER and PR levels were decreased by 80% and 60%, respectively. In female adult rats, although the same rate of decrease in ER levels was observed (75%), the decrease in PR levels was more strong (95%) (Figures 2A and 2B).

Administration of progesterone, LH and danazole during puberty inhibited the ALDH3 inducibility in both sexes later in adult life. Estradiol provoked a decrease of ALDH3 inducibility that was not found to be statistically significant. In contrast, hydrocortisone and cyproterone enhanced the ALDH3 inducibility only in males whereas tamoxifen increased ALDH3 inducibility in males or in females. Finally, testosterone had no effect in males or in females (Figures 3A and 3B).

3.2. *In Vitro* Studies

In the pituitary cell culture system there was a tendency toward a decrease of LH release from pituitary cells incubated for 24 h with BP. On the contrary, BP did not affect FSH release in these cells (Table 2).

Table 1. Serum levels of LH and FSH after administration of BP in adult female rats

Treatment	FSH (ng/ml)	LH (ng/ml)
Control	20.1 ± 9.8	12.77 ± 1.5
BP	10.71 ± 1.66#	4.05 ± 0.11

#P.01

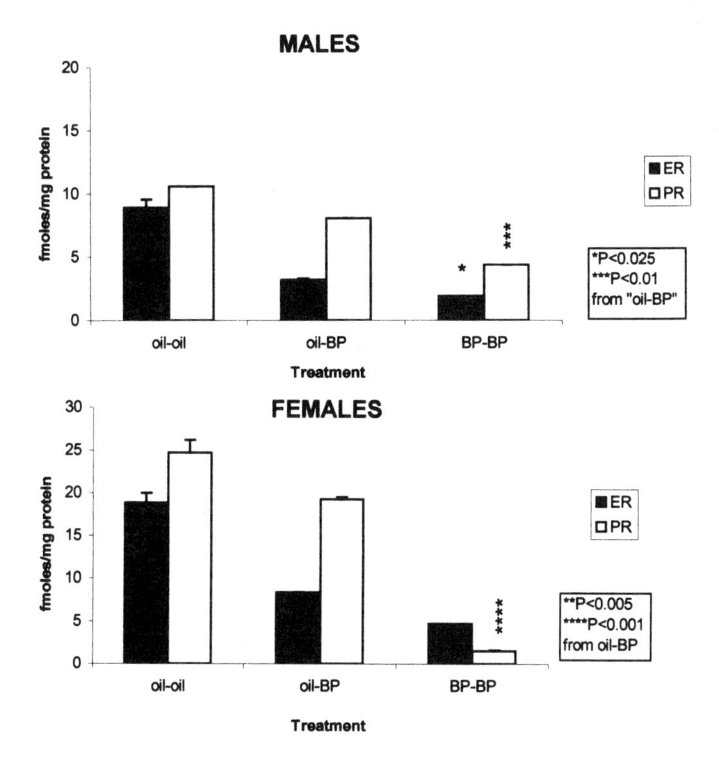

Figure 2. Effects of BP administration on the cytosolic levels of hepatic estrogen (ER) and progesterone (PR) receptors in male (A) and in female (B) rats. For further comments, see legend to Figure 1.

4. DISCUSSION

This report shows for the first time that administration of BP and/or steroids during puberty affect ALDH3 induction by BP later in adult life, indicating that sexual differentiation may be a regulatory mechanism involved in this induction process. The steroids tested here can be classified into three distinct groups based on their ability to regulate the ALDH3 inducibility. The first include progesterone, LH, danazole and estradiol that act like inhibitors in both males and females. The second consists of hydrocortisone, cyproterone and tamoxifen that enhance ALDH3 inducibility. The third group includes testosterone that did not affect the induction of this enzyme. Interestingly, BP administration at puberty inhibited ALDH3 inducibility later in adult life in both sexes. Several mechanisms may explain these alterations in ALDH3 induction. A possible interaction between nuclear steroid receptors and the transcriptional machinery involved in the up-regulation of [Ah] genes by AHR-ligands has been suggested (Thomsen, 1994). Hepatic ALDH3 induction was found to be diminished when 3-methylcholanthrene was administered in adult rats together with any of the following: estradiol, progesterone, hydrocortisol, diethylstilbestrol or tamoxifen (Karageorgou, 1995). Supporting this interaction, PAHs decrease the mouse and rat ER levels via an AHR-dependent mechanism (Romkes et al., 1987; Lin et al., 1991). It has also been found that AHR mediates the down-regulation of ER by TCDD (Harris, 1990). In addition, progesterone has been shown to exert similar effects on ER (Leavitt, 1985). Thus, it is possible that BP either acts as an antiestrogen or has a progestational activity, even when administered at puberty.

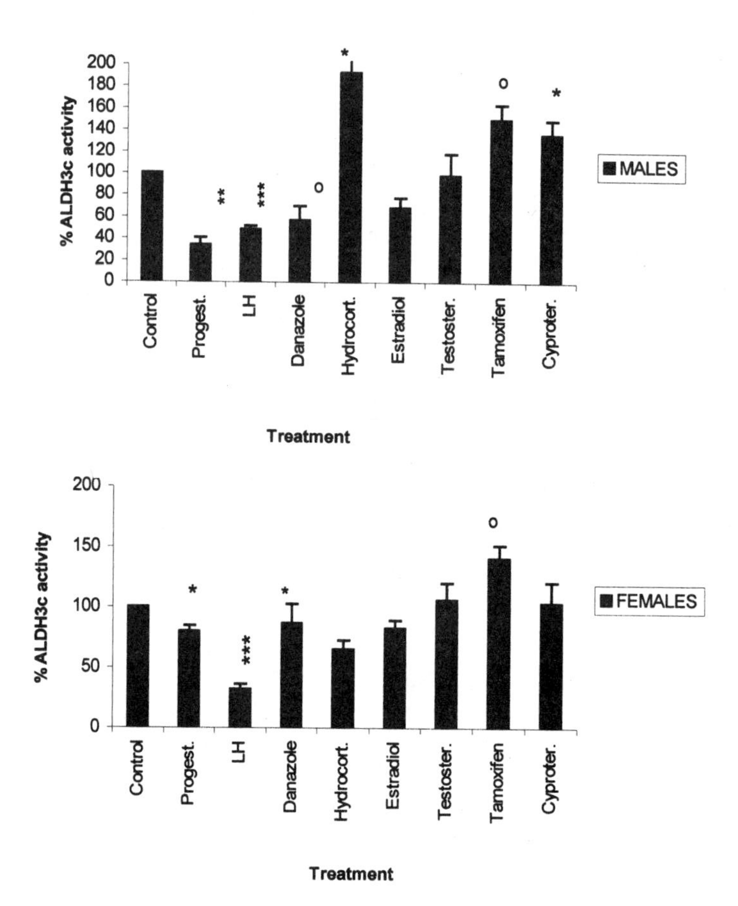

Figure 3. ALDH3 activity (%) after treatment of rats with steroids during puberty. (A) Males; (B) Females. * $P < 0.025$; ** $P < 0.005$; *** $P < 0.01$; **** $P < 0.0001$; ø $P < 0.05$ statistically significant from control group.

The results are somehow different when the ER agonist, estradiol, and the ER antagonist, tamoxifen, were administered during puberty. As discussed earlier, both estradiol and tamoxifen have been shown to have a strong inhibitory effect on ALDH3 induction when were administered together with PAHs in adult rats (Karageorgou, 1995). Based on our results, although estradiol decreases ALDH3 activity, tamoxifen administration at pu-

Table 2. Gonadotrophins release from pituitary cells after incubation with BP

Treatment	FSH (ng/ml)	LH (ng/ml)
DMEM	30.6 ± 0.3	14.4 ± 2.7
DMEM/DMSO	27.1 ± 1.1	26.5 ± 2.7
BP (0.1 μM)	–	24.9 ± 5.1
BP (0.5 μM)	–	12.5 ± 2.4##
BP (1.0 μM)	–	14.9 ± 1.2##
BP (2.0 μM)	28.5 ± 0.4	–
BP (5.0 μM)	–	15.9 ± 2.3##
BP (10.0 μM)	29.1 ± 1.0	–
BP (20.0 μM)	28.2 ± 1.0	–

##P.05 from DMEM/DMSO group.

berty up-regulates the ALDH3 induction in adult life. It appears that the state of activation of the ER is important for the regulation of ALDH3 induction. It is also possible that BP and steroids regulate ALDH3 activity by acting via the pituitary gland. Anterior pituitary is responsible for the synthesis and secretion of gonadotrophins (LH and FSH), thyroid-stimulating hormone (TSH), adrenocortical-stimulating hormone (ACTH), prolactin and growth hormone (GH). Hypophysectomy has been shown to influence the hepatic metabolism of steroids (Kato, 1993) and the regulation of rat hepatic drug metabolizing enzymes (Finnen, 1984). In our experiments, administration of LH and an inhibitor of the release of gonadotrophins by pituitary, danazole, during puberty provoked a decrease of the ALDH3 inducibility by BP in adult rats. These apparently controversial results indicate first that deviation from the normal pituitary function is deteriorating the expression of hepatic ALDH3, and, second that the liver is a target organ of LH. In the present work, BP was shown to decrease gonadotrophin serum levels in female rats, probably by acting as an inhibitor of the release of these hormones from pituitary. Our results obtained from the in vitro studies support such a hypothesis. A slight decrease in the LH was found in pituitary cells after treatment with BP.

In conclusion, our results show that administration of BP and steroids during puberty can "imprint" the induction of by BP in adult life. The exact mechanism of this regulation is unknown, but probably hepatic estrogen receptors and pituitary gland could mediate the effects of BP and steroids on ALDH3 activity. It would be very interesting if such a regulation existed for the other members of the [Ah] gene battery, especially in CYP1A1 and CYP1A2 that are known to generate carcinogenic BP-metabolites. These experiments are currently in progress in our laboratory.

ACKNOWLEDGMENTS

We are grateful to Dr. Aglaia Pappa for her assistance on the *in vitro* studies.

REFERENCES

Byrne, B., Fowler, P.A., Fraser, M., Culler, M. and Templeton, A., (1995), GnSAF bioactivity in serum from superovulated women is not blocked by inhibin antibody, Biol. Reprod., 52, 88–95.

Clark, J.H., Williams, M., Upchurch, S., Eriksson, H., Helton, E. and Markaverish, B.M., (1982), Effects of estradiol-17beta on nuclear occupancy of the estrogen receptor, stimulation of nuclear type II sites and uterine growth, J. Steroid Biochem., 16, 323–328.

Finnen, M.J. and Hassal, K.A., (1984), Effects of hypophysectomy on sex differences in the induction of hepatic drug-metabolizing enzymes in the rat, J. Pharmacol., Exp. Ther., 229, 250–254.

Hankinson, O., (1995), The Aryl hydrocarbon receptor complex, Annu. Rev. Pharmacol. Toxicol., 35, 307–340.

Harris, M., Zacharewski, T. and Safe, S., (1990), Effects of 2,3,7,8-tetrachlodibenzo-p-dioxin and related compounds on the occupied nuclear estrogen receptor in MCF-7 human breast cancer cells, Cancer Res., 50, 3579–3584.

Karageorgou, M., Vasiliou, V., Nebert, D.W. and Marselos M., (1995), Ligands of four receptors in the nuclear steroid7thyroid hormone superfamily inhibit induction of rat cytosolic aldehyde dehydrogenase-3 (ALDH3) by 3-methylcholanthrene, Biochem. Pharmacol., 50, 2113–2117.

Kato, R., (1974), Sex related differences in drug metabolism, Drug Metab. Rev., 3, 1–32.

Kato, R., and Yamazoe, Y., (1993), Hormonal regulation of cytochrome P450 in rat liver, Hanbook Exp. Pharmacol., 105, 447–459

Leavitt, W.W., Evans, R.W., Okulicz, W.C., MacDonald, R.G., Henory, W.J. and Robidoux, W.F., (1982), Progesterone inhibition of nuclear estrogen receptor, In: Hormone Antagonists, Agarwal M.K. ed., de Gruyter, Berlin, 213–232.

Lin, F.H., Stohs, S.J., Birnbaum, L.S., Clark, G., Lucier, G.W. and Goldstein, J.A., (1991), The effects of 2,3,7,8-Tetrachloro-p-dioxin (TCDD) on the hepatic estrogen and glucocorticoid receptors in congenic strains of Ah responsive and Ah nonresponsive C57BL/6J mice, Toxicol. Appl. Pharmacol., 108, 129–139.

Marselos, M., (1976), Genetic variation of drug metabolizing enzymes in the Wistar rat, Acta Pharmacol. Toxicol., 39, 186–197.

Nebert, D.W., (1989), The Ah locus: Genetic differences in toxicity, cancer, mutation and birth defects, Crit. Rev. Toxicol., 20, 153–174.

Nebert, D.V., Puga, A. and Vasiliou V., (1993), Role of the Ah receptor and the dioxin-inducible [Ah] gene battery in toxicity, cancer and signal transduction, Ann. N. Y. Acad. Sci., 685, 624–640.

Romkes, M., Piskorska-Pliszczynska, J., and Safe S., (1987), Effects of 2,3,7,8-Tetrachlorodibenzo-p-dioxin on hepatic and uterine estrogen receptor levels in rat, Toxicol. Appl. Pharmacol., 87, 306–314.

Skett, P., (1988), Biochemical basis of sex differences in drug metabolism, Pharmac. Ther., 38, 269–304.

Skett, P. and Paterson, P., (1985), Sex differences in the effects of microsomal enzyme inducers on hepatic phase I drug metabolism in the rat, Biochem. Pharmacol., 34, 3533–3536.

Thomsen, J.S., Wang, X., Hines, R.N. and Safe S., (1994), Introduction of a functional human estrogen receptor restores the function of the Ah receptor in the human breast carcinoma cell line MDA-MBA-231, The Toxicologist, 13, 34

Williams, G.M. and Weisburger, J.H., (1986), Chemical carcinogens, In: Casarett, L.J. and Doull J. (Eds.), Toxicology, New York, MacMillan, 99–173.

18

EFFECTS OF TAMOXIFEN AND TOREMIFENE ON ALDH1 AND ALDH3 IN HUMAN RETINAL PIGMENT EPITHELIAL CELLS AND RAT LIVER

Periklis Pappas,[1] Panayiotis Stephanou,[1] Marianthi Sotiropoulou,[1] Carol Murphy,[1] Lotta Salminen,[2] and Marios Marselos[1]

[1]Department of Pharmacology
Medical School
University of Ioannina
451 10 Ioannina, Greece
[2]Department of Ophthalmology
Medical School
University of Tampere
335 21 Tampere, Finland

1. INTRODUCTION

Aldehyde dehydrogenase is a NAD(P)-dependent enzyme with wide distribution virtually in all animal tissues (Vasiliou and Marselos, 1989; Lindahl, 1992). Different constitutively expressed ALDHs are found in liver, stomach, brain, kidney, skin and eye (Vasiliou and Marselos, 1989; Pappas et al., 1997). In general, ALDHs are located especially in organs with a high content of epithelial cells. An increased interest has been shown over the last years for aldehyde dehydrogenase-3 (ALDH3) activity in the cornea where much higher constitutive specific activity is detected compared to the liver cells (Boesch et al., 1996). High levels of ALDH3 activity occurs in the cornea from baboon, cow, human, opossum, pig and sheep (King and Holmes, 1997). The same study reports also the presence of ALDH1 as the 1–2% of human lens soluble protein. Furthermore, retina and retinal pigment epithelial (RPE) cells have been shown to play an important protective role for the photosensitive cells of the eye against photo- and chemical toxicity.

Tamoxifen, a synthetic antiestrogen widely used for the treatment of breast cancer, is also being used for the prevention of this type of cancer among women at high risk for the disease. Focused attention has been given to potential adverse effects of long-term tamoxifen use, including the possibility of ocular toxicity (Pavlidis et al., 1992). Toremifene is a triphenylethylene antiestrogen with significant antitumor activity. It is structurally very similar to tamoxifen. Both drugs undergo extensive hepatic metabolism.

Enzymology and Molecular Biology of Carbonyl Metabolism 7
edited by Weiner *et al.* Kluwer Academic / Plenum Publishers, New York, 1999.

In the present study, the *in vitro* activities of ALDH1 and ALDH3 were studied in human RPE cells after exposure to these two novel anti-estrogenic drugs. In addition, the effects of a classical inducer of drug metabolism on the ALDH enzyme activity of RPE cells were also examined. Finally, acute and subchronic *in vivo* experiments were carried out with both tamoxifen and toremifene in female Wistar rats. Measurements of drug metabolizing enzymes were performed in whole eye extracts and in hepatic subcellular fractions.

2. MATERIALS AND METHODS

The citrate salts of both tamoxifen and toremifene were provided by ORION Corporation Ltd., Turku, Finland; benzo[a]pyrene (BaP) and MTT (3-[4,5-dimethylthiazol-2-yl]-2,5-diphenyltetrazolium bromide) obtained by Sigma Chemical Co. (St. Louis, Mo, USA), and D407 cell line was supplied by Dr. R. C. Hunt (Univ. of S. Carolina School of Medicine, Columbia, SC 29208). All other chemicals, reagents and supplies were purchased from commercial sources. For the in vivo experiments, female Albino rats (250–300 g) of the Wistar/Mol/Io/RR substrain (Marselos 1976) from the Animal Center of the University of Ioannina.

2.1. Drug Exposure Protocol

In vitro: D407 RPE cells were harvested and seeded into 75 cm^2 dishes (5% CO_2 and 37°C); when they were 70% confluent, different concentrations of tamoxifen or toremifene (0.25–20.0 μM f.c., in DMSO) and BaP (1.0–50.0 μM f.c., in DMSO) were added to the cells (0.1% DMSO in medium f.c.). For MTT test 40,000 cells were plated in each well allowed to grow for 24 hrs. After this period they were exposed to different concentrations of drugs for 24 hrs (tam and tor) or 1 and 24 hrs (BaP).

In vivo: Both drugs were administered as suspensions in olive oil, by intraperitoneal injections, according to the following protocols: *i. acute treatment* (40 mg/kg bw, once), *ii. weekly treatment* (20 mg/kg bw, for 7 consecutive days), and *iii. monthly treatment* (20 mg/kg bw, 5 doses/week for 4 weeks). Liver and whole eyes were selected 24 hrs after ending animal treatment, and treated as described previously (Pappas et al., 1997).

2.2. Preparation of the in Vitro Samples

D407 RPE cells were scraped, washed twice with ice-cold saline, resuspended in 1.0 ml of assay buffer and after sonication (6 times of 10 sec; on ice) samples were centrifuged at 14,000 rpm for 20 min. Supernatants were collected and used the same day for biochemical measurements.

2.3. Preparation of the Tissue

To measure the ALDH1, ALDH3 and GST enzyme activities in rat liver after acute and subchronic treatment with either tamoxifen or toremifene, rats were decapitated and livers were removed, homogenized and centrifuged at 105,000 g for 60 min in order to collect the cytosolic and microsomal fraction. All preparations were performed at 4°C.

2.4. Assays for Biochemical Measurements

ALDH1 and 3 enzyme activities were measured as already described (Marselos et al., 1987; Vasiliou and Marselos 1989) with small modifications. For total glutathione-S-transferases (GSTs) enzyme activity, 1-chloro-2,4-dinitrobenzene was used as substrate (Habig et al., 1974). The protein concentration of both in vivo and in vitro samples was determined according to the BioRad method. For each of the above measurements cells from two 75 cm^2 dishes were used. The MTT test was used in order to measure cell viability and cell proliferation, and performed according to Mosmann (1983). The linearity of the MTT assay, concerning the relationship between cell number and absorption was examined (data not shown); the number of 40,000 cells/well was chosen for the assays. Student's two-tailed *t*-test was used for the statistical analysis of the results. All data are expressed as means ± S.E.

3. RESULTS AND DISCUSSION

Retinal pigment epithelial (RPE) cells function in the formation of blood-retinal barrier, control the access of blood-borne molecules to the neural retina and participate in the metabolism of retinoids (Bridges 1984). For the present study, D407 human RPE cell line was used. It is an immortal cell line, cloned from a primary culture of human retinal pigment epithelial cells. D407 cells have a doubling time of 24hrs, remarkable retention of differentiated function, but lack the ability to synthesize pigment (Hunt and Davis, 1990). D407 cells are also referred to as the only cell line which express the cellular retinaldehyde binding protein and retinol dehydrogenase activity (Davis et al. 1995; Dunn et al., 1996). Although the metabolic capacity of RPE cells has been shown to be high, little is known about the expression of major drug metabolizing enzymes (DME).

In the first group of experiments, in addition to studies for basic enzyme activity, the effects of different concentrations of BaP, a classical DME-inducer, on ALDH3 and ALDH1 enzyme activities were examined. After 24 hrs of exposure, RPE cells produced a statistically significant reduction of both ALDH1 and ALDH3 activities (Figure 1a). The decreased enzyme activities, as a function of time, were verified ($P<0.02$) after exposure to BaP (Figure 1b). However, after 48 and 72 hrs of exposure RPE had apparently recovered and the activities were normalized. The possible inhibition of cellular proliferation and cell viability by different BaP concentrations were examined by using the MTT test (Figure 2); our data showed that the RPE exposure to BaP for 1 and 24 hrs, does not affect the viability and proliferation rate of the cells.

Tamoxifen and toremifene are known non steroidal anti-estrogenic drugs. Although this is not their only action, both drugs bind to estrogen receptors (ER). TCDD and polycyclic aromatic hydrocarbons, like BaP, are known to decrease levels of ER both in intact animals and in cell culture lines (De Vito et al., 1992). It has been shown by our group (Karageorgou et al., 1995) that tamoxifen inhibits ALDH3 induction by 3-methylcholanthrene (3-MC), suggesting a possible role for tamoxifen in the ALDH3 induction process, which is known to be Aryl Hydrocarbon Receptor-dependent. The synergistic effects of different members of nuclear steroid/thyroid hormone receptor superfamily in gene regulation are also known (Kurokawa et al., 1994; Leng et al., 1994). The two next series of experiments show some data about the effects of tamoxifen and toremifene on the basal enzyme activity of ALDH1, ALDH3 and GSTs, in two different systems. RPE cells were exposed to four different concentrations of the two drugs (from 0.25 to 20.0 μM). After 24 hrs both ALDH1 and ALDH3 enzyme activities, and also GSTs, were not affected by tamoxifen, even at the highest dose (Figure 3a). The same results were found with toremifene, a compound which differs from tamoxifen in the substitution of a chlorine atom for a hydrogen atom in the ethyl group (Figure 3b); the concentration of 50.0 μM was

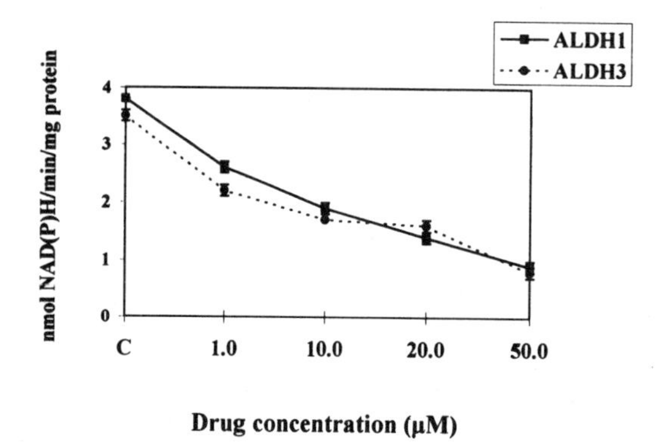

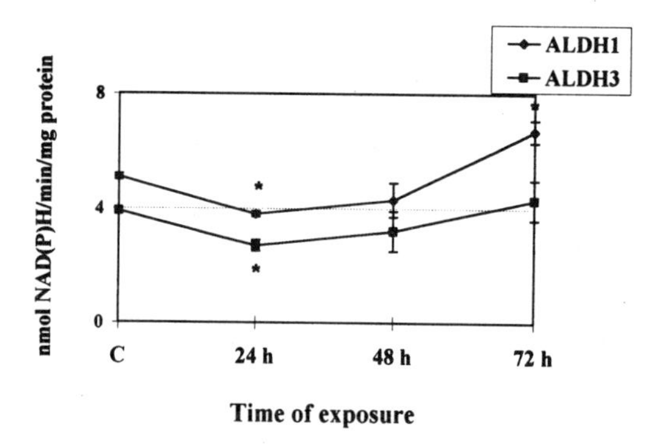

Figure 1. Effects of BaP on ALDH1 and ALDH3 enzyme activities of RPE cells: (**a**) *dose response*; exposure time: 24 hrs. (**b**) *time response*; final concentration of BaP: 1.0 μM.

found to be toxic for both drugs. The presence of 0.1% DMSO which was used as drug solvent did not affect the enzyme activities compared to untreated cells.

The results were different when cells were exposed to tamoxifen for a longer time period; enzyme activities were found to be decreased at 20.0 μM (Figure 4). The cytotoxic effect of drug treatment which is expressed by the inhibition of cellular proliferation, is shown in Figure 5. Tamoxifen and toremifene had the same range of inhibition concentration (IC50) for cell viability and cell proliferation which is actually present at an earlier stage than the inhibition concentration for enzyme activities (Figures 3a, 3b). Only if the

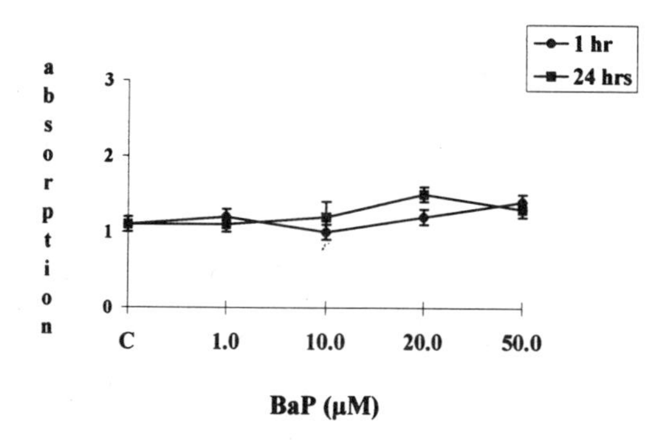

Figure 2. MTT cleavage by RPE cells after exposure to BaP for 1 h and 24 hrs.

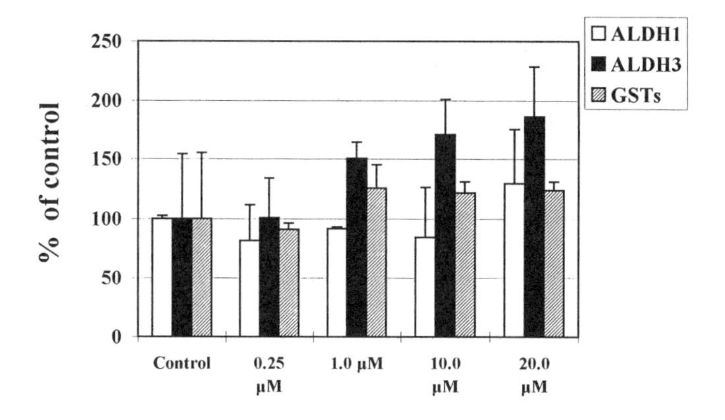

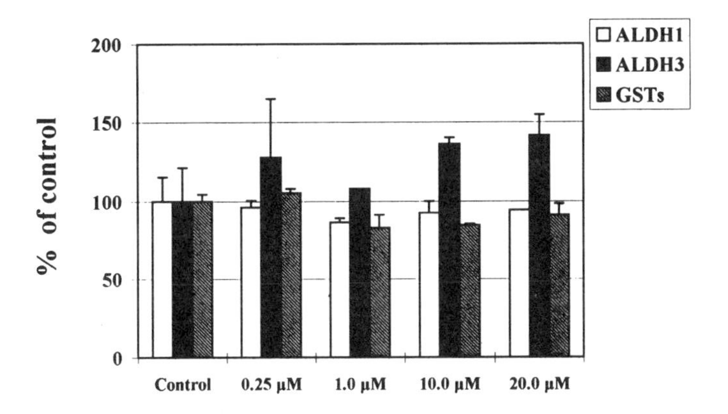

Figure 3. ALDH1, ALDH3 and GSTs enzyme activities after exposure of RPE cells to: (**a**) *tamoxifen*, and (**b**) *toremifene*; the exposure time was 24 hrs, for both drugs.

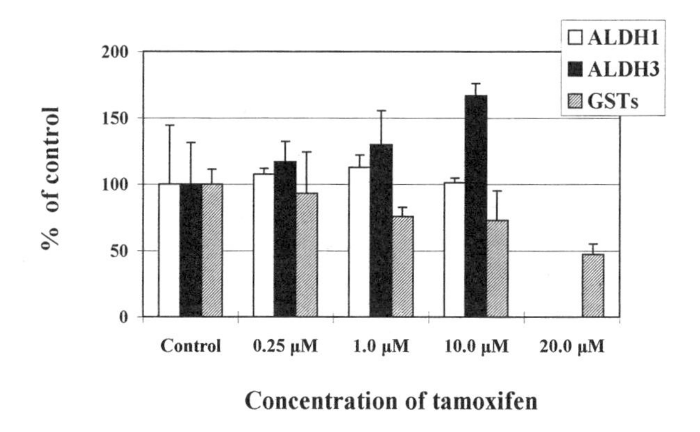

Figure 4. Effects of tamoxifen on ALDH1, ALDH3 and total GSTs enzyme activities of RPE cells. The exposure time was 4 days.

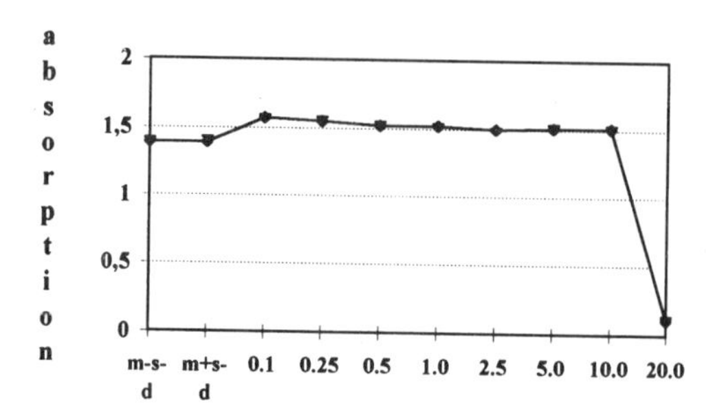

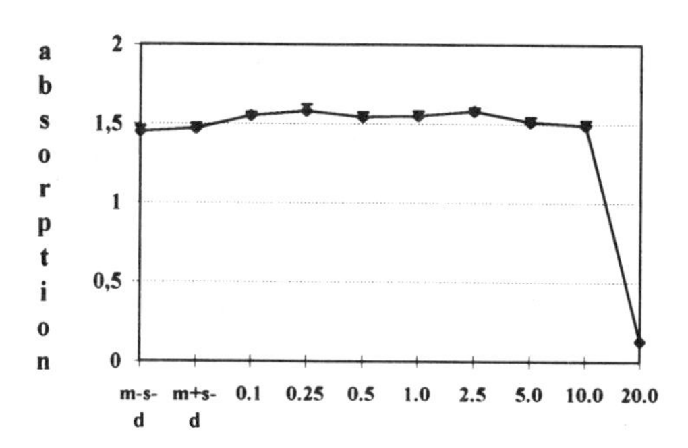

Figure 5. MTT cleavage by RPE cells after exposure either to (a) tamoxifen, or (b) toremifene; the exposure time was 24 hrs, for both drugs.

exposure time for tamoxifen is increased to 4 days, the IC50 for enzyme activities will be the same as for cellular proliferation. It appears the decreased enzyme activities at high drug doses are due to cytotoxicity.

Finally, the acute and subchronic effects of tamoxifen and toremifene were studied in rat liver and eye, in vivo. It has been already shown that under normal conditions or following phenobarbital treatment the ALDH3/ALDH1 ratio is lower than 1.0 and after BaP or 3-MC treatment the ratio becomes greater than 1.0 (Marselos and Vasiliou, 1991). This criterion has been used in order to classify new compounds as phenobarbital- or aromatic hydrocarbon-type of inducers. In the present study, although an acute dose of both drugs

Table 1. Effects of acute and subchronic treatment of tamoxifen and toremifene on ALDH1 and ALDH3c enzyme activities in rat liver

	ALDH1 (nmoles of NADH/min/mg protein)			ALDH3 (nmoles of NADPH/min/mg protein)		
	Acute (40mg/kg)	Weekly (20mg/kg)	Monthly (20mg/kg)	Acute (40mg/kg)	Weekly (20mg/kg)	Monthly (20mg/kg)
Control	20.0 ± 1.1	19.9 ± 1.7	20.0 ± 0.6	3.0 ± 0.7	3.3 ± 0.3	3.2 ± 0.1
Tamoxifen	21.3 ± 1.1	25.7 ± 1.6***	28.6 ± 1.1*	3.4 ± 0.2	3.2 ± 0.3	4.5 ± 0.2**
Toremifene	21.3 ± 1.5	26.4 ± 1.0***	27.9 ± 0.8*	4.1 ± 0.5	3.3 ± 0.4	3.6 ± 0.2

*Means statistically different from control groups, at: *P<0.001, **P<0.0025, ***P<0.025.

did not affect the hepatic enzyme activities, weekly treatment produced a statistically significant increase of ALDH1 (Table 1). The enzyme activity was induced at the same levels when a monthly protocol of administration was applied. The data in Table 1 shows that both tamoxifen and toremifene are phenobarbital-type of inducers. ALDH3 was increased by tamoxifen only after monthly treatment, indicating that a longer treatment period is more effective. Total GST activity, measured by the rate of chlorodinitrobenzene conjugation, was not affected (data not shown).

In conclusion, the D407 RPE cell line expresses both ALDH1 and both ALDH3, as far as enzyme activity is concerned. The inhibition of expression of both ALDH activities in RPE cells occurs at concentrations where cell viability and cell proliferation are unaltered. We also report here for the first time that tamoxifen and also toremifene are phenobarbital-type inducers for hepatic aldehyde dehydrogenases, since they meet the criterion for the ratio of ALDH3/ALDH1 to be lower than 1.0 (Marselos and Vasiliou, 1991).

ACKNOWLEDGMENTS

The authors wish to thank both ORION Corporation Ltd., Turku, Finland, and personally Eero Mantyla, for providing tamoxifen and toremifene. This work was supported by a EU grant (BIOMED II BMH4-CT79-2324).

REFERENCES

Boesch, J.S., Lee, C. and Lindahl, R.G.: Constitutive expression of class 3 aldehyde dehydrogenase in cultured rat corneal epithelium. J. Biol. Chem. 271 (1996) 5150–5157.

Bridges, C.D.B.: Retinoids in photosensitive systems. In: The retinoids, Vol 2. Sporn, M.B., Roberts, A.B. and Goodman, D.S., (Eds.), Academic Press, Orlando, FL, (1984) 126–176.

Davis, A., Bernstein, P., Bok, D., Turner, J., Nachtigal, M. and Hunt, R.: A human retinal pigment epithelial cell line that retains epithelial characteristics after prolonged culture. Invest. Ophthalmol. Vis. Sci. 36 (1995) 955–964.

De Vito, M.J., Thomas, T., Martin, E., Umbreit, T. and Gallo, M.: Anti-estrogenic action of 2,3,7,8-tetrachlorodibenzo-p-dioxin: tissue-specific regulation of estrogen receptor in CD1 mice. Toxicol Appl. Pharmacol. 113 (1992) 284–292.

Dunn, K.C., Aotaki-Keen, A.E., Putkey, F.R. and Hjelmeland, L.M.: ARPE-19, a human retinal pigment epithelial cell line with differentiated properties. Exp. Eye Res. 62 (1996) 155–169.

Habig, W.H., Pabst, M.J. and Jakoby, W.B.: Glutathione S-transferases. The first enzymatic step in mercapturic acid formation. J. Biol. Chem. 249 (1974) 7130–7139.

Hunt, R.C. and Davis, A.A.: Altered expression of keratin and vimentin in human retinal pigment epithelial cells in vivo and in vitro. J. Cellular Physiol. 145 (1990) 187–199.

Karageorgou, M., Vasiliou, V., Nebert, D.W. and Marselos, M.: Ligands of four receptors in the nuclear steroid/thyroid hormone superfamily inhibit induction of rat cytosolic aldehyde dehydrogenase-3 (ALDH3c) by 3-methylcholanthrene. Biochem. Pharmacol. 50 (1995) 2113–2117.

King, G. and Holmes, R.: Human corneal and lens aldehyde dehydrogenases. Adv. Exp. Med. Biol. 414 (1997) 19–27.

Kurokawa, R., DiRenzo, J., Boehm, M., Sugarman, J., Gloss, B., Rosenfeld, M.G., Heyman, R.A. and Glass, C.K.: Regulation of retinoid signalling by receptor polarity and allosteric control of ligand binding. Nature 371 (1994) 528–531.

Leng, X., Blanco, J., Tsai, S.Y., Ozato, K., O'Malley B.W. and Tsai, M.J.: Mechanisms for synergistic activation of thyroid hormone receptor and retinoid X receptor on different response elements. J. Biol. Chem. 269 (1994) 31436–31442.

Lindahl, R.: Aldehyde dehydrogenases and their role in carcinogenesis. CRC Crit. Rev. Biochem. Mol. Biol. 27 (1992) 283–335.

Marselos, M.: Genetic variation of drug metabolizing enzymes in the Wistar rat. Acta Pharmacol. Toxicol. 39 (1976) 186–197.

Marselos, M., Strom, S.C. and Michalopoulos G.: Effect of phenobarbital and 3-methylcholanthrene on the aldehyde dehydrogenase activity in cultures of HepG2 cells and normal human hepatocytes. Chem-Biol. Inter. 62 (1987) 75–88.

Marselos, M. and Vasiliou, V.: Effect of various chemicals on the aldehyde dehydrogenase activity of the rat liver cytosol. Chem-Biol. Interact. 79 (1991) 79–89.

Mosmann T.: Rapid colorimetric assay for cellular growth and survival: Application to proliferation and cytotoxicity assays. J. Immun. Methods 65 (1983) 55–63.

Pappas, P., Stephanou, P., Karamanakos, P., Vasiliou, V. and Marselos, M.: Ontogenesis and expression of ALDH activity in the skin and the eye of the rat. Adv. Exp. Med. Biol. 414 (1997) 73–80.

Pavlidis, N.A., Petris, C., Briassoulis, E., Klouvas, G., Psilas, C., Rempapis, J. and Petroutsos, G.: Clear evidence that long-term, low-dose tamoxifen treatment can induce ocular toxicity. A prospective study of 63 patients. Cancer 69 (1992) 2961–2964.

Vasiliou, V., and Marselos, M.: Tissue distribution of inducible aldehyde dehydrogenase activity in the rat after treatment with phenobarbital or methylcholanthrene. Pharmacol. Toxicol. 64 (1989) 39–42.

19

NEGATIVE REGULATION OF RAT HEPATIC *ALDEHYDE DEHYDROGENASE 3* BY GLUCOCORTICOIDS

Ronald Lindahl,[1] Gong-Hua Xiao,[2] K. Cameron Falkner,[2] and Russell A. Prough[2]

[1]Departments of Biochemistry and Molecular Biology
University of South Dakota School of Medicine
Vermillion, South Dakota 57069
[2]Departments of Biochemistry and Molecular Biology
University of Louisville School of Medicine
Louisville, Kentucky 40292

1. INTRODUCTION

The aldehyde dehydroghenases are a family of proteins whose function is to catalyze the oxidation of a wide variety of aliphatic and aromatic aldehydes to their corresponding carboxylic acids and are classified according to their sequence similarity (Lindahl, 1994). One family member, aldehyde dehydrogenase 3, is induced by polycyclic aromatic hydrocarbons or chlorinated aromatic compounds, such as TCDD (Dunn et al., 1988). These compounds are ligands for the *Ah* receptor, which functions as a transcription factor. In concert with a dimerization partner, known as ARNT (*Ah* receptor nuclear translocator), *Ah* receptor binds to a canonical consensus sequence (TNGCGTC), denoted *Ah*RE. Binding facilitates gene transactivation (Whitlock et al., 1996). Two AhRE core sequences have been identified in the 5' flanking region of the rat ALDH3 gene (Takimoto et al., 1994; Xie et al., 1996).

Previous studies in our laboratories (Prough et al., 1996, 1997; Xiao et al., 1997) have shown that *ALDH3* is also negatively regulated by both glucocorticoids and cAMP in primary rat hepatocyte cultures. The hypothesis of the present study is that down-regulation of *ALDH3* expression by glucocorticoids (GC) occurs at the transcriptional level and involves the binding of glucocorticoid receptor (GR) to its canonical consensus element in the 5' flanking region of the gene. Here we describe the effects of glucocorticoids on *ALDH3* expression in primary rat hepatocytes and in *ALDH3*-promoter transfected HepG2 cells.

Enzymology and Molecular Biology of Carbonyl Metabolism 7
edited by Weiner *et al.* Kluwer Academic / Plenum Publishers, New York, 1999.

2. RESULTS AND DISCUSSION

2.1. Glucocorticoid-Mediated Negative Regulation of *ALDH3* in Rat Hepatocytes

To examine whether the negative regulation of *ALDH3* expression was mediated by the GC receptor, we initially tested the concentration-dependence of the repression phenomenon in primary hepatocytes. As anticipated, ALDH3 enzyme activity was induced 4.6-fold by the inclusion of 50 uM 1,2 benzanthracene (BA) in the culture media. Addition of the synthetic glucocorticoid dexamethasone (DEX) caused a decline in BA-inducible ALDH3 enzyme activity in the concentration range between 10^{-6} and 10^{-12} M with significant decreases (>50%) in activity being observed at concentrations of 10^{-8} or greater. No consistent effect of DEX on the very low of basal activity of ALDH in hepatocytes was observed. Consistent with our earlier work (Prough et al., 1996; Xiao et al., 1997), DEX increased the functional activity (2 to 3-fold) and transcript levels of *CYP1A1*, another gene regulated by both the *Ah* receptor and glucocorticoids (data not shown).

Treatment of rat hepatocytes with BA caused a 14-fold induction of *ALDH3* mRNA levels (Figure 1). Administration of DEX suppressed the BA-dependent induction of this gene by 75%. Administration of the glucocorticoid antagonist RU 38 486 significantly blunted the DEX-dependent repression of BA-induced *ALDH3* expression. Taken together, the dose dependence and antagonist studies are consistent with the involvement of glucocorticoid receptor and support the conclusion that the effects of GC on *ALDH3* expression occurs at the pretranslational level.

To test the hypothesis that GR-dependent negative regulation of *ALDH3* might occur through a second transcription factor, we tested the cycloheximide sensitivity of the GC

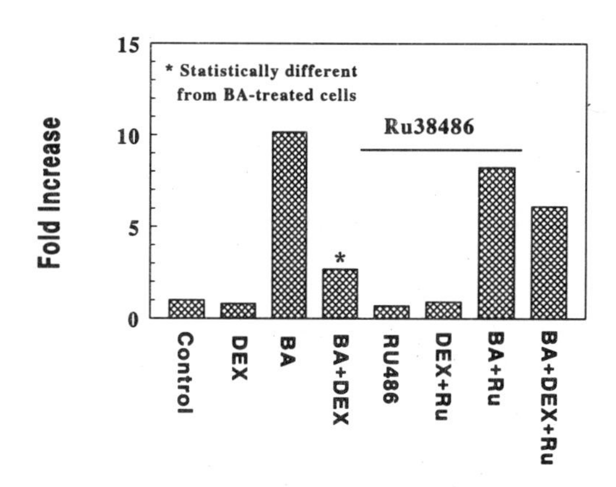

Figure 1. Effect of GC and the GC antagonist RU 38 486 on *ALDH3* transcript levels. Primary rat hepatocytes were cultured as described (Xiao et al., 1997). Twenty-four hours after plating the media was changed and cells were either treated with DMSO, 50 uM 1,2-benzanthracene in DMSO, 1 uM dexamethasone in DMSO, 10 uM RU 38 486 or combinations of the three agents for 72 hours. Cells were harvested, mRNA prepared and separated as described (Xiao et al., 1997). The mRNA was probed with labeled fragments from the cDNA for ALDH3 or GAPDH. The complexes were recorded using x-ray film and quantitated using a Bio-Rad 620 video densitometer. *ALDH3* mRNA levels were normalized to the level of *GAPDH* message. The data shown is a representative experiment. When the results for three experiments were pooled, induction of *ALDH3* mRNA by BA was significantly different from controls ($p<0.01$). For three experiments, RU 38 486 had little or no effect on BA-induction of *ALDH3* mRNA ($p>0.05$), but had a significant effect on DEX repression of BA-induction ($p<0.05$).

repression of BA-induced *ALDH3* expression. Administration of cycloheximide to rat hepatocyte cultures (5 ug/ml for 24 hr) completely blunted the GC effect and actually resulted in a glucocorticoid-dependent superinduction of *ALDH3* transcripts compared to BA plus DEX-treated hepatocytes (changed from a 75% suppression of BA-induced levels in the absence of CHX to a 290% increase in it's presence). This concentration of cycloheximide had a small effect on the GC-dependent potentiation of BA-dependent induction of *CYP1A1*. The cycloheximide-mediated glucocorticoid-dependent superinduction of *ALDH3* strongly suggests that the GC-dependent negative regulation of this gene is in part caused by interaction of GRs with a second, labile transcription factor.

2.2. Glucocorticoid-Mediated Negative Regulation of Reporter Gene Activity in ALDH3-Promoter Transfected HepG2 Cells

To identify regions of the *ALDH3* 5' flanking region that are responsible for GC-mediated negative regulation, a series of CAT reporter gene assays were performed. As anticipated, BA-induced expression of pALDH3.5CAT (containing the proximal 3.5 Kb of the rat ALDH3 5' flanking region) increased approximately 20-fold when this reporter construct was transiently transfected into HepG2 cells (Figure 2). Dexamethasone suppressed the BA-dependent induction of pALDH3.5CAT approximately 50–60%. This is in good agreement with the data from both *ALDH3* mRNA and enzyme activities in primary hepatocyte cultures, supporting our hypothesis that GC suppress the BA-dependent induction of *ALDH3* at the transcriptional level.

RU 38 486 had no effect on either the basal or BA-induced levels of pALDH3.5CAT expression. However, CAT activity of cells treated with BA and RU 38 486 or BA and

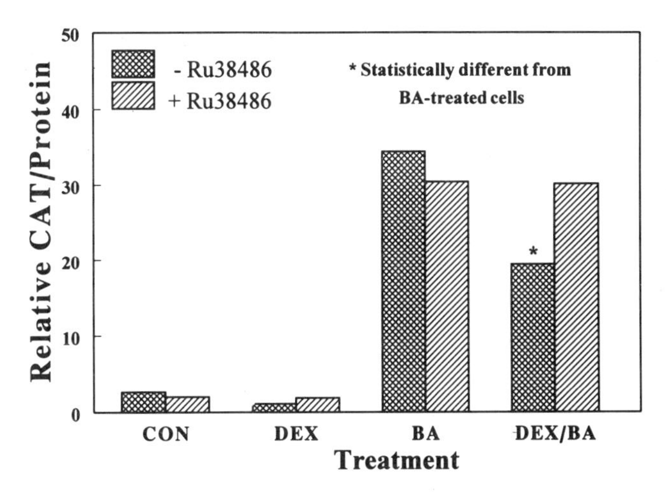

Figure 2. Effects of GC and the GC antagonist RU 38 486 on CAT activity of HepG2 cells transfected with pALDH3.5CAT. CAT and β-galactosidase assays were performed on lysates of HepG2 cells that had been transiently transfected with pALDH3.5CAT and expression plasmids for β-galactosidase (pCMVβ) and human glucocorticoid receptor (pRSVGR) and then treated with either 50 uM 1,2-benzanthracene, 1 uM dexamethasone, 10 uM RU 38 486 or combinations of these compounds for 24 hours. The normalized CAT activity is the % conversion of chloramphenicol to its acetyl derivative relative to β-galactosidase activity and is the mean of three flasks +/- S.D. *, significantly different from CAT activity of control cells ($p<0.001$); **, significantly different from CAT activity of BA-treated cells ($p<0.005$).

DEX and RU 38 486 were identical, confirming RU 38 486 is an effective antagonist of the DEX-dependent suppression of *ALDH3*, most likely acting through GR.

By sequence analysis of the 5' flanking region of rat *ALDH3*, we identified an imperfect palindromic glucocorticoid response element (pGRE) at approximately -920 bp relative to the transcription start site. Deletion of the PstI fragment containing the pGRE (-1057 to -380 bp) from pALDH3.5CAT had dramatic effects on both the BA-dependent induction of CAT activity and the DEX-dependent suppression of induction (Figure 3). Most striking was the markedly reduced BA-dependent induction of CAT activity in pGHX1. BA caused only a 4-fold increase in CAT activity relative to the 20-fold induction observed with the parent construct. Basal activity was nearly identical between the two constructs. A sequence with identity to the canonical *Ah*RE is also located within the deleted PstI fragment (at -690 to -685 bp). These results are consistent with this *Ah*RE being actively involved in determining the magnitude of the xenobiotic induction of *ALDH3* in cooperation with the *Ah*RE located at approximately -3,500 bp. Co-treatment of cells with DEX had no effect on the BA-dependent CAT induction of pGHX1, indicating that sequences critical for the GC effect are located between -1057 and -390. This is consistent with the pGRE at -920 bp being involved in the GC regulation of *ALDH3*.

To directly test whether glucocorticoids act through the pGRE at -920 bp, we used a PCR-based strategy to generate products from pALDH3.5CAT that were subcloned in pCRII plasmids to replace portions of the important PstI fragment. These products contained either a mutated (AGGACA to ACGGCA) or wild type sequence of the pGRE. The PCR products were synthesized to span the region between -930 to -374 in the 5' flanking region of *ALDH3*. The rationale was to produce plasmids that had full BA-dependent inducibility by retaining both the *AhRE* sequences at -690 and -3,500 bp. Thus, when the PCR products are reintroduced in pGHX1, the resulting plasmids lack sequences (bp -1057 to -930) immediately 5' to the pGRE. Both plasmids. pGHX2 (wild type GRE) and pGHX3 (mutant GRE) were induced approximately 20 fold by BA, nearly identical to that observed with pALDH3.5CAT. When both BA and DEX were coadministered, only pGHX2 exhibited reduced CAT activity (50–60% reduced). Dexamethasone failed to sig-

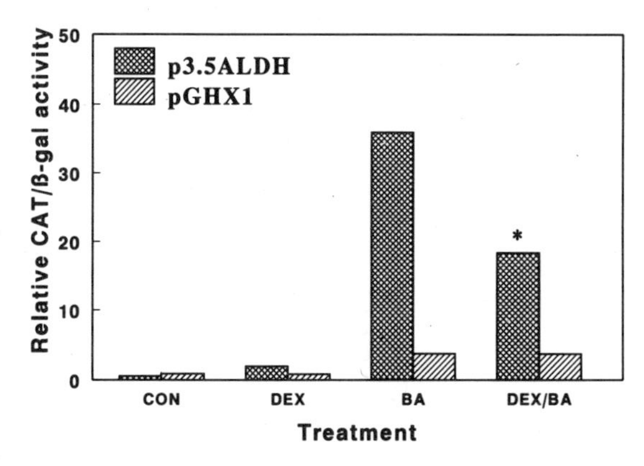

Figure 3. Effects of *ALDH3* 5' flanking region deletion on CAT activity of transfected HepG2 cells. CAT and β-galactosidase assays were performed on lysates of HepG2 cells that had been transiently transfected with either pALDH3.5CAT or pGHX1 (which carries the PstI fragment-deleted ALDH3.5 5' flanking region). Cells were also transfected with both pCMVβ and pRSVGR and treated and as described in the legend to Figure 2. *Significantly different from CAT activity of control cells (p<0.002); **Significantly different from CAT activity of BA-treated cells (p<0.003).

nificantly repress CAT expression in pGHX3, which contains the mutated GRE. Thus, the palindromic GRE located at -920 bp is functionally involved in the negative regulation of the aldehyde dehydrogenase 3 gene.

Since pGHX2 and pGHX3 lack the element responsible for the cAMP-mediated negative regulation of *ALDH3* (-1057 to -930) the negative effects of GC and cAMP on *ALDH3* expression are clearly due to separate cis-acting elements. However, both pGHX2 and pGHX3 contain the two NF1 sites implicated by our earlier work as involved in the down regulation of basal *ALDH3* expression (Xie et al., 1997). These NF1 sites are immediately 3' to the GRE.

To test whether the pGRE sequence was capable of binding the GC receptor, we performed electrophoretic mobility shift assays. A specific DNA-protein complex was formed between pGRE oligonucleotides and rat liver nuclear extracts. Formation of this complex could be prevented by inclusion of excess oligonucleotides identical to either the *ALDH3* pGRE or the pGRE derived from the Mouse Mammary Tumor Virus Long Terminal Repeat. Complex formation was less effectively blocked by an oligonucleotide containing the mutated pGRE and not effectively at all by an unrelated oligonucleotide, namely the *Ah*RE from *CYP1A1*. Complex formation with an *ALDH3* pGRE-containing oligonucleotide was much greater in nuclear extracts from HepG2-GR4 cells than parental HepG2 cells, a cell line which does not express GR. HepG2-GR4 cells are stably transfected with an expression vector for the human glucocorticoid receptor. This result is consistent with GR binding to the *ALDH3* pGRE. Finally, inclusion of antibodies specific for human GR blocked formation of the specific DNA-protein complex between HepG2-GR4 extracts and an ALDH3 pGRE oligonucleotide.

3. SUMMARY

The expression of the aldehyde dehydrogenase 3 gene is known to be controlled by multiple regulatory processes. In liver, inducible expression appears to be mediated by two *Ah*RE sequences which allow regulation of this gene by xenobiotic compounds which are ligands for the *Ah* receptor (Takimoto et al., 1994; this work). Constitutive expression of *ALDH3* in tissues such as the cornea also involves the -3,500 region which contains an *Ah*RE (Boesch et al., 1996; Boesch et al, 1998). However, the constellation of transcription factors which appear to interact with the *Ah*RE in constitutively expressing corneal cells does not include either the *Ah* receptor nor the prototypical ARNT protein (Boesch et al., 1998). For both inducible and constitutive *ALDH3* expression the more distal 5' flanking region sequences appear to interact with more proximal regulatory elements. Of particular interest is the region near -1 kb which includes the GC (-930 to -910) and cAMP (-1057 to -991) responsive elements as well as the 2 NF1 sites (-916 to -815), all of which appear to act as negative modulators of ALDH3 expression. A second putative *ALDH3* negative regulatory region lies even more distal than -3,500 bp. To date, this region has been little studied, but appears to be involved in regulating both inducible and constitutive *ALDH3* expression. This region may also be responsible for some of the tissue-specificity of *ALDH3* expression.

With respect to the work described here, in both isolated hepatocytes and HepG2 cells, no consistent negative regulation by glucocorticoids was observed in the basal expression of *ALDH3*. This indicates that the mechanism of GC-mediated negative regulation involves direct interference with ALDH3 gene activation mediated by the *Ah* receptor. Our results suggest a complex interplay between multiple transcription factors, including the GC and *Ah* receptors, regulates the hepatic expression of the ALDH3 gene. Active recruitment of transcription factors needed for gene transactivation, amelioration of the actions of negative regulatory trans-acting factors or cis-acting elements and/or chromatin remodeling may be required for achieve proper regulation of the aldehyde dehydrogenase 3 gene.

ACKNOWLEDGMENTS

This research was supported in part by USPHS Grants CA21103 (RL) and ES04244 (RAP).

REFERENCES

Boesch, J.S., Lee, C., and Lindahl, R., (1996) Constitutive expression of class 3 aldehyde dehydrogenase in cultured rat corneal epithelium. *J. Biol. Chem.* **271**, 5150–5157.

Boesch, J.S., Miskimins, R., Miskimins, W.K., and Lindahl, R., (1998) Class 3 aldehyde dehydrogenase gene: Basis of constitutive expression in corneal epithelium. *Molec. Cell. Biol.*, Submitted.

Dunn, T.J., Lindahl, R., and Pitot, H.C., (1988) Differential gene expression in response to 2,3,7,8-tetrachlorodibenzo-p-dioxin (TCDD): noncoordinate regulation of a TCDD-induced aldehyde dehydrogenase and cytochrome P450c in the rat. *J. Biol. Chem.* **263**, 10878–10886.

Lindahl, R., (1994) Mammalian aldehyde dehydrogenases: regulation of gene expression. *Alcohol & Alcoholism* Supp **2**, 147–154.

Prough, R.A., Xiao, G.-H., Pinaire, J.A., and Falkner, K. C., (1996) Hormonal regulation of xenobiotic drug metabolizing enzymes. *FASEB J.* **10**, 1369–2377.

Prough, R.A., Falkner, K.C., Xiao, G.-H., and Lindahl, R., (1997) Regulation of rat ALDH-3 by hepatic protein kinases and glucocorticoids. In *Enzymology and Molecular Biology of Carbonyl Metabolism 6*, Adv. Exptl. Biol. Med. **414**, 29–36.

Takimoto, K., Lindahl, R., Dunn, T.J. and Pitot, H.C., (1994) Structure of the 5' flanking region of the class 3 aldehyde dehydrogenase gene in the rat. *Arch. Biochem. Biophys.* **312**, 539–546.

Whitlock, J.P.Jr., Okino, S.T., Dong, L., Ko, H.P., Clarke-Katzenberg, R., Ma, Q., and Li, H., (1996) Induction of cytochrome P4501A1: a model for analyzing mammalian gene transcription. *FASEB J.* **10**, 809–818.

Xiao, G.-H., Falkner, K.C., Xie, Y., Lindahl, R., and Prough, R.A., (1997) cAMP-deponent negative regulation of rat aldehyde dehydrogenase 3 gene expression. *J. Biol. Chem.* **272**, 3238–3245.

Xie, Y.Q., Takimoto, K., Pitot, H.C., Miskimins, W.K., and Lindahl, R., (1996) Characterization of the rat class 3 aldehyde dehydrogenase gene promoter. *Nucl. Acids Res.* **24**, 4185–4195.

20

MODULATION OF CLASS 3 ALDEHYDE DEHYDROGENASE GENE EXPRESSION

An Eye Opening Experience

Maureen Burton,[1] Richard Reisdorph,[1] Russell Prough,[2] and Ronald Lindahl[1]

[1]Departments of Biochemistry and Molecular Biology
University of South Dakota School of Medicine
Vermillion, South Dakota 57069
[2]Departments of Biochemistry and Molecular Biology
University of Louisville School of Medicine
Louisville, Kentucky 40292

1. INTRODUCTION

Class 3 aldehyde dehydrogenase (ALDH3) is an enzyme that efficiently converts a variety of endogenous as well as exogenous aldehydes to their corresponding carboxylic acids. For some time now the tissue distribution, protein and gene sequence of ALDH3 have been known. A low Km for medium length aliphatic as well as aromatic aldehydes is the mark of ALDH3. The ALDH3 monomer is 453 amino acids long, and unlike other mammalian ALDHs, the native protein is a homodimer. Recently, the ALDH3 crystal structure has been determined and, arguably, nearly understood. *ALDH3* is expressed differentially in mammalian tissues, being constitutively expressed in cornea, stomach, lung and urinary tract, and only inducible by xenobiotics in liver; it is also found in certain cancer cells.

The intricacies of obtaining and maintaining a particular level of gene expression remain obscure for most genes and *ALDH3* is no exception. Analysis of the ALDH3 gene identifies it as one of the "11 exon" aldehyde dehydrogenases. *ALDH3* is approximately 10 kb in length with an additional 10 kb of 5' flanking sequence that has been isolated (Asman, *et al.* 1993). The proximal promoter region of the rat *ALDH3* gene has been characterized and found to contain CAAT and TATA boxes; two Sp1 sites appear to promote basal levels of transcription and 2 NF1 sites repress it (Xie, *et al.* 1996). Beyond the promoter region there are two strong inhibitory regions separated by a strong enhancer. A xenobiotic response element (XRE), located about –3 kb from the transcriptional start site and in the en-

Enzymology and Molecular Biology of Carbonyl Metabolism 7
edited by Weiner *et al.* Kluwer Academic / Plenum Publishers, New York, 1999.

hancer segment, is important for expression in all *ALDH3*-expressing cells (Takimoto, *et al.* 1994; Boesch, *et al.*). In cornea, a complex of transcription factors appear to be involved at the XRE, whereas the Aryl hydrocarbon receptor (AhR) and Aryl hydrocarbon receptor nuclear translocator (ARNT) are the transcription factors bound during *ALDH3* induction in the liver. Interestingly, it appears that different tissues activate this cis-acting xenobiotic response element using different trans-acting factors (Boesch, *et al.*). Numerous other putative response elements also exist in the 5' flanking region of *ALDH3*. An important question still unanswered is whether they are all capable of modulating *ALDH3* gene expression and how these multiple response elements interact. In this chapter we will discuss some of the response elements we have found that may contribute to the differential expression pattern of the mammalian ALDH3 gene observed among tissues.

2. REGULATION OF *ALDH3* EXPRESSION

2.1. Glucocorticoids

We have examined the effects of dexamethasone (Dex), a glucocorticoid receptor agonist, on ALDH3 enzyme activity and *ALDH3* gene expression in primary cultures of rat corneal epithelial cells. Enzyme assays show that Dex dose-dependently increases ALDH3 activity with 1 μM Dex resulting in a 2-fold increase in ALDH3 activity. Western blot analyses corroborates the dose-dependent activity increase, indicating that elevated ALDH3 protein levels are responsible for the increased activity. The increase in both these parameters is inhibited by the receptor antagonist RU38,486. Together, this suggests that regulation of *ALDH3* in corneal epithelium occurs at the level of transcription or translation. Northern blot analysis demonstrates that *ALDH3* mRNA levels are also upregulated by Dex. Thus, it is clear that the Dex-induced upregulation of *ALDH3* occurs at the level of transcription. This is perhaps not a surprising result as the 5' flanking region of the gene contains at least one glucocorticoid response element (GRE) palindrome and several GRE half-sites. Reporter gene analyses using increasing lengths of the *ALDH3* 5' flanking region show that the GRE at –1 kb is active in corneal epithelial cells. The reporter gene construct containing 1.2 kb upstream of the transcription start site is capable of doubling CAT activity in the presence of 1 μM Dex. Several GRE half-sites also exist throughout the 5' flanking region, but the significance/importance of these sites is still unclear. However, they may serve as negative regulators of the palindromic responsiveness since augmentation of reporter gene expression seems to be inhibited with increasing lengths of the *ALDH3* 5' flanking region. Thus the emerging picture indicates that glucocorticoids, acting through the glucocorticoid receptor and GRE, upregulate ALDH3 gene expression which leads to increased ALDH3 enzyme activity in corneal epithelial cells.

Quite contrary to what happens in the cornea, xenobiotic-induced, hepatic *ALDH3* expression is downregulated by Dex at similar concentrations. This also occurs at the level of transcription as demonstrated by activity, Western, Northern, and reporter gene analyses (Falkner, *et al.* 1998). It was established in these cells that the –1 kb 5' flanking region is responsible for the downregulation. The major difference between corneal and liver expression of *ALDH3* is that corneal expression is constitutive while liver expression occurs only when induced. So ultimately it would appear that the differences between expression and regulation in these two cell types lies not only in the transcription factors present but also in the interaction of the various activated elements, whether they are inhibitory or stimulatory.

2.2. Protein Kinases A and C

It is well established that phosphorylation of proteins can alter their functionality. Also well established is that post-translational modification of the ALDH3 polypeptide is not a mechanism of ALDH3 regulation (Lindahl, 1992). However, it has recently been shown that phosphorylation events can alter levels of *ALDH3* enzyme activity and that this effect is likely to occur via phosphorylation of transcription factors critical to *ALDH3* expression (Xiao, *et al.* 1997; Burton, *et al.* 1997). Several aspects of transcription factor function can be altered by phosphorylation: nuclear translocation, DNA binding affinity and transactivation. Since a constellation of transcription factors is involved in ALDH3 expression and regulation, the effects of protein phosphorylation are also likely to be complex.

Utilizing protein kinase A (PKA) inhibitors and activators, we have shown that PKA is a negative regulator of inducible ALDH3 expression in rat hepatocytes. This was true at the level of ALDH3 enzyme activity, protein content, mRNA, and CAT reporter gene activity. PKA appears to be affecting a transcription factor(s) which bind(s) right around –1 kb. Similarly, preliminary data suggest that constitutive *ALDH3* expression in cornea is attenuated by PKA (Burton, *et al.* 1997). There are several transcription factors necessary for *ALDH3* expression that may be affected by PKA. First is Sp1, whose DNA binding affinity is increased by PKA-mediated phosphorylation (Rohlff, *et al.* 1997). Second is CREB, for which there is a putative binding site near –1 kb. Third is HNF4, at least in corneal cells, which exhibits decreased binding affinity when phosphorylated by PKA (Viollet, *et al.* 1997). Since HNF4 is part of the transcription factor complex at the XRE in corneal epithelial cells and not liver, inhibition of PKA would not produce as prominent a change in corneal cells as that seen with induced expression. Clearly, the PKA-sensitive transcription factors involved in modulation of *ALDH3* expression differ in each cell type and thus phosphorylation by PKA affects expression in a different manner in each tissue.

Protein kinase C (PKC) is also involved in inducible and constitutive *ALDH3* expression. We have also demonstrated that PKC augments both constitutive and xenobiotic-induced *ALDH3* expression. One of the transcription factors believed to be common to both mechanisms of *ALDH3* expression is ARNT. PKC phosphorylation is considered crucial to transactivation in the Ah/ARNT pathway by activating ARNT. It is also possible that other critical transcription factors are also affected by PKC phosphorylation, whether it is direct or indirect. These data in concert have led us to develop the following working model of ALDH3 gene regulation.

2.3. Hypoxia

In response to decreased oxygen levels, cells will alter expression of certain genes in order to compensate for the resultant stresses and demands placed on them. Generally, hypoxia results in increased expression of glycolytic enzymes, angiogenic factors, and proteins associated with oxygen transport. Because it has recently come to light that ARNT is involved in modulating gene expression during hypoxia (Wang *et al.*, 1995) and we know that ARNT is involved in ALDH3 expression, we have begun to study the effect of hypoxia on *ALDH3* expression. Corneal epithelial *ALDH3* activity is reduced by nearly 40% of control when rat corneal cultures are exposed to 1% oxygen (hypoxia) for 72 hours. Additionally, there is an even greater decrease in ALDH3 protein and message levels. *ALDH3* transcript levels are approximately 18-fold lower in hypoxic cells. Importantly, ALDH3 protein has a half-life of 96 hours, which accounts for the seemingly greater effect of hypoxic

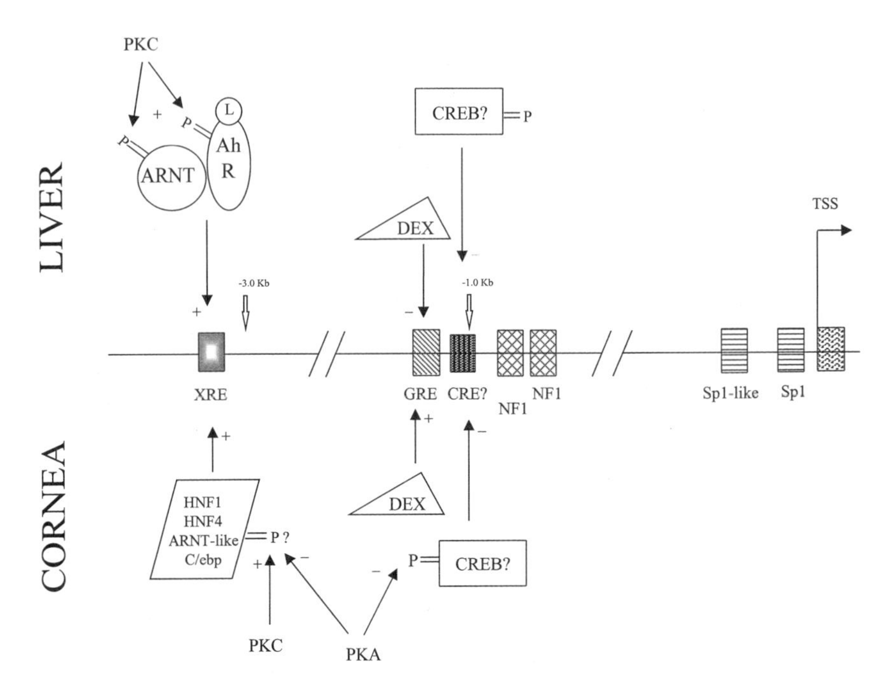

Figure 1. Current model of rat ALDH3 gene regulation in liver and cornea. Terms not described in text: TSS, transcription start site; CRE, cAMP response element; CREB, cAMP response element binding protein; HNF, hepatocyte nuclear factor; C/ebp, CCAAT/enhancer binding protein); L, ligand; P, phosphorylated.

treatment on *ALDH3*message levels than on ALDH3 activity. Rat H4-II-EC3 hepatoma cells (EC3) do not constitutively express *ALDH3*, but it can be induced in these cells by treatment with 3-methylcholanthrene. Treatment of EC3 cells for 48 hours with hypoxia in the presence of 3-MC caused a significant suppression of 3-MC mediated induction of ALDH3 enzyme activity, protein level and transcript. Rat HTC hepatoma cells (HTC) are a line which do constitutively express *ALDH3*, in addition to supporting 3-MC inducible expression of *ALDH3*. However, hypoxia was ineffective at altering either constitutive or induced *ALDH3* expression in HTC cells at any time tested. Thus, hypoxia is yet another signaling system that is capable of altering *ALDH3* expression, although not in all cell types.

The mechanism by which hypoxia modulates ALDH3 expression appears to differ from that of most other hypoxia-responsive genes. All the hypoxia-responsive genes currently described in the literature are upregulated by hypoxia (Semenza *et al.* 1992, 1996; Levy, *et al.* 1995). It has, however, been shown that ARNT will preferentially bind hypoxia inducible factor 1α (HIF-1α) over the AhR (Gradin, *et al.* 1996). This would effectively sequester ARNT from binding response elements other than the hypoxia response element (HRE), where the HIF-1α/ARNT dimer activates transcription. If such is the case in cornea and EC3 cells, then downregulation of both constitutive and inducible *ALDH3* expression is an indirect effect of hypoxia caused by sequestration of ARNT. However, a direct effect, whereby HIF-1α/ARNT binds to the 5' flanking region of *ALDH3* resulting in downregulation, can not be ruled out since there are HREs found far upstream in this region. The fact that *ALDH3* expression in HTC cells is resistant to hypoxia suggests that ARNT is not a critical player, or else is not a limiting factor, in modulation of *ALDH3* expression in this cell type.

3. CONCLUSIONS

Although the ALDH3 gene sequence has been studied for years, we are only beginning to understand the forces controlling its varied expression pattern. We now have some clues as to which signals affect ALDH3 transcription. We have demonstrated that at least 3 known signaling mechanisms (hypoxia, phosphorylation, glucocorticoid) can alter both constitutive and induced *ALDH3* expression. Further investigation is underway to determine exactly how each of these pathways impinge on *ALDH3* transcription. By regulating various crucial transcription factors through a multitude of pathways, cells can balance just how much ALDH3 is or is not present in a cell. This regulation is probably based on the level of need for the detoxifying enzyme. Clearly, not only is any one regulator of gene expression important, but also the interplay between all the regulators of a gene found in a given cell.

ACKNOWLEDGMENTS

This research is supported by NIH grant CA-21103 to R.L. RU38,486 was provided by Research Biochemicals International as part of the Chemical Synthesis Program of the National Institute of Mental Health, Contract N01MH30003.

REFERENCES

Asman, D.C., Takimoto, K., Pitot, H.C., Dunn, T.J. and Lindahl, R.: Organization and characterization of the rat class 3 aldehyde dehydrogenase gene. J. Biol. Chem. 268 (1993) 12530–12536.

Boesch, J., Miskimins, R., Miskimins, W.K.and Lindahl, R.: Identification of elements responsible for constitutive expression of the rat aldehyde dehydrogenase-3 gene in cornea. Arch. Biochem. Biophys. (submitted)

Burton, M.B., Rivera-Hopkins, D. and Lindahl, R.: Regualation of aldehyde dehydrogenase-3 expression by cAMP and glucocorticoids in corneal epithelium. Faseb J. Abstr. 11 (1997) A944.

Falkner, K.C., Xiao, G.-H., Pinaire, J.A., Blaze, M.J., Pendleton, M.L., Lindahl, R. and Prough, R.A.: The negative regulation of the rat aldehyde dehydrogenase 3 gene by glucocorticoids: involvement of an imperfect palindromic response element. Mol. Pharmacol. (submitted).

Gradin, K., McGuire, J., Wenger, R.H., Kvietikova, I., Whitelaw, M.L., Togtgard, R., Tora, L., Gassmann, M. and Poellinger, L.: Functional interference between hypoxia and dioxin signal transduction pathways: competition for recruitment of the Arnt transcription factor. Mol. Cell. Biol. 16 (1996) 5221–5231.

Levy, P.A., Levy, N.S., Wegner, S. and Goldberg, M.A.: Transcriptional regulation of the rat vascular endothelial growth factor gene by hypoxia. J. Biol. Chem. 270 (1995) 13333–13340.

Lindahl, R.: Aldehyde dehydrogenaseand their role in carcinogenesis. Crit. Rev. Biochem. Mol. Biol. 27 (1992) 283–335.

Rohlff, C., Ahmed, S., Borellini, F., Lei, J. and Glazer, R.I.: Modulation of transcription factor Sp1 by cAMP-dependent protein kinase. J. Biol. Chem. 272 (1997) 21137–41.

Semenza, G.L. and Wang, G.L: A Nuclear Factor Induced by Hypoxia via De Novo Protein Synthesis Binds to the Human Erythropoietin Gene Enhancer at a Site Required for Transcriptional Activation. Mol. Cell. Biol. 12 (1992) 5447–5454.

Semenza, G.L., Jiang, B., Leung, S.W., Passantino, R., Concrdet, J., Maire, P. and Giallongo, A: Hypoxia response elements in the aldolase A, enolase 1, and lactate dehydrogenase A gene promotors contain essential binding sites for hypoxia-inducible factor 1. J. Biol. Chem. 272 (1996) 32529–32537.

Takimoto, K., Lindahl, R., Dunn, T.J. and Pitot, H.C.: Structure of the 5' flanking region of class 3 aldehyde dehydrogenase in the rat. Arch. Biochem. Biophys. 312 (1994) 538–546.

Viollet, B., Kahn, A. and Raymondjean, M.: Protein Kinase A-dependent phosphorylation modulates DNA-binding activity of hepatocyte nuclear factor 4. Mol. Cell. Biol. 17 (1997) 4208–4219.

Wang, G.L., Jiang, B., Rue, E.A. and Semenza, G.L.: Hypoxia-inducible factor 1 is a basic-helix-loop-helix-PAS heterodimer regulated by cellular O_2 tension. Proc. Natl. Acad. Sci. 92 (1995) 5510–5514.

Xiao, G.-H., Falkner, K.C., Xie, Y., Lindahl, R.G., and Prough, R.A.: cAMP-dependent negative regulation of rat aldehyde dehydrogenase class 3 gene expression. J. Biol. Chem. 272 (1997) 3238–3245.

Xie, Y.Q., Takimoto, K., Pitot, H.C., Miskimins, W.K. and Lindahl, R.: Characterization of the rat class 3 aldehyde dehydrogenase gene promotor. Nucleic Acids Research 24 (1996) 4185–4191.

21

SUSCEPTIBILITY OF HEPATOMA CELLS TO LIPID PEROXIDATION AND ADAPTATION OF ALDH 3C ACTIVITY TO IRON-INDUCED OXIDATIVE STRESS

Margherita Ferro,[1] Anna Maria Bassi,[1] Susanna Penco,[1] Giuliana Muzio,[2] and Rosa A. Canuto[2]

[1]Department of Experimental Medicine
University of Genoa
Via L. B. Alberti 2
16132 Genova, Italy
[2]Department of Clinical and Biological Sciences
St. Louis Gonzaga Hospital
University of Turin
10043 Orbassano, Italy

1. INTRODUCTION

Free radical iron-induced oxidative stress causes membrane lipid peroxidation and consequent cell damage (Bacon et al., 1985), a process that, in rat liver, is significantly lower in tumor than normal tissue (Dianzani et al., 1984, 1986; Borrello et al., 1985). Factors contributing to this difference include a reduction in cytochrome P450 (an enzyme system implicated in the production of initiating and propagating radicals), a reduction in the peroxidizable substrate arachidonic acid (ARA), an increase in the lipid soluble antioxidant α-tocopherol (Borrello et al., 1985; Cheeseman et al., 1988).

Hepatoma cells can be enriched by externally added ARA, restoring their susceptibility to oxidative stress (Canuto et al., 1991). After exposure to 0.5 mM ARA, hepatoma cells become more sensitive to iron-induced peroxidation and less viable, thus confirming the toxic potential of the peroxidative process (Canuto et al., 1995). The sensitivity to lipid peroxidation and its consequences appears to be correlated with the aldehyde-metabolizing ability of the different hepatoma cell populations (Canuto et al., 1996).

In this study, two rat hepatoma cell lines, MH1C1 and 7777, were chosen for their different phenotypes as regards cytochrome P450 and class 3 aldehyde dehydrogenase

Enzymology and Molecular Biology of Carbonyl Metabolism 7
edited by Weiner *et al.* Kluwer Academic / Plenum Publishers, New York, 1999.

(ALDH3; Canuto et al., 1996) This was to assess the effective roles of cytochrome P450 and ARA in the susceptibility of hepatoma cells to iron-stimulated lipid peroxidation. Cells were therefore enriched with a concentration of ARA low enough to leave them fully viable prior to exposure to pro-oxidants. Parallel series of cell cultures were exposed to the classic cytochrome P450 inducer beta-nafthoflavone (BNF) or to ARA plus BNF. Since the cells were still viable after application of the pro-oxidant stimulus, they were again exposed to the usual growth medium, so that one major enzyme activity involved in the detoxication of lipid peroxidation-derived aldehydes in these cells, i.e. ALDH 3, could be detected.

2. MATERIALS AND METHODS

2.1. Culture Conditions

MH1C1 and 7777 rat hepatoma cells were routinely grown as monolayers, using DMEM/F12 medium supplemented with 7% fetal calf serum and 2 mM glutamine. Cells were seeded in 24- or 6-well plates, to determine lipid peroxidation levels and for enzyme analyses, respectively.

2.2. Treatments

Cell cultures, pre-treated or not for the last 48 hours of culture time with 15 μM BNF, were enriched or not with ARA (10 nmoles/10^6 cells) for the last 24 hours of culture time. They were then washed twice with low-calcium Hank's solution (150 μM) and exposed for one hour to a pro-oxidant system [mixtures of $FeSO_4$/histidine (Fe^{2+}/His, 3/30 mM), or $FeSO_4$/ascorbic acid (Fe/Asc, 0.1/0.5 mM), or $Fe_2(SO_4)_3$/ADP (Fe/ADP, 0.1/2.5 mM)]. After pro-oxidation, one group of samples was directly analyzed as described below; another group was rinsed twice with complete medium, then refed with culture medium for further 24 hours, and analyzed as described below. Each experimental condition, repeated in triplicate, was checked for viability.

2.3. Assessment of Total Fatty Acid Composition in Membrane Phospholipids

The percentage content of each fatty acid was evaluated in membrane phospholipids extracted from MH1C1 and 7777 hepatoma cell lines, exposed or not to arachidonic acid, following the protocol described elsewhere (Canuto et al., 1991).

2.4. Lipid Peroxidation

Lipid peroxidation in MH1C1 and 7777 cells was assessed directly in the wells by measuring malondialdehyde (MDA) production by a spectrofluorometric thiobarbituric acid (TBA) assay (Iguchi et al., 1993).

2.5. Viability Tests

The monolayers were exposed for 3 hr to a neutral red dye-containing medium (NRU test, Borenfreund and Puerner, 1985), whereas the protein content (TPC test) was quantified directly in the well using Coomassie blue (Bradford, 1976). For the MTT test the cultures were exposed to tetrazolium salt and processed as described by Mossman et al. (1983).

2.6. Enzyme Activities

7-Ethoxyresorufin *O*-deethylase activity (EROD) was evaluated by exposing the monolayers to the substrate dissolved in culture medium; after one hour, the resorufin-conjugates were hydrolyzed and extracted with methanol. The fluorescence of resorufin was compared to that of a standard curve (Donato et al., 1994).

ALDH 3 activity was measured on cell homogenates by monitoring the changes in A_{340} due to NADPH production during oxidation of the substrate (2.5 mM benzaldehyde) (Canuto et al., 1983).

2.7. Statistical Analysis

Results are arithmetic means ± SD or SEM of five to seven independent experiments; statistical analysis was by ANOVA test followed by the Tukey Kramer test (GraphPad INSTAT).

3. RESULTS

3.1. Effects of Arachidonic Acid Supplementation on MH1C1 and 7777 Hepatoma Cell Lines

Since in previous experiments arachidonic acid had been seen to be cytotoxic at certain concentrations, a dose-response curve was deduced from the results of the NRU test, expressed on a per-cell-number rather than a molarity basis (Figure 1). We believe this is the most correct way to express the concentration of a possible toxicant, since the same final concentration might affect a small cell population much more strongly than a larger one. The concentration corresponding to 10 nmoles/10^6 was chosen as the highest non-cytotoxic concentration for this study.

Despite the low dosage, the arachidonic acid content increased in membrane phospholipids after 24 h exposure 6-fold and 4-fold in MH1C1 and 7777 cells, respectively (Table 1).

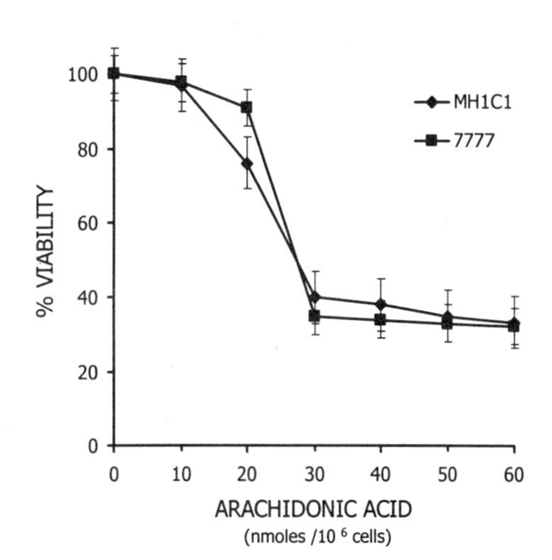

Figure 1. Viability of MH1C1 and 7777 hepatoma cells after exposure to 0-60 nmoles/10^6 cells of arachidonic acid for 24 hours. Values are percentages of viable cells evaluated by NRU test and are the mean ± SEM of 3 separate experiments.

Table 1. Percentage of total fatty acid extracted from membrane phospholipids of MH1C1 and 7777 cells, enriched or not with arachidonic acid

Fatty Acid	MH1C1		7777	
	Unenriched	ARA-enriched	Unenriched	ARA-enriched
14:0	4.22 ± 0.3	3.24 ± 0.5	5.17 ± 0.1	3.64 ± 0.1
16:0	19.67 ± 0.9	21.87 ± 1.2	20.11 ± 1.9	24.23 ± 1.6
16:1	6.28 ± 0.2	6.65 ± 0.3	6.37 ± 0.6	8.49 ± 0.7
18:0	10.70 ± 0.5	16.78 ± 1.0	21.57 ± 1.6	21.82 ± 1.8
18:2	9.57 ± 0.4	12.57 ± 0.8	17.59 ± 0.8	7.58 ± 0.2
20:4	1.19 ± 0.3	7.31 ± 0.3	0.84 ± 0.2	4.84 ± 0.5

Percentages of the total fatty acid content in phospholipids extracted after 24 h exposure to arachidonic acid (10 nmoles/10^6 cells); values are means ± SD of three determinations.

3.2. Effects of Beta-Nafthoflavone Treatment on Enzyme Activities in MH1C1 and 7777 cells

Treatment with BNF, a classical cytochrome P450 inducer, increased EROD acticity 9-fold and 11-fold in MH1C1 and 7777 cells, respectively (Table 2). However, the final enzymatic activity in 7777 cells remained lower than that of MH1C1. ALDH 3 was induced 4-fold and 3-fold in MH1C1 and 7777 cells, respectively, but the level of ALDH 3 remained very low in the latter, although some induction had occurred (2.6 ± 0.2 nmol/min/mg prot).

3.3. Effect of Treatments on Cell Viability

Three end-points were evaluated to assess cell viability after each experimental procedure, and in all conditions cells were viable. The slight loss of viability after Fe/His treatment and NRU test might be an artefact (Table 3).

Table 2. Induction of 7-ethoxyresorufin *O*-deethylase (EROD) and aldehyde dehydrogenase Class 3 (ALDH 3) activities by beta-nafthoflavone (BNF)

ENZYMATIC ACTIVITY	MH1C1		7777	
	Basal	BNF	Basal	BNF
EROD pmol/min/mg prot	23.6 ± 0.5	189 ± 3.4* (9)	3.7 ± 0.7	40.2 ± 2.3* (11)
ALDH 3 nmol/min/mg prot	36.5 ± 1.3	135 ± 2.3* (4)	1.0 ± 0.1	2.6 ± 0.2 (3)

Values are expressed as pmoles of resorufin, and nmoles of NADPH, produced min^{-1} mg^{-1} protein, for EROD and ALDH 3 respectively . The results are means ± SEM of 5-7 experiments.
* = $p<0.001$ treated vs untreated cultures.

Table 3. Viability in MH1C1 and 7777 cells after exposure to pre-treatments and pro-oxidants as described in Materials and Methods

Pro-oxidant	Pre-treatment	MH1C1 *NRU*	MH1C1 *TPC*	MH1C1 *MTT*	7777 *NRU*	7777 *TPC*	7777 *MTT*
None	None	100 ± 0.0	100 ± 0.0	100 ± 0.0	100 ± 0.0	100 ± 0.0	100 ± 0.0
	ARA	97 ± 2.1	95 ± 1.6	97 ± 1.9	99 ± 0.4	96 ± 2.0	100 ± 0.2
	BNF	98 ± 1.0	98 ± 1.3	100 ± 0.3	98 ± 0.7	97 ± 1.8	99 ± 0.2
	ARA + BNF	99 ± 0.5	98 ± 1.4	100 ± 0.5	98 ± 0.5	96 ± 1.5	99 ± 1.0
Fe/His	None	87 ± 3.4	96 ± 3.1	99 ± 2.4	89 ± 1.8	98 ± 0.6	98 ± 1.0
	ARA	87 ± 1.8	95 ± 2.3	99 ± 1.0	87 ± 1.2	97 ± 0.9	98 ± 2.0
	BNF	99 ± 0.2	99 ± 0.4	99 ± 0.1	88 ± 2.0	97 ± 1.1	97 ± 1.8
	ARA + BNF	89 ± 1.3	98 ± 1.6	97 ± 1.5	87 ± 2.3	96 ± 1.0	95 ± 2.1
Fe/Asc	None	98 ± 1.0	99 ± 0.3	96 ± 2.6	98 ± 1.9	97 ± 1.2	99 ± 0.9
	ARA	95 ± 1.8	99 ± 0.8	93 ± 2.6	98 ± 1.2	96 ± 0.8	98 ± 1.4
	BNF	94 ± 2.5	99 ± 0.2	97 ± 1.2	97 ± 0.9	96 ± 0.3	98 ± 0.5
	ARA + BNF	94 ± 1.0	99 ± 0.1	97 ± 1.6	96 ± 1.3	95 ± 0.9	97 ± 1.2
Fe/ADP	None	96 ± 2.5	98 ± 0.8	97 ± 1.9	96 ± 2.3	97 ± 1.0	97 ± 1.9
	ARA	97 ± 1.2	98 ± 0.9	95 ± 2.1	96 ± 0.9	98 ± 1.8	95 ± 2.1
	BNF	98 ± 1.8	98 ± 0.5	98 ± 1.4	96 ± 1.3	97 ± 2.4	94 ± 1.8
	ARA + BNF	98 ± 1.5	98 ± 0.3	94 ± 2.5	94 ± 1.2	96 ± 1.5	94 ± 2.4

Values are expressed as percentages of viable cultures compared to controls and are means ± SEM of 5-7 separate experiments.
NRU = test of neutral red uptake; TPC = total protein content ; MTT = test of tetrazolium salt.

3.4. Effect of the Pre-Treatments on Susceptibility of MH1C1 and 7777 Cells to Lipid Peroxidation

Figure 2 shows the effect on lipid peroxidation, evaluated as MDA production, in stimulated and unstimulated MH1C1 cells; low levels of ARA enrichment produced a significant increase of MDA production with each stimulant, even in the absence of pro-oxi-

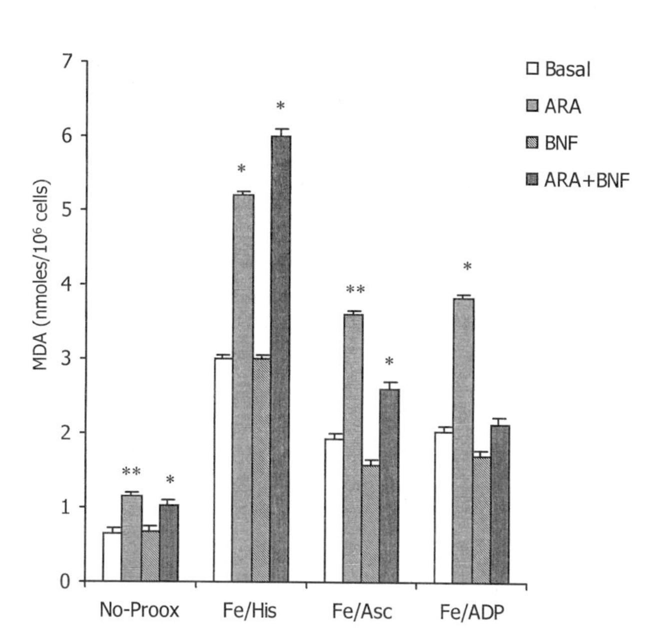

Figure 2. MDA production in MH1C1 cells after exposure to various pro-oxidant stimuli. Values are means ± SEM of 7–10 separate experiments. *,** = $p < 0.01$ and 0.001 vs. untreated cultures (ANOVA followed by Tukey Kramer test).

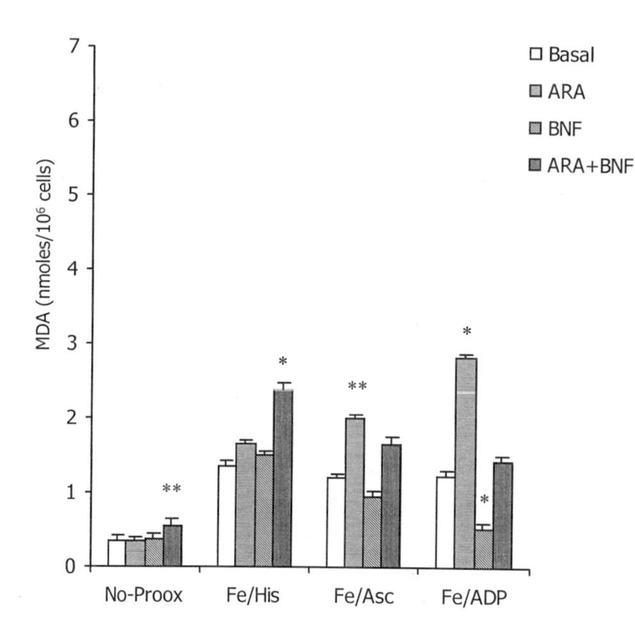

Figure 3. MDA production in 7777 cells after exposure to various pro-oxidant stimuli. Values are means ± SEM of 7–10 separate experiments. *,** = $p < 0.01$ and 0.001 vs. untreated cultures (ANOVA followed by Tukey Kramer test).

dants. By contrast, induction of cytochrome p450 (cells treated with BNF alone) failed to increase lipid peroxidation in MH1C1 cells. However, the MDA level increased when BNF pre-treatment was associated with ARA enrichment with two of the pro-oxidants (Fe/His and Fe/Asc) and in basal conditions. Unexpectedly, with Fe/His, ARA-induced lipid peroxidation was potentiated by BNF (12%).

The capacity of 7777 cells to undergo lipid peroxidation was much lower than that of MH1C1 cells, as shown in Figure 3; however, the pattern of inducibility of mda production was similar to that of MH1C1 cells, maximal MDA levels being found with ARA alone, or with ARA + BNF .

3.5. Effects of Oxidative Stress on ALDH 3 activity

Figure 4 shows the results on ALDH 3 activity in MH1C1 cell line; at the end of pre-treatment with BNF and ARA+BNF, and before exposure to Fe/His, ALDH 3 activity was increased 3-fold and 5-fold, respectively. After 1 h exposure to Fe/His, the two induced activities had decreased by 30%, whereas the non-induced activities (basal and ARA pre-treatment) had decreased by 25%. Refeeding the cells with fresh medium produced a sharp increase in activity, especially in the absence of ara enrichment: the increase of activity was much less marked in ARA and ARA+BNF conditions (Figure 4, lower curves).

When the 7777 cell line was exposed to the same pre-treatment and pro-oxidant, the very low ALDH 3 activity of these cells did not change significantly (data not shown).

4. DISCUSSION

As already shown (Canuto et al., 1995; 1996), hepatoma cells enriched with arachidonic acid increase their sensitivity to pro-oxidant agents, with consequent loss of viabil-

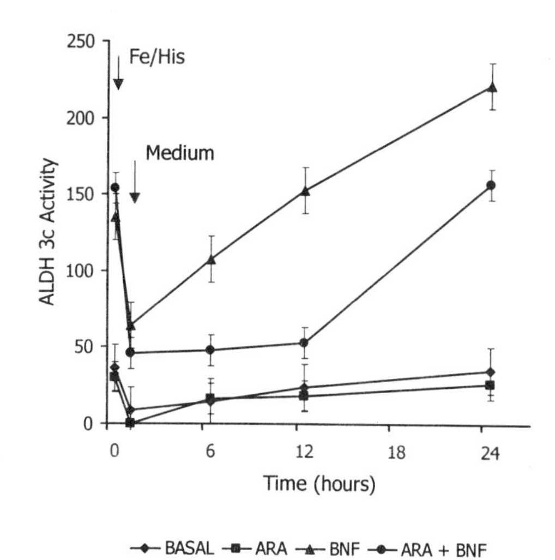

Figure 4. ALDH 3 activity in MH1C1 cells exposed or not to Fe/His for 1 hour and after 6–12–24 hours of return to standard culture conditions. Values are expressed as nmoles of NADPH produced min^{-1} mg^{-1} protein.

ity. Whereas in the previous studies ARA doses were 125 nmoles/10^6 cells or above, in this study we used 10 nmoles/10^6 cells; as a consequence the cells contained less arachidonic acid and were still viable after pro-oxidant treatment. The loss of viability thus clearly depends on the entity of the lipid peroxidation phenomenon.

Even if ara dose was low, it produced a significant increase in mda production (Figures 2 and 3), thus confirming the role of this fatty acid in the well-known resistance of hepatoma cells to oxidative stress.

The full role of cytochrome p450 is not easy to understand, since it may function in several ways: by generating metabolic intermediates, by initiating oxygen radicals, through peroxidase activity, etc. Studies with purified isozymes indicate that different isozymes of p450 influence microsomal lipid peroxidation in specific ways (Ohmori et al., 1993). However, a recent study by Lambert et al. (1996), using microsomes obtained from recombinant cell lines expressing specific p450 isoforms showed that the rate of peroxidation depends on the total level of p450, and not on any specific isozyme.

In contrast with the effect shown by ara, which influenced lipid peroxidation at a very low concentration , bnf, which markedly induced the CYP1A1-dependent erod activity (Table 2), did not affect the peroxidation of unenriched cells. The low peroxidability of cells after BNF treatment might be due to induction of ARA metabolism, limiting the availability of this substrate for lipid peroxidation. However, p450 induction exacerbates the rate of peroxidation when cells are enriched with ara (Figures 2 and 3), even with fe/asc, which is considered "non-enzymic", as pro-oxidant (Koster et al., 1980). It is possible that, in these experimental conditions cytochrome p450 acts through its peroxidase activity, catalyzing the breakdown of looh (Bast and Haenen, 1984). When cells were stimulated with fe/his, the effect of bnf was even greater than that of ARA, thus suggesting a specific role of CYP1A1 in the mechanism of action of this pro-oxidant.

l-histidine is known to markedly sensitize cells to oxidative attack by mechanisms still largely unknown. It has been suggested that it might modulate the impact of oxidative stress in mammalian cells, with the formation of dna double strand breaks (Cantoni et al., 1994). The fe/his association has been used to induce lipid peroxidation in normal and transformed liver cells to investigate the role of 4-hydroxy-2-nonenal (4-HNE) in cell proliferation (Hammer et al., 1997). At present we have no good explanation for the enhancement of ara-mediated effects by bnf in fe/his stimulated cells. Iron-induced lipid peroxidation strongly affected ALDH 3 activity, but only in MH1C1 cells, and not in 7777 cells, which were practically devoid of this enzyme.

It is known that many different aldehydes are generated during lipid peroxidation, and that 4-HNE is one of the chief products. The end-product aldehydes easily react with proteins and inactivate enzymes with toxic effects (for a review see Esterbauer et al., 1990). They are normally metabolized through various pathways, the aldehyde dehydrogenases and the glutathione S-transferases being the most important in normal cells. In particular, ALDH 3 is involved in the oxidation of lipoperoxidation aldehydes in tumor cells (reviewed by Lindahl, 1992). The early disappearance of ALDH 3 after an oxidative stimulus is thus a potentially toxic event, since the cell loses an important detoxication system. MH1C1 cells are able to adapt to the situation by increasing enzyme activity immediately after the challenge, which is a great advantage for cells exposed to mild oxidative stresses. Further studies are needed to elucidate the mechanisms responsible for this adaptation.

REFERENCES

Bacon B.R., Tavill A.S., Brittenham G.M., Parck C.H., and Recknagel R.O., 1985, in: "Free Radicals in Liver Injury" (Poli G., Cheeseman K.H., Dianzani M.U., Slater T.F., eds.), pp. 49–57, IRL Press, Oxford.

Bast A. and Haenen G.R.M.M., 1984, Cytochrome P_{450} and glutathione: what is the significance of their interrelationship in lipid peroxidation?, *Trends Biochem. Sci.*, 9, 510–513.

Borenfreund E. and Puerner J.A., 1985, Toxicity determined *in vitro* by morphological alterations and neutral red absorbtion, *Toxicol. Letts.*, 24: 119–124.

Borrello S., Minotti G., Palombini G., Grattagliano A., and Galeotti T., 1985, Superoxide-dependent lipid peroxidation and vitamine E content of microsomes from hepatomas with different growth rates, *Arch. Biochem. Biophys.*, 238: 588–595.

Bradford M.M., 1976, A rapid and sensitive method for the quantitation of microgram quantities of protein utilizing the principle of protein-dye binding, *Anal. Biochem.*, 72: 248–254.

Cantoni O., Sestili P., Brandi G., and Cattabeni F., 1994, The L-histidine-mediated enhancement of hydrogen peroxide-induced cytotoxicity is a general response in cultured mammalian cell lines and is always associated with the formation of DNA double strand breaks. *FEBS Letts.*, 353: 75–78.

Canuto R.A., Garcea R., Biocca M.E., Pascale R., Pirisi L., and Feo F., 1983, The subcellular distribution and properties of aldhyde dehydrogenase of hepatoma AH-130, *Eur. J., Cancer. Clin.. Oncol.*, 19:389–400.

Canuto R.A., Muzio G., Biocca M.E., and Dianzani M.U., 1991, Lipid peroxidation in rat AH-130 hepatoma cells enriched *in vitro* with arachidonic acid, *Cancer Res.*, 51: 4603–4608.

Canuto R.A., Muzio G., Bassi A.M., Maggiora M., Leonarduzzi G., Lindahl R., Dianzani M.U., and Ferro M., 1995, Enrichment with arachidonic acid increases the sensitivity of hepatoma cells to the cytotoxic effects of oxidative stress, *Free Rad. Biol. Med.*, 18: 287–293.

Canuto R.A., Ferro M., Maggiora M., Federa R., Brossa O., Bassi A.M., Lindahl R., and Muzio G., 1996, In hepatoma cell lines restored lipid peroxidation affects cell viability inversely to aldehyde metabolizing enzyme activity. In: "Enzymology and Molecular Biology of Carbonyl Metabolism 6", (Weiner et al. eds), pp. 113–122, Plenum Press, New York.

Cheeseman K.H., Emery S., Maddix S.P., Slater T.F., Burton G.W., and Ingold K.U., 1988, Studies on lipid peroxidation in normal and tumor tissues, *Biochem. J.*, 250: 247–252.

Dianzani M.U., Canuto R.A., Rossi M.A., Poli G., Garcea R., Biocca M.E., Cecchini G., Ferro M., and Bassi A.M., 1984, Further experiments on lipid peroxidation in transplanted and experimental hepatomas, *Toxicol. Pathol.* 12: 189–199.

Dianzani M.U., Poli G., Canuto R.A., Rossi M.E., Biasi F., Cecchini G., Muzio G., Ferro M., and Esterbauer H., 1986, New data on kinetics of lipid peroxidation in experimental hepatomas and preneoplastic nodules, *Toxicol. Pathol.*, 14: 404–410.

Donato T., Bassi A.M., Gòmez-Lechòn M.J., Penco S., Herrero E., Adamo D., Castell J.V., and Ferro M., 1994, Evaluation of the xenobiotic biotransformation capability of six rodent hepatoma cell lines in comparison with rat hepatocytes. *In vitro Cell Dev. Biol.* 30A: 574–580.

Esterbauer H., Zollner H., and Schaur R.J., 1990, Aldehydes formed by lipid peroxidation:mechanisms of formation, occurence and determination, in: "Membrane Lipid Oxidation", Vol.1, (Vigo Pelfrey C., ed.) Boca Raton. FL.

Hammer A., Ferro M., Tillian H.M:, Tatzber F., Zollner H., Schauenstein E., and Schaur R.J., 1997, Effects of oxidative stress by iron on 4-hydroxynonenal formation and proliferative activity in hepatomas of different degrees of differentiation, *Free Rad. Biol. Med.*, 23: 26–33.

Iguchi H., Kojo S., and Ikeda M., 1993, Lipid peroxidation and disintegration of the cell membrane structure in cultures of of rat lung fibroblasts treated with asbestos, *J. Appl. Toxicol.*, 13: 269–275.

Koster J.F., and Slee R.G., 1980, Lipid peroxidation of rat liver microsomes, *Biochem., Biophys. Acta,* 620: 489–499.

Lambert N., Chambers S.J., Plumb G.W., and Williamson G., 1996, Human cytochrome P_{450} are pro-oxidants in iron/ascorbate-initiated microsomal lipid peroxidation, *Free Rad. Res.*, 24: 177–185.

Lindahl R., 1992, Aldehyde dehydrogenases and their role in carcinogenesis, *Critical Rev. Biochem. Mol. Biol.,* 27: 283–335.

Mosmann T., 1983, Rapid colorimetric assay for cellular growth and survival: application to proliferation and cytotoxicity assays, *J. Immunol. Meth.*, 65: 55–63.

Ohmori S., Misaizu T., Nakamura T., Takano N., Kitagawa H., and Kitada M., 1993, Differential role of lipid peroxidation between rat P_{450} 1A1 and P_{450} 1A2, *Biochem. Pharmacol.*, 46: 55–60.

THE LACK OF AHD4 INDUCTION BY TCDD IN CORNEAL CELLS MAY INVOLVE TISSUE-SPECIFIC REGULATORY PROTEINS

Vasilis Vasiliou and Teyen Shiao

Molecular Toxicology and Environmental Health Sciences
Department of Pharmaceutical Sciences
University of Colorado Health Sciences Center
4200 East Ninth Avenue
Denver, Colorado 80262

1. INTRODUCTION

The AHD4 enzyme is encoded by a single gene which is a member of the mouse aromatic hydrocarbon [*Ah*] battery. The latter contains at least six genes that are coordinately induced in liver cells by 2,3,7,8-tetrachlorodibenzo-p-dioxin (TCDD) (Nebert et al., 1993). These include two P450 genes, ***Cyp1a1*** and ***Cyp1a2***, and four Phase II genes, NA(P)H:menadione oxidoreductase (***Nqo1***); aldehyde dehydrogenase (***Ahd4***); UDP glucuronosyltransferase (***Ugt1*a6***); and glutathione transferase (***Gsta1***). Induction of all six [*Ah*] genes by TCDD requires both the Ah receptor (AHR) and Ah receptor nuclear translocator protein (ARNT), and operates via nuclear translocation followed by binding to ***a***romatic ***h***ydrocarbon ***r***esponse ***e***lements (AhREs). Increased expression of the Phase II genes serves as a defense mechanism against oxidative stress caused by endogenous and exogenous stimuli (Shertzer et al., 1995; Vasiliou et al., 1995). This up-regulation occurs via the ***e***lectro***p***hile ***r***esponsive ***e***lement (EpRE; Jaiswal, 1994). AhREs and EpRE are present and functional in the 5' flanking region of the mouse ***Ahd4*** gene (Vasiliou et al., 1997).

In several mammalian species, corneal epithelium expresses high constitutive AHD4 levels, which alone may represent 10–40% of the total soluble corneal protein (Downes and Holmes, 1993; Boesch et al., 1996). Constitutive AHD4 enzymatic activities and/or mRNA levels have also been found in rat stomach, urinary bladder, lung and skin (Dunn et al., 1988; Vasiliou and Marselos, 1989; Pappas et al., 1997). Differential inducibility of the ALDH3 (rat ortholog of AHD4) was found in various rat tissues after treatment with TCDD or 3-methylcholanthrene (Dunn et al., 1988; Vasiliou and Marselos, 1989). However, TCDD did not increase constitutive AHD4 mRNA levels in mouse eye and stomach

Enzymology and Molecular Biology of Carbonyl Metabolism 7
edited by Weiner *et al.* Kluwer Academic / Plenum Publishers, New York, 1999.

(Marks-Hull et al., 1997). Hence, TCDD inducibility and tissue-specific constitutive expression appear to be two independent regulatory mechanisms for ***Ahd4*** gene expression.

In this chapter, we present evidence that constitutive and TCDD-inducible ***Ahd4*** gene expression are two distinct mechanisms of regulation that may not occur in the same tissue. Constitutive expression appear to involve tissue-specific transcription factors that recognize the same DNA motifs in the 5' flanking region of this gene as xenobiotic-activated cellular receptors.

2. MATERIALS AND METHODS

2.1. Cell Cultures

Three immortalized corneal epithelial cell lines, human (HCE), rat (RatCE) and rabbit (RabbitCE) were cultured in supplemented hormone epithelial medium (SHEM) (Araki et al., 1993; 1995). The human breast cancer MCF-7 cell line was cultured in Dulbecco's modified Eagle's medium (DMEM), supplemented with 10% fetal calf serum. The mouse hepatoma (Hepa-1) wild-type (*wt*) cell culture (Bernard et al., 1973) was routinely grown in DMEM supplemented with 5% fetal calf serum. All cells were cultured in a 37°C incubator injected with 5% CO_2.

2.2. Functional Analysis

Constructs containing 3.2 and 1.6 kb 5' flanking region of the mouse ***Ahd4*** and human ***NQO1*** gene (respectively) fused to luciferase reporter gene were used. In addition, luciferase chimeras containing either two AhRE copies of the mouse ***Cyp1a1*** gene (AhRE.Luc; Hoffer et al., 1996), or one copy of the EpRE of the mouse ***Ahd4*** gene (EpRE.Luc) were also used. Reporter plasmids were transiently transfected into the cell lines by the calcium phosphate procedure (Parker and Stark, 1979). Luciferase gene expression, in the presence and absence of 10 nm TCDD, was monitored by measuring luciferase activity. The pGL3.basic (promotorless) and pGL3.promoter (SV40 promoter) luciferase plasmids (PROMEGA) were used as negative and positive controls, respectively. Cells were co-transfected with Rous Sarcoma Virus β-Gal plasmid in each case to permit normalization of transfection efficiency through the measurement of β-galactosidase activity (Guarente, 1983).

2.3. RNA Extraction and Northern Blot Analysis

Total RNA from mouse tissue samples was extracted by the acid guanidinium thiocyanate method (Chomczynski and Sacchi 1987). RNA was separated by formaldehyde-agarose gel electrophoresis, transferred to Nytran membranes and UV cross-linked. Radioactively labeled cDNA probes were prepared by random priming (Sambrook et al., 1989) using [α-^{32}P]dCTP as the labeled precursor. Prehybridization and hybridization were carried out at 52°C in a solution containing 50% deionized formamide, 6X SSC, 2.5X Denhardt's solution, 0.5% sodium dodecylsulfate (SDS), and denatured salmon sperm DNA (0.1 mg/mL). After hybridization for 16–20 h, filters were washed twice under progressively stringent conditions and exposed to X-ray film at -70°C with intensifying screens for 24–48 h. Mouse AHD4 cDNA was used as probe (Vasiliou et al., 1993). Levels of ethidium bromide stained 18S rRNA and 28S rRNA were used to standardize the amount of RNA loaded per lane.

3. RESULTS AND DISCUSSION

3.1. Lack of *Ahd4* Inducibility in Corneal Epithelial Cells

Transient transfection experiments using luciferase chimeras containing 3.2 and 1.6 kb of 5' flanking region of the mouse ***Ahd4*** and the human ***NQO1*** gene (respectively) showed that both ***Ahd4*** and ***NQO1*** 5' flanking regions can drive luciferase expression in all three corneal epithelial cell lines. The magnitude of basal luciferase activity using either AHD4 or NQO1 chimeric plasmid was found in the order of HCE < RatCE < RabbitCE. When the AHD4 reporter chimera was used, a significant increase of the luciferase activity was found in Hepa-1 but not in corneal epithelial cells (Table 1). The latter is consistent with our previous studies, which showed that AHD4 mRNA levels are constitutively expressed but not further induced by TCDD in mouse cornea (Marks-Hull et al., 1997). The 3.2 kb 5' flanking region of the mouse ***Ahd4*** gene contains seven AhREs, of which four are functional and involved in the up-regulation of this gene by TCDD in liver cells (Vasiliou et al., 1997). Similarly, one functional AhRE is present in the 1.6 kb of the human NQO1 gene (Jaiswal, 1991). Western blot analysis showed that both AHR and ARNT proteins, which are necessary for TCDD-induction, are present in all three cell lines (data not shown). In addition, both transcription factors appear to be functional in terms of TCDD inducibility, since luciferase activity was increased after TCDD treatment in all corneal cell lines transfected with the NQO1 reporter chimera (Table 1). The lack of AHD4 inducibility in corneal epithelial cells led us to speculate that this gene might be under the control of cornea-specific *cis-* and/or *trans-* acting factor(s) which prevent the activated AHR-ARNT complex from interacting with the AhRE binding sites.

3.2. Promoter and Enhancer Activities in the 5' Flanking Region of the *Ahd4* Gene

Transient transfection experiments with the RabbitCE cell line showed that there are two areas in the 5' flanking region of the ***Ahd4*** gene with promoter/enhancer activities. The first lies within the first 1.1 kb and shows strong promoter activity, as indicated by an 8-fold higher luciferase activity in transfected cells, as compared with that of the SV40 promoter

Table 1. Transient expression of AHD4–3.2.Luc and NQO1.Luc recombinant plasmids in Hepa-1, HCE, RatCE and RabbitCE cells

	AHD4-3.2.Luc		NQO1.Luc	
Cell line	Control	TCDD	Control	TCDD
Hepa-1	+	++++++	+	++++
HCE	+	+	+	++++
RatCE	++	++	+	+++
RabbitCE	++++	++++	++	++++

The AHD4-3.2.Luc and NQO1.Luc chimeric plasmids contain 3.2 and 1.6 kb 5' flanking region of the mouse *Ahd4* and human *NQO1* gene (respectively) upstream of the luciferase reporter gene. The chimeric plasmids were transiently transfected, and their expression in the absence or presence of 10 nM TCDD (18 h) was monitored by measuring the luciferase activity. RSV-βGal plasmid was co-transfected in each case to normalize the transfection efficiency. The pGL3.Basic (promotorless) and pGL3.Promoter (containing the minimal SV40 promoter) luciferase plasmids were used as negative and positive controls, respectively.

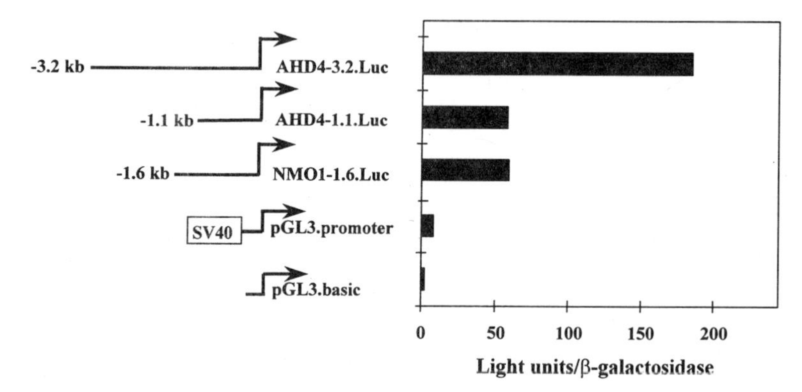

Figure 1. Transient expression of AHD4 and NQO1 luciferase recombinant plasmids in RabbitCE cell cultures. The AHD4 and NQO1 chimeric plasmid DNAs are diagrammed at *left*. The numbers on the left of the constructs indicate the distance from the transcription start site. The pGL3.promoter and pGL3.basic luciferase plasmids (diagrammed also at *left*) were used as negative and positive controls respectively. The RSV-β-Gal plasmid was co-transfected in each to normalize the transfection efficiency. Data presented as mean values of four experiments and expressed as light units per β-galactosidase activity are shown at *right*. The standard deviations of four experiments were always less than 15% of the means.

driven luciferase plasmid (Figure 1). The region between 3.2 kb and 1.1 kb caused a further increase in luciferase activity relative to to that found with the 1.1 kb luciferase chimera. It would appear that this area (-3.2 to -1.1kb) of the ***Ahd4*** gene contains *trans*-acting element(s) that may play a central role in the cornea-specific ***Ahd4*** expression.

3.3. Is the AHR Involved in the Cornea-Specific *Ahd4* Expression?

Several transcription factor binding sites have been identified within the 3.2 kb 5' flanking region. These include seven AhREs located at -93, -377, -393, -879, -1195 -1667 and -2597, one EpRE at -742, four AP-1 sites at -290, -713, -3005, and -3072, and two NF-κB sites at -638 and -2546, a C/EBPβ site at -361, and a C/EBPα site was found at -2700. These are summarized in Figure 2.

The high luciferase activities found in RabbitCE cells transfected with the AHD4 chimeras and the multiple AhREs present in the 5' flanking region raises the question as to whether transcription factors bound to these elements are responsible for cornea-specific ***Ahd4*** expression. To address this possibility, HCE, RabbitCE, MCF-7 and Hepa-1 cell lines were transfected with the AHD4–3.2.Luc, AHD4–1.1.Luc, AhRE.Luc and EpRE.Luc. RabbitCE cells transfected with the AHD4–3.2.Luc and/or AhRE.Luc plasmids were found to have the highest luciferase activities (Figure 3). The AhRE increased dramatically the ability of the heterologous SV40-promoter to drive luciferase expression only in RabbitCE cells. On the contrary, the EpRE sequence did not cause any increase in the RabbitCE. These data indicate that AhRE binding motifs are involved in the constitutive luciferase expression in these cells.

Is the AHR activated with its unknown yet endogenous ligand involved in the constitutive tissue-specific ***Ahd4*** expression? A molecular connection between the AHR signaling and retinoic acid homeostasis was recently suggested in mice (Andreola et al., 1997). ALDH1, which is involved in the formation of retinoic acid, was found to be markedly decreased in the AHR null mice. In contrast, ALDH1 expression was unaffected by

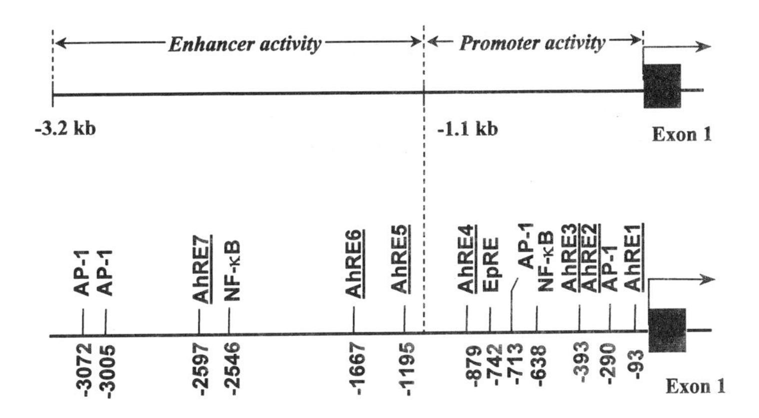

Figure 2. The two domains of the 3.2 kb 5' flaking region of the *Ahd4* gene that appear to possess promoter and enhancer activities in RabbitCE cell cultures. The lengths of the domains are given at *top*, and the transcription factor binding sites, either functional or putative, are shown *below*. The AhREs are underlined.

TCDD treatment in wild type mice, indicating that AHR is not directly involved in the down-regulation of this gene (Andreola et al., 1997). To elucidate whether AHR plays a central role in the tissue-specific constitutive expression of *Ahd4* gene, we have examined AHD4 mRNA levels in the eye, stomach and liver of wild type and AHR-null mice (Fernandez-Salguero et al., 1995). AHD4 mRNA levels were found in remarkably high levels in the eye and stomach, but not in liver, of knockout homozygotes [*Ahr(-/-)*], heterozygotes *[Ahr(-/+)*] and wild type mice [*Ahr(+/+)*] (Table 2). These data argue against the possibility of the AHR being responsible in the tissue-specific *Ahd4* expression.

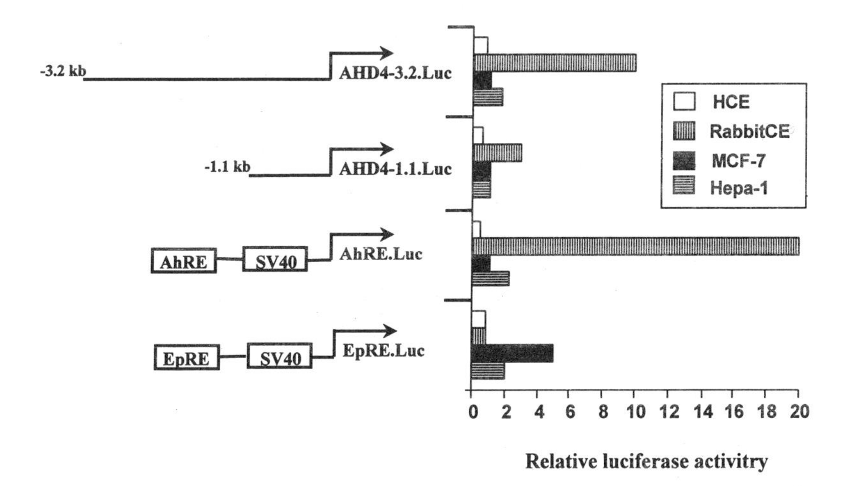

Figure 3. Transient expression of AHD4–3.2.Luc, AHD4–1.1Luc, AhARE.Luc and EpRE.Luc recombinant plasmids in HCE, RabbitCE, MCF-7 and Hepa-1 cell lines. The luciferase chimeric plasmid DNAs are diagrammed at *left*. The numbers on the left of the constructs indicate the distance from the transcription start site. The RSV-β-Gal plasmid was co-transfected in each to normalize the transfection efficiency. Relative luciferase activities are shown at *right*. The mean activities of all groups are compared to the respective luciferase activity found in each cell line transfected with pGL3.promoter plasmid (containing the minimal SV40 promoter), which represents 1 arbitrary unit. The pGL3.basic luciferase plasmid was used as negative controls. The standard deviations of three experiments in each cell line were always less than 18% of the means.

Table 2. AHD4 mRNA levels in the eye, stomach, and liver of *Ahr*(–/–), *Ahr*(–/+) and *Ahr* (+/+) mice

Mouse line	Eye	Stomach	Liver
Ahr (–/–)	++++	++	0
Ahr (–/+)	++++	++	0
Ahr (+/+)	++++	++	0

RNA isolation and hybridization conditions are described in Materials and Methods. Mouse AHD4 full length cDNA was used as probe.

3.4. Pax-6 May Be Involved in the Lack of *Ahd4* Inducibility by TCDD in Corneal Cells

Although both AHR and ARNT are present in the corneal epithelial cells, as described earlier, the AhREs of the ***Ahd4*** gene do not respond to TCDD-inducibility. The observation that the AhRE of the human ***NQO1*** gene is functional in corneal cells led us to speculate that other nuclear regulatory proteins are bound to the AhREs of the ***Ahd4*** gene. Pax-6 is a transcription factor which contains both a paired domain and homeodomain and regulates lens determination and crystallin gene expression (Cvekl and Piatigorsky, 1996). Besides lens, Pax-6 is also expressed at significant levels in corneal epithelium (Koroma et al., 1997). The core DNA sequence that interacts with the paired domain of Pax-6 consists of at least 20 bp, which is larger than most recognition motifs of DNA-binding proteins(Cvekl and Piatigorsky, 1996; Gopal-Srivastava et al., 1996). Alignment of the Pax-6 consensus binding site with the AhRE sequences of the ***Ahd4*** gene revealed a 55–75% overlap between these two responsive elements (Table 3). More precisely, the entire 5 bp core sequence of the AhRE motif (***GCGTG***; Hankinson, 1995) is found within the binding core of the Pax-6 (Table 3). Such an overlap offers an explanation for the lack of functionality of the AhREs in relation to TCDD inducibility in the corneal epithelium. It is possible that Pax-6, as a corneal regulatory protein, is bound to these DNA motifs in the upstream region of the ***Ahd4*** gene; thus the TCDD-activated AHR-ARNT transcriptional complex does not transactivate ***Ahd4*** gene expression. The involvement of Pax-6 in cornea-specific ***Ahd4*** expression is currently under investigation in our laboratory.

Table 3. Alignment of the AhRE motifs present in the 3.2 kb 5'flanking region of the ***Ahd4*** gene with the Pax-6 consensus sequence

AhREs in the 5' flanking region of the **Ahd4** gene	**Pax-6 concensus*** A G GA T	Matching nucleotides
	ANNTTCACGCTTCANTTCNC	
–101 AhRE1 →	tTCgTCACGCgaacTTcCTg	(12/20)
–386 AhRE2 →	gCTTaCACGCTgtAGacCTT	(14/20)
–388 AhRE3 ←	gCAgTCACGCAgttGTGgGG	(13/20)
–873 AhRE4 ←	cTGcaCACGCcatgGcTCAg	(11/20)
–1204 AhRE5 →	tGTggCACGCTgttATcCTT	(13/20)
–1676 AhRE6 →	ATAcaCACGCccCATcaCTa	(13/20)
–2592 AhRE7 ←	ACATgCACGCTcCcGgGAAg	(15/20)

*This consensus sequence is from (Cvekl and Piatigorsky, 1996 and references therein).

4. CONCLUSIONS

The lack of ***Ahd4*** inducibility by TCDD in corneal cells, along with the overlap in the sequences of the AhRE and Pax-6 DNA binding motifs, suggest that at least two distinct tissue-specific regulatory protein complexes may bind to the same DNA sequences in the upstream region of this gene. It would appear that protein(s) involved in the tissue-specific constitutive expression of the ***Ahd4*** gene have higher binding activities than the TCDD-activated AHR-ARNT complex.

ACKNOWLEDGMENTS

We thank Frank J. Gonzalez for providing tissues of the ***Ahr*** knockout mouse line and David Thompson for a critical reading of this manuscript. This work was supported by NEI Grant 1R29 EY11490.

REFERENCES

Andreola, F., Fernandez-Salguero, P.M., Chiantore, M.V., Petkovich, M.P., Gonzalez, F.J., and De, L.L. (1997) Aryl hydrocarbon receptor knockout mice (AHR-/-) exhibit liver retinoid accumulation and reduced retinoic acid metabolism. *Cancer Res.* **57,** 2835–2838.

Araki-Sasaki, K., Ohashi, Y., Sasabe, T., Hayashi, K., Watanabe, H., Tano, Y., and Handa, H. (1995) An SV40-immortalized human corneal epithelial cell line and its characterization. *Inv.Ophthalmol.Vis.Sci.* **36,** 614–621.

Araki, K., Ohashi, Y., Sasabe, T., Kinoshita, S., Hayashi, K., Yang, X.Z., Hosaka, Y., Aizawa, S., and Handa, H. (1993) Immortalization of rabbit corneal epithelial cells by a recombinant SV40-adenovirus vector. *Invest.Ophthalmol.Vis.Sci.* **34,** 2665–2671.

Bernard, H.P., Darlington, G.J., and Ruddle, F.H. (1973) Expression of liver phenotypes in cultured mouse hepatoma cells: synthesis and secretion of serum albumin. *Dev.Biol.* **35,** 83–96.

Boesch, J.S., Lee, C., and Lindahl, R.G. (1996) Constitutive expression of class 3 aldehyde dehydrogenase in cultured rat corneal epithelium. *J.Biol.Chem.* **271,** 5150–5157.

Chomczynski, P. and Sacchi, N. (1987) Single-step method of RNA isolation by acid guanidium thiocyanate-phenol-chloroform extraction. *Anal.Biochem.* **162,** 156–159.

Cvekl, A. and Piatigorsky, J. (1996) Lens development and crystallin gene expression: many roles for Pax-6. [Review] [72 refs]. *Bioessays* **18,** 621–630.

Downes, J.E. and Holmes, R.S. (1993) Purification and properties of murine corneal aldehyde dehydrogenase. *Biochemistry &.Molecular.Biology.International.* **30,** 525–535.

Dunn, T.J., Lindahl, R., and Pitot, H.C. (1988) Differential gene expression in response to 2,3,7,8- tetrachlorodibenzo-p-dioxin (TCDD). Noncoordinate regulation of a TCDD-induced aldehyde dehydrogenase and cytochrome P-450c in the rat. *J.Biol.Chem.* **263,** 10878–10886.

Fernandez-Salguero, P., Pineau, T., Hilbert, D.M., McPhail, T., Lee, S.S., Kimura, S., Nebert, D.W., Rudikoff, S., Ward, J.M., and Gonzalez, F.J. (1995) Immune system impairment and hepatic fibrosis in mice lacking the dioxin-binding Ah receptor. *Science* **268,** 722–726.

Gopal-Srivastava, R., Cvekl, A., and Piatigorsky, J. (1996) Pax-6 and alphaB-crystallin/small heat shock protein gene regulation in the murine lens. Interaction with the lens-specific regions, LSR1 and LSR2. *J.Biol.Chem.* **271,** 23029–23036.

Guarente, L. (1983) Yeast promoters and lac Z fusions designed to study expression of cloned genes in yeast. *Methods in Enzymol.* **101,** 181–191.

Hankinson, O. (1995) The aryl hydrocarbon receptor complex. *Annu.Rev.Pharmacol.Toxicol.* **35,** 307–340.

Hoffer, A., Chang, C.Y., and Puga, A. (1996) Dioxin induces transcription of fos and jun genes by Ah receptor-dependent and -independent pathways. *Toxicol.Appl.Pharmacol.* **141,** 238–247.

Jaiswal, A.K .(1991) Human NAD(P)H:Quinone oxydoreductase (NQO1) gene structure and induction by dioxin. *Biochemistry* **30,** 10647–10653.

Jaiswal, A.K. (1994) Antioxidant response element. *Biochem.Pharmacol.* **48,** 439–444.

Koroma, B.M., Yang, J.M., and Sundin, O.H. (1997) The Pax-6 homeobox gene is expressed throughout the corneal and conjunctival epithelia. *Invest.Ophthalmol.Vis.Sci.* **38,** 108–120.

Marks-Hull, H., Shiao, T.Y., Araki-Sasaki, K., Traver, R., and Vasiliou, V. (1997) Expression of ALDH3 and NMO1 in human corneal epithelial and breast adenocarcinoma cells. *Adv.Exp.Med.Biol.* **414,** 59–68.

Nebert, D.W., Puga, A., and Vasiliou, V. (1993) Role of the Ah receptor and dioxin inducible [*Ah*] gene battery in toxicity, cancer, and signal transduction. *Ann.New York Acad.Sci.* **685,** 624–640.

Pappas, P., Stephanou, P., Vasiliou, V., Karamanakos, P., and Marselos, M. (1997) Ontogenesis and expression of ALDH activity in the skin and the eye of the rat. *Adv.Exp.Med.Biol.* **414,** 73–80.

Parker, B. and Stark, G. (1979) Regulation of simian virus 40 transcription: Sensitive analysis of the RNA species present early in infections by virous or viral DNA. J. Virol. **31,** 360–369.

Sambrook, J., Fritsch, E.F., and Maniatis, T. (1989) Molecular Cloning: A Laboratory Manual, Cold Spring Harbor Laboratory Press, Cold Spring Harbor, New York.

Shertzer, H.G., Vasiliou, V., Liu, R.M., Tabor, M.W., and Nebert, D.W. (1995) Enzyme induction by L-buthionine (S,R)-sulfoximine in cultured mouse hepatoma cells. *Chem.Res.Toxicol.* **8,** 431–436.

Vasiliou, V. and Marselos, M. (1989) Tissue distribution of inducible aldehyde dehydrogenase activity in the rat after treatment with phenobarbital or methylcholanthrene. *Pharmacol.Toxicol.* **64,** 39–42.

Vasiliou, V., Puga, A., Chang, C.Y., Tabor, M.W., and Nebert, D.W. (1995) Interaction between the Ah receptor and proteins binding to the AP-1-like electrophile response element (EpRE) during murine phase II [Ah] battery gene expression. *Biochem.Pharmacol.* **50,** 2057–2068.

Vasiliou, V., Reuter, S.F., Kozak, C.A., and Nebert, D.W. (1993) Mouse dioxin-inducible cytosolic aldehyde dehydrogenase-3: AHD4 cDNA sequence, genetic mapping, and differences in mRNA levels. *Pharmacogenetics.* **3,** 281–290.

Vasiliou, V., Reuter, S.F., Shiao, T.Y., Puga, A., and Nebert, D.W. (1997) Mouse dioxin-inducible Ahd4 gene. Structure of the 5' flanking region and transcriptional regulation. *Adv.Exp.Med.Biol.* **414,** 37–46.

23

HUMAN CORNEAL AND LENS ALDEHYDE DEHYDROGENASES

Localization and Function(s) of Ocular ALDH1 and ALDH3 Isozymes

Gordon King,[1] Lawrie Hirst,[2] and Roger Holmes[3*]

[1]School of Science
Griffith University
Qld 4111, Australia
[2]Department of Surgery
University of Queensland
Princess Alexandra Hospital
Brisbane Qld 4102, Australia
[3]The University of Newcastle
Callaghan NSW 2308, Australia

1. INTRODUCTION

The mammalian cornea and lens play an important role in protecting the eye from ultra-violet radiation (UVR) induced damage, particularly the photosensitive retina, by the absorption of UVR in the 290–320 nm (UV-B) and 320–400 nm (UV-A) wavelength ranges, respectively (Boettner and Wolters, 1962; Zigman, 1983). The responsible compounds in the cornea have not, as yet, been conclusively established, however, a number of studies have supported a major role for proteins, particularly soluble proteins, in corneal UV-B absorption, especially in the lower part of this range between 290–300 nm (Cogan and Kinsey, 1946; Mitchell and Cenedella, 1995). In contrast, photoreception of UV-A radiation in the mammalian lens is achieved by a variety of chromophores, the best characterized being the water soluble kynurenine derivatives (see Zigman, 1985).

Aldehyde dehydrogenases (ALDHs) are now recognised as a complex gene family, including a group of NAD-dependent ALDH (EC 1.2.1.3) isozymes, which are responsible

* To whom correspondence should be addressed.

for the metabolism of a variety of biological aldehydes in the body (see MacKerrell et al, 1986; Algar and Holmes, 1986). Two of these isozymes, ALDH1 and ALDH3, have been reported in very high levels in human anterior eye tissues (Holmes, 1988; King and Holmes, 1993; 1997). In the human cornea, ALDH1 and ALDH3 constitute approximately 3 and 5 percent respectively, of soluble protein, whereas ALDH1 constitutes 1–2 percent of soluble protein in the human lens (King and Holmes, 1998). It has been proposed that these high levels of ALDH protein and activity may assist in protecting the cornea and lens from UV-induced damage, by detoxifying peroxidic aldehydes generated following UVR absorption, as well as performing a more direct role in the absorption of UV-B radiation (King and Holmes, 1993, 1997, 1998).

In this present investigation, we report on the cellular distribution of ALDH1 and ALDH3 in the human cornea and lens, and provide immunohistochemical evidence that these isozymes are predominantly localized within corneal and lens epithelial cells. In addition, we describe a similar study of the distribution of albumin within these human ocular tissues, since this has been recently reported as the major soluble protein of the human cornea (Zhu and Crouch, 1992; King, 1997).

2. MATERIALS AND METHODS

2.1. Tissue Sources and Extraction

Human corneas and lenses were obtained from the Queensland Eye Bank and stored frozen at -70°C until used.

2.2. Antibodies

Rabbit antihuman corneal ALDH3 polyclonal antibodies were produced from purified human corneal ALDH3 (King and Holmes, 1993). Rabbit antihuman albumin antibodies were purchased from Dako Corporation, Hercules, Denmark. Rabbit antihuman liver ALDH1 polyclonal antibodies were gifts from Drs A.Yoshida, L. Hsu and A. Shibuya of the Beckman Research Institute of the City of Hope, Los Angeles, USA, and prepared as previously described (Ikawa et al., 1983). These antihuman ALDH polyclonal antibodies have been previously shown to be specific for the respective human corneal and lens ALDH1 and ALDH3 isozymes, with no cross-reactivity being observed for the alternative isozyme under the conditions used (King and Holmes, 1998).

2.3. Tissue Sectioning and Immunohistochemical Staining

Human corneas and lenses were fixed in phosphate-buffered 4 percent paraformaldehyde in isotonic sodium chloride, and equilibrated in phosphate-buffered 30 percent sucrose at 4° C. The samples were set in Tissue TEK and sectioned at -20° C in a Minitome cryostat. The tissue sections were then processed for immunohistochemical staining. Reaction with the primary antibodies was performed at room temperature using antibody diluted into 5 percent w/v Dutch Jug milk powder in Tris buffered saline (pH 7.5; 0.9 percent (w/v) in sodium chloride) for one hour. The antibodies were used at 1:150 dilution. Reaction with the secondary antibody was also at room temperature with a 1:3000 dilution of the anti-G in 5 percent Dutch Jug milk powder in Tris buffered saline. Washes were performed following treatments with both the primary and secondary antibodies using 5 percent w/v Dutch Jug milk powder in Tris buffered saline (pH 7.5; 0.9 % (w/v) sodium chloride).

Development of the antibody reactions was performed using the o-tolidine based procedure (Scopsi and Larsson, 1986).

3. RESULTS

3.1. Immunohistochemical Localization of Human Corneal ALDH3, ALDH1, and Albumin

Figure 1 illustrates paraformaldehyde-fixed sections of a human cornea reacted with either rabbit antihuman corneal ALDH3 antibodies, rabbit antihuman serum albumin antibodies or without treatment with the primary antibodies, with the reactions visualised using the horseradish peroxidase method with o-tolidine as the substrate. The antibodies against human corneal ALDH3 reacted intensely with the corneal epithelium (Ep), with reduced staining in or near the endothelial cells (En), and with little reactivity being observed for the stromal region (S). Intense staining was observed for both the epithelial and endothelial human corneal cells, using antibodies against human serum albumin, and again with little reactivity for the stromal region. The control section was essentially free of any histochemical stain. These results support the presence of high levels of ALDH3 and albumin in human corneal epithelial cells. In addition, human corneal endothelial cells apparently contain very high levels of albumin and significant levels of ALDH3.

Similar results were observed for human corneal sections when antibodies were used against human liver ALDH1 (Figure 2), although the staining appeared to be restricted to corneal epithelial cells, with little or no detectable activity being observed for corneal endothelial cells. These results are supportive of a high level of ALDH1 protein within human corneal epithelial cells, with little or no ALDH1 being present in other regions of the human cornea.

3.2. Immunohistochemical Localization of Human Lens ALDH1, ALDH3, and Albumin

Figure 3 illustrates an immunohistochemical stain of a section of human lens, using rabbit antihuman liver ALDH1 antibodies or without treatment with the primary antibodies. The lens section was taken from the anterior of the lens, as is demonstrated by the presence of the epithelium beneath the lens capsule. This result supports the major presence of ALDH1 protein within the epithelium cells of the human lens, as well as in the lens fibres of the outer cortex. No detectable stain was observed in the other regions of the lens cortex.

Whole cryostat sections of paraformaldehyde-fixed samples of the human lens were reacted with rabbit antihuman corneal ALDH3 antibodies, rabbit antihuman serum albumin antibodies, or without treatment with the primary antibodies, and visualised for antigen cross-reactivity. ALDH3 and albumin were apparently stained in the epithelial cells of the human lens and in the lens fibres of the outer cortex. The inset for the upper plate of Figure 4 shows immunohistochemical staining for ALDH3 protein within the epithelial cells lining the lens capsule.

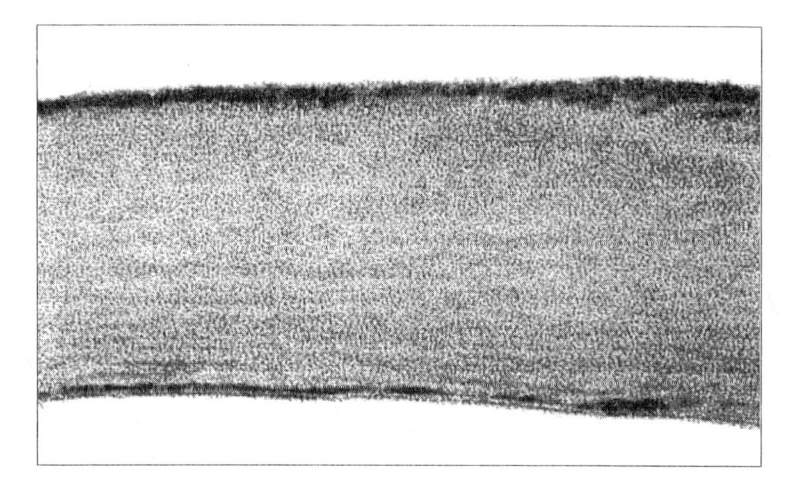

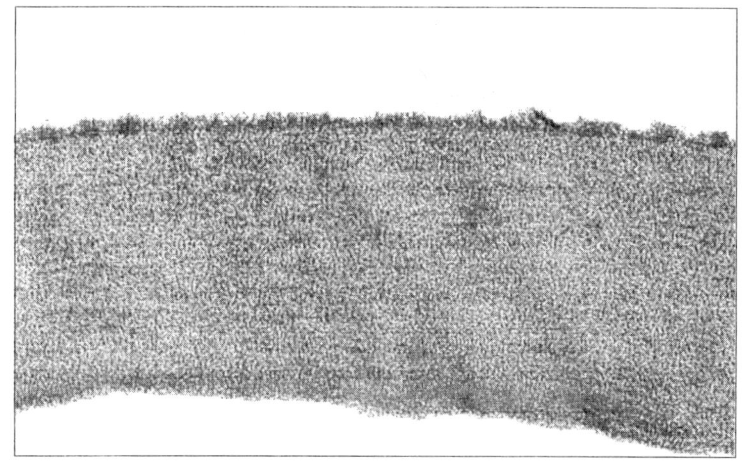

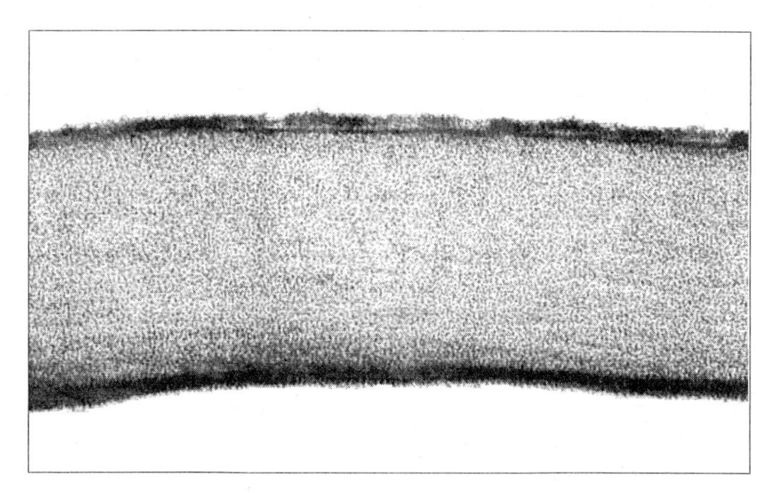

Figure 1. 25-micron tissue sections of paraformaldehyde-fixed samples of a human cornea reacted with antihuman corneal ALDH3 polyclonal antibodies (upper figure); a control sample not exposed to the primary antibody (middle figure); and a sample reacted with antihuman serum albumin polyclonal antibodies (lower figure). The antibody reactions were visualised using the horse preroxidase method with o-tolidine as the substrate.

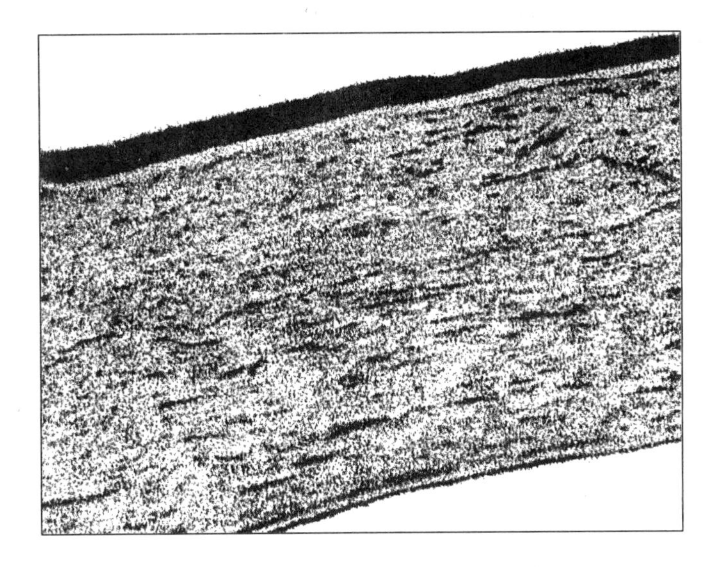

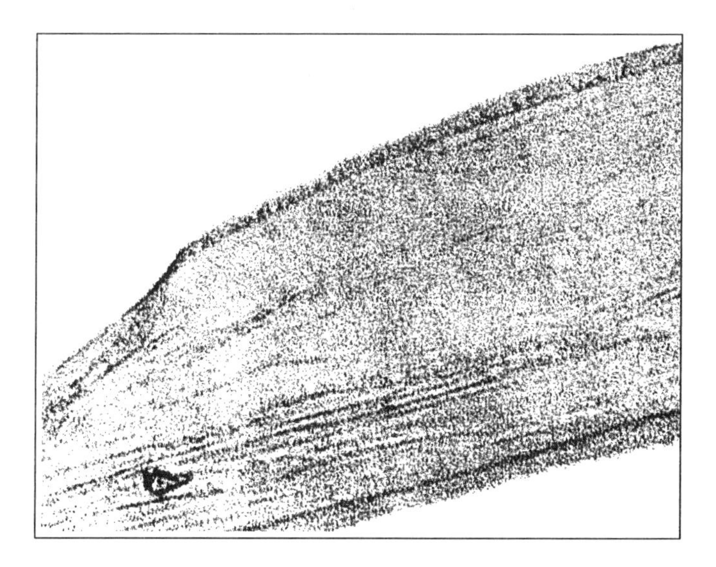

Figure 2. 10-micron tissue sections of paraformaldehyde-fixed samples of a human cornea reacted with antihuman liver ALDH1 polyclonal antibodies (upper figure); and a control section processed without reaction with primary antibody (lower figure). The antibody reaction was visualised as described for Figure 1.

4. DISCUSSION

The results of this study provide evidence that the major soluble proteins of the human cornea, namely albumin (King, 1997), and the ALDH isozymes, ALDH3 (King and Holmes, 1993) and ALDH1 (King and Holmes, 1997, 1998), are present in high levels within the epithelial cells of the human cornea. In addition, these immunohistochemical studies suggest that albumin and ALDH3 are also present within the endothelial cell layer of the human cornea, and that the stroma has little or no albumin and ALDH present. This latter observation may reflect the relatively acellular nature of the stroma within the cornea.

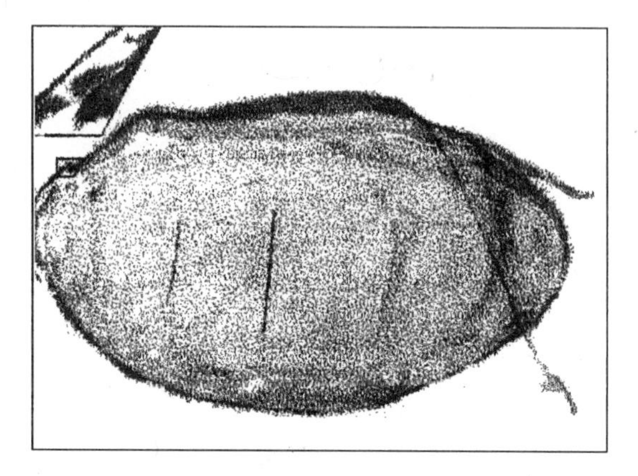

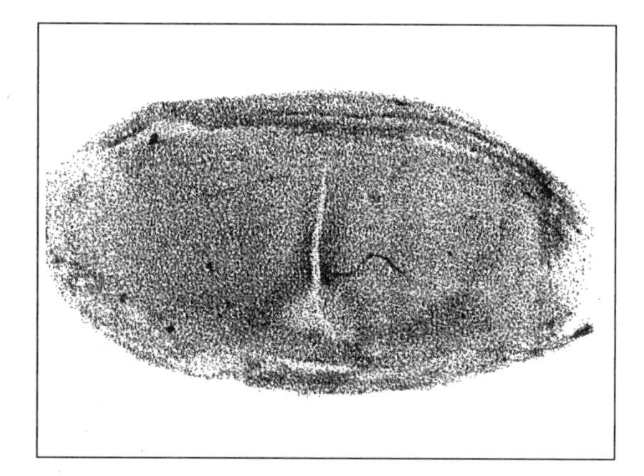

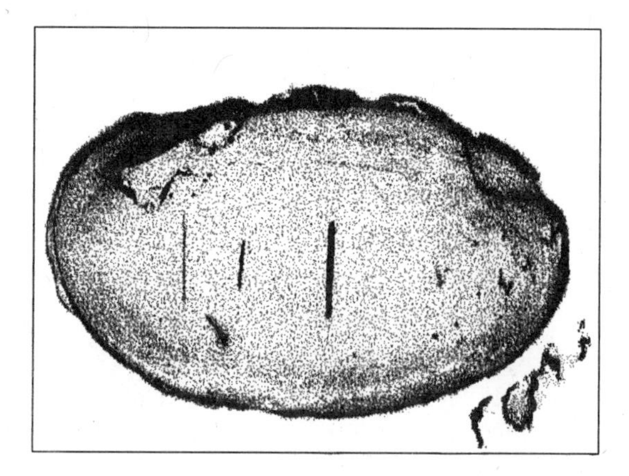

Figure 3. 10-micron tissue sections of paraformaldehyde-fixed samples of a human lens reacted with antihuman liver ALDH1 polyclonal antibodies (upper figure); and a control section processed without reaction with primary antibody (middle figure). The antibody reaction was visualised as desribed for Figure 1.

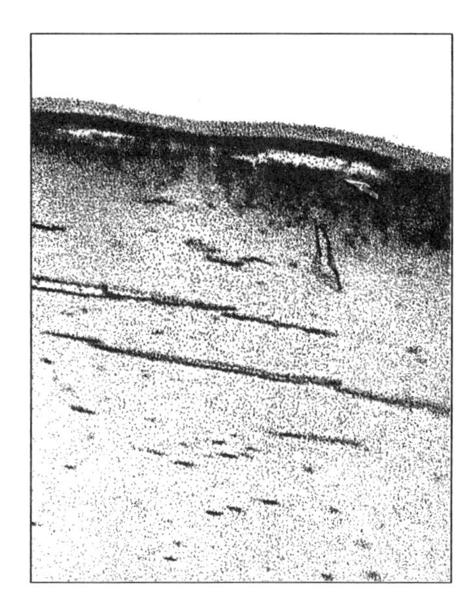

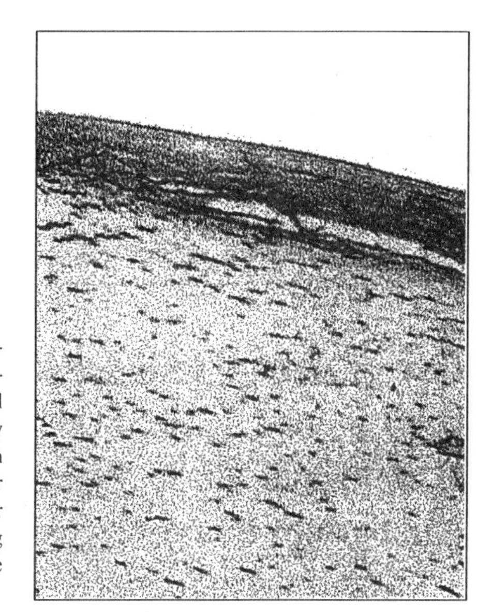

Figure 4. 25-micron tissue sections of paraformaldehyde-fixed samples of a human lens reacted with antihuman corneal ALDH3 polyclonal antibodies (upper figure); a control section processed without reaction with primary antibody (middle figure); and a sample reacted with antihuman serum albumin polyclonal antibodies (lower figure). The inset for the upper figure shows a higher level of magnification for epithelial cells attached to the lens capsule stained following treatment with antihuman corneal ALDH3 antibodies. The antibody reactions were visualised as described for Figure 1.

The earlier studies of Gondhowiardjo et al. (1991; 1993) reported the major presence of ALDH activity within human corneal epithelial extracts, with lower levels in stromal and endothelial extracts, as well as in epithelial extracts from keratoconus corneas. Studies of the regional distribution of ALDH activity within corneas of several mammalian species, including baboon, pig, cattle and sheep, have also shown a high level of activity within the corneal epithelial cells in each case (Downes et al., 1992). Moreover, an investigation of the constitutive expression of rat ALDH3 in cultured corneal epithelial cells has proposed a light inducible-light maintenance mechanism of gene regulation (Boesch et al., 1996). In addition, a recent study has investigated the expression of ALDH3 within mouse corneas using a reporter gene expression method, and demonstrated an exceptionally high level of expression within mouse corneal epithelial cells, with little ALDH3 expression within the keratocytes or corneal endothelial cells (Kays and Piatigorsky, 1997).

Zhu and Crouch (1992) have reported that albumin is the major soluble protein of the rat and human corneas. The latter finding has been recently confirmed by King (1997). The high levels for albumin and for the ALDH isozymes, ALDH3 and ALDH1, within corneal epithelial cells are presumably indicative of a major function or range of functions for these proteins. Given that these cells are the first to be transversed by light entering the eye, and taking account of the previous evidence for a role of corneal soluble proteins in UV-B absorption (Cogan and Kinsey, 1946; Mitchell and Cenedella, 1995), the major presence of albumin, ALDH3 and ALDH1 within the epithelial cells may contribute strongly to the absorption of UV-B by the cornea. To reflect this role, the latter workers have suggested the name ëabsorbinsí for the major soluble proteins of mammalian corneal epithelial cells.

Additional roles for corneal ALDHs, as major soluble proteins or as enzymes with catalytic activities, have been proposed (see King and Holmes, 1998), including: (1) ALDH3 serving as a corneal "crystallin", in a similar way to that described for major soluble proteins in the lens, by maintaining the structural integrity and transparency of the cornea (Cooper et al., 1993); (2) detoxifying peroxidic aldehydes, such as malondialdehyde and medium chain aliphatic aldehydes, generated following UV-B absorption and lipid peroxidation (Holmes and VandeBerg, 1986; Holmes, 1998; Evces and Lindahl, 1989; Abedinia et al., 1990; Algar et al., 1990; Downes and Holmes, 1993; King and Holmes, 1993, 1997); (3) detoxifying free radicals, also generated following lipid peroxidation, through reaction with ALDH sulphydryl groups (Uma etal, 1996); and (4) in the absorption of upper UV-B radiation (300–320 nm) by way of E-NAD(H) complexes, which absorb strongly across this range (King and Holmes, 1997).

Zhu and Crouch (1992) have suggested, on the basis of their *in vitro* studies on the oxidation of corneal proteins with hydrogen peroxide, that corneal albumin, which is the most prevalent soluble protein, may serve a role as an anti-oxidant, scavenging hydrogen peroxide and other oxidising species, and thereby preventing any further oxidative damage within the cornea.

Soluble proteins present at high levels (10% or more of total soluble protein) in the ocular lens of a range of phyla have been termed lens crystallins (see Piatigorsky and Wistow, 1991).

Although human lens ALDHs do not reach these very high levels of protein concentration within the lens overall, the current immunohistochemical studies (Figures 3 and 4) demonstrate that ALDH1 is concentrated within the lens epithelial cell layer, and may therefore serve as a ëcrystallin í protein for the lens epithelial cells. Previous studies have shown that some lens crystallins are metabolic enzymes, which have been recruited in a taxon-specific manner to serve a structural role. Recruitment of soluble proteins as lens crystallins has been reported under two different mechanisms: (1) by gene duplications, as for the α-crystallins and β/γ crystallins, for which single and multiple gene duplications have occurred, respectively; and (2) by gene sharing, for the taxon-specific enzyme crystallins, where a gene for an enzyme has been recruited for a high level of lens expression (see Piatigorsky and Wistow, 1991; Wistow and Kim, 1991; Tomarev and Piatigorsky, 1996).

A number of studies have reported on the induction of UV-A (~360 nm) (Zigman and Vaughan, 1974; Zuclich and Connolly, 1976) and UV-B (290–340 nm) cataracts in experimental animals (Pitts et al, 1977; Jose and Pitts, 1985). These reports supported an hypothesis that UV radiation is a causative factor in human cataracts and in lens ageing. More recently, Stuart et al., (1993) have reported evidence that the UV-B waveband (300–315 nm) is more damaging to the lens than the UV-A waveband (315–400). In addition, Hightower et al (1994) have investigated the effects of UV-B upon cultured rabbit lenses, and reported a 4-fold increase in several primary lipid peroxidation products within membranes of the lens. These results suggest that UV-B radiation may induce cataract formation in the lens, and that lipid peroxidation may contribute to this process. Residual UV-B radiation passing through the cornea and other anterior eye tissue and fluid would have to transverse the lens epithelial cell layer prior to entry into the lens. It is our hypothesis that ALDH1 and albumin, as the major soluble proteins within this cell layer, may perform a similar role to those proposed for the major corneal epithelial cell soluble proteins. Thus, lens epithelial cells which contain very high levels of ALDH1 and albumin, provide a further barrier to the entry of UV-B radiation within the major cortex region of the lens. This latter region contains the normally transparent lens fibre cells and the nucleus of the lens, which are subject to radiation induced damage and cataract formation (Zigman, 1985). The absence of ALDH1, ALDH3 and albumin protein within the lens fibre cells and lens nucleus demonstrates that these proteins have not been recruited as major proteins for the lens overall.

In summary, these studies have demonstrated that the major soluble proteins of the human cornea, namely albumin, ALDH3 and ALDH1 are predominantly localized within the epithelial cells, and may perform a variety of functions in protecting the cornea against UV-B induced tissue damage, namely by the direct absorption of UV-B radiation, and/or by the detoxification of UVR induced lipid peroxidic aldehyde by-products. In addition, a study of the distribution of these proteins within the human lens has demonstrated that these proteins are also localized within the epithelial cell layer, and may serve a similar range of functions to those proposed for human corneal albumin, ALDH1 and ALDH3.

ACKNOWLEDGMENTS

The human eye samples were kindly supplied by Dr. Peter Madden of the Queensland Eye Bank. We acknowledge the expert assistance of Ms. Josephine Bancroft with eye tissue sectioning and immunohistochemical staining procedures. This research was supported by a grant from the National Health and Medical Research Council of Australia (94/1098).

REFERENCES

Abedinia, M., Pain, T., Algar, E.M. and Holmes, R.S.: Bovine corneal aldehyde dehydrogenase: the major soluble corneal protein with a possible dual protective role for the eye. Exp. Eye Res. 51 (1990) 419–426.

Algar, E.M. and Holmes, R.S.: Kinetic properties of murine liver aldehyde dehydrogenases. In Weiner, H. and Flynn, T.G. (Eds), Enzymology and Molecular Biology of Carbonyl Metabolism 2. Alan R. Liss, N.Y., 1989 pp 93–103.

Algar, E.M., Abedinia, M., VandeBerg, J.L. and Holmes, R.S.: Purification and properties of baboon aldehyde dehydrogenase: proposed UVR protection role. In Weiner, H., Crabb, D.W. and Flynn, T.G. (Eds), Enzymology and Molecular Biology of Carbonyl Metabolism 3. Plenum Press, N.Y., 1990 pp 53–60.

Boesch, J.S., Lee, C. and Lindahl, R.G.: Constitutive expression of class 3 aldehyde dehydrogenase in cultured rat corneal epithelium. J. Biol. Chem. 271 (1996) 13011–13018.

Boettner E.A. and Wolters J.R.: Transmittance of the ocular media. Invest. Ophthalmol. 1 (1962) 776–783.

Cogan, D.G. and Kinsey, V.E.: Action spectrum of keratitis produced by ultraviolet radiation. Arch. Ophthalmol. 55 (1946) 670–677.

Cooper D.L., Isola N.R., Stevenson K. and Baptist E.W.: Members of the ALDH gene family are lens and corneal crystallins. In Weiner, H., Crabb, D.W., and Flynn, G.F. (Eds), Enzymology and Molecular Biology of Carbonyl Metabolism 4. Plenum Press, New York, 1993, pp 169–179.

Downes, J.E. and Holmes, R.S.: Purification and properties of murine aldehyde dehydrogenase. Biochem. Mol. Biol. Intl. 30 (1993) 525–535.
Downes, J.E., VandeBerg, J.L., Hubbard, G.B. and Holmes, R.S.: Regional distribution of mammalian corneal aldehyde dehydrogenase and alcohol dehydrogenase. Cornea 11 (1992) 560–566.
Evces, S. and Lindahl, R.: Characterization of rat cornea aldehyde dehydrogenase. Arch. Biochem. Biophys. 274 (1989) 518–524.
Gondhowiardgo, T.D., van Haeringen, N.J., Hoekzema, R., Pels, L. and Kijlstra, A.: Detection of aldehyde dehydrogenase activity in human corneal extracts. Curr. Eye Res. 10 (1991) 1001–1007.
Gondhowiardgo, T.D., van Haeringen, N.J., Volker-Dieben, H.J., Beekhuis, H.W., Kok, J.H.C., van Rij, G., Pels, L. and Kijlstra, A.: Analysis of corneal aldehyde dehydrogenase patterns in pathologic corneas. Cornea 12 (1993) 146–154.
Hightower, K., McGready, J. and Borchman, D.: Membrane damage in UV-irradiated lenses. Photochem. Photobiol. 59 (1994) 485–490.
Holmes, R.S. and VandeBerg, J.L.: Ocular NAD-dependent alcohol dehydrogenase and aldehyde dehydrogenase in the baboon. Exp. Eye Res. 43 (1986) 383–396.
Holmes, R.S.: Alcohol dehydrogenases and aldehyde dehydrogenases of anterior eye tissues from humans and other mammals. In Kuriyama, K., Takada, A. and Ishii, H. (Eds.), Biomedical and Social Aspects of Alcohol and Alcoholism. Elsevier Science Publishers, Amsterdam (1988), pp 51–57.
Ikawa, M., Wang, G. and Yoshida, A.: Isolation and characterization of aldehyde dehydrogenase isozymes from usual and atypical human livers. J. Biol. Chem. 258 (1983) 6282–6287.
Jose, J.G. and Pitts, D.G.: Wavelength dependency of cataracts in albino mice following chronic exposure. Exp. Eye Res. 41 (1985) 545–563.
Kays, T.W. and Piatigorsky, J.: Aldehyde dehydrogenase class 3 expression: identification of a cornea-preferred gene promoter in transgenic mice. Proc. Natl. Acad. Sci. USA 94 (1997) 13594–13599.
King, G.J.: Aldehyde dehydrogenases of the human eye. PhD Thesis. 1997. Griffith University, Brisbane, Australia.
King, G. and Holmes, R.S.: Human corneal aldehyde dehydrogenase: purification, kinetic characterization and phenotypic variation. Biochem. Mol. Biol. Intl. 31 (1993) 49–63.
King, G. and Holmes, R.S.: Human corneal and lens aldehyde dehydrogenases. Purification and properties of human lens ALDH1 and differential expression as major soluble proteins in human lens (ALDH1) and cornea (ALDH3). In Weiner.H., Crabb, D.W. and Flynn, T.G. (Eds), Enzymology and Molecular Biology of Carbonyl Metabolism 6. Plenum Press, N.Y., 1997, pp:19–27.
King, G. and Holmes, R.S.: Human ocular aldehyde dehydrogenase isozymes: distribution and properties as major soluble proteins in cornea and lens. J. Exp. Zool. (1998) in press.
MacKerrell, A.D., Blatter, E.E. and Pietruszko, R.: Human aldehyde dehydrogenase: kinetic identification of the isozyme for which biogenic aldehydes and acetaldehyde compete. Alcoholism: Clin. Exp. Res. 10 (1986) 266–270.
Mitchell, J. and Cenedella, R.J.: Quantification of ultraviolet light-absorbing fractions of the cornea. Cornea 14 (1995) 266–272.
Piatigorsky, J. and Wistow, G.: The recruitment of crystallins: new functions precede gene duplication. Science 252 (1991) 1078–1079.
Pitts, D.G., Cullen, A.P. and Hacker, P.D.: The ocular effects of ultraviolet radiation from 295 to 365 nm. Invest. Ophthalmol. & Visual Sci. 16 (1977) 932–939.
Scopsi, L. and Larsson, L.-I.: Increased sensitivity in peroxidase immunochemistry: A comparative study of a number of peroxidase visualisation methods employing a model system. J. Histochem. 84 (1986) 221–230.
Stuart, D.D., Cullen, J.G. and Doughty, M.J.: Optical effects of UV-A and UV-B radiation on the cultured bovine lens. Curr. Eye Res. 13 (1994) 371–376.
Tomarev, S.I. and Piatigorsky, J.: Lens crystallins of invertebrates: diversity and recruitment from detoxification enzymes and novel proteins. Eur. J. Biochem. 235 (1996) 449–465.
Uma, L., Hariharan, J., Sharma, Y. and Balasubramanian, D.: Corneal aldehyde dehydrogenase displays antioxidant properties. Exp. Eye Res. 63 (1996) 117–120.
Wistow, G. and Kim, H.: Lens proteins expression in mammals: taxon-specificity and the recruitment of crystallins. J. Mol. Evol. 32 (1991) 262–269.
Zhu, L. and Crouch, R.K.: Albumin in the cornea is oxidized by hydrogen peroxide. Cornea 11 (1992) 567–572.
Zigman S.: The role of sunlight in human cataract formation. Surv. Ophthalmol. 27 (1983) 317- 326.
Zigman, S.: Photobiology of the lens. In Maisel, H. (Ed.) The Ocular Lens. Marcel Dekker Inc., 1985, pp301–347.
Zigman, S. and Vaughan, T.: Near-ultraviolet light effects on the lenses and retinas of mice. Invest. Opthalmol. 13 (1974) 462–465.
Zuclich, S, and Connolly, J.S.: Ocular damage induced by near-ultraviolet laser radiation. Invest. Ophthalmol. 15 (1976) 760–764.

24

THE ROLE OF RETINOID METABOLISM BY ALCOHOL AND ALDEHYDE DEHYDROGENASES IN DIFFERENTIATION OF CULTURED NEURONAL CELLS

Treena J. Blythe,[1] Mark L. Grimes,[1] and Kathryn E. Kitson[2]

[1]Institute of Molecular Biosciences
[2]Institute of Food, Nutrition and Human Health
Massey University
Palmerston North, New Zealand

1. INTRODUCTION

Retinoic acid is a morphogen that plays important roles during fetal growth in specification of head and limb bud development (Hofman and Eichele, 1994), and probably also in regulation of neuronal development (Holder and Maden, 1992). Exposure of the developing fetus to ethanol can cause a range of deformities and neurological abnormalities, leading to conditions known in humans as fetal alcohol syndrome (FAS) and alcohol related neurodevelopmental disorder (ARND) (Sampson et al., 1997). Retinoic acid is formed from retinol (vitamin A) via oxidation first to retinal, catalysed by isozymes of alcohol dehydrogenase (ADH) or the short chain dehydrogenase/reductase family. Retinal is then oxidised to retinoic acid by aldehyde dehydrogenase (AlDH), retinal dehydrogenase or microsomal cytochrome P450 1A1 or 1A2 (Duester, 1996). Retinoic acid has its effect on fetal development through binding to nuclear receptors and forming transcription factors which then regulate gene expression (De Luca, 1991; Chambon, 1996). It has been postulated that one of the mechanisms leading to the development of FAS or ARND may be inhibition of retinoic acid production caused by competitive inhibition of alcohol and/or aldehyde dehydrogenases during ethanol metabolism (Duester, 1991, Deltour et al., 1996). The present study was initiated to see whether cultured neuronal cells could be used as a model system in which to study the effects of ethanol and acetaldehyde on retinoid-regulated neuronal cell differentiation. A human neuroblastoma cell line, SH-SY5Y, was chosen because it had previously been shown to be responsive to retinoic acid (Pahlman et al., 1984).

Enzymology and Molecular Biology of Carbonyl Metabolism 7
edited by Weiner *et al.* Kluwer Academic / Plenum Publishers, New York, 1999.

2. MATERIAL AND METHODS

Human neuroblastoma SH-SY5Y cells (obtained from M. Israel, Dept. of Neurology, University of California, San Francisco) were grown in RPMI 1640 medium, supplemented with 10% (v/v) fetal calf serum. To assess differentiation, confluent cells were re-plated onto coverslips on fresh plates, grown for 48 hours, then effectors (retinal, retinoic acid, ethanol, acetaldehyde, disulfiram) were added at required concentrations, using dimethylformamide (DMF) as a solvent. Control plates had either no addition, or solvent only. Cells were then viewed at 0, 24 or 48 hours using a Zeiss Axioskop microscope to assess cell morphology, and the percentage of differentiated cells was calculated. A differentiated cell was defined as having a neurite of length at least 1.5 times that of the cell body. To test for expression of Class 1 AlDH protein, SH-SY5Y cells were harvested, lysed, and the cytosolic contents separated using SDS polyacrylamide gel electrophoresis. The proteins were then electroblotted onto nitrocellulose membrane and probed with sheep Class 1 AlDH antibodies, with purified sheep and human Class 1 AlDH protein used as controls.

3. RESULTS AND DISCUSSION

3.1. Presence of Class 1 Aldehyde Dehydrogenase in Cultured SH-SY5Y Cells

As an initial step, we investigated whether cultured SH-SY5Y cells grown under the experimental conditions of this study did produce Class 1 AlDH, as described in Methods. A clear reaction to the Class 1 AlDH antibody was seen in the cell extracts, at a position that coincided exactly with that of standards of purified sheep or human Class 1 AlDH protein. Attempts were made to extract and assay active enzyme from the cultured cells, with the aim of characterising the enzyme with respect to substrate specificity. However, sufficient cell material was not available to extract useable quantities of active enzyme, so further characterisation was not possible.

3.2. Differentiation of Cultured Neuronal Cells in Response to Retinal and Retinoic Acid

When SH-SY5Y cells are cultured in the absence of effectors, they form clumps of cells that are rounded in appearance with numerous small filopodia. When treated with appropriate effectors, these cells undergo differentiation which leads to the cells separating from each other and becoming bipolar in form, with processes of up to twice the length of the cell body. As expected from previously published work (Pahlman et al., 1984), there was a significant increase in the percentage of differentiated cells in the presence of 1 μM all-*trans*-retinoic acid, as compared to cells grown with no additions or with solvent (DMF) alone (Figure 1). In addition, the presence of 1 μM all-*trans*-retinal also led to a significant increase in cell differentiation at both 24 and 48 hours (Figure 1), although the percentage of differentiated cells (about 20%) was less than that seen with retinoic acid (about 30%).

These data suggest that the cells metabolise added all-*trans*-retinal to retinoic acid, a reaction that is likely to be catalysed by Class 1 AlDH which has a K_m for retinal in the

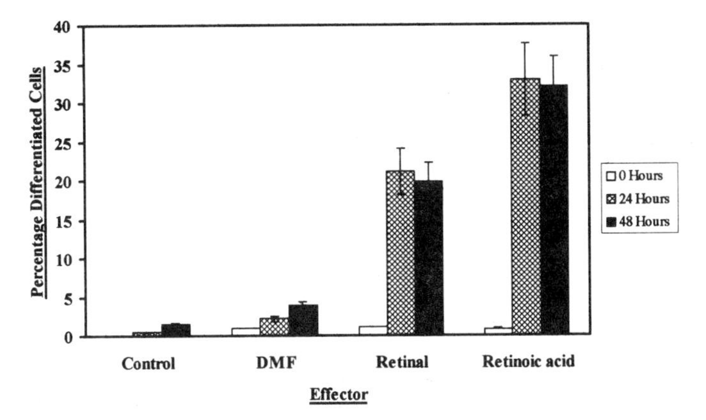

Figure 1. Effects of retinal and retinoic acid on cell differentiation. Cells were grown as described in the Methods section, and exposed to 1 μM concentrations of either all-*trans*- retinoic acid or all-*trans*-retinal as indicated. Control preparations had either no addition or solvent (DMF) only. Cell morphology was quantified by counting the total number of cells (250–300 in each case) and the number of differentiated cells and expressing the number of differentiated cells as a percentage of the total. Results are the mean of 5 experiments ± SEM.

sub-micromolar range (Kitson and Blythe, this volume). It is understandable that the net effect of added retinal at a concentration of 1 μM would not be as great as that of retinoic acid at the same concentration, since some of the added retinal is likely to be reduced to retinol by alcohol dehydrogenase in the cells.

3.3. Effect of All-*Trans*-Retinol and 9-*Cis*-Retinal on Cell Differentiation

The results of the experiment above confirm that the SY5Y cells respond to both all-*tran*- retinoic acid and all-*trans*-retinal, and the results suggest that some retinal might be converted back to retinol in the cells. This is an indication that enzymes which convert retinol to retinal are likely to be present, since the ADH enzymes catalyse reversible reactions, and suggests that the cells should also differentiate in response to addition of retinol. This is confirmed by the results shown in Figure 2. The addition of 1 μM all-*trans*-retinol gave much the same degree of cell differentiation as did all-*trans*-retinal.

In addition to oxidising all-*trans*-retinal, Class 1 AlDH also efficiently oxidises 9-*cis*-retinal. The results in Figure 2 indicate that 9-*cis*-retinal promotes cell differentiation with the same efficiency as the all-*trans* isomer. These results indicate that the SY5Y cell line provides an excellent experimental tool for studying the differentiation of neuronal cells in response to retinoids, and provides the opportunity to study factors that may influence the interconversion of retinol and retinal and the formation of retinoic acid from retinal.

3.4. The Influence of Ethanol, Acetaldehyde, and Disulfiram on Retinal-Induced Cell Differentiation

Some preliminary experiments were carried out to assess the possible effects of ethanol and acetaldehyde on cell differentiation promoted by retinal. It is reasonably well established that ethanol can inhibit the ADH-catalysed conversion of retinol to retinal (Deltour et al., 1996). It could be assumed that acetaldehyde, either added directly to cell

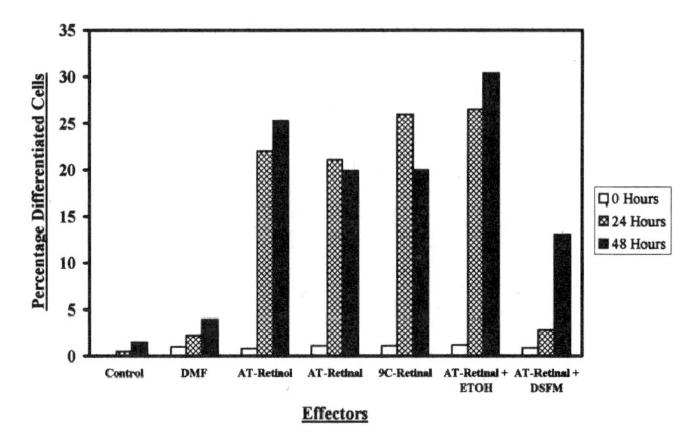

Figure 2. The effects of retinol, 9-*cis*-retinal, ethanol and disulfiram on cell differentiation. Cells were grown and cell morphology quantified as described for Figure 1. The concentration of retinoids was 1 μM, ethanol 50 μM and disulfiram 10 μM. Results are an average of two experiments.

cultures or formed as a result of ethanol metabolism, might inhibit the oxidation of retinal to retinoic acid catalysed by AlDH. In these experiments the direct addition of acetaldehyde (50 μM) caused the cultured neuronal cells to die. There were no viable cells present at any time intervals in acetaldehyde-treated samples. The addition of ethanol (50 μM) in addition to all-*trans*-retinal, however, did not decrease the amount of cell differentiation (Figure 2). This was an unexpected result, as it had been predicted that acetaldehyde formed from the added ethanol might inhibit the oxidation of retinal to retinoic acid by AlDH, and thereby inhibit cell differentiation. A further test with the addition of disulfiram (10 μM), which is a potent inactivator of Class 1 AlDH, indicated that at 24 hours, all-*trans*-retinal-stimulated differentiation was completely prevented by this inhibitor (Figure 2). This supports the involvement of Class 1 AlDH in the metabolism of retinal in these cells. At 48 hours, some differentiation was apparent. By this stage, all added disulfiram would probably have been reduced or metabolised, and there would have been time for protein synthesis to occur to replace some of the inactivated AlDH. These results, although preliminary, indicate that cultured SH-SY5Y cells provide a method for studying the effects of substrates and inhibitors of ADH and AlDH on retinoic-acid induced neuronal cell differentiation.

4. CONCLUSION

There are three possible mechanisms that can be suggested to explain the fact that ethanol did not inhibit retinal-induced cell differentiation. Firstly, it is possible that acetaldehyde does not compete effectively with retinal as a substrate for Class 1 AlDH. Secondly, it is possible that ethanol, or acetaldehyde produced from ethanol, causes an induction of AlDH, so that when the added ethanol has been metabolised and no further acetaldehyde is formed, the net result is an increase in the conversion of retinal to retinoic acid. A third explanation is that the presence of ethanol prevents ADH-catalysed reduction of retinal to retinol, so that more retinal is available for conversion to retinoic acid. The latter two possibilities provide intriguing alternatives to the theory that the effect of ethanol is to inhibit retinoic acid formation; perhaps it is an excess formation of retinoic acid that leads to the changes in fetal development that are caused by ethanol. Indeed, the alterations in fetal development seen with ethanol are, in some respects, similar to those seen with excess retinol/vitamin A in the diet. This was pointed out in one of the first papers suggesting a link between fetal alcohol syndrome and the effects of ethanol on retinoid metabolism (Duester, 1991). In addition, another report has shown that while ethanol inhibited the oxidation of retinal to retinoic acid in liver extracts, the formation of retinoic acid from retinal in rat conceptal cytosol was increased by ethanol (Chen et al., 1996). The fact that different tissues respond to ethanol differently, and may use different enzymes to oxidise retinal to retinoic acid (Chen et al., 1995), suggests an alternative hypothesis to explain developmental changes caused by maternal alcohol intake. It is possible that during alcohol intake by the mother, the developing fetus is exposed to large fluctuations in retinoic acid concentration depending on the particular tissue and time after ethanol intake. Retinoic acid concentrations might be lower than normal when ethanol is present at high concentrations and prevents retinol oxidation by ADH. The concentration might fluctuate to higher than normal as induced enzyme activity and accumulated retinal cause a high rate of retinoic acid production once the ethanol has been metabolised. For a morphogen for which the correct developmental effects are dependent on precise intracellular concentration gradients (Hofman and Eichele, 1994), such fluctuations in concentration would have the potential to cause dramatic changes in fetal development.

Further studies using SH-SY5Y cells would provide an excellent means by which to investigate whether ethanol leads to an increase, a decrease, or to large fluctuations in retinoic acid concentrations in neuronal cells. Such studies would be greatly assisted by development of a direct assay method for retinol, retinal and retinoic acid that is sensitive enough to detect the amounts present in harvested cells. An assay based on previously-published methods (Napoli and Race, 1990; Martini and Murray, 1994) was developed in this study, but the extraction of retinoids from samples was not sufficiently reproducible at the very low concentrations required to allow accurate measurement of retinoid concentrations in the cells.

ACKNOWLEDGMENTS

We wish to acknowledge funding support from: The Neurological Foundation of New Zealand, The New Zealand Lottery Grants Board, The Palmerston North Medical Research Foundation, and a Massey University Postgraduate Scholarship (T. J. Blythe).

REFERENCES

Chambon, P. (1996) A decade of molecular biology of retinoic acid receptors, *FASEB Journal* **10**, 940–954.

Chen, H., Namkung, M.J. and Juchau, M.R. (1995) Biotransformation of all-*trans*-retinol and all-*trans*-retinal to all-*trans*-retinoic acid in rat conceptal homogenates, *Biochem. Pharmacol.* **50**, 1257–1264.

Chen, H., Namkung, M.J. and Juchau, M.R. (1996) Effects of ethanol on biotransformation of all-*trans*-retinol and all-*trans*-retinal to all-*trans*-retinoic acid in rat conceptal cytosol, *Alc. Clin. Exp. Res* **20**, 942–947.

Deltour, L., Ang, H.L. and Duester, G. (1996) Ethanol inhibition of retinoic acid synthesis as a potential mechanism for fetal alcohol syndrome, *FASEB Journal* **10**, 1050–1057.

De Luca, L.M. (1991) Retinoids and their receptors in differentiation, embryogenesis, and neoplasia, *FASEB Journal* **5**, 2924–2932.

Duester, G. (1991) Human liver alcohol dehydrogenase gene expression. Retinoic acid homeostasis and fetal alcohol syndrome. In Watson, R.R. (Ed.), *Drug and Alcohol Abuse Reviews, Vol. 2: Liver Pathology and Alcohol*, Humana Press, Inc., pp. 375–402.

Duester, G. (1996) Involvement of alcohol dehydrogenase, short-chain dehydrogenase /reductase, aldehyde dehydrogenase, and cytochrome P450 in the control of retinoid signalling by activation of retinoic acid synthesis, *Biochemistry* **35**, 12221–12227.

Hofman, C. and Eichele, G. (1994) Retinoids in development. In Sporn, M.B., Roberts, A.B. and Goodman, D.S. (Eds), *The Retinoids: Biology, Chemistry and Medicine*, 2nd Edition, Raven Press, Ltd., New York, pp. 387–441.

Holder, N. and Maden, M. (1992) Mice with half a mind, *Nature* **360**, 708.

Martini, R. and Murray, M. (1994) Retinal dehydrogenation and retinoic acid 4-hydroxylation in rat hepatic microsomes: Developmental studies and effect of foreign compounds on the activities, *Biochem. Pharmacol.* **47**, 905–909.

Napoli, J.L. and Race, K.R. (1990) Microsomes convert retinol and retinal into retinoic acid and interfere in the conversions catalysed by cytosol, *Biochim. Biophys. Acta* **1034**, 228–232.

Pahlman, S., Ruusala, A.-I., Abrahamsson, L., Mattsson, M.E.K. and Esscher, T. (1984) Retinoic acid-induced differentiation of cultured human neuroblastoma cells: comparison with phorbolester-induced differentiation, *Cell Differnetiation*, **14**, 135–144.

Sampson, P.D., Streissguth, A.P., Bookstein, F.L., Little, R.E., Clarren, S.K., DeHaene, P., Hanson, J.W. and Graham, J.M. (1997) Incidence of fetal alcohol syndrome and prevalence of alcohol-related neurodevelopmental disorder, *Teratology* **56**, 317–326.

25

METABOLISM OF RETINALDEHYDE BY HUMAN LIVER AND KIDNEY

Wojciech Ambroziak, Gonzalo Izaguirre, Darryl Abriola, Ming-Kai Chern, and Regina Pietruszko

Center of Alcohol Studies and Department of Molecular Biology and Biochemistry
Rutgers The State University of New Jersey
Piscataway, New Jersey 08854-8001

1. INTRODUCTION

In mammalian organisms retinoids and their derivatives are important in regulation of diverse physiological functions. Retinoic acid, the most potent of naturally occurring retinoids, only recently has been recognized as a major hormone in cell differentiation and development (Gudas, 1994). In mammals biosynthesis of retinoids proceeds via central or excentric cleavage of carotene to retinaldehyde followed by its reduction to retinol or oxidation to retinoic acid (Goodman and Huang, 1965; Olson and Hyashi, 1965; Wang et al., 1996b).

Ethanol was shown to inhibit retinaldehyde oxidation *in vitro,* and ethanol treatment of mouse embryos has been demonstrated to reduce endogenous retinoic acid levels (Deltour et al., 1996). Chronic alcohol consumption and alcoholism lead to defects observed in vitamin A deficiency (Duester, 1996). Fetal alcohol syndrome may also result from retinoic acid deficiency.

It has been shown that retinoic acid formation is inhibited by citral (Connor and Smit, 1987) and disulfiram (McCaffery et al., 1992), both known aldehyde dehydrogenase inhibitors, and that such inhibition produced abnormalities in developing embryos. It was also reported that NAD-dependent aldehyde dehydrogenase was able to metabolize retinaldehyde (Dockham et al., 1992; Yoshida et al., 1992). Aldehyde dehydrogenases, claimed to be "specific" towards *all trans* retinaldehyde, were recently purified and cloned from rat and mouse tissues (Wang et al., 1996a; Zhao et al., 1996). Both showed structural similarity to NAD-linked aldehyde dehydrogenase (EC 1.2.1.3) with broad substrate specificity. An attempt was made during this investigation to identify the enzyme that metabolizes retinaldehyde to reti-

Enzymology and Molecular Biology of Carbonyl Metabolism 7
edited by Weiner *et al.* Kluwer Academic / Plenum Publishers, New York, 1999.

noic acid in mature human liver and kidney. The results presented demonstrate involvement of aldehyde dehydrogenase with broad substrate specificity in retinoic acid metabolism.

2. MATERIALS AND METHODS

Human livers and kidneys obtained from autopsy by NDRI (Philadelphia, PA) were used during this investigation. Purification procedures, conducted under nitrogen and followed by activity with retinaldehyde and propionaldehyde, used methods previously described (Hempel et al., 1982), employing consecutive chromatographic steps on CM-Sephadex, DEAE-Sephadex and 5'AMP Sepharose 4B columns. The separation method of E1 and E2 isozymes using a 5'AMP column is new and is being prepared for publication. During purification we carefully assayed proteins that did not attach or firmly attached to all columns for ALDH activity. Aldehyde dehydrogenase activity was measured in 0.1M sodium pyrophosphate buffer, containing 500 μM NAD, 1 mM propionaldehyde and 1mM EDTA at pH 9.0; an assay using 0.2 mM octanaldehyde, instead of propionaldehyde, was also employed in the same conditions. Retinaldehyde activity (incubation conditions see Results and Discussion) was determined by HPLC on Waters μ Bondapak C_{18} column with isocratic elution of acetonitrile and 1% ammonium acetate (80:20% v/v) at 1.25 ml/min flow rate and 340 nm detection. All work was done under gold lights. Kinetic studies were done in 0.05M Tris/HCl buffer, pH 7.6 or 0.1M Tris/glycine buffer, pH 9.0; Km and V_{max} values were calculated by non-linear regression by Lineweaver-Burk (1934) procedure. Protein concentration was determined by the microbiuret procedure (Goa, 1953). Electrophoretic procedures were employed for determination of enzyme homogeneity, native and subunit molecular weights, and pI values.

3. RESULTS AND DISCUSSION

3.1. Assay System for Retinaldehyde Activity Used during This Investigation

Before starting purification, a spectrophotometric activity assay measuring retinaldehyde disappearance at 425 nm was attempted. During the first purification attempt it was discovered that the assay method was unreliable, due to retinoid-binding proteins copurifying with aldehyde dehydrogenase. It was decided, therefore, to use low enzyme concentrations during incubation and the HPLC procedure and to determine retinoic acid formed.

Preliminary investigation of the relationship of retinaldehyde to propionaldehyde activity was done to determine the amount of enzyme to be used in incubations with retinaldehyde as substrate. All experiments were done under nitrogen and gold lights. Incubations were done at 25°C in 0.1 M Tris-glycine buffer, pH 9.0, containing 0.5 mM NAD, 20 μM *all trans* retinaldehyde and 1 mM EDTA in 1 ml total volume. The reaction was initiated after 5 min preincubation at 25°C by addition of retinaldehyde dissolved in 10 μl ethanol. Total enzyme activity was adjusted to obtain linear rates up to 40 min incubation and to assure that conversion of substrate at 20 min incubation was not greater than 30%. After 10–20 min the reaction was terminated by rapid freezing in dry ice/ethanol and stored, if necessary, at -70°C before determination of retinoic acid formed via HPLC.

Table 1. Propionaldehyde and retinaldehyde activity of human liver homogenates

Liver source	IEF ID	Propionaldehyde μmol/min/g	Retinaldehyde μmol/min/g	Retinaldehyde as propionaldehyde ratio
F (C)	E1, E2	0.58	0.023	4.0
F (C)	E1, E2	0.77	0.048	6.2
F (C)	E1, E2	1.09	0.061	5.6
F (O)	E1 –	0.14	0.016	11.4
M (O)	E1 –	0.36	0.052	14.4
F (O)	E1 –	0.55	0.075	13.6
F (O)	E1, E2	1.05	0.060	5.7

Data shown are per gram of liver (wet weight). F = female; M = male; C = Caucasian; O = Oriental. Aldehyde dehydrogenase isozymes were identified by isoelectric focusing (IEF) and staining for activity with propionaldehyde as substrate. Results are mean of three determinations of propionaldehyde activity and two determinations of retinaldehyde activity.

3.2. Homogeneous Human Liver Aldehyde Dehydrogenase Isozymes and Their Retinaldehyde Activity

When incubations were done in the presence of 0.1% Triton X-100 or Tween-80 strong inhibition of retinaldehyde oxidizing activity was seen with the E1 isozyme. Apparent Km values for E1 and retinaldehyde in the presence of Triton X-100 were in the range 2–6 μM and V_{max} was only 2–6 nmol/min/mg. In the absence of detergents V_{max} increased by an order of magnitude and substrate saturation of E1 and E2 occurred in the range of 5–8 μM retinaldehyde; no substrate inhibition was observed up to 20 μM retinaldehyde. Km and V_{max} values (determined at 0.02–1 μM retinaldehyde) were 60–100 nM and 30–50 nmol/min/mg for the E1 isozyme and 100–120 nM and ca. 2 nmol/min/mg for the E2 isozyme. This is in agreement with estimated Km of human liver E1 isozyme of 60 nM, reported by Yoshida et al., (1992). However, the retinaldehyde concentration used in kinetic studies affected calculated values of Km and V_{max}; the higher the retinaldehyde concentration, the higher the calculated Km and V_{max} values. The reason for this is unknown, but problems with solubility of retinaldehyde may be an important factor contributing to the observed effects.

Activity of the E2 isozyme with retinaldehyde was also measured in the presence of Mg^{2+} ions (150μM). Magnesium stimulated retinaldehyde oxidizing activity showing that this activity belonged to the E2 isozyme. IEF gel electrophoresis showed that E2 was homogeneous and that E2 retinaldehyde activity was not due to E1 contamination (Table 1). These results are in agreement with those of Chen et al., (1994) who found that cloned ALDH2 from the mouse metabolized retinaldehyde. Formation of retinoic acid from retinaldehyde in the presence of the E3 isozyme was also established by HPLC. The velocity was, however, extremely low, preventing kinetic constants determination. Thus, all three isozymes recognized retinaldehyde as substrate, but metabolized it at low velocity; even with the E1 isozyme the velocity of retinaldehyde dehydrogenation was only 10% of that with acetaldehyde.

3.3. Measurement of Activity with Propionaldehyde and Retinaldehyde in Human Liver Homogenates

Measurement of NAD-dependent propionaldehyde and retinaldehyde oxidizing activity in homogenates from different human livers (Caucasian and Oriental) are shown in Table 1.

Table 2. Summary of purification of aldehyde dehydrogenases and of retinaldehyde dehydrogenase from three human livers and kidneys

	Propionaldehyde yield % ± SD		Retinaldehyde yield % ± SD		Ret/Prop ratio × 100 ± SD		
Purification step	Liver	Kidney	Liver	Kidney	Liver	Kidney	IEF ID**
1. Homogenate	100	100	100	100	4.9 ± 2.1	5.1 ± 3.1	E1, E2
2. Dialysis	91 ± 6	88 ± 6	92 ± 6	94 ± 3.5	4.9 ± 2.1	5.3 ± 3.1	E1, E2
3. CM-sephadex (loading and elution)	81 ± 6	81 ± 7	84 ± 8	82 ± 6	4.9 ± 2.1	5.1 ± 3.2	E1, E2
4. CM-sephadex wash (1M NaCl, pH 8.0)	trace	trace	NAD-independent activity eluted				–
5. DEAE-sephadex loading	2.1		6.7		8.0		E1
6. DEAE-sephadex wash (buffer)	7.4 ± 4	2 ± 0.6	20 ± 10	13 ± 4	16.9 ± 11.4	24 ± 1.6	E1
7. DEAE-sephadex gradient elution	68 ± 13	64*	59 ± 12	68*	4.4 ± 1.7	5.6 ± 3	E1, E2
8. DEAE-sephadex (1M NaCl, pH 8.0)	Inactive		Inactive				–
9. 5'AMP sepharose 4B loading and wash	10.8		0.3		0.03		E3
10. 5'AMP sepharose 4B elution: (a) (1M NaCl, pH 6.0)	40 ± 9	36.5 ± 7	0.8 ± 0.3	0.7 ± 0.4	0.1 ± 0	0.1 ± 0	E2
(b) (0,1M phosphate, pH 8.0, 1 mg/ml NAD)	19 ± 6	17.5 ± 8	39 ± 11	51 ± 20	11.0 ± 2.6	10.4 ± 1.6	E1
11. 5'AMP sepharose 4B(1.0M NaCl + 0.3 mg/ml NADH, pH 8 Wash)	0.9		5.7		17.1		E1
6A. 5'AMP sepharose 4B of step 6 elution NAD:	1.1		24		54		E1
6B. 5'AMP sepharose 4B (1.0 M NaCl, pH 8.0 + 0.3 mg/ml NADH Wash)	0.4		8.6		48		E1

Results are mean ±SD for n = 3 (*n = 2) purifications; values from a single purification are presented without adnotation. Total activity liver: propionaldehyde = 1.23 ± 0.41 μmol/min/gram of liver; retinaldehyde = 0.060 ± 0.03 μmol/min/gram of liver. Total activity kidney: propionaldehyde = 0.33 ± 0.13 μmpl/min/gram of kidney; retinaldehyde = 0.016 ± 0.005 μmol/min/gram of kidney. ** Isozyme composition was identified by isoelectric focusing followed by development of activity with propionaldehyde and retinaldehyde as substrates.

In Oriental livers missing the E2 isozyme (IEF evaluation), a much higher retinaldehyde/propionaldehyde activity ratio was observed than in the Caucasian livers or in the Oriental liver containing both E1 and E2 isozymes (Table 1). These data indicated that the assay used for measurement of activity with retinaldehyde worked well in liver homogenates.

3.4. Purification of Retinal Dehydrogenases from Caucasian Human Livers and Kidneys

The procedure employed for purification was the same as that previously used for aldehyde dehydrogenase purification (Hempel et al., 1982). It offered an advantage of knowledge of the behavior of E1, E2 and E3 isozymes on the chromatographic columns selected. Three purifications were done from human liver and three from kidney; the results are summarized in Table 2. The purification was also followed by the isoelectric focusing procedure and gel staining with propionaldehyde and retinaldehyde as substrates. During the entire purification, from both liver and kidney, no activity bands other than E1 were visualized with retinaldehyde as substrate. It can be also seen (Table 2) that retinaldehyde activity copurifies with aldehyde dehydrogenase.

The majority of NAD-dependent retinaldehyde activity of the human liver was eluted from 5'AMP Sepharose 4B with 1mg/ml of NAD in 0.1M sodium phosphate buffer, pH 8.0; the conditions of elution of the E1 isozyme (see step 10 (b) in Table 2). Although retinaldehyde/propionaldehyde activity ratio was constant throughout the elution peak, suggesting homogeneity, further attempts were made to separate retinaldehyde from propionaldehyde oxidizing activity using an FPLC procedure with a Mono P (Pharmacia) column (not shown in Table 2) as described by Kurys et al., (1989). The attempts of alteration of retinaldehyde/propionaldehyde ratio were, however, unsuccessful. Following elution of the E1 isozyme, further elution of the 5'AMP column was attempted with 0.3mg/ml NADH in 0.1M sodium phosphate buffer, pH 8.0 containing 1M NaCl and 1mM EDTA (step 11, Table 2). Some enzyme with retinaldehyde oxidizing activity was eluted in this fraction but by several procedures it was identified as the E1 isozyme. Its retinaldehyde/propionaldehyde activity ratio was the same as that of the E1 isozyme. Tight binding of the E1 isozyme to 5'AMP-Sepharose 4B may be due to the absence of affinity bead uniformity. Thus it has been concluded that the majority of retinaldehyde oxidizing activity in the human liver and kidney was associated with the E1 isozyme. Chromatographic profiles with propionaldehyde, octanal and retinaldehyde as substrates of kidney homogenates were identical to those of liver. Both isozymes, E1 and E2, from kidney were indistinguishable from those from liver in pI, native and subunit MW and Km for acetaldehyde.

During purification on DEAE-Sephadex a small loss of propionaldehyde and larger losses of retinaldehyde activity occurred during column washing (step 6 in Table 2). This fraction was active with both propionaldehyde and retinaldehyde but the activity with *all trans* retinaldehyde was higher, resulting in a retinaldehyde/propionaldehyde ratio 1.5 - 2 times as high as that with the E1 isozyme. Rechromatography of this material on 5'AMP (steps 6A and 6B in Table 2) increased this ratio to about 5 times. Measurement of the octanal/propionaldehyde activity ratio, the Km for acetaldehyde and the susceptibility to inhibition by diethylaminobenzaldehyde and disulfiram indicated that it was the E1 isozyme. Isoelectric focusing gels with either propionaldehyde or retinaldehyde also showed the same single activity band which corresponded to the E1 isozyme. No other activity band was detected with retinaldehyde. The same fractions also contained a large amount of small molecular weight protein (MW = ca 15, 000) which could be cellular retinoid binding protein (Ong and Chytil, 1978).

All chromatographic columns employed for enzyme purification from both liver and kidney, were additionally eluted with high salt (1M NaCl + 1 mg/ml NAD at pH 8.0) to see if NAD-dependent retinaldehyde oxidizing activity could be detected separate from aldehyde dehydrogenase activity; these attempts were, however, unsuccessful in all cases. NAD-independent retinaldehyde oxidizing activity (presumably, aldehyde oxidase) was retained on CM-Sephadex column, which is the first of the three columns employed for purification. It was eluted (see step 4 of Table 2) from CM-Sephadex column by high salt concentrations and was present in both liver and kidney homogenates. It constituted less than 10% of total retinaldehyde oxidizing activity, using the same assay as that used for dehydrogenase. With the exception of the enzyme with properties like E1 but higher retinaldehyde/propionaldehyde activity ratio, (steps 6, 6A and 6B in Table 2), which requires further investigation, it can be stated with confidence that the majority of NAD-linked retinaldehyde oxidizing activity is associated with aldehyde dehydrogenase of broad substrate specificity in both human liver and kidney. Thus, in the mature liver and kidney the same enzymes that metabolize acetaldehyde also metabolize retinaldehyde.

ACKNOWLEDGMENTS

Financial support of USPHS NIAAA Grant 1RO1 AA00186 to RP is acknowledged.

REFERENCES

Chen, M., Achkar, C., and Gudas, L.J. (1994) Enzymic conversion of retinaldehyde to retinoic acid by cloned murine cytosolic and mitochondrial aldehyde dehydrogenases. Molec. Pharmacol. 46, 88–96.

Connor, M.J., and Smit, M.H. (1987) Terminal group oxidation by mouse epidermis, inhibition *in vivo* and *in vitro*. Biochem. J. 244, 489–492.

Deltour, L., Ang, H.-L., and Duester, G. (1996) Ethanol inhibition of retinoic acid synthesis as a potential mechanism for fetal alcohol syndrome. FASEB J. 10:1050–1057.

Dockham, A.F., Lee, M.-O., and Sladek, N.E. (1992) Identification of human liver aldehyde dehydrogenases that catalyze the oxidation of aldophosphamide and retinaldehyde. Biochem. Pharmacol. 43, 2453–2469.

Duester, G. (1996) Involvement of alcohol dehydrogenase, short chain dehydrogenase/reductase, aldehyde dehydrogenase and cytochrome P450 in the control of retinoid signalling by activation of retinoic acid synthesis. Biochemistry USA 35, 12221–12227.

Goa, J. (1953) A microbiuret method for protein determination. Determination of total protein in cerebrospinal fluid. Scand. J. Clin. Lab. Invest. 5, 218–222.

Goodman, D.S., and Huang, H.S. (1965) Biosynthesis of vitamin A with rat intestinal enzymes. Science 149, 879–880.

Gudas, L.J. (1994) Retinoids and vertebrate development (minireview). J. Biol. Chem. 260, 15399–15402.

Hempel, J.D., Reed, D. M., and Pietruszko, R. (1982) Human aldehyde dehydrogenase: improved purification procedure and comparison of homogeneous isozymes E1 and E2. Alcoholism: Clin. Expl. Res. 6, 417–425.

Kurys, G., Ambroziak, W., and Pietruszko, R. (1989) Human Aldehyde Dehydrogenase. Purification and characterization of a third isozyme with low Km for γ-aminobutyraldehyde. J. Biol. Chem. 264, 4715–4721.

Lineveaver, H., and Burk, D.J. (1934) The determination of enzyme dissociation constants. J. Amer. Chem. Soc. 56:658–667.

McCaffery, P., Lee, M.-O., Wagner, M.A., Sladek, N.E., and Drager, U.C. (1992) Asymmetrical retinoic acid synthesis in the dorsoventral axis of the retina. Development 115, 371–382.

Olson, J.A., and Hyashi, O. (1965) The enzymatic cleavage of beta-carotene into vitamin A by soluble enzymes of rat liver and intestine. Proc. Natl. Acad. Sci., USA 54, 1364–1370.

Ong, D.E., and Chytil, F. (1978) cellular retinol-binding protein from rat liver: purification and characterization. J. Biol. Chem. 253, 828–832.

Wang, X., Penzes, P., and Napoli, J.L. (1996a) Cloning of a CDNA encoding an aldehyde dehydrogenase and its expression in *Escherichia coli*. J. Biol. Chem. 271, 16288–16293.

Wang, X.D., Russell, R.M., Liu, C., Stickel, F., Smith, D.E., and Krinsky, N.L. (1996b) ß-Oxidation in rabbit liver *in vitro* and in the perfused ferret liver contributes to retinoic acid biosynthesis from ß-apocarotenoic acids. J. Biol. Chem. 271, 26490–26498.

Yoshida, A., Hsu, L.C., and Dave, V. (1992) Retinal oxidation activity and biological role of human cytosolic aldehyde dehydrogenase. Enzyme 46, 239–244.

Zhao, D., McCaffery, P., Ivins, K.J., Neve, R.L., Hogan, P., Chin, W.W., and Drager, U.C. (1996) Molecular identification of a major retinoic-acid-synthesising enzyme, a retinaldehyde specific dehydrogenase. Eur. J. Biochem. 240, 15–22.

26

THE HUNT FOR A RETINAL-SPECIFIC ALDEHYDE DEHYDROGENASE IN SHEEP LIVER

Kathryn E. Kitson[1] and Treena J. Blythe[2]

[1]Institute of Food, Nutrition, and Human Health
[2]Institute of Molecular Biosciences
Massey University
Palmerston North, New Zealand

1. INTRODUCTION

During the past 7 years, a number of retinal-specific aldehyde dehydrogenases have been identified in rat (RalDH1, el Akawi and Napoli, 1994; Posch *et al.*, 1992; RalDH2, Wang *et al.*, 1996; RALDH-1, Bhat *et al.*, 1995; Labreque *et al.*, 1993; Labreque *et al.*, 1995) and mouse (RALDH-2; Zhao *et al.*, 1996) tissues. In higher mammals such as horse, sheep and human, the major Class 1 aldehyde dehydrogenase (AlDH) is known to oxidise retinal, but it has not been clearly established whether this is the major AlDH that oxidises retinal or whether there is another retinal-specific enzyme as well. The Class 1 AlDH in higher mammals is not specific for retinal; it readily oxidises a wide range of other aldehydes (Klyosov, 1996). For human Class 1 AlDH, previous kinetic studies have yielded varying estimates of the K_m for all-*trans*- retinal. Using a spectrophotometric method giving a single reaction progress curve, Yoshida *et al.* (1992) reported a K_m of 0.06 μM, and using a similar method Klyosov (1996) reported a K_m of 1.1 μM. This difference is significant in considering the possible importance of the Class 1 AlDH in retinal metabolism *in vivo*, since cellular retinal concentrations are likely to be about 0.1 μM (Yoshida *et al.*, 1992). The purpose of the current study was two-fold: firstly, to investigate the possibility that retinal-specific AlDHs are present in sheep as well as in rat and mouse tissues, and secondly, to carry out more accurate kinetic characterisation of the Class 1 AlDHs from sheep and human with retinal as a substrate.

Enzymology and Molecular Biology of Carbonyl Metabolism 7
edited by Weiner *et al.* Kluwer Academic / Plenum Publishers, New York, 1999.

2. MATERIALS AND METHODS

Sheep liver Class 1 AlDH (sAlDH) was purified as previously described (Kitson and Kitson, 1994), and human recombinant Class 1 AlDH (hAlDH) was prepared following cloning and expression in *E. coli* as previously described (Jones *et al.*, 1995). Human cellular retinol binding protein (CRBP) with a C-terminal 6-histidine tag was expressed in E. coli then purified by affinity chromatography using a HiTrap® (Pharmacia) chelating column loaded with nickel ions. Isoelectric focusing gels (Ampholine PAG plates, Pharmacia, pH range 3.5–9.5) were run on an LKB Multiphor apparatus. Activity staining for AlDH was carried out using phenazine methosulphate and nitro-blue tetrazolium, as previously described (Agnew *et al.*, 1981). Routine enzyme assays were performed on a Varian Cary double beam UV-visible spectrophotometer, and fluorimetric assays on a Perkin-Elmer LS50B luminescence spectrophotometer.

3. RESULTS AND DISCUSSION

3.1. Dehydrogenase Activity in Tissue Extracts Using Retinal as a Substrate

Extracts from sheep liver were activity stained following isoelectric focusing using a range of aldehyde substrates and NAD^+ as a cofactor. With acetaldehyde, activity bands for both Class 1 and Class 2 AlDHs could be clearly detected, with isoelectric points at about pH 5.2 and pH 5.4–5.8 respectively, as expected (Agnew *et al.*, 1981). With retinal as a substrate very little activity was observed in the pH range expected for the Class 1 and 2 isoenzymes. There was, however, a marked activity band observed with retinal at an isoelectric point of pH 8.3–8.5. Since retinal was known to be a substrate with a K_m in the sub-micromolar range for the purified Class 1 sAlDH, it was surprising that this enzyme did not show an activity band with retinal as a substrate in the activity stain. It seemed possible that the hydrophobic retinal was not able to diffuse into the gel matrix, and in an attempt to improve this stain, enzyme was blotted from the polyacrylamide gel on to a nitrocellulose sheet. Staining of the nitrocellulose-adsorbed enzyme produced some reaction with Class 1 AlDH with retinal as a substrate, but the activity stain at an isoelectric point of pH 8.5–8.5 was more marked. This activity stain was not present in the absence of NAD^+, and was present with *p*-carboxybenzaldehyde, 3,5-diiodosalicylaldehyde, and in some experiments acetaldehyde, but not with citral. It appeared possible that this protein was an AlDH with a very high affinity, but not absolute specificity, for retinal, and this possibility appeared to be reinforced by the similarity of the isoelectric point to that of a recently identified retinal-specific AlDH from rat tissue (pI ~ 8.5; Labreque *et al.*, 1993; Posch *et al.*, 1992).

3.2. A Retinal Dehydrogenase or a Novel Alcohol Dehydrogenase?

The pI 8.5 protein identified by activity staining was partially purified, and found to have a native molecular weight of 42–45 kDa, and the same molecular weight from SDS gel electrophoresis, which suggested a monomeric protein. It was subsequently found, however, that the protein had high alcohol dehydrogenase (ADH) activity with ethanol as a substrate, and that the apparent activity in the presence of retinal was due to traces of ethanol in the

NAD^+ being used. The fact that this reaction did not occur with some aldehyde substrates was probably explained by product inhibition of the ADH reaction. The variable reaction with acetaldehyde present was probably due to varying product inhibition depending on changes in the relative ratios of ethanol and acetaldehyde in the activity stain over time. Since it appeared that the ADH might be a novel protein, further attempts were made to characterise it. The protein proved to be very difficult to purify, as it did not appear to elute successfully from gel filtration or dye columns of any type. Preparative isoelectric focusing followed by preparative SDS polyacrylamide gel electrophoresis finally yielded some pure protein. The protein was N-terminally blocked, so that N-terminal sequencing of the intact protein was not possible. An amino acid composition was obtained, and submitted to two database searching programs, with no exact match being found. Some internal sequence was obtained from a tricine gel separation of a peptide band from a limited tryptic digest, but this sequence was ambiguous as there appeared to be two peptides of identical molecular weight present in the band. Submission of all possible combinations of sequence from these peptides to a database search, however, did not yield any close matches. In terms of substrate specificity, the enzyme showed a high preference for ethanol, low activity with n-propanol, n-butanol, n-petanol, n-hexanol and 2-propanol and no activity with methanol, sorbitol or glucose. Tissue extracts of kidney, heart, spleen and brain showed no activity. This protein therefore appears to be a novel monomeric ADH with relatively high specificity for ethanol. It is presumably mammalian in origin, although it is possible, particularly in a ruminant animal such as the sheep, that a microbial enzyme from the digestive tract might be found in the liver (Popper and Schaffer, 1957).

3.3. Activity of Class 1 sAlDH and hAlDH with Retinal as a Substrate

Previous studies have shown that Class 1 hAlDH has a low K_m for retinal, but as indicated in the introduction there is variation in published values. In this study a fluorimetric method was used to allow determination of the K_m of sAlDH and hAlDH for isomers of retinal by a conventional kinetic approach. The sensitivity of the fluorimeter allowed use of sufficiently dilute enzyme solutions to permit linear reaction rates to be determined at retinal concentrations as low as 0.02 μM. All kinetic determinations were carried out using 50 mM phosphate buffer, pH 7.4, at 25°C, with 0.2 mM NAD^+. Using this approach, the K_m values of sAlDH and hAlDH for all-*trans*- and 9-*cis*-retinal were determined to be in the sub-micromolar range (Table 1). In addition, a K_m value for sAlDH with all-*trans*-retinal was determined using the single reaction progress curve method used by Yoshida *et al.* (1992). This value was close to that determined using the fluorimetric method (Table 1). The K_m for all-*trans*-retinal for hAlDH determined in this study was essentially identical to that determined by Yoshida *et al.* (1992), and much lower than that published by Klyosov (1996).

In addition to using free all-*trans*-retinal as a substrate, AlDH activity was studied in the presence of all-*trans*-retinal bound to cellular retinol binding protein (CRBP). For both sAlDH and hAlDH, the K_m in the presence of CRBP-bound all-*trans*-retinal was higher than in the presence of free retinal, but was still in the sub-micromolar range (Table 1).

The data presented in Table 1 indicate that both sAlDH and hAlDH have K_m values for 9-*cis*-retinal, all-*trans*-retinal and CRBP-bound all-*trans*-retinal that suggest that these enzymes would contribute to conversion of these forms of retinal to retinoic acid under physiological conditions. Although the Class 1 AlDHs have broad substrate specificity, there are few substrates for which these enzymes have been reported to have such low K_m values as those seen with retinal. The reported K_m values for acetaldehyde for Class 1 AlDH range from 22 to about 400 μM (reviewed by Rashovetsky *et al.*, 1994), which sug-

Table 1. K_m values for Class 1 AlDH with retinal substrates

Enzyme	Substrate	K_m (μM)	Reference
Class 1 sAlDH	All-*trans*-retinal	0.14 ± 0.013	This study
(fluorimetric)	9-*cis*-retinal	0.14 ± 0.011	"
	CRBP-all-*trans*-retinal	0.92 ± 0.20	"
(single reaction)	All-*trans*-retinal	0.24 ± 0.08	"
Class 1 hAlDH	All-*trans*-retinal	0.058 ± 0.008	"
(fluorimetric)	9-*cis*-retinal	0.060 ± 0.008	"
	CRBP-all-*trans*-retinal	0.22 ± 0.035	"
(single reaction)	All-*trans*-retinal	0.060	Yoshida *et al.*, 1992
(single reaction)	All-*trans*-retinal	1.10 ± 0.02	Klyosov, 1996
RalDH1	All-*trans*-retinal	1.6 ($K_{0.5}$)	El Akawi & Napoli, 1994
	9-*cis*-retinal	5.2 ($K_{0.5}$)	"
RALDH-1	All-*trans*-retinal	8-10	Labrecque *et al.*, 1993
RalDH2	All-*trans*-retinal	0.4 ± 0.1	Wang *et al.*, 1996
	CRBP-all-*trans*-retinal	0.2 ± 0.06	"
RALDH-2	All-*trans*-retinal	–	Zhao *et al.*, 1996
	9-*cis*-retinal	–	"

Kinetic parameters for this study were determined as described in the text. As a comparison, a range of values from the literature for hAlDH and rat retinal dehydrogenases have also been included.

gests, since acetaldehyde concentrations during ethanol metabolism are usually in the low micromolar range (Lindros, 1989), that the primary role of the Class 1 enzyme is certainly not acetaldehyde metabolism. On the basis of kinetic data, it could be suggested that Class 1 sAlDH and hAlDH are primarily retinal dehydrogenases.

3.4. Sequence Comparisons of Aldehyde Dehydrogenases

In order to analyse the relationship between AlDHs from different species, and assess the degree of relationship between retinal-metabolising AlDH isoenzymes, a multiple sequence alignment was performed of 32 enzymes, including all the known retinal-oxidising forms (Blythe, 1997). From this alignment, generated using the program 'Pileup', a dendrogram was prepared which indicates the difference between sequences in the alignment (Figure 1). Table 2 summarises the sources of the information that was used to compile the sequence alignment and dendrogram. The overall pattern obtained is consistent with current classification based on functional data, with the Class 1, 2 and 3 enzymes each clustering together, and the Class 1 and 2 enzymes being more similar to each other than to the Class 3 enzymes.

The sequence alignment showed that the retinal-specific AlDHs from rat and mouse form two closely-aligned pairs. RalDH1 and RALDH-1 (both from rat) are identical apart from residues 99 (arginine in RalDH1 and cysteine in RALDH-1) and 169 (glutamate in RalDH1 and asparagine in RALDH-1). Asparagine 169 is completely conserved in all other 31 enzymes and forms part of the oxyanion hole, and position 99 is an arginine in all Class 1 and 2 enzymes, suggesting that the differences between RalDH1 and RALDH-1 may be due to one sequencing error in each case. The fact that different kinetic mechanisms have been reported for these two enzymes (el Akawi and Napoli, 1994; Labrecque *et al.*, 1993; Labrecque *et al.*, 1995) may reflect different experimental conditions, rather than a genuine difference in the enzymes. RalDH2 (rat) and RALDH-2 (mouse) differ by one residue only; residue 353 is isoleucine in RalDH2 and valine in RALDH-2. The isoenzymes are presumably the same enzyme isolated from two different species.

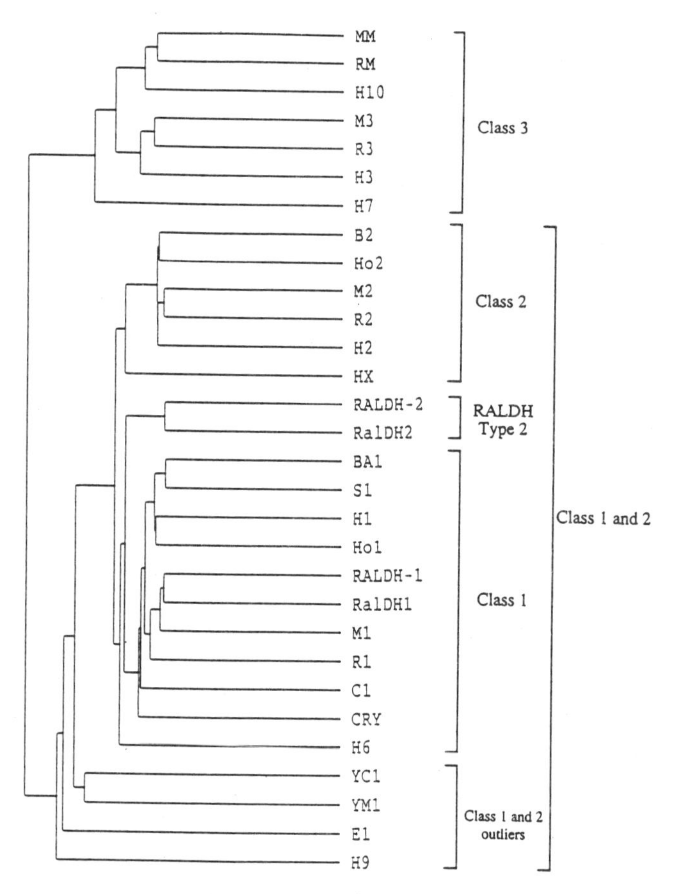

Figure 1. A dendrogram showing clustering relationships between aldehyde dehydrogenase enzymes. Thirty-two aldehyde dehydrogenase sequences were aligned using the GCG program 'pileup'. The above diagram is a pictorial representation of the clustering relationships used in the construction of a sequence alignment (Blythe, 1997). The diagram represents similarity between primary amino acid sequences. The enzymes listed and sources of information are identified in Table 2.

From the dendrogram (Figure 1) it can be seen that RalDH1 and RALDH-1 show close sequence similarity with rat, mouse, bovine, ovine, equine and human Class 1 AlDHs. RalDH2 and RALDH-2 also cluster with the Class 1 enzymes, but show more sequence differences suggesting a distinct enzyme form, which perhaps fits with the tissue and substrate specificity reported for RALDH-2. This enzyme does not accept acetaldehyde, citral, propanal or benzaldehyde as substrate, and is found mainly in the testis, with low levels only in liver, brain, heart, lung and kidney (Wang *et al.*, 1996).

The sequence alignment also includes a number of recently-cloned AlDH genes: AlDH6 (H6), AlDH7 (H7), AlDH9 (H9), AlDH10 (H10) and AlDHX (HX). H7 and H10 (fatty aldehyde dehydrogenase) cluster with the Class 3 enzymes, H9 (γ-aminobutyraldehyde-oxidising AlDH) is an outlier in the Class 1 and 2 cluster, HX (human stomach) clusters with Class 2, and H6 (human saliva) with Class 1. Another gene, AlDH 8, has also been cloned, but was omitted from the alignment as the cDNA encodes only 385 amino acids, and is therefore probably not a complete transcript. None of these novel hAlDH sequences clusters closely with the retinal-oxidising isozymes; it is unlikely that any other retinal-specific isozyme exists that has not been detected already.

Table 2. Identification of enzymes and sources of information used in sequence alignment

Name	Identification	Databank accession number	Number of amino acids	Reference
BA1	Bovine Class 1 from amacrine cells (eye)	L36128 (GenBank)	501	Saari *et al.*, 1995
Ho1	Horse Class 1	P15437 (SwissProt)	500	Von Bahr-Lindstrom *et al.*, 1984
E1	*E.Coli* Class 1	P23883 (SwissProt)	494	Heim and Strehler, 1991
YC1	Yeast Class 1 (*S.Cerivisiae*)	U56604 (GenBank)	501	Weiner *et al.*, Direct Submission (cytosolic)
C1	Chicken Class 1	P27463 (SwissProt)	509	Godbout, 1992
M1	Mouse Class 1	P24549 (SwissProt)	500	Rongnoparut and Weaver, 1991
R1	Rat Class 1	P13601 (SwissProt)	500	Dunn *et al.*, 1989
S1	Sheep liver Class 1	P51977 (SwissProt)	500	Stayner and Tweedie, 1995
H1	Human liver Class 1	P00352 (SwissProt)	500	Hsu *et al.*, 1989
RALDH-1	Rat kidney retinal-specific Class 1, type 1	P51647 (SwissProt)	501	Bhat *et al.*, 1995
RalDH1	Rat liver retinal-specific Class 1, type 1	U79118 (GenBank)	501	Penzes *et al*, 1997
RALDH-2	Mouse retinal-specific, type 2	X99273 (GenBank)	499	Zhao *et al.*, 1996
RalDH2	Rat retinal-specific, type 2	U60063 (GenBank)	499	Wang *et al.*, 1996
B2	Bovine liver Class 2	P20000 (SwissProt)	520	Guan and Weiner, 1990
Ho2	Horse Class 2	P12762 (SwissProt)	500	Johanssen *et al.*, 1988
YM1	Yeast Class 2 (*S.Cerivisiae*)	U56605 (GenBank)	519	Weiner *et al.*, Direct Submission (mitoch.)
M2	Mouse Class 2	P47738 (SwissProt)	519	Chang and Yoshida, 1994
R2	Rat liver Class 2	P11884 (SwissProt)	519	Fares *et al.*, 1989
H2	Human Class 2	P05091 (SwissProt)	517	Hsu *et al.*, 1988
CRY	Short eared elephant shrew crystallin (eye)	U03906 (GenBank)	501	Graham *et al.*, 1996
H6	Human saliva precursor	A55684 (PIR)	512	Hsu *et al.*, 1994a
HX	Human stomach	M63967 (GenBank)	453	Hsu and Chang, 1991
YM3	Yeast mitochondrial 3 precursor	P40047 (SwissProt)	519	Dietrich and Davis, 1994
M3	Mouse Class 3 (dioxin inducible)	P47739 (SwissProt)	453	Vasiliou *et al.*, 1993
R3	Rat Class 3 (Tumour-associated)	P11883 (SwissProt)	452	Jones *et al.*, 1988
H3	Human Class 3	P30838 (SwissProt)	453	Schuuring *et al.*, 1992
MM	Mouse liver microsomal (dioxin inducible)	U14390 (GenBank)	484	Nebert *et al.*, 1994
RM	Rat liver microsomal	A41028 (PIR)	484	Miyauchi *et al.*, 1991
H4	Human 4-aminobutyraldehyde oxidising	P47895 (SwissProt)	462	Pietruszko, 1993, Direct Submission
H9	Human γ-aminobutyraldehyde oxidising	P49189 (SwissProt)	493	Lin *et al.*, 1996
H7	Human stomach AlDH7	P43353 (SwissProt)	468	Hsu *et al.*, 1994b
H10	Human liver fatty aldehyde dehydrogenase	U46689 (GenBank)	485	Chang and Yoshida, 1997

4. CONCLUSIONS

From the isoelectric focusing and activity staining experiments performed, it can be concluded that the only retinal dehydrogenase activity in sheep liver is that due to Class 1 AlDH. The additional activity band that appeared to be a retinal dehydrogenase was actually an alcohol dehydrogenase. While a minor retinal-specific activity might not have been detected in these experiments, it is unlikely that an enzyme contributing a lot of retinal dehydrogenase activity would have been missed. Kinetic analyses show similar K_m values, in the low micromolar range, for both sAlDH and hAlDH with 9-*cis* and all-*trans*-retinal as substrates. Both enzymes also recognise CRBP-bound all-*trans*-retinal as a substrate. The K_m for the sheep enzyme was increased more than that for the human enzyme by the use of CRBP-bound retinal; this possibly reflects the fact that the CRBP used was human. If AlDH recognises specific groups on CRBP with which it binds in accepting the retinal substrate from the carrier protein, this recognition may not be as good for the sheep enzyme with human CRBP. Nevertheless, both enzymes clearly readily oxidise CRBP-bound retinal, so that if, as is likely, a significant proportion of retinal occurs in this form *in vivo*, the Class 1 AlDHs are not precluded from being producers of retinoic acid as a result.

The dendrogram produced from sequence alignment of 32 AlDHs shows that the 'retinal dehydrogenases' from rat and mouse tissue cluster closely with the Class 1 AlDHs. In particular, the type 1 retinal dehydrogenases show very close similarity with the Class 1 ovine, bovine, horse and human AlDHs. The K_m values reported in this study for Class 1 sAlDH and hAlDH isozymes with free or CRBP-bound retinal are as low as, or lower than, those reported for the rat retinal dehydrogenases (see Table 1). Given the current kinetic evidence, and the close sequence similarities, serious consideration should be given to classifying the major Class 1 enzymes from sheep, horse, cow and human as retinal dehydrogenases, in recognition of what is likely to be a primary physiological role for these enzymes. In terms of studying the role of Class 1 AlDHs in retinoic acid production during fetal development, it is clear that rat and mouse models will probably not reflect the human situation as closely as will sheep, horse or cow models. The close similarity in kinetic parameters between the sheep and human Class 1 AlDHs indicates that the sheep enzyme, which can be prepared easily in relatively large amounts, can provide a useful alternative to the human in studies on aldehyde oxidation.

The recent elucidation of the tertiary structure of the Class 1 AlDH from sheep liver has allowed comparison of the residues in the substrate entrance tunnel of this enzyme with those of the Class 2 enzyme for which the structure was recently reported (Steinmetz *et al.*, 1997). Such comparison reveals differences in amino acid side chains and in the tertiary structure within the entrance tunnel that can explain the difference in affinity for retinoid substrates for these two enzymes (Moore *et al.*, this volume). With solving of further structures, classification of AlDHs as retinal-specific may be able to be made on a structural as well as a kinetic basis.

ACKNOWLEDGMENTS

We wish to thank Associate Professor M.J. Hardman, Dr. M.L. Patchett, J. Wyatt and E. Loughnane (Massey University) for samples of hAlDH, Professor H. Weiner (Purdue University, U.S.A.) for cDNA for hAlDH, Professor J. Findlay (Leeds University, U.K.) for CRBP cDNA, Dr. S. Lott for assistance with CRBP preparation and Associate Professor T.M. Kitson for assistance with kinetic analyses. We wish to acknowledge funding support from: The Neurological Foundation of New Zealand, The New Zealand Lottery Grants Board, The Palmerston North Medical Research Foundation, and a Massey University Postgraduate Scholarship (T.J. Blythe).

REFERENCES

Agnew, K.E.M., Bennett, A.F., Crow, K.E., Greenway, R.M., Blackwell, L.F. and Buckley, P.D. (1981) A reinvestigation of the purity, isoelectric points and some kinetic properties of the aldehyde dehydrogenases from sheep liver, *Eur. J. Biochem.* **119**, 79–84.

Bhat, P.V., Labreque, J., Boutin, J.M., Lacroix, A. and Yoshida, A. (1995) Cloning of a cDNA encoding rat aldehyde dehydrogenase with high activity for retinal oxidation, *Gene* **166**, 303–306.

Blythe, T.M. (1997) Studies on proteins involved in retinoid and alcohol metabolism, PhD Thesis, Massey University, Palmerston North, New Zealand.

Chang, C. and Yoshida, A. (1994) Cloning and characterisation of the gene encoding mouse mitochondrial aldehyde dehydrogenase, *Gene* **148**, 331–336.

Chang, C. and Yoshida, A. (1997) Human fatty aldehyde dehydrogenase gene (ALDH10): organisation and tissue-dependent expression, *Genomics* **40**, 80–85.

Dietrich, F.S. and Davis, R.W. (1994) Direct Submission (YM3).

Dunn, T.J., Koleske, A.J., Lindahl, R., and Pitot, H.C. (1989) Phenobarbital-inducible aldehyde dehydrogenase in the rat. cDNA sequence and regulation of the mRNA by phenobarbital in responsive rats, *J. Biol. Chem.* **264**, 13057–13065.

el Akawi, Z. and Napoli, J.L. (1994) Rat liver cytosolic retinal dehydrogenase: comparison of 13-*cis*, 9-*cis*, and all-*trans* retinal as substrates and effects of cellular retinoid binding proteins and retinoic acid on activity, *Biochemistry* **33**, 1938–43.

Fares, J., Guan, K. and Weiner, H. (1989) Primary structure of rat and bovine liver mitochondrial aldehyde dehydrogenases deduced from cDNA sequences, *Eur. J. Biochem.* **180**, 67–74.

Godbout, R. (1992) High levels of aldehyde dehydrogenase transcripts in the undifferentiated chick retina, *Exp. Eye Res.* **54**, 297–305.

Graham, C., Hodin, J. and Wistow, G. (1996) A retinaldehyde dehydrogenase as a srtuctural protein in a mammalian eye lens. Gen recruitment of eta-crystallin, *J. Biol. Chem.* **271**, 15623–15628.

Guan, K. and Weiner, H. (1990) Sequence of the precursor of bovine liver mitochondrial aldehyde dehydrogenase as determined from its cDNA, its gene, and its functionality, *Arch. Biochem. Biophys.* **277**, 351–360.

Heim, R. and Strehler, E.E. (1991) Cloning and E. coli gene encoding a protein remarkably similar to mammalian aldehyde dehydrogenases, *Gene* **99**, 15–23.

Hsu, L.C. and Chang, W.-C. (1991) Cloning and characterisation of a new functional aldehyde dehydrogenase gene, *J. Biol. Chem.* **266**,12257–12265.

Hsu, L.C., Bendel, R.E. and Yoshida, A. (1988) Genomic structure of the human mitochondrial aldehyde dehydrogenase gene, *Genomics* **2**, 57–65.

Hsu, L.C., Chang, W.-C. and Yoshida, A. (1989) Genomic structure of the human cytosolic aldehyde dehydrogenase gene, *Genomics* **5**, 857–865.

Hsu, L.C., Chang, W.C., Hiraoka, L. and Hsieh, C.L. (1994a) Molecular cloning, genome organisation, and chromosomal localisation of an additional human aldehyde dehydrogenase gene, ALDH6, *Genomics* **24**, 333–341.

Hsu, L.C., Chang, W.-C. and Yoshida, A. (1994b) Cloning of a cDNA encoding human ALDH7, a new member of the aldehyde dehydrogenase family, *Gene* **151**, 285–289.

Johanssen, J., von Bahr-Lindstrom, H., Jeck, R., Woenckhaus, C. and Jornvall, H. (1988) Mitochondrial aldehyde dehydrogenase from horse liver. Correlations of the same species variants for both the cytosolic and mitochondrial forms of an enzyme, *Eur. J. Biochem.* **172**, 527–533.

Jones, D.E., Brennan, M.D., Hempel, J. and Lindahl, R. (1988) Cloning and complete nucleotide sequence of a full length cDNA encoding a catalytically functional tumor-associated aldehyde dehydrogenase *Proc. Nat. Acad. Sci. (USA)* **85**, 1782–1786.

Jones, K.M., Kitson, T.M., Kitson, K.E., Hardman, M.J. and Tweedie, J.W. (1995) Human Class 1 aldehyde dehydrogenase. Expression and site-directed mutagenesis In Enzymology and Molecular Biology of Carbonyl Metabolism 5, Weiner, H., Holmes, R.S. and Wermuth, B. (Eds), Plenum Press, New York, pp. 17–23.

Kitson, T.M. and Kitson, K.E. (1994) Probing the active site of cytoplasmic aldehyde dehydrogenase with a chromophoric reporter group, *Biochem. J.* **300**, 25–30.

Klyosov, A.A. (1996) Kinetics and specificity of human liver aldehyde dehydrogenases toward aliphatic, aromatic, and fused polycyclic aldehydes, *Biochemistry* **35**, 4457–4467.

Labreque, J., Bhat, P.V. and Lacroix, A. (1993) Purification and partial characterisation of a rat kidney aldehyde dehydrogenase that oxidises retinal to retinoic acid, *Biochem. Cell Biol.* **71**, 85–89.

Labreque, J., Dumas, F., Lacroix, A. and Bhat, P.V. (1995) A novel isoenzyme of aldehyde dehydrogenase specifically involved in the biosynthesis of 9-*cis* and all-*trans* retinoic acid, *Biochem. J.* **305**, 681–684.

Lin, S.W., Chen, J.C., Hsu, L.C., Hsieh, C.L. and Yoshida, A. (1996) Human gamma-aminobutyraldehyde dehydrogenase ALDH9: cDNA sequence, genomic organisation, polymorphism, chromosomal localisation, and tissue expression, *Genomics* **34**, 376–380.

Lindros, K.O. (1989) Human blood acetaldehyde in Human Metabolism of Alcohol, Vol. II, Crow, K.E. and Batt, R.D., Eds, CRC Press, Boca Raton, Florida, pp.177–192.

Miyauchi, M., Makasi, R., Taketawi, S., Yamamoto, A., Akayama, M. and Tashiro, V. (1991) Molecular cloning, sequencing and expression of cDNA for rat liver microsomal aldehyde dehydrogenase, *J. Biol. Chem.* **266**, 19536–19542.

Nebert, E.W., Kozak, C.A., Lindahl, R. and Vasiliou, V. (1994) Mouse microsomal class 3 aldehyde dehydrogenase:AHD3 cDNA sequence, chromosomal mapping and dioxin inducibility, *Toxicologist* **14**, 410.

Penzes, P., Wang, X.S., Sperkova, Z. and Napoli, J.L. (1997) Cloning of a rat cDNA encoding retinal dehydrogenase isozyme type 1 and its expression in E. coli, *Gene* **191**, 176–172.

Popper, H. and Schaffner, F. (1957) Liver: Structure and Function, McGraw-Hill, New York, p99.

Posch, K.C., Burns, R.D. and Napoli, J.L. (1992) Biosynthesis of all-*trans* retinoic acid from retinal. Recognition of retinal bound to cellular retinol binding protein (type 1) as substrate by a purified cytosolic dehydrogenase, *J. Biol. Chem.* **267**, 19676–82.

Rashovetsky, L.G., Maret, W. and Klyosov, A.A. (1994) Human liver aldehyde dehydrogenases: a new method of purification of the major mitochondrial and cytosolic enzymes and re-evaluation of their kinetic properties, *Biochim. Biophys. Acta* **1205**, 301–307.

Rongnoparut, P. and Weaver, S. (1991) Isolation and characterisation of a cytosolic aldehyde dehydrogenase-encoding cDNA from mouse liver, *Gene* **101**, 261–265.

Saari, J.C., Champer, R.J., Asson-Batres, M.A., Garwin, G.G., Huang, J., Crabb, J.W. and Milam, A.H. (1995) Characterisation and localisation of an aldehyde dehydrogenase to amacrine cells of bovine retina, *Visual Neuroscience* **12**, 263–272.

Schuuring, E.M., Verhoeven, E., Eckey, R., Vos, H.L. and Michalides, R.J. (1992) Direct Submission (H3).

Stayner, C.K. and Tweedie, J.W. (1995) Cloning and characterisation of the cDNA for sheep liver cytosolic aldehyde dehydrogenase, *Adv. Exp. Med. Biol.* **372**, 61–66.

Steinmetz, C.G., Xie, P., Weiner, H. and Hurley, T.D. (1997) Structure of a mitochondrial aldehyde dehydrogenase: the genetic component of ethanol aversion, *Structure* **5**, 701–711.

Vasiliou, V., Reuter, S.F., Kozak, C.A. and Nebert, D.W. (1993) Mouse dioxin- inducible cytosolic ALDH3: AHD4 cDNA sequence, genetic mapping, and differences in mRNA levels, *Pharmacogenetics* **3**, 281–290.

von Bahr-Lindstrom, H., Hempel, J. and Jornvall, H. (1984) The cytoplasmic isoenzyme of horse liver aldehyde dehydrogenase. Relationship to the corresponding human isoenzyme, *Eur. J. Biochem.* **141**, 37–42.

Wang, X., Penzes, P. and Napoli, J.L. (1996) Cloning of a cDNA encoding an aldehyde dehydrogenase and its expression in Escherichia coli. Recognition of retinal as a substrate, *J. Biol. Chem.* **271**, 16288–16293.

Yoshida, A., Hsu, L. C. and Dave, V. (1992) Retinal oxidation activity and biological role of human cytosolic aldehyde dehydrogenase, *Enzyme*, **46**, 239–244.

Zhao, D., McCaffery, P., Ivins, K.J., Neve, R.L., Hogan, P., Chin, W.W. and Drager, U. (1996) Molecular identification of a major retinoic-acid synthesizing enzyme, a retinaldehyde-specific dehydrogenase, *Eur. J. Biochem.* **240**, 15–22.

27

CARDIAC METABOLISM OF ENALS

Aruni Bhatnagar,[1,2] Sanjay Srivastava,[1] Li-Fei Wang,[1] Animesh Chandra,[1] Naseem H. Ansari,[1] and Satish K. Srivastava[1]

[1]Department of Human Biological Chemistry and Genetics
[2]Department of Physiology and Biophysics
University of Texas Medical Branch
Galveston, Texas 77555-1067

1. INTRODUCTION

Recent evidence suggests that in addition to their well studied toxic effects, reactive oxygen species (ROS) are also essential mediators of cell growth, differentiation and apoptosis (Sen and Packer, 1996; Lander, 1997). Nonetheless, the mechanisms by which ROS alter cell function remain obscure. Since free radicals derived from oxygen are generally short-lived and react predominantly at their site of generation, it is likely that their metabolic as well as the toxic effects are mediated, in part, by their metastable products, particularly those derived from the peroxidation of membrane lipids. Peroxidation of unsaturated fatty acids generates a variety of metastable compounds of which aldehydes are the most abundant end-products (Esterbauer *et al.*, 1991; Witz, 1983). In comparison only minor amounts of ketones, epoxides, hydrocarbons, alcohols and acids are formed (Grosch, 1987). Therefore, in cellular systems, aldehyde burden may be an important consequence of lipid peroxidation.

Of the several aldehydes generated during lipid peroxidation, the contribution of unsaturated aldehydes (enals) is particularly important, since these are potent electrophiles and react avidly with most cell constituents. Moreover, unlike their radical precursors, enals are long-lived, and thus, can amplify disturbances due to local radical events. Enhanced enal burden due to increased lipid peroxidation is evident under several pathological conditions such as ischemia-reperfusion (Lucas and Szweda, 1998), atherosclerosis (Yla-Herttuala *et al.*, 1989), Alzheimer's (Sayre *et al.*, 1997) and Parkinson (Yoritaka *et al.*, 1996) disease, where enal-derived lesions could contribute to the sequalae of clinically documented symptoms.

While extensive research efforts have been directed towards understanding the mechanisms by which radical and non-radical ROS are scavenged and detoxified, little is known in regard to the cellular metabolism of lipid-derived enals. However, elucidation of

Enzymology and Molecular Biology of Carbonyl Metabolism 7
edited by Weiner *et al.* Kluwer Academic / Plenum Publishers, New York, 1999.

such metabolism is essential to identify conditions that lead to the accumulation of aldehyde-modified proteins and aldehyde-DNA adducts. While aldehyde-modified proteins are thought to contribute to tissue dysfunction and disease, the formation of aldehyde-DNA adducts have been linked to spontaneous carcinogenesis and aging (Nath and Chung, 1994). Furthermore, enals similar to those generated by lipid peroxidation are either present in several foods, drugs, and pollutants, or are generated during their metabolism (Esterbauer *et al.* 1991, Witz, 1989). Thus, understanding the metabolism of lipid peroxidation-derived enals may provide additional insights into the processes that detoxify several apparently unrelated toxicants.

To understand the metabolic roles of lipid-derived aldehydes and to assess their contribution to the cellular effects of ROS, we have recently begun a series of studies related to the metabolism of these compounds. Using 4-hydroxy *trans*-2- nonenal (HNE) as a model aldehyde, we have found that in isolated perfused rat heart unsaturated aldehydes such as HNE are metabolized by several distinct biochemical pathways (Srivastava *et al.*, 1998 a & b). However, the metabolism of HNE in perfused hearts is likely to be due to several cell types. Thus, to identify the biochemical pathways of enal metabolism specific to the working myocardium we examined HNE metabolism in adult isolated cardiac myocytes. Also, as compared to the perfused heart, with isolated myocytes, more detailed and mechanistic studies are possible. The biochemical pathways of enal metabolism may be of significance under conditions when the cardiac myocytes are exposed to oxidative stress such as that due to ischemia-reperfusion and free radical generating drugs (e.g. adriamycin)

2. MATERIALS AND METHODS

The HNE was synthesized as its dimethyl acetal starting from dimethyl acetal of fumaraldehyde as described before (Chandra and Srivastava, 1997). For the synthesis of 4-^{3}H-HNE the dimethyl acetal of HNE was oxidized to the 4-ketoderivative using polymer supported chromic acid as an oxidizing agent. The ketone was reduced to the dimethyl acetal of HNE by using tritiated $NaBH_4$ and 4-^{3}H-HNE obtained by acid hydrolysis was purified on HPLC. The 4-^{3}H-HNE thus synthesized had a specific activity of 175 mCi/mmole. Myocytes were isolated from adult guinea-pig hearts as described before (Bhatnagar, 1995, 1997). The myocytes were stored in "KB" solution at 4°C. Before each experiment, the cells were suspended in modified Krebs-Henseleit solution containing (in mM): NaCl, 118; KCl, 4.7; $MgCl_2$, 1.25; $CaCl_2$, 3.0; KH_2PO_4, 1.25; EDTA, 0.5; $NaHCO_3$, 25; glucose, 10; pH 7.4, equilibrated with a gas mixture of 95% O_2 + 5% CO_2, maintained at 37°C.

Synthesized standards and metabolites of HNE were seperated by HPLC using a Rainin $ODS_{c\text{-}18}$ column equilibrated wtih 0.1% trifluoroacetic acid (TFA) at a flow rate of 1 ml/min. The compounds were eluted using a gradient consisting of solvent A (0.1 % aqueous TFA) and solvent B (0.1% TFA in 60 % acetonitrile) at a flow rate of 1 ml/min. The gradient was established such that B reached 16.6 % in 20 min, 41.5% in 35 min and then held at 41.5 % for an additional 30 min. B reached from 41.5 % to100 % in 10 min and was held at this value for 20 min. In this system, the conjugates GS-HNE and GS-DHN eluted at a retention time (RT) of 44 min (24.9 % acetonitrile), and the RT of DHN, HNA and HNE standards were: 53, 58 and 65 min, respectively (Srivastava *et al.*, 1998b). The identity of the glutathione conjugates was established by ESI-MS as described before (Srivastava *et al.*, 1998b).

3. RESULTS AND DISCUSSION

Approximately 70 % of the isolated cardiac myocytes displayed rod-shaped morphology, and excluded tryphan blue. The myocytes were incubated in 1 ml KH buffer bubbled with 95 % O_2 and 5% CO_2 for 30 min in a shaking water bath for buffer. After equilibration, 50 nmoles with ^{3}H-HNE were added to the incubation mixture. After 30 min at 37 °C the cells were harvested by centrifugation in an eppendorf tube for 10 min at 1, 000 rpm. The pellet was saved and the supernatant was further cleaned by centrifuation at 14, 000 rpm for 20 min. The radioactivity in the pellet was extracted with 96 % acetonitrile and 4 % acetic acid. The HNE metabolites in the cell extract as well as the supernatant were separated on HPLC as decribed under Materials and Methods. One ml fractions were collected and assayed for radioactivity. The total radioactivity recovered in the myocyte pellet was 4 to 5 % of the administered dose. The supernatent contained 70 to 80 % of the radioactivity. These results suggest that HNE is rapidly metabolized by isolated cardiac myocytes and most of the metabolites are extruded in the external medium. The high capacity of the myocytes to metabolize HNE indicates that under oxidative stress, most of the HNE (and related aldehydes) generated by lipid peroxidation will be rapidly metabolized into other products and extruded from the cells. Therefore, direct measurements of tissue aldehydes and their protein adducts may grossly underestimate the extent of lipid peroxidation as well as the generation of lipid-derived enals.

As shown in Figure 1, radioactivity in the supernatant was separated into 4 distinct peaks (I-V). Peak V displayed high absorbance at 224 nm and co-eluted with externally added reagent HNE. This peak was, therefore, assigned to HNE. It represented minimal radioactivity; indicating that most of the HNE was metabolized under the present experimental conditions. Peak II accounted for 5 to 10 % of the added radioactivity and co-eluted with externally added reagent DHN. About 37 % of the radioactivity was recovered in peak III. Since the retention time of this peak corresponded to that of the synthesized reagent HNA, it was assigned to the oxidation product of HNE. A small peak of radioactivity eluted with a retention time of 63 min (peak III). However, the chemical identity of this peak could not be established.

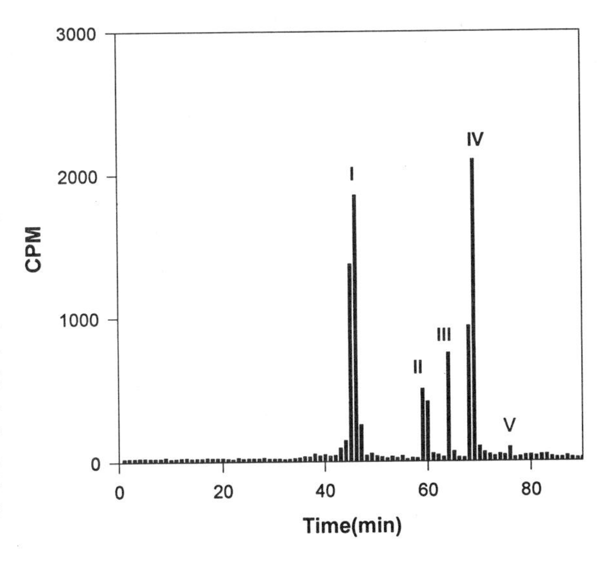

Figure 1. HPLC separation of the metabolites of HNE. Cardiac myocytes were isolated from adult guinea pig ventricles and incubated with 50 nmoles of [^{3}H]-HNE for 30 min. The myocytes were pelleted by low speed centrifugation and the radioactivity in the supernatant was separated by reverse phase HPLC using an ODS C_{18} column. Radioactivity in the HPLC eluate was measured in 1-ml fractions. The peaks containing radioactivity corresponded to the retention time of the synthesized standards of glutathione conjugates (peak I), DHN (peak II), HNA(peak IV) and unmetabolized HNE (peak V). The identity of peak III could not be established.

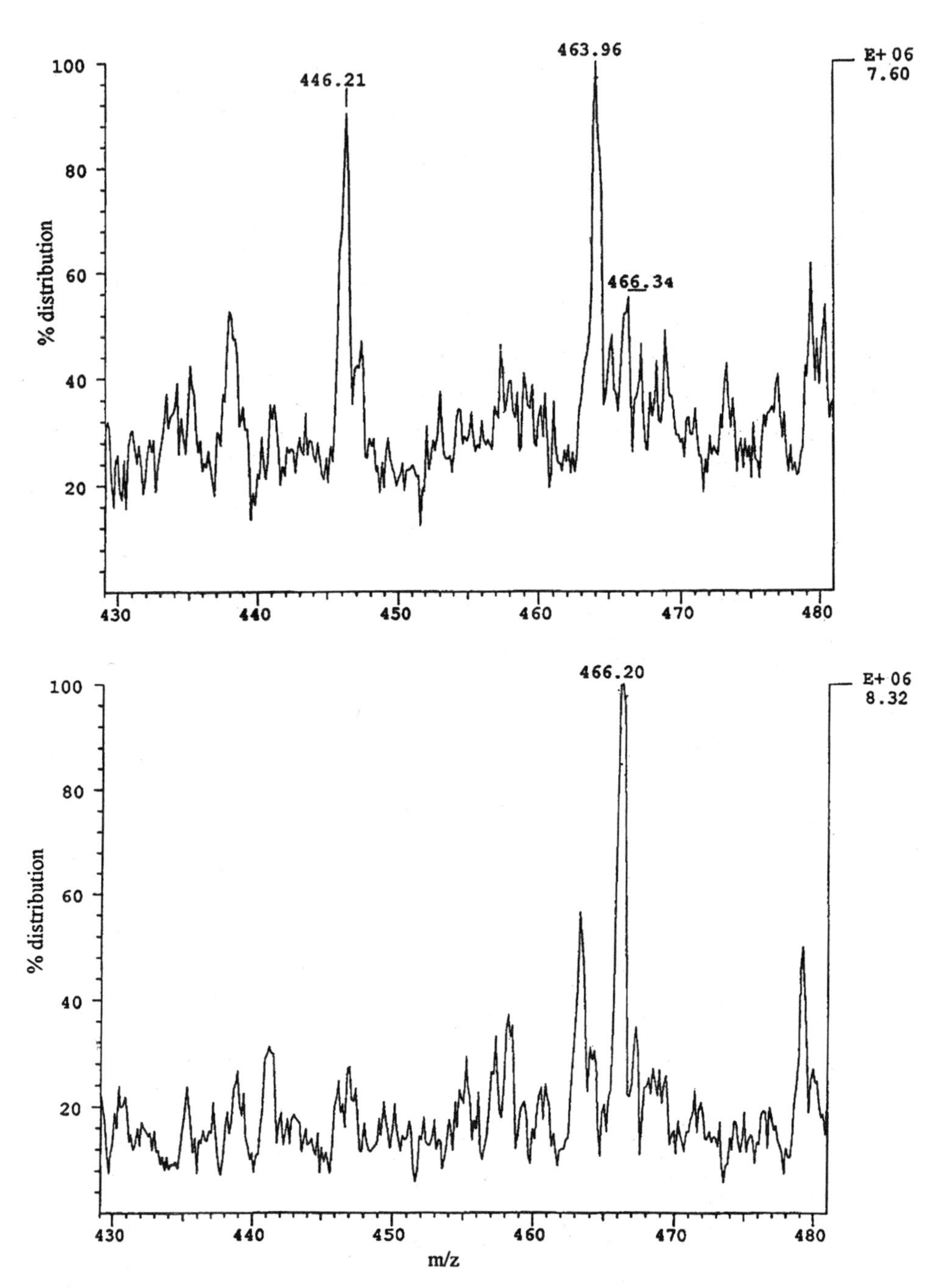

Figure 2. Electrospray mass spectra of the metabolites of HNE. Isolated cardiac myocytes were incubated with [^{3}H]-HNE and the metabolites recovered in the supernatant (upper panel) or the tissue extract (lower panel) were separated by HPLC as described in the text. Pooled fractions corresponding to peak I (see Figure 1) were collected and injected into electrospray. Peaks at m/z of 464 and 466 were identified to be due to GS-HNE and GS-DHN respectively. The peak with a m/z ratio of 446 is due to in source fragmentation of GS-HNE.

Peak I contained the polar metabolites of HNE and corresponded to 44 % of the total radioactivity. The retention time of peak I was identical to that obtained with the synthesized gluathione conjuates of HNE and its corresponding alcohol (GS-HNE and GS-DHN). In order to further establish the identity of this peak, the eluate corresponding to peak I was collected and analyzed by ESI-MS. As shown in Figure 2, two distinct peaks were observed which corresponded to the molecular mass of 464 and 466 Da, which is identical to the molecular mass of the expected metabolites GS-HNE and GS-DHN, respectively. The ratio of the 464/466 peaks, which corresponds to the ratio of the oxidized and reduced forms of the conjugate was 0.41 (Table 2), indicating that 60 % of the extruded conjugate was GS-HNE.

A similar distribution of metabolites was observed when the cell extract was examined. However, as shown in Figure 2 the 464/466 ratio in the cells was 5.0 which is markedly different from that observed in the supernatant (0.41). This dramatic difference suggests that at least in isolated cardiac myocytes, most of the reduced form of the conjugate (GS-DHN) is retained in the cells, whereas the oxidized form of the conjugate (GS-HNE) is preferentially extruded.

To examine the biochemical pathways responsible for HNE metabolism, the myocytes were incubated in the presence of inhibitors of the aldose reductase and aldehyde dehydrogenase. In the presence of 0.4 mM sorbinil, a significant decrease in the per cent of HNE converted to GS-DHN was observed. The ratio 464/466 in the supernatant was decreased from 0.41 to 0.16, whereas in the cell extract, the ratio decreased from 5.0 to 1.8. Nonetheless, no major change in the distribution pattern of other metabolites was observed either in the supernatant or the cell extract (Table 1), indicating that sorbinil selectively inhibits the conversion of GS-HNE to GS-DHN. These results indicate that the formation of GS-DHN in the cardiac myocytes is catalyzed by aldose reductase. In agreement with this conclusion, we have recently demonstrated that homogenous cardiac aldose reductase efficiently catalyzes the reduction of GS-HNE *in vitro* (Srivastava *et al.*, 1998a). Thus, aldose reductase, an enzyme previously thought to be involved in the metabolism of glucose (Bhatnagar and Srivastava, 1992), may also be an important participant in cardiac metabolism of enals.

Table 1. Percent distribution of HNE metabolites in the supernatant and the extract of isolated cardiac myocytes incubated with [^{3}H]-HNE

Treatment	Peak I (GS-conjugates)	Peak II (DHN)	Peak III (?)	Peak IV HNA
Supernatant				
HNE	44	11	8	37
+ sorbinil	47	5	7	41
+ cyanamide	62	12	12	14
+ sorbinil and cyanamide	75	7	8	10
Cell extract				
HNE	39	8	10	43
+ sorbinil	41	5	9	45
+ cyanamide	61	8	12	19
+ cyanamide and sorbinil	58	6	14	22

Cardiac myocytes were incubated with 50 nmoles of ^{3}H-HNE for 30 min and the radiolabeled metabolites were extracted and identified as described in the text. The percent distribution was calculated on the basis of the total radioactivity recovered in peaks I-IV. No unmetabolized HNE (peak V) was detected.

Table 2. ESI-MS analysis of the glutathione conjugates recovered from isolated cardiac mycoytes incubated with HNE

Treatment	Intensity at m/z		Intensity ratio for 464/466
	464	466	
Supernatant			
+ HNE	5.11E+06	2.09E+06	0.41
+ sorbinil	1.06E+07	1.69E+06	0.16
+ cyanamide	9.08E+06	1.82E+06	0.20
+ sorbinil and cynamide	6.90E+06	1.29E+06	0.19
Cell extract			
+ HNE	1.43E+06	7.15E+06	5.00
+ sorbinil	1.87E+06	4.50E+06	1.8
+ cynamide	1.96 E+06	9.30E+06	4.7
+ sorbinil and cyanamide	2.58E+06	4.23E+06	1.6

Isolated cardiac myocytes were incubated with 50 nmoles of [^{3}H]-HNE and the metabolites were extracted from the myocytes as well as the supernatant. The metabolites were separated on HPLC and the peak corresponding to the glutathione conjugates were collected, pooled and analyzed by ESI-MS. The m/z values of 464 and 466 were identified to be due to GS-HNE and GS-DHN respectively.

As expected, inhibition of aldehyde dehydrogenase by cyanamide, led to a decrease in the HPLC peak corresponding to HNA. These results identify aldehyde dehydrogenase catalyzed oxidation as a significant component of HNE metabolism. Interestingly, inhibition of aldehyde dehydrogenase led to a corresponding increase in the formation of glutathione conjugates, such that in the presence of cynamide the gluathione conjugates represented 60 to 70 % of the glutathione conjugates in the supernatant as well as the cell extract. This, suggests that during HNE metabolism, glutathione conjugation directly competes with aldehyde dehydrogenase catalyzed oxidation, such that inhibition of acid formation enhances the efficiency of conjugation. Moreover, the cynamide-induced increase in conjugate formation persisted even when aldose reductase was inhibited (Table 1), indicating that inhibition of aldose reductase does not alter the extent of HNE metabolized via glutathione conjugation. However, as suggested by our ESI-MS results (Table 2) the aldose reductase-mediated pathway does determine the ratio of the reduced to the oxidized conjugate and thus regulate the rate of the extent to which the conjugate is extruded from the cardiac myocytes.

In conclusion, the results of our present study show that HNE is rapidly metabolized in cardiac myocytes. The major pathways for HNE metabolism are similar to those recognized in other tissues such as the liver (Esterbauer *et al.,* 1991). However, in contrast to the rat heart (Srivastava *et al.*, 1998b), the guinea pig cardiac myocytes, show a greater extent of HNE metabolism via glutathione conjugation. The conjugate once formed appears to be rapidly extruded possibly via the GSSG transporter, or other transporters of glutathione conjugates such as the multi drug resistance protein. The kinetic properties and the identity of the GS-HNE transporter in the heart are currently under investigation in our laboratory.

ACKNOWLEDGMENTS

This work was supported in part by the NIH grants HL 59378 and DK36118.

REFERENCES

Bhatnagar, A. (1995) Electrophysiological effects of 4-hydroxynonenal, an aldehydic product of lipid peroxidation, on isolated rat ventricular myocytes. *Circ. Res.* **76**, 293–304.

Bhatnagar, A. (1997) Contribution of ATP to oxidative stress-induced changes in the action potential of isolated cardiac myocytes. *Am. J. Physiol.* **272**, H1598-H1608.

Bhatnagar, A. and Srivastava, S.K. (1992) Aldose reductase: Congenial and injurious profiles of an enigmatic enzyme. *Biochem. Med. Metabol. Biol.* **48**, 91–121.

Chandra, A. and Srivastava, S.K. (1997) A synthesis of 4-hydroxy-2-trans-nonenal and 4-(^{3}H) 4- hydroxy-2-trans-nonenal. *Lipids* **32**, 779–782.

Esterbauer, H., Schaur, R.J. and Zollner, H. (1991) Chemistry and biochemistry of 4-hydroxynonenal, malonaldehyde and related aldehydes. *Free Radic Biol Med.* **11**, 81–128.

Grosch, W. (1987) Reactions of hydroperoxides - products of low molecular weight. In Autoxidation of Unsaturated Lipids. (H.W.-S. Chan Ed.) Academic Press, London, pp 95–139.

Lander, H.M. (1997) An essential role of free radicals and derived species in signal transduction. *FASEB J.* **11**, 118–124.

Lucas, D.T. and Szweda, L.I. (1998) Cardiac reperfusion injury: aging, lipid peroxidation, and mitochondrial dysfunction. *Proc. Natl. Acad. Sci.* **95**, 510–514.

Nath, R.G. and Chung, F-L. (1994) Detectionof exocyclic 1,N^2-propanodeoxyguanosine adducts as common DNA lesions in rodents and humans. *Proc. Natl. Acad. Sci.* **91**, 7491–7495.

Sayre, L.M., Zelasko, D.A., Harris, P.L., Perry, G., Salomon, R.G. and Smith, M.A. (1997) 4-Hydroxynonenal-derived advanced lipid peroxidation end products are increased in Alzheimer's disease. *J. Neurochem.* **68**, 2092–2097.

Sen, C.K. and Packer, L. (1996) Antioxidant and redox regulation of gene transcription. *FASEB J.* **10**, 709–720.

Srivastava, S., Chandra, A., Ansari, N.H., Srivastava, S.K. and Bhatnagar, A. (1998a) Identification of cardiac oxidoreductase(s) involved in the metabolism of the lipid peroxidation derived aldehyde - 4-hydoxynonenal. *Biochem. J.* **329**, 469–475.

Srivastava, S., Chandra, A., Wang, L.-F., Seifert, W.E., DaGue, B.B., Ansari, N.H., Srivastava, S.K. and Bhatnagar, A. (1998b) Metabolism of the lipid peroxidation product, 4-hydoxy-*trans*-2- nonenal, in isolated perfused rat heart. *J. Biol. Chem.***273**, 10893–10900.

Witz, G. (1989) Biological interactions of alpha, beta-unsaturated aldehydes. *Free Radic. Biol. Med.* **7**, 333–49.

Yla-Herttuala, S., Palinski, W., Rosenfeld, M.E., Parthasarathy, S., Carew, T.E., Butler, S., Witztum J.L. and Steinberg D. (1989) Evidence for the presence of oxidatively modified low density lipoprotein in atherosclerotic lesions of rabbit and man. *J. Clin. Invest.* **84**, 1086–1095.

Yoritaka, A., Hattori, N., Uchida, K., Tanaka, M., Stadtman, E.R. and Mizuno, Y. (1996) Immunohistochemical detection of 4-hydroxynonenal protein adducts in Parkinson disease. *Proc. Natl. Acad. Sci.* **93**, 2696–2701.

OXIDATION OF ETHANOL TO ACETALDEHYDE IN BRAIN AND THE POSSIBLE BEHAVIORAL CONSEQUENCES

Sergey M. Zimatkin,[1,2] Anton V. Liopo,[1] and Richard A. Deitrich[2]

[1]Institute of Biochemistry
Academy of Sciences of Belarus, Grodno, Belarus
Alcohol Research Center
[2]University of Colorado
Denver, Colorado

1. INTRODUCTION

One of the oldest theories of ethanol's action is that the first metabolite, acetaldehyde, is responsible for some of the actions in the central nervous system (Hunt, 1996; Hashimoto et al. 1989; Tan et al. 1993; Bergamaschi et al. 1988; Zimatkin and Deitrich, 1997; Thadani and Truitt, 1977; Collins et al. 1988; Heap et al. 1995). This hypothesis fell into disfavor for a number of reasons. The first was that following ethanol ingestion, the blood levels of acetaldehyde are extremely low, provided that the artifactual formation of acetaldehyde is accounted for (Sippel, 1974; Westcott et al. 1980; Sippel and Eriksson, 1975; Tabakoff et al. 1976; Zimatkin and Pronko, 1995). The levels in normal humans are nearly undetectable in the blood, of the order of 1 μM. The second problem was that even if the blood acetaldehyde levels were significant, the molecule does not seem to be able to penetrate the blood brain barrier because of the presence of aldehyde dehydrogenase in blood vessels. Substantial blood levels were required before acetaldehyde appeared in the brain (Westcott et al. 1980; Tabakoff et al. 1976; Sippel and Eriksson, 1975; Sippel, 1974; Sippel, 1974). A third issue was that one could inhibit the oxidation of ethanol to acetaldehyde with pyrazole but intoxication still ensued. Indeed the use of pyrazole was critical in the vapor chamber method of Goldstein where ethanol metabolism was slowed in order to physically addict mice to ethanol (Goldstein and Pal, 1971).

One idea that received some attention earlier was that the brain itself might oxidize ethanol (Tabakoff and Gelpke, 1976; Zimatkin and Deitrich, 1997; Sinet et al. 1980; Cohen et al. 1980). However, studies of perfused brain with ^{14}C labeled ethanol failed to reveal any accumulation of any labeled metabolic products whereas both ^{14}C labeled

Enzymology and Molecular Biology of Carbonyl Metabolism 7
edited by Weiner *et al.* Kluwer Academic / Plenum Publishers, New York, 1999.

acetaldehyde and acetate labeled numerous amino acids and carbon dioxide (Mukherji et al. 1975). Numerous attempts to directly demonstrate oxidation of ethanol to acetaldehyde were unsuccessful, primarily because, in the presence of iron, ethanol undergoes oxidation to acetaldehyde and in brain tissue this was of such a magnitude so as to mask any enzymatic production of ethanol (Aragon, Baker and Deitrich, unpublished observations). This problem was solved in two ways. One method uses a very effective iron chelator, desferrioximine while the other method is to perfuse the brain with ice cold buffer and remove all hemoglobin before homogenization. These methods allowed the demonstration of a significant oxidation of ethanol to acetaldehyde in brain tissue in vitro (Gill et al. 1992; Aragon et al. 1992; Aragon et al. 1991; Aragon et al. 1991; Aragon et al. 1992). The oxidation seemed to be primarily at the hands of catalase, and indeed a number of indirect experiments that earlier suggested that the this was likely. These findings have now been replicated (Hamby-Mason et al. 1997).

The significance of these findings is that it is not necessary to postulate that acetaldehyde enters the brain from the blood now that it has been shown that it can be formed in situ. Thus, even though there may be little blood acetaldehyde, or if there is, little of it penetrates into the brain itself. However, in a back-door attack, acetaldehyde may be present by virtue of its production in the brain by catalase and perhaps other enzymes.

2. METHODS

In this study we have used mice and rats from a number of sources. 1. Rats from the randomally outbred stock at the Institute of Biochemistry Academy of Sciences of Belarus. 2. HAS (high alcohol sensitive), LAS (low alcohol sensitive, short-sleeping) and CAS (control alcohol sensitive, intermediate-sleeping) rats selectively bred for initial sensitivity to ethanol from the Alcohol Research Center at the University of Colorado. 3. Genetically heterogeneous mice, short sleep and long sleep mice selectively bred for sensitivity to ethanol obtained from the Institute for Behavioral Genetics at the University of Colorado. 4. Acatalasemic mice, superoxide overexpressing and C3H mice from the Webb-Waring Institute of the University of Colorado.

To test the sensitivity of mice and rats to the hypnotic effects, ethanol was administrated in intraperitoneally, and the duration of alcohol-induced sleep was measured by the interval from the loss to recovery of righting response (sleep time). At the moment of recovery blood was taken from the retroorbital sinus for determination of ethanol and acetaldehyde. Brains were taken for determination of the rates of acetaldehyde accumulation from HS mice 5 days after testing and from HS rats 2 weeks following testing for sleep time.

2.1. In Vitro Assays

To remove erythrocyte catalase activity from the brain, the animals (180–200 gm) were perfused under anaesthesia through the heart by 300–400 ml (30–40 ml/min) of ice-cold isotonic saline. The brains were quickly excised, rinsed in saline, placed on filter paper on ice and the hypothalamus, cerebellum, basal ganglia, brain hemispheres and brain stem were taken and stored in liquid nitrogen. Brain samples were weighed and 10% homogenates prepared in ice-cold 0.1 M potassium phosphate buffer (pH 7.4), containing 0.1% Triton, using a homogenizer with glass mortar and Teflon pestle. All homogenates were stored at 0° and examined the same day.

Varying amounts of brain tissue were incubated for periods from 5 to 60 min at 37° in sealed clear 15 ml vials. The incubation medium (1.25 ml) included 0.1 M potassium phosphate buffer (pH 7.4), 10 mM glucose and various concentrations of ethanol in the presence or absence of the catalase inhibitor 3-amino-1,2,4-triazole (8 mM, added to the incubation medium 20 min before ethanol). The reaction was stopped by an ice-cold mixture (1.0 ml) of perchloric acid (1.6 M), thiourea (13 mM), semicarbazide (3 mM), EDTA (0.4%) and 1 mM 1-propanol as internal standard added through the stopper with a needle and syringe. This treatment prevents further formation of acetaldehyde and in preliminary experiments gave the best signal to noise ratio. The vials were stored at 4° for 60 min. The acetaldehyde content of the gaseous phase of each vial was measured by a head-space gas chromatography procedure as follows. The flasks were incubated at 65° for 20 min, and 2 ml of the head-space was injected into Model 3700 gas chromatograph with a flame ionization detector. A 200 cm, 2 mm glass column of Poropak Q mesh 80/100 was used with inlet and detector temperatures at 125° and 150°, respectively, and a nitrogen flow rate of 30 ml/min. Under these conditions, the retention time for acetaldehyde was 1.8 min, for ethanol, 3.0 min, and for propanol (internal standard), 3.8 min. Relative peak heights were determined by comparison with standards prepared by the addition of known amounts of acetaldehyde to "zero-time" controls. To establish the blank, each experimental sample was accompanied by a control which included the same homogenate and reagents, except for ethanol, and treated in the same manner. Blanks containing ethanol, but stopped before incubation were also used. The possible nonenzymatic production of acetaldehyde in the course of incubation with ethanol was also examined in heated brain homogenates.

2.2. Catalase Assay

The homogenates were assayed for catalase activity by the spectrophotometric method of Aebi (1984). Aliquots of the homogenates equivalent to 50 mg of wet tissue were incubated for 3 min at 25° in the incubation medium containing 10 mM potassium phosphate buffer (pH 7.0) and 0.5 M hydrogen peroxide. The catalase activity was determined by H_2O_2 elimination measured spectrophotometrically at 240 nm. The enzyme activities were calculated using a millimolar extinction coefficient of 0.0436 mM^{-1} cm^{-1}. One unit is equal to 1 μM H_2O_2/min/mg protein, equal to 0.5 μM/min/mg protein, if determined with an oxygen electrode.

Protein was determined by the method of Lowry et al (1951) using bovine serum albumin as the standard.

3. RESULTS AND DISCUSSION

The ability to detect acetaldehyde at all in brain homogenates requires that ethanol be metabolized to acetaldehyde at a rate greater than its rate of removal by aldehyde dehydrogenase or other mechanisms, including diffusion into the blood (Figure 1). Since the blood already contains acetaldehyde arising from the liver oxidation of acetaldehyde, this may not be a source of removal of much acetaldehyde from the brain.

We have studied the distribution of this activity in different brain areas (Table 1). An important feature of these data is that while not strictly linear for 60 minutes in all brain areas, acetaldehyde continues to accumulate, indicating that the mechanisms for its removal are slower than its rate of formation. This renders somewhat moot the discussions about adequacy of the source of hydrogen peroxide, since it is apparent that this is not a serious impediment to the oxidation of ethanol in the brain.

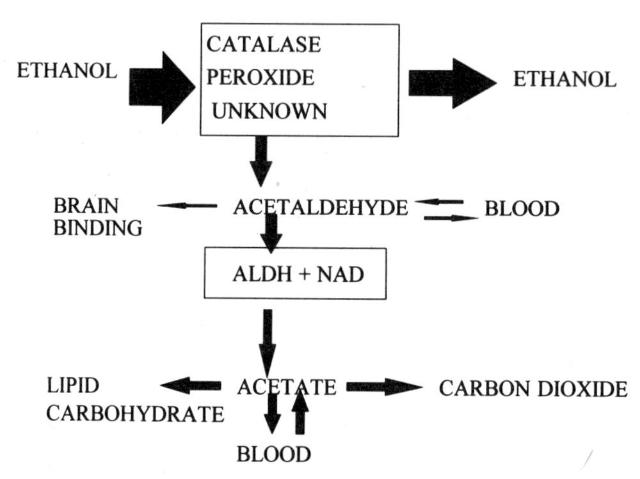

Figure 1. Scheme of possible routes of ethanol metabolism in brain. Arrows are not to scale.

In Figure 2 we illustrate the inhibition of catalase by aminotriazole in these brain areas as compared to the inhibition of the accumulation of acetaldehyde in these brain areas. Two aspects are apparent. First aminotriazole is not equally effective in all brain areas in inhibition of catalase, but secondly, the inhibition of acetaldehyde accumulation is significantly less in each brain area, indicating that there is a second source of acetaldehyde which is not inhibited by the catalase inhibitor. We also found that acatalsemic mice metabolized ethanol nearly as efficiently as did the control C3H mice (Table 2). Earlier experiments (Gill et al, 1992) found that neither pyrazole nor metyrapone inhibited the accumulation of acetaldehyde, indicating that it is unlikely that alcohol dehydrogenase or cytochrome P450 are responsible. The best evidence for this will come from use of CYP 450 2E1 and ADH knockout mice.

We have also investigated the correlation between the behavioral effects of ethanol and the metabolism of ethanol in the brain. As seen in Table 3, there is a relationship between the ethanol sensitivity of a number of animal lines that differ in the hypnotic effects of ethanol and the amount of acetaldehyde which accumulates in the brain in in vitro studies. This includes both mice and rats. For the heterogeneous rats and mice, the sleep times are divided arbitrarily and the brain ethanol metabolism tested several weeks later. For the selected lines, the animals genetically selected to have short or long sleep times were used without behavioral testing.

Table 1. Effect of incubation time on acetaldehyde accumulation in brain regions (nmoles/mg protein)

	Time min				
Brain region	5	15	30	45	60
Hemispheres	1.5±0.4	2.4±0.7	3.7±1.2	4.6±1.4	5.1±1.3
Stem	0.4±0.07	1.3±0.1	2.6±0.4	3.2±0.3	3.7±1.4
Cerebellum	1.5±0.1	1.8±0.4	3.6±0.7	5.4±1.2	7.2±1.7
Whole	0.8±0.2	1.8±0.3	3.3±0.4	4.4±0.6	5.3±0.8

Each data point is mean± SEM of 5 animals. 50mM ethanol

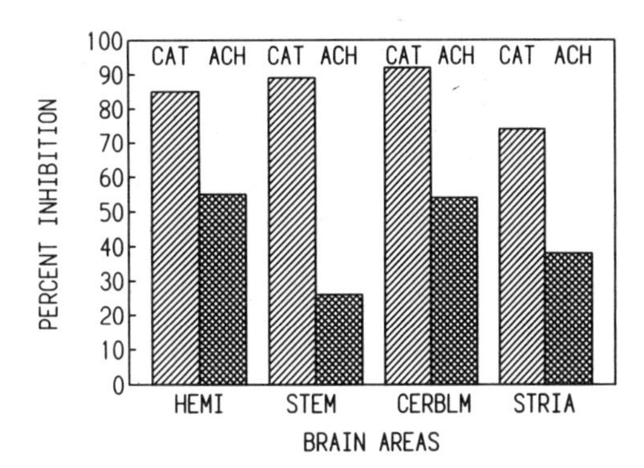

Figure 2. Relationship of the inhibition of catalase to the inhibition of acetaldehyde by aminotriazole in various brain areas. Hemi, cereberal hemispheres; Stem, brain stem; Cerblm, cerebellum; Stria, stiatum.

Table 2. Ethanol sensitivity, acetaldehyde accumulation and catalase activity in brains of mice with different catalase content

Group (n)	Sleep time (min)	BAC RRR mM	Acetaldehyde nm/H/mg p	Catalase (um O2/min/mg p)
C3H (20)	70.6 ± 6.8	96.3 ± 1.9	12 ± .8	1.9 ± .2
Acat (33)	96 ± 9.3	93.3 ± 1.13	6.7 ± .5	0.1 ± 0.01
SODO (47)	90 ± 4.4	86.5 ± 1.11	7.3 ± .8	1.9 ± .2

C3H: C3H inbred mice as controls for acatalasemic mice (Acat). SODO: Superoxide overexpressing mice. Values are ± SEM.

Table 3. Acetaldehyde accumulation in brains of animals differing in ethanol sensitivity

Phenotype	LAS/CAS/HAS rats	HS mice	HS rats	SS/LS mice	C3H mice
Short sleep	0.63 ± .14	0.67 ± .18	0.32 ± .03	0.6 ± .18	0.83 ± .075
Intermed.	0.72 ± .12	0.96 ± .08	0.44 ± .05	–	0.99 ± .15
Long sleep	1.23 ± .17*#	1.65 ± .21*#	0.48 ± .05*	1.41 ± .2*	1.26 ± .12*

Values are nmoles/mgprotein/5min with 50mM Ethanol
*Significantly different from Short Sleep
#Significantly different from Intermediate

In summary we have confirmed the oxidation of ethanol to acetaldehyde in the brains of mice and rats and have extended this to various areas of the brain. We have shown that catalase cannot be solely responsible for the action since nearly complete inhibition of catalase by aminotriazole results in vastly different levels of inhibition of acetaldehyde accumulation. In addition brains from acatalsemic mice still produce acetaldehyde from ethanol. There appears to be a correlation between the capacity to produce acetaldehyde from ethanol and the behavioral sensitivity of a variety of mice and rats to the hypnotic effects of ethanol. The possible role of acetaldehyde in the effects of ethanol in the brain should continue to be investigated.

ACKNOWLEDGMENTS

Supported by a Grant from CRDF #112100, AA 03527. AA00093 and AA010598.

REFERENCES

Aebi, H. (1984) Catalase in vitro. Methods. Enzymol. *105*: 121–126.

Aragon, C.M.G., F. Rogan, and Z. Amit (1992) Ethanol metabolism in rat brain homogenates by a catalase-H_2O_2 system. Biochem. Pharmacol. *44*: 93–98.

Aragon, C.M.G., L.M. Stotland, and Z. Amit (1991) Studies on ethanol-brain catalase interaction: Evidence for central ethanol oxidation. Alc.: Clin. Exp. Res. *15*: 165–169.

Bergamaschi, S., S. Govoni, R.A. Rius, and M. Trabucchi (1988) Acute ethanol and acetaldehyde administration produce similar effects on L-type calcium channels in rat brain. Alcohol. *5*: 337–340.

Cohen, G., P.M. Sinet, and R. Heikkila (1980) Ethanol Oxidation by Rat Brain in Vivo. Alc.: Clin. Exp. Res. *4*,: 366–370.

Collins, M.A., N.S. Ung-Chun, and K. Raikoff (1988) Stable acetaldehyde adducts with brain proteins. Alc. Alcoholism *23*: 31

Gill, K., J.F. Menez, D. Lucas, and R.A. Deitrich (1992) Enzymatic production of acetaldehyde from ethanol in rat brain tissue. Alc. : Clin. Exp. Res. *16*: 910–915.

Goldstein, D.B. and N. Pal (1971) Alcohol dependence produced in mice by inhalation of ethanol; grading the withdrawal reaction. Science *172*: 288–290.

Hamby-Mason, R., J.J. Chen, S. Schenker, A. Perez, and G.I. Henderson (1997) Catalase mediates acetaldehyde formation from ethanol in fetal and neonatal rat brain. Alcohol. Clin. Exp. Res. *21*: 1063–1072.

Hashimoto, T., T. Ueha, T. Kuriyama, M. Katsura, and K. Kuriyama (1989) Acetaldehyde-induced alterations in metabolism of monoamines in mouse brain. Alc. Alcoholism *24*: 91–99.

Heap, L., R.J. Ward, C. Abiaka, D. Dexter, M. Lawlor, O. Pratt, A. Thomson, K. Shaw, and T.J. Peters (1995) The influence of brain acetaldehyde on oxidative status, dopamine metabolism and visual discrimination task. Biochem. Pharmacol. *50*: 263–270.

Hunt, W.A. (1996) Role of acetaldehyde in the actions of ethanol on the brain - A review. Alc. *13*: 147–151.

Jääskeläinen, I.P., E. Pekkonen, K. Alho, J.D. Sinclair, P. Sillanaukee, and R. Näätänen (1995) Dose-related effect of alcohol on mismatch negativity and reaction time performance. Alc. *12*: 491–495.

Lowry, O.H., Rosenbrough, N.J., Farr, A.L., and Randall, R.J., (1951) Protein measurement with aht Folin phenol reagent. J. Biol. Chem. 193:L 265–275)

Mukherji, B., Y. Kashiki, H. Ohyanagi, and Sloviter (1975) Metabolism of ethanol and acetaldehyde by the isolated perfused rat brain. J. Neurochem. *24*: 841–843.

Sinet, P.M., R.E. Heikkila, and G. Cohen (1980) Hydrogen Peroxide Production by Rat Brain In Vivo. J. Neurochem. *34*: 1421–1428.

Sippel, H.W. (1974) The acetaldehyde content in rat brain during ethanol metabolism. J Neurochem *23*: 451–452.

Sippel, H.W. and C.J.P. Eriksson (1975) The acetaldehyde content in rat brain during ethanol oxidation. In *The Role of Acetaldehyde in the Actions of Ethanol, Satellite Symposium 6th Int. Congr. Pharmac.* Vol.23, K.O. Lindros and C.J.P. Eriksson, eds., pp. 149–157, The Finnish Foundation for Alcohol Studies, Helsinki.

Tabakoff, B., R.A. Anderson, and R.F. Ritzmann (1976) Brain acetaldehyde after ethanol administration. Biochem. Pharmacol. *25*: 1305–1309.

Tabakoff, B. and C.G. Gelpke (1976) Alcohol and aldehyde metabolism in brain. In *Biochemical Pharmacology of Ethanol*, E. Majchrowicz, ed., pp. 141–164, Plenum Publishing Corporation, New York.

Tan, X., A.F. Castoldi, L. Manzo, and L.G. Costa (1993) Interaction of ethanol with muscarinic receptor-stimulated phosphoinositide metabolism during the brain growth spurt in the rat: Role of acetaldehyde. Neurosci. Lett. *156*: 13–16.

Thadani, P.V. and E.B. Truitt (1977) Effect of acute ethanol or acetaldehyde administration on the uptake, release, metabolism and turnover of norepinephrine in the brain. Biochem. Pharm. *26*: 1147–1150.

Westcott, J.Y., H. Weiner, J. Schultz, and R.D. Myers (1980) In vivo acetaldehyde in the brain of the rat treated with ethanol. Biochem. Pharmacol. *29*: 411–417.

Zimatkin, S.M. and R.A. Deitrich (1997) Ethanol metabolism in the brain. Addiction Biology *2*: 387–392.

Zimatkin, S.M. and P.S. Pronko (1995) Levels of ethanol and acetaldehyde and morphological disturbances in brain following administration of alcohol and ALDH inhibitors. In *International Association of Forensic Science*, B. Jacob and W. Bonte, eds., pp. 1–4, Verlag, Berlin.

CLONING AND EXPRESSION OF A cDNA ENCODING A CONSTITUTIVELY EXPRESSED RAT LIVER CYTOSOLIC ALDEHYDE DEHYDROGENASE

Eva C. Kathmann and James J. Lipsky

Department of Pharmacology
Mayo Clinic and Foundation
Rochester, Minnesota 55905

1. INTRODUCTION

Aldehyde dehydrogenase (ALDH) is a ubiquitous enzyme expressed in nearly all mammalian tissues and subcellular organelles. Rat liver contains multiple isoforms of ALDH, including the following for which cDNAs have been cloned: phenobarbital (PB)-inducible cytosolic ALDH1 (Dunn et al., 1989), mitochondrial ALDH2 (Farres et al., 1989), microsomal ALDH3 (Miyauchi et al., 1991), and tumor-associated cytosolic ALDH3 (Jones et al., 1988). However, a constitutively expressed rat liver cytosolic ALDH1 has been difficult to characterize based on confusion in the literature as to whether or not this ALDH isoform even exists (Tottmar, et al., 1973; Lindahl and Evces, 1984; Berger and Weiner, 1977; Tank et al., 1981; Truesdale-Mahoney et al., 1981; Cao et al., 1989). In order to study and characterize this ALDH1 isoform, we have cloned and expressed a cDNA encoding a rat liver constitutively expressed cytosolic ALDH1.

2. MATERIALS AND METHODS

2.1. cDNA Cloning

The cDNA encoding a rat liver constitutively expressed ALDH1 was cloned using a PCR-based method (Kathmann and Lipsky, 1997). Primers were designed based on regions of conservation in five mammalian cytosolic ALDH1 cDNA sequences. An 853 nucleotide segment of the open reading frame was amplified and sequenced from a Sprague

Dawley rat liver cDNA library (Stratagene). From the sequence obtained, gene specific primers were designed to perform 5'- and 3'-RACE (rapid amplification of cDNA ends). The entire open reading frame and the 5'- and 3'-UTR sequences were obtained from Sprague Dawley rat liver Marathon-Ready cDNA (Clontech). Finally, primers specific for the 5'- and 3'-UTRs were designed to amplify the full-length cDNA from the Marathon-Ready cDNA.

2.2. Tissue Distribution

Northern blot analysis was performed to determine the tissue distribution of constitutively expressed cytosolic ALDH1 mRNA in rat. Since constitutive cytosolic ALDH1 and PB-inducible ALDH1 are 89% identical at the nucleotide level, specific probes were prepared to differentiate between the two mRNAs. A probe specific for the constitutive cytosolic ALDH1 3'-UTR corresponding to nucleotides 1660 to 1905 (numbering is relative to the adenosine of the translation initiation ATG start codon) and a probe specific for the PB-inducible ALDH1 corresponding to nucleotides 1496 to 1884 were prepared. The two ALDH1 sequences share only 79% identity in this region of their 3'-UTRs. A Sprague Dawley rat Multiple Tissue Northern Blot (Clontech) containing approximately 2 μg of purified poly (A) RNA per lane was probed with α^{32}P-dCTP randomly-labeled cDNA probes.

2.3. Protein Expression

The rat liver constitutively expressed cytosolic ALDH1 cDNA open reading frame was cloned into the pET-14b expression vector (Novagen) and transformed into *E. coli* BL21(DE3)pLysS. The recombinant protein contained a 6-residue histidine tag upstream of a thrombin cleavage site at the amino terminus. The protein was purified over a nickel column and the histidine tag was removed by thrombin cleavage subsequent to purification.

2.4. Western Blot Analysis

Antiserum was raised in rabbits (Cocalico Biologicals) against peptides corresponding to rat liver constitutively expressed ALDH1 residues 5–16 and rat liver mitochondrial ALDH2 residues 3–14. Chemiluminescent detection was performed with the ECL Western Blotting Analysis System (Amersham Life Science).

3. RESULTS AND DISCUSSION

A cDNA encoding a constitutively expressed rat liver cytosolic ALDH1 was cloned using a PCR-based method (Kathmann and Lipsky, 1997). The cDNA contained 28 nucleotides of the 5'-UTR, a 1503 nucleotide open reading frame, and three different length 3'-UTRs (239, 504 and 564 nucleotides) differing only in the position of the poly (A) tail. The cDNA encoded a peptide of 501 amino acids with a calculated molecular mass of 54,458 daltons. The deduced amino acid sequence of rat liver constitutive cytosolic ALDH1 was 99.6, 96.4, 89.8, and 86.6 % identical to rat kidney, mouse liver, rat liver PB-inducible, and human liver cytosolic ALDH1 sequences, respectively. The rat liver and kidney cytosolic (Bhat et al., 1995) ALDH1 proteins differ by only two amino acids (residue 100 is arginine in liver while cysteine in kidney and residue 170 is asparagine in liver

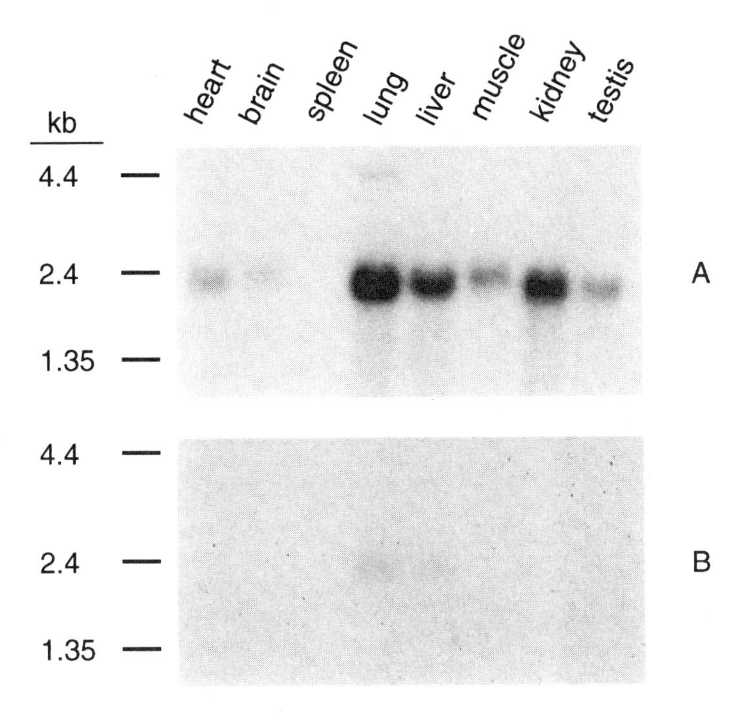

Figure 1. Northern blot analysis with a rat constitutive cytosolic ALDH1 specific probe (A). The blot was stripped and reprobed with a rat phenobarbital-inducible cytosolic ALDH1 specific probe (B). Panel A was exposed for 20 hours while panel B was exposed for 93 hours.

while glutamate in kidney), thus they appear to be the same gene product expressed in two different tissues.

Northern blot analysis indicated that constitutively expressed cytosolic ALDH1 mRNA was highly expressed in lung, kidney and liver, moderately expressed in skeletal muscle and testis, and weakly expressed in heart and brain of untreated Sprague Dawley rats (Figure 1). In comparison, PB-inducible cytosolic ALDH1 mRNA was very weakly expressed only in lung and liver of untreated Sprague Dawley rats (Figure 1). This is in contrast to a previous report indicating that untreated rats contain PB-inducible ALDH1 mRNA in lung, kidney, brain, colon, heart, and small intestine (Dunn et al., 1989). However, the probe used for Northern blot analysis in the previous report contained the entire open reading frame of PB-inducible ALDH1 which we now know is about 90% identical to constitutive cytosolic ALDH1. Therefore, the mRNA detected in untreated rat tissues in the previous report may have been constitutively expressed cytosolic ALDH1 mRNA rather than PB-inducible cytosolic ALDH1 mRNA.

Rat liver constitutive cytosolic ALDH1 was expressed with a histidine tag at the amino terminus in *E. coli*. The purified recombinant protein had a subunit molecular weight of 55 kDa as determined by SDS-PAGE. ALDH activity of the recombinant protein was measured spectrophotometrically at 340 nm as the rate of formation of NADH. At 24°C and pH 8.8, the measured K_m for acetaldehyde was 47 μM which is in the range expected for a class 1 ALDH. Antiserum raised against the amino terminus of rat liver cytosolic ALDH1 recognized the recombinant constitutive cytosolic ALDH1 protein, while antiserum raised against the amino terminus of rat liver mitochondrial ALDH2 did not (Figure 2).

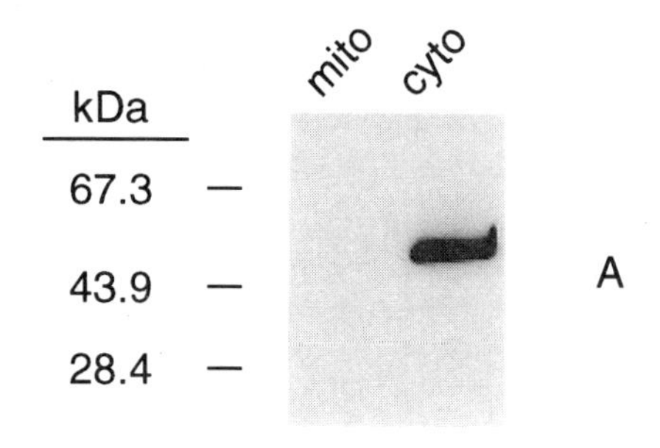

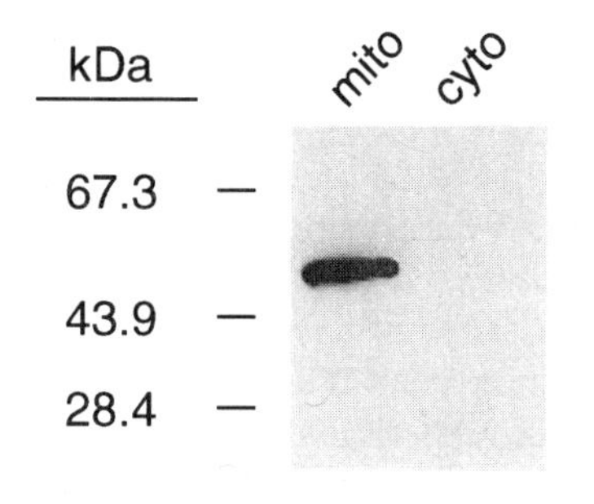

Figure 2. Western blot analysis of recombinant rat liver constitutive cytosolic ALDH1 and recombinant rat liver mitochondrial ALDH2 probed with antiserum raised against the amino terminus of rat liver constitutive cytosolic ALDH1 (A). The blot was stripped and reprobed with antiserum raised against the amino terminus of rat liver mitochondrial ALDH2 (B).

In conclusion, we have cloned and expressed a cDNA which, based on identity with other cytosolic ALDH1 sequences, kinetic analysis, and Western blot analysis, encodes a constitutively expressed rat liver cytosolic ALDH1. The recombinant protein will be useful to further study and characterize this rat liver ALDH isozyme.

ACKNOWLEDGMENTS

This work was supported by grants NIH AA-09543 and FDT-000-886. We thank Dr. Henry Weiner for his gift of rat liver mitochondrial ALDH2 cDNA from which recombinant protein was produced.

REFERENCES

Berger, D. and Weiner, H., 1977, Relationship between alcohol preference and biogenic aldehyde metabolizing enzymes in rats, *Biochem. Pharmacol.* **26**, 841–846.

Bhat, P. V., Labrecque, J. Boutin, J.-M., Lacroix, A. and Yoshida, A., 1995, Cloning of a cDNA encoding rat aldehyde dehydrogenase with high activity for retinal oxidation, *Gene* **166**, 303–306.

Cao, Q.-N., Tu, G.-C., and Weiner, H., 1989, Presence of cytosolic aldehyde dehydrogenase isozymes in adult and fetal rat liver, *Biochem. Pharmacol.* **38**, 77–83.

Dunn, T. J., Koleske, A. J., Lindahl, R. and Pitot, H. C., 1989, Phenobarbital-inducible aldehyde dehydrogenase in the rat, *J. Biol. Chem.* **264**, 13057–13065.

Farres, J., Guan, K. L. and Weiner, H., 1989, Primary structures of rat and bovine liver mitochondrial aldehyde dehydrogenases deduced from cDNA sequences, *Eur. J. Biochem.* **180**, 67–74.

Jones, D. E., Brennan, M. D., Hempel, J. and Lindahl, R., 1988, Cloning and complete nucleotide sequence of a full-length cDNA encoding a catalytically functional tumor-associated aldehyde dehydrogenase, *Proc. Natl. Acad. Sci. U.S.A.* **85**, 1782–1786.

Kathmann, E. C. and Lipsky, J. J., 1997, Cloning of a cDNA encoding a constitutively expressed rat liver cytosolic aldehyde dehydrogenase, *Bioch. Bioph. Res. Comm.* **236**, 527–531.

Lindahl, R. and Evces, S., 1984, Comparative subcellular distribution of aldehyde dehydrogenase in rat, mouse and rabbit liver, *Biochem. Pharmacol.* **33**, 3383–3389.

Miyauchi, K., Masaki, R., Taketani, S., Yamamoto, A., Akayama, M. and Tashiro, Y., 1991, Molecular cloning, sequencing and expression of cDNA for rat liver microsomal aldehyde dehydrogenase, *J. Biol. Chem.* **266**, 19536–19542.

Tank, A. W., Weiner, H. and Thurman, J. A., 1981, Enzymology and subcellular localization of aldehyde oxidation in rat liver, *Biochem. Pharmacol.* **30**, 3265–3275.

Tottmar, S. O. C., Pettersson, H. and Kiessling, K.-H., 1973, The subcellular distribution and properties of aldehyde dehydrogenases in rat liver, *Biochem. J.* **135**, 577–586.

Truesdale-Mahoney, N. Doolittle, D. P. and Weiner, H., 1981, Genetic basis for the polymorphism of rat liver cytosolic aldehyde dehydrogenase, *Biochem. Genet.* **19**, 1275–1282.

30

THE ROLES OF ACETALDEHYDE DEHYDROGENASES IN *SACCHAROMYCES CEREVISIAE*

Wayne Tessier, Mark Dickinson, and Melvin Midgley

Department of Biological Sciences
University of Hull
Cottingham Road
Hull HU6 7RX, United Kingdom

1. INTRODUCTION

This work was undertaken to try to understand the routes that are used for synthesis of acetyl-CoA in *Saccharomyces cerevisae* under a variety of nutritional conditions. It is a topic of interest in connection with the production of flavour compounds in beer and lager. Distinct physiological roles have been proposed for the Mg^{2+} activated, $NADP^+$-dependent acetaldehyde dehydrogenase (Mg-ACDH) and the K^+-activated $NAD^+/NADP^+$ dependent mitochondrial enzyme (K-ACDH) (Llorente & de Castro 1977). The former enzyme was thought to function in acetate formation from acetaldehyde derived from pyruvate through the action of pyruvate decarboxylase. The latter enzyme has been implicated in acetate formation from acetaldehyde derived from ethanol. On this basis, disruption mutants for each of the enzymes would be expected to have markedly different potential when grown on 5% glucose, where mitochondrial function is severely repressed and on 1% ethanol where mitochondrial function is developed.

Just after our experimental work was completed a study with similar aims was published by Wang *et al.* 1998. In some ways this was a more detailed work involving the use of single, double and triple mutations of aldehyde dehydrogenase genes. There appear to be some differences in the growth characteristics of some mutants in the two studies, but our main conclusions about the identity of genes coding for Mg^{2+}-activated and K^+-activated acetaldehyde dehydrogenases and the function of these enzymes are in broad agreement with those of Wang *et al.* (1998).

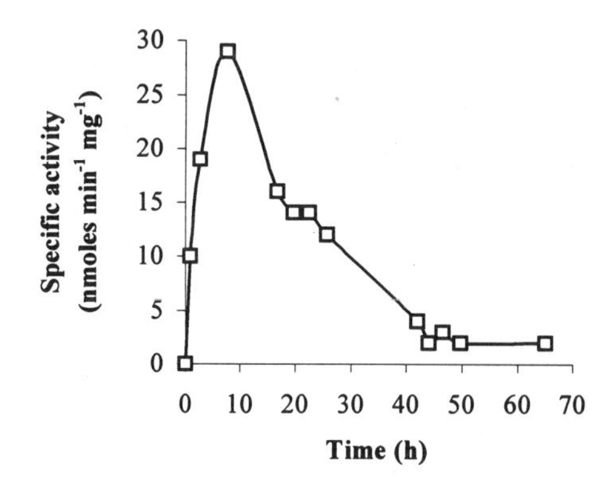

Figure 1. Variation of the specific activity of Mg-ACDH during a commercial pilot scale lager fermentation.

2. RESULTS

2.1. Mg-ACDH in Brewing

Figure 1 shows the variation in the specific activity of Mg-ACDH in the course of a lager brew with a commercial brewing strain (*Saccharomyces cerevisiae* BBII). The rapid rise and fall has been reproduced on a laboratory scale (1 litre) with this yeast and with a number of other yeasts including YPH499 which was the isogenic haploid of the wild type of the yeast used for the deletion work (see below). It was clear that the activity of the enzyme was under tight regulatory control. A number of experiments have shown that the burst of activity was primarily in response to the presence of glucose in the medium and not to the level of dissolved oxygen. It was also noted that the burst of activity always preceded the phase of rapid growth. These observations seemed to support the idea that Mg-ACDH was used in processing acetaldehyde derived from pyruvate and was a key factor in preparing the cell for active growth.

Table 1. ACDH activities in YPH499 and mutant strains

	Specific activity [nmoles/min/mg protein]	
Yeast strain	Mg-ACDH	K-ACDH
YPH499 (*ALD6:ALD7*)	74 ± 2	93 ± 14
RY124 (*ald6*)	0	41
RY125 (*ald6*)	0	49
RY126 (*ald6*)	0	49
RY127 (*ald6*)	0	45
RY270 (*ald7*)	74	0
RY271 (*ald7*)	74	0
RY272 (*ald7*)	66	0
RY273 (*ald7*)	77	0
RY274 (*ald7*)	74	0
RY275 (*ald7*)	70	0

All cells were grown at 30°C on complex medium containing 0.5% glucose.

Table 2. Growth rates for YPH499 and mutant strains at 30°C

	Maximum specific growth rate (h^1)		
Culture conditions	YPH499 (*ALD6*)	RY124 (*ald6*)	RY 125 (*ald6*)
YNB 0.5% glucose	0.29	0.16	0.17
YNB 5% glucose	0.29	0.14	0.16
YEP 0.5% glucose	0.32	0.12	not tested
YEP 5% glucose	0.3	0.1	not tested

For mutants RY270-275 (*ald7*) specific growth rates were all in the range 0.3-0.33 h^{-1}
YNB, yeast nitrogen base; YEP, yeast extract peptone

2.2. Deletion of Mg-ACDH

The gene (*ALD6*) encoding Mg-ACDH was identified by a TFASTA search of Gen Bank (release 93.0) after purification of the enzyme (Dickinson, 1996) and establishing the N-terminal sequence. Following gene disruption and transformation four mutant strains of YPH501 were obtained (Meaden *et al.* 1997) all completely lacking in Mg-ACDH (Table 1). Growth studies (Table 2) showed that growth on glucose was impaired with specific growth rates reduced to a half or a third of wild type levels on 5% or 0.5% glucose in shake-flask or fermenter-grown cultures. Evidently the mutation was not completely inhibitory for growth on glucose. We concluded that, although Mg-ACDH was clearly important for glucose metabolism, there were compensating reaction(s) which could provide acetate for acetyl-CoA production. One possible route was that acetaldehyde accumulating in the cytosol could enter the mitochondria for oxidation by K-ACDH followed by export of the acetate back to the cytosol. Table 1 shows that the K-ACDH was active in these mutants, although interestingly, at somewhat reduced levels from the wild type. In addition, however, we made the finding that the repressibility of the K-ACDH on change from 0.5% to 5% glucose was much less marked (Table 3) in the Mg-ACDH disrupted mutants. This seemed to imply a mechanism to meet the increased need for K-ACDH in the absence of Mg-ACDH under conditions where Mg-ACDH was normally active and where K-ACDH was severely repressed because mitochondrial function was undeveloped.

A surprising result with the Mg-ACDH mutants was that all were very severely impaired in growth on 1% ethanol. Growth rates could not be determined, but cell masses at the end of the incubation period were of the order of 1% of those of the wild type. Since K-ACDH was present and the role of this enzyme was thought to be in ethanol assimila-

Table 3. ACDH in YPH499 and RY124 (*ald6*) under different growth conditions

		Specific activity (nmoles/min/mg)		
Strain	Culture conditions	Mg-ACDH	K-ACDH	K-ACDH % repression
YPH499 (*ALD6*)	0.5% glucose	54	29	
YPH499 (*ALD6*)	5% glucose	62	3	90
RY124 (*ald6*)	0.5% glucose	0	28	
RY124 (*ald6*)	5% glucose	0	21	25

Cultures were grown at 30°C in shake flasks in complex media.
Essentially the same results were also obtained with minimal medium.
RY125, 126 and 127 behaved in a similar way to RY124.

tion, the lack of growth on ethanol seemed difficult to account for. All but one of the mutants proved capable of switching from ADH I (high Km, constitutive) to ADH II (low Km ethanol induced) on transfer from 5% glucose to 1% ethanol. The wild type of course did this successfully.

2.3. K-ACDH in Brewing

The specific activity of K-ACDH did not vary dramatically in brewing type experiments like that of Mg-ACDH shown in Figure 1. With brewing yeasts, or with YPH499 the cultures were largely anaerobic after 1 hour and the K-ACDH activity drifted up from a specific activity of 20 mu/mg to that of 40 mu/mg over a period of 40 hours. We did not expect this enzyme to be active throughout a fermentation because of its repressibilty by anaerobiosis. We suggest that it may have been present to compensate for the Mg-ACDH, which was absent after 48 h.

2.4. Deletion of K-ACDH

The gene (*ALD7*) encoding the enzyme was identified by a TFASTA search of GenBank after purification of K^+ ACDH from a number of yeasts and determining the N-terminal sequence(s) (Tessier *et al.* 1998). At the time it was surprising to find that the gene sequence bore no resemblance to that given by Saigal *et al* (1991) which was supposed to be that of the mitochondrial ACDH. That claim has now been withdrawn, however, and the gene responsible has been shown to be the same as we have reported (Wang *et al*, 1998).

Gene disruption and transformation produced 6 mutant strains of YPH501 (Tessier *et al.* 1998) all completely lacking K-ACDH activity (Table 1). The Mg-ACDH in these mutants was at normal wild type levels. All the mutants grew at the same rate as the wild type on 5% and 0.5% glucose (Table 2). Growth was similarly unaffected on 2% sucrose, 1% acetate or 2% glycerol. Growth on 1% ethanol by all mutants was extremely poor with one mutant RY 270 taking 110 hours to increase cell number 4-fold. (The wild type had a doubling time of 9.7 h). The mutants' viability was not affected by exposure to 1% ethanol or 50 mM acetaldehyde, so the lack of growth was due to be a genuine inability to utilise ethanol effectively.

3. CONCLUSION

The work with the K-ACDH mutants supports the idea (Llorente and de Castro 1977) that K-ACDH primary role is in metabolising acetaldehyde derived from ethanol. Thus growth on glucose was completely unaffected by gene disruption and growth on ethanol was almost non-existent.

The results with the Mg-ACDH mutants were more ambiguous. Thus, growth on glucose was strongly reduced, but not to the very low-levels expected if this enzyme was essential for the disposal of acetaldehyde derived through pyruvate from glucose. Also, the inability of these mutants to grow effectively on ethanol was a surprise in view of the role proposed for the enzyme in glucose metabolism. However, the change in the repressibility and levels of K-ACDH in the mutants all suggested that other changes had occurred. This makes a simple interpretation of the growth characterisation of the Mg-ACDH mutants difficult.

ACKNOWLEDGMENTS

WT acknowledges receipt of the Henry Mitchell Memorial Scholarship from the Institute of Brewing. We are also grateful to Bass Brewers plc and in particular Dr. C. A. Boulton for provision of brewing strains and facilities.

REFERENCES

Dickinson, F. M. (1996). Purification and some properties of the Mg^{2+}-activated cytosolic aldehyde dehydrogenase of *Saccharomyces cerevisiae*. *Biochem. J.* 315, 393–399.

Llorente, N. and de Castro, I. N. Physiological role of yeasts $NAD(P)^+$ and $NADP^+$ linked aldehyde dehydrogenases. *Rev. Esp. Fisiol.* 33, 135–142.

Meaden, P. G., Dickinson, F. M., Mifsud, A., Tessier, W. D., Westwater, J., Bussey, H. and Midgley, M. (1997). The *ALD6* gene of *Saccharomyces cerevisiae* encodes a cytosolic, Mg^{2+}-activated acetaldehyde dehydrogenase. *Yeast.* 13, 1319–1327.

Saigal, D., Cunningham, S. J., FarrËs, J. and Weiner, H. (1991). Molecular cloning of the mitochondrial aldehyde dehydrogenase of *Saccharomyces cerevisiae* by genetic complementation. *J. Bacteriol.* 173, 3199–3208.

Tessier, W. D., Meaden, P. G., Dickinson, F. M. and Midgley, M. (1998). Identification and disruption of the gene encoding the K^+-activated acetaldehyde dehydrogenase of *Saccharomyces cerevisiae*. *F.E.M.S Letters* (In Press).

Wang, X., Mann, C. J., Ni, Y. B. L. and Weiner, H. (1998). Molecular cloning, characterisation and potential roles of cytosolic and mitochondrial aldehyde dehydrogenases in ethanol metabolism in *Saccharomyces cerevisiae*. *J. Bacteriol.* 180, 822–830.

31

CHARACTERIZATION OF AN ALDEHYDE DEHYDROGENASE GENE FRAGMENT FROM MUNG BEAN (*VIGNA RADIATA*) USING THE POLYMERASE CHAIN REACTION

Anatoley G. Ponomarev,[1] Victoria V. Bubyakina,[1*] Tatyana D. Tatarinova,[1] and Sergey M. Zelenin[2]

[1]Institute of Biological Problems of Cryolithozone
Siberian Branch
Russian Academy of Sciences
Prospekt Lenina 41
Yakutsk 677007, Russia
[2]Institute of Bioorganic Chemistry
Siberian Branch
Russian Academy of Sciences
Lavrentieva 8
Novosibirsk 630090, Russia

1. INTRODUCTION

NAD(P)$^+$-dependent aldehyde dehydrogenases of higher plants have not been studied well enough at the moment as compared to those of mammals. The betaine aldehyde dehydrogenase gene was obtained from rice (Nacamyra et al., 1997). There is some information on the structure of spinach, sugar beet, sorghum, burley cDNAs (Rhodes and Hanson, 1993; Wood et al., 1996), non-phosphorylating glyceraldehyde 3-phosphate dehydrogenase cDNAs from pea and maize (Habenicht et al., 1994), maize RF2 protein cDNA (about 60% identity to mammalian mitochondrial aldehyde dehydrogenases) (Cui et al., 1996), tobacco mitochondrial aldehyde dehydrogenase cDNA (op den Camp and Kuhlemeier, 1997).

* To whom correspondence should be addressed.

Enzymology and Molecular Biology of Carbonyl Metabolism 7
edited by Weiner *et al.* Kluwer Academic / Plenum Publishers, New York, 1999.

Further information on the structure of the genes and enzymes of aldehyde dehydrogenases from higher plants will provide advanced study of their important functions in plant metabolism. Betain aldehyde dehydrogenase is involved in the biosynthesis of glycine betaine which provides tolerance of certain higher plants to dry and saline environments (Rhodes and Hanson, 1993). Some hypotheses suggest that mitochondrial aldehyde dehydrogenase from plants may be implicated in the biosynthesis of indoleacetic acid (phytohormone auxin) and biosynthesis of glycine (Davies, 1960; Asker and Davies, 1985).

The main objectives of the paper were the search and identification of the fragments of Vigna radiata aldehyde dehydrogenase genes using a PCR-based approach.

2. MATERIALS AND METHODS

2.1. PCR

Total DNA was isolated from the roots of 2 day old seedlings of Vigna radiata using ultracentrifugation in a CsCl gradient. This DNA was used as a template in PCR amplification using primers U3 (upper) and L1 (lower). The sequences of these primers have been synthesised using regions of amino acid sequences which were conserved among *Aspergillus niger* aldehyde dehydrogenase, *Aspergillus nidullans* aldehyde dehydrogenase, spinach and sugar beet betain aldehyde dehydrogenases. They are as follows [amino acid residue numbers correspond to the human classes 1 and 2 aldehyde dehydrogenases (Hsu et al., 1985)]:

```
     177                            184
U3   Trp Lys  Ile  Ala Pro Ala    Leu Ala
5'   TGGAAAATTGCTCCAGCA/TCTTGC   3'
     404                     398
L1   Val  Pro  Gly  Phe  Ile  Glu  Glu
5'   GACAGGGCCAAAAATTTCC/TTC  3'
```

PCR was performed using Taq polymerase and the annealing temperature from 66°C to 46°C. An amplified U3/L1 fragment was cloned into pGEM-7Zf(+) (Promega).

2.2. Sequence Analysis

The sequences were determined with the help of the ABI Prism Dye Terminator Cycle Sequencing Ready Reaction Kit (Perkin Elmer) with AmpliTaq DNA polymerase. The U3/L1 fragment sequence was determined entirely on both strands and from two independent PCR amplifications. The alignment of nucleotide sequences was carried out with the help of BLASTN 1. 4. 9. MP program (Altschul et al., 1990).

3. RESULTS AND DISCUSSION

A number of amplified products (from 700-bp to 1700–bp) has been obtained in the result of PCR on Vigna radiata total DNA using primers U3 and L1 based on the conserved regions of the aldehyde dehydrogenase amino acid sequences. The sequence of the smallest amplified product named as U3/L1L (Figure 1) was compared to the sequences in the GenBank, EMBL, DDBJ, PDB databases. It is homologious to the genes of the aldehyde dehydrogenase superfamily.

```
Ala Gly Asn Cys Ile Val Leu Lys Pro Ala Glu Gln Thr Pro Ala Ser Ile Leu    18
T GCG GGT AAC TGT ATT GTT TTA AAA CCT GCT GAA CAG ACC CCT GCG AGC ATT TTG
  Val Leu Ile Glu Leu Ile Gln Asp Leu Leu Pro Pro Gly Ile Leu Asn Ile Val   36
  GTT TTA ATT GAA CTC ATT CAA GAC TTA CTA CCA CCC GGT ATT TTA AAC ATC GTC
  Asn Gly Tyr Gly Val Glu Val Gly Arg Pro Leu Ala Thr Asn Pro Arg Ile Ala   54
  AAC GGT TAT GGT GTT GAA GTG GGT CGT CCA CTC GCA ACC AAT CCA CGT ATT GCT
  Lys Ile Ala Phe Thr Gly Ser Thr Ala Val Gly Gln Met Ile Met Gln Tyr Ala   72
  AAA ATT GCA TTC ACT GGC TCT ACT GCT GTG GGT CAA ATG ATC ATG CAA TAT GCC
  Thr Glu Asn Val Ile Pro Val Thr Leu Glu Leu Gly Gly Lys Ser Pro Asn Ile   90
  ACT GAA AAC GTA ATT CCT GTA ACG TTG GAA TTA GGT GGT AAG TCT CCA AAC ATC
  Phe Phe Glu Asp Ile Met Asp Gln Asp Asn Glu Phe Leu Asp Lys Ala Leu Glu   108
  TTC TTT GAA GAC ATT ATG GAT CAA GAC AAT GAG TTC TTA GAC AAA GCC CTT GAA
  Val Leu Pro Cys Leu Gln Leu Asn Gln Gly Glu Ile Cys Thr Cys Pro Ser Arg   126
  GTT TTG CCA TGT TTG CAA CTT AAC CAA GGT GAA ATT TGT ACA TGT CCA TCA CGC
  Ala Leu Val Gln Gly Ser Ile Ala Asp Lys Phe Leu Glu Met Ala Val Glu Arg   144
  GCT TTA GTT CAA GGA AGT ATT GCA GAT AAA TTC CTT GAA ATG GCA GTT GAA CGT
  Val Lys Arg Ile Lys Thr Gly His Pro Leu Asp Thr Glu Thr Met Ile Gly Ala   162
  GTG AAA CGC ATT AAA ACG GGT CAC CCG CTC GAT ACA GAA ACG ATG ATT GGT GCG
  Gln Ala Ser Leu Glu Gln Gln Glu Lys Ile Leu Gly Cys Ile Ala Thr Gly Arg   180
  CAA GCT TCA CTT GAA CAA CAA GAG AAA ATT TTG GGT TGT ATC GCA ACA GGT CGC
  Gly Glu Gly Ala Gln Val Leu Thr Gly Gly Ser Glu Arg Asn Glu Val Gly Ser   198
  GGT GAA GGT GCA CAA GTA CTG ACA GGT GGT AGT GAG CGT AAT GAA GTG GGT TCT
  Gly Phe Tyr Ile Glu Pro Thr Ile Phe Lys Gly Thr Asn Asp Met Lys Ile Phe Gln 217
  GGT TTC TAT ATT GAA CCA ACC ATT TTC AAA GGC ACC AAT GAC ATG AAA ATT TTC CAA
```

Figure 1. Nucleotide sequence and deduced amino acid sequence of the *Vigna radiata* aldehyde dehydrogenase gene U3/L1L fragment. Figures on the right side show the numbers of amino acid residues from the beginning of the fragment.

The deduced amino acid sequence of the *Vigna radiata* aldehyde dehydrogenase gene fragment coding for 217 amino acid residues shows about 45% sequence identity to classes 1 and 2 mammalian aldehyde dehydrogenases, aldehyde dehydrogenases from *Aspergillus niger* and *Aspergillus nidulans*, betain aldehyde dehydrogenase from spinach and high identity (about 65%) to aldehyde dehydrogenases from *Vibrio cholerae*, *Rhodococcus* sp., *Alcaligenes eutrophus*. Its comparative analysis has been performed with 16 aldehyde dehydrogenase amino acid sequences (Hempel et al., 1993). The alignment with 5 sequences is shown in Figure 2. All the positions of amino acid residues set in the group of invariant residues (shown in bold) coincide. The Cys position (index 366) corresponds to the important Cys302 of classes 1 and 2 mammalian aldehyde dehydrogenases (Hempel et al., 1982; von Bahr – Lindsrom et al., 1985; Pietruszko et al., 1991). In the group of nearly invariant residues (underlined) most of the amino acid residues coincide. In the positions corresponding to Phe (index 359) and Gln (index 364) there have been revealed Gln and Glu. The correspondence of the Glu position (index 327) and the Glu268 of human classes 1 and 2 aldehyde dehydrogenases (Abriola et al., 1987; Pietruszko et al., 1991, 1993) has been found. And in the group of invariant similar and similar amino acid residues (indicated by # and *, respectively) belonging to the conservative changes of the types Thr / Ser, Arg / Lys and Leu / Ile / Val / Met most of the positions coincide as well.

The alignment portion mostly resembling the expected consensus nucleotide binding site Gly - X - Gly - X - X – Gly (Wierenga and Hol, 1983; Hempel et al., 1993) is represented in the Vigna radiata aldehyde dehydrogenase as the -gly - ser - thr - ala - val – gly- sequence (indexes 297-302) with a treonine in place of the second glycine residue. The consensus -val - thr - leu - Glu –leu - gly - gly - lys - ser – pro- sequence (indexes 324–333) (Hempel et al., 1993) is present in the Vigna radiata aldehyde dehydrogenase sequence. The alignment for the portion displaying the active site motif (indexes 359 – 368)

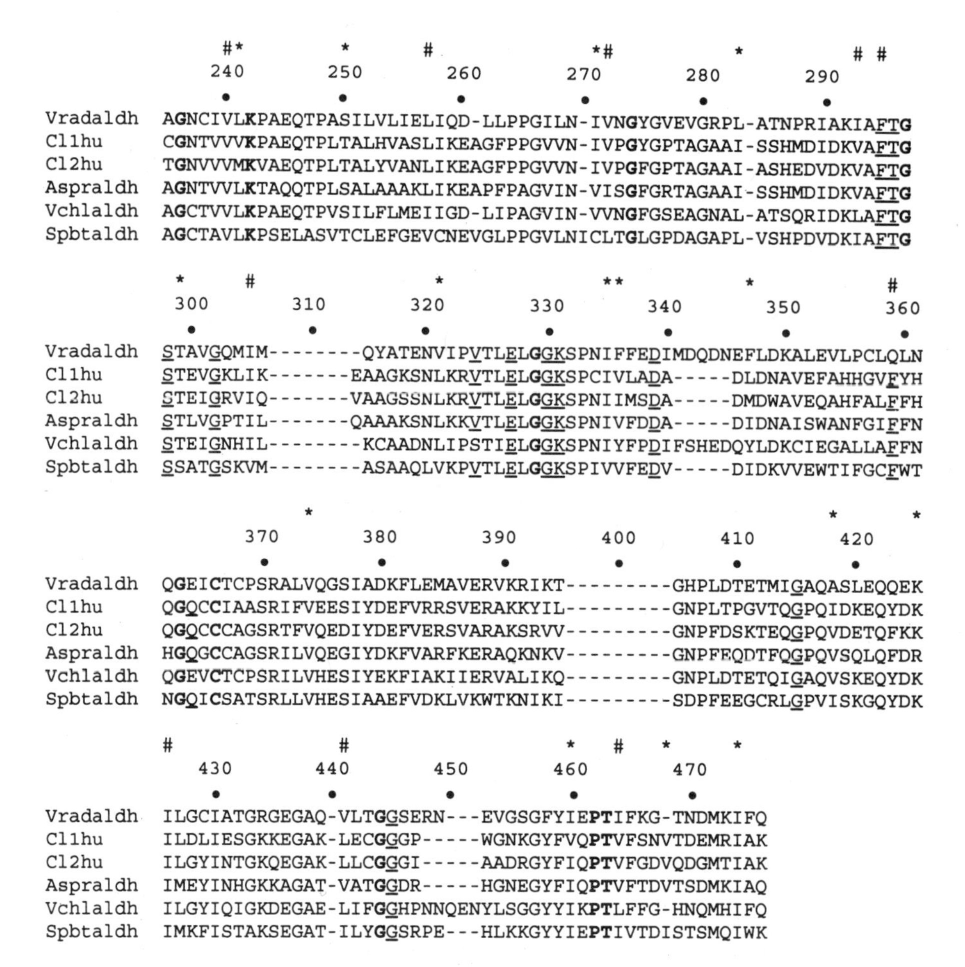

Figure 2. Alignment of deduced amino acid sequence of the *Vigna radiata* aldehyde dehydrogenase gene fragment with 16 aldehyde dehydrogenase sequences (Hempel et al., 1993). Sequences are: C11hu from human liver cytosol (class 1) (Hempel et al., 1984); C12hu from human liver mitochondria (class 2) (Hempel et al., 1985); Aspraldh from *Aspergillus nidulans* (Pickett et al., 1987); Vchlaldh from *Vibrio cholerae* (Parsot and Mekalanos, 1991); Spbtaldh from spinach (a betaine aldehyde dehydrogenase) (Weretilnyk and Hanson, 1990). Vradaldh: the *Vigna radiata* aldehyde dehydrogenase (this study). Figures on the top line show index numbers (Hempel et al., 1993). The other explanations are given in the text.

shows that the invariant Cys (index 366) and Gly (index 363) is found in the *Vigna radiata* aldehyde dehydrogenase sequence.

The deduced amino acid sequence of the *Vigna radiata* aldehyde dehydrogenase fragment gene represents a central amino acid portion of the unknown aldehyde dehydrogenase from *Vigna radiata* which shows high identity (about 65%) to NAD^{+}-dependent aldehyde dehydrogenases from *Vibrio cholerae* (Parsot and Mekalanos, 1991), *Rhodococcus* sp., (Nagy et al., 1995), *Alcaligenes eutrophus* (Priefert et al., 1992). On this basis we can assume a common evolutionary origin of the Vigna radiata aldehyde dehydrogenase eukaryotic gene and three genes from prokaryotes. This assumption may be confirmed only after having studied the kinetic and biochemical properties of the Vigna radiata aldehyde dehydrogenase and the full structure of its gene.

ACKNOWLEDGMENTS

This research was supported by grant of the Russian Basic Research Foundation.

REFERENCES

Abriola, D. P., Fields, R., Stein, S., MacKerell, A. D. and Pietruszko, R. (1987) Active site of human liver aldehyde dehydrogenase, *Biochemistry* 26, 5679–5684.

Altschul, S. F., Gish, W., Miller, W., Myers, E. W., Lipman, D. J. (1990) Basic local alignment search tool, *J. Mol. Biol.* 215, 403–410.

Asker, H., Davies, D. D. (1985) Mitochondrial aldehyde dehydrogenase from higher plants, *Phytochemistry* 24, 689–693.

Cui, X., Wise, R. P., Schnable, P. S. (1996) The rf2 nuclear restorer gene of male–sterile T-ytoplasm maize, *Science* 272, 1334–1336.

Davies, D. D. (1960) The purification and properties of glycolaldehyde dehydrogenase, *J. Exp. Botany* 11, 289–294.

Habenicht, A., Hellman, U., Cerff, R. (1994) Non-hosphorylating GAPDH of higher plants is a member of the aldehyde dehydrogenase superfamily with no sequence homology to phosphorylating GAPDH, *J. Mol. Biol.* 237, 165–171.

Hempel, J., Kaiser, R. and Jornvall, H. (1985) Mitochondrial aldehyde dehydrogenase from human liver: Primary structure, differences in relation to the cytosolic enzyme, and functional correlations, *Eur. J. Biochem.*153, 13–28.

Hempel, J., Nicholas, H. and Lindahl, R. (1993) Aldehyde dehydrogenases: Widespread structural and functional diversity within a shared framework, *Protein Science 2*, 1890–1900.

Hempel, J., Pietruszko, R., Fietzek, P. and Jornvall, H. (1982) Identification of a segment containing a reactive cysteine residue in human liver cytoplasmic aldehyde dehydrogenase (isoenzyme E1), *Biochemistry* 21, 6834–6838.

Hempel, J., von Bahr - Lindstrom, H. and Jornvall, H. (1984) Aldehyde dehydrogenase from human liver. Primary structure of the cytoplasmic isoenzyme, *Eur. J. Biochem.* 141, 21–35.

Hsu, L. C., Tani, K., Fujiyoshi, T., Kurachi, K. and Yoshida, A. (1985) Cloning of a cDNAs for human aldehyde dehydrogenases 1 and 2 (cDNA expression library / synthetic oligodeoxynucleotide probe / isozymes), *Proc. Nat. Acad. Sci. USA* 82, 3771–3775.

Nagy, I., Schoofs, G., Compernolle, F., Proost, P., Vanderleyden, J. and De Mot, R. (1995) Degradation of the thiocarbamate herbicide EPTC (S-thyl dipropylcarbamothioate) and biosafening by *Rhodococcus* sp. strain N186 / 21 involve an inducible cytochrome P-50 system and aldehyde dehydrogenase, *J. Bacteriology* 177, 676–687.

Nakamura, T., Yokota, S., Muramoto, Y., Tsutsui, K., Oguri, Y., Fukui, K., Takabe, T. (1997) Expression of a betaine aldehyde dehydrogenase gene in rice, a glycinebetaine nonaccumulator, and possible localization of its protein in peroxisomes, *Plant J.* 11, 1115–1120.

op den Camp, R. G., Kuhlemeier, C. (1997) Aldehyde dehydrogenase in tobacco pollen, *Plant. Mol. Biol.* 35, 355–365.

Parsot, C. and Mekalanos, J. J. (1991) Expression of the *Vibrio cholerae* gene encoding aldehyde dehydrogenase is under control of ToxR, the cholerae toxin transcriptional activator, *J. Bacteriology* 173, 2842–2851.

Pickett, M., Gwynne, D. I., Buxton, F. P., Elliot, R., Davies, R. W., Lockington, R. A., Scazzocchio, C. and Sealy-Lewis, H. M. (1987) Cloning and characterization of the aldA gene of *Aspergillus nidulans, Gene* 51, 217–226.

Pietruszko, R., Abriola, D. P., Blatter, E. E., Mukerjee, N. (1993) Aldehyde dehydrogenase: aldehyde dehydrogenation and ester hydrolysis, *Adv. Exp. Med. Biol.* 328, 221–230.

Pietruszko, R., Blatter, E. E., Abriola, D. P. and Prestwich, G. (1991) Localization of cysteine 302 at the active site of aldehyde dehydrogenase, *Adv. Exp. Med. Biol.* 284, 19 – 30.

Priefert, H., Kruger, N., Jendrossek, D., Scmidt, B. and Steinbuchel, A. (1992) Identification and molecular characterization of the gene coding for acetaldehyde dehydrogenase II (acoD) of *Alcaligenes eutrophus, J. Bacteriology* 174, 899–907.

Rhodes, D., Hanson, A. D. (1993) Quaternary ammonium and tertiary sulfonuim compounds in higher plants, *Annu. Rev. Plant Physiol. Plant Mol. Biol.* 44, 357–384.

von Bahr - Lindstrom, H., Jeck, R., Woenckhaus, C., Sohn, S., Hempel, J. and Jornvall, H. (1985) Characterization of the coenzyme-binding site of liver aldehyde dehydrogenase: Differential reactivity of coenzyme analogs, *Biochemistry* 24, 5847–5851.

Weretilnyk, E. A. and Hanson, A. D. (1990) Molecular cloning of a plant betaine-aldehyde dehydrogenase , an enzyme implicated in adaptation to salinity and drought, *Proc. Natl. Acad. Sci. USA* 87, 2745–2749.

Wierenga, R. K. and Hol, W. G. J. (1983) Predicted nucleotide-binding properties of p21 and its cancer-associated variant, *Nature* 302, 842–844.

Wood, A. J., Saneoka, H., Rhodes, D., Joly, R. J., Goldsbrough, P. B. (1996) Betaine aldehyde dehydrogenase in sorghum, *Plant Physiology* 110, 1301–1308.

32

ALDEHYDE DEHYDROGENASE GENE SUPERFAMILY

The 1998 Update

Thomas L. Ziegler and Vasilis Vasiliou

Molecular Toxicology and Environmental Health Sciences
Department of Pharmaceutical Sciences
University of Colorado Health Sciences Center
Denver, Colorado 80262

1. INTRODUCTION

Aldehyde dehydrogenases (ALDHs) represent a group of $NAD(P)^{+}$-dependent enzymes, which have similar primary structures and oxidize a wide spectrum of endogenous and exogenous aldehydes (Lindahl, 1992; Vasiliou et al., 1995). Several ALDHs display broad substrate specificities, oxidizing a variety of both aliphatic and aromatic aldehydes, whereas other forms possess narrower substrate preferences. Based on substrate specificities, ALDH enzymes are broadly categorized as: **a**. *semialdehyde dehydrogenases,* including hydroxymuconic semialdehyde dehydrogenase (E.C. 1.2.1.32), Escherichia coli (EC 1.2.1.16) and mammalian (EC 1.2.1.24) succinate-semialdehyde dehydrogenase, glutamate semialdehyde dehydrogenase (E.C. 1.2.1.41), aspartate semialdehyde dehydrogenase (E.C. 1.2.1.11), 2-amino-adipate-6-semialdehyde dehydrogenase (E.C. 1.2.1.31) and metylmalonate-semialdehyde dehydrogenase (E.C. 1.2.1.27), **b.** *nonspecific ALDHs* (E.C.1.2.1.3), **c.** ***other ALDHs*** including betaine dehydrogenase (E.C. 1.2.1.8), non phosphorylating glyceraldehyde 3-phosphate dehydrogenase (E.C. 1.2.1.9), phenylacetaldehyde dehydrogenase (EC 1.2.1.39), lactaldehyde dehydrogenase (EC 1.2.1.22), and **d.** ***ALDH-like proteins*** which represent certain other protein sequences, containing either complete or almost complete ALDH sequences. These include the 10-formyltetrahydrofolate dehydrogenase (E.C. 1.5.1.6), Δ^{1}-pyrroline-5-carboxylate dehydrogenase (EC 1.5.1.12), antiquitin, a human 56-kDa androgen-binding protein, and the crystallins.

Enzymology and Molecular Biology of Carbonyl Metabolism 7
edited by Weiner *et al.* Kluwer Academic / Plenum Publishers, New York, 1999.

2. THE ALDH GENE SUPERFAMILY

The organization of proteins into superfamilies was first introduced by M. Dayhoff in 1976. A gene superfamily is "a cluster of evolutionarily related sequences" (Dayhoff, 1976). Each superfamily consists of homologous gene families, which are clusters of genes from different genomes that include both orthologs and paralogs (Tatusov et al., 1997). Orthologs are genes in different species that evolved from a common ancestor by separation, whereas paralogous genes are products of gene duplication events within the same genome. In the last five years it has become evident that ALDHs represent a large gene superfamily. In 1996, a total of 65 cDNAs/genes encoding ALDH enzymes had been isolated and sequenced from various sources including bacteria, plants, fungi, birds and mammals (Vasiliou, 1997). The number of ALDH genes has increased significantly the last two years. Besides the cloning of ALDH genes by various laboratories, many new ALDH genes have been identified by the genome sequencing projects. At latest count (June 1998), in bacteria, plants, yeast, and animals there are 164 distinct cDNAs/genes whose protein products contain the ALDH "signature sequence"(Hempel et al., 1993), and, thus, are regarded as members of the same superfamily. These sequences include four archaebacterial genes, 71 eubacterial, and 89 eukaryotic sequences (Table 1). In this report, we list all ALDH cDNA/gene sequences retrieved from GenBank, including accession numbers and functions of the gene product, if known. For human, mouse, and yeast

Table 1. Summary of ALDH genes. As of June 1998, the number of public entries of ALDH genes in GeneBank is 164. Species with more than two genes are listed

Taxa	Species*		Number of genes	
Archaea				**4**
Eubacteria				**71**
	Acetobacter europaeus	2		
	Aquifex aeolicus	2		
	Bacillus subtilis	12		
	Deinococcus radiodurans	2		
	Escherichia coli	14		
	Mycobacterium tuberculosis	7		
	Pseudomonas putida	5		
	Pseudomonas sp.	3		
	Pseudomonas aeruginosa	2		
	Rhizobium sp.	2		
	Synechocystis sp.	3		
Eukaryota				**89**
Fungi			16	
	Saccharomyces cerevisiae	13		
Plants			18	
	Oryza sativa	2		
	Pisum sativum	2		
	Sorghum bicolor	2		
	Zea mays	2		
Animals			55	
	Bos taurus	4		
	Caenorhabditis elegans	8		
	Homo sapiens	17		
	Mus musculus	5		
	Rattus norvegicus	10		

*, Species are listed in alphabetical order.

genes, the chromosomal location is also given. Due to the space limitation, references for all ALDH sequences are not included; however, they can be found in Vasiliou's Homepage in the World Wide Web at http://www.uchsc.edu/sp/sp/alcdbase/aldhcov.html.

2.1. Occurrence of ALDH Genes in Archaea

Excitingly, four ALDH genes have been described in archaea to date (Table 2). The *T. tenax* ***TTGAPN*** gene encodes a NAD-dependent glyceraldehyde-3-phosphate dehydrogenase enzyme (G3PDH), which shows no similarity with the phosphorylating G3PDHs, but it is very similar to non-phosphorylating G3PDHs (Brunner et al., 1998). The latter enzymes represent a separate ancient branch within the ALDH superfamily (Habenicht et al., 1997). One gene was found in *Haloferax volcanii* and encodes a protein with high similarity to an ALDH gene in *Pseudomonas putida* (L23214, Table 3). The third and fourth genes, isolated from *Methanococcus jannaschii* and *Methanobacterium*, encode NADP-dependent G3PDHs, which are similar to *Rhizobium sp.* glutamate dehydrogenase (Z68203, Table 3).

2.2. ALDH Genes in Eubacteria

A total of 71 genes has been sequenced in eubacteria, of which fourteen are from *Escherichia coli*, twelve from *Bacillus subtilis*, seven from *Mycobacterium tuberculosis*, five from *Pseudomonas putida*, six from *Synechocystis sp.*, two from each one of *Acetobacter europaeus, Aquifex aeolicus, Deinococcus radiodurans, Pseudomonas aeruginosa*, and *Rhizobium sp.*, and, finally, one from each one of *Acetobacter polyoxogenes, Agrobacterium radiobacter, Alcaligenes eutrophus, Alteromonas sp., Azotobacter vinelandii, Bacillus stearothermophilus, Clostridium acetobutylicum, Comamonas testosteroni, Haemophilus influenzae, Mycoplasma capricolum, Pseudomonas oleovorans, Rhodococcus erythropolis, Sinorhizobium meliloti, Streptococcus mutans, Synechococcus, Vibrio cholerae,* and *Vibrio harveyi* (Table 3).

2.3. Fungal ALDH Genes

Sixteen fungal genes encode ALDH peptides. From these sequences, one is from *Cladosporium herbarum*, one from *Pichia angusta*, one from *Schizosaccharomyces pombe*, and thirteen from *Saccharomyces cerevisiae* (Table 4).

2.4. ALDH Genes in Plants

A total of 18 cDNA or genes has been sequenced from plants. Two ALDHs have been sequenced in each of *Oryza sativa*, *Pisum sativum*, *Sorghum bicolor*, and *Zea mays*,

Table 2. Archaea ***ALDH*** genes

Gene name or locus	GenBank	Species	Function
HVALDH	U95374	*Haloferax volcanii*	ALDH
MJNADP-G3PDH	U67581	*Methanococcus jannaschii*	NADP-G3PDH
MTH978	AE000871	*Methanobacterium thermoautotrophicum*	NADP-G3PDH
TTGAPN	Y10625	*Thermoproteus tenax*	G3PDH

G3PDH, glyceraldehyde-3-phosphate dehydrogenase; NADP-G3PDH, NADP-dependent glyceraldehyde-3-phosphate dehydrogenase.

Table 3. Eubacterial *ALDH* genes

Gene name or locus	GenBank	Species	Function/substrate
aldF	Y08696	*Acetobacter europaeus*	ALDH/?
aldH	Y08696	*Acetobacter europaeus*	ALDH/?
abcaldH	D00521	*Acetobacter polyoxogenes*	ALDH/?
aldA	X95394	*Agrobacterium radiobacter*	ALDH/?
acoD	M74003	*Alcaligenes eutrophus*	ALDH/Acetaldehyde
olgA	AB009654	*Alteromonas sp.*	ALDH/?
aldH1	AE000677	*Aquifex aeolicus*	ALDH/?
aldH2	AE000679	*Aquifex aeolicus*	ALDH/?
aldh	U94420	*Azotobacter vinelandii*	ALDH/?
aldhT	D13846	*Bacillus stearothermophilus*	ALDH/?
aldX	AB005554	*Bacillus subtilis*	ALDH/?
ycbD	D30808	*Bacillus subtilis*	ALDH/?
gbsA	U47861	*Bacillus subtilis*	ALDH/Betaine
mmsA	AL021958	*Bacillus subtilis*	ALDH/Methylmalonate semialdehyde
rocA	Z99123	*Bacillus subtilis*	Δ^1P5CDH
ycgN	D50453	*Bacillus subtilis*	Δ^1P5CDH
ycnH	D50453	*Bacillus subtilis*	ALDH/Succinic semialdehyde
ywdH	Z99123	*Bacillus subtilis*	ALDH/?
iolA	Z99124	*Bacillus subtilis*	ALDH/Methylmalonate semialdehyde
YfmT	D86418	*Bacillus subtilis*	ALDH/Benzaldehyde
aldY	Z99123	*Bacillus subtilis*	ALDH/?
dhaS	AF027868	*Bacillus subtilis*	ALDH/Acetaldehyde
tsaD	U32622	*Comamonas testosteroni*	ALDH/4-Carboxybenzaldehyde
aaD	L14817	*Clostridium acetobutylicum*	ALDH/?
ssdA	AB003475	*Deinococcus radiodurans*	ALDH/Succinic semialdehyde
aldA	AB003475	*Deinococcus radiodurans*	ALDH/?
putA	AE000203	*Escherichia coli*	Proline dehydrogenase
adhE	AE000222	*Escherichia coli*	ALDH/Acetaldehyde
betB	AE000138	*Escherichia coli*	ALDH/Betaine
aldH	AE000228	*Escherichia coli*	ALDH/?
ECAE000235	AE000235	*Escherichia coli*	ALDH/?
ECAE000241	AE000241	*Escherichia coli*	ALDH/?
ECAE000269	AE000269	*Escherichia coli*	ALDH/?
eutE	AE000332	*Escherichia coli*	ALDH/?
aldA	AE000239	*Escherichia coli*	ALDH/Lactaldehyde
ECAE000236	AE000236	*Escherichia coli*	ALDH/?
ECAE000250	AE000250	*Escherichia coli*	ALDH/?
aldB	AE000436	*Escherichia coli*	ALDH/Betaine
gabD	AE000351	*Escherichia coli*	ALDH/Succinic semialdehyde
hpcC	X81322	*Escherichia coli*	ALDH/5-Carboxymethyl-2-hydroxymuconic semialdehyde
aldH	U32731	*Haemophilus influenzae*	ALDH/?
mc137	Z33096	*Mycoplasma capricolum*	ALDH/?
mtcy369.13	Z80226	*Mycobacterium tuberculosis*	ALDH/Betaine
mtv038	AL021933	*Mycobacterium tuberculosis*	ALDH/?
mtci5.21	Z92770	*Mycobacterium tuberculosis*	ALDH/?
mtcy71.33	Z92771	*Mycobacterium tuberculosis*	ALDH/?
mtcy08d5.18	Z92669	*Mycobacterium tuberculosis*	ALDH/L-Sorbosone
mtv003.04c	AL008883	*Mycobacterium tuberculosis*	ALDH/Phenylacetaldehyde
mtcY04C12.16	Z81360	*Mycobacterium tuberculosis*	ALDH/Succinic semialdehyde
aruD	AF011922	*Pseudomonas aeruginosa*	ALDH/Succinic semialdehyde
mmsA	M84911	*Pseudomonas aeruginosa*	ALDH/Methylmalonate semialdehyde
alkH	X65936	*Pseudomonas oleovorans*	ALDH/?
todI	U09250	*Pseudomonas putida*	ALDH/?
cymC	U24215	*Pseudomonas putida*	ALDH/p-Cumic aldehyde

(*continued*)

Table 3. Eubacterial *ALDH* genes

Gene name or locus	GenBank	Species	Function/substrate
psepadA	L23214	*Pseudomonas putida*	ALDH/p-Hydroxybenzaldehyde
dmpC	X52805	*Pseudomonas putida*	ALDH/2-Hydroxymuconic semialdehyde
xylG	M64747	*Pseudomonas putida*	ALDH/2-Hydroxymuconic semialdehyde
phnG	AF023839	*Pseudomonas sp.*	ALDH/4-Carboxy-2-hydroxymuconate-6-semialdehyde
terpE	M91440	*Pseudomonas sp.*	ALDH/?
doxF	M60405	*Pseudomonas sp.*	ALDH/Salicylaldehyde
glutamate DH	Z68203	*Rhizobium sp.*	Glutamate dehydrogenase
y4uC	AE000099	*Rhizobium sp.*	ALDH/?
thcA	U17129	*Rhodococcus erythropolis*	ALDH/?
betB	U39940	*Sinorhizobium meliloti*	ALDH/Betaine
aldH	D32049	*Synechococcus*	ALDH/?
Δ^1-P5CD	D90913	*Synechocystis sp.*	Δ^1P5CDH
ssdH	D63999	*Synechocystis sp.*	ALDH/Succinic semialdehyde
aldH	D64004	*Synechocystis sp.*	ALDH/?
aldA	AF034434	*Vibrio cholerae*	ALDH/?
aldH	U39638	*Vibrio harveyi*	ALDH/Fatty aldehydes

Δ^1P5CDH, Δ^1-pyrroline-5-carboxylate dehydrogenase; NPG3PDH, glyceraldehyde-3-phosphate dehydrogenase. The question mark denotes unknown substrate.

and one in each of *Amaranthus hypochondriacus*, *Atriplex hortensis*, *Aspergillus niger*, *Beta vulgaris*, *Brassica napus*, *Emericella nidulans*, *Hordeum vulgare*, *Nicotiana plumbaginifolia*, *Nicotiana tabacum*, and *Spinacia oleracea* (Table 5). Three of these genes encode non-phosphorylating glyceraldehyde-3-phosphate dehydrogenases (NPG3PDH), seven betaine ALDHs (BALDH), one methylmalonate ALDH (MMSDH), four non-specific or uncharacterized yet ALDHs, and two codes for the turgor 26g protein. The latter seem to be the ortholog of the human antiquitin (ATQ1). Finally, one gene encodes a putative ALDH polypeptide known also as restorer factor 2.

Table 4. Fungal *ALDH* genes

Gene name or locus	GenBank	Species	Chromosome	Function
CHALDH	X78228	*Cladosporium herbarum*		ALDH
PAALDH	U40996	*Pichia angusta*		ALDH
SPALDH	AL022299	*Schizosaccharomyces pombe*		ALDH
SCPUT1	Z73314	*Saccharomyces cerevisiae*	XII	Δ^1P5CDH
SCPUT2	U00062	*Saccharomyces cerevisiae*	VIII	Δ^1P5CDH
SCALDH1	U56604	*Saccharomyces cerevisiae*	XVI	ALDH
SCALD2	Z49705	*Saccharomyces cerevisiae*	XIII	ALDH
SCALD3	Z49705 Z71257	*Saccharomyces cerevisiae*	XIII	ALDH
SCALDH5	U56605	*Saccharomyces cerevisiae*	V	ALDH
SCALD6	U39205 U00094	*Saccharomyces cerevisiae*	XVI	ALDH
YOR374W	Z75282 Y13140	*Saccharomyces cerevisiae*	XV	ALDH
SCE3612	U18814	*Saccharomyces cerevisiae*	V	ALDH
SCSSDH	Z35875 Y13134	*Saccharomyces cerevisiae*	II	SSDH
SCH8179	U00062	*Saccharomyces cerevisiae*	VIII	ALDH
SC9718	Z49702 Z71257	*Saccharomyces cerevisiae*	XIII	ALDH
HOM2	X15649	*Saccharomyces cerevisiae*	?	ASDH

Δ^1P5CDH, Δ-1-pyroline 5-carboxylate dehydrogenase; SSDH succinic semialdehyde dehydrogenase; ASDH, aspartate semialdehyde dehydrogenase.

Table 5. Plant *ALDH* genes

Gene name or locus	GenBank	Species	Function/substrate
AHYBADH	AF017150 AHAF000132	*Amaranthus hypochondriacus*	ALDH/Betaine
ANALDH	M32351	*Aspergillus niger*	ALDH
AHBADH	X69770	*Atriplex hortensis*	ALDH/Betaine
BVBADH	X58463	*Beta vulgaris*	ALDH/Betaine
BNBTG26	S77096	*Brassica napus*	ATQ1
ENALDH	M16197	*Emericella nidulans*	ALDH
HVBADH	D26448	*Hordeum vulgare*	ALDH/Betaine
NPGAPN	U87848	*Nicotiana plumbaginifolia*	NPG3PDH
NTALDH	Y09876	*Nicotiana tabacum*	ALDH
OSBADH	AB001348	*Oryza sativa*	ALDH/Betaine
OSMMSDH	AF045770	*Oryza sativa*	MMSDH
PSCC26G	X54359	*Pisum sativum*	26g turgor protein
PSGAPN	X75327	*Pisum sativum*	NPG3PDH
SBBADH	U12196 U12195	*Sorghum bicolor*	ALDH/Betaine
SBALDH	U87982	*Sorghum bicolor*	ALDH
SPBADH	M31480 U69142	*Spinacia oleracea*	ALDH/Betaine
ZMGAPN	X75326	*Zea mays*	NPG3PDH
ZMRF2	U43082	*Zea mays*	ALDH/restorer factor 2*

*, Plant ortholog of the mammalian ALDH2; NPG3PDH, non phosphorylating glyceraldehyde 3-phosphate dehydrogenase

2.5. ALDH Genes in Invertebrates

Fourteen ALDH genes have been described in invertebrate species, of which eight are derived from *Caenorhabditis elegans* and one from each of *Enchytraeus buchholzi, Entamoeba histolytica, and Leishmania tarentolae* (Table 6). The remaining three genes are of particular interest because, one is the first *Drosophila melanogaster* ALDH sequence, and the two others are the *Octopus dofleini* and *Ommastrephes sloanei* orthologs of the mammalian ***ω-crystallin*** (Table 6).

2.6. ALDH Genes in Vertebrates

To date, 40 have been sequenced in mammalian and one in non-mammalian (chicken) species (Table 7). The 40 mammalian ALDH cDNAs/genes include ten from rat, five from mice, four from bovine, one from horse, and one from sheep. Mammalian ***η-crystallin*** has been isolated and sequenced from *Macroscelides proboscideus* and *Elephantulus edwardi* (shrews). The remaining seventeen vertebrate ALDHs have been described in human (Table 8). Sixteen and four ***ALDH*** genes have been mapped in the human and mouse genome, respectively (Table 7 and 8).

3. CLOSING REMARKS

How large is the ALDH gene superfamily? The explosion in molecular biology during the past decade, along with the release of complete or partially complete genome sequences of various species, has resulted in the discovery of many unexpected additional

Table 6. *ALDH* genes in invertebrates

Gene name or locus	GenBank	Species	Function/substrate
CET05H4	AF016452	*Caenorhabditis elegans*	ALDH/?
CEF56D12	AF016672	*Caenorhabditis elegans*	ALDH/?
CEC54D1	U46673	*Caenorhabditis elegans*	ALDH/?
CEF42G9	U00051	*Caenorhabditis elegans*	ALDH/Betaine
CEF54D8	U12966	*Caenorhabditis elegans*	ALDH/?
CET08B1	AF039039	*Caenorhabditis elegans*	ALDH/?
CEF45H10	Z81538	*Caenorhabditis elegans*	ALDH/Succinic semialdehyde
CEF01F1	U13070	*Caenorhabditis elegans*	ATQ1
drmmsdh	AL009147	*Drosophila melanogaster*	ALDH/Methylmalonate-semialdehyde
EBALDH	X95396	*Enchytraeus buchholzi*	ALDH/?
EHALDH1	L05667	*Entamoeba histolytica*	ALDH/Fatty and aromatic aldehydes
LTALDDE	Z31698	*Leishmania tarentolae*	ALDH/?
ODCRST	L06902	*Octopus dofleini*	ω-*crystallin*
OSCRST	L06903	*Ommastrephes sloanei*	ω-*crystallin*

ATQ1, antiqutin; the question mark denotes unknown substrate.

genes in almost every gene superfamily. Near 100 newly discovered ALDH genes have been deposited in the GenBank between June 1995 and June 1998. The cytochrome P450 gene superfamily, one of the largest superfamilies, consists of at least 1,000 genes, of which 90 have been described in bacteria (Nelson et al., 1996; for more details on P450 genes see David Nelson's Homepage at http://drnelson.utmem.edu/nelsonhomepage.html).

Table 7. *ALDH* genes in vertebrates[a]

Gene name or locus	GenBank	Species	Function/substrate
ALDH1	L36128	*Bos Taurus*	ALDH/Retinal
ALDH2	S03565	*Bos Taurus*	ALDH/Aliphatic aldehydes
ALDH5	S61045	*Bos Taurus*	ALDH/?
ALDH3	M37384	*Bos Taurus*	ALDH/Aromatic aldehydes
EECRYST	U02483	*Elephantulus edwardi*	η-***crystallin***/Retinal
ALDH1	S00364	*Equus Caballus*	ALDH/Aliphatic aldehydes
GGALDH1	X58869	*Gallus gallus*	ALDH/Retinal
MPCRYST	U03906	*Macroscelides proboscideus*	***h-crystallin***/Retinal
Aldh1	M74570	*Mus Musculus*	ALDH/Aliphatic aldehydes
Aldh1-like	U96401	*Mus Musculus*	ALDH/?
Aldh10	U14390	*Mus Musculus*	ALDH/Fatty and aromatic aldehydes
Aldh3	U12785	*Mus Musculus*	ALDH/Fatty and aromatic aldehydes
Aldh2	U07235	*Mus Musculus*	ALDH/Acetaldehyde
ALDH1	U12761	*Ovis aries*	ALDH/ Aliphatic aldehydes
PB-ALDH	M23995	*Rattus norvegicus*	ALDH/Phenylacetaldehyde
ALDH1A	L42009 U79118	*Rattus norvegicus*	ALDH/Retinal
ALDH1B	AF001897	*Rattus norvegicus*	ALDH/?
RALDH1C	U60063 X99273	*Rattus norvegicus*	ALDH/Retinal
ALDH2	X14977	*Rattus norvegicus*	ALDH/Acetaldehyde
ALDH3	J03637	*Rattus norvegicus*	ALDH/Aromatic aldehydes
ALDH3m	M73714	*Rattus norvegicus*	ALDH/Fatty aldehydes
SSDH	L34820	*Rattus norvegicus*	ALDH/Succinate semialdehyde
FOLATEDH	M59861	*Rattus norvegicus*	ALDH/10-Formyltetrahydrofolate
MMSDH	M93401	*Rattus norvegicus*	ALDH/Methylmalonate-semialdehyde

[a]Human genes are listed in Table 8.

Table 8. Human *ALDH* genes

Gene name or locus	GenBank	Chromosome	Function/substrate
ALDH1	J04748	9q21	ALDH/Retinal
ALDH2	K03001 Y00109	12q24	ALDH/acetaldehyde
ALDH3	M74542 M77477	17p11.2	ALDH/Aromatic alehydes
ALDH4	U24266	1	ALDH/glutamic γ-semialdehyde
ALDH5	M63967	9p13	ALDH/Aliphatic aldehydes
ALDH6	U07919	15q26	ALDH/Aliphatic aldehydes
ALDH7	U10868	11q13	ALDH/Aliphatic and aromatic aldehydes
ALDH8	U37519	11q13	ALDH/?
ALDH9	X75425 U34252	1q22-24	ALDH/γ–Aminobutyraldehyde
ALDH10	L47162 U46689	17p11.2	ALDH/Fatty and aromatic aldehydes
SSDH	L34821	6p22	ALDH/Succinic semialdehyde
ATQ1	S74728	5q31	Antiquitin
ATQL1	AF002693	5q14	ATQ/Pseudogene?
ATQL3	AF002694	7q36	ATQ/Pseudogene?
ATQL4	AF002695	10q21	ATQ/Pseudogene?
MMSDH	M93405 ?		ALDH/Methylmalonate semialdehyde
AL021939	AL021939	6q24.1-25.1	ALDH/?

To date, the ALDH superfamily contains 71 bacterial ALDH genes, a number that is approximately the same of that found in the P450 superfamily. The complete genome of the *Escherichia coli* (4,653,831 bp) contains 4,283 protein encoding genes of which 15 are ALDHs, or 0.35% of the total genome. Similarly, the *Saccharomyces cerevisiae* genome (2,068 Kbp) encodes 5,932 proteins, of which 14 are ALDHs, or 0.23% of the total genome. As of July 1998, the number of public cDNA Expressed Sequence Tags (ESTs) is 1,757,810 of which 61% are human. There are also 348,551 mouse ESTs, meaning that 81% of all ESTs in the dbEST database are either mouse or human. The majority of human genes are probably represented here now, and the number that are not will be shrinking steadily over time. A recent search of the EST database for ALDH sequences has resulted in 2,330 documents, or 0.13% of the total EST database. Based on these numbers and on the fact that the estimated number of human genes ranges between 60,000 and 100,000 (Antequera and Bird, 1993), we predict that the number of individual ALDH genes in any mammalian species might range from 20 to 30. Because of the growing number of newly discovered genes that are members of the ALDH superfamily, a unified systematic nomenclature system for this superfamily has become a necessity. Such a nomenclature for the eukaryotic ALDHs was discussed at the 9th International Workshop on the "Enzymology and Molecular Biology of Carbonyl Metabolism," and soon will be available to all colleagues in the field (Vasiliou et al., manuscript in preparation).

ACKNOWLEDGMENTS

We thank our colleagues, especially Daniel W. Nebert and Ronald Lindahl, for valuable discussions and a critical reading of this manuscript. This work was supported by NEI Grant 1R29 EY11490 (VV). T. L. Ziegler was supported by NIAAA Training Grant 5T32AA07464.

REFERENCES

Antequera, F. and Bird, A. (1993) Number of CpG islands and genes in human and mouse. *Proc. Natl. Acad. Sci. U.S.A.* **90,** 11995–11999.

Brunner, N.A., Brinkmann, H., Siebers, B., and Hensel, R. (1998) NAD+-dependent glyceraldehyde-3-phosphate dehydrogenase from Thermoproteus tenax. The first identified archaeal member of the aldehyde dehydrogenase superfamily is a glycolytic enzyme with unusual regulatory properties. *J.Biol.Chem.* **273,** 6149–6156.

Dayhoff M.O. (1976) The origin and evolution of protein superfamilies. *Fed.Proc.* **35,** 2132–2138.

Fields, C., Adams, M.D., White O., and Venter, J.C. (1998) How many genes in the human genome? *Nature Genet.* **7,** 345–346.

Habenicht, A., Quesada, A., and Cerff, R. (1997) Sequence of the non-phosphorylating glyceraldehyde-3-phosphate dehydrogenase from Nicotiana plumbaginifolia and phylogenetic origin of the gene family. *Gene* **198,** 237–243.

Hempel, J., Nicholas, H., and Lindahl, R. (1993) Aldehyde dehydrogenases: widespread structural and functional diversity within a shared framework. *Protein Sci.* **2,** 1890–1900.

Nelson, D.R., Koymans, L., Kamataki, T., Stegeman, J.J., Feyereisen, R., Waxman, DJ, Waterman, M.R., Gotoh, O., Coon, M.J., Estabrook, R.W., Gunsalus, I.C., and Nebert, DW. (1996) P450 superfamily: update on new sequences, gene mapping, accession numbers and nomenclature. *Pharmacogenetics* **6,** 1–42.

Tatusov, R.L., Koonin, E.V., and Lipman, D.J. (1997) A genomic perspective on protein families. *Science* **278,** 631–637.

Vasiliou, V. (1997) Aldehyde dehydrogenase genes. *Adv.Exp.Med.Biol.* **414,** 595–600.

Vasiliou, V., Weiner, H., Marselos, M., and Nebert, D. W. (1995) Aldehyde dehydrogenase genes: classification based on evolution, structure and regulation. *Eur. J. Drug. Metabol. Pharmacokinet.* **20,** 53–64.

33

HUMAN ALCOHOL DEHYDROGENASE FAMILY

Functional Classification, Ethanol/Retinol Metabolism, and Medical Implications

Shih-Jiun Yin,[1] Chih-Li Han,[1] An-I Lee,[1] and Chew-Wun Wu[2]

[1]Department of Biochemistry
National Defense Medical Center
[2]Department of Surgery
Veterans General Hospital
Taipei, Taiwan, Republic of China

1. INTRODUCTION

Alcohol dehydrogenase (ADH), a NAD^+-dependent zinc-containing dimeric enzyme, functions as a rate-limiting step in the mammalian ethanol metabolism (Edenberg and Bosron, 1997; Yin, 1996; Crabb et al., 1987). Primarily based on the homology of primary structure and also on the electrophoretic mobility, the Michaelis constants for ethanol and the sensitivity to pyrazole inhibition, human ADH family members have been categorized into five classes (Jornvall, this volume; Jornvall et al., 1997; Jornvall and Hoog, 1995; Vallee and Bazzone, 1983). The intraclass and interclass sequence similarities at the amino acid/nucleotide level are approximately 90% and 60%, respectively. To date seven ADH genes have been identified in humans. The *ADH3*, *ADH2* and *ADH1* (in closely tandem array) as well as the *ADH4*, *ADH5*, and *ADH7* have been mapped on chromosome 4q21–q25 (Smith, 1986; Yoshida et al., 1991; Yokoyama et al., 1996), suggesting that they belong to the family resulting from gene duplications and diversifications. *ADH1* through *ADH5* and *ADH7* encode α, β, γ, π, χ, and μ (or denoted σ) polypeptides, respectively (Smith, 1986; Yoshida et al., 1991; Jornvall and Hoog, 1995). The *ADH6*-encoding subunit has not yet been designated a Greek letter (Yasunami et al., 1991).

Class I ADHs are composed of α-, β-, and γ-subunits having low K_m values (<5 mM) for ethanol oxidation and class II ππ and class IV μμ have intermediate K_m (~30 mM). Class III χχ is not saturable with ethanol and virtually functions as glutathione-dependent formaldehyde dehydrogenase (Koivusalo et al., 1989; Holmquist and Vallee, 1991). Three allelic variations occur at the gene locus of *ADH2*, viz., *ADH2*1*, *ADH2*2*, and *ADH2*3* which encode the subunits of β_1, β_2, and β_3, respectively; two variants occur at the *ADH3*

Enzymology and Molecular Biology of Carbonyl Metabolism 7
edited by Weiner *et al.* Kluwer Academic / Plenum Publishers, New York, 1999.

locus, *ADH3*1* and *ADH3*2*, coding for γ_1 and γ_2, respectively. The distribution of these allele frequencies varies among racial populations (Smith, 1986; Yoshida et al., 1991; Yin, 1996). The kinetic properties of the resultant allelozymes differ significantly owing to single amino acid substitutions in the coenzyme binding domain of the enzymes (Edenberg and Bosron, 1997). The allelic variations of the *ADH2* and *ADH3* appear to influence the susceptibility to alcoholism and alcoholic liver cirrhosis in Asians (Thomasson et al., 1991; Chao et al., 1994; Higuchi et al., 1995).

Human ADH exhibits tissue-specific expression (Yin, 1996; Yin et al., 1997a). Class I ADHs are detected in the liver, kidney, lung, and the mucosa of stomach and lower digestive tract. Class II $\pi\pi$ appears solely in the liver. Class III $\chi\chi$ is ubiquitously expressed. Class IV $\mu\mu$ displays a limited distribution to the mucosa of upper digestive tract and stomach. The class I α-, β-, and γ-ADHs and class II enzyme, and the class I $\gamma\gamma$ and class IV enzyme, are the major forms involved in the metabolism of ethanol in the liver and stomach, respectively (Yin et al., 1997a,b; Han et al., 1998). This paper reviews recent progress in the enzymology of human ADH family with emphasis on the significance for functional classification, ethanol and retinol metabolism, as well as the implications for first-pass metabolism of alcohol and for etiology of fetal alcohol syndrome, alcohol-related cancer, skin disease and sterility.

2. FUNCTIONAL CLASSIFICATION

Results of initial velocity as well as product and dead-end inhibition studies of the class I, II, and IV ADHs are consistent with an Ordered Bi Bi mechanism (Han and Yin, 1998; Hurley et al., 1990; Yin et al., 1984; Bosron et al., 1979, 1983; Dubied et al., 1977). Coenzyme binds with the enzyme first and the product coenzyme releases last in the catalytic cycle as shown in Scheme 1, where E represents enzyme; A, NAD^+; B, ethanol; P, acetaldehyde; Q, NADH. Steady-state and stopped-flow kinetic studies as well as the deuterium isotope effects indicate that dissociation of NADH is the rate-limiting step for ethanol oxidation in the human ADH family (Han and Yin, 1998; Stone et al., 1993; Burnell et al., 1989; Yin et al., 1984; Bosron et al., 1979, 1983). Similar kinetic mechanism and rate-limiting step in catalysis have been previously reported for the class I ADH from horse (Theorell and Chance, 1951; Wratten and Cleland, 1963; Plapp et al., 1973; Dworschack and Plapp, 1977) and rat (Crabb et al., 1983).

Isoelectric point, the Michaelis and catalytic constants for ethanol, and the inhibition constants of 4-methylpyrazole are key physical and kinetic features in the functional classification of human ADH family. Class I enzymes belong to the high isoelectric-point forms, i.e. pI = 9.1–9.8 for $\alpha\alpha$, $\beta_2\beta_2$, $\gamma_2\gamma_2$, $\beta_1\beta_1$, and $\gamma_1\gamma_1$ (in increasing order). Class III $\chi\chi$ (6.0) is the low pI form and class II $\pi\pi$ (8.7) and class IV $\mu\mu$ (8.5) are the intermediate pI

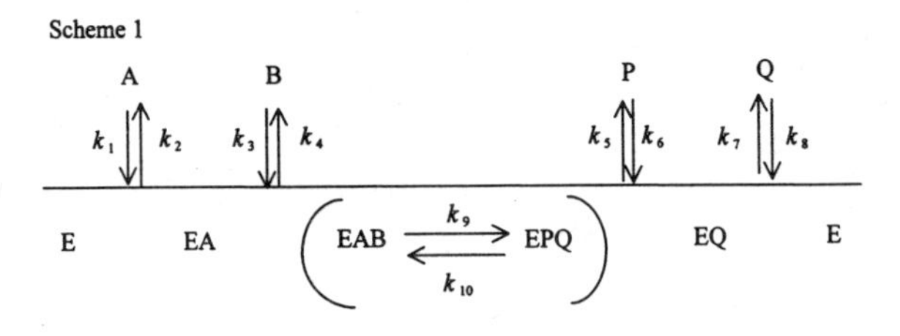

Scheme 1.

forms. This classification of ADH family based on the isoelectric point values is generally in agreement with the amino acid composition of the respective enzyme forms.

Featured kinetic constants for ethanol oxidation of ADH family are summarized in Table 1. Class I $\beta_1\beta_1$, $\beta_2\beta_2$, $\gamma_1\gamma_1$ and $\gamma_2\gamma_2$ are the low Michaelis-constant forms (K_m <1 mM). Class I $\beta_3\beta_3$ (36 mM), class II $\pi\pi$ (36 mM) and class IV $\mu\mu$ (27 mM) are the high K_m forms; class I $\alpha\alpha$ (4.4 mM), the intermediate K_m form. The smaller K_m values may reflect a higher affinity with the substrate. This inference is compatible with the findings from X-ray crystallographic studies that the active site of $\beta_1\beta_1$ is narrow and restrictive near the catalytic zinc atom and can bind small substrates such as ethanol in relatively few conformations, thus increasing the number of productive encounters between enzyme and ethanol (Hurley et al., 1994) and that $\alpha\alpha$ exhibits a more spacious active site near the catalytic zinc atom due to the substitution of Phe93, in the $\beta_1\beta_1$, by Ala (Hurley et al., 1997). In the similar way, the inability to saturate the class III $\chi\chi$ with ethanol as a substrate can be explained by the enormous size of this enzyme's substrate binding site as revealed by the X-ray structure (Yang et al., 1997).

Class I $\beta_2\beta_2$ and $\beta_3\beta_3$, and class IV $\mu\mu$ are the high catalytic-constant forms, while class I $\beta_1\beta_1$ is the low k_{cat} form. The other class I isoenzymes and class II $\pi\pi$ are the intermediate k_{cat} forms. The striking contrasts in turnover numbers for the $\beta\beta$ allelozymes are solely attributed to a single amino acid substitution, i.e. weak base His47 in $\beta_2\beta_2$ and Cys369 in $\beta_3\beta_3$, for both the strong base Arg in $\beta_1\beta_1$. These can be explained by the fact that the side chains of the above two amino acid residues interact with the negatively charged pyrophosphate bridge of the dinucleotide coenzyme, thus affecting the rate-limiting dissociation of the binary E–NADH complex (Hurley et al., 1994; Davis et al., 1996). Class I enzymes are highly sensitive to the competitive inhibition of 4-methylpyrazole with K_i values 1.1 μM or less, while class III $\chi\chi$ is virtually insensitive to the inhibition. Class II $\pi\pi$ (K_i = 500 μM) and class IV $\mu\mu$ (190 μM) are the forms with intermediate 4-methylprazole sensitivity. This differential sensitivity to the inhibition by pyrazole derivatives appears to be due to a different active site geometry near the catalytic zinc atom and the nicotinamide ring of the coenzyme as revealed by the X-ray structures as well as computer modeling of the $\beta\beta$ allelozymes, $\mu\mu$ and $\chi\chi$ (Hurley et al., 1994; Davis et al., 1996; Xie et al., 1997).

Table 1. Featured kinetic constants for ethanol oxidation in the functional classification of human ADH family[a]

Class	Form	$K_{m,EtOH}$ (mM)	k_{cat} (min^{-1})	$K_{i,4\text{-}MePy}$ (μM)
I	$\alpha\alpha$	4.4	23	0.53
	$\beta_1\beta_1$	0.019	3.6	0.51
	$\beta_2\beta_2$	0.93	250	0.34
	$\beta_3\beta_3$	36	320	1.1
	$\gamma_1\gamma_1$	1.0	87	0.10
	$\gamma_2\gamma_2$	0.63	35	–
II	$\pi\pi$	36	30	500
III	$\chi\chi$	–	–	–
IV	$\mu\mu$	27	1300	190

[a]From Han and Yin (1998), Bosron et al. (1983) and Burnell et al. (1989); $S_{0.5}$ value for $\gamma_1\gamma_1$ and $\gamma_2\gamma_2$; values for $\chi\chi$ too large to be determined.
$K_{i,4\text{-}MePy}$ is the inhibition constant for 4-methyl pyrazole.

3. ETHANOL METABOLISM

Under saturating concentrations of both ethanol and NAD^+, the catalytic efficiencies of ADH family are reflected on the k_{cat} values (Table 1). Therefore class IV μμ and class I $\beta_2\beta_2$, $\beta_3\beta_3$ are among the most efficient forms for oxidation of ethanol. At low concentrations of coenzyme or substrate, the catalytic efficiency can be expressed as V_1/K_a, V_1/K_b, V_2/K_p or V_2/K_q, a bimolecular rate constant for enzyme and coenzyme or for enzyme and substrate (Han and Yin, 1998). V_1 and V_2 are the maximal velocity (turnover number in this case) for ethanol oxidation and acetaldehyde reduction, respectively. K_a, K_b, K_p, and K_q are the Michaelis constants for NAD^+, ethanol, acetaldehyde, and NADH, respectively. At low ethanol concentrations, class I $\beta_1\beta_1$ (190 $mM^{-1}min^{-1}$) and $\beta_2\beta_2$ (270 $mM^{-1}min^{-1}$) are among the most efficient forms. The former is mainly due to the low K_m value and the latter mainly due to the high k_{cat} value. For overall catalytic efficiency at low concentrations of both NAD^+ and ethanol, i.e. $V_1/K_{ia}K_b$, a termolecular rate constant, $\beta_1\beta_1$ (3300 $mM^{-2}min^{-1}$) stands out the most efficient form (Han and Yin, 1998). This is attributable to the lowest K_{ia} and K_b values of the enzyme in the ADH family. K_{ia} is a product inhibition constant, also a dissociation constant, of the enzyme and NAD^+.

The control of metabolic flux of ethanol through ADH is best quantified using a steady-state rate equation for the Ordered Bi Bi system (Equation 1). The rate equation describes the effects of enzyme content, the kinetic properties of the enzyme, and the concentrations of substrates and products on the *in vivo* activity, or velocity, of the enzyme (Crabb et al., 1987).

$$v = \frac{V_1V_2\left([A][B]-[P][Q]/K_{eq}\right)}{\left(\begin{array}{l} V_2K_{ia}K_b + V_2K_b[A] + V_2K_a[B] + V_1K_q[P]/K_{eq} + V_1K_p[Q]/K_{eq} + \\ V_2[A][B] + V_1K_q[A][P]/K_{eq}K_{ia} + V_1[P][Q]/K_{eq} + V_2K_a[B][Q]/K_{iq} \\ +V_2[A][B][P]/K_{ip} + V_1[B][P][Q]/K_{ib}K_{eq} \end{array}\right)} \quad (1)$$

Table 2 shows kinetic simulations of the *in vivo* steady-state activity of ADH family, assuming the cellular concentrations of ethanol (10 mM), acetaldehyde (10 μM), NAD^+ (0.5 mM), and NADH (1.5 μM) (Han and Yin, 1998; Crabb et al., 1983). Class I $\beta_2\beta_2$ and class IV μμ appear to be the most active forms *in vivo*. The tissue contents of the ADH family are not included in the calculation because of the lack of relevant human data.

Table 2. Kinetic simulations of steady-state ethanol-oxidizing activity of ADH family[a]

		Rate (min^{-1}) at acetaldehyde concentrations			Inhibition (%) versus that of 10 μM acetaldehyde	
Class	Form	10 μM	100 μM	500 μM	100 μM	500 μM
I	αα	12	8.6	0.43	29	96
	$\beta_1\beta_1$	2.8	2.3	0.97	18	65
	$\beta_2\beta_2$	130	73	8.4	45	94
II	ππ	3.7	3.3	1.8	10	51
IV	μμ	130	110	31	18	76

[a] *In vivo* ADH activity was calculated from the Ordered Bi Bi rate equation using the obtained kinetic constants and assuming the substrate concentrations of ethanol (10 mM) and NAD^+ (0.5 mM) and the product concentrations of NADH (1.5 μM) and acetaldehyde (10, 100, or 500 μM) (Han and Yin, 1998).

Among Asians, there is a functional polymorphism of the mitochondrial aldehyde dehydrogenase (*ALDH2*) gene with the variant allele (*ALDH2*2*) encoding a protein subunit that confers low activity to the tetrameric enzyme in a dominant fashion (Weiner et al., 1997; Xiao et al., 1997). Pharmacokinetic studies using an ethanol dose of 0.2 g/kg body weight of Chinese men with different *ALDH2* genotypes indicate that the area under the blood ethanol concentration–time curve (AUC) for the variant homozygous *ALDH2*2/*2* subjects was 2.6 fold greater than that for the usual *ALDH2*1/*1* homozygotes (Table 3). The greater AUC implies a slower ethanol elimination. Peak blood acetaldehyde levels reached 75 μM in the variant homozygotes (cf. 1 μM in the usual homozygotes). It would be reasonable to estimate that in liver the concentrations of acetaldehyde may be more than doubled the blood value, e.g. 150–300 μM, in the *ALDH2*2/*2* individuals. To assess the effects of high acetaldehyde level on the *in vivo* activity of ADH, the steady-state concentrations of acetaldehyde at 100 μM and 500 μM are also used for kinetic simulation and compared with that of the 10 μM acetaldehyde (Table 2). Roughly 50% product inhibition of the ethanol-oxidizing activity in the ADH family can be obtained by such high acetaldehyde levels. This may partly account for the slower alcohol elimination seen in the variant homozygous *ALDH2*2/*2* subjects (Peng et al., 1998).

It should be pointed out that, in addition to the kinetic properties of the respective enzyme forms and the *in situ* steady-state concentrations of the substrates and products, the tissue expression pattern as well as the contents of the ADH forms also need to be considered to accurately assess the *in vivo* contribution of the ADH family to ethanol metabolism.

4. RETINOL METABOLISM

There are multiple pathways involved in the metabolism of retinol (Duester, 1996; Napoli, 1996). Based on the findings of coenzyme preference, the role of cellular retinol-binding proteins in retinoid storage, and the spatiotemporal pattern of gene expression in rodent embryo, it has been proposed that ADH may participate as a rate-limiting step in the synthesis of retinoic acid from retinol and that cellular retinol-binding protein-dependent short-chain dehydrogenase/reductase (SDR) may function in the conversion of retinal to retinol for storage (Duester, 1996). Retinoic acid, a ligand controlling a nuclear receptor signaling pathway, plays a key role in the regulation of embryonic development, spermatogenesis and epithelial differentiation (Chambon, 1996; Mangelsdorf and Evans, 1995).

Human ADHs, except for the class III enzyme, are active toward the oxidation of retinol (Han et al., 1998; Allali-Hassani et al., 1998; Yang et al., 1994). K_m values for all-

Table 3. Pharmacokinetic parameters of blood ethanol and acetaldehyde in men with different *ALDH2* allelotypes

	Genotype		
Parameter	*ALDH2*1/*1*	ALDH2*1/*2	*ALDH2*2/*2*
Peak ethanol (mM)	2.16 ± 0.38	3.07 ± 0.23	4.08 ± 0.49[a]
AUC for ethanol (mM × h)	1.91 ± 0.34	3.68 ± 0.41	4.98 ± 0.66[b]
Peak acetaldehyde (μM)	1.0 ± 0.4	23.6 ± 1.3[c]	75.4 ± 10.6[b,d]
AUC for acetaldehyde (μM × h)	0.4 ± 0.2	19.1 ± 2.4	89.2 ± 9.7[c,e]

All subjects had the genotype of *ADH2*2/*2* and *ADH3*1/*1*. Ethanol dose was 0.2 g/kg. [a]$P < 0.05$ vs. *ALDH2*1/*1*; [b]$P < 0.01$ vs. *ALDH2*1/*1*; [c]$P < 0.001$ vs. *ALDH2*1/*1*; [d]$P < 0.05$ vs. *ALDH2*1/*2*; [e]$P < 0.001$ vs. *ALDH2*1/*2* (Peng et al., 1998).

trans-retinol for the class I αα, $\beta_1\beta_1$, $\beta_2\beta_2$, $\gamma_1\gamma_1$, class II ππ, and class IV μμ are 36, 30, 96, 74, 7.3, and 27 μM, respectively, and the k_{cat} values are 11, 1.2, 17, 7.2, 3.3, and 120 min^{-1}, respectively (Han et al., 1998). Relative values of the catalytic efficiency (k_{cat}/K_m) for retinol to that for ethanol for αα, $\beta_1\beta_1$, $\beta_2\beta_2$, $\gamma_1\gamma_1$, ππ, and μμ are 65, 0.24, 0.60, 2.9, 650, and 83, respectively. Thus at low substrate concentrations, class I αα, class II ππ and class IV μμ catalyze the oxidation of retinol much more efficiently than ethanol. This is largely due to the tremendously lower Michaelis constants for retinol for these three ADH forms.

The catalytic efficiency for retinol for class IV μμ is 10–110 fold greater than those for the other ADH forms (Han et al., 1998). This is consistent with the findings from X-ray structure and model building studies that the deletion at position 117 widens a key area near the entrance to the substrate binding site of μμ and allows very long chain alcohols, such as retinol, to bind in an extended conformation (Kedishvili et al., 1995; Xie et al., 1997). Thus, the high efficiency of the class IV enzyme for retinol oxidation appears to be due to an enlarged active site which permits retinol to bind in a productive, low energy, conformation.

5. MEDICAL IMPLICATIONS

5.1. First-Pass Metabolism

First-pass metabolism, i.e., the difference between the quantity of ethanol that reaches the systemic circulation by the intravenous route and the quantity that entered by the oral dose, has recently been an issue of extensive interest and intense controversy because of its medical and legal implications in the bioavailability of alcohol and the intoxication, drug/alcohol interaction, and alcohol-induced organ injury (Gentry et al., 1994; Levitt, 1994; Ammon et al., 1996). To assess the effective potential of human ADH family to metabolize first-passed ethanol, it is critical to measure the total enzyme activity as well as the magnitude of activity increase in the liver and stomach corresponding to the alcohol concentration difference between the organ and systemic blood during the period of first-pass metabolism.

Recent reports on the expression pattern and activity of ADH family in human tissues indicate that there is no significant gender and age difference in the gastric ethanol-oxidizing activity and that the total organ activity of the stomach appears to be negligibly low (0.36%) relative to that of the liver (Yin et al., 1997a,b). Kinetic studies show class I αα, $\beta_1\beta_1$, $\beta_2\beta_2$, and class II ππ exhibit substrate inhibition under high ethanol concentrations with K_i values over 250–720 mM, while class IV μμ still conforms to the simple Michaelis-Menten kinetics (Han et al., 1998). On the other hand, class I $\gamma_1\gamma_1$ displays negative cooperativity under wide range of ethanol concentrations (Bosron et al., 1983; Han et al., 1998). Kinetic simulations of the contribution by ADH family to the first-pass metabolism reveal that ππ and μμ exhibit the greatest activity increase corresponding to concentration difference of 33 mM vs. 3 mM ethanol and 500 mM vs. 3 mM ethanol, which simulate the possible *in vivo* situations for the liver and stomach, respectively, to the peripheral circulation (Table 4).

Therefore, several lines of evidence including the tissue expression pattern, the kinetic properties of ADH family, the organ ethanol-oxidizing activity, as well as the kinetic simulation under *in vivo* conditions, support the concept that first-pass metabolism of alcohol in humans may occur mainly in the liver through the class II ππ ADH.

Table 4. Kinetic simulations of contribution by human ADH family to first-pass metabolism of ethanol[a]

Class	Form	$K_{i,EtOH}$ (mM)	Activity change (%)	
			33 mM vs. 3 mM	500 mM vs. 3 mM
I	$\alpha\alpha$	720	120	57
	$\beta_1\beta_1$	620	−3.9	−44
	$\beta_2\beta_2$	340	14	−48
	$\gamma_1\gamma_1$	–	44	43
II	$\pi\pi$	250	460	290
III	$\chi\chi$	–	–	–
IV	$\mu\mu$	–	450	850

[a]For $\alpha\alpha$, $\beta_1\beta_1$, $\beta_2\beta_2$, and $\pi\pi$, activity change = $100\times\{VS_1/[K_m + S_1 + (S_1^2/K_i)] \div VS_2/[K_m + S_2 + (S_2^2/K_i)]-1\}$, where S_1 and S_2 are the high and low concentrations of ethanol, respectively; for $\mu\mu$, activity change = $100\times[VS_1/(K_m + S_1) \div VS_2/(K_m + S_2)-1]$; for $\gamma_1\gamma_1$, activity change directly estimated from the ethanol saturation curve; not active with $\chi\chi$ up to 250 mM ethanol (Han et al., 1998).

5.2. Fetal Alcohol Syndrome and Alcohol-Related Disease

It has been postulated that inhibition of retinoic acid by ethanol at the ADH step underlies the etiology of fetal alcohol syndrome, a birth defect characterized by craniofacial, limb, and brain malformations (Duester, 1991). Recent reports on the expression patterns of ADH and the inhibition by ethanol of retinoic acid synthesis in mouse embryos support this hypothesis (Ang et al., 1996; Deltour et al., 1996; Ang and Duester, 1997; Duester, this volume). Ethanol acts as a competitive inhibitor against retinol oxidation by human ADH (Han et al., 1998; Allali-Hassani et al., 1998). The K_i values of ethanol range from 37 μM to 11 mM in increasing order for $\beta_1\beta_1$, $\beta_2\beta_2$, $\gamma_1\gamma_1$, $\alpha\alpha$, $\pi\pi$, and $\mu\mu$ (Han et al., 1998). Results of the kinetic simulation indicate that the oxidation of retinol, a rate-limiting step in retinoic acid synthesis, can be blocked 80–100% through the ADH-linked pathway by physiologically attainable concentrations of ethanol in heavy drinkers (Table 5). It should be stressed that this disturbance of retinoid homeostasis may be significant even in a social drinking setting. Therefore, women during first four weeks after gestation without noticeable signs may be vulnerable to ethanol teratogenicity in social or binge drinking. The possibility of the risk has been substantiated by a mouse model study that human-like fetal alcohol syndrome could be produced in embryos

Table 5. Kinetic simulations of inhibition of human ADH family by ethanol for oxidation of all-*trans*-retinol[a]

Class	Form	$K_{i,EtOH}$ (mM)	Inhibition (%) by ethanol at		
			5 mM	20 mM	50 mM
I	$\alpha\alpha$	3.8	51	81	91
	$\beta_1\beta_1$	0.037	99.0	99.7	99.9
	$\beta_2\beta_2$	0.31	94	98	99
	$\gamma_1\gamma_1$	0.43	91	98	99
II	$\pi\pi$	6.9	23	55	78
III	$\chi\chi$[b]	–	–	–	–
IV	$\mu\mu$	11	25	57	76

[a]Inhibition = $100 \times \{1-VS/[K_m(1 + I/K_i) + S] \div VS/(K_m + S)\}$, where S and I represent the concentrations of retinol (10 μM) and ethanol, respectively. [b]$\chi\chi$ not active with up to 100 μM retinol (Han et al., 1998).

by just two doses of ethanol (peak blood levels reached 42–47 mM) administered to pregnant mice during the gastrulation stage of embryogenesis (Sulik et al., 1981).

Retinoic acid also functions to maintain differentiation of epithelia and spermatogenesis in adult vertebrates (Chambon, 1996; Mangelsdorf and Evans, 1995; Duester, 1996). It was proposed that inhibition of retinol oxidation by ethanol may be responsible for the testicular atrophy and aspermatogenesis commonly seen in male chronic alcoholics (Van Thiel et al., 1974). Skin diseases, such as psoriasis, have been associated with heavy drinking (Higgins and du Vivier, 1992). The kinetic simulations shown in Table 5 suggest that retinoic acid signaling in spermatogenesis and keratinocyte differentiation may be significantly interfered with by ethanol through ADH pathways. Epidemiologic studies have demonstrated that alcohol consumption is a risk factor for development of the oral, esophageal and colorectal cancers (Blot, 1992). Recent reports of the tissue expression pattern and kinetic simulation support a possible tumorigenic mechanism that during heavy drinking, in addition to acetaldehyde cytotoxicity, retinoid signaling needed for maintaining epithelial differentiation may be interrupted by ethanol through the class IV μμ or class I γγ pathways (Han et al., 1998; Yin et al., 1997a).

6. CONCLUSIONS

Classification of human ADH family can be based on the gene structure and chromosome location, the homology of primary structure, as well as the feature of catalytic property. Both the primary structural and functional characteristics of the ADHs should be interpreted on the basis of the three dimensional structure of the enzyme molecules. Results of the kinetic mechanism-based simulation under the possible *in vivo* conditions indicate that liver ethanol-oxidizing activity can be significantly inhibited by the product acetaldehyde in Asian individuals with the variant *ALDH2*2* alleles following ingestion of alcohol, that class II ππ ADH in liver appears to be responsible for the first-pass metabolism, and that the pathogenesis of fetal alcohol syndrome, alcohol-related cancer, skin disease and male sterility may be in part due to perturbation of retinoid signaling by ethanol via ADH pathways.

ACKNOWLEDGMENTS

This work was supported by grants from the National Science Council (NSC88-2316-B016-002), the National Health Research Institutes (DOH87-HR-612), and the Ministry of Defense, Taiwan, Republic of China.

REFERENCES

Allali-Hassani, A., Peralba, J. M., Martras, S., Farres, J. and Pares, X. (1998) Retinoids, ω-hydroxyfatty acids and cytotoxic aldehydes as physiological substrates, and H_2-receptor antagonists as pharmacological inhibitors, of human class IV alcohol dehydrogenase, *FEBS Lett.* **426**, 362–366.

Ammon, E., Schafer, C., Hofmann, U. and Klotz, U. (1996) Disposition and first-pass metabolism of ethanol in humans: Is it gastric or hepatic and does it depend on gender? *Clin. Pharmacol. Ther.* **59**, 503–513.

Ang, H. L., Deltour, L., Hayamizu, T. F., Zgombic-Knight, M. and Duester, G. (1996) Retinoic acid synthesis in mouse embryos during gastrulation and craniofacial development linked to class IV alcohol dehydrogenase gene expression, *J. Biol. Chem.* **271**, 9526–9534.

Ang, H. L. and Duester, G. (1997) Initiation of retinoid signaling in primitive streak mouse embryos: Spatiotemporal expression patterns of receptors and metabolic enzymes for ligand synthesis, *Dev. Dyn.* **208**, 536–543.

Blot, W. J. (1992) Alcohol and cancer, *Cancer Res.* **52** (*Suppl.*), 2119s–2123s.

Bosron, W. F., Li, T.-K., Dafeldecker, W. P. and Vallee, B. L. (1979) Human liver π-alcohol dehydrogenase: Kinetic and molecular properties, *Biochemisty* **18**, 1101–1105.

Bosron, W. F., Magnes, L. J. and Li, T.-K. (1983) Kinetic and electrophoretic properties of native and recombined isoenzymes of human liver alcohol dehydrogenase, *Biochemistry* **22**, 1852–1857.

Burnell, J. C., Li, T.-K. and Bosron, W. F. (1989) Purification and steady-state kinetic characterization of human liver $\beta_3\beta_3$ alcohol dehydrogenase, *Biochemistry* **28**, 6810–6815.

Chambon, P. (1996) A decade of molecular biology of retinoic acid receptors, *FASEB J.* **10**, 940–954.

Chao, Y.-C., Liou, S.-R., Chung, Y.-Y., Tang, H.-S., Hsu, C.-T., Li, T.-K. and Yin, S.-J.(1994) Polymorphism of alcohol and aldehyde dehydrogenase genes and alcoholic cirrhosis in Chinese patients, *Hepatology* **19**, 360–366.

Crabb, W. D., Bosron, W. F. and Li, T.-K. (1983) Steady-state kinetic properties of purified rat liver alcohol dehydrogenase: Application to predicting alcohol elimination rates *in vivo*, *Arch. Biochem. Biophys.* **224**, 299–309.

Crabb, D. W., Bosron, W. F. and Li, T.-K. (1987) Ethanol metabolism, *Pharmac. Ther.* **34**, 59–73.

Davis, G. J., Bosron, W. F., Stone, C. L., Owusu-Dekyi, K. and Hurley, T. D. (1996) X-ray structure of human $\beta_3\beta_3$ alcohol dehydrogenase: The contribution of ionic interactions to coenzyme binding, *J. Biol. Chem.* **271**, 17057–17061.

Deltour, L., Ang, H. L. and Duester, G. (1996) Ethanol inhibition of retinoic acid synthesis as a potential mechanism for fetal alcohol syndrome, *FASEB J.* **10**, 1050–1057.

Dubied, A., von Wartburg, J.-P., Bohlken, D. P. and Plapp, B. V. (1977) Characterization and kinetics of native and chemically activated human liver alchol dehydrogenase, *J. Biol. Chem.* **252**, 1464–1470.

Duester, G. (1991) A hypothetical mechanism for fetal alcohol syndrome involving ethanol inhibition of retinoic acid synthesis at the alcohol dehydrogenase step, *Alcohol. Clin. Exp. Res.* **15**, 568–572.

Duester, G. (1996) Involvement of alcohol dehydrogenase, short-chain dehydrogenase/reductase, aldehyde dehydrogenase, and cytochrome P450 in the control of retinoid signaling by activation of retinoic acid synthesis, *Biochemistry* **35**, 12221–12227.

Dworschack R. T. and Plapp B. V. (1977) Kinetics or native and activated isoenzymes of horse liver alcohol dehydrogenase, *Biochemistry* **16**, 111–116.

Edenberg, H. J. and Bosron, W. F. (1997) Alcohol dehydrogenases. In Guengerich, F. P. (Ed.), Comprehensive Toxicology. Pergamon, New York, Vol. 3, pp.119–131.

Gentry, R. T., Baraona, E. and Lieber, C. S. (1994) Agonist: Gastric first pass metabolism of alcohol, *J. Lab. Clin. Med.* **123**, 21–26.

Han, C.-L., Liao, C.-S., Wu, C.-W., Hwong, C.-L., Lee, A.-R. and Yin, S.-J. (1998) Contribution to first-pass metabolism of ethanol and inhibition by ethanol for retinol oxidation in human alcohol dehydrogenase family. Implications for etiology of fetal alcohol syndrome and alcohol-related diseases, *Eur. J. Biochem.* **254**, 25–31.

Han, C.-L. and Yin, S.-J. (1998) Steady-state kinetic mechanism of human ADH family: Implications for functional classification and ethanol metabolism, manuscript in preparation.

Higgins, E. M. and du Vivier, A. W. P. (1992) Alcohol and the skin, *Alcohol Alcohol.* **27**, 595–602.

Higuchi, S., Matsushita, S., Murayama, M., Takagi, S. and Hayashida, M. (1995) Alcohol and aldehyde dehydrogenase polymorphisms and the risk for alcoholism, *Am. J. Psychiatry* **152**, 1219–1221.

Holmquist, B. and Vallee, B. L. (1991) Human liver class III alcohol and glutathione dependent formaldehyde dehydrogenase are the same enzyme, *Biochem. Biophys. Res. Commun.* **178**, 1371–1377.

Hurley, T. D., Edenberg, H. J. and Bosron, W. F. (1990) Expression and kinetic characterization of variants of human $\beta_1\beta_1$ alcohol dehydrogenase containing substitutions at amino acid 47, *J. Biol. Chem.* **265**, 16366–16372.

Hurley, T. D., Bosron, W. F., Stone, C. L. and Amzel, L. M. (1994) Structures of three human β alcohol dehydrogenase variants. Correlations with their functional differences, *J. Mol. Biol.* **239**, 415–429.

Hurley, T. D., Steinmetz, C. G., Xie, P. and Yang, Z.-N. (1997) Three-dimensional structures of human alcohol dehydrogenase isoenzymes reveal the molecular basis for their functional diversity, *Adv. Exp. Med. Biol.* **414**, 291–302.

Jornvall, H. and Hoog, J.-O. (1995) Nomenclature of alcohol dehydrogenases, *Alcohol Alcohol.* **30**, 153–161.

Jornvall, H., Shafqat, J., El-Ahmad, M., Hjelmqvist, L., Persson, B. and Danielsson, O. (1997) Alcohol dehydrogenase variability. Evolutionary and functional conclusions from characterization of further variants, *Adv. Exp. Med. Biol.* **414**, 281–289.

Kedishvili, N. Y., Bosron, W. F., Stone, C. L., Hurley, T. D., Peggs, C. F., Thomasson, H. R., Popov, K. M., Carr, L. G., Edenberg, H. J. and Li, T.-K. (1995) Expression and kinetic characterization of recombinant human stomach alcohol dehydrogenase. Active-site amino acid sequence explains substrate specificity compared with liver isozymes, *J. Biol. Chem.* **270**, 3625–3630.

Koivusalo, M., Baumann, M. and Uotila, L. (1989) Evidence for the identity of glutathione-dependent formaldehyde dehydrogenase and class III alcohol dehydrogenase, *FEBS Lett.* **257**, 105–109.

Levitt, M. D. (1994) Antagonist: The case against first-pass metabolism of ethanol in the stomach, *J. Lab. Clin. Med.* **123**, 28–31.

Mangelsdorf, D. J. and Evans, R. M. (1995) The RXR heterodimers and orphan receptors, *Cell* **83**, 841–850.

Napoli, J. L. (1996) Retinoic acid biosynthesis and metabolism, *FASEB J.* **10**, 993–1001.

Peng, G.-S., Wang, M.-F., Chen, C.-Y., Luu, S.-U., Chou, H.-C., Li, T.-K. and Yin, S.-J. (1998) Alcohol metabolism, cardiovascular hemodynamic effects and subjective perceptions in men with different mitochondrial aldehyde dehydrogenase allelotypes: Mechanism for full protection against alcoholism by homozygosity of the *ALDH2*2* allele, submitted.

Plapp, B. V., Brooks, R. L. and Shore, J. D. (1973) Horse liver alcohol dehydrogenase. Amino groups and rate-limiting steps in catalysis, *J. Biol. Chem.* **248**, 3470–3475.

Smith, M. (1986) Genetics of human alcohol and aldehyde dehydrogenase, *Adv. Hum. Genet.* **15**, 249–290.

Stone, C. L., Bosron, W. F. and Dunn, M. F. (1993) Amino acid substitutions at position 47 of human $\beta_1\beta_1$ and $\beta_2\beta_2$ alcohol dehydrogenase affect hydride transfer and coenzyme dissociation rate constants, *J. Biol. Chem.* **268**, 892–899.

Sulik, K. K., Johnston, M. C. and Webb, M. A. (1981) Fetal alcohol syndrome: Embryogenesis in a mouse model, *Science* **214**, 936–938.

Theorell, H. and Chance, B. (1951) Studies on liver alcohol dehydrogenase. II. The kinetics of the compound of horse liver alcohol dehydrogenase and reduced diphosphopyridine nucleotide, *Acta Chem. Scand.* **5**, 1127–1144.

Thomasson, H. R., Edenberg, H. J., Crabb, D. W., Mai, X.-L., Jerome, R. E., Li, T.-K., Wang, S.-P., Lin, Y.-T., Lu, R.-B. and Yin, S.-J. (1991) Alcohol and aldehyde dehydrogenase genotypes and alcoholism in Chinese men, *Am. J. Hum. Genet.* **48**, 677–681.

Vallee, B. L. and Bazzone, T. J. (1983) Isozymes of human liver alcohol dehydrogenase, *Isozymes: Curr. Top. Biol. Med. Res.* **8**, 219–244.

Van Thiel, D. H., Gavaler, J. and Lester, R. (1974) Ethanol inhibition of vitamin A metabolism in the testes: Possible mechanism for sterility in alcoholics, *Science* **186**, 941–942.

Weiner, H., Sheikh, S., Zhou, J. and Wang, X. (1997) Subunit interactions in mammalian liver aldehyde dehydrogenases, *Adv. Exp. Med. Biol.* **414**, 181–185.

Wratten, C. C. and Cleland, W. W. (1963) Product inhibition studies on yeast and liver alcohol dehydrogenases, *Biochemistry* **2**, 935–941.

Xiao, Q., Weiner, H. and Crabb, D. (1997) Studies on the dominant negative effect of the *ALDH2*2* allele, *Adv. Exp. Med. Biol.* **414**, 187–194.

Xie, P., Parsons, S. H., Speckhard, D. C., Bosron, W. F. and Hurley, T. D. (1997) X-ray structure of human class IV $\sigma\sigma$ alcohol dehydrogenase. Structural basis for substrate specificity, *J. Biol. Chem.* **272**, 18558–18563.

Yang, Z.-N., Davis, G. J., Hurley, T. D., Stone, C. L., Li, T.-K. and Bosron, W. F. (1994) Catalytic efficiency of human alcohol dehydrogenases for retinol oxidation and retinal reduction, *Alcohol. Clin. Exp. Res.* **18**, 587–591.

Yang, Z.-N., Bosron, W. F. and Hurley, T. D. (1997) Structure of human $\chi\chi$ alcohol dehydrogenase: A glutathione-dependent formaldehyde dehydrogenase, *J. Mol. Biol.* **265**, 330–343.

Yasunami, M., Chen, C.-S. and Yoshida, A. (1991) A human alcohol dehydrogenase gene (*ADH6*) encoding an additional class of isozyme, *Proc. Natl. Acad. Sci. USA* **88**, 7610–7614.

Yin, S.-J., Bosron, W. F., Magnes, L. J. and Li, T.-K.(1984) Human liver alcohol dehydrogenase: Purification and kinetic characterization of the $\beta_2\beta_2$, $\beta_2\beta_1$, $\alpha\beta_2$ and $\beta_2\gamma_1$ "Oriental" isoenzymes, *Biochemistry* **23**, 5847–5853.

Yin, S.-J. (1996) Alcohol dehydrogenase: Enzymology and metabolism. In Saunders, J. B. and Whitfield, J. B. (Eds.), The Biology of Alcohol Problems. Elsevier Science, Oxford, pp. 113–119.

Yin, S.-J., Liao, C.-S., Wu, C.-W., Li, T.-T., Chen, L.-L., Lai, C.-L. and Tsao, T.-Y. (1997a) Human stomach alcohol and aldehyde dehydrogenases: Comparison of expression pattern and activities in alimentary tract, *Gastroenterology* **112**, 766–775.

Yin, S.-J., Han, C.-L., Liao, C.-S. and Wu, C.-W. (1997b) Expression, activities, and kinetic mechanism of human stomach alcohol dehydrogenase. Inference for first-pass metabolism of ethanol in mammals, *Adv. Exp. Med. Biol.* **414**, 347–355.

Yokoyama, H., Baraona, E. and Lieber, C. S. (1996) Molecular cloning and chromosomal localization of the *ADH7* gene encoding human class IV (σ) ADH, Genomics **31**, 243–245.

Yoshida, A., Hsu, L. C. and Yasunami, M. (1991) Genetics of human alcohol-metabolizing enzymes, *Prog. Nucl. Acid Res. Mol. Biol.* **40**, 255–287.

34

DYNAMICS IN ALCOHOL DEHYDROGENASE ELUCIDATED FROM CRYSTALLOGRAPHIC INVESTIGATIONS

Ramaswamy S

Department of Molecular Biology
Swedish University of Agricultural Sciences
Box 590
Biomedical Centre, Uppsala, Sweden

1. INTRODUCTION

A major criticism of x-ray crystallographic studies of proteins is that they do not provide dynamic information. Crystallographic data provides time-averaged and space-averaged models. However, entries in the protein data bank contain both co-ordinates as well as temperature factors. These temperature factors are normally modeled isotropically and are taken to signify the thermal motion of the given atom; large temperature factors show that the positions of those atoms are less accurately determined.

Protein function must involve dynamics and motion. The motions can be as large as a domain closure such as observed in alcohol dehydrogenase (ADH) or as small as the vibrations of individual atoms in the active site where reactions take place. With the recent developments in synchrotron radiation, cryo-cooling and computing, it is possible to extract some dynamic information from high-resolution protein crystallographic studies.

Dynamic information from small molecule crystallographic studies is routinely extracted for both rigid body motions of parts of their structure and atomic vibrations. Methods of quantifying rigid body motions from crystallographic data are discussed in detail in the International Tables of Crystallography, Volume 4. The broad area of study is called thermal motion analysis. Carroll Johnson, when receiving the 1997 ACA Buerger award, described this well by calling it "The Fossil Footprints of Restless Atoms".

Due to the importance of intramolecular mobility for the biological activity of proteins, the subject has been addressed several times. Most of these studies have been on lysozyme, ribonuclease A and crambin (Artymuik et al., 1979; Sternberg et al., 1979; Frauenfelder et al., 1979; Clarage et al., 1992; Perez et al., 1996; Harata et al., 1998, Kuriyan et al., 1991), and have relied on conventional methods of extracting dynamic informa-

Enzymology and Molecular Biology of Carbonyl Metabolism 7
edited by Weiner *et al.* Kluwer Academic / Plenum Publishers, New York, 1999.

tion. An analysis of 'normal modes' of protein vibrations has been done on both pancreatic trypsin inhibitor (Diamond, 1990) and lysozyme (Kidera and Go, 1990). The use of these two structures has essentially been due to the availability of high-resolution crystallographic data. These methods gain more relevance today as the number of protein structures being determined at higher resolution increases (Dauter et al., 1995). As more such studies are being carried out the amount of information that becomes available will enable parameterisation of the motion of atoms in protein molecules. These can then be used in theoretical methods to calculate these motions from protein structures that are determined at much lower resolutions. In this article I discuss thermal motion analysis of alcohol dehydrogenase from the near atomic resolution data of a ternary complex of ADH and combine it with the extra information available from the large number of ADH structures to understand motion from domain movement to proton transfer.

2. PROTEIN MOLECULES IN CRYSTALS

The temperature factor determined from crystallography has information about motion of the protein in the crystalline state; one has to separate the various kinds of motions in order to make chemical sense of them. These can be classified in an hierarchical fashion as:

- The motion (disorder) of the entire content of the asymmetric unit
- The motion of the various independent pieces in the asymmetric unit
- The motion of various domains as rigid groups
- The motions of various smaller rigid groups such as the plane of the phenyl ring
- The individual vibrations of the atoms.

Though each of the individual components can be determined (Schomaker and Trueblood, 1968), the correlations between these components generally cannot, as such information is lost during data collection. It is necessary to understand what causes each of these motions in order to be able to separate the ones caused by the process of crystallisation from those that can be of chemical and biological relevance. Also, one must be aware that temperature factors in low resolution structures act as error sinks.

3. METHODS

Refinement of crystallographic co-ordinates with TLS (Translation, Libration and S-the correlation matrix) parameters and individual anisotropic temperature factors are well established (Driessen et al., 1989). The aim of this section is to summarize the relevant theory and terminology.

Crystallographic data is the measurement of intensities of reflections that are converted into structure factors, Fs. These observed values are called Fobs. However once a structure is determined and one has positions of atoms in the asymmetric unit, one can calculate structure factors (Fcalc). A crystallographic refinement process involves minimizing the difference between Fobs and Fcalc. This process adjusts the various parameters that contribute to Fcalc in order to minimise the difference. As this involves solving linear equations the number of equations is equal to the number of reflections and the number of variables to be determined is the number of parameters refined. In most protein crystallographic refinements, the co-ordinates of the atomic positions are refined (3 parameters:

x, y, z). With higher resolution data, there are more observed reflections and hence one can refine more parameters, such as group temperature factors, individual isotropic temperature factors, and at atomic resolution, individual anisotropic temperature factors. During anisotropic refinement the number of parameters refined for every atom is nine (3 position + 6 temperature factors).

As mentioned earlier, atoms in real crystals are not point atoms. Their deviation from being point atoms is parameterised as the temperature factor. The temperature factor can be modeled in different ways: individual isotropic temperature factors, where it is assumed that the atom's movement is isotropic in space or anisotropic when the probability of finding the atom around the given position is not the same in the three directions. Modeling group temperature factors, where one assumes that a group of atoms is moving as a rigid body is chemically reasonable. This kind of modeling allows us either to model a single isotropic temperature factor for the entire group (one extra parameter), or model it as anisotropic (6 extra parameters).

Modeling motions of rigid groups can also be done differently if one understands the various degrees of freedom that exist for a rigid body to move. The rigid body motion of molecules in crystals can be represented as coming from three tensors: the libration (or rotation tensor) - L that is symmetric, a translation tensor - T - that is symmetric, and an S tensor that describes the average quadratic correlation of translation and libration and is not symmetric (Schomaker and Trueblood, 1968). This method of modeling would involve addition of 20 parameters to a rigid group.

Crystallographic refinement can thus be carried out where the parameterisation can be done in various ways; the computer program RESTRAIN (Driessen et al., 1989) allows one to do this. This program was used to refine the structures reported in this analysis after refinement converged and the structures were reported. The structure was refined with atoms having 3 co-ordinate variables and 1 isotropic temperature factor. Each rigid group's TLS tensors were also refined simultaneously. In order to be consistent, the high resolution structure of the ternary complex of ADH with pentafluorobenzyl alcohol was refined with isotropic temperature factors when TLS refinement of the protein was carried out for the various rigid groups. The program TLSANL (Driessen et al., 1989) was used to analyse the refined TLS tensors.

4. RESULTS

4.1. Motion of the Entire Asymmetric Unit

Mosaicity in crystals occurs due to the small but random breakage of periodicity. This breakage of periodicity can be attributed to the random rotation and translation of molecules, suggesting that protein molecules in crystals can rotate and move. Horse liver alcohol dehydrogenase crystallises in a variety of crystal forms. One of the crystal forms for the holo enzyme is P21 with cell dimensions of 51.1Å, 180.7Å, 44.3Å and β=108.0° when data are collected at 4o C. However, when the crystals are frozen, the cell along the c-axis almost doubles; a slight rotation of the molecule causes this doubling. The cell dimensions of the frozen form are 49.8Å, 179.3Å, 86.4Å and β=105.9°. Freezing changes the orientation of the molecules, suggesting that protein molecules in crystals have a tendency to rotate. The TLS parameters for the entire protein being treated as one unit can be refined along with the other parameters. This is an increase of only 20 parameters in refinement. Hence, such a refinement can be done even at reasonably medium resolution such as 2.5 Å. This may be considered equivalent to refinement of an overall anisotropic temperature factor. This information can also be extracted from the individual anisotropic temperature factors as the common motion exhibited by all the atoms in the asymmetric unit from high-resolution refinement of structures. In the case of the ternary complex of horse liver ADH with NAD and pentafluorobenzyl alcohol, the atoms have a mean square translation of 0.77 Å2, while there is a very negligible rotational component.

4.2. Motion of Large Parts of Asymmetric Units and Domains

Horse liver alcohol dehydrogenase is a classic example of an enzyme that undergoes domain motion. The active enzyme is a dimer and each subunit is made up of a catalytic domain and a coenzyme binding domain. When the structures of the apo and the holo enzyme were determined, it was observed that there is a domain motion that closes the domains towards each other and covers the NAD which binds in between the two domains (Elkund et al., 1981). This motion has been described as a sliding motion (Colonna-Cesari et al., 1986.). The structure determination of the cod liver class III alcohol dehydrogenase (Ramaswamy et al., 1996) revealed the 2 subunits having 2 different degrees of domain motion with respect to holo and apo-horse liver enzyme. One of the subunits was closed and closer to the horse liver holo-enzyme and the other was open and closer to the apo enzyme. Thus it can be considered that closure of the horse liver apo form to the holo form goes through the two states exhibited by the cod liver enzyme in its crystal structure. These are visible in Figure 1. Another significant change in this motion is the conformation of a loop (between residues 293 and 300, horse liver enzyme numbering). A double

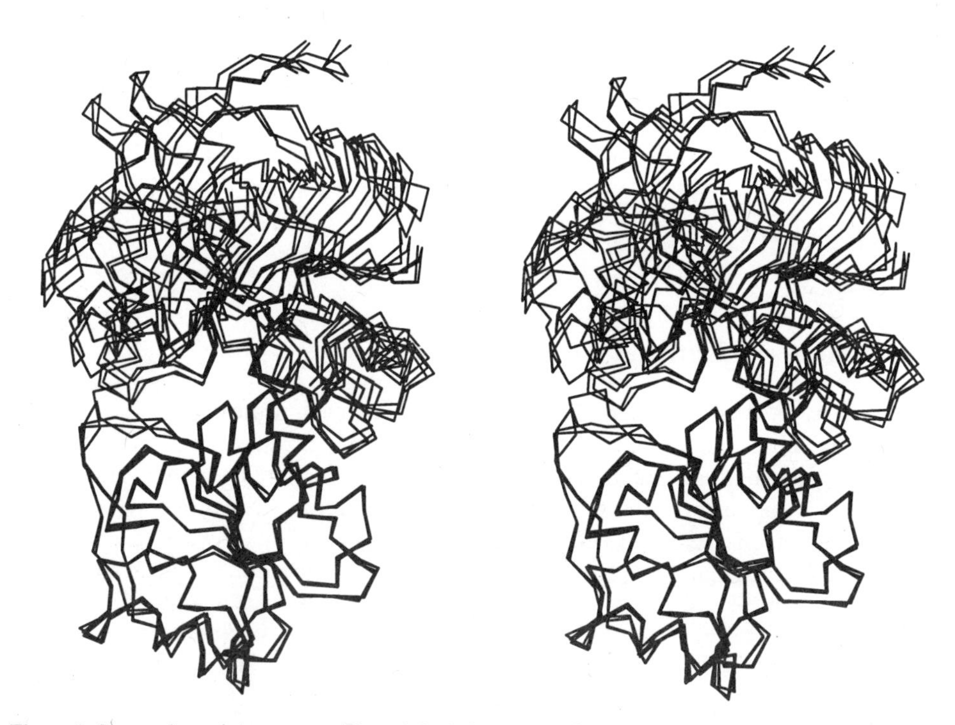

Figure 1. Stereo view of the superposition of the holo and apo forms of horse liver ADH as well as the two subunits of the cod liver ADH. The coenzyme domains are superposed to show the maximum variability in the catalytic domain. The four different states are clearly visible.

mutant enzyme, G293A and P239T, has much higher activity than the wild type liver ADH (D.-H. Park and B. V. Plapp, unpublished). These two residues are in the loop and this prevents this "AT" enzyme from changing to the holo conformation. We have now determined by X-ray crystallography the structures of both the apo and holo forms of the mutant enzyme, and both adopt an open conformation (unpublished results).

This calculation was done for the holo enzyme, the 2 subunits of the cod enzyme and the AT mutant holo enzyme, which is similar to the apo form of the horse liver enzyme. From the four available structures, the enzyme was divided into six rigid components: the catalytic domain (1–174), the helix connecting the 2 domains (174–187), the co-enzyme binding domain (188–292), the loop region (293–300), the rest of the coenzyme binding region (300–318) and the folded back part of the catalytic domain. As before, one can refine the TLS parameters for each of these components. These TLS components can then be visualised if we consider the libration and translation parts independently as two ellipsoids.

The most interesting result is that of the helix in the hinge region. The helix in the holo enzyme has a rotation vibration along the helix axis (Figure 2a). However, when a similar analysis was done on the two cod ADH subunits, which are open to different extents (Figure 2b and 2c), and the holo form of the AT mutant (Figure 2d), which has an apo enzyme conformation, it was observed that the ellipsoid representing the rotation vibrations longest axis moved from being perpendicular to the helix axis to being almost parallel. The magnitude of the translation ellipsoid is larger in the direction perpendicular to the longest axis of the librational ellipsoid in the completely open structure. The direction of motion required for the closure of the domains corresponds to the largest radius of the translation and libration ellipsoids. The magnitudes of the axis are also much larger in the completely open form. Such an effect is anticipated as the holo form is probably slightly more stable than the apo form due to the presence of NAD^+ which makes H-bonds with both of the protein domains holding it closed.

4.3. Motions of Side Chains and Local Conformational Flexibility

Horse liver alcohol dehydrogenase is a good model for studying flexibility of side chains. There are a large number of structures of ADH solved with alcohol analogues, aldehyde analogues as well as with other compounds. The structure of the ternary complex of ADH with NAD^+ and pentafluorobenzyl alcohol shows that the aromatic ring was bound in a position that was shifted from the position of the ring in the complex with bromobenzyl alcohol (Ramaswamy et al., 1994). Though the rings adopted different positions, the methylene carbons and the oxygens were bound in almost exactly the same places. This shows that alcohols can bind in different ways in the active site of ADH. The structure determination of the ternary complexes of ADH with N-formylpiperidine and N-cyclohexylformamide revealed that these two aldehyde analogues also had the rings bound in different orientations (almost perpendicular to each other), while the oxygens and the carbons of the carbonyl groups superimposed well (Ramaswamy et al., 1997). The structure determination of complexes with the two isomers of 3-butylthiolane 1-oxides bound to ADH and NADH revealed more surprising results (Cho et al., 1997). Here the 2 isomers of the same compound bound very differently in the active site. All these different binding modes are associated with simultaneous changes in the conformations of the side chains of the amino acids in the substrate binding pocket. This is illustrated in Figure 3a.

One of the problems with drug design based on crystal structures has been the inability of the programs to predict movements of side chains that enable the designed drugs

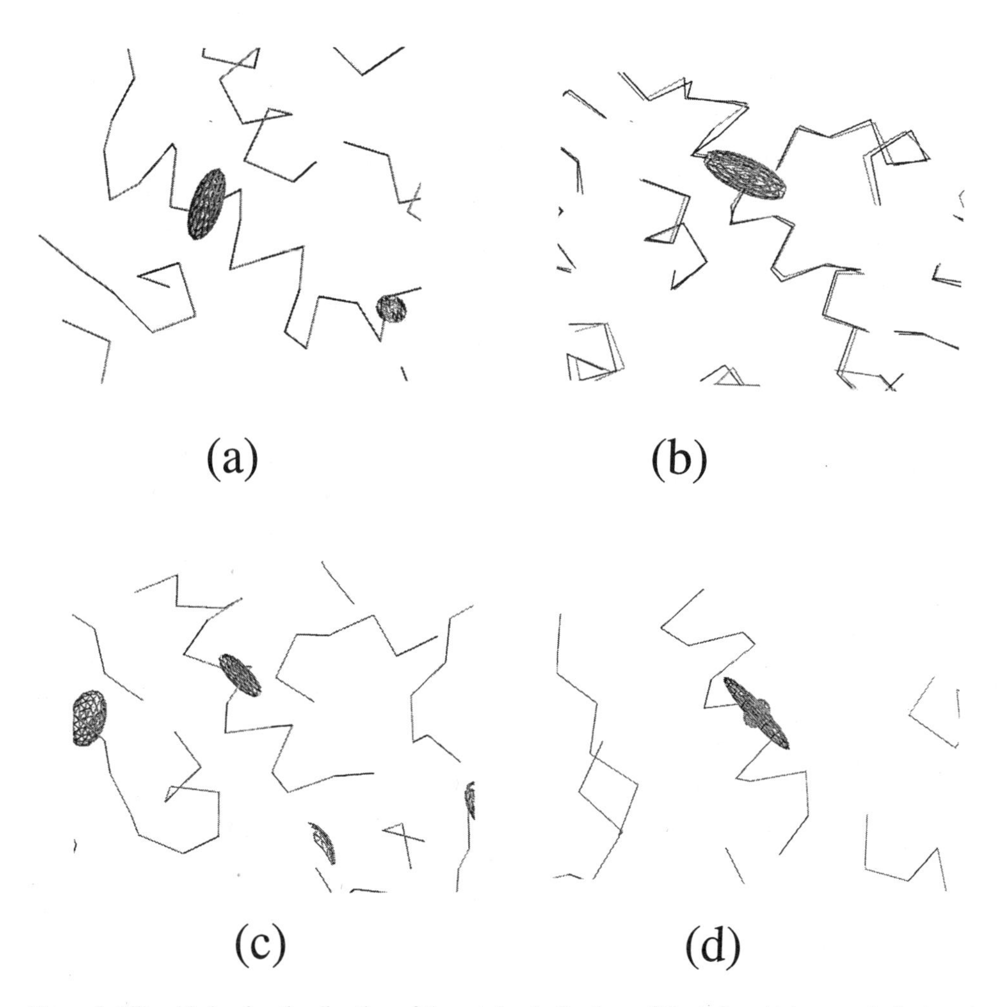

Figure 2. Ellipsoid showing the direction of the rotational vibrations of the helix which connects the two domains: a) holo form of horse liver enzyme, b) subunit 1 of the cod liver enzyme, c) subunit 2 of the cod liver enzyme and d) the double mutant G293A and P295T of horse liver enzyme that stays in the apo conformation.

to bind differently than expected based on the substrate pocket conformation. If one could predict possible movements of the residues and incorporate these possibilities in the design programs then the programs would become much more powerful. One way of obtaining this information is to determine a large number of structures of these complexes and gather an ensemble of structures that have these variations. The ideal method would be to be able to extract this information from a single experiment. Though all possible modes of flexibility cannot be determined (like possible changes in rotamer distributions) it is possible to predict some by looking at the anisotropic vibration of the atoms in the side chains from well-refined atomic resolution structures. Comparison of Figures 3(a) and 3(b) suggests that it might be possible to guess the possible movements of the proline and valine in the substrate pocket of ADH. For comparison the thermal ellipsoids of Phe-93 are also presented. For this information to be meaningfully extracted, one has to correct the anisotropic temperature factors for other features. In order to make sense of this one has to subtract the contribution from the overall TLS of the entire asymmetric unit as well as that of

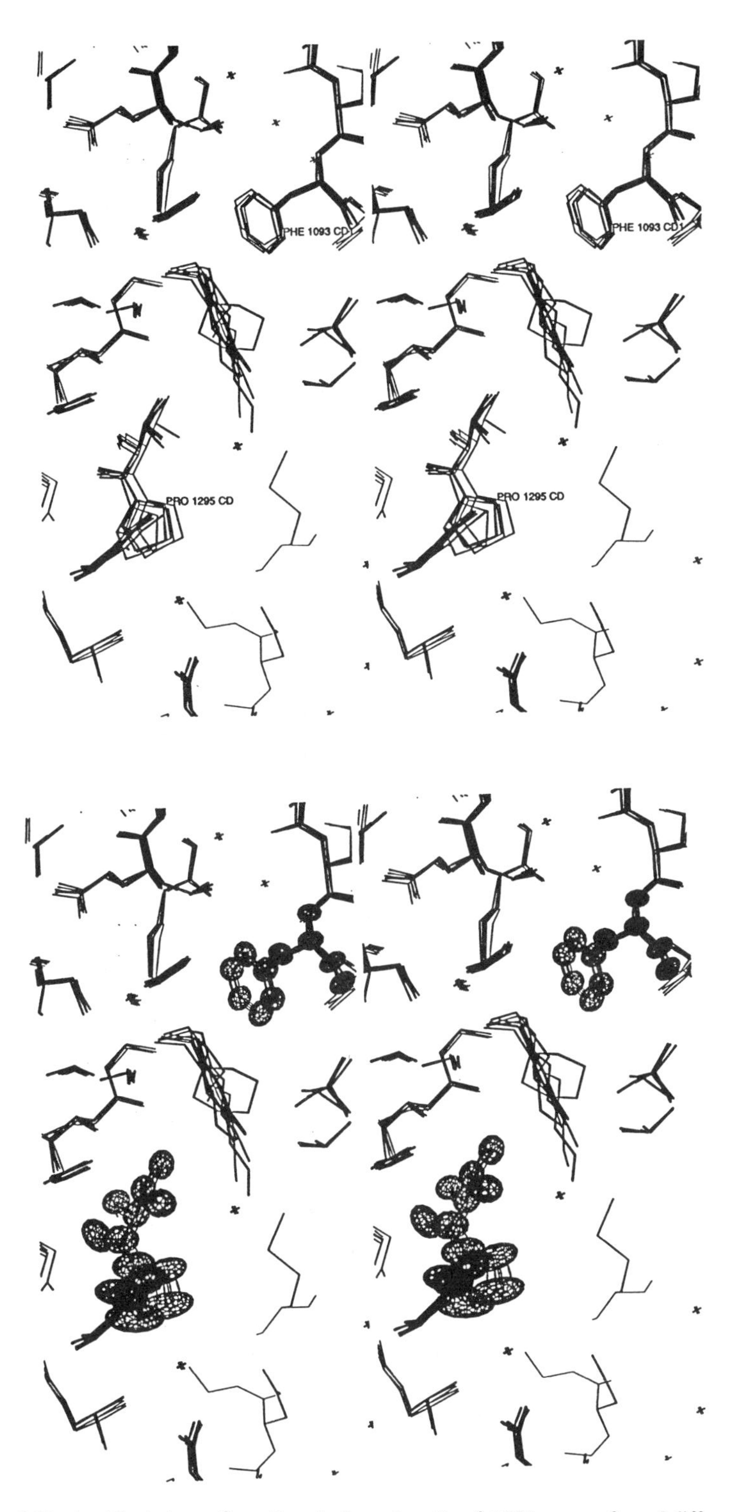

Figure 3. Flexibility in side chain conformations in the active site of ADH as seen from 6 different structures of horse liver ADH ternary complexes: a) superposed structures and b) superposed structures and the 50% thermal ellipsoid plot of the refined 1.25Å resolution ternary complex.

the rigid group to which these residues belong. It is possible to calculate the TLS effect from the anisotropic temperature factor in the program TLSANL.

4.4. Individual Anisotropic Vibrations of Atoms and Implications for Enzyme Action

Interpretation and extraction of individual anisotropic motion of atoms in a protein requires well-refined atomic resolution structures of proteins. The anisotropic temperature factors represent root mean square deviations of atomic positions. If one assumes that this RMS is a result of a Gaussian distribution, then one can calculate ellipsoids where the lengths of the axis correspond to the different probability values. We collected 1.25 Å data on a ternary complex of ADH with pentafluorobenzyl alcohol and NAD^+ and refinement is in progress. It has been well established that the transfer of the proton from the carbonyl carbon to the C4 atom of the NAD involves quantum mechanical tunnelling (Bahnson et al., 1997). The pro-R hydrogen of the methylene carbon points towards the C4 of the

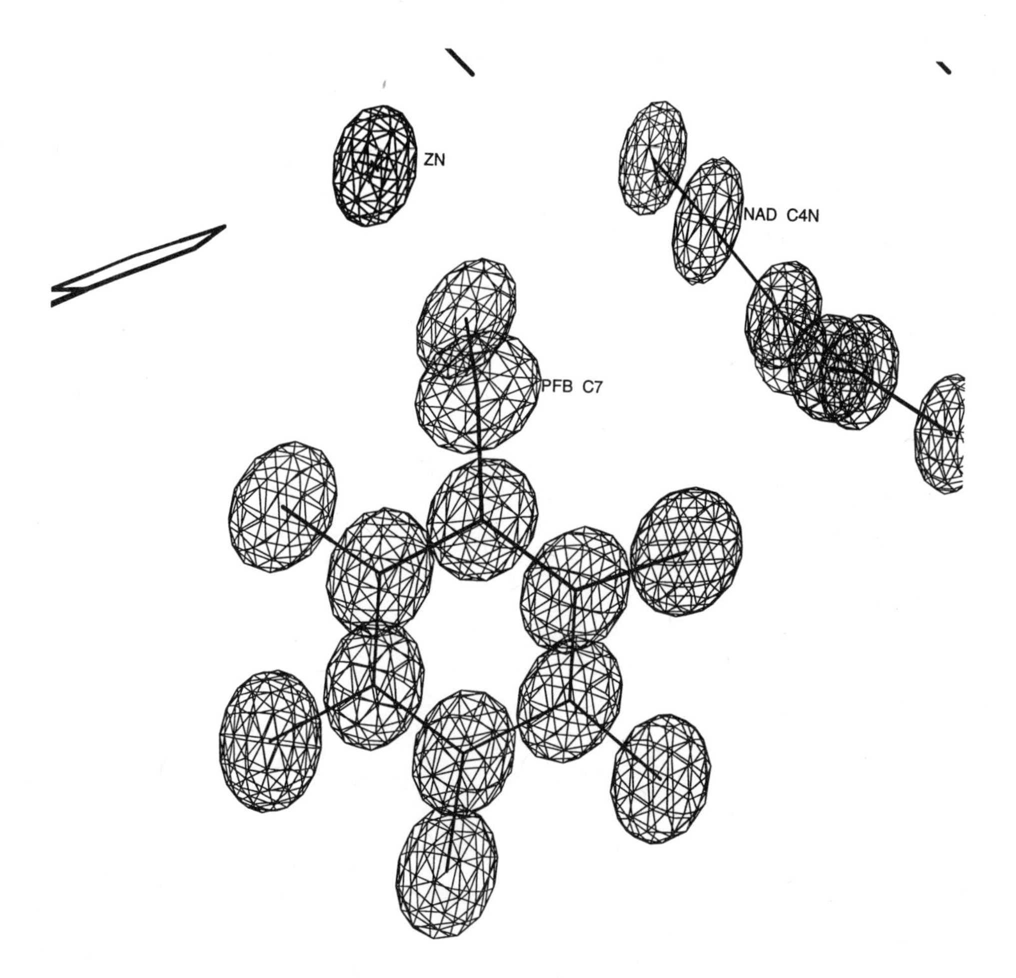

Figure 4. Thermal ellipsoid plot (50% probability) of the pentafluorobenzyl alcohol and NAD. This shows that the ellipsoids of the C4 atom of the NAD and the methylene carbon of pentafluorobenzyl alcohol (PFB) point towards each other.

NAD. Refinement of the individual anisotropic temperature factors of the atoms using shelx96 (Sheldrick and Schneider, 1996) and displaying their thermal ellipsoids (using the program Xtalview) suggests that the thermal ellipsoids of these atoms point towards each other (Figure 4). The volumes are calculated at 50% probability. It is not possible from crystallography to determine whether or not these motions are correlated. However, one can calculate this from normal mode refinement (Kidera and Go, 1992). Also, recent developments in methods of full matrix refinement, such as Eigen value analysis of the second order derivative of the least squares matrix may reveal the extent of correlation (Ten Eyck, personal communication). These calculations have not been done. However, the fact that the thermal ellipsoids point towards each other suggests that one can calculate probabilities when the distance between the two carbon atoms will be less than 2.6 Å for hydrogen transfer to take place. Whether or not this can lead to sensible calculations for the rate of hydrogen transfer still needs to be seen.

In conclusion, one can say that a large amount of dynamic information is present in crystallographic data. Careful refinement and extraction of this information is possible with current methods and theory. It definitely seems possible to extend these methods and theories to utilise them for extracting information like hydrogen transfer rates. The information of probable directions of motion as an indicator of flexibility, when introduced into molecular modeling and drug design programs will enable better use of the results obtained from a crystallographic endeavour.

ACKNOWLEDGMENTS

I thank Devapriya Choudhury, Enrique Carredano, Hans Eklund and Bryce V. Plapp for useful discussions; Ian Tickle, Birkbeck College, for answering questions on RESTRAIN and thermal motion analysis; and the Swedish Agriculture and Forestry research foundation for travel funding.

REFERENCES

Artymuik, P. J., Blake, C. C. F., Grace, D. E. P., Oatley, S. J., Phillips, D. C., and Sternberg, M. J. E. (1979) Crystallographic Studies of the Dynamic Properties of Lysozymes, *Nature* **280**, 563–568.

Bahnson, B. J., Colby, T. D., Chin, J. K., Goldstein, B. M., and Klinman, J. P. (1997) A Link between Protein Structure and Enzyme Catalyzed Hydrogen Tunneling, *Proc. Natl. Acad. Sci. USA.* **94**,12797–12802.

Cho, H., Ramaswamy, S., and Plapp, B. V. (1997) Flexibility of Liver Alcohol Dehydrogenase in Stereoselective Binding of 3-Butylthiolane 1-Oxides, *Biochemistry* **36**, 382–389.

Clarage, J. B., Clarage, M. S., Phillips, D. C., Sweet, R. M., and Casper, D. L. D. (1992) Correlation of Atomic Movements in Lysozyme Crystals, *Proteins* **12**, 145–157.

Colonna-Cesari, F., Perahia, D., Karplus, M., Eklund, H., Brändén, C. I., and Tapia, O., (1986) Interdomain Motion in Liver Alcohol Dehydrogenase—Structural and Energetic Analysis of the Hinge Bending Mode, *J. Biol. Chem.* **261**, 15273–15280.

Dauter, Z., Lamzin, V. S., and Wilson, K. S. (1995) Proteins at Atomic Resolution, *Current Opinion in Structural Biology* **5**, 784–790.

Diamond, R. (1990) On the Use of Normal Modes in Thermal Parameter Refinement: Theory and Application to the Bovine Pancreatic Trypsin Inhibitor, Acta *Crystallogr.* **A46**, 425–435.

Driessen, H., Haneef, M. I. J., Harris, G. W., Howlin, B., Khan, G., and Moss, D. S. (1989) RESTRAIN: Restrained Structure Factor Least Squares Refinement Program for Macromolecules, *J. Appl Cryst.* **22**, 510–516.

Eklund, H., Samama, J-P., Wallen, L., Bränden, C-I., Åkeson, Å., and Jones, T.A. (1981) Structure of a triclinic ternary complex of Horse Liver Alcohol Dehydrogenase at 2.9Å resolution, *J. Mol.Biol.* **146**, 561–587.

Frauenfelder, H., Petsko, G. A., and Tsernoglou, D. (1979) Temperature Dependent X-Ray Diffraction as a Probe of Protein Structural Dynamics, *Nature* **280**, 558–563.

Harata, K., Abe, Y., and Muraki, M. (1998) Full-Matrix Least-Squares Refinement of Lysozymes and Analysis of Anisotropic Thermal Motion, *PROTEINS: Structure, Function, and Genetics* **30**, 232–243.

Kidera, A., and Go, N. (1990) Refinement of Protein Dynamic Structure: Normal Mode Refinement, *Proc. Nat. Acad. Sci. U.S.A.* **87**, 3718–3722.

Kidera, A., and Go, N. (1992) Normal Mode Refinement: Crystallographic Refinement of Protein Dynamic Structure, *J. Mol. Biol.* **225**, 457–475.

Kuriyan, J., Ösapay, K., Burley, S. K., Brunger, A T., Hendrickson, W. A., and Karpuls, M. (1991), Exploration of disorder in protein structures by X-ray restrained molecular dynamics, *PROTEINS: Structure, Function and Genetics*. **10**, 340–358.

Perez, J., Faure, P., and Benout, J-P. (1996) Molecular Rigid Body Displacements in a Tetragonal Lysozyme Crystal Confirmed by X-Ray Diffuse Scattering, *Acta Crystallogr.* **D52**, 722–729.

Ramaswamy, S., El-Ahmad, M. Danielsson, O., Jörnvall, H., and Eklund, H. (1996) Crystal Structure of Cod Liver Class I Alcohol dehydrogenase: Substrate Pocket and Structurally Variable Loops, *Protein Science* **5**, 663–671.

Ramaswamy, S., Eklund, H., and Plapp, B. V. (1994) Structures of Horse Liver Alcohol Dehydrogenase Complexed with NAD^+ and Substituted Benzyl Alcohols, *Biochemistry* **33**, 5230–5237.

Ramaswamy, S., Scholze. M., and Plapp, B. V. (1997) Binding of Formamides to Liver Alcohol Dehydrogenase, *Biochemistry* **36**, 3522–3527.

Schomaker, V., and Trueblood, K. N. (1968) On the Rigid-Body Motion of Molecules in Crystals, *Acta Crystallogr.* **B24**, 63–76.

Sheldrick G. M., and Schneider T. R. (1997) "SHELXL: High Resolution Refinement", *Methods Enzymol.* **277**, 319–343.

Sternberg, M. J. E., Grace, D. E. P., and Phillips, D.C. (1979) Dynamic Information from Protein Crystallography. An Analysis of Temperature Factors from the Refinement of the Hen egg-white lysozyme structure, *J. Mol. Biol.* **130**, 231–253.

35

STUDIES ON VARIANTS OF ALCOHOL DEHYDROGENASES AND ITS DOMAINS

Jawed Shafqat, Jan-Olov Höög, Lars Hjelmqvist, Udo Oppermann, Carlos Ibanez, and Hans Jörnvall

Department of Medical Biochemistry and Biophysics
Karolinska Institutet
S-171 77 Stockholm, Sweden

1. INTRODUCTION

Zinc-containing medium-chain alcohol dehydrogenases/reductases (MDR) are ubiquitous (Danielsson et al., 1994), constituting a group of enzymes with great multiplicity. Natural variants of the family serve to illustrate functional properties, but can also serve as tools for studying evolution of novel forms. Formation of families, enzymes, classes and isozymes reflects these aspects. Class III/glutathione-dependent formaldehyde dehydrogenase, with its constant properties in vertebrates (Jörnvall and Höög, 1995; Hjelmqvist et al., 1995), invertebrates (Danielsson et al., 1994; Kaiser et al., 1993), plants (Shafqat et al., 1996; Martinez et al., 1996), fungi (Sasnaukas et al., 1992; Fernandez et al., 1995), prokaryotes (Gutheil et al., 1992; Ras et al., 1995), and archaeons (Ammendola, 1992) has a distant origin. Class I, with ethanol dehydrogenase activity, has a duplicatory origin from class III (Danielsson and Jörnvall, 1992; Jörnvall, 1994), as has also the ethanol active ADH in plants (Shafqat et al., 1996). Different regions of the molecules have been assessed for various functional and structural properties, like binding of substrate and coenzyme (Eklund, 1984; Hurley, 1991), metals (Eklund et al., 1976), other subunits (Persson et al., 1993; Danielsson et al., 1994), and further aspects.

During our studies of the MDR enzymes, we expressed domain-shuffled recombinant ADH in *E. coli*, and then detected ethanol dehydrogenase activity not derived from the recombinant protein, but native to *E. coli*. The bacterial enzyme was purified and shown to be an ethanol dehydrogenase, related to the *Zymomonas mobilis* type of enzyme (Keshav et al., 1990). However, it also had several properties related to those of the class I vertebrate enzyme, like a sensitivity toward pyrazole and a lack of glutathione-dependent formaldehyde dehydrogenase activity. Remarkably, it was found to be inducible with ethanol, at least in the strain (TG1) first studied, and this probably explains the late discovery of the enzyme. However, it was later also found as an open reading frame when the *E. coli* genome was completed (Blattner et al., 1997, SWISS-PROT entry ADHP_ECOLI).

Enzymology and Molecular Biology of Carbonyl Metabolism 7
edited by Weiner *et al.* Kluwer Academic / Plenum Publishers, New York, 1999.

The domain-shuffled enzyme that we also produced was composed of inactive chimeric molecules with properties different from both parent forms, showing that subunit folding and interactions are sensitive to gross domain changes.

2. MATERIALS AND METHODS

E. coli K12 (strain TG1) containing plasmid pKK223–2 (Amp^r) was grown in LB medium with 50 µg ampicillin, 0.02% glucose and 17.2 mM ethanol for 16 h. The cells were pelleted by centrifugation at 4000 × g for 10 min, resuspended in 10 mM Tris-Cl, pH 8.0, 1 mM dithiothreitol, sonicated, and recentrifuged at 17000 × g for 50 min. The supernatant obtained was applied to DEAE-Sepharose fast flow in the resuspension buffer, and enzyme activity was eluted with a gradient of 0–0.5 M NaCl in the same buffer. The active fraction was dialysed against 20 mM bis-Tris-Cl, pH 7.1, and further fractionated on High-Load Q-Sepharose 16/10 in the same buffer. After elution with a gradient of 0–0.5 M NaCl and subsequent dialysis against the next starting buffer, the active fraction was chromatographed twice on Mono-Q HR 5/5, first in 20 mM Tris-Cl, pH 8.0, and secondly in bis-Tris-Cl, pH 7.1. It was then eluted with 0–0.5 M NaCl in the same buffers.

The entire coding sequences corresponding to the human γ (class I) and χ (class III) enzyme subunits were ligated with the Blue script vector (Amp^r) and transformed into MC 106 cells. Single stranded DNA was isolated by superinfection with helper phages R 408. Site directed mutagenesis of the positive strand was carried out by synthesis of two sets of primers for both γ and χ at the domain borders, introducing *Bsi*W I and *Nar* I cleavage sites. After annealing of primers and synthesis of the second strand, the vector was transformed into MC 106 cells. The clones having both mutations were selected. Domain shufflings were carried out by cleaving the region corresponding to the domain borders with *Bsi*W I and *Nar* I from their respective vectors. The fragments were then shuffled and ligated to give χ-γ-χ and γ-χ-γ subunits, respectively. DNA sequence analysis was carried out with the Sequenase kit (Amersham) or the T7 sequencing™ kit (Pharmacia Biotech).

The domain shuffled constructs were transformed into expression vector pKK223–2 (Amp^r) (Pharmacia Biotech). In later preparations these constructs were amplified by PCR to introduce *Bam*H 1 and *Nde* 1sites at the 5' and 3' ends, respectively, and ligated with pET 15b (Amp^r), resulting after expression in an N-terminal His-tag sequence with an integrated thrombin cleavage site or a pET 27b ($Kana^r$) vector introducing after expression an N-terminal *pelB* signal sequence for potential periplasmic localization.

Recombinant χ-γ-χ and γ-χ-γ constructs were over-expressed in TG1 cells with the pKK223–2 vector and in BL21 cells with the pET vectors. The cells were grown in LB medium and 1 mM IPTG was added as soon as the O.D. of the culture reached 0.6 at 600 nm. The harvested cells were lysed by sonication and were centrifuged at 15000 x g for 20 min. The supernatant was loaded on a His-bind resin (Novagen), and after a washing step the column was eluted with 0.5 M NaCl, 1 M imidazole, 20 mM Tris-Cl, pH 7.9. In order to prevent inactivation, the eluted enzyme was immediately subjected to buffer change to 20 mM Tris-Cl, pH 7.9, 100 mM NaCl, 1 mM DTT, employing a Pharmacia PD 10 column. The His-tag was removed by incubation of the sample with thrombin in the same buffer containing 2.5 mM $CaCl_2$ for 16 h at 4°C. Final purification was achieved by Mono-Q FPLC in 20 mM Tris-Cl, pH 8.0, and elution with a linear gradient of 0–0.5 M NaCl.

Purities were evaluated by polyacrylamide gel electrophoresis in native or SDS-containing gels and by isoelectric focusing at pH 3–9, followed by staining with Coomassie brilliant blue for protein and Nitro blue tetrazolium/phenazine methosulphate for activity.

Molecular mass estimates were carried out on Superdex-200 FPLC column in 20 mM Tris-Cl, pH 8.0, 100 mM NaCl, 1 mM DTT.

Enzyme activities were determined at 25°C in 0.1 M glycine/NaOH, pH 10, with ethanol and octanol for class I, and in 0.1 M sodium pyrophosphate, pH 8.0, with glutathione-conjugated formaldehyde and 12-hydroxy-dodecanoic acid for class III, using Beckman DU68 or Hitachi 3000 spectrophotometers.

Amino acid compositions were determined on a Pharmacia AlphaPlus analyzer after acid hydrolysis in 6 M HCl for 24 h at 110°C. Sequence analysis was performed with Milligen Prosequencer 6625 or ABI 477 instruments. Peptides were generated by proteolytic digestion with Lys-C, Asp-N, and Glu-C proteases (Boehringer Mannheim) at 37°C overnight with enzyme:substrate ratios of 1:25–1:100 in 0.1 M ammonium bicarbonate, pH 8.1, with 2 M urea, and by CNBr cleavage in 70% formic acid for 24 h at room temperature. Peptides obtained were purified by reverse phase HPLC on Vydac C18 (2.4mm x 25 cm) with a gradient of acetonitrile in 0.1% trifluroacetic acid. Sequence comparison and calculation of phylogenetic trees utilized the program CLUSTAL W (Thompson et al., 1994), with bootstrap analysis for evaluation of confidence limits (Felsenstein, 1985).

CD spectroscopy was carried out on an AVIV model 62DS circular dichroism spectropolarimeter (Aviv, Lakewood) equilibrated with a dual syringe titration device. Purified proteins were analyzed in 10 mM Tris-Cl, pH 8.0, 100 mM NaCl and 1 mM DTT, both with and without NAD^+ at a concentration of 0.5 mg/ml by measurement of the ellipticity as a function of the wavelength at 0.5 nm increments between 260 and 190 nm at 4°C.

3. RESULTS

The ethanol-active *E. coli* enzyme was isolated from strain TG1 during expression of domain-shuffled recombinant ADH. The enzyme activity was found to be native to *E. coli*, as TG1 cells without recombinant ADH also express the same activity. Purification was achieved by a four-step chromatographic procedure. The purified enzyme showed a homogeneous band on SDS/polyacrylamide gel electrophoresis, with an estimated subunit molecular mass of 35–38 kDa. The enzyme was found to be highly active towards ethanol (k_{cat} 4600 min^{-1}) with a specific activity of 43 U/mg. K_m values for ethanol and octanol at pH 10 were determined to be 0.7 mM and 1.2 mM, respectively. The inhibition with pyrazole and 4-methylpyrazole showed K_i values of 0.2 μM and 44 μM (Table 1), respectively, and was found to be non-competitive with ethanol, which is in contrast to the inhibition of the traditional class I enzymes. The enzyme was found to be inducible with ethanol. This was discovered, when soon after finding the enzyme, it could suddenly not be detected in some preparations. On checking all conditions, we then realized that the initial prepara-

Table 1. Enzyme parameters for the inducible, ethanol-active *E. coli* enzyme at pH 10

Enzyme	K_m ethanol	K_m octanol	k_{cat} ethanol	K_i pyrazole	K_i 4-methylpyrazole
E. coli (ethanol-active)	0.7 mM	1.2 mM	4600 min^{-1}	0.2 μM	44 μM
E. coli, class III	>3000 mM	>1 mM	–	–	–
Human, class I	1.2 mM	0.013 mM	35 min^{-1}	2.6 μM	0.3 μM

For comparison, corresponding values for the class III *E. coli* enzyme (Gutheil, et al. 1992) and the human ethanol-active class I $\beta_1\beta_1$ form are also shown (Wagner, et al. 1983).

Table 2. Ethanol induction of the ethanol-active *E. coli* strain K12 TG1 alcohol dehydrogenase[a]

Ethanol, medium (mM)	Ethanol activity (U/ml)	GSH/HCHO activity (U/ml)
0	0.48	1.4
1.7	0.92	1.4
3.4	1.18	1.7
17.2	2.03	1.7
34	1.52	1.8
172	1.42	1.8

[a]The first column gives the concentration of ethanol added to the cultures.
Activities with ethanol and with glutathione/formaldehyde (GSH/HCHO) were measured in the crude lysates after centrifugation and resuspension in 10 mM Tris-Cl, pH 8.0, 1 mM dithioerythritol.

tions contained ampicillin that had been dissolved in ethanol/water. Once we added ethanol in the growth medium, we rediscovered the enzyme. Table 2 shows a titration of induction conditions, establishing that about 17 mM ethanol is required in the growth medium for maximal enzyme production in the present strain. However, the enzyme later was found without any induction in strain BL21.

The primary structure, determined by sequence analysis of peptides from Lys-C, Asp-N, Glu-C and CNBr digests, showed a 336-residue polypeptide chain (Figure 1) (Shafqat et al., 1998), in complete agreement with one of the open reading frames in the *E. coli* genome. The enzyme has apparently not been purified before and is a clear MDR-type protein. It exhibits a low identity (29%) with human class I, somewhat higher identity (30%) with *E. coli* class III, but is quite similar (79%) with an already determind structure from another bacterium, *Zymomonas mobilis* (Figure 1).

The domain-shuffled construct χ-γ-χ resulted in a fusion protein with the pET 15b vector. The protein could be purified on His-binding resin but was found to be inactive before and after thrombin cleavage of the His-tag. It was unstable and denatured completely after three days. However in the presence of NAD^+, it could be detected for three more days. It had a different folding pattern as revealed by CD spectroscopy and a low tendency to dimerization as judged by gel permeation FPLC, which suggested that about 80% was present as monomer. The other domain-shuffled construct γ-χ-γ, in vectors pKK223–2 or pET 15b/27b, did not give any protein expression.

4. DISCUSSION

The ethanol-active and ethanol-inducible alcohol dehydrogenase characterized from *Escherichia coli* is clearly an MDR-type of enzyme. The results open discussion on several points: ethanol inducibility, origin and function of alcohol dehydrogenase, substrate specificity, and gene detection from the genome project.

Figure 1. Alignment of the *E. coli* ethanol-active enzyme now characterized with the corresponding *Z. mobilis*, *S. solfataricus*, yeast, and human class I form β. Residues identical in all five forms are given against a black background, those in four forms against a stippled background.

```
E. coli, inducible     MKAAVVTKDHHVDVTYKTLRSLKHGEALLKMECCGVCHTDLHVKNG-------DFGDKTG--VILGH
Z. mobilis 1           MKAAVITKDHTIEVKDTKLRPLKYGEALLEMEYCGVCHTDLHVKNG-------DFGDETG--RITGH
S. solfataricus 1      MRAVRLVEIGKPLSLQEIGVPKPKGPQVLIKVEAAGVCHSDVHMRQGRFGNLRIVEDLGVKLPVTLGH
Yeast                  SIPETQKGVIFYESHGKLEHKDIPVPKPKANELLINVKYSGVCHTDLHAWHG-------DWPLPVKLPLVGGH
Human, class Iβ        STAGKVIKCKAAVLWEVKKPFSIEDVEVAPPKAYEVRIKMVAVGICRTDDHVVSG-------NLVTPLP--VILGH

EGIGVVAEVGPGVTSLKPGDRASVAWFYEGCGHCEYCNSGNETLCRSVKNAG--------------------YSVDG
EGIGIVKQVGEGVTSLKAGDRASVAWFFKGCGHCEYCVSGNETLCRNVENAG--------------------YTVDG
EIAGKIEEVGDEVVGYSKGD-LVAVNPWQGEGNCYYCRIGEEHLCDSPRWLG--------------------INFDG
EGAGVVVGMGENVKGWKIGDYAGIKWLNGSCMACEYCELGNESNCPHADLSG--------------------YTHDG
EAAGIVESVGEGVTTVKPGD-KVIPLFTPQCGKCRVCKNPESNYCLKNDLGNPRGTLQDGTRRFTCRGKPIHHFLGTS

GMAEECIVVADYAVKVPDGLDSAAASSITCAGVTTYKAVKLSKIR-PGQWIAIYGLGGLGNLALQYAKNVFNAKVIAI
AMAEECIVVADYSVKVPDGLDPAVASSITCAGVTTYKAVKVSQIQ-PGQWLAIYGLGGLGNLALQYAKNVFNAKVIAI
AYAEYVIVPHYKYMYKLRRLNAVEAAPLTCSGITTYRAVRKASLDPTKTLLVVGAGGGLGTMAVQIAKAVSGATIIGV
SFQQYATADAVQAAHIPQGTDLAQVAPILCAGITVYKALKSANLMAGHWVAISGAAGGLGSLAVQYAKAMG-YRVLGI
TFSQYTVVDENAVAKIDAASPLEKVCLIGCGFSTGYGSAVNVAKVTPGSTCAVFGLGGVGLSAVMGCKAAGAARIIAV

DVNDEQLKLATEMGADLAINS--HTEDAAKIVQEKTGG--AHAAVVTAVAKAAFNSAVDAVRAGGRVVAVGLPPES-M
DVNDEQLAFAKELGADMVINP--KNEDAAKIIQEKVGG--AHATVVTAVAKSAFNSAVEAIRAGGRVVAVGLPPEK-M
DVREEAVEAAKRAGADYVINASMQDPLAEIRRITESKG--VDAVIDLNNSEKTLSVYPKALAKQGKYVMVGLFGAD-L
DGGEGKEELFRSIGGEVFIDFTKEKDIVGAVLKATDGG--AHGVINVSVSEAAIEASTRYVRANGTTVLVGMPAGAKC
DINKDKFAKAKELGATECINPQDYKKPIQEVLKEMTDGGVDFSFEVIGRLDTMMASLLCCHEACGTSVIVGVPPASQN

SLDIPRLVLDGIEVVGSLVGTR------QDLTEAFQFAAEGKVVPKVALRPLADINTIFTEMEEGKIRGRMVIDFRH
DLSIPRLVLDGIEVLGSLVGTR------EDLKEAFQFAAEGKVKPKVTKRKVEEINQIFDEMEHGKFTGRMVVDFTHH
HYHAPLITLSEIQFVGSLVGNQS-----DFLGIMRLAEAGKVKPMITKTMKLEEANEAIDNLENFKAIGRQVLIP
CSDVFNQVVKSISIVGSYVGNR------ADTREALDFFARGLVKSPIKVVGLSTLPEIYEKMEKGQIVGRYVVDTSK
LSINPMLLLTGRTWKGAVYGGFKSKEGIPKLVADFMAKKFSLDALITHVLPFEKINEGFDLLHSG-KSIRTVLTF
```

The inducibility shows a new regulatory element at the genetic level, now accessible to analysis. However, the activity differs widely between different *E. coli* strains, like BL21, where the activity was detected in low amounts without induction. The finding also illustrates that care needs to be taken in expression of recombinant alcohol dehydrogenases in *E. coli*, as many of them have been expressed apparently without notable interference of the present *E. coli* enzyme. This also applies to the most closely related *Zymomonas* enzyme, which has been expressed in *E. coli* (Keshav et al., 1990). This illustrates the capacity of prokaryotes to utilize induction and amplification to secure proper metabolism.

The structure determined reflects the origin of alcohol dehydrogenases. Class III has already been suggested as an ancestral form, and this has been consistent with the finding of only this type in all living forms. However, ethanol-active forms have been produced repeatedly through a number of gene duplications in many life forms. Thus, separate duplications producing ethanol-active forms have been traced in vertebrates (Danielsson et al., 1994), plants (Shafqat et al., 1996), yeasts (Jörnvall et al., 1996), and now also prokaryotes. The novel enzyme illustrates the repeated and separate origin of ethanol-active lines in different life forms, and does not contradict the conclusion of class III as of ancient origin. As noted before, only marine life forms appear to lack ethanol-active alcohol dehydrogenase (Kaiser et al, 1993). Apparently, they do not need ethanol-active alcohol dehydrogenase for elimination of low- molecular-weight alcohols which can presumably be done directly to the sea instead. This difference further supports the central role of ethanol-active alcohol dehydrogenase in detoxifications (Jörnvall et al., 1996).

Regarding the development of substrate specificity, the present *E. coli* enzyme like the closely related *Zymomonas* enzyme, has gap segments corresponding to the human enzyme, influencing a part of the substrate pocket, making exact equivalents of the human positions 140 and 141difficult to predict. Similar gap segment has also been observed in sorbitol dehydrogenases (Luque et al., 1998), ζ-crystallins (Borras et al., 1989) and mycothiol-dependent alcohol dehydrogenase (Norin et al., 1997). The different details but the overall similar patterns regarding the active site residues produce a related set of substrate binding and co-enzyme interacting residues, compatible with the specificity and development of the different enzyme lines of alcohol dehydrogenases. The gap segments are also close to the four-Cys binding site of the structural zinc atom and compatible with the low zinc stoichiometry of the *Zymomonas* enzyme (Wills et al., 1981) compared to that of the mammalian ADH.

Regarding genome projects, it is notable that ADH exists in multiple forms. Initially, ADH was hardly known from prokaryotes, and the class III enzyme rather than any ethanol-active type was the first form characterized (Gutheil et al., 1992) until the complete *E. coli* genome recently established the presence of 14 MDR alcohol dehydrogenase type of enzymes. One of these new forms has now also been characterized at the protein and enzyme level, and has been established to be a typical ethanol-active alcohol dehydrogenase, linking ADH activity in *E. coli* with that of most other life forms (Figure 2).

Finally the significance of domain interactions was now studied in human class III and class I domain-shuffled forms. Expression of χ-γ-χ in pET 15b resulted in a recombinant protein, which lacked alcohol dehydrogenase activity and showed differences in folding pattern toward the native human class III enzyme. It also showed low tendency of dimerization with apparently <10% found in dimers, supporting NAD^+ binding as a stabilizing factor like in the mammalian MDR alcohol dehydrogenases. However, the lack of activity, low stability and low dimerization of χ-γ-χ, and the absence of detectable γ-χ-γ, clearly demonstrate that domain organization is complex and of importance for native alcohol dehydrogenases.

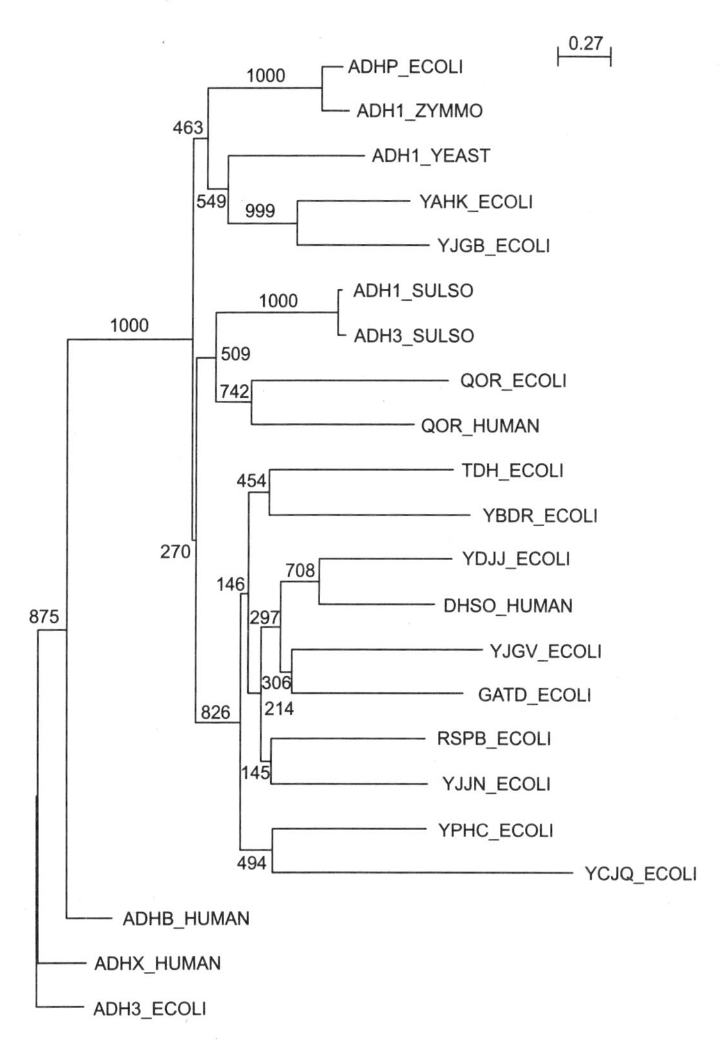

Figure 2. Relationship of the *E. coli* ethanol-active alcohol dehydrogenase now studied (ADHP_ECOLI) towards the other 13 MDR forms (other abbreviations ending in ECOLI) previously known or recently deduced from the *E. coli* genome, corresponding *Zymomonas* (ADH1_ZYMMO) and *Sulfolobus* (ADH1_SULSO and ADH3_SULSO) alcohol dehydrogenases, the human forms of the other enzymes now included in the figure are; sorbitol dehydrogenase (DHSO), quinone oxidoreductase (QOR) class I (ADHB) and III (ADHX) alcohol dehydrogenases. The phylogenetic tree was calculated with the program CLUSTAL W. Positions with gaps were excluded and corrections were made for multiple substitutions. Numbers indicate results from bootstrap analysis for evaluation of conficence limits (1000 bootstrap replicates).

ACKNOWLEDGMENTS

This work was supported by grants from the Swedish Medical Research Council (projects 03X-3532) and EC Biotechnology grant EC (97-2123).

REFERENCES

Ammendola, S., Raia, C.A., Caruso, C., Camardella, L., D'Auria, S., Rosa, M.D. and Rossi, M. (1992) Thermostable NAD$^+$-dependent alcohol dehydrogenase from *Sulfolobus solfataricus*: gene and protein sequence determination and relationship to other alcohol dehydrogenases. Biochemistry 31, 12514–12523.

Blattner, F.R, plunkett, G., Block, C.A., Perna, N.T., Burland, V., Riley, M., Colladovides, J., Glasner, J.D., Rode, C.K., Mayhew, G.F., Gregor, J., Davis, N.W., Krikpatric, H.A., Goeden, M.A., Rose, D.J., Mau, B., and Shao, Y. (1997) The complete genome sequence of *Escherichia coli* K-12. Science 277, 1453–1471

Borrás, T., Persson, B. and Jörnvall, H. (1989) Eye lens ζ-crystallin relationships to the family of "long-chain" alcohol/polyol dehydrogenases. Protein trimming and conservation of stable parts. Biochemistry 28, 6133–6139.

Danielsson, O., Atrian, S., Luque, T., Hjelmqvist, L., Gonzàlez-Duarte, R. and Jörnvall, H. (1994) Fundamental molecular differences between alcohol dehydrogenase classes. Proc. Natl. Acad. Sci. USA, 91, 4980–4984.

Danielsson, O., Eklund, H. and Jörnvall, H. (1992) Major piscine liver alcohol dehydrogenase has class-mixed properties in relation to mammalian alcohol dehydrogenases of classes I and III. Biochemistry 31, 3751–3759.

Danielsson, O and Jörnvall, H. (1992) "Enzymogenesis": classical liver alcohol dehydrogenase origin from the glutathione-dependent formaldehyde dehydrogenase line. Proc. Natl. Acad. Sci. USA, 89, 9247–9251.

Eklund, H., Nordström, B., Zeppezaner, E., Söderlund, G., Ohlsson, I., Boiwe, T., Söderberg, B-O., Tapia, O., Bränden, C-I. and Åkeson, Å. (1976) Three dimensional structure of horse liver alcohol dehydrogenase at 2.4Å resolution. J.Mol.Biol. 102, 27–59.

Eklund, H., Samama, J.-P. and Jones, T.A. (1984) Crystallographic investigations of nicotinamide adenine dinucleotide binding to horse liver alcohol dehydrogenase. Biochemistry 23, 5982–5996.

Felsenstein, J. (1985) Confidence limits on phylogenies: An approach using the bootstrap. Evolution. 39, 783–789.

Fernández, M.R., Biosca, J.A., Norin, A., Jörnvall, H. and Parés, X. (1995) Class III alcohol dehydrogenase from *Saccharomyces cerevisiae*: Structural and enzymatic features differ towards the human/mammalian forms in a manner consistent with functional needs in formaldehyde detoxication. FEBS Lett. 370, 23–26.

Gutheil, W.G., Holmquist, B. and Vallee, B.L. (1992) Purification, characterization, and partial sequence of the glutathione-dependent formaldehyde dehydrogenase from *Escherichia coli*: a class III alcohol dehydrogenase. Biochemistry 31, 475–481.

Hjelmqvist, L., Shafqat, J., Siddiqi, A.R. and Jörnvall, H. (1995) Alcohol dehydrogenase of class III: consistent patterns of structural and functional conservation in relation to class I and other proteins. FEBS Lett. 373, 212–216.

Höög, J.-O., Weis, M., Hedén, L.-O., Zeppezauer, M., Jörnvall, H. and von Bahr-Lindström, H. (1987) Expression in *Escherichia coli* of active human alcohol dehydrogenase lacking N-terminal acetylation. Biosci. Rep. 7, 969–974.

Hurley, T.D., Bosron, W.F., Hamilton, J.A. and Amzel, L.M. (1991) Structure of human $\beta_1\beta_1$ alcohol dehydrogenase: catalytic effects of non-active-site substitutions. Proc. Natl. Acad. Sci. USA. 88, 8149–8153.

Jörnvall, H. (1994) The alcohol dehydrogenase system. In: Toward a Molecular Basis of Alcohol Use and Abuse (Jansson, B., Jörnvall, H., Rydberg, U., Terenius, L. and Vallee, B.L. eds) pp. 221–229, Birkhäuser, Basel.

Jörnvall, H. and Höög, J.-O. (1995) Nomenclature of alcohol dehydrogenases. Alcohol and Alcoholism 30, 153–161.

Jörnvall, H., Shafqat, J., El-ahmad, M., Hjelmqvist, L., Persson, B. and Danielsson, O. (1996) Alcohol dehydrogenase variability: Evolutionary and functional conclusions from characterization of further variants. Enzymology and molecular Biology of Carbonyl Metabolism 6. Edited by Weiner *et al.* Plenum press, New York. 281–289

Kaiser, R.,Fernández, M.R., Parés, X. and Jörnvall, H. (1993). Origin of the human alcohol dehydrogenase system: Implications from the structure and properties of the octopus protein. Proc. Natl. Acad. Sci. USA, 90, 11222–11226

Keshav, K.F., Yomano, L.P., An, H. and Ingram, L.O. (1990) Cloning of the *Zymomonas mobilis* structural gene encoding alcohol dehydrogenase I (adhA): sequence comparison and expression in *Escherichia coli*. J. Bacteriol. 172, 2491–2497.

Luque, T., Hjelmqvist, L., Marfany, G., Danielsson, O., El-Ahmad, M., Persson, B., Jörnvall, H. and Gonzàlez-Duarte, R. (1998) Sorbitol dehydrogenase of *Drosophila*: gene, protein and expression data show a two-gene system. J. Biol. Chem., submitted.

Martínez, M.C., Achkor, H., Persson, B., Fernández, M.R., Shafqat, J., Farrés, J., Jörnvall, H. and Parés, X. (1996) *Arabidopsis* formaldehyde dehydrogenase. Eur. J. Biochem. 241, 849–857.

Norin, A., van Ophem, P.W., Piersma, S., Duine, J.A. and Jörnvall, H. (1997) Factor-dependent formaldehyde dehydrogenase: A prokaryotic enzyme linking different eukaryotic dehydrogenase lines. Eur. J. Biochem. 248, 282–289.

Persson, B., Bergman, T., Keung, W.M., Waldenström,U., Holmqvist, B. and Jörnvall, H. (1993) Basic features of class I alcohol dehydrogenase: variable and constant segments coordinating by inter-class and intra-class variability. Conclusions from characterization of alligator enzyme. Eur. J. Biochem. 216, 49–56.

Ras, J., Van Ophem, P.W., Reijnders, W.N.M., Van Spanning, R.J.M., Duine, J.A., Stouthamer, A.H. and Harms, N. (1995) Isolation, sequencing, and mutagenesis of the gene encoding NAD- and glutathione-dependent formaldehyde dehydrogenase (GD-FALDH) from *Paracoccus denitrificans*, in which GD-FALDH is essential for methylotrophic growth. J. Bacteriol. 177, 247–251.

Sasnaukas, K., Jomantienė, R., Januska, A., Lebedienė, E., Lebedys, J. and Janulaitis, A. (1992) Cloning and analysis of a *Candida maltosa* gene which confers resistance to formaldehyde in *Saccharomyces cerevisiae*. Gene 122, 207–211.

Shafqat, J., El-Ahmad, M., Danielsson, O., Martínez, M.C., Persson, B., Parés, X. and Jörnvall, H. (1996) Pea formaldehyde-active class III alcohol dehydrogenase: Common derivation of the plant and animal forms but not of the corresponding ethanol-active forms (classes I and P). Proc. Natl. Acad. Sci. USA, 93, 5595–5599.

Shafqat, J., Höög, J-O., Hjelmqvist, L., Oppermann, U., Ibanez, C. and Jörnvall, H. (1998) An ethanol-active and ethanol-inducible *E. coli* alcohol dehydrogenase. Structural and enzymatic relationships to the eukaryotic protein forms. Eur.J.Biochem. submitted.

Thompson, J.D., Higgins, D.G. and Gibson, T.J. (1994) CLUSTAL W: improving the sensitivity of progressive multiple sequence alignment through sequence weighting, position-specific gap penalties and weight matrix choice. Nucl. Acid Res. 22, 4673–4680.

Wagner, F.W., Burger, A.R. and Vallee, B.L. (1983) Kinetic properties of human liver alcohol dehydrogenase: oxidation of alcohols by class I isoenzymes. Biochemistry 22, 1857–1863.

Wills, C., Kratofil, P., Londo, D. and Martin, T. (1981) Characterization of the two alcohol dehydrogenases of *Zymomonas mobilis*. Arch. Biochem. Biophys. 210, 775–785.

36

UNCOMPETITIVE INHIBITORS OF ALCOHOL DEHYDROGENASES

Bryce V. Plapp,[1] Vijay K. Chadha,[1] Kevin G. Leidal,[1] Heeyeong Cho,[1] Michael Scholze,[1] John F. Schindler,[1] Kristine B. Berst,[1] and Ramaswamy S[2]

[1]Department of Biochemistry
The University of Iowa
Iowa City, Iowa 52242
[2]Department of Molecular Biology
Swedish University of Agricultural Sciences
Biomedical Center, S-751 24 Uppsala, Sweden

1. INTRODUCTION

Selective inhibitors of alcohol dehydrogenases could be useful for prevention of poisoning due to metabolism of alcohols, such as methanol or ethylene glycol, that lead to toxic products (Jacobsen and McMartin, 1997). Good inhibitors could also be used to study the physiological functions of the various isoenzymes of alcohol dehydrogenase and for therapeutic intervention after the metabolic roles of the enzymes are established. Although 4-methylpyrazole is a potent inhibitor of some of the liver alcohol dehydrogenases, it is not very effective against all of the human isoenzymes, and it is a competitive inhibitor against alcohol, which makes it less effective when the concentration of substrate alcohol is increased. Thus, we have been designing, synthesizing, and evaluating potentially specific inhibitors that are uncompetitive against varied concentrations of alcohols. The research has led to some selective inhibitors of human alcohol dehydrogenases and structure-function information about the specificities of the enzymes.

As analogues of the aldehyde substrates, amides and formamides bind preferentially to the enzyme-NADH complex and uncompetitively inhibit alcohol oxidation (Winer and Theorell, 1960; Porter et al., 1976; Freudenreich et al., 1984; Ramaswamy et al., 1997; Schindler et al., 1998). Sulfoxides are analogues of the ketone substrates, products of the oxidation of secondary alcohols, and also bind to the enzyme-NADH complex (Chadha et al., 1983; Cho et al., 1997; Cho and Plapp, 1998). Amides and sulfoxides are good inhibitors of alcohol metabolism in experimental animals (Lester and Benson, 1970; Porter et al., 1976; Delmas et al., 1983; Chadha et al., 1983, 1985; Plapp et al., 1984). The struc-

Enzymology and Molecular Biology of Carbonyl Metabolism 7
edited by Weiner *et al.* Kluwer Academic / Plenum Publishers, New York, 1999.

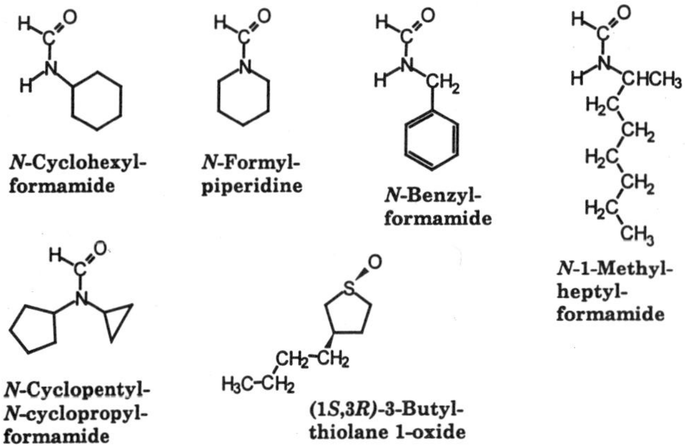

N-Cyclohexyl-formamide **N-Formyl-piperidine** **N-Benzyl-formamide** **N-1-Methyl-heptyl-formamide** **N-Cyclopentyl-N-cyclopropyl-formamide** **(1S,3R)-3-Butyl-thiolane 1-oxide**

Figure 1. Uncompetitive inhibitors of alcohol dehydrogenases. The formamides can exist in the cis or trans configuration, and *N*-1-methylheptylformamide has an unspecified chiral center. The 3-butylthiolane 1-oxide has four isomers, due to the chirality of the sulfur and carbon 3; the isomer shown is the most potent inhibitor of horse liver alcohol dehydrogenase (Cho et al., 1997).

tures of the compounds that are especially interesting and are found to be potent or selective inhibitors of human alcohol dehydrogenases are shown in Figure 1. The purpose of this article is to review the preparation, evaluation, and utility of these compounds.

2. EXPERIMENTAL METHODS

Crystalline horse liver alcohol dehydrogenase (*Eq*ADH), NAD^+ and NADH were purchased from Boehringer Mannheim. Clones for the expression of human alcohol dehydrogenases (*Hs*ADH) α, β_1, and σ were obtained from Dr. Thomas D. Hurley (Indiana University School of Medicine), and the vectors for the γ_2 and π enzymes were from Dr. Jan-Olov Höög (Karolinska Institutet). The isoenzymes of *Hs*ADH were purified essentially as described by Stone et al. (1993) and Hurley et al. (1997). *N*-cyclohexylformamide and *N*-formylpiperidine were obtained from Aldrich Chemical Company, and other formamides were prepared by reaction of the corresponding amino compound with formic acid and purification by distillation or chromatography (Schindler et al., 1998). 3-Butylthiolane 1-oxide was prepared from 2-n-butylsuccinic acid as the mixture of all four stereoisomers (Chadha et al., 1985).

The inhibition constants were determined by initial velocity studies in 46 mM sodium phosphate and 0.25 mM EDTA buffer, pH 7.0, at 25 ºC by measuring the change in absorbance at 340 nm due to formation or utilization of NADH. The coenzyme concentration was held constant at 0.2 mM NADH or 2.0 mM NAD^+, and the concentration of a substrate appropriate for the specificity of each enzyme was varied. The concentrations of inhibitor were varied over at least a 3-fold range, and 20 to 25 different assay conditions, with duplicates, were tested. The inhibition data were fitted to the equations for competitive or uncompetitive inhibition with the appropriate computer programs (Cleland, 1979). The K_i values were not dependent on the substrate used, and the inhibition constants determined for the forward or reverse reactions were about (within 2-fold) the same, which is expected if coenzyme dissociation is rate-limiting in the ordered mechanism. The formamides are uncompetitive inhibitors against alcohols and competitive against aldehydes or ketones indicating that the formamides bind to the enzyme-NADH complex. K_i values determined at pH 7 were similar to values obtained at pH 8.

Crystals of the complexes of the horse liver alcohol dehydrogenase were prepared by dialysis of 1 ml of 10 mg/ml enzyme in washed Spectra/por tubing against 10 ml of 50 mM ammonium *N*-[tris(hydroxymethyl]-2-aminoethanesulfonate (pH 7.0) at 5 °C containing 0.7 mM NADH and 14 mM *N*-formylpiperidine or 0.7 mM NAD^+ and 8 mM 2,3,4,5,6-pentafluorobenzyl alcohol (Ramaswamy et al., 1994, 1997). Crystals formed after stepwise addition of 2-methyl-2,4-pentanediol to a concentration of about 10% to the outer dialyzate, and the final concentration was brought to 25%. X-Ray data were collected to 2.5 Å for the formamide complex and to 2.1 Å for the alcohol complex. The structures were solved by molecular replacement and refined to *R* values of 20.5% and 18.3%, respectively. Molecular modeling was carried out using the computer program O (Jones et al., 1991). Inhibitors were built and minimized in SYBYL (Tripos Associates).

3. RESULTS AND DISCUSSION

As shown in Figure 2, *N*-cyclohexylformamide is an uncompetitive inhibitor against varied concentrations of ethanol and a competitive inhibitor against acetaldehyde with the horse liver alcohol dehydrogenase, EE isoenzyme (*Eq*ADH). The K_i values (intercept or slope inhibition constants) were about 8 μM for either experiment. Other formamides showed the same patterns, but some sulfoxides were uncompetitive against the alcohol and noncompetitive against the aldehyde.

The simplest kinetic explanation for the inhibition results is based on the Ordered Bi Bi mechanism established for the horse liver enzyme, shown in Figure 3. An inhibitor that binds only to the enzyme-NADH complex (dissociation constant K_J) would produce the pattern of inhibition shown in Figure 2. If the inhibitor binds to both the enzyme-NAD^+ and enzyme-NADH complexes (K_I and K_J), the inhibition patterns would be noncompetitive (slope and intercept effects on the double reciprocal plots). In general, the formamides and sulfoxides appear to most closely resemble the carbonyl substrates of the dehydrogenases and bind preferentially to the enzyme-NADH complex. Some of the inhibitors may also bind weakly to the enzyme-NAD^+ complex.

The binding of formamides to *Eq*ADH was studied by x-ray crystallography in order to determine how the enzyme-NADH complex binds aldehyde analogs. Crystallography of complexes with aldehyde substrates is not feasible since the enzyme will reduce the alde-

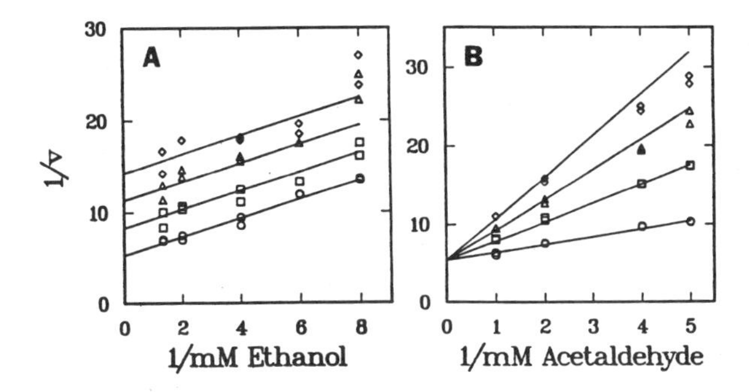

Figure 2. Inhibition of horse liver alcohol dehydrogenase by *N*-cyclohexylformamide. The inhibition was studied in 33 mM sodium phosphate buffer and 0.25 mM EDTA at pH 8 and 25 °C by measuring the change in of absorbance due to NADH. A, Uncompetitive inhibition with concentrations of 0 μM (○), 10 μM (□), 15 μM (Δ), 20 μM (◇) against varied concentrations of ethanol and a saturating concentration of 2 mM NAD^+. B, Competitive inhibition with concentrations of 0 μM (○), 10 μM (□), 20 μM (Δ), 30 μM (◇) against varied concentrations of acetaldehyde at 0.1 mM NADH. Data were fitted with the appropriate programs of Cleland (1979).

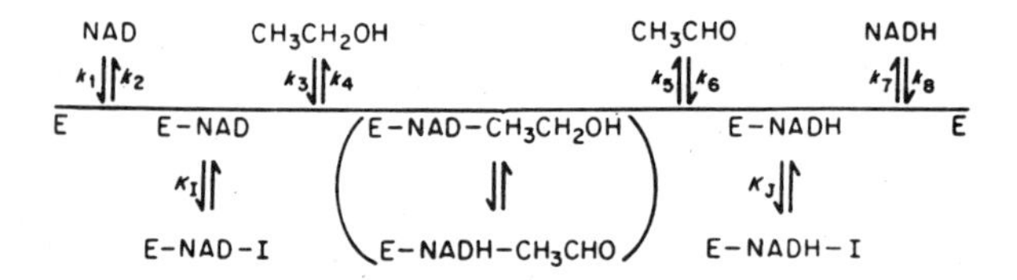

Figure 3. Ordered Bi Bi mechanism of alcohol dehydrogenase. Binding of a dead-end inhibitor that is an analog of the aldehyde substrate to the enzyme-NADH complex (K_J) produces uncompetitive inhibition against varied concentrations of substrate alcohol and competitive inhibition against varied concentrations of aldehyde.

hyde to the alcohol. When the enzyme is saturated with NAD^+ and benzyl alcohol or NADH and benzaldehyde, the "on-enzyme equilibrium" favors NAD^+ and alcohol by 4 to 1 (Shearer et al., 1993). Crystallography shows that the enzyme-NAD^+ complex binds *p*-bromobenzyl alcohol (a good substrate) and 2,3,4,5,6-pentafluorobenzyl alcohol (a good inhibitor that is not a substrate due to the electron-withdrawing fluorines) in slightly different positions, with the benzene rings in the same plane, but shifted. The oxygens of the alcohols are bound to the catalytic zinc in the same position, but the methylene carbons are in different positions. The binding of the fluoroalcohol appears to mimic closely the structure expected for a Michaelis complex where the reactants are poised for a direct transfer of the *pro-R* hydrogen from the methylene carbon of the alcohol to the *re* face of C4 of NAD^+ (Ramaswamy et al., 1994).

Thus, it is of interest to compare the binding of *N*-formylpiperidine (a benzaldehyde analog) with the enzyme-NADH complex to the binding of the pentafluorobenzyl alcohol with the enzyme-NAD^+ complex, as these two complexes could represent the two ground states for the interconversion of ternary complexes. Figure 4 shows a superpositioning of the structures of the two complexes. The atoms of the protein and coenzyme are in very similar positions in the two complexes, and the oxygens of the substrate analog bind the catalytic zinc and form hydrogen bonds to the hydroxyl group of Ser-48. The exocyclic carbon (what would be the reactive carbon in a substrate) of the substrate analogues are also in similar positions, and oriented as would be expected for direct hydrogen transfer to or from the nicotinamide ring of the coenzyme. The distances between the exocyclic carbons and C4 of the nicotinamide ring are 3.4 to 4.0 Å in the two complexes, which would be appropriate for ground state structures with direct hydrogen transfer. During the chemical reaction, the C-C distance is calculated to be 2.6–2.9 Å in the transition state, which can occur because of the protein dynamics (Ramaswamy, 1998). It may be noted, however, that the six-membered rings of the inhibitors are rotated into different positions. This may be due to different interactions of the rings with the amino acid side chains or to the fixed, planar geometry of the formamide as compared to the rotatable bonds in the alcohol. In any case, the results show that there is some flexibility in binding of ligands, and that structural data are required to define the positions.

Structures of *Eq*ADH complexed with NADH and two isomers (1*S*,3*R* and 1*S*,3*S*) of 3-butylthiolane 1-oxide have also been determined at high resolution (Cho et al., 1997). In both cases, the oxygen of the sulfoxide is ligated to the catalytic zinc and hydrogen bonded to the hydroxyl group of Ser-48, and the position of the tetramethylene ring is the same. The butyl side chains bind in different positions, however, due to conformational flexibility of the amino acid side chains, showing that the active site adapts to optimize the binding interactions. The binding of tetramethylene sulfoxide is especially good as compared to other sulfoxides (Chadha et al., 1983), which may be ascribed to the good fit into

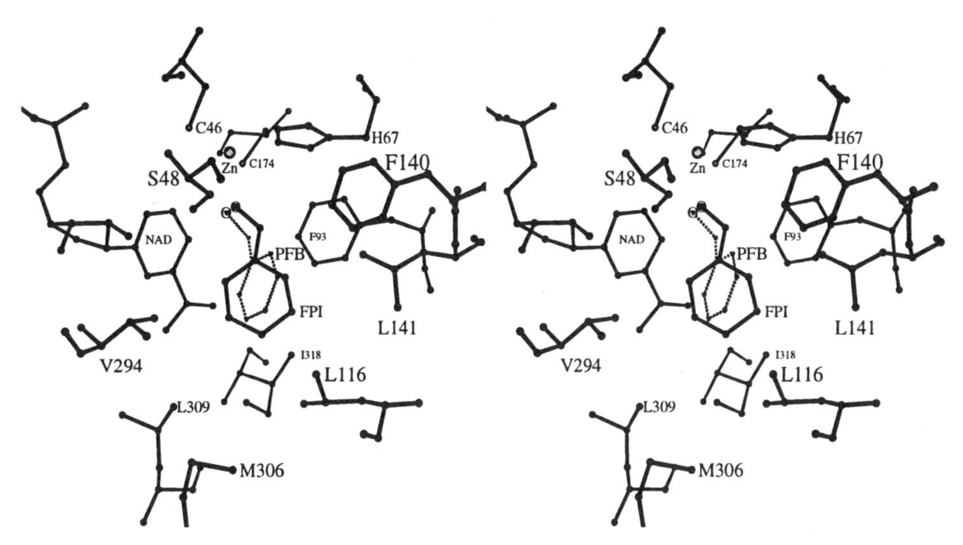

Figure 4. Comparison of the binding modes of *N*-formylpiperidine (*FPI*) and 2,3,4,5,6-pentafluorobenzyl alcohol (*PFB*) to horse liver alcohol dehydrogenase complexed with coenzyme, NADH or NAD^+, respectively. The inhibitors represent analogues of good substrates of the enzyme, benzaldehyde and benzyl alcohol. Coordinates were taken from Brookhaven Protein Data Bank entries 1LDE and 1HLD. The protein atoms are in essentially the same positions in the two complexes.

the substrate binding pocket and to some contribution from a cation-π interaction of the sulfur with the benzene ring of Phe-93 (Cho and Plapp, 1998).

Specific inhibitors of the human alcohol dehydrogenases (*Hs*ADH) could be useful for preventing metabolism of particular substrates by the various isoenzymes. Five isoenzymes (α, β, γ, π, and σ) have significant activity on a variety of alcohols (Edenberg and Bosron, 1997). The substrate specificities differ as the isoenzymes have different amino acid residues in their active sites (Table 1). Since the alcohol dehydrogenases have homologous structures (Hurley et al., 1994; Xie et al., 1997), selective inhibitors could be designed by rational approaches, but because of the uncertainties in modeling due to conformational flexibility, empirical testing is required. Thus, we prepared and tested 30 different compounds (Cho and Plapp, 1998; Schindler et al., 1998) and found some potent, selective inhibitors, indicated in bold font in Table 2. The inhibition results can be partially explained in terms of the known three-dimensional structures of the enzymes and their complexes. Structures for the complexes of the inhibitors with the human enzymes have not yet been determined.

Horse and human α, β and γ enzymes have similar affinities for *N*-cyclohexylformamide, whereas the π and σ enzymes have much lower affinity. *N*-cyclohexylmethylformamide was a good inhibitor of all of the enzymes except for *Hs*ADH σ. These compounds could be used as general inhibitors of ethanol metabolism. Tighter binding could be obtained with more specific inhibitors.

*Hs*ADH α is inhibited strongly and selectively by the disubstituted formamide, with *N*-cyclopentyl and *N*-cyclobutyl substituents. Molecular modeling suggests that this inhibitor would bind with favorable contacts between the cyclobutyl group and Ala-93 and Ile-318. The α enzyme is unusual in having a small aliphatic residue at position 93, in contrast to the aromatic residue in the other enzymes, and thus there is space for two large

Table 1. Amino acid residues that could directly contact substrates and inhibitors in the substrate binding sites of horse and human alcohol dehydrogenases[a]

Amino acid	*Eq* E	*Hs* α	*Hs* β_1	*Hs* γ_2	*Hs* π	*Hs* σ
48	S	T	T	S	T	T
57	L	M	L	L	F	M
93	F	A	F	F	Y	F
116	L	V	L	L	L	I
140	F	F	F	F	F	F
141	L	L	L	V	F	M
294	V	V	V	V	V	V
306[b]	M	M	M	M	E	M
309[b]	L	L	L	L	I	F
318	I	I	V	I	F	V

[a]From Sun and Plapp (1992) and Satre et al. (1994). Abbreviations: *Hs*ADH, human (*Homo sapiens*) alcohol dehydrogenase; *Eq*ADH, horse (*Equus caballus*) ADH.
[b]Residues provided by the second subunit.

substituents. The other enzymes have much weaker affinity for the disubstituted formamides, but it is interesting that the γ_2 enzyme binds *N*-cyclopentyl-*N*-cyclopropylformamide about as well as it binds the monosubstituted compounds.

*Hs*ADH β shows strong affinities for *N*-benzylformamide and *N*-heptylformamide, which indicates that different substituents can optimize interactions. Modeling suggests that the benzyl and heptyl groups would make favorable contacts with residues Val-318 and Leu-309. Binding of *N*-1-methylheptylformamide is not as good, probably because of steric interference with Thr-48.

*Hs*ADH γ_2 binds *N*-1-methylheptylformamide particularly well, which probably reflects favorable interactions of the 1-methyl group with Ser-48. Other branched chain derivatives are better inhibitors than the linear alkyl compounds. The three-dimensional structure of *Hs*ADH γ_2 is not known, but the amino acid residues at the active site are similar to those of *Eq*ADH. Both of these enzymes have Ser-48, rather than Thr-48, and there should be more space to accommodate the branched compounds. This structural feature is also apparent in the exceptionally good binding of 3-butylthiolane 1-oxide, which does not fit well when Thr-48 is present. The good binding of *N*-cyclopentyl-*N*-cyclopropylformamide to *Hs*ADH γ_2 suggests that there is more room in the active site than in the horse enzyme, and it will be important to determine the structure.

Table 2. Selective inhibitors for alcohol dehydrogenases[a]

	K_i (μM)					
Inhibitor	*Eq* E	*Hs* α	*Hs* β_1	*Hs* γ_2	*Hs* π	*Hs* σ
N-cyclohexylformamide	8.7	2.3	3.4	5.2	84	380
N-cyclohexylmethyl-formamide	12	3.6	1.0	5.8	7.1	47
N-benzylformamide	***0.74***	31	***0.33***	4.9	110	11
N-n-heptylformamide	3.0	3.6	*0.33*	12	11	***0.74***
N-1-methylheptyl-formamide	5.4	7.0	1.7	***0.41***	40	100
N-cyclopentyl-*N*-cyclopropyl-formamide	110	***0.88***	4600	4.6	96	930
N-cyclopentyl-*N*-cyclobutyl-formamide	8600	***0.36***	10000	47	360	1100
3-butylthiolane 1-oxide	***0.56***	73	120	***0.30***	110	530

[a] Inhibition constants were determined against varied concentrations of appropriate substrates for each enzyme.

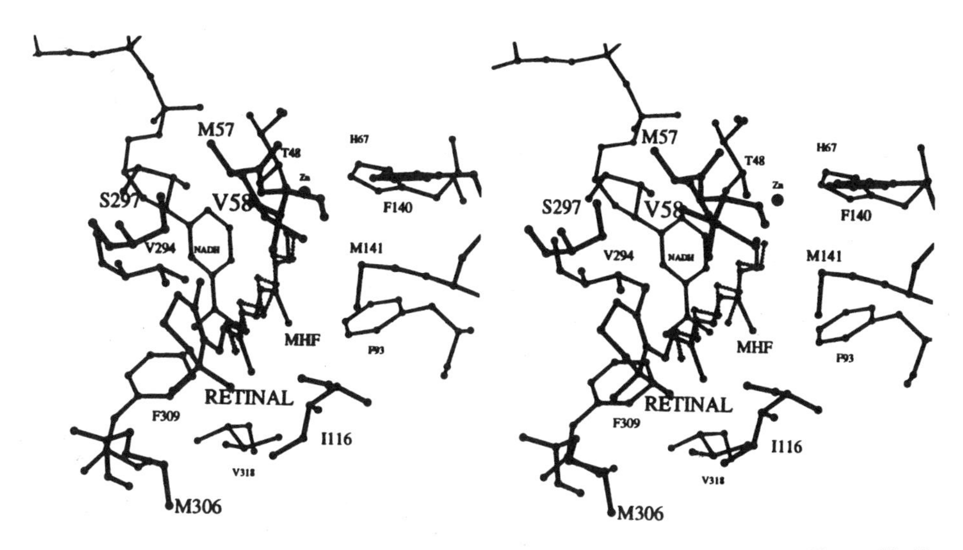

Figure 5. Modeling of the binding of *all-trans*-retinal (solid lines) and *cis-N*-1(*R*)-methylheptylformamide (dotted lines) into the active site of the human alcohol dehydrogenase σ isoenzyme. Coordinates for the enzyme were taken from Brookhaven Data Bank entry 1AGN (Xie et al., 1997). The positions of the ligands were chosen so that the distance between the oxygen of the ligand and the catalytic zinc was about 2.2 Å, the *re* face of the carbonyl group of the ligand was directed toward the nicotinamide ring in the orientation observed for complexes of formamides with horse liver alcohol dehydrogenase (Ramaswamy et al., 1997), and the aliphatic moiety was not in close contact with protein side chains. The side chain of Met-141 was rotated slightly to accommodate the ligands, but otherwise the protein side chains are the same as those found in the crystalline complex.

*Hs*ADH π has relatively low affinity for most of the formamides, but modeling with *N*-cyclohexylformamide does not indicate any steric conflicts between the cyclohexyl ring and residues in the active site. The best inhibitor is *N*-cyclohexylmethylformamide. It is surprising that this enzyme does not bind *N*-benzylformamide better, as the active site has four phenylalanine residues.

*Hs*ADH σ is most strongly inhibited by *N*-heptylformamide, which could make several hydrophobic contacts in the active site. Met-57, Met-141 and Phe-309 in the σ enzyme narrow the binding pocket as compared with the β and γ enzymes, which have smaller residues at these positions. However, the β enzyme also binds *N*-heptylformamide tightly. The greatly decreased affinity of the σ enzyme for *N*-1-methylheptylformamide as compared to that for *N*-heptylformamide is surprising as the σ enzyme has good catalytic activity with *all-trans*-retinol. Modeling of *all-trans*-retinal and *cis-N*-1(*R*)-methylheptylformamide suggests that both molecules could fit well into the active site (Figure 5).

It is interesting that the small differences in the active sites of the human enzymes can result in large differences in the binding of the various inhibitors. Clearly, favorable hydrophobic interactions, offset by unfavorable steric effects, account for the differences, but quantitative analysis remains for the future. Three dimensional structures must be determined in order to see which interactions exist.

The new inhibitors have not been tested in humans, but short chain formamides, amides and tetramethylene sulfoxides effectively inhibit ethanol metabolism in rats (Porter et al., 1976; Chadha et al., 1983, 1985; Plapp et al., 1984). The specificities of the inhibitors need to be tested with the purified rat and mouse alcohol dehydrogenases and with other enzymes that can bind carbonyl substrates, such as the aldehyde dehydrogenases. Toxicities of the formamides need further evaluation. The LD_{50} for *N*-cyclohexylformamide is 2.5 mmoles/kg (Schindler et al., 1998), and that for 3-butylthiolane 1-oxide is 1.4 mmoles/kg (Chadha et al., 1983). Toxicity was not observed for the doses used in the metabolic studies (Porter et al., 1976; Delmas et al., 1983). The effectiveness of inhibition will also depend upon the rates of metabolism of the inhibitors. Thiolane 1-oxide was eliminated in rats with first-order kinetics with a half-time of 4.5 h, and isovaleramide was eliminated with zero order kinetics, with a rate of 1.1 mmol kg^{-1} h^{-1} (Chadha et al., 1983). It appears that these compounds would be useful inhibitors of alcohol metabolism. They have already been used to show that alcohol dehydrogenase is a rate-limiting factor in the elimination of ethanol (Plapp et al., 1984).

ACKNOWLEDGMENTS

This work was supported by United States Public Health Service Grant AA00279. We thank Dr. Thomas D. Hurley at the Indiana University Medical School, Indianapolis, and Dr. Jan-Olov Höög, Karolinska Institutet, Stockholm, for the expression vectors for the human alcohol dehydrogenases.

REFERENCES

Chadha, V. K., Leidal, K. G., and Plapp, B. V. (1983) Inhibition by carboxamides and sulfoxides of liver alcohol dehydrogenase and ethanol metabolism, *J. Med. Chem.* **26**, 916–922.

Chadha, V. K., Leidal, K. G., and Plapp, B. V. (1985) Inhibition of liver alcohol dehydrogenase and ethanol metabolism by 3-substituted thiolane 1-oxides, *J. Med. Chem.* **28**, 36–40.

Cho, H., Ramaswamy, S., and Plapp, B. V. (1997) Flexibility of liver alcohol dehydrogenase in stereoselective binding of 3-butylthiolane 1-oxides, *Biochemistry* **36**, 382–389.

Cho, H., and Plapp, B. V. (1998) Specificity of alcohol dehydrogenases for sulfoxides, *Biochemistry* **37**, 4482–4489.

Cleland, W. W. (1979) Statistical analysis of enzyme kinetic data, *Methods Enzymol.* **63**, 103–138.

Delmas, C., de Saint Blanquat, G., Freudenreich, C., and Biellmann, J.-F. (1983) New inhibitors of alcohol dehydrogenase: Studies in vivo and in vitro in the rat, *Alcohol. Clin. Exp. Res.* **7**, 264–270.

Edenberg, H. J., and Bosron, W. F. (1997) Alcohol Dehydrogenases, in *Comprehensive Toxicology, Vol. 3 Biotransformation*, Guengerich, F. P., Ed., Pergamon Press, New York, pp. 119–131.

Freudenreich, C., Samama, J.-P., and Biellmann, J.-F. (1984) Design of inhibitors from the three-dimensional structure of alcohol dehydrogenase. Chemical synthesis and enzymatic properties, *J. Am. Chem. Soc.* **106**, 3344–3353.

Hurley, T. D., Bosron, W. F., Stone, C. L., and Amzel, L. A. (1994) Structures of three human β alcohol dehydrogenase variants. Correlations with their functional differences, *J. Mol. Biol.* **239**, 415–429.

Hurley, T. D., Steinmetz, C. G., Xie, P., and Yang, Z.-N. (1997) Three-dimensional structures of human alcohol dehydrogenase enzymes reveal the molecular basis for their functional diversity, in *Enzymology and Molecular Biology of Carbonyl Metabolism 6*, Weiner, H., Lindahl, R., Crabb, D. W., Flynn, T. G., Eds., Plenum Press, New York, pp. 291–302.

Jacobsen, D., and McMartin, K. E. (1997) Antidotes for methanol and ethylene glycol poisoning, *Clinical Toxicology* **35**, 127–143.

Jones, T. A., Zou, J. Y., Cowan, S. W., and Kjeldgaard, M. (1991) Improved methods for binding protein models in electron density maps and the location of errors in these models, *Acta Crystallogr.* **A47**, 110–119.

Kedishvili, N. Y., Bosron, W. F., Stone, C. L., Hurley, T. D., Peggs, C. F., Thomasson, H. R., Popov, K. M., Carr, L. G., Edenberg, H. J., and Li, T.-K. (1995) Expression and kinetic characterization of recombinant human stomach alcohol dehydrogenase. Active-site amino acid sequence explains substrate specificity compared with liver enzymes, *J. Biol. Chem.* **270**, 3625–3630.

Lester, D., and Benson, G. D. (1970) Alcohol oxidation in rats inhibited by pyrazole, oximes and amides, *Science* **169**, 282–284.

Plapp, B. V., Leidal, K. G., Smith, R. K., and Murch, B. P. (1984) Kinetics of inhibition of ethanol metabolism in rats and the rate-limiting role of alcohol dehydrogenase, *Arch. Biochem. Biophys.* **230**, 30–38.

Porter, C. C., Titus, D. C., and DeFelice, M. J. (1976) Liver alcohol dehydrogenase inhibition by fatty acid amides, *N*-alkylformamides and monoalkylureas, *Life Sciences* **18**, 953–959.

Ramaswamy, S., Scholze, M., and Plapp, B. V. (1997) Binding of formamides to liver alcohol dehydrogenase, *Biochemistry* **36**, 3522–3527.

Ramaswamy, S. (1998) Dynamics in alcohol dehydrogenase elucidated from crystallographic investigations, *this volume.*

Satre, M. A., •gombiæ-Knight, M., and Duester, G. (1994) The complete structure of human class IV alcohol dehydrogenase (retinol dehydrogenase) determined from the ADH7 gene, *J. Biol. Chem.* **269**, 15606–15612.

Schindler, J. F., Berst, K. B., and Plapp, B. V. (1998) Inhibition of human alcohol dehydrogenases by formamides, *J. Med. Chem.* **41**, 1696–1701.

Shearer, G. L., Kim, K., Lee, K. M., Wang, C. K., and Plapp, B. V. (1993) Alternative pathways and reactions of benzyl alcohol and benzaldehyde with horse liver alcohol dehydrogenase, *Biochemistry* **32**, 11186–11194.

Stone, C. L., Bosron, W. F., and Dunn, M. F. (1993) Amino acid substitutions at position 47 of human $\beta_1\beta_1$ and $\beta_2\beta_2$ alcohol dehydrogenase affect hydride transfer and coenzyme dissociation rate constants, *J. Biol. Chem.* **268**, 892–899.

Sun, H.-W., and Plapp, B. V. (1992) Progressive sequence alignment and molecular evolution of the Zn-containing alcohol dehydrogenase family, *J. Mol. Evol.* **34**, 522–535.

Winer, A. D., and Theorell, H. (1960) Dissociation constants of ternary complexes of fatty acids and fatty acid amides with horse liver alcohol dehydrogenase-coenzyme complexes, *Acta Chem. Scand.* **14**, 1729–1742.

Xie, P., Parsons, S. H., Speckhard, D. C., Bosron, W. F., and Hurley, T. D. (1997) X-ray structure for human class IV σσ alcohol dehydrogenase. Structural basis for substrate specificity, *J. Biol. Chem.* **272**, 18558–18563.

37

TANDEM MASS SPECTROMETRY OF ALCOHOL DEHYDROGENASE AND RELATED BIOMOLECULES

William J. Griffiths

Department of Medical Biochemistry and Biophysics
Karolinska Institutet
S-17177 Stockholm, Sweden

1. INTRODUCTION

In the last ten years it has become possible to routinely analyse molecules of the size of alcohol dehydrogenase (ADH) by mass spectrometry. This has come about primarily as a result of the introduction of two new ionisation methods, namely matrix assisted laser desorption ionisation (MALDI) (Karas and Hillenkamp, 1988) and electrospray (ES).

Since the introduction of ES (Yamashita and Fenn, 1984) as a method of forming gas phase ions from solution, it has become possible using mass spectrometry to analyse a wide range of biomolecules. Both proteins and peptides are readily ionised by ES, giving spectra which characteristically show a coherent series of peaks, where successive peaks differ in their charge state by one unit (Fenn et al., 1989). Equine liver ADH was one of the first large molecules analysed by ES (Meng et al., 1988). The spectrum was recorded under denaturing conditions and a series of peaks ranging from $[M+46H]^{46+}$ to $[M+32H]^{32+}$ was observed. Yeast ADH is a molecule which has been studied using ES on a wide range of mass spectrometers, including Fourier transform ion cyclotron resonance (Speir et al., 1995), quadrupole and magnetic sector (Loo, 1995) instruments. In each of these studies molecular weights were determined with a reasonable degree of accuracy (M_r 36774 Da from sequence, e.g. 36771 ± 6 Da in experiment (Loo, 1995)), but higher mass components were always noted. Loo has investigated multimeric complexes of ADH. He observed the non-covalently bound dimeric subunit complex of equine liver ADH (M_r ~ 80000 Da) and the tetrameric complex of yeast ADH (M_r ~ 147000 Da) by recording ES spectra under non-denaturing conditions. Under these conditions dimeric yeast ADH complexes were not observed, consistent with the solution-phase characteristics of the protein. However, under acidic conditions abundant dimer species were observed, in agreement with the observation of other workers (Kelly et al., 1991).

Enzymology and Molecular Biology of Carbonyl Metabolism 7
edited by Weiner *et al.* Kluwer Academic / Plenum Publishers, New York, 1999.

Mass spectrometry is now being used routinely in the identification of proteins, particularly those obtained from polyacrylamide gels, via data base search methods. In 1993, five laboratories independently developed computer algorithms to use peptide mass maps to search data-bases and so identify proteins (Henzel et al., 1993, Pappin et al., 1993, Yates et al., 1993, James et al., 1993, Mann et al., 1993). A mass map is the mass spectrum of an enzymatic digest of a protein. Provided a sufficient number of peptide ions are observed in the mass spectrum, and the protein is not heavily modified, and that there are not more than two proteins present, a match can generally be found. Tandem mass spectrometry (MS/MS) data is now also being used with database search methods to identify proteins (Yates et al., 1996).

In a MS/MS experiment, the ion of interest is selected by the first mass analysis section (MS1) of a tandem mass spectrometer, induced to fragment, usually by collision with an inert gas, and the resultant fragment ions are mass analysed by a second mass analyser (MS2). From the product ion spectrum obtained, the structure of the precursor ion can be determined. The ion of interest may, for example, be a protonated peptide eluting from a high performance liquid chromatography column separation of an enzymatic digest, or alternatively be a protonated peptide ion from a crude unseparated digest. In either case MS1 selects the ion of interest so that it is selectively fragmented and identified. The fragment ion spectrum can either be interpreted manually (which, although time consuming, is not difficult for the experienced worker) or fed into a data base searching programme. If the protein from which the peptide is derived is in the data base a sequence will be suggested.

With the advent of low flow-rate ES (Gale and Smith, 1993), micro-ES (Emmett and Capriloi, 1994), and particularly nano-ES (Wilm and Mann, 1994) mass spectrometry has become the most sensitive method for sequencing peptides and identifying proteins via data base searches.

2. EXPERIMENTAL

Bakers yeast ADH in citrate buffer was from Sigma-Aldrich Sweden AB. Recombinant rat sorbitol dehydogenase (SDH) was a kind gift from Christina Kaiser.

All spectra were recorded on an AutoSpec-OATOFFPD (Micromass, Manchester, UK) high resolution tandem mass spectrometer. The instrument is fitted with an ES interface equipped with both nano-ES and micro-ES probes. In general sample spectra were recorded using the nano-ES probe, and calibration spectra with the micro-ES probe. Protein mass spectra were recorded over the *m/z* range 3000 (or 8000)–400 (1000) at a rate of 10 s/decade with an instrument resolution of 1500 (10 % valley definition). Spectra of enzymatic digests were recorded over the *m/z* range 2400–100 at a rate of 10 s/decade at an instrument resolution of 5000 (10 % valley definition). Tandem mass spectra were recorded by selecting the precursor ions of interest using the double focusing sectors of the instrument (MS1), focusing them into the fourth field free region collision cell containing xenon gas and pulsing the resultant fragment ions into the orthogonal acceleration time-of-flight (OATOF) mass analyser (MS2).

3. RESULTS AND DISCUSSION

Shown in Figure 1 is a nano-ES mass spectrum of an acidic aqueous solution (pH 2) of bakers yeast ADH which has been carboxymethylated. A broad range of charge states are observed ranging from $[M+25H]^{25+}$–$[M+45H]^{45+}$, this envelope is shifted to lower charge

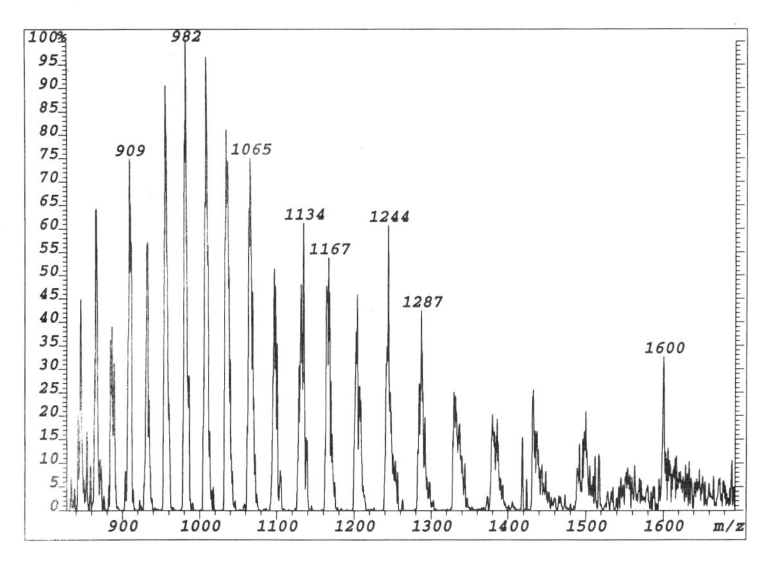

Figure 1. Nano-ES spectrum of carboxymethylated yeast ADH (5 pmol/μL in 20 % acetic acid, citrate buffer).

states by raising the pH of the solution. As can be deduced from the broadness of the peaks in the mass spectrum, the sample is heterogeneous. The major components of the mixture were found to have molecular weights of 37142 ± 8, 37172 ± 8, 37199 ± 8, 37228 ± 8 and 37286 ± 10 Da. The high error limits are a consequence of the heterogeneity of the sample. This may be a result of incomplete carboxymethylation, and/or, as others have noted (Loo, 1995; Speir et al., 1995) heterogeneity in Sigma-Aldrich preparations of yeast ADH.

Shown in Figures 2 and 3 is a nano-ES mass spectrum of rat SDH (2 pmol/μl) in an aqueous solution of 10 mM ammonium bicarbonate, 20 % acetic acid. Some non-covalent interactions are maintained under these condition. There is evidence for the presence of zinc bound SDH dimers (M_r 76295 ± 8, 76361 ± 8 Da) and tetramers (M_r ~152700), although monomers SDH (M_r 38047 + 8, 38109 ± 8, 38148 ± 10 Da) are most abundant. Although SDH in solution-phase is believed to be a tetramer, in acidic solution an abundant dimer species is apparently stable. The present results parallel those obtained by others (Loo, 1995; Kelly et al, 1991) who obtained evidence for abundant dimer species of yeast ADH in acidic solutions.

As was mentioned in the introduction, computer algorithms exist which use peptide mass maps to search protein data bases and so identify proteins. This is illustrated here for a trypsin digest of a yeast protein whose molecular weight is approximately 37 kDa. Shown in Figure 4 is an ES mass spectrum of the tryptic digest. Ten ions were initially selected and their monoisotopic masses fed into MS-Fit 2.0.4 search programme provided over the internet (http://prospector.ucsf.edu/cgi-bin/msfit.ex) by the UCSF Mass Spectrometry Facility. No species was specified and the complete data base (NCBInr.07.15.98) searched, allowing a peptide mass tolerance of ± 0.2 Da and three missed cleavages. The top hit was alcohol dehydrogenase 1 from Saccharomyces cerevisiar (Bakers yeast) with 6/10 matches corresponding to 26% coverage. A further 12 masses were fed into the programme resulting in 18/22 matches and an increased coverage of 52%.

To further confirm the identity of the digested protein a selection of "matched" peptides was analysed by tandem mass spectrometry. In each case the MS/MS spectrum recorded was compatible with the sequence provided by the data base search.

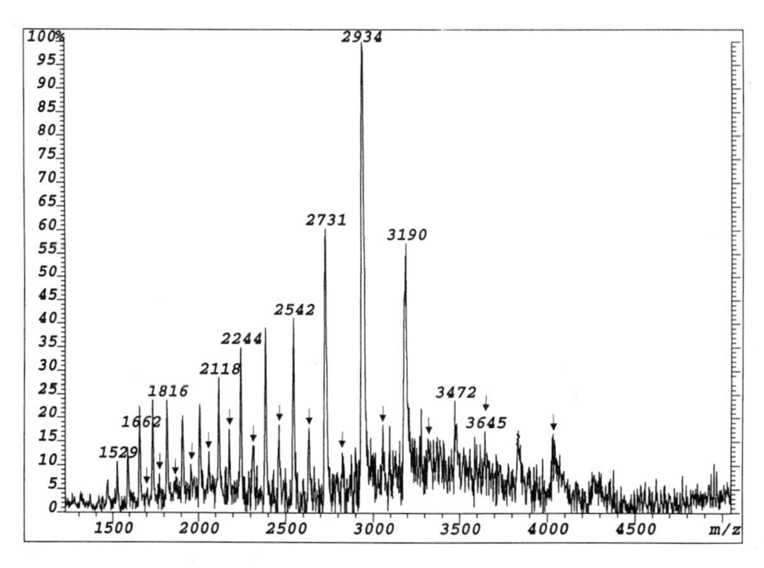

Figure 2. Nano-ES spectrum of rat SDH (2 pmol/μL in 10 mM ammonium bicarbonate, 20 % acetic acid). Single headed arrows represent odd charged state dimer ions.

4. CONCLUSION

Mass spectrometry has now become an accepted part of the protein chemist's armamentarium. Not only can peptide and protein molecular weights be determined accurately, but proteins can be identified by peptide mass mapping, and subunit and metal non-covalent interactions investigated.

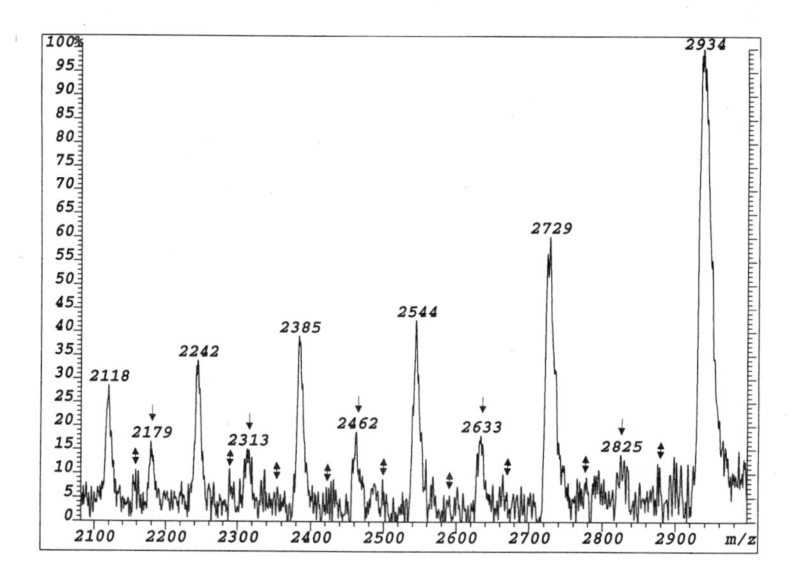

Figure 3. Close-up of the *m/z* region 2100–3000 of Fig. 2. Single headed arrows represent odd charged state dimer ions, double headed arrows represent odd charged state tetramer ions.

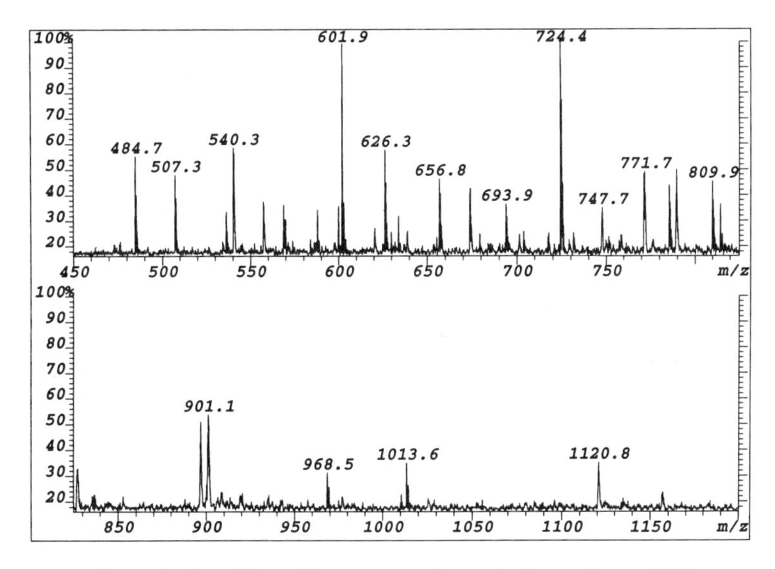

Figure 4. Nano-ES peptide mass map of a tryptic digest of yeast ADH.

ACKNOWLEDGMENTS

All members of the Karolinska Institute mass spectrometry group are warmly acknowledged for their co-operation. Professors Jan Sjövall and Hans Jörnvall are thanked for encouraging and maintaining mass spectrometry at the Institute. The service engineers of Micromass AB are thanked for their excellent work. This work was supported by the Swedish Medical Research Council (project no. K98-03X-12551-01A), the research funds of the Karolinska Institute and Stiftelsen Lars Hjertas Minne.

REFERENCES

Emmett, M. R. and Caprioli, R. M., 1994, Micro-electrospray mass spectrometry: ultra-high-sensitivity analysis of peptides and proteins. *J. Am. Soc. Mass Spectrom.*, **5**, 605–613.

Fenn, J. B., Mann, M., Meng, C. K., Wong, S. F. and Whitehouse, C. M., 1989, Electrospray ionization for mass spectrometry of large biomolecules. *Science*, **246**, 64–714.

Gale, D. C. and Smith, R. D., 1993, Small volume and low flow-rate electrospray ionization mass spectrometry of aqueous samples. *Rapid Commun. Mass Spectrom.*, **7**, 1017–1021.

James, P., Quadroni, M., Carafoli, E. and Gonnett, G., 1993, Protein identification by mass profile fingerprinting. *Biochem. Biophys. Res. Com.*, **195**, 5–8.

Karas, M. and Hillenkamp, F., 1988, Laser desorption ionization of proteins with molecular masses exceeding 10 000 daltons. *Anal. Chem.*, **60**, 1299–2301.

Kelly, M. A., Murphy, C. M. and Fenselau, C. C., 1991, Evaluation of a vestec electrospray source and a phrasor high energy dynode detector on a Hewlett Packard quadrupole mass spectrometer. In: *Proceedings of the 39th Annual Conference on Mass Spectrometry and Allied Topics, Nashville, TN, June 1991.* ASMS, East Lansing, MI, 136–137.

Loo, J. A., 1994, Observation of large subunit protein complexes by electrospray ionization mass spectrometry. *J. Mass Spectrom.*, **30**, 180 -183.

Mann, M., Højrup, P. and Roepstorff, P., 1993, Use of mass spectrometric information for the identification of proteins in sequence databases. *Biolog. Mass Spectrom.*, **22**, 338.

Meng, C. K., Mann, M. and Fenn, J. B., 1988, Electrospray ionization of some polypeptides and small proteins. In: *Proceedings of the 36th Annual Conference on Mass Spectrometry and Allied Topics, San Francisco, CA, June 1988.* ASMS, East Lansing, MI, 711–772.

Pappin, D. J. C., Højrup, P. and Bleasby, A. J., 1993, Rapid identification of proteins by peptide-mass fingerprinting. *Current Biology*, **3**, 327.

Speir, J. P., Senko, M. W., Little, D. P., Loo, J. A. and McLafferty, F. W., 1995, High resolution tandem mass spectra of 36–67 kDa proteins. *J. Mass Spectrom.*, **30**, 39–42.

Wilm, M. and Mann, M., 1996, Analytical properties of the nanoelectrospray ion source. *Anal. Chem.*, **68**,1–8.

Yamashita, M. and Fenn, J. B., 1984, Electrospray ion source. Another variation on the free-jet theme. *J. Phys. Chem.*, **88**, 4451–4459.

Yates, J. R., Speicher, S., Griffin, P. R. and Hunkapiller, T., 1993, Peptide mass maps: a highly informative approach to protein identification. *Anal. Biochem.*, **214**, 397–408.

Yates, J. R., Eng, J. K., Clauser, K. R. and Burlingame, A. L., 1996, Search of sequence data bases with uninterprited high energy collision-induced dissociation spectra of peptides. *J. Am. Soc. Mass Spectrom.*, **7**, 1089–1098.

38

FUNCTION OF ALCOHOL DEHYDROGENASE AND ALDEHYDE DEHYDROGENASE GENE FAMILIES IN RETINOID SIGNALING

Gregg Duester

Gene Regulation Program
Burnham Institute
10901 N. Torrey Pines Road
La Jolla, California 92037

1. INTRODUCTION

Vitamin A (retinol) must be metabolized to retinoic acid in order to fulfill its roles in vertebrate growth and development (Blaner and Olson, 1994; Maden, 1994; Hofmann and Eichele, 1994). Retinoic acid mediates retinoid signaling by binding to and modulating the transcriptional regulatory properties of nuclear retinoic acid receptors (Kastner et al., 1994; Mangelsdorf et al., 1994). Retinoic acid reporter assays of individual mouse embryos indicate that retinoic acid is undetectable during the initial stage of gastrulation, but that it is easily detectable later in gastrulation and neurulation (Rossant et al., 1991; Ang et al., 1996a). Retinoic acid receptors are expressed prior to detection of retinoic acid in mouse embryos (Dollé et al., 1990; Ruberte et al., 1991; Ang and Duester, 1997), indicating that a major factor governing retinoid signaling is the induction of endogenous retinoic acid synthesis during embryogenesis. To unravel this layer of biological control in the retinoid signaling pathway an understanding of the enzymes involved in retinoic acid synthesis is needed.

Enzymes able to metabolize retinol to retinoic acid *in vitro* have been identified as the same enzymes which metabolize ethanol to acetic acid (Figure 1). Retinol is first oxidized to an inactive intermediate, retinal, by members of the alcohol dehydrogenase (ADH) family, followed by oxidation of retinal to the active ligand retinoic acid by members of the aldehyde dehydrogenase (ALDH) family (Duester, 1996) Thus, it is likely that some members of these enzyme families have a degree of selectivity for retinoids that allows them to contribute to the production of retinoic acid. However, conclusive evidence that any of these enzymes function in retinoid metabolism *in vivo* has up until now been lacking.

Enzymology and Molecular Biology of Carbonyl Metabolism 7
edited by Weiner *et al.* Kluwer Academic / Plenum Publishers, New York, 1999.

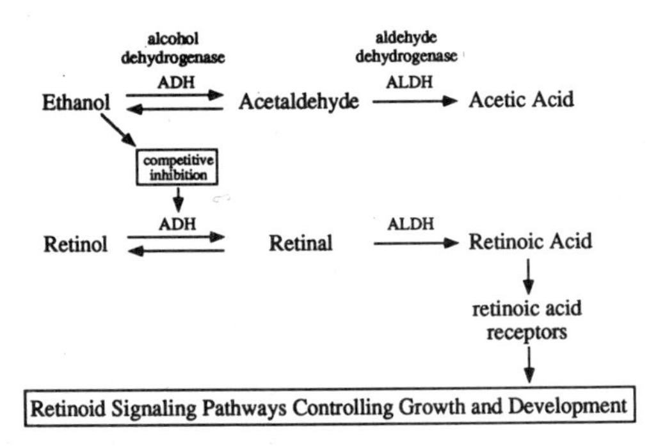

Figure 1. Enzymes involved in the metabolism of ethanol and retinol. Both ethanol and retinol can be oxidized *in vitro* by some forms of ADH. Likewise, both acetaldehyde and retinal can be oxidized *in vitro* by some forms of ALDH. The production of retinoic acid by these two enzyme families provides the ligand for nuclear retinoic acid receptors that control a retinoid signaling pathway involved in growth and development of vertebrate animals. Oxidation of retinol by ADH is competitively inhibited by ethanol leading to reduced retinoic acid synthesis; this forms the basis of an hypothesis for the mechanism of fetal alcohol syndrome (Deltour et al., 1996).

2. MOUSE ADH AND ALDH FAMILIES

In order to examine the potential physiological roles of ADH and ALDH in retinoid metabolism and retinoid signaling, studies in this laboratory have focused upon the mouse which is amenable to genetic and embryological studies. The presently known members of the mouse ADH and ALDH families are shown (Figure 2). Seven distinct classes of vertebrate ADH have been discovered, with five of these forms presently identified in the human and three forms identified in the mouse (Zgombic-Knight et al., 1995). Of the seven classes, four are known to function as retinol dehydrogenases *in vitro* (Duester, 1996). Among these, ADH1 (class I ADH) and ADH4 (class IV ADH) are conserved in all mammalian lines examined including the mouse, and both catalyze retinol oxidation *in vitro* (Connor and Smit, 1987; Boleda et al., 1993; Yang et al., 1994). A third form of ADH conserved in the mouse is class III ADH (ADH3) which does not utilize retinol as a substrate (Boleda et al., 1993; Yang et al., 1994).

The human ALDH family is composed of more than ten distinct classes, with five forms presently identified in the mouse, two of which function as retinal dehydrogenases *in vitro* (Yoshida et al., 1998). Cytosolic class I aldehyde dehydrogenase (ALDH1) from liver has long been known to have high activity for the oxidation of retinal to retinoic acid *in vitro* (Futterman, 1962; Lee et al., 1991; McCaffery et al., 1992). Mitochondrial class 2 ALDH (ALDH2) has been reported to have very weak activity for retinal oxidation (Dockham et al., 1992; Chen et al., 1994a). Cytosolic class 3 ALDH (ALDH3) found at high levels in the stomach (Vasiliou et al., 1995) as well as microsomal ALDH (ALDH10) found at high levels in the liver (Vasiliou et al., 1996) have been identified in the mouse, but neither of these enzymes has activity for retinal (Lee et al., 1991). An additional form of cytosolic ALDH called RALDH-2 in the mouse (Zhao et al., 1996), or RalDH(II) in the rat (Wang et al., 1996), or ALDH11 in the human (Yoshida et al., 1998), has high activity with retinal and is most closely related to ALDH1.

Studies on the expression patterns for retinoid-active ADHs and ALDHs (i.e. ADH1, ADH4, ALDH1, and ALDH11) provide further evidence supporting a function in retinoid metabolism. Using either immunohistochemistry or in situ hybridization, mouse ADH1

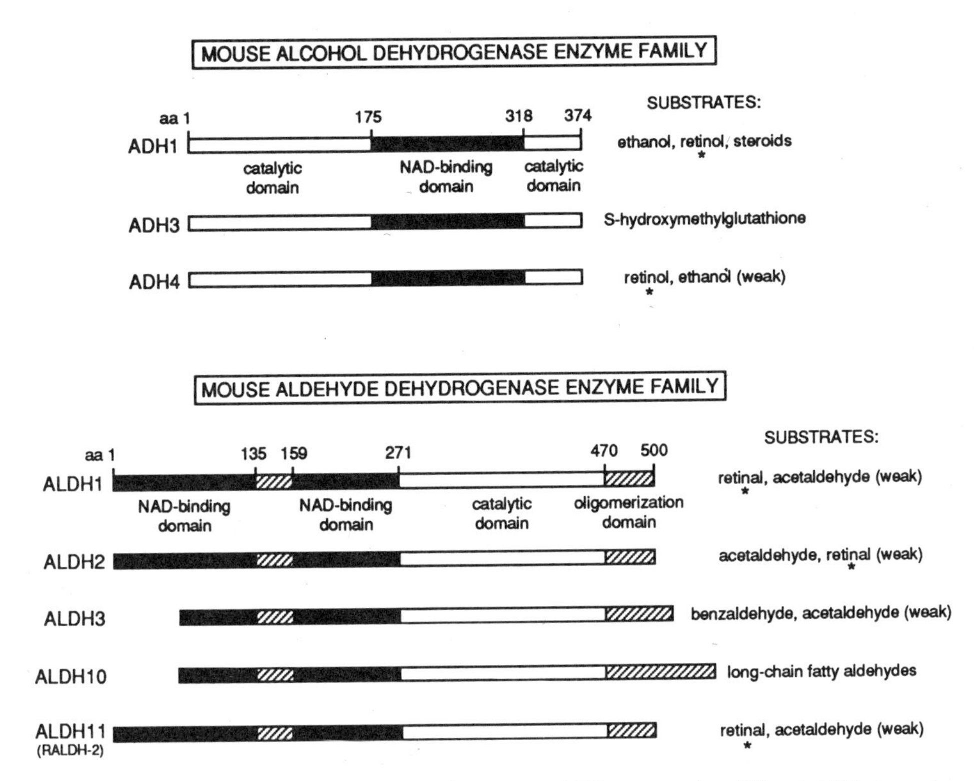

Figure 2. The mouse ADH and ALDH families. Three mouse ADHs representing different ADH classes have been identified. Crystal structures for horse and human ADHs indicate subunits of approximately 374 amino acid residues with the NAD-binding domain flanked on either side by residues contributing to the catalytic domain (Eklund et al., 1976; Hurley et al., 1991; Yang et al., 1997; Xie et al., 1997). Five mouse ALDHs are known with subunits ranging from approximately 450–500 amino acid residues in length. The crystal structures for rat and bovine ALDHs indicate a bipartite NAD-binding domain located on the N-terminal end, followed by the catalytic domain, and an oligomerization domain which has contribution from both the C-terminus as well as a region separating the two halves of the NAD-binding domain (Liu et al., 1997; Steinmetz et al., 1997). Some representative alcohol or aldehyde substrates are indicated for each form of ADH or ALDH.

and ADH4 have been localized in numerous retinoid-responsive epithelia including the epidermis (Haselbeck et al., 1997b), testis (Deltour et al., 1997), digestive tract (Haselbeck and Duester, 1997), genitourinary tract (Ang et al., 1996b), and brain/craniofacial tissues of neural tube stage embryos (Ang et al., 1996a; Ang and Duester, 1997). This positive correlation provides evidence that ADH1 and ADH4 may function to provide retinoic acid locally for retinoid signaling in these tissues. In addition, ADH1 and ADH4 and significant levels of retinoic acid are found in the adult and embryonic adrenal glands (Haselbeck et al., 1997a). Thus, another function of ADH may be to produce retinoic acid as an endocrine hormone in the adrenal gland. In adult tissues the expression pattern for mouse ALDH1 mimics that seen for ADH1 and ADH4 (unpublished data), and for mouse ALDH11 expression is observed mostly in the testis (Zhao et al., 1996) and other reproductive tract tissues (unpublished data). During mouse embryogenesis, the initial expression of both ALDH1 and ALDH11, as well as ADH4, correlates temporally with the initial detection of retinoic acid during gastrulation (Ang and Duester, 1997; Niederreither et al., 1997). Thus, at this time in development it appears that there is an induction of both en-

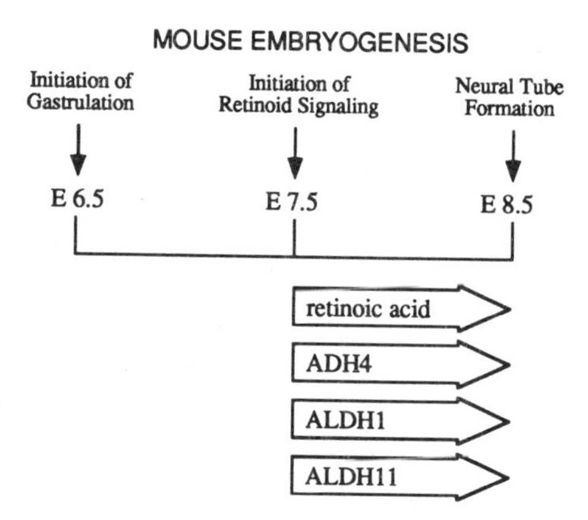

Figure 3. Retinoic acid synthesis during mouse embryogenesis. At 6.5 days of embryogenesis (E6.5) the mouse embryo begins to gastrulate, but no retinoic acid is yet detectable suggesting that retinoid signaling has not yet commenced. At E7.5 retinoic acid is now detectable, plus this is the stage when ADH4, ALDH1, and ALDH11 expression is first detectable. At stage E8.5 the formation of the neural tube and central nervous system is well underway, a process that requires retinoic acid in mammals (Dickman et al., 1997), birds (Maden et al., 1996), and amphibians (Blumberg et al., 1997).

zymes needed to convert retinol to retinoic acid. This provides circumstantial evidence that ADH and ALDH function in retinoid signaling *in vivo* (Figure 3).

3. USE OF *XENOPUS* TO STUDY RETINOIC ACID SYNTHETIC ENZYMES

Embryos of the frog *Xenopus laevis* have proven useful for studies on the retinoid signaling pathway. Retinoic acid and retinoid receptors are found in *Xenopus* embryos (Chen et al., 1994b; Creech Kraft et al., 1994a; Sharpe and Goldstone, 1997). Administration of excess retinoic acid to *Xenopus* embryos during gastrulation is highly teratogenic resulting in anteroposterior truncation and central nervous system defects (Durston et al., 1989; Sive et al., 1990). The vitamin A precursors retinol and retinal are both present in *Xenopus* eggs and embryos, with retinal being more abundant (Creech Kraft et al., 1994b; Blumberg et al., 1996). Retinal in *Xenopus* eggs exists as a Schiff base linkage between its reactive aldehyde group and the yolk storage protein lipovitellin 1 (Irie et al., 1991). Metabolism of retinol and retinal to retinoic acid has been demonstrated in *Xenopus* embryos (Kraft et al., 1995), and has been shown to be sensitive to the ALDH inhibitor citral (Schuh et al., 1993). Discussed below is the use of *Xenopus* embryos to examine the potential function of ALDH1 as a retinoic acid-synthetic enzyme *in vivo* (H.L. Ang and G. Duester, manuscript submitted).

In order to establish an assay for retinoic acid during *Xenopus* development, embryos at various stages were individually assayed for the presence of endogenous retinoic acid using the F9-RARE-*lacZ* reporter cell bioassay (Wagner et al., 1992). This assay monitors diffusion of retinoic acid from *Xenopus* embryo explants into the surrounding monolayer of reporter cells. The reporter cells are mouse F9 embryonal carcinoma cells stably transfected with an indicator gene composed of *lacZ* (encoding beta-galactosidase) controlled by a retinoic acid response element. Retinoic acid diffusing from the *Xenopus* embryos was detected by the induction of beta-galactosidase activity in the reporter cell monolayer at the site where the embryo was placed for incubation. Examination of the temporal appearance of endogenous retinoic acid during development demonstrated that unfertilized eggs had no detectable retinoic acid, and that following fertilization there was still no retinoic acid detection in embryos at the early cleavage stages, nor in morulas, stage 8 blastulas, or initial gastrulas at stage 10 (Table 1). Retinoic acid was first detected, albeit weakly, at stage 11 (large yolk plug gastrula) in 92% of the embryos, and subsequently a higher level of retinoic acid detection was observed in 87% of stage 12 embryos (medium yolk plug gastrula), and in 100% of embryos examined at stage 15 (neural fold) and thereafter (Table 1).

The above findings indicate that there exists a time window from fertilization through blastula stage 8 during which *Xenopus* embryos have no detectable endogenous retinoic acid and can thus be manipulated and assayed for effects on premature retinoic acid synthesis with no background detection. We took advantage of this property of *Xenopus* by performing overexpression studies in which embryos were injected with several different vertebrate ALDH1 mRNAs including the *Xenopus* homolog which was cloned. Embryos were microinjected at the 2–8 cell stages with mouse, chick, or *Xenopus* ALDH1 mRNAs then allowed to grow to stage 8 at which time they were subjected to the retinoic acid bioassay. High levels of retinoic acid were detected in ALDH1-injected stage 8 blastulas whereas control-injected *Xenopus* embryos had no detection at this stage (Table 2). This indicates that ALDH1 overexpression induces premature synthesis of retinoic acid *in vivo*. When ALDH1-injected embryos were allowed to develop to the tadpole stage most were observed to have undergone teratogenesis resembling that induced by retinoic acid treatment, i.e. truncation of the anteroposterior axis (data not shown). These studies provide the first evidence that ALDH is involved in retinoic acid synthesis and retinoid signaling *in vivo*. Also, this property is conserved in ALDH1 from amphibian to mammal, further indicating that retinoid metabolism is an important function for this enzyme.

Interestingly, injection of mouse ADH4 mRNA had no effect on retinoic acid detection either when injected alone or in combination with ALDH1 mRNA. This finding is

Table 1. Bioassay detection of endogenous retinoic acid in individual *Xenopus* embryos during development

Stage	% embryos with retinoic acid detection
Unfertilized egg	0
2-cell (stage 2)	0
8-cell (stage 4)	0
Morula (stage 6.5)	0
Blastula (stage 8)	0
Initial gastrula (stage 10)	0
Large yolk plug (stage 11)	92
Medium yolk plug (stage 12)	87
Neural fold (stage 15)	100
Neural groove (stage 18)	100
Neural tube (stage 19)	100

In this bioassay, embryos were incubated as explants on top of a monolayer of F9-RARE-*lacZ* reporter cells. Retinoic acid diffusing from the embryo to the reporter cells was detected *in situ* by induction of beta-galactosidase activity in the surrounding reporter cells. Embryos did not develop any farther once placed on the reporter cells.

Table 2. Premature retinoic acid synthesis induced by overexpression of ALDH1 in *Xenopus*

Injection	% embryos with retinoic acid detection
Mouse ALDH1 mRNA	69
Mouse ALDH1 mRNA + Ro 41-5253	0
Chick ALDH1 mRNA	82
Chick ALDH1 mRNA + Ro 41-5253	0
Xenopus ALDH1 mRNA	100
Xenopus ALDH1 mRNA + Ro 41-5253	0
Mouse ADH4 mRNA	0
Water-injected	0
Uninjected	0

Embryos were injected at the 2-8 cell stage, then incubated until they reached stage 8 (blastula). At this stage they were assayed for retinoic acid detection using the F9-RARE-*lacZ* bioassay in the presence or absence of the retinoic acid receptor inhibitor Ro 41-5253.

more understandable when one considers that large amounts of the substrate retinal are present in the yolk of *Xenopus* eggs, as discussed above, indicating that there is no need to convert retinol to retinal during the early stages of *Xenopus* development. In *Xenopus* embryos the rate-limiting step for retinoic acid synthesis is apparently conversion of preexisting retinal to retinoic acid as opposed to conversion of retinol to retinal which occurs in mammalian embryos (Chen et al., 1995). A role for ADH in *Xenopus* retinoic acid synthesis is still suspected, however, based upon the recent identification of the *Xenopus* homologs of ADH1 and ADH4, and the discovery that ADH1 is highly expressed in the maternal liver where retinol is converted to retinal for linkage to vitellogenin (I. Hoffmann, H. L. Ang, and G. Duester, manuscript submitted).

4. CONCLUSIONS

Analysis of ADHs and ALDHs in vertebrate embryos has provided evidence that both of these enzyme families participate in retinoid metabolism to produce the active ligand retinoic acid needed for retinoid signaling pathways controlling growth and development. Some key findings are summarized below:

1. Retinoic acid is first detectable in vertebrate embryonic development just after the initiation of gastrulation. This is true in *Xenopus* embryos (described herein), chick embryos (unpublished data), as well as mouse embryos (Rossant et al., 1991; Ang et al., 1996a). In quail embryos subjected to vitamin A deficiency, embryogenesis appears to proceed normally up until gastrulation, but is then followed by defects in the cardiovascular and central nervous systems (Maden et al., 1996). Blockage of retinoic acid receptor activity in *Xenopus* embryos leads to central nervous system defects (Blumberg et al., 1997). This suggests that retinoid signaling does not play a role in very early embryonic development, but becomes important just after gastrulation initiates when several tissues begin to differentiate.
2. ADHs and ALDHs can metabolize retinoids *in vitro*, and both are expressed during embryonic development. Expression of mouse ADH4 (Ang et al., 1996a),

mouse ALDH1 (Ang and Duester, 1997), and mouse ALDH11 (Niederreither et al., 1997) coincides with the first detection of retinoic acid during embryogenesis. At the beginning of gastrulation during stage E6.5 of mouse embryogenesis, all three of these enzymes and retinoic acid are absent, but at stage E7.5 the mRNAs for all three of these enzymes are detectable and retinoic acid is now observed. This provides circumstantial evidence that the enzymes cooperate to catalyze both steps of retinoic acid synthesis *in vivo*.

3. *Xenopus* appears to be an excellent system to examine the role of ALDH in retinoic acid synthesis since this organism stores large amounts of retinal in the egg (Irie et al., 1991), and does not begin endogenous conversion of retinal to retinoic acid until after gastrulation begins (described herein). ALDH1 overexpression in *Xenopus* embryos results in premature induction of embryonic retinoic acid synthesis, and this function is conserved in amphibian, avian, and mammalian forms of the enzyme (described herein). This is the first example of a functional assay demonstrating that ALDH participates in retinoic acid synthesis *in vivo*.

ACKNOWLEDGMENTS

Thanks to H.L. Ang, L. Deltour, M. Foglio, R.J. Haselbeck, and I. Hoffmann for stimulating discussions. This work was supported by NIH grants AA07261 and AA09731.

REFERENCES

Ang, H.L., Deltour, L., Hayamizu, T.F., Zgombic-Knight, M., and Duester, G. (1996a) Retinoic acid synthesis in mouse embryos during gastrulation and craniofacial development linked to class IV alcohol dehydrogenase gene expression, *J. Biol. Chem.* **271**, 9526–9534.

Ang, H.L., Deltour, L., Zgombic-Knight, M., Wagner, M.A., and Duester, G. (1996b) Expression patterns of class I and class IV alcohol dehydrogenase genes in developing epithelia suggest a role for alcohol dehydrogenase in local retinoic acid synthesis, *Alcohol. Clin. Exp. Res.* **20**, 1050–1064.

Ang, H.L. and Duester, G. (1997) Initiation of retinoid signaling in primitive streak mouse embryos: Spatiotemporal expression patterns of receptors and metabolic enzymes for ligand synthesis, *Dev. Dyn.* **208**, 536–543.

Blaner, W.S. and Olson, J.A. (1994) Retinol and retinoic acid metabolism, in "*The Retinoids: Biology, Chemistry, and Medicine, 2nd edition*," M.B. Sporn, A.B. Roberts, and D.S. Goodman, eds., Raven Press, Ltd., New York, p. 229.

Blumberg, B., Bolado, J.,Jr., Derguini, F., Craig, A.G., Moreno, T.A., Chakravarti, D., Heyman, R.A., Buck, J., and Evans, R.M. (1996) Novel retinoic acid receptor ligands in *Xenopus* embryos, *Proc. Natl. Acad. Sci. USA* **93**, 4873–4878.

Blumberg, B., Bolado, J.,Jr., Moreno, T.A., Kintner, C., Evans, R.M., and Papalopulu, N. (1997) An essential role for retinoid signaling in anteroposterior neural patterning, *Development* **124**, 373–379.

Boleda, M.D., Saubi, N., Farrés, J., and Parés, X. (1993) Physiological substrates for rat alcohol dehydrogenase classes: Aldehydes of lipid peroxidation, omega-hydroxyfatty acids, and retinoids, *Arch. Biochem. Biophys.* **307**, 85–90.

Chen, H., Namkung, M.J., and Juchau, M.R. (1995) Biotransformation of all-*trans*-retinol and all-*trans*-retinal to all-trans-retinoic acid in rat conceptal homogenates, *Biochem. Pharmacol.* **50**, 1257–1264.

Chen, M., Achkar, C., and Gudas, L.J. (1994a) Enzymatic conversion of retinaldehyde to retinoic acid by cloned murine cytosolic and mitochondrial aldehyde dehydrogenases, *Mol. Pharmacol.* **46**, 88–96.

Chen, Y., Huang, L., and Solursh, M. (1994b) A concentration gradient of retinoids in the early *Xenopus laevis* embryo, *Dev. Biol.* **161**, 70–76.

Connor, M.J. and Smit, M.H. (1987) Terminal-group oxidation of retinol by mouse epidermis: Inhibition *in vitro* and *in vivo*, *Biochem. J.* **244**, 489–492.

Creech Kraft, J., Schuh, T., Juchau, M., and Kimelman, D. (1994a) The retinoid X receptor ligand, 9-*cis*-retinoic acid, is a potential regulator of early *Xenopus* development, *Proc. Natl. Acad. Sci. USA* **91**, 3067–3071.

Creech Kraft, J., Schuh, T., Juchau, M.R., and Kimelman, D. (1994b) Temporal distribution, localization and metabolism of all-*trans*-retinol, didehydroretinol and all-*trans*-retinal during *Xenopus* development, *Biochem. J.* **301**, 111–119.

Deltour, L., Ang, H.L., and Duester, G. (1996) Ethanol inhibition of retinoic acid synthesis as a potential mechanism for fetal alcohol syndrome, *FASEB J.* **10**, 1050–1057.

Deltour, L., Haselbeck, R.J., Ang, H.L., and Duester, G. (1997) Localization of class I and class IV alcohol dehydrogenases in mouse testis and epididymis: Potential retinol dehydrogenases for endogenous retinoic acid synthesis, *Biol. Reprod.* **56**, 102–109.

Dickman, E.D., Thaller, C., and Smith, S.M. (1997) Temporally-regulated retinoic acid depletion produces specific neural crest, ocular and nervous system defects, *Development* **124**, 3111–3121.

Dockham, P.A., Lee, M.-O., and Sladek, N.E. (1992) Identification of human liver aldehyde dehydrogenases that catalyze the oxidation of aldophosphamide and retinaldehyde, *Biochem. Pharmacol.* **43**, 2453–2469.

Dollé, P., Ruberte, E., Leroy, P., Morriss-Kay, G., and Chambon, P. (1990) Retinoic acid receptors and cellular retinoid binding proteins. I. A systematic study of their differential pattern of transcription during mouse organogenesis, *Development* **110**, 1133–1151.

Duester, G. (1996) Involvement of alcohol dehydrogenase, short-chain dehydrogenase/reductase, aldehyde dehydrogenase, and cytochrome P450 in the control of retinoid signaling by activation of retinoic acid synthesis, *Biochemistry* **35**, 12221–12227.

Durston, A.J., Timmermans, J.P.M., Hage, W.J., Hendriks, H.F.J., De Vries, N.J., Heideveld, M., and Nieuwkoop, P.D. (1989) Retinoic acid causes an anteroposterior transformation in the developing central nervous system, *Nature* **340**, 140–144.

Eklund, H., Nordstrom, B., Zeppezauer, E., Soderlund, G., Ohlsson, I., Boiwe, T., Soderberg, B.-O., Tapia, O., and Branden, C.-I. (1976) Three-dimensional structure of horse liver alcohol dehydrogenase at 2.4 Å resolution, *J. Mol. Biol.* **102**, 27–59.

Futterman, S. (1962) Enzymatic oxidation of vitamin A aldehyde to vitamin A acid, *J. Biol. Chem.* **237**, 677–680.

Haselbeck, R.J., Ang, H.L., Deltour, L., and Duester, G. (1997a) Retinoic acid and alcohol/retinol dehydrogenase in the mouse adrenal gland: A potential endocrine source of retinoic acid during development, *Endocrinology* **138**, 3035–3041.

Haselbeck, R.J., Ang, H.L., and Duester, G. (1997b) Class IV alcohol/retinol dehydrogenase localization in epidermal basal layer: Potential site of retinoic acid synthesis during skin development, *Dev. Dyn.* **208**, 447–453.

Haselbeck, R.J. and Duester, G. (1997) Regional restriction of alcohol/retinol dehydrogenases along the mouse gastrointestinal epithelium, *Alcohol. Clin. Exp. Res.* **21**, 1484–1490.

Hofmann, C. and Eichele, G. (1994) Retinoids in development, in "*The Retinoids: Biology, Chemistry, and Medicine, 2nd Edition*," M.B. Sporn, A.B. Roberts, and D.S. Goodman, eds., Raven Press, Ltd., New York, p. 387.

Hurley, T.D., Bosron, W.F., Hamilton, J.A., and Amzel, L.M. (1991) Structure of human beta1 alcohol dehydrogenase: catalytic effects of non-active-site substitutions, *Proc. Natl. Acad. Sci. USA* **88**, 8149–8153.

Irie, T., Azuma, M., and Seki, T. (1991) The retinal and 3-dehydroretinal in *Xenopus laevis* eggs are bound to lipovitellin 1 by a Schiff base linkage, *Zool. Sci.* **8**, 855–863.

Kastner, P., Chambon, P., and Leid, M. (1994) Role of nuclear retinoic acid receptors in the regulation of gene expression, in "*Vitamin A in Health and Disease*," R. Blomhoff, ed.,Marcel Dekker, Inc., New York, p. 189.

Kraft, J.C., Kimelman, D., and Juchau, M.R. (1995) *Xenopus laevis*: A model system for the study of embryonic retinoid metabolism: I. Embryonic metabolism of 9-*cis*- and all-*trans*-retinals and retinols to their corresponding acid forms, *Drug Metab. Dispos.* **23**, 72–82.

Lee, M.-O., Manthey, C.L., and Sladek, N.E. (1991) Identification of mouse liver aldehyde dehydrogenases that catalyze the oxidation of retinaldehyde to retinoic acid, *Biochem. Pharmacol.* **42**, 1279–1285.

Liu, Z.J., Sun, Y.J., Rose, J., Chung, Y.J., Hsiao, C.D., Chang, W.R., Kuo, I., Perozich, J., Lindahl, R., Hempel, J., and Wang, B.C. (1997) The first structure of an aldehyde dehydrogenase reveals novel interactions between NAD and the Rossmann fold, *Nature Struct. Biol.* **4**, 317–326.

Maden, M. (1994) Role of retinoids in embryonic development, in "*Vitamin A in Health and Disease*," R. Blomhoff, ed.,Marcel Dekker, Inc., New York, p. 289.

Maden, M., Gale, E., Kostetskii, I., and Zile, M.H. (1996) Vitamin A-deficient quail embryos have half a hindbrain and other neural defects, *Curr. Biol.* **6**, 417–426.

Mangelsdorf, D.J., Umesono, K., and Evans, R.M. (1994) The retinoid receptors, in "*The Retinoids: Biology, Chemistry, and Medicine, 2nd Edition*," M.B. Sporn, A.B. Roberts, and D.S. Goodman, eds., Raven Press, Ltd., New York, p. 319.

McCaffery, P., Lee, M.-O., Wagner, M.A., Sladek, N.E., and Dräger, U.C. (1992) Asymmetrical retinoic acid synthesis in the dorsoventral axis of the retina, *Development* **115**, 371–382.

Niederreither, K., McCaffery, P., Dräger, U.C., Chambon, P., and Dollé, P. (1997) Restricted expression and retinoic acid-induced downregulation of the retinaldehyde dehydrogenase type 2 (RALDH-2) gene during mouse development, *Mech. Dev.* **62**, 67–78.

Rossant, J., Zirngibl, R., Cado, D., Shago, M., and Giguère, V. (1991) Expression of a retinoic acid response element-*hsplacZ* transgene defines specific domains of transcriptional activity during mouse embryogenesis, *Genes Dev.* **5**, 1333–1344.

Ruberte, E., Dolle, P., Chambon, P., and Morriss-Kay, G. (1991) Retinoic acid receptors and cellular retinoid binding proteins. II. Their differential pattern of transcription during early morphogenesis in mouse embryos, *Development* **111**, 45–60.

Schuh, T.J., Hall, B.L., Creech Kraft, J., Privalsky, M.L., and Kimelman, D. (1993) v-erbA and citral reduce the teratogenic effects of all-*trans* retinoic acid and retinol, respectively, in *Xenopus* embryogenesis, *Development* **119**, 785–798.

Sharpe, C.R. and Goldstone, K. (1997) Retinoid receptors promote primary neurogenesis in *Xenopus*, *Development* **124**, 515–523.

Sive, H.L., Draper, B.W., Harland, R.M., and Weintraub, H. (1990) Identification of a retinoic acid-sensitive period during primary axis formation in *Xenopus laevis*, *Genes Dev.* **4**, 932–942.

Steinmetz, C.G., Xie, P.G., Weiner, H., and Hurley, T.D. (1997) Structure of mitochondrial aldehyde dehydrogenase: The genetic component of ethanol aversion, *Structure* **5**, 701–711.

Vasiliou, V., Reuter, S.F., Kozak, C.A., and Nebert, D.W. (1995) Mouse class 3 aldehyde dehydrogenases, *Adv. Exp. Med. Biol.* **372**, 151–158.

Vasiliou, V., Kozak, C.A., Lindahl, R., and Nebert, D.W. (1996) Mouse microsomal class 3 aldehyde dehydrogenase: AHD3 cDNA sequence, inducibility by dioxin and clofibrate, and genetic mapping, *DNA Cell Biol.* **15**, 235–245.

Wagner, M., Han, B., and Jessell, T.M. (1992) Regional differences in retinoid release from embryonic neural tissue detected by an in vitro reporter assay, *Development* **116**, 55–66.

Wang, X.S., Penzes, P., and Napoli, J.L. (1996) Cloning of a cDNA encoding an aldehyde dehydrogenase and its expression in *Escherichia coli* - Recognition of retinal as substrate, *J. Biol. Chem.* **271**, 16288–16293.

Xie, P.G., Parsons, S.H., Speckhard, D.C., Bosron, W.F., and Hurley, T.D. (1997) X-ray structure of human class IV sigma alcohol dehydrogenase - Structural basis for substrate specificity, *J. Biol. Chem.* **272**, 18558–18563.

Yang, Z.-N., Davis, G.J., Hurley, T.D., Stone, C.L., Li, T.-K., and Bosron, W.F. (1994) Catalytic efficiency of human alcohol dehydrogenases for retinol oxidation and retinal reduction, *Alcohol. Clin. Exp. Res.* **18**, 587–591.

Yang, Z.N., Bosron, W.F., and Hurley, T.D. (1997) Structure of human chi alcohol dehydrogenase: A glutathione-dependent formaldehyde dehydrogenase, *J. Mol. Biol.* **265**, 330–343.

Yoshida, A., Rzhetsky, A., Hsu, L.C., and Chang, C. (1998) Human aldehyde dehydrogenase gene family, *Eur. J. Biochem.* **251**, 549–557.

Zgombic-Knight, M., Ang, H.L., Foglio, M.H., and Duester, G. (1995) Cloning of the mouse class IV alcohol dehydrogenase (retinol dehydrogenase) cDNA and tissue-specific expression patterns of the murine ADH gene family, *J. Biol. Chem.* **270**, 10868–10877.

Zhao, D., McCaffery, P., Ivins, K.J., Neve, R.L., Hogan, P., Chin, W.W., and Dräger, U.C. (1996) Molecular identification of a major retinoic-acid-synthesizing enzyme, a retinaldehyde-specific dehydrogenase, *Eur. J. Biochem.* **240**, 15–22.

PRIMARY ROLE OF ALCOHOL DEHYDROGENASE PATHWAY IN ACUTE ETHANOL-INDUCED IMPAIRMENT OF PROTEIN KINASE C-DEPENDENT SIGNALING SYSTEM

Cinzia Domenicotti,[1] Dimitri Paola,[1] Antonella Vitali,[1] Mariapaola Nitti,[1] Damiano Cottalasso,[1] Giuseppe Poli,[2] Maria Adelaide Pronzato,[1] and Umberto Maria Marinari[1]

[1]Department of Experimental Medicine
Section of General Pathology
University of Genoa
via L. B. Alberti 2, 16132 Genoa, Italy
[2]Department of Experimental Medicine and Oncology
Section of General Pathology
University of Turin
C.so Raffaello 30, 10125 Turin, Italy

1. INTRODUCTION

Ethanol is primarily metabolized by oxidation and the three enzymatic systems currently considered to contribute to alcohol metabolism are alcohol dehydrogenase (ADH), cytochrome P-450 and peroxisomal catalase (Lieber et al., 1970; Thurman et al., 1972). It is generally accepted that in many species alcohol oxidation is primarily catalyzed by ADH and that accounts for over 60% of ethanol metabolism. The contribution of other metabolic pathways to ethanol metabolism has been estimated by utilizing specific inhibitors (in particular 4-methylpyrazole, an inhibitor of ADH).

Moreover, the experimental use of ADH deficient deer mice have represented an useful model to evaluate the role of different ethanol metabolic pathways and to investigate the effect of 4-methylpyrazole (4MP) in the absence of ADH (Handler et al., 1988). 4MP administered to deer mice lacking hepatic cytosolic ADH has been found to act on catalase-dependent alcohol metabolism by limiting the production of hydrogen peroxide (H_2O_2).

Enzymology and Molecular Biology of Carbonyl Metabolism 7
edited by Weiner *et al.* Kluwer Academic / Plenum Publishers, New York, 1999.

Bradford and collaborators (1993) have demonstrated that in the absence of ADH, the effect of 4MP is due to competitive inhibition of fatty-acyl CoA-synthetase, resulting in a reduced supply of fatty acids for peroxisomal oxidation of alcohols (methanol, ethanol) by catalase-H_2O_2. Therefore 4MP is not a specific inhibitor of ADH but it is also able to inhibit the catalase pathway acting on fatty-acyl CoA-synthetase.

Ethanol both directly and indirectly by its metabolism affects the intracellular signal transduction pathways (Hoek et al., 1990; Hoek et al., 1992). A direct effect of alcohol on protein kinase C (PKC), an enzyme playing a key role in the signal transduction system, has been reported (Slater et al., 1993). PKC is a family of serine-threonine specific kinases structurally correlated and variously distributed (Nishizuka, 1986; Dekker et al., 1994). These phosphorylating enzymes which are involved in many cellular responses have been classified into three groups (Newton, 1995; Nishizuka, 1995). Classic PKCs (α, βI, βII, γ) are calcium and phospholipid dependent whereas novel PKCs (δ, ε, η, θ and μ) do not require calcium for activation. In addition, atypical (ζ, λ) PKCs are unresponsive to calcium and phorbol esters.

PKC activity may be modulated by oxidants and products of lipid peroxidation which induce selective modifications of cysteine residues present within the regulatory and catalytic domains (Gopalakrishna et al., 1987; Gopalakrishna et al., 1989; Gopalakrishna et al., 1997). PKC contains a hydrophobic binding pocket for alcohols and anesthetics whose effects on PKC-mediated processes depend on the location of the enzyme, the type of activator and on isoform specific differences. Ethanol and aliphatic alcohols at least up to hexanol, have recently been demonstrated to inhibit PKC activity (Slater et al., 1997). In addition, acute exposure to ethanol induced an increase in PKC activity in human lymphocytes and epidermal keratinocytes (De Petrillo et al., 1993; Fanò et al., 1993). Moreover, chronic ethanol treatment caused an increase in PKC activity in neuroblastoma × glioma hybrid cells, NG108–15 and in PC12 cells (Messing et al., 1991; Virmani et al., 1992).

Considering the different reports obtained within this context, our first aim was to evaluate the effects of acute exposure to ethanol on the enzymatic activity, protein levels and translocation to the membrane of novel PKCs present in rat hepatocytes. Our second aim was to identify by means of 4MP pre-treatment the effective involvement of different ethanol metabolism pathways in alterations of the PKC-dependent signaling system.

2. MATERIALS AND METHODS

2.1. Chemicals

Ethanol was supplied by Merck (Darmstadt, Germany). [^{32}P] ATP (specific activity 3,000 Ci/mmol) and hyperfilm ECL were supplied by Amersham International (Amersham, Buckinghamshire, UK); rabbit polyclonal antibodies specific for δ and ε PKC isoforms were supplied by Santa Cruz Biotechnology (Santa Cruz, USA); protein A-sepharose and dithiothreitol (DTT) were supplied by Pharmacia Biotech (Uppsala, Sweden); all other reagents were from Sigma Chemicals Co. (St. Louis, MO, USA).

2.2. Isolation and Incubation of Rat Hepatocytes

Rat hepatocytes were isolated by collagenase perfusion as described by Poli (1979). Some cellular suspensions (5×10^6 cells/ml) were pre-treated with 200 μM 4-methylpyrazole (4MP) for 15 min.

Two different experimental systems were employed : in the first, control and 4MP pretreated hepatocytes were exposed to 20 mM ethanol for 60 min. ; in the second, lysed hepatocytes were placed in a standard stock solution, containing 0.15 M KCl, 0.1 M tris-HCl pH 7.4, and an appropriate NADPH-generating system: 0.1 M glucose 6-phosphate, 0.245 mM $NADP^+$ and 8.4 I.U. of glucose 6-phosphate dehydrogenase, to maintain ethanol metabolism. Incubation was for 60 min at 37°C and ethanol was added to the cell-free system to achieve a final concentration of 20 mM. Some samples of the cell-free system were supplemented with 1mM palmitoyl CoA and catalase at a concentration of 40 I.U./mg protein. The reactions were terminated by adding 10% TCA and the samples were kept on ice for at least 30 min.

2.3. Glutathione Assay

Reduced glutathione (GSH) determination in the hepatocytes was carried out as described by Fariss and Reed (1987). An aliquot of cellular suspensions (3.5×10^6 cells/ml) was layered over a dibutyl phtalate oil layer, which was in turn layered over 10% perchloric acid containing 1mM bathophenanthroline disulphonic acid, and centrifuged for 1 min. at 13,000 × g.

100 mM iodoacetic acid in m-cresol purple was then added to a portion of the acid extract with 15 mM γ-glutamyl-glutamate. Finally, 1-fluoro-2,4-dinitrobenzene (1% v/v in ethanol) was added and the solution stored in the dark at 4°C until it was analysed by high-performance liquid chromatography (HPLC) analysis.

2.4. Malondialdehyde (MDA) Assay

MDA was assayed following the HPLC method of Young and Trimble (1991). The thiobarbituric acid (TBA) reaction was carried out by mixing 250 μl of 1.22 M phosphoric acid, 450 μl of HPLC grade water, 50 μl of control or ethanol-exposed cellular suspensions (5×10^6 cells/ml) treated with 0.2 M sodium acetate buffer pH 5 and 2,6-Di-tert-butyl-4-methylphenol (BHT) and 250 μl of 0.44 M TBA. The reaction mixture was incubated in a boiling water bath for 1h in sealed glass tubes and then cooled to 4°C on ice. Ten minutes or less before injection onto the column, all samples were neutralized with 40 μl of 1 M NaOH plus 360 μl of HPLC grade methanol. The TBA-MDA adducts were measured by an isocratic HPLC method using fluorimetric detection (excitation 532 nm, emission 553 nm).

2.5. Separation of Cytosolic and Membrane Fractions

Hepatocytes were lysed by sonication in 2 ml of 10 mM N-2-hydroxyethyl piperazine-N-N-2 ethane sulfonic acid (HEPES) buffer pH 7.5, containing 0.25 M sucrose, 5 mM EDTA, 10 mM mercaptoethanol, 2 mM phenylmethylsulphonyl fluoride (PMSF) and 1 mM leupeptin, and centrifuged at 100,000 × g for 30 min. The soluble fraction was collected; the membrane pellet was treated with the lysing buffer, containing 0.2% Triton X-100, for 20 min on ice and then centrifuged at 100,000 × g for 30 min, yielding the particulate fraction. Cytosolic and membrane fractions were subjected to novel PKC activity assay and to immunoblot analysis.

2.6. Novel PKC Activity Assay by Histone Phosphorylation

Novel PKC isoforms (δ and ε) from cytosol and membrane fractions from 100 μg of protein samples were immunoprecipitated with specific antibodies and protein A sepharose. The immunoprecipitates were washed three times in a PKC buffer (10 mM tris HCl, 150 mM NaCl, 10 mM $MgCl_2$ and 0.5 mM DTT). The activity assay of novel isoforms was performed by adding 40 μl of PKC buffer containing 0.1 mM ATP, [γ^{32}P]ATP (2μCi per sample), 1 μg of phosphatidylserine, 0.4 μg of dioleylglycerol, and 10 μg of histone H1 as substrate (Pessino et al., 1995; Monks et al., 1997).

The reaction was continued for 10 min at 30°C, then stopped by addition of Laemmli sample buffer. After centrifugation, the reaction mixtures were loaded onto 12.5% SDS-polyacrilamide gel and blotted on nitro-cellulose membrane, which was than exposed to an autoradiographic film for 24h. Relative intensity of phosphorylated substrates was measured by densitometric scanning of autoradiographs.

2.7. Immunoblot Analysis

Cytosol (20 μg of total proteins) and membrane (40 μg of total proteins) fractions were denaturated in Laemmli buffer and then subjected to 8% SDS-polyacrilamide gel electrophoresis, followed by electroblotting onto a nitro-cellulose membrane. Immunodetection was performed using specific PKC isoenzyme antibodies. After incubation with the second antibody labelled with horseradish peroxidase, the immunoblots were detected with an enhanced chemiluminesce (ECL) system. The film employed exhibited a linear response to the light produced by ECL which was proportional to the protein content of samples.

2.8. Statistical Analysis

Each experiment was performed at least in quadruplicate. Results are expressed as means ± S.D. for each set of conditions. Significance of parametric difference among groups was evaluated with one-way ANOVA and Dunnett's test for multiple comparisons.

3. RESULTS

3.1. Glutathione and Malondialdehyde Levels in Rat Hepatocytes Pre-Treated with 4MP and Exposed to Ethanol

Our experimental system was represented by isolated rat hepatocytes pre-treated with 200 μM 4MP for 15 min and then exposed to 20 mM ethanol for 60 min. The oxidative state of alcohol-treated cells was analysed by measuring the intracellular glutathione (GSH) level which was found to be depleted by 50%. Ethanol-induced lowering of thiol antioxidant was completely prevented by cell pre-incubation with 4MP (Table 1).

Although the oxidative state of hepatocytes was affected by ethanol treatment, the production of MDA, as index of lipid peroxidation, appeared only slightly modified; in fact the MDA level was 25% increased in rat hepatocytes exposed to ethanol. Pre-treatment of liver cells with 4MP prevented the MDA formation, demonstrating that GSH depletion induced an imbalance of the physiological antioxidative defense system rather than a real lipoperoxidative event (Table 1).

3.2. Ethanol-Induced Effects on Activity and Protein Level Expression of Novel PKCs Present in Rat Hepatocytes

Under these experimental conditions, the activity and the protein level of novel isoforms were evaluated both in the cytosol and in the membrane fractions of rat hepatocytes.

Table 1. GSH and MDA levels in rat hepatocytes pre-treated with 200 μM 4MP for 15 min and exposed to 20 mM ethanol for 60 min

	GSH (nmol/10^6 cells)	MDA (nmol/10^6 cells)
Control	12.3 ± 1.8	0.091 ± 0.009
200 μM 4MP	12.7 ± 2.6	0.087 ± 0.007
20 mM EtOH	7.4 ± 0.4★	0.126 ± 0.013★
200 μM 4MP + 20 mM EtOH	12.1 ± 1.8●	0.087 ± 0.01●

GSH and MDA concentrations are expressed as nmol/mg protein and represent the mean ± S.D. of four separate experiments.
★ $p < 0.01$ vs. control ● $p < 0.01$ vs. 20 mM EtOH

A 40% decrease in δ PKC activity was found in the cytosol of 20 mM ethanol-treated cells and a similar degree of inactivation was also detected in the novel isoenzyme associated to the membrane. Pre-incubation of hepatocytes with 200 μM 4MP did not affect δ PKC activity in control cells, and restored the cytosolic and particulate enzymatic activity inhibited by 20 mM ethanol (Figure 1).

Ethanol-induced inactivation of δ PKC was accompanied by a 40% reduction of immunoreactive protein level in both the subcellular compartments. Also in this case, 4MP pre-treatment prevented PKC modifications induced by alcohol metabolism (Figure 1).

The behaviour of ε PKC, the other component of novel PKC isoforms present in rat hepatocytes behaved similarly in that cytosolic and particulate kinase activities were depressed by 40% after alcohol treatment and the corresponding protein levels were also de-

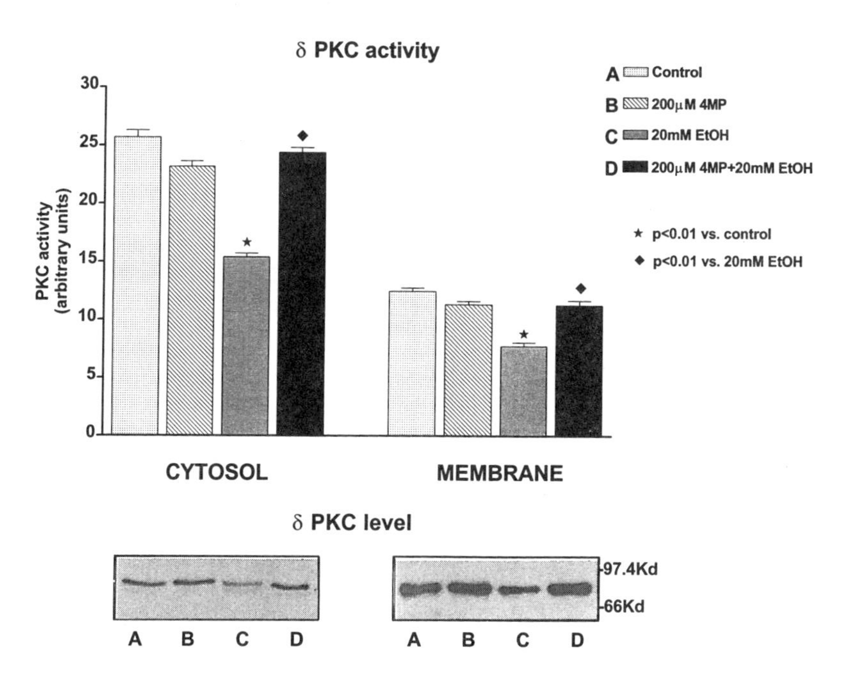

Figure 1. Effect of ethanol metabolism on the activity and the protein level of δ PKC in the cytosol and in the membrane of rat hepatocytes. The graph reports the arbitrary units of δ isoform activity expressed as mean ± S.D. of four separate experiments. The corresponding protein level was revealed by Western blot analysis and the immunoblots shown originate from one representative experiment.

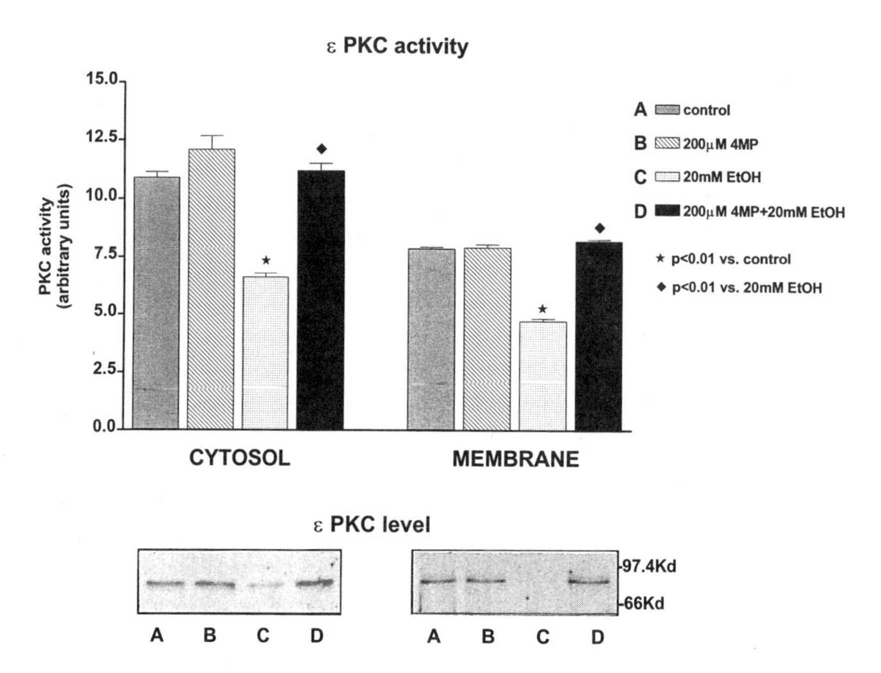

Figure 2. Effect of ethanol metabolism on the activity and the protein level of ε PKC in the cytosol and in the membrane of rat hepatocytes. The graph reports the arbitrary units of ε isoform activity expressed as mean ± S.D. of four separate experiments. The corresponding protein level was revealed by Western blot analysis and the immunoblots shown originate from one representative experiment.

creased. 4MP pre-incubation was able to prevent ethanol-induced changes in the functional activity and in the protein content of ε PKC (Figure 2).

3.3. Protective Effect Induced by 4MP Is Not Due to Its Competitive Inhibition of Fatty Acyl CoA Synthetase which Leads to the Decreased Hydrogen Peroxide (H_2O_2) and Related Reactive Oxygen Species (ROS) Production

The observed 4MP-induced prevention of PKC inactivation and loss after acute ethanol exposure might be due to a decrease in acetaldehyde production and/or a decrease in H_2O_2 and other related ROS generation during peroxisomal oxidative metabolism.

To investigate the putative involvement of these oxidant species in PKC isoform modifications, an experimental system represented by lysed hepatocytes was employed. The results obtained in intact cells were confirmed in the hepatocyte lysate treated with 20 mM ethanol: also in the cell-free system δ and ε PKCs showed a marked loss of functional activity, accompanied by a significant decrease in their immunoreactive protein levels. 4MP pre-treatment almost completely prevented ethanol-induced changes in the activity and protein level of novel isoenzymes (Figure 3).

The addition of catalase (40 I.U./mg protein) itself was found not to affect the activity or protein content of novel PKC isoforms (Figure 3, lane E), nor did it prevent the PKC changes consequent to acute ethanol treatment (Figure 3, lane F). Further, supplementation of 20 mM ethanol-treated hepatocyte lysate with 1mM palmitoyl CoA, which has

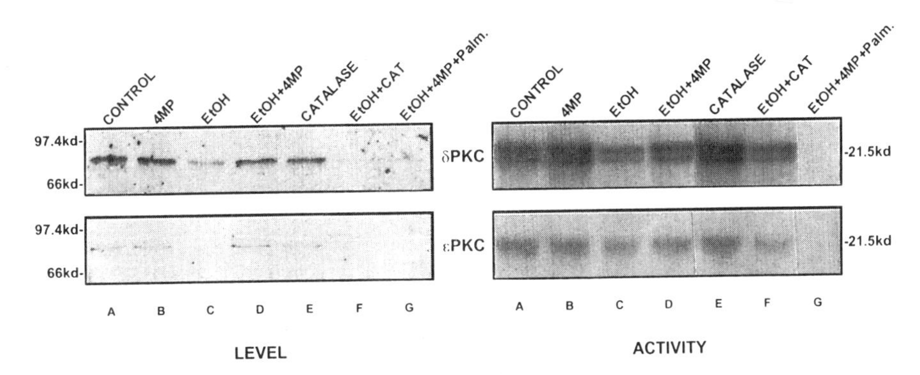

Figure 3. Role of ADH and catalase-dependent ethanol metabolism in the inactivation and loss of novel PKC isoforms. Protein level of novel PKCs (δ and ε) present in the hepatocyte lysate was revealed by Western blot analysis (left side). Novel PKC isoform activity has been assayed by the immunoprecipitation technique: the autoradiographic bands show the phosphorylation of H1 histone as substrate (right side). Lane A : untreated; lane B : 200 μM 4MP; lane C : 20 mM ethanol; lane D : 200 μM 4MP + 20 mM ethanol; lane E : catalase (40 I.U./mg protein); lane F : 20 mM ethanol + catalase (40 I.U./mg protein); lane G : 200 μM 4MP + 20 mM ethanol + 1 mM palmitoyl CoA.

been shown to stimulate peroxisomal H_2O_2 production and ethanol metabolism, induced almost complete inactivation of novel PKC isoenzymes, accompanied, as revealed by Western Blot analysis, by the disappearance of their protein levels. This phenomenon was observed also in the conditions of 4MP pre-treatment (Figure 3, lane G).

4. DISCUSSION

Ethanol induces significant perturbations of intracellular signal transduction pathways in consequence of a direct effect by ethanol and/or its metabolism. Many reports have produced evidence that PKC isoenzymes may be considered as molecular and cellular targets for the action of ethanol in particular during the cellular tolerance mechanisms to chronic alcohol exposure (Coe et al., 1996).

According to previous results (Domenicotti et al., 1996), the aim of our recent studies was to investigate the effects induced by ethanol metabolism on thePKC-dependent signaling system, looking for a possible correlation between the impairment of this central component of signal transduction network and the pathogenesis of alcoholic liver injury.

In the present report, a negative effect of alcohol metabolism on the enzymatic activity and protein level expression of novel PKC isoforms was demonstrated by using 4MP, an inhibitor of ADH. 4MP has been demonstrated to inhibit both ADH and fatty acyl CoA synthetase, a key enzyme involved in the supply of substrates for peroxisomal oxidation of alcohols (Thurman et al., 1993). Therefore, the observed prevention of ethanol-induced effects on novel PKC isoenzymes by 4MP, might be due to a decrease in acetaldehyde production and/or a decrease in H_2O_2 and ROS generated from peroxisomal oxidative metabolism.

In lysed hepatocytes exposed to ethanol, the addition of catalase has been found unable to prevent ethanol-induced PKC modifications, demonstrating that H_2O_2 does not play a relevant role in the observed phenomenon. Moreover, supplementation of cell-free system with palmitoyl CoA, shown to stimulate peroxisomal H_2O_2 production, induced almost complete inactivation and loss of novel PKC isoenzymes, also in the conditions of 4MP pre-treatment.

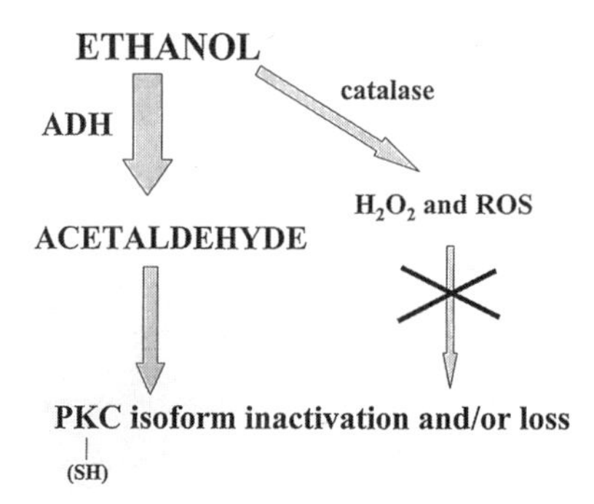

Figure 4. Primary role of ADH pathway in acute ethanol-induced impairment of PKC-dependent signaling system.

Both lines of experimental evidence are against the involvement of catalase pathway and consequently of H_2O_2 and ROS in the mechanisms of ethanol-induced impairment on novel PKCs. Thus, ethanol primarily metabolized by ADH leads to a production of acetaldehyde which rather than aldehydic products of lipid peroxidation appears to be a major responsible for PKC isoform inactivation and loss (Figure 4).

PKC may be particularly sensitive to acetaldehyde since the enzyme has thiol-rich regions within regulatory and/or catalytic domains. In fact, acetaldehyde has been demonstrated to covalently bind to proteins and GSH in consequence of its high affinity for sulphydryl groups (Chiarpotto et al., 1995; Grattagliano et al., 1996).

In conclusion, a single overdose of ethanol can interfere with the regulation of hepatocyte PKC presumably through acetaldehyde-mediated partial proteolytic degradation of novel isoforms, more sensitive than classic PKCs to protein fragmentation (Chen et al., 1995; Domenicotti et al., 1998). These results indirectly confirm the primary role of ADH pathway in ethanol metabolism and suggest a possible involvement of ethanol-induced impairment of PKC-dependent signaling system and, in particular, of novel isoenzymes in the pathogenesis of alcoholic liver injury.

Protein phosphorylation in fact is a fundamental process in regulating cellular transport processes of proteins and lipoproteins from the endoplasmic reticulum through the Golgi apparatus; different lines of evidence have pointed out a role of PKCs in the vesicular transport (DeMatteis et al., 1993; Buccione et al., 1996). Moreover, a stimulation of PKC activity might accelerate the intracellular exocytic transport and, consequently, the observed inactivation of PKC by ethanol treatment could play a role in the impairment of Golgi vesicle formation and lipoprotein secretion, a crucial step in the pathogenesis of fatty liver.

Further studies are required to understand the role of PKC changes in ethanol-induced impairment of hepatic protein and lipoprotein secretion analysing the protein content and the activity of PKC isoforms located at liver Golgi apparatus and involved in the regulation of secretion, following acute ethanol intoxication.

ACKNOWLEDGMENTS

This work was supported by CNR (96-04995-ST74) and MURST 40% (Patologia da radicali liberi e degli equilibri redox) and 60%. We wish to thank Mr. Giuseppe Maloberti and Mr. Giuseppe Catalano for their technical assistance.

REFERENCES

Bradford, B.U., Forman, D.T. and Thurman, R.G. (1993) 4-Methylpyrazole inhibits fatty acyl coenzyme synthetase and diminishes catalase-dependent alcohol metabolism: has the contribution of alcohol dehydrogenase to alcohol metabolism been previously overestimated? *Mol. Pharmacology* 43, 115–119.

Buccione, M., Bannykh, S., Santone, I., Baldassarre, M., Facchiano, F., Bozzi, Y., Di Tullio, G., Mironov, A., Luini, A. and DeMatteis, M.A.. (1996) Regulation of constitutive exocytic transport by membrane receptors. A biochemical and morphometric study. *J. Biol. Chem.* 271, 3523–3533.

Chen, C.C., Cheng, C.S., Chang, J. and Huang, H.C. (1995) Differential correlation between translocation and down-regulation of conventional and new protein kinase C isozyme in C_6 glioma cells. *J. Neurochem.* 64, 818–824.

Chiarpotto, E., Biasi, F., Scavazza, A., Camandola, S., Aragno, M., Tamagno, E., Danni, O., Dianzani, M.U. and Poli, G. (1995) Acetaldehyde involvement in ethanol-induced potentiation of rat hepatocyte damage due to the carcinogen 1,2-dibromoethane. *Alcohol Alcohol.* 30, 721–728.

Coe, I.R., Yao, L., Diamond, I. and Gordon, A.S. (1996) The role of protein kinase C in cellular tolerance to ethanol. *J. Biol. Chem.* 271, 29468–29472.

Dekker, L.V. and Parker, P.J. (1994) Protein kinase C-a question of specificity. *Trends Biochem. Sci.* 19, 73–77.

De Matteis, M.A., Santini, G., Kahn, R.A., Di Tullio, G. and Luini, A. (1993) Receptor and protein kinase C-mediated regulation of ARF binding to the Golgi complex. *Nature* 364, 818–821.

De Petrillo, P.B. and Caspar, S.L. (1993) Ethanol exposure increases total protein kinase C activity in human lymphocytes. *Alcohol. Clin. Exp. Res.* 17, 351–354.

Domenicotti, C., Paola, D., Lamedica, A., Ricciarelli, R., Chiarpotto, E., Marinari, U.M., Poli, G., Melloni, E. and Pronzato, M.A. (1996) Effects of ethanol metabolism on PKC activity in isolated rat hepatocytes. *Chem. Biol. Interact.* 100, 155–163.

Domenicotti, C., Paola, D., Vitali, A., Nitti, M., Cottalasso, D., Melloni, E., Poli, G., Marinari, U.M. and Pronzato, M.A. (1998) Mechanisms of inactivation of hepatocyte protein kinase C isoforms following acute ethanol treatment. *Free Radic. Biol. Med.* 25, in press.

Fanò, G., Belia, S., Mariggiò, M.A., Antonica, A., Agea, E. and Spinozzi, F. (1993) Alteration of membrane transductive mechanisms induced by ethanol in human lymphocyte cultures. *Cell. Signal.* 5, 139–143.

Fariss, M.W. and Reed, D.J. (1987) High-performance liquid chromatography of thiols and disulfides: dinitrophenol derivatives. *Methods Enzymol.* 143, 101–109.

Gopalakrishna, R. and Anderson, W.B. (1987) Susceptibility of protein kinase C to oxidative inactivation: loss of both phosphotransferase activity and phorbol diester binding. *FEBS Lett.* 225, 233–237.

Gopalakrishna, R. and Anderson, W.B. (1989) Ca^{2+} and phospholipid-independent activation of protein kinase C by selective oxidative modification of regulatory domain. *Proc. Natl. Acad. Sci. USA* 86, 6758–6762.

Gopalakrishna, R., Gundimeda, U. and Chen, Z.H. (1997) Cancer-preventive selenocompounds induce a specific redox modification of cysteine-rich regions in Ca^{2+}dependent isoenzymes of protein kinase C. *Arch. Biochem. Biophys.* 348, 25–36.

Grattagliano, I., Vendemiale, G., Sabbà, C., Buonamico, P. and Altomare, E. (1996) Oxidation of circulating proteins in alcoholics: role of acetaldehyde and xanthine oxidase. *J. Hepatol.* 25, 28–36.

Handler, J.A. and Thurman, R.G. (1988) Catalase-dependent ethanol oxidation in perfused rat liver. Requirement for fatty-acid-stimulated H_2O_2 production by peroxisomes. *Eur. J. Biochem.* 176, 477–484.

Hoek, J.B. and Rubin, E. (1990) Alcohol and membrane-associated signal transduction. *Alcohol Alcohol.* 25, 143–156.

Hoek, J.B., Thomas, A.P., Rooney, T.A., Higashi, K. and Rubin, E. (1992) Ethanol and signal transduction in the liver. *FASEB J.* 6, 2386–2396.

Lieber, C.S. and DeCarli, L.M. (1970) Hepatic microsomal ethanol-oxidizing system. In vitro characteristics and adaptive properties in vivo. *J. Biol. Chem.* 245, 2505–2512.

Messing, R.O., Peterson P.J. and Heinrich, C.J. (1991) Chronic ethanol exposure increases levels of protein kinase C δ and ε and protein kinase C-mediated phosphorylation in cultured neural cells. *J. Biol. Chem.* 266, 23428–23432.

Monks, C.R.F., Kupfer, H., Tamir, I., Barlow, A. and Kupfer, A. (1997) Selective modulation of protein kinase C-θ during T-cell activation. *Nature* 385, 83–86.

Newton, A.C. (1995) Protein kinase C: structure, function and regulation. *J. Biol. Chem.* 270, 28495–28498.

Nishizuka, Y. (1986) Studies and perspectives of protein kinase C. *Science* 233, 305–312.

Nishizuka, Y. (1995) Protein kinase 5: protein kinase C and lipid signalling for sustained cellular responses. *FASEB J.* 9, 484–496.

Pessino, A., Passalacqua, M., Sparatore, B., Patrone, M., Melloni, E. and Pontremoli, S. (1995) Antisense oligodeoxynucleotide inhibition of δ protein kinase C expression accelerates induced differentiation of murine erythroleukaemia cells. *Biochem. J.* 312, 549–554.

Poli, G., Gravela, E., Albano, E. and Dianzani, M.U. (1979) Studies on fatty liver with isolated hepatocytes. II-The action of carbon tetrachloride on lipid peroxidation, protein, and triglyceride synthesis and secretion. *Exp. Mol. Pathol.* 30, 116–127.

Slater, S.J., Cox, K.J.A., Lombardi, J.V., Ho, C., Kelly, M.B., Rubin, E. and Stubbs, C.D. Inhibition of protein kinase C by alcohols and anesthetics. *Nature* 364, 82–84.

Slater, S.J., Kelly, M.B., Larkin, J.D., Ho, C., Mazurek, A., Taddeo, F.J., Yeager, M.D. and Stubbs, C.D. Interaction of alcohols and anesthetics with protein kinase C α. *J. Biol. Chem.* 272, 6167–6173.

Thurman, R.G., Ley, H.G. and Scholz, R. (1972) Hepatic microsomal ethanol oxidation. Hydroperoxide formation and the role of catalase. *Eur. J. Biochem.* 25, 420–428.

Virmani, M. and Ahluwalia, B. (1992) Biphasic protein kinase C translocation in PC12 cells in response to short-term and long-term ethanol exposure. *Alcohol Alcohol.* 27, 393–401.

Young, I.S. and Trimble, E.R. (1991) Measurement of malondialdehyde in plasma by high performance liquid chromatography with fluorimetric detection. *Ann. Clin. Biochem.* 28, 504–508.

40

CLASS II ALCOHOL DEHYDROGENASE

A Suggested Function in Aldehyde Reduction

Jan-Olov Höög, Stefan Svensson, Patrik Strömberg, and Margareta Brandt

Department of Medical Biochemistry and Biophysics
Karolinska Institutet
S-171 77 Stockholm, Sweden

1. INTRODUCTION

Enzymes are defined from their catalytic activity and generally they have a specific function in cell metabolism. Mammalian alcohol dehydrogenase (ADH) was early isolated and classified (Vallee and Bazzone, 1983) and the function was stated as either to metabolize ethanol into acetaldehyde or to produce ethanol. In the mammals and especially in humans a large number of different ADHs have been identified, as different enzymes (classes), as different isozymes and as allelic forms (Jörnvall et al., this volume). For the different classes, solely class III ADH, glutathione-dependent formaldehyde dehydrogenase, has a specific function in the turnover of formaldehyde (Uotila and Koivusalo, 1986). Class I ADH, the main ethanol metabolizing enzyme in the liver, probably has an additional function in the metabolism of steroids and bile acids (human class I γγ; Marschall et al., 1998) The extrahepatically expressed class IV ADH has been ascribed a function in the turnover of retinoids (Duester, 1998).

The class II ADH was identified in human liver long ago (Li et al., 1977) and the enzyme was isolated as a high K_m enzyme in the metabolism of ethanol, but until recently it has not been characterized thoroughly from other species. With the today characterized enzymes from ostrich (Hjelmqvist et al., 1995) and rabbit (Svensson et al., 1998) it is obvious that the origin was before the vertebrate radiation, thus this class of ADH is as widespread as the other characterized classes. The human class II enzyme has been shown to have activity in the metabolism of norepinephrine, serotonin and 4-hydroxyalkenals (Mårdh et al., 1986; Consalvi et al., 1986; Sellin et al., 1990). The enzyme has further been postulated to be regulated by thyroid hormones (Mårdh et al., 1987), to reduce p-nitrosophenol (Maskos and Winston, 1994) and to be the most active hepatic ADH in retinol metabolism (Yang et al., 1994). Some ADHs have dismutase activity, i.e., formation of both an alcohol and an acid from the corresponding aldehyde. This has been shown also

Enzymology and Molecular Biology of Carbonyl Metabolism 7
edited by Weiner *et al.* Kluwer Academic / Plenum Publishers, New York, 1999.

for class II ADH, although to a less extent as compared to the class I γγ isozyme (Svensson et al., 1996). From tissue distribution studies of the human and rat enzymes the class II ADH is mainly detected in the liver, but the rat enzyme shows high mRNA levels in the duodenum while the human enzyme only shows trace amounts in the intestinal tract (Estonius et al., 1993, 1996).

So far, no specific function for the class II ADHs has been found but the enzyme can participate in the metabolism of a large number of alcohols and aldehydes. With the recently cloned and expressed rodent forms (Höög, 1995; Höög and Svensson, 1997; unpublished) new insight into the class II ADHs function become available. The class II ADH has alcohol dehydrogenase activity, dismutase activity and capability to reduce quinone compounds (Figure 1).

2. MATERIALS AND METHODS

2.1. Isolation of cDNAs

cDNAs coding for class II ADHs from human, rat and rabbit have been isolated earlier (Höög et al, 1987; Höög, 1995; Svensson, 1998) and now a cDNA coding for mouse class II ADH was isolated from an adaptor-ligated double-stranded liver cDNA library (Marathon-Ready cDNA, Clontech), by PCR amplification utilizing *Pfu* polymerase (Stratagene). Multiple sequence alignments of the earlier characterized class II ADH sequences were performed to identify regions with a high degree of positional identity for design of PCR primers. The full-length cDNA was obtained by using the rapid amplification of cDNA ends technique (RACE). DNA sequence analyses were performed on both strands with sequence specific primers using T7 DNA polymerase (Pharmacia Biotech), [α-^{35}S]dATP (Amersham) and alkali-denatured plasmids. Obtained DNA sequences were analyzed with the GCG computer program package (Genetics Computer Group, Univ. of Wisconsin) and compared with EMBL data banks.

2.2. Expression of Recombinant Protein

cDNAs coding for the different class II ADHs were subcloned into pET expression vectors (pET3d, rat and rabbit; pET12b, human and rabbit; pET29, mouse; Novagen). Prior to subcloning into the expression plasmids the coding parts of the cDNAs were PCR amplified with *Pfu* polymerase (Stratagene) to introduce suitable restriction enzyme sites.

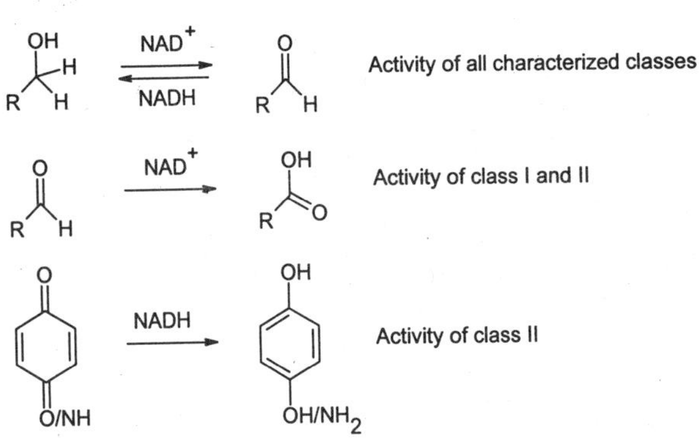

Figure 1. Reactions catalyzed by mammalian ADHs. The upper reaction is the reversible dehydrogenase reaction, the middle is the conversion of an aldehyde into an acid in the dismutase function of ADH and the bottom is the reduction of quinones.

Recombinant proteins were expressed in 1 l cultures using an *E. coli lacI*^{q} strain, BL21 (DE3). The cells were grown at 29°C to circumvent inclusion bodies and 0.8 mM isopropyl- β-thiogalactopyranoside was added at an OD_{595} of 1.0. The cells were harvested after another 4 h of growth, disrupted in 0.3 mM dithiothreitol, 10 mM sodium-phosphate, pH 7.5, sonicated intermittently and centrifuged at 48,000 × *g*. The supernatant was applied on a DEAE-cellulose column (DE-52, Whatman) and the void volume, containing recombinant ADH, was applied to an AMP-Sepharose column (Pharmacia Biotech). A final purification step was performed on a Resource Q column/FPLC (Pharmacia Biotech). Recombinant proteins were analyzed for purity by SDS-polyacrylamide gel electrophoresis. Protein concentrations were determined with the Bio-Rad protein assay kit standardized with bovine serum albumin.

Enzymatic activities were determined spectrophotometrically at 340 nm using a Hitachi 3000 spectrophotometer to follow NADH formation in 0.1 M glycine-NaOH, pH 10, 25°C or in sodium phosphate, pH 7.5, 25°C. Concentrations used for NAD^+ and NADH were 2.4 mM and 0.8 mM, respectively, and the absorption coefficient for NADH 6.22×10^3 $M^{-1}cm^{-1}$ was used. Substrates of analysis grade were used without further purification, except acetaldehyde that was distilled prior to use. A weighted non-linear regression analysis program, was used to fit all lines to data points and to calculate kinetic parameters.

3. RESULTS

cDNAs coding for class II ADH have been cloned from human, rat, rabbit and mouse. The addition of an isolated cDNA coding for the mouse class II ADH shows that this enzyme has a wide-spread distribution in different species as ADH of other classes. Comparison of the different amino acid sequences for class II ADHs, including the enzyme from ostrich, further shows that this class is a highly divergent class within the ADH family, but the two rodent forms shows a high degree of positional identity, 92.3%. This is in contrast to the other relationships of class II ADHs that show identities of about 75% and the two isozymic forms from rabbit that are 88.4% identical (Table 1).

From the alignment of the characterized class II ADHs (Figure 2) it is seen that the different enzymes differ in polypeptide chain length as well, the rodent forms translate into 376 amino acid residues, the rabbit forms into 378 and the human and ostrich enzymes into 379. This makes the class II chains slightly longer than the other ADH forms. The rodent forms have deletions in two regions, a one residue deletion at position 57 and a two residue deletion after position 302. The rabbit enzymes have one deletion after position 327. In common, all class II ADHs have an insertion of four residues around position

Table 1. Positional identities at protein level between the different characterized class II ADHs. The value for the two rodent forms is indicated in bold

	Mouse	Rat	Rabbit A	Rabbit B	Ostrich
Human	72.3	72.6	78.0	76.7	69.6
Mouse	–	**92.3**	70.4	71.1	65.3
Rat		–	70.4	70.1	66.1
Rabbit A			–	88.4	68.4
Rabbit B				–	67.7

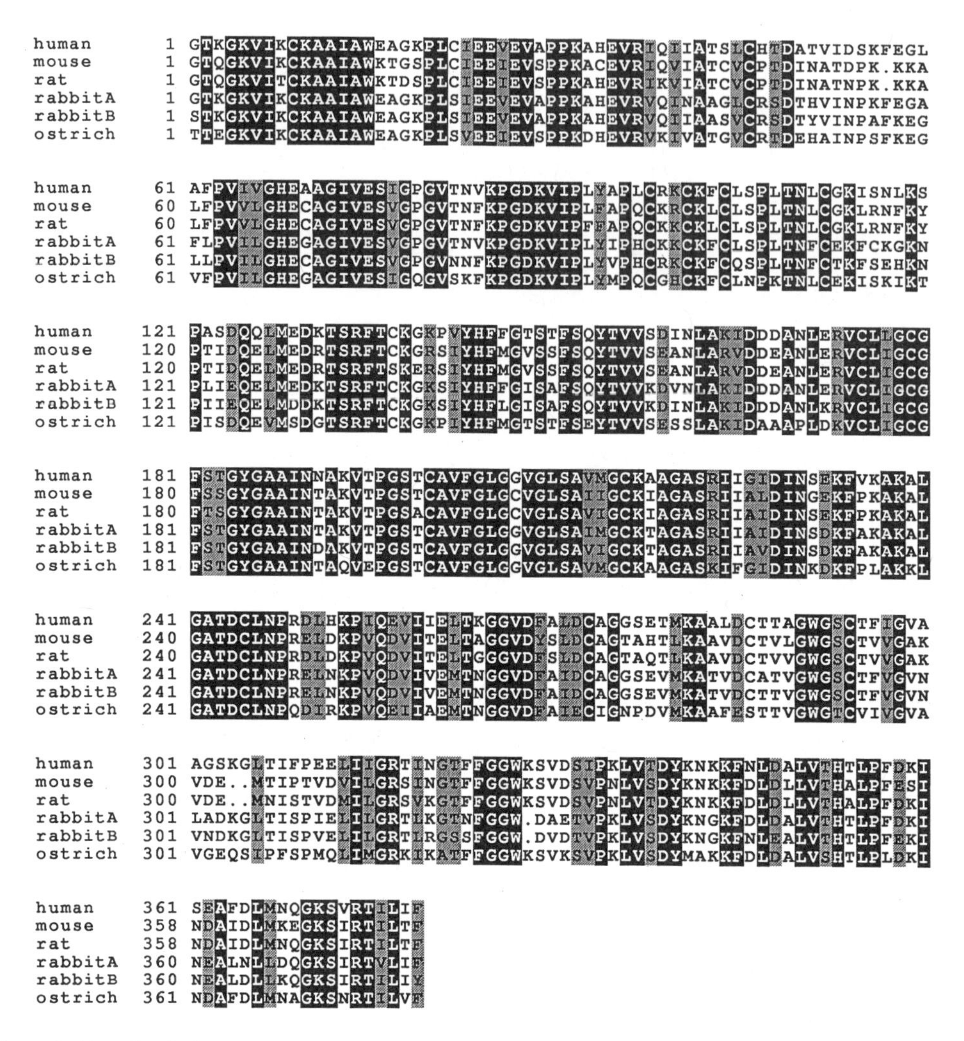

Figure 2. An updated alignment of class II ADH amino acid sequences. From top to bottom: human, mouse, rat, rabbit isozyme A, rabbit isozyme B, and ostrich. Conserved residues in all species are inverted in black. Stippled residues indicates conserved exchanges. While (.) denotes deletions. The numbering to the left of every sequence refers to the individual numbering also indicating the difference in polypeptide lengths.

120 as compared to all other classes of ADH. So far, this is the only characteristic at the amino acid level that distinguish class II from the other classes of ADHs.

The rodent structures show active sites that differ at several positions as compared to other ADHs. Worthwhile to notice is Pro at position 47 and Asn at position 51 (Figure 3). Furthermore, the highly conserved Val at position 298 (class I numbering) is exchanged for Ala in the rodent forms and the conserved Ser178 in all vertebrate ADHs is in the rodent class II exchanged for Thr. All these exchanges and the deletions place the rodent class II enzyme in a subgroup of class II.

All mammalian class II ADH were expressed in *E. coli* as recombinant proteins and purified to homogeneity in a three step purification scheme with ion-exchange and affinity chromatography. Specific activities were determined in glycine-NaOH, pH 10, for the five

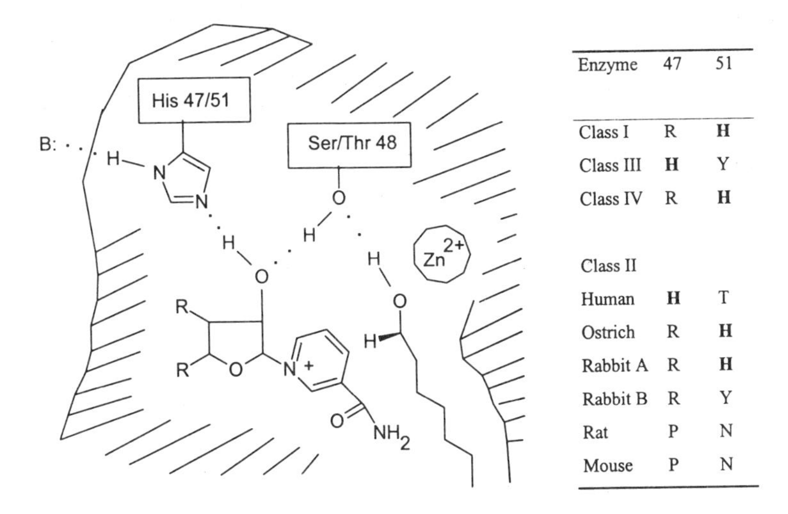

Figure 3. Schematic representation of the active site in ADH. The suggested proton shuttle is here given as substrate, Ser/Thr48, NAD^+ and a His at either position 47 or 51. The alternatives of residues at these latter positions in different ADHs are given to the right. The suggestion here is that a basic amino acid residue (His) is essential at either position 47 or position 51 for ethanol dehydrogenase activity.

different expressed enzymes. The obtained results are: 1.1 U/mg for the human class II enzyme, 0.6 U/mg for the rabbit A isozyme, 0.07 U/mg for the rabbit B isozyme, 0.001 U/mg for the rat enzyme and 0.0003 U/mg for the mouse enzyme. The kinetic constant determined at pH 7.5 differed as well between the species (Table 2). The catalytic efficiency for ethanol (k_{cat}/K_m) differs from 2.9 $min^{-1}mM^{-1}$ for the rabbit A isozyme to almost no activity for the rodent enzymes. The rodent forms have low dehydrogenase activity, but show fairly high activity for reduction of benzoquinones (Table 2).

4. DISCUSSION

Class II alcohol dehydrogenase from mammals, ADH2 in the new nomenclature system (Duester at al., 1998), has recently been identified, cloned and expressed from rat,

Table 2. Comparison of specific activities for different mammalian class II ADHs at pH 7.5, given as k_{cat}/K_m ($mM^{-1}min^{-1}$)

	Substrate				
Species	Ethanol	Acetaldehyde	Benzylalcohol	Benzaldehyde	Benzoquinone
Human	1.4	80[a]	79000[b]	45000[a]	9900
Mouse	NA	1.3	0.21	170	2250
Rat	NA	0.41	ND	25	ND
Rabbit A	2.9	41	700	ND	225
Rabbit B	0.24	6.2	147	ND	400
Ostrich[c]	51	ND	ND	ND	ND

NA - no detectable activity that could be translated into a k_{cat}/K_m value, ND - not determined.
[a]pH 7.0 (Deetz et al., 1984). [b]pH 10.0 (Ditlow et al., 1984). [c]ph 7.0 (Hjelmqvist et al., 1995).

rabbit and mouse (Höög, 1995; Svensson et al., 1998; unpublished). Together with the earlier cloned human class II ADH (Höög et al., 1987) the results show that the human class II and the corresponding rabbit enzymes have similar characteristics. This is in contrast to the two rodent forms that differ widely both at the structural and functional levels.

Previous attempts to isolate a class II ADH from mouse have been unsuccessful, it has therefore been indicated that the mouse either lacks genes for class II or possesses very different forms. With the now isolated cDNA coding for a mouse class II ADH it is shown that this class also is present in the mouse, and thereby the class II ADH shows a species distribution similar to other ADHs. However, the mouse form and the earlier isolated rat class II ADH are very different from the human one which can explain why these enzymes have been overlooked, probably due to their lack of ethanol oxidation activity (Table 2).

Class II ADH was first isolated from human liver as a high K_m enzyme (Li et al., 1977), but the high K_m is not a general theme for class II ADH. In contrast, recombinantly expressed human class II ADH does not show a high K_m for ethanol, the K_m values for the recombinantly expressed and the liver isolated enzymes are 8 mM and 34 mM, respectively. This can be due to different buffer conditions, it has been shown that the class II activity is sensitive for the salt concentration (Davis et al., 1988; unpublished), or different allelic forms (Höög and Svensson, 1997). Today, the only identified theme in common for class II ADH is an insertion of four amino acid residues around position 120. Comparison of residues lining the active site pocket in the different class II ADHs indicate that the kinetic values will differ widely between the species, as they obviously do. Position 47, shown to interact with the pyrophosphate in the coenzyme (Eklund et al., 1982), is occupied by His, Pro and Arg in the human, rodent and rabbit enzymes, respectively. Pro at this position will prevent a charge interaction. The rodent enzymes has further an exchange at position 51, Asn, that is not found in any other ADH. It was early suggested that His51, in the class I ADH, will participate in a proton relay system, essential for catalysis (Eklund et al., 1982). Probably a His at position 47 can replace the effect of His51 in the proton relay system (cf. Figure 3). In the structurally determined class III ADH it is suggested that His47 can participate in such a system (Yang et al., 1997), and His47 can probably have a dual function, both as coenzyme binding and proton relay. This can explain the low activity for the rodent enzymes and the rabbit B isozyme (forms without His at positions 47 and 51). The ostrich class II ADH has similar kinetic characteristics as the class I enzymes which is, in some respect, expected from the amino acid residues (Figure 3; Hjelmqvist et al., 1995). The typical His51 is present and the enzyme also show a class I-like K_m value for ethanol. In contrast, the rabbit A isozyme, with His at position 51, has a K_m value similar to that of human class II ADH, but the rabbit B isozyme with Thr at this position is a fairly bad ethanol dehydrogenase. The B isozyme is in one sense a transition form, without the suggested essential His, and absence of dehydrogenase activity for longer alcohols than pentanol (Svensson et al., 1998). The latter rabbit isozyme is further a low activity form but show a fairly high specific activity for benzoquinone reduction (Table 2). All results from the different class II ADH strongly indicates that a His is essential at position 47 or 51 for an efficient alcohol oxidation. However, the benzoquinone reduction activity seems to be independent of a His at any of these positions. The specific activity for benzoquinones are about 1000-fold higher for all class II ADHs as compared to alcohol dehydrogenase activity.

The dehydrogenase activity, independent of substrate, is very low for the rodent class II enzymes, explaining the late identification of the mouse variant. The activities for these two enzymes are roughly 100 fold lower as compared to the human enzyme and makes it unlikely that the dehydrogenase activity of class II ADH has any importance in rodents. In fact, the only reaction which is efficiently catalyzed by the rodent forms of class II ADH is the reduction of benzoquinones. Performing site-directed mutagenesis on the mouse class II ADH, exchanging His for Pro, partly restored the ethanol dehydrogenase activity while the capacity to reduce benzoquinones were more or less unaffected. This further strengthens the importance of a His at position 47 or 51 to obtain a proton relay system. K_m values for the coenzyme were in a μM value for NAD^+ and value less than 0.5 μM for NADH. However, the ratio of coenzymes in the cell favor NAD^+ but the K_m for NADH and the cell concentration, around 0.5 μM (Bücher, 1970), shows that reduction of aldehydes is possible during physiological conditions. This fine tuned activity together with the conservation of residues that suppress the alcohol dehydrogenase activity suggest a function for class II ADHs as reductases of various quinones.

Interestingly, the rodent class II ADHs show a very high degree of positional identity, 92.3%, indicating a specific function for the rodent forms. In comparison, the two rabbit isozymic forms are only 88.4% identical and the highly conserved, both at structural and functional level, class III ADH is 96% identical. All results taken together, both the sequence data and the determined kinetic constants divide the class II ADHs into two subgroups, one group active with ethanol and one non-ethanol active. The ethanol active group is represented by the human form and the non-ethanol active group consists of the rodent class II enzymes. The rabbit A isozyme is clearly a human type of class II while the B isozyme is close to the rodent forms in characteristics. The rabbit is the only species investigated that shows isozymic forms for class II ADH and thereby has enzymes that can be placed in both subgroups of class II. The rodent type of class II ADH can be seen as aldehyde reductase-like enzyme instead of alcohol dehydrogenases. In conclusion, these results indicate that the reductase activity in the benzoquinone reduction is a probable physiological function for class II ADH.

ACKNOWLEDGMENTS

This work was supported by grants from the Swedish Medical Research Council, the Swedish Alcohol Research Fund, and the Karolinska Institutet.

REFERENCES

Bücher, T. (1970) In: *Pyridine nucleotide dependent dehydrogenases* (Sund, H., ed.) Springer Verlag.

Consalvi, V., Mårdh, G. and Vallee, B.L. (1986) Human alcohol dehydrogenases and serotonin metabolism, *Biochem. Biophys. Res. Commun.* **139,** 1009–1016.

Davis, G.J., Carr, L.G., Hurley, T.D., Li, T.-K. and Bosron, W.F. (1994) Comparative roles of histidine 51 in human $\beta_1\beta_1$ and threonine in ππ alcohol dehydrogenases, *Arch. Biochem. Biophys.* **311,** 307–312.

Deetz, J.S., Luehr, C.A., Vallee, B.L. (1984) Human liver alcohol dehydrogenase isozymes: Reduction of aldehyde and ketones, *Biochemistry* **23,** 6822–6828.

Ditlow, C.C., Holmquist, B., Morelock, M.M. and Vallee, B.L.(1984) Physical and enzymatic properties of a class II alcohol dehydrogenase isozyme of human liver: π-ADH, *Biochemistry* **23,** 6363–6368.

Duester G. (1998) Alcohol dehydrogenase as a critical mediator of retinoic acid synthesis from vitamin A in the mouse embryo, *J. Nutrition.* **128,** 459S-462S.

Duester, G., Farrés, J., Felder, M.R., Holmes, R., Höög, J.-O., Parés, X., Plapp, B.V., Yin, S.-J. and Jörnvall, H. (1998) Recommended nomenclature for the vertebrate alcohol dehydrogenase gene family, Submitted to *FEBS Lett.*

Eklund, H., Plapp, B.V., Samama, J.-P. and Brändén, C.-I. (1992) Binding of substrate in a ternary complex of horse liver alcohol dehydrogenase, *J. Biol. Chem.* **257,** 14349–14358.

Estonius, M., Danielsson, O., Karlsson, C., Persson, H., Jörnvall, H. and Höög, J.-O. (1993) Distribution of alcohol and sorbitol dehydrogenases: Assessment of mRNAs in rat tissues, *Eur. J. Biochem.* **215,** 497–503.

Estonius, M., Svensson, S. and Höög, J.-O. (1996) Alcohol dehydrogenase in human tissues. Localisation of transcripts coding for five classes of the enzyme, *FEBS Lett.* **397,** 338–342.

Hjelmqvist, L., Estonius, M. and Jörnvall, H. (1995) The vertebrate alcohol dehydrogenase system: Variable class II type form elucidates separate stages of enzymogenesis, *Proc. Natl. Acad. Sci. USA* **92,** 10905–10909.

Höög, J.-O. (1995) Cloning and characterization of a novel rat alcohol dehydrogenase of class II type, *FEBS Lett.* **368,** 445–448.

Höög, J.-O., von Bahr-Lindström, H., Hedén, L.-O., Holmquist, B., Larsson, K., Hempel, J., Vallee, B.L. and Jörnvall, H. (1987) Structure of the class II enzyme of human liver alcohol dehydrogenase. Combined cDNA and protein sequence determination of the π subunit, *Biochemistry* **26,** 1926–1932.

Höög, J.-O. and Svensson, S. (1997) Mammalian class II alcohol dehydrogenase—a highly variable enzyme, In: *Enzymology and molecular biology of carbonyl metabolism 6* (Weiner, H., Lindahl, R., Crabb, D.W. and Flynn, T.G., eds.) Plenum Press, pp 303–311.

Li, T.-K., Bosron, W.F., Dafeldecker, W.P., Lange, L.G. and Vallee, B.L. (1977) Isolation of Π-alcohol dehydrogenase of human liver: Is it a determinant of alcoholism? *Proc. Natl. Acad. Sci. USA* **74,** 4378- 4381.

Marschall, H.-U., Oppermann, U.C.T., Svensson, S., Nordling, E., Persson, B., Höög, J.-O and Jörnvall, H. (1998) Human liver alcohol dehydrogenase γγ isozyme: the sole cytosolic 3β-hydroxysteroid dehydrogenase of iso-bile acids, *Biochemistry*, in press.

Maskos, Z. and Winston, G.W. (1994) Mechanism of p-nitrosophenol reduction catalyzed by horse liver and human π-alcohol dehydrogenase (ADH), *J. Biol. Chem.* **269,** 31579–31584.

Mårdh, G., Dingely, A.L., Auld, D.S. and Vallee, B.L. (1986) Human class II (π) alcohol dehydrogenase has a redox-specific function in norepinephrine metabolism, *Proc. Natl. Acad. Sci. USA* **83,** 8908–8912.

Mårdh, G., Auld, D.S. and Vallee, B.L. (1987) Thyroid hormones selectively modulate human alcohol dehydrogenase isozyme catalyzed ethanol metabolism, *Biochemistry* **26,** 7585–7588.

Parés, X., Cederlund, E., Moreno, A., Hjelmqvist, L., Farrés, J. and Jörnvall, H. (1994) Mammalian class IV alcohol dehydrogenase (stomach alcohol dehydrogenase): Structure, origin and correlation with enzymology, *Proc. Natl. Acad. Sci. USA* **91,** 1893–1897.

Sellin, S., Holmquist, B., Mannervik, B. and Vallee, B.L. (1990) Oxidation and reduction of 4-hydroxyalkenals catalyzed by isozymes of human alcohol dehydrogenase, *Biochemistry* **30,** 2514–2518.

Svensson, S., Lundsjö, A., Cronholm, T. and Höög, J.-O. (1996) Aldehyde dismutase activity of human liver alcohol dehydrogenase, *FEBS Lett.* **394,** 217–220.

Svensson, S., Hedberg, J.J. and Höög, J.-O. (1998) Structural and functional divergence of class II alcohol dehydrogenase. Cloning and characterization of rabbit liver isoforms of the enzyme, *Eur. J. Biochem.* **251,** 236–243.

Uotila, L. and Koivusalo, M. (1989) Glutathione-dependent oxidoreductases:formaldehyde dehydrogenase, In: *Glutathione: Chemical, biochemical and medical aspects, part A* (Dolphin, D., Poulson, R. and Avramovic, O., eds.) John Wiley and sons, pp. 517–551.

Vallee, B.L. and Bazzone, T.J. (1983) Isozymes of human liver alcohol dehydrogenase, *Isozyme* **8,** 219–244.

Yang, Z.N., Davies, G.J., Hurley, T.D., Stone, C.L., Li, T.-K. and Bosron, W.F. (1994) Catalytic efficiency of human alcohol dehydrogenases for retinol oxidation and retinal reduction, *Alcohol Clin. Exp. Res.* **18,** 587–591.

Yang, Z.N., Bosron, W.F. and Hurley, T.D. (1997) Structure of human χχalcohol dehydrogenase: A glutathione-dependent formaldehyde dehydrogenase, *J. Mol. Biol.* **265,** 330–343.

ZINC BINDING CHARACTERISTICS OF THE SYNTHETIC PEPTIDE CORRESPONDING TO THE STRUCTURAL ZINC SITE OF HORSE LIVER ALCOHOL DEHYDROGENASE

Tomas Bergman,[1] Carina Palmberg,[1] Hans Jörnvall,[1] David S. Auld,[2] and Bert L. Vallee[2]

[1]Department of Medical Biochemistry and Biophysics
Karolinska Institutet
S-171 77 Stockholm, Sweden
[2]Center for Biochemical and Biophysical Sciences and Medicine
Harvard Medical School
Boston, Massachusetts 02115

1. INTRODUCTION

Medium-chain dehydrogenases/reductases of the liver alcohol dehydrogenase type are zinc metalloenzymes (Vallee and Hoch, 1957; Åkeson, 1964; Drum et al., 1969), with two zinc atoms per subunit, one catalytic at the active site and one structural at a site influencing subunit interactions (Sytkowski and Vallee, 1976; Brändén et al., 1975). In horse liver alcohol dehydrogenase and in all other mammalian liver forms, the structural zinc atom is liganded by four closely spaced Cys residues, at positions 97, 100, 103, and 111 (Brändén et al., 1975; Vallee and Auld, 1990). The mechanism by which this zinc atom maintains its structural role is largely unknown. To probe the metal-binding characteristics of the structural zinc site of alcohol dehydrogenase, we have analyzed zinc-binding to a synthetic replica of the protein segment containing the four Cys residues and covering residues 93–115 of the parent molecule.

Previously we have shown this peptide to mimick the metal-binding properties of the intact enzyme in cobalt-substitution experiments (Bergman et al., 1992; Bergman et al., 1993; Gheorghe et al., 1995) and we have now continued this study and investigated experimental approaches to determine the zinc-binding constant. Titration of the peptide/metal complex with a metallochromic chelator is potentially a useful technique and we now report on tests with 4-(2-pyridylazo)resorcinol (cf. Hunt et al., 1985) to analyze the zinc site.

Enzymology and Molecular Biology of Carbonyl Metabolism 7
edited by Weiner *et al.* Kluwer Academic / Plenum Publishers, New York, 1999.

Phe-Thr-Pro-Gln-Cys-Gly-Lys-Cys-Arg-Val-Cys-Lys-His-Pro-Glu-Gly-Asn-Phe-Cys-Leu-Lys-Asn-Asp
93 115

Figure 1. Amino acid sequence of the zinc-binding synthetic peptide corresponding to residues 93–115 of horse liver alcohol dehydrogenase. The four Cys ligands are underlined.

2. EXPERIMENTAL

Peptide synthesis was carried out with side-chain-protected tertiary butyloxycarbonyl amino acid derivatives. Final purification was by preparative reverse-phase HPLC and peptide integrity was checked by amino acid analysis and mass spectrometry.

In the zinc-binding studies, Hepes buffer (20 mM, pH 7.5) was used after pretreatment with diphenylthiocarbazone (dithizone) to remove traces of metal ions (Holmquist, 1988). The Cys-containing peptides were reduced with dithiothreitol (DTT) and stored at -70°C until used. Before zinc incubation the DTT was removed via exclusion chromatography on a Bio-Gel P4 column (BioRad) and the reduced peptides were collected under anaerobic conditions. Zinc was added to fractions and the excess unbound zinc was separated from the peptide/zinc complex via another exclusion chromatography (Bio-Gel P4). The zinc-binding stoichiometry was evaluated by atomic absorption spectrophotometry and amino acid analysis of the elution fractions. The metallochromic chelator 4-(2-pyridylazo)resorcinol (PAR) was tested for determination of the zinc-binding constant via extraction of zinc from the metal-saturated peptide by measuring the absorbance at 500 nm for the $Zn(PAR)_2$ complex.

3. RESULTS AND DISCUSSION

A zinc-binding peptide corresponding to the loop around the structural zinc atom of horse liver alcohol dehydrogenase was synthesized for metal affinity studies. The 23-residue peptide covers the protein segment between residues 93–115 with Cys zinc ligands at positions 97, 100, 103, and 111 in the protein (Figure 1).

Zinc was added to the reduced peptide followed by anaerobic exclusion chromatography and measurement of zinc-binding. As previously reported (Bergman et al., 1992), zinc is bound at a 1:1 zinc/peptide ratio which is also the stoichiometry found in the protein. To further characterize the zinc-binding properties of the peptide replica we needed a method to determine the zinc-binding affinity of the peptide. Metallochromic chelators have been used to quantify zinc in biological fluids and for monitoring zinc release from proteins (Pollák and Kubáñ, 1979; Hunt et al., 1985). We tested the metallochromic chelator PAR (Figure 2) as a competitor to the peptide for its bound zinc.

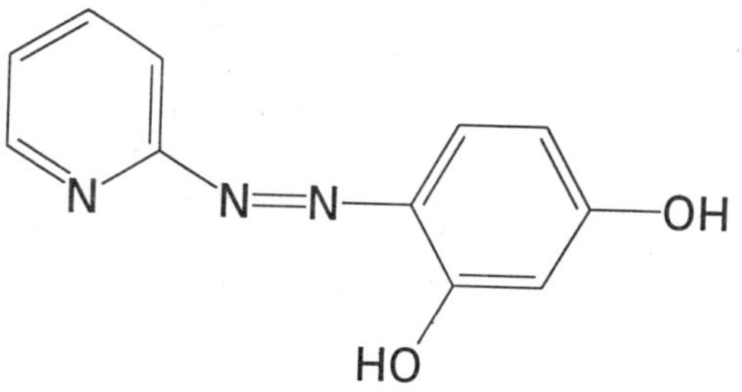

Figure 2. Structure of 4-(2-pyridylazo)resorcinol (PAR). The oxygen at position 3 and the azo-nitrogen at position 4 of the resorcinol ring together with the nitrogen of the pyridyl moiety have been suggested to constitute a terdentate ligand to zinc with the three donor atoms in a plane (cf. Iwamoto, 1961).

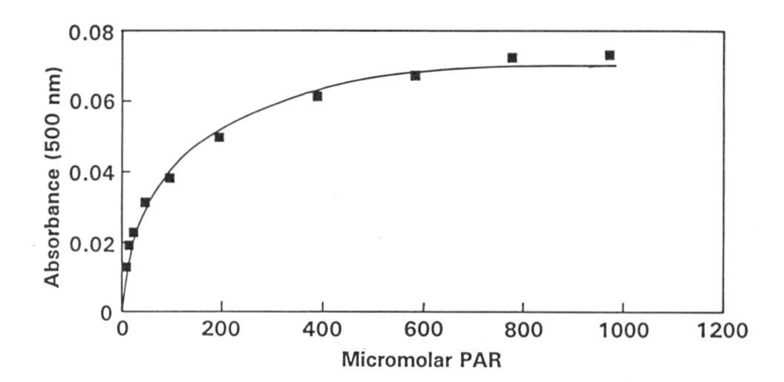

Figure 3. PAR-titration of the peptide/Zn complex (1 μM) of the 23-residue replica of the zinc-binding protein segment. The corresponding zinc-binding constant is 6.7×10^9 M^{-1}.

The saturated peptide/zinc complex is exposed to increasing concentrations of PAR which extracts and chelates peptide-bound zinc to form a $Zn(PAR)_2$ complex whose concentration is measured by its absorbance at 500 nm. At sufficiently high concentrations of PAR all peptide-bound zinc is transferred to PAR and the absorbance at 500 nm becomes constant. The peptide zinc-binding constant can be calculated from the corresponding graph correlating absorbance at 500 nm with PAR concentration (Figure 3).

The complete extraction of the zinc bound to the peptide replica of the protein segment requires a PAR concentration that is 1000-fold higher than that of the peptide concentration. The corresponding zinc-binding constant is of the order 10^9 M^{-1} and separate determinations of different peptide/zinc preparations are highly reproducible. The peptide/zinc complex appears fairly stable towards oxidation and anaerobicity during titration is not critical.

In conclusion, zinc-binding to a model peptide corresponding to the structural zinc site of alcohol dehydrogenase indicates that the metallochromic chelator PAR is a good candidate for direct titration experiments aiming at determination of zinc-binding constants. It should be possible to detect significant differences in the zinc affinity of coordination sites that differ in nature and number of ligands and intervening residues by means of PAR titration. This will lead to a better understanding of the forces governing zinc-binding in proteins (cf. Vallee and Auld, 1990) which is important for design of synthetic enzymes.

ACKNOWLEDGMENTS

This work was supported by grants from the Swedish Medical Research Council (Projects 13X-10832 and 13X-3532), the Swedish Cancer Society (Project 1806), the Endowment for Research in Human Biology, Inc., the Emil and Wera Cornell Foundation and the Magn. Bergvall Foundation.

REFERENCES

Bergman, T., Jörnvall, H., Holmquist, B., and Vallee, B.L., 1992, A synthetic peptide encompassing the binding site of the second zinc atom (the "structural" zinc) of alcohol dehydrogenase, *Eur. J. Biochem.*, 205:467–470.

Bergman, T., Jörnvall, H., Härd, T., Holmquist, B., and Vallee, B.L., 1993, A synthetic approach to analysis of the structural zinc site of alcohol dehydrogenase, *in*: Enzymology and Molecular Biology of Carbonyl Metabolism 4, Weiner, H., Crabb, D.W., and Flynn, T.G., eds., Plenum Press, New York, pp. 419–428.

Brändén, C.-I., Jörnvall, H., Eklund, H., and Furugren, B., 1975, Alcohol dehydrogenases, *in*: The Enzymes 11, Boyer, P.D., ed., Academic Press, New York, pp. 103–190.

Drum, D.E., Li, T.-K., and Vallee, B.L., 1969, Zinc isotope exchange in horse liver alcohol dehydrogenase, *Biochemistry*, 8:3792–3797.

Gheorghe, M. T., Lindh, I., Griffiths, W.J., Sjövall, J., and Bergman, T., 1995, Analytical approaches to alcohol dehydrogenase structures, *in*: Enzymology and Molecular Biology of Carbonyl Metabolism 5, Weiner, H., Holmes, R.S., and Wermuth, B., eds., Plenum Press, New York, pp. 417–426.

Holmquist, B., 1988, Elimination of adventitious metals, *Methods Enzymol.*, 158:6–12.

Hunt, J.B., Neece, S.H., and Ginsburg, A., 1985, The use of 4-(2-pyridylazo)resorcinol in studies of zinc release from *Escherichia coli* aspartate transcarbamoylase, *Anal. Biochem.*, 146:150–157.

Iwamoto, T., 1961, Acid-base property and metal chelate formation of 4-(2-pyridylazo)-resorcinol, *Bull. Chem. Soc. Japan*, 34:605–610.

Pollák, M., and Kubáň, V., 1979, Comparison of spectrophotometric methods of determination of zinc (II) in biological material and study of its complex formation reactions with 4-(2-pyridylazo)resorcinol, *Collection Czechoslov. Chem. Commun.*, 44:725–741.

Sytkowski, A.J., and Vallee, B.L., 1976, Chemical reactivities of catalytic and noncatalytic zinc or cobalt atoms of horse liver alcohol dehydrogenase: differentiation by their thermodynamic and kinetic properties, *Proc. Natl. Acad. Sci. USA*, 73:344–348.

Vallee, B.L., and Auld, D.S., 1990, Zinc coordination, function, and structure of zinc enzymes and other proteins, Biochemistry, 29:5647–5659.

Vallee, B.L., and Hoch, F.L., 1957, Zinc in horse liver alcohol dehydrogenase, *J. Biol. Chem.*, 225:185–195.

Åkeson, Å., 1964, On the zinc content of horse liver alcohol dehydrogenase, *Biochem. Biophys. Res. Commun.*, 17:211–214.

42

AMPHIBIAN ALCOHOL DEHYDROGENASE

Purification and Characterization of Classes I and III from *Rana perezi*

Josep Maria Peralba, Bernat Crosas, Susana E. Martínez, Pere Julià,
Jaume Farrés, and Xavier Parés

Department of Biochemistry and Molecular Biology
Faculty of Sciences
Universitat Autònoma de Barcelona
08193 Bellaterra, Spain

1. INTRODUCTION

Alcohol dehydrogenase (ADH) catalyses the reversible interconversion of a variety of alcohols and their corresponding aldehydes and ketones, and is widely distributed in organisms. It has been detected in all animals, in plants, and eukaryotic and prokaryotic microorganisms. In vertebrates the ADH system is complex, with at least seven different enzymatic classes (Jörnvall and Höög, 1995, Kedishvili et al., 1997). The study in submammalian species is of interest to understand the relationship between the classes found in mammals and the evolutionary origins of the present structures. Moreover, since the enzyme exhibits a wide substrate specificity, the comparative study of the enzymes from distant species may provide valuable information on the function of the enzyme, and how it may change among groups of vertebrates to adapt to different physiological needs.

A number of works exist on the properties of avian (Nussrallah et al., 1989, Kaiser et al., 1990, Estonius et al., 1990, Estonius et al., 1994, Hjelmqvist et al., 1995), reptilian (Persson et al., 1993, Hjelmqvist et al., 1996) and piscine (Danielson et al., 1992, Danielsson and Jörnvall, 1992, Ramaswamy et al., 1996, Danielsson et al., 1996) alcohol dehydrogenases. The information on amphibian ADH is, however, still scarce. In a previous publication we reported the amino acid sequence of the major liver ADH from the frog *Rana perezi* (Cederlund et al., 1991). It consisted of a class I enzyme, homologous to the class I forms present in all other groups of vertebrates. The analysis of the primary structure of the amphibian class I ADH gave the first estimate for the time of emergence of the class I type from the class III type, which was positioned to roughly the time of the vertebrate divergence (Cederlund et al., 1991).

Enzymology and Molecular Biology of Carbonyl Metabolism 7
edited by Weiner *et al.* Kluwer Academic / Plenum Publishers, New York, 1999.

In the present work we have studied the kinetic constants and substrate specificity of the previously sequenced amphibian class I enzyme. Moreover, we have purified and characterized the class III ADH from the same organism (*Rana perezi*) and we have recognized the existence of a different ADH form present in the frog stomach.

2. EXPERIMENTAL PROCEDURES

2.1. Animals and Enzyme Purification

Rana perezi, a green frog common in Northeastern Spain, was used. The animals were sacrificed by decapitation, and their organs were dissected and stored at -80°C.

Prior to analysis, the organs were thawed, cut in small pieces and homogenized in 10 mM Tris/HCl, 0.5 mM dithiothreitol, pH 8.0 (1:1, w/v). The homogenates were centrifuged at 4°C and supernatants were used for ADH activity and electrophoretic analysis. Total protein was determined by the Coomassie Blue method (Bradford, 1976), using bovine serum albumin as a standard.

Typically, class I and class III enzymes were purified from about 80 frog livers (97 g). Homogenate supernatant was applied to a DEAE-Sepharose column (2.5 × 27 cm) equilibrated with the homogenation buffer, and the enzymes were eluted by a linear gradient of 0–150 mM NaCl. Fractions showing class I or class III activities were pooled and concentrated separately. Both enzymes were finally purified by application to an AMP-Sepharose column (1×25 cm) equilibrated with 10 mM sodium phosphate, 0.5 mM dithiothreitol, pH 7.4. The purified enzyme was eluted by a linear gradient of 0–66 μM NADH in 0.1 M Tris/HCl, 0.5 mM dithiothreitol, pH 8.2 (for class I ADH) or pH 7.6 (for class III ADH).

2.2. Enzyme Assays

ADH activity was determined spectrophotometrically in 0.1 M glycine/NaOH, pH 10.0, or in 0.1 M sodium phosphate, pH 7.5, at 25°C, by measuring the change in absorbance at 340 nm. The activity in partially purified fractions was measured at pH 10.0 and the specific conditions (standard assay) for each enzyme were: 5 mM ethanol and 1.2 mM NAD^+ for class I enzyme and 1 mM octanol and 2.4 mM NAD^+ for class III ADH. Kinetic results were processed by using the data analysis program Enzfitter. The constants shown represent the mean of at least three determinations performed with different preparations of enzyme.

2.3. Electrophoresis

Electrophoresis on starch gel was performed as described earlier (Moreno and Parés, 1991). Gel slices were stained for ADH activity by using ethanol or 2-buten-1-ol (0.1 M) as the substrate.

Isoelectric focusing was performed on a Miniprotean instrument (Bio-Rad, Richmond, USA), at 4°C, for a pH range of 3–10. The staining for enzyme activity on polyacrylamide gel was performed as described for starch gel electrophoresis.

Analysis of the purity and determination of the subunit molecular mass of the frog enzymes were performed by electrophoresis on SDS/polyacrylamide gel and staining by the silver-stain technique.

3. RESULTS AND DISCUSSION

3.1. Starch Gel Electrophoresis and Activity Analysis of Homogenates from Frog Tissues

The starch gel electrophoresis of homogenates from frog liver, stomach and eyes (Figure 1) demonstrated the existence of three enzyme types, unevenly distributed. The class III ADH is the most anodic form and is detected in all tissues studied, using pentanol or 2-buten-1-ol as substrates. The class III band is clearly observed in liver, but it is very faint in other tissues. The class I ADH shows several bands with a slower mobility than class III and is the major liver enzyme, being revealed with ethanol, pentanol or 2-buten-1-ol. A stomach ADH, also detected in a moderate amount in intestine but not in the liver or in the eye, has a cathodic mobility and exhibits a diffuse strong band with 2-buten-1-ol, or pentanol as substrates. The characterization of this form will be reported elsewhere (Peralba et al., 1998).

Liver class I shows several bands, and the pattern changes from one animal to another, suggesting the existence of different isozymes and polymorphism (Figure 1). Structural heterogeneity was, however, not detected in a purified preparation (Cederlund et al., 1991), although polymorphism due to the existence of several genes and alleles has been suggested for the major liver ADH of the amphibian *Xenopus laevis* (Wesolowski and Lyerla, 1979 and 1983). In any case, the purified enzyme used for kinetic characterization (see below) contained a mixture of forms.

3.2. Purification of Frog Alcohol Dehydrogenase Enzymes

Class I and class III ADHs were purified from the same preparation of frog liver by extraction in aqueous media, ion-exchange chromatography on DEAE-Sepharose, and affinity chromatography on AMP-Sepharose. From 97 g of frog liver, 7.8 mg of a class I enzyme, with a specific activity of 0.5 U/mg, and 0.74 mg of a class III enzyme, with a specific activity of 2.8 U/mg, were obtained. The yields were 23 % for class I ADH and 24 % for class III ADH (Tables 1 and 2). Class I is a major protein in the frog liver, representing about 1% of the total soluble protein (Table 1). Class III is much less abundant (Table 2).

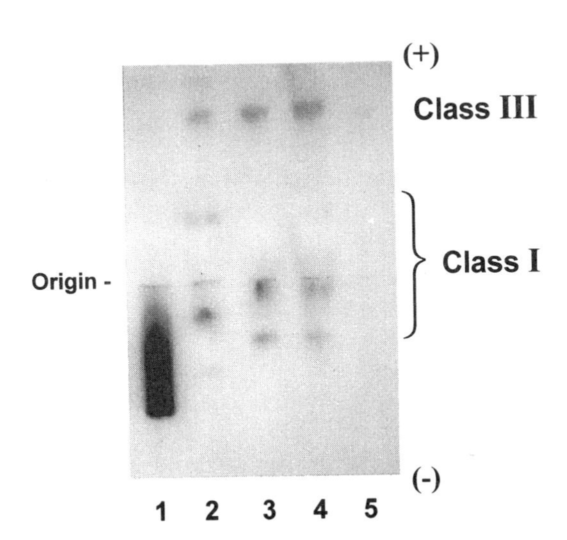

Figure 1. Starch gel electrophoresis of frog tissue homogenates. The electrophoresis was performed at pH 7.6 and the gel was stained with 2-buten-1-ol. Lane 1, stomach, showing the distinct ADH form; lanes 2–4, liver from different animals; lane 5, whole eye.

Table 1. Purification of frog liver class I ADH

Step	Protein (mg)	Total activity (U)	Specific activity (U/mg)	Purification (-fold)	Yield (%)
Supernatant	3977	18.22	0.005	1	100
DEAE-Sepharose	213	10.05	0.047	9	55
AMP-Sepharose	7.8	4.23	0.5	110	23

About 80 livers, 97 g, were used. Activity was measured at pH 10.0, with 5 mM ethanol and 1.2 mM NAD^+.

3.3. Enzymatic Characterization

3.3.1. Molecular Properties. SDS/polyacrylamide gel electrophoresis indicated a very similar subunit molecular mass for the two frog ADH enzymes. From the mobility of known molecular mass standards, an apparent subunit Mr of 41000 for class I ADH, and 42000 for class III ADH were estimated.

Isoelectric focusing of frog class I ADH yielded multiple bands with an isoelectric point of 6.8–7.4, while class III exhibited a sharp single band with pI of 6.2.

3.3.2. Kinetic Analysis. 3.3.2.1. Kinetic Properties of Amphibian Class I ADH. At 1.2 mM cofactor concentration the enzyme shows 12 times more activity with NAD^+ than with $NADP^+$, demonstrating the NAD^+ dependence of the amphibian class I. The pH-activity profile with 1 mM ethanol as the substrate for the oxidation reaction shows an optimum at pH 10.0. The aldehyde reduction activity with 0.7 mM acetaldehyde shows a maximum value at pH 6.7.

The Km values for class I ADH (Table 3) are in general low for both the alcohols and the aldehydes, although with the highest Km value, the enzyme can oxidize methanol. At pH 10.0, Km values decrease as the chain length of the alcohols increases up to pentanol. Km values for octanol and 12-hydroxydodecanoic acid are higher than that for pentanol. Benzyl alcohol is the best alcohol substrate because of the lowest Km. The comparatively large Km value with cyclohexanol suggests that the enzyme uses preferentially primary alcohols. The similar kcat values for most alcohols suggest a common limiting step for the overall reaction, probably NADH dissociation. The kinetic properties for alcohols at pH 7.5 are similar to those at pH 10.0, except for the lower kcat values.

As for many other vertebrate ADHs, the frog class I exhibits higher efficiency with aldehydes than with alcohols at pH 7.5, mostly because of the about 30-fold increase in kcat. Kinetics with aldehydes confirm the preference of the enzyme for medium-chain and aromatic substrates, and that primary alcohols and the corresponding aldehydes are better substrates than the secondary compounds. All substrates, alcohols and aldehydes, show inhibition by excess of substrate at concentrations of about 10×Km.

Table 2. Purification of frog liver class III ADH

Step	Protein (mg)	Total activity (U)	Specific activity (U/mg)	Purification (-fold)	Yield (%)
Supernatant	3977	8.8	0.002	1	100
DEAE-Sepharose	574	5.2	0.009	4.5	60
AMP-Sepharose	0.74	2.1	2.8	1400	24

Purification was performed from the same liver preparation as that for class I ADH (Table 1). Activity was measured at pH 10.0, with 1 mM octanol and 2.4 mM NAD^+. Activity in the supernatant was measured in the presence of 1 mM 4-methylpyrazole to inhibit Class I ADH.

Table 3. Kinetic constants of frog class I ADH

Substrate	pH	K_m (μM)	k_{cat} (min^{-1})	k_{cat}/K_m ($min^{-1} \cdot mM^{-1}$)
Methanol	10	5000	13	2.6
Ethanol	10	90	43	480
1-Butanol	10	15	26	1700
1-Pentanol	10	5	25	5000
1-Octanol	10	14	28	2000
12-Hydroxydodecanoic acid	10	15	26	1700
Benzylalcohol	10	3	22	7300
Cyclohexanol	10	220	27	120
2-buten-1-ol	10	12	46	3800
NAD^+	10	100		
Ethanol	7.5	100	17	170
1-Butanol	7.5	11	11	1000
1-Octanol	7.5	16	7	440
NAD^+	7.5	30		
Acetaldehyde	7.5	73	565	7740
Butanal	7.5	5	484	96800
Hexanal	7.5	8	714	89250
Octanal	7.5	12	378	31500
m-Nitrobenzaldehyde	7.5	0.6	183	305000
Benzaldehyde	7.5	9	358	39800
Cyclohexanone	7.5	4500	45	10
K_i pyrazole (pH 10)	0.15 μM			

Km values for NAD^+ and NADH were determined with 1 mM ethanol and 0.7 mM acetaldehyde respectively. Coenzyme concentrations for the forward and reverse reactions were 1.2 mM NAD^+ and 0.05 mM NADH, respectively.

The inhibition constant for pyrazole of frog class I ADH was determined at pH 10.0, as indicated in Table 3. The amphibian enzyme, similarly to its mammalian counterpart, is strongly inhibited by pyrazole, in a competitive manner.

The wide substrate specificity, the low Km values for most substrates, and the very low Ki for pyrazole (Table 3) confirm that the major liver enzyme is of class I type, as previously concluded from the primary structure studies (Cederlund et al., 1991). The structural features also help to understand the basis of the remarkably low Km values of the frog enzyme. The substrate-binding pocket of the amphibian enzyme is extremely hydrophobic. Thus, when compared to the γ1 subunit of the human class I ADH, the frog class I has Leu instead of Tyr at position 110 and two larger residues, Ile instead of Val at oposition 141 and Leu instead of Val at position 294. (The importance of residue 294 on the Km values has been also recognized for the class IV enzymes by Farrés et al., 1994). The overall effect of these changes is an extremely hydrophobic and somewhat space-restricted substrate-binding site (Cederlund et al., 1991). The resulting lower amount of water molecules to be displaced by the substrate (Julià et al., 1988), and the fewer possibilities of unproductive encounters with the substrate (Xie et al., 1997) would explain the tight binding of substrates and inhibitors to the amphibian ADH.

3.3.2.2. Kinetic Properties of Amphibian Class III ADH. With a cofactor concentration of 2.4 mM, the enzyme is 7-fold more active with NAD^+ than with $NADP^+$, demonstrating the NAD dependence of amphibian class III. The optimal pH for formaldehyde oxidation in the presence of glutathione is 8.2, while activity at pH 7.5 is only 40% that of the optimal value.

Typically for a class III enzyme, short-chain alcohols do not saturate the enzyme, although Km and kcat values can be obtained for medium and long-chain compounds (Table 4). However, catalytic efficiency is 100 times higher for S-hydroxymethylglutathione (the adduct formed by spontaneous reaction between glutathione and formaldehyde), unambiguously demonstrating the specificity of the enzyme towards this substrate. An additional property, and also a class III feature, is the very high Ki values for pyrazole derivatives (Table 4), indicating a lack of inhibition at the usual pyrazole concentrations. A property unique for the class III ADH is the activation of the alcohol oxidation by carboxylic acids, first demonstrated for the human enzyme (Moulis et al., 1991). The frog enzyme exhibits also a similar behavior. Thus, 1 mM octanoic acid produces a 2.3-fold activation of ethanol (0.5 M) oxidation, but octanoic acid is a competitive inhibitor (Ki=0.8 mM) of octanol oxidation.

Class III ADH has been fully characterized from different species: vertebrates (mammals, reptiles), invertebrates (*Drosophila*, octopus), plants and microorganisms (yeast and bacteria). Comparison of the kinetic constants clearly reveals that two class III types exist: a multicellular organism type with low Km (1–6 μM) and low kcat (200–900 min^{-1}), and a unicellular type (yeast and bacteria) with both high Km (40–90 μM) and kcat (5000–9000 min^{-1}). These two kinetically different types may represent adaptations to intracellular or environmental concentrations of formaldehyde, potentially higher in microorganisms than in multicellular organisms (Fernández et al., 1995 and 1996). The amphibian class III characteristics follow the expected pattern, with low Km (1.6 μM) for S-hydroxymethylglutathione and relatively low kcat (386 min^{-1}).

3.4. Functional and Evolutionary Relationships

Class I ADH has been detected in the liver of all vertebrates, from bony fishes to mammals. In contrast, class III is the only member of the medium-chain ADH family in invertebrates or the cyclostome (Fernández et al., 1993, Danielsson et al., 1994). General

Table 4. Kinetic constants of frog class III ADH

Substrate	pH	K_m (μM)	k_{cat} (min^{-1})	k_{cat}/K_m ($min^{-1} \cdot mM^{-1}$)
Ethanol	10	n.s.	–	–
1-Butanol	10	n.s.	–	–
1-Pentanol	10	37000	390	11
1-Octanol	10	800	450	560
12-Hydroxydodecanoic acid	10	170	524	3080
16-Hydroxyhexadecanoic acid	10	30	55	1830
NAD^+	10	240		
Octanal	7.5	2000	374	190
NADH	7.5	10		
S-Hydroxymethylglutathione	8.0	1.6	386	241250
K_i pyrazole (pH 10)	40 mM			
K_i 4-methylpyrazole (pH 10)	20 mM			

Km values for NAD^+ and NADH were determined with 1 mM octanol and 1 mM octanal respectively. Coenzyme concentrations for the forward and reverse reactions were 2.4 mM NAD^+ and 0.1 mM NADH.
n. s., the enzyme did not reach saturation even at high concentrations of substrate.

kinetic properties of class I are similar in all groups of vertebrates, with low Km for ethanol (in the low millimolar range), and low Ki for pyrazole (in the low micromolar range). The kinetic properties, the wide substrate specificity, and the hepatic localization suggest a general role in the detoxication of ingested alcohols, or those produced by the intestinal flora. The activity of class I with endogenous compounds of physiological interest, such as retinoids, hydroxysteroids and aldehydes derived from lipid peroxidation gives the possibility of a more specific function for class I (Boleda et al., 1993). In this regard the activity of the amphibian class I with retinoids is only moderate because of the low kcat values (unpublished results).

The results on the properties of the amphibian class III ADH are consistent with the general features known for this enzyme. Class III is the most ancestral ADH form, present in all animals, and with little changes in structure and function among species, reflecting an essential role in formaldehyde elimination. Km and kcat values for the frog enzyme towards S-hydroxymethylglutathione are very similar to those of other vertebrate groups, but also to those of invertebrates or plants, thus defining the multicellular organism kinetic type for class III. In contrast Km and kcat values are higher for the unicellular organisms.

4. CONCLUSIONS

Three alcohol dehydrogenase types have been detected in tissues from the frog *Rana perezi*: a class I ADH found in liver, which exhibits electrophoretic heterogeneity; a class III ADH observed in all organs analyzed, and a stomach type, which is the major ADH in the gastric tissue and it is absent in liver. Classes I and III have been purified to homogeneity and characterized. Class I shows the general properties of this class found in all vertebrate groups. The remarkably low Km values shown by the frog enzyme are consistent with a more hydrophobic and space-restricted substrate binding site demonstrated by the primary structure. The molecular and kinetic properties of amphibian class III sustain the notion of a very conserved enzyme in both structure and function for all vertebrates.

ACKNOWLEDGMENTS

This work was supported by grants from the Commission of the European Union (BIO4-CT97-2123), Comissionat per a Universitats i Recerca (1997SGR 00040), and Dirección General de Enseñanza Superior (PM96-0069).

REFERENCES

Boleda, M.D., Saubi, N., Farrés, J., and Parés, X., 1993, Physiological substrates for rat alcohol dehydrogenase classes: Aldehydes of lipid peroxidation, ω-hydroxyfatty acids, and retinoids, *Arch. Biochem. Biophys.* **307**, 85–90.

Bradford, M.M., 1976, A rapid and sensitive method for the quantitation of microgram quantities of protein utilizing the principle of protein-dye binding, *Anal. Biochem.* **72P**, 248–254.

Cederlund, E., Peralba, J.M., Parés, X., and Jörnvall, H., 1991, Amphibian alcohol dehydrogenase, the major frog liver enzyme. Relationships to other forms and assessment of an early gene duplication separating vertebrate class I and class III alcohol dehydrogenases, *Biochemistry* **30**, 2811–2816.

Danielsson, O., Eklund, H., and Jörnvall, H., 1992, The major piscine liver alcohol dehydrogenase has class-mixed properties in relation to mammal alcohol dehydrogenases of class I and III, *Biochemistry* **31**, 3751–3759.

Danielsson, O., and Jörnvall, H., 1992, "Enzymogenesis": Classical liver alcohol dehydrogenase origin from the glutathione-dependent formaldehyde dehydrogenase line, *Proc. Natl. Acad. Sci. USA* **89**, 9247–9251.

Danielsson, O., Shafqat, J., Estonius, M. and Jörnvall, H., 1994, Alcohol dehydrogenase class III contrasted to class I: Characterization of the cyclostome enzyme, the existence of multiple forms as for the human enzyme, and distant cross-species hybridization. *Eur. J. Biochem.* **225**, 1081–1088.

Danielsson, O., Shafqat, J., Estonius, M., El-Ahmad, M., and Jörnvall, H., 1996, Isozyme multiplicity with anomalous dimer patterns in a class III alcohol dehydrogenase. Effects on the activity and quaternary structure of residue exchanges at "nonfunctional" sites in a native protein, *Biochemistry* **35**, 14561–14568.

Estonius, M., Karlsson, C., Fox, E.A., Höög, J.-O., Holmquist, B., Vallee, B.L., Davidson, W.S. and Jörnvall, H., 1990, Avian alcohol dehydrogenase: The chicken liver enzyme. Primary structure, cDNA cloning, and relationships to other alcohol dehydrogenases, *Eur. J. Biochem.* **194**, 593–602.

Estonius, M., Hjelmqvist, L. and Jörnvall, H., 1994, Diversity of vertebrate class I alcohol dehydrogenase. Mammalian and non-mammalian enzyme functions correlated through the structure of a ratite enzyme, *Eur. J. Biochem.* **224**, 373–378.

Farrés, J., Moreno, A., Crosas, B., Peralba, J.M., Allali-Hassani, A., Hjelmqvist, L., Jörnvall, H., and Parés, X., 1994, Alcohol dehydrogenase of class IV (ss-ADH) from human stomach. cDNA sequence and structure/function relationships, *Eur. J. Biochem.* **224**, 549–557.

Fernández, M.R., Jörnvall, H., Moreno, A., Kaiser, R., and Parés, X., 1993, Cephalopod alcohol dehydrogenase: purification and enzymatic characterization, *FEBS Lett.* **328**, 235–238.

Fernández, M.R., Biosca, J.A., Norin, A., Jörnvall, H., and Parés, X., 1995, Class III alcohol dehydrogenase from *Saccharomyces cerevisiae*: Structural and enzymatic features differ toward the human/mammalian forms in a manner consistent with functional needs in formaldehyde detoxication, *FEBS Lett.* **370**, 23–26.

Fernández, M.R., Biosca, J.A., Martínez, M.C., Achkor, H., Farrés, J., and Parés, X., 1997, Formaldehyde dehydrogenase from yeast and plant. Implications for the general functional and structural significance of class III alcohol dehydrogenase, *Adv. Exp. Med. Biol.* **414**, 373–381.

Hjelmqvist, H., Metsis, M., Persson, H., Höög, J.-O., McLennan, J. anf Jörnvall, H., 1995, Alcohol dehydrogenase of class I: kiwi liver enzyme, parallel evolution in separate vertebrate lines, and correlation with 12S rRNA patterns, *FEBS Lett.* **367**, 306–310.

Hjelmqvist, L., Shafqat, J., Siddiqi, A.R. and Jörnvall, H., 1996, Linking of isozyme and class variability patterns in the emergence of novel alcohol dehydrogenase functions: Characterization of isozymes in *Uromastix hardwickii. Eur. J. Biochem.* **236**, 563–570.

Jörnvall, H., and Höög, J.-O., 1995, Nomenclature of alcohol dehydrogenases, *Alcohol Alcohol.* **30**, 153–161.

Julià, P., Parés, X., and Jörnvall, H., 1988, Rat liver alcohol dehydrogenase of class III. Primary structure, functional consequences and relationships to other alcohol dehydrogenases, *Eur. J. Biochem.* **172**, 73–83.

Kaiser, R., Nussrallah, B.A., Dam, R., Wagner, F.W., and Jörnvall, H., 1990, Avian alcohol dehydrogenase. Characterization of the quail enzyme, functional interpretations, and relationships to the different classes of mammalian alcohol dehydrogenase, *Biochemistry* **29**, 8365–8371.

Kedishvili, N.Y., Gough, W.H., Chernoff, E.A.G., Hurley, T.D., Stone, C.L., Bowman, K.D., Popov, K.M., Bosron, W.F., and Li, T.-K. (1997) cDNA sequence and catalytic properties of a chick embryo alcohol dehydrogenase that oxidizes retinol and 3β,α-Hydroxysteroids, *J. Biol. Chem.* **272**, 7494–7500.

Moreno, A., and Parés, X., 1991, Purification and characterization of a new alcohol dehydrogenase from human stomach, *J. Biol. Chem.* **266**, 1128–1133.

Moulis, J.-M., Holmquist, B., and Vallee, B.L., 1991, Hydrophobic anion activation of human liver $\chi\chi$ alcohol dehydrogenase, *Biochemistry* **30**, 5743–5749.

Nussrallah, B., Dam, R. and Wagner, F.W., 1989, Characterization of *Coturnix* quail liver alcohol dehydrogenase enzymes. *Biochemistry* **28**, 6245–6251.

Peralba, J.M., Cederlund, E., Crosas, B., Moreno, A., Julià, P., Martínez, S.E., Persson, B., Farrés, J., Parés, X., and Jörnvall, H., 1998, Structural and enzymatic properties of a gastric NADP(H)-dependent and retinal-active alcohol dehydrogenase (submitted).

Persson, B., Bergman, T., Keung, W.-M., Waldenström, U., Holmquist, B., Vallee, B.L., and Jörnvall, H., 1993, Basic features of class I alcohol dehydrogenase. Variable and constant segments coordinated by inter-class and intra-class variability. Conclusions from characterization of the alligator enzyme, *Eur. J. Biochem.* **216**, 49–56.

Ramaswamy, S., El-Ahmad, M., Danielsson, O., Jörnvall, H., and Eklund, H., 1996, Crystal structure of cod liver class I alcohol dehydrogenase: Substrate pocket and structurally variable segments, *Protein Science* **5**, 663–671.

Wesolowski, M.H., and Lyerla, T.A., 1979, The developmental appearance of hexokinase and alcohol dehydrogenase in the *Xenopus laevis*, *J. Exp. Zool.* **210**, 211–220.

Wesolowski, M.H. and Lyerla, T.A., 1983, Alcohol dehydrogenase isozymes in the clawed frog, *Xenopus laevis*. *Biochem. Genet.* **21**, 1003–1017.

Xie, P., Parsons, S.H., Speckhard, D.C., Bosron, W.F., and Hurley T.D., 1997, X-ray structure of human class IV $\sigma\sigma$ alcohol dehydrogenase. Structural basis for substrate specificity, *J. Biol. Chem.* **30**,18558–18563.

43

EXTRACELLULAR ACIDIFICATION: A NOVEL DETECTION SYSTEM FOR LIGAND/RECEPTOR INTERACTIONS

Demonstration with Bioactive Peptides and CHO or Pancreatic β Cells, but of Possible Interest for Tracing Putative Receptors in Ethanol Metabolism

Valentina Bonetto, Elo Eriste, Madis Metsis, and Rannar Sillard

Department of Medical Biochemistry and Biophysics
Karolinska Institute
S-17177 Stockholm, Sweden

1. INTRODUCTION

Microphysiometry is a technique that measures changes in the metabolic activity of cells (acid secretion) as a response to extracellular stimuli (McConnell et al., 1992). Cellular metabolism in response to the stimulation leads to consumption of glucose, synthesis and use of ATP, and production of several acidic metabolites such as lactic acid and bicarbonate. They are finally excreted from the cell by passive and active transport systems. The resulting change in the metabolic rate leads usually to a measurable increase in acid excretion from the cell, but may in a few cases lead to a decrease. Measurements of extracellular acidification rates (ECAR) can be carried out using an instrument with a silicon pH sensor, Cytosensor, and can trace when cell surface receptors have been stimulated by various signalling substances, such as hormones, neurotransmitters or growth factors.

Since this may be of interest in many systems influenced by ethanol metabolism, including perhaps the search for ethanol receptors (Wick et al., 1998), we here describe the system and its use with bioactive peptides and specific cells.

For this purpose, we have examined Chinese hamster ovary cells (CHO) and pancreatic β cell clones (HIT-T15 and RINm5F) as model systems. Transiently and stably transfected CHO cells are useful as hosts for receptor protein expression and are widely used in receptor studies (Fehmann et al., 1994; Vilardaga et al., 1997). Similarly, the pancreatic β cells express a large number of different receptors (Lagny-Pourmir et al., 1989; Thermos et

Enzymology and Molecular Biology of Carbonyl Metabolism 7
edited by Weiner *et al.* Kluwer Academic / Plenum Publishers, New York, 1999.

al., 1990; Moens et al., 1996; Inagaki et al., 1996) and also exhibit evidence for the presence of orphan G-protein coupled receptors (Blache et al., 1998). To evaluate the applicability of this system, the behavior of several individual hormonal substances and bioactive peptides, including peptide mixtures from tissue extracts was studied with the Cytosensor.

The results indicate variable involvement of signal transduction pathways to different G-protein coupled receptors in the same cell line and make it possible to draw conclusions for applicability in screening experiments for novel endogenous receptor ligands.

2. MATERIALS AND METHODS

2.1. Materials

Cell culture media were from Gibco and adenosine, AMP, ADP, ATP, serotonin, dopamine, adrenaline and carbachol from Sigma. All peptides were isolated from pig intestine (Bonetto et al., 1995). Porcine brain extracts were prepared and fractionated as described (Sillard et al., 1992).

2.2. Cells and Cytosensor Assays

All cells were grown in DMEM with 10% fetal bovine serum, containing penicillin and streptomycin. CHO cells were passaged every 2–3 days at a ratio of 1:5 using trypsin/EDTA. HIT-T15 and RINm5F cells were passaged every 6–7 days at the same ratio. The cells were detached with trypsin/EDTA, plated out into the capsule cups and assayed with the Cytosensor after 2 (CHO) or 5 (β cells) days of recovery period. The Cytosensor was run with DMEM without the addition of fetal bovine serum. The cells were in contact with the samples for 5 min during 3 pump cycles. A 2-min standard pump cycle was used. The cells were allowed to recover for at least 40 min before the next sample was applied.

2.3. Transfection with VIP Receptor

CHO cells were plated at a density 5×10^6 cells per 10 cm cell culture dish. Next day the cells were transfected with different ratios of DNA:lipofectin (Gibco) according to the recommendations of the manufacturer. After 14 h, cells were collected with 2 mM EDTA/PBS, washed with medium and plated into capsules for measurements on the subsequent day. Transfection efficiences were monitored by β-galatosidase activity in parallel transfections.

3. RESULTS AND DISCUSSION

3.1. Responses in CHO Cells

A series of purinergic ligands, adenosine, AMP, ADP and ATP, were assayed with CHO cells (Figure 1). Adenosine caused responses in both directions—either towards an increase or a decrease of extracellular acidification depending on the concentration. At 0.1 nM and 1 nM concentrations, adenosine did not evoke any significant response. At 10 nM the substance produced an increase in the extracellular acidification at about 20% over the basal level. However, at 100 nM and 1 μM concentrations a marked decrease in acidification rate was observed, which was particularly pronounced at the highest concentration

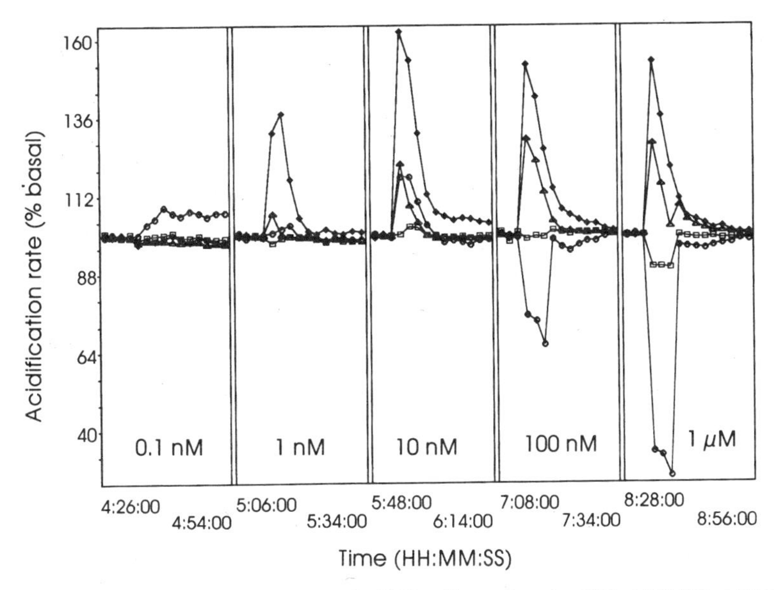

Figure 1. Extracellular acidification rate responses in CHO cells to adenosine (○), AMP (□), ADP (Δ), and ATP (◆). In each panel the samples were applied for 5 minutes within 3 pump cycles followed by a 40 min recovery interval and by an identical period with 1 nM ATP as a standard (not shown).

used, where approximately a 70% reduction of the basal acidification rate was detected. Such behavior is quite unique for this type of cellular response to a receptor ligand. For adenosine the response was also biphasic, with a strongly negative rapid drop in the metabolic rate followed by a slow and less pronounced second negative response phase.

AMP did not cause an acidification increase, but at 1 μM concentration a decrease in the acidification rate of about 10% was also observed. ADP however, evoked an extracellular acidification increase already at 1 nM, and a biphasic response at the highest concentration. The response of CHO cells to ATP was the strongest in this series of receptor ligands. Like with the other substances used, no response was observed at 0.1 nM ATP. However, at 1 nM concentration, the response reached up to 50% of the maximal response elicited by 10 nM ATP. The ECAR value at 1 nM for ATP was about the same as that for ADP at a 100 times higher concentration. The ATP response reached maximum at 10 nM and at higher concentrations a desensitization effect was observed. Thus, the ECAR value was less at 100 nM and at 1 μM ATP than that at 10 nM. A dramatic reduction in peak values of ECAR was detected, when 1 nM ATP was injected after 10, 100 or 1000 nM concentrations. However, such desensitization behavior was not detectable with other substances. The results indicate that purinergic receptors are present in CHO cells. The responses caused by ATP are powerful, indicating that the receptors are mainly of the P2 type, and that they are strongly coupled to the cellular metabolism in CHO cells.

Several other commonly occurring hormonal substances or their derivatives, such as carbachol, serotonin, dopamine, and adrenaline were also assayed with CHO cells (Figure 2). The amplitudes, durations, and patterns of the responses observed with these ligands were all different. The strongest response, reaching up to 145% of basal level was obtained with carbachol at 1 μM concentration with a half-maximal value at close to 10 nM. No significant response was detected at 0.1 nM. However, serotonin showed a signifi-

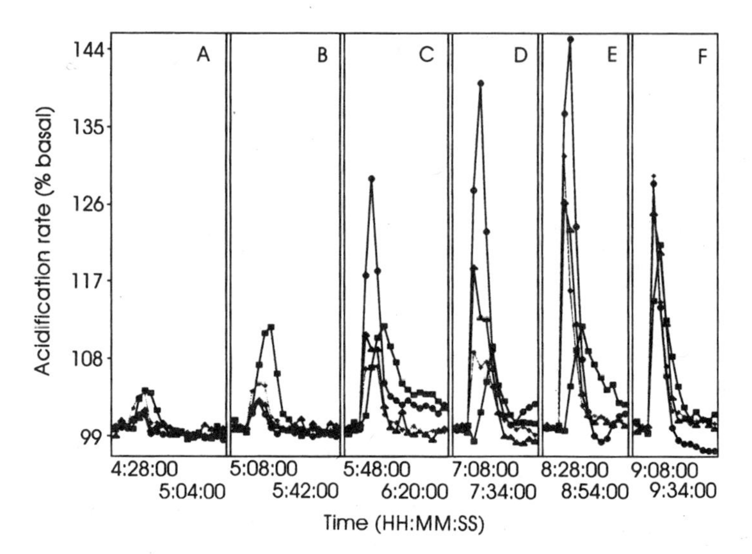

Figure 2. Extracellular acidification rate responses in CHO cells to carbachol (●), serotonin (■), dopamine (▲), and adrenaline (◆) (A - D). (A) All compounds at 0.1 nM; (B) all compounds at 1 nM; (C) 10 nM carbachol and serotonin, 100 nM dopamine and adrenaline; (D) 100 nM carbachol and serotonin, 1 μM dopamine and adrenaline; (E) 1 μM carbachol (●), 1 μM serotonin (■), 1 nM ATP (▲ and ◆); (F) 1 nM ATP as a standard at all positions. Each assay included a 40 min recovery period.

cant effect, about 50% of maximum, already at that low concentration. At the same time, the maximal value for the serotonin effect never reached above 112% of basal, but was of the longest duration, and over 10 nM the effect was also biphasic.

Similarly, effects of dopamine and adrenaline were weak, reaching up to approximately 20% and 10% above basal values, respectively, at the highest concentrations used (1 μM). The duration of the response was much shorter than that for serotonin. Moreover, none of these pharmacological agents caused significant desensitization of CHO cells like 1 nM ATP did. These results indicate the presence of cholinergic receptors in CHO cells and a strong coupling of such receptors to acid secretion. The three other hormones evoked only weak responses, suggesting a weak coupling or, more likely, low levels of expression of the corresponding receptors in these cells. Despite the background levels, CHO cells are widely used for pharmacological characterization of subtypes of receptors for all the hormones (Pedersen et al., 1994; Richards and van Giersbergen, 1995).

In a further experimental set, CHO cells were transiently transfected with VIP receptors and the resulting cells tested with VIP and other peptides (Figure 3). In general, the total response of transiently transfected cells was low, reaching about 10% above the basal values with 1 μM VIP. Most likely this indicates that only a fraction of the cells was transfected successfully. However, the cells responded to VIP and its related peptides with an increase in ECAR, indicating that the receptor was correctly inserted into the plasma membrane. Among the ligands tested, VIP evoked the strongest response, while the response to 1 μM secretin reached only 5% above the basal level.

Chromatographic fractions from pig brain known to contain VIP and related peptides such as PHI and variant forms of VIP, were also found to increase ECAR values dose-dependently. The response to a VIP-containing fraction at 0.01 mg/ml was lower than that at 0.05 mg/ml. However, the unrelated gastrointestinal peptide galanin did not

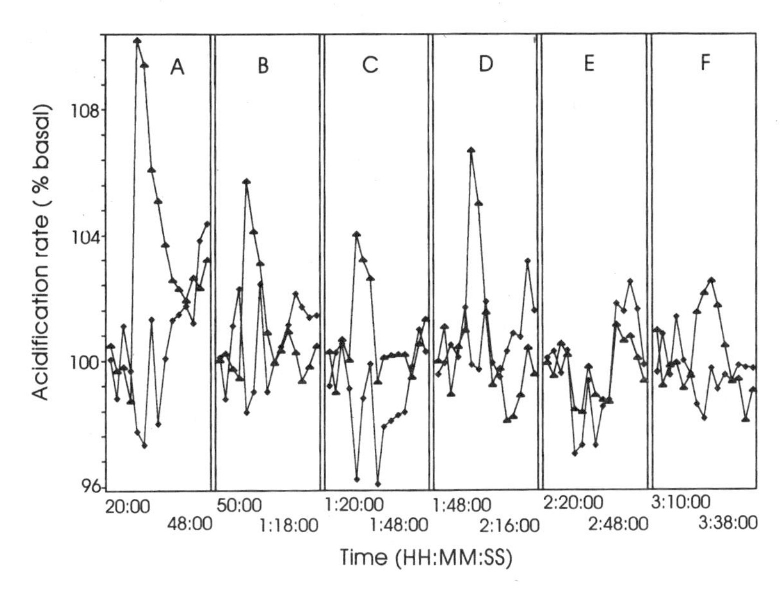

Figure 3. Acidification rate responses in nontransfected CHO cells (◆) and CHO cells transiently transfected with VIP receptor (▲). (A) 1 μM VIP, (B) 1 μM secretin, (C) a VIP-containing chromatographic fraction from pig brain extract at 0.01 mg/ml, (D) the same fraction as in B at 0.05 mg/ml, (E) 1 μM galanin, (F) a PHI-containing fraction from pig brain at 0.02 mg/ml.

evoke any ECAR, showing that the effects caused by VIP were specific for VIP receptor. A peptide fraction known to contain the VIP-related PHI elicited only 2% increase in ECAR. Such response is at the noise level of the measurements, but following it over a time-frame of 30 min and comparing it with traces obtained with nontransfected cells, it is possible to detect a specific response. However, as mentioned, the nontransfected cells did not respond to any of the peptide hormones tested, VIP, secretin, PHI, galanin or CCK showing that the plasma membrane of CHO cells lack appreciable amounts of receptors for these components. The positive responses with transfected cells, however, suggest that ligands in such complex mixtures as tissue extracts can be successfully detected and their purification monitored. This is of interest in receptor for metabolic regulation, including these of relevance in ethanol metabolism.

3.2. Responses in Pancreatic Beta Cell Clones HIT-T15 and RINm5F

VIP, GIP, galanin, CCK and daintain were found to strongly activate cultivated pancreatic β cells in the Cytosensor. The extracellular acidification response in HIT-T15 cells to increasing concentrations of GIP was dose-dependent (Figure 4), with maximal response at 140% of a basal value, at a micromolar range, with an EC50 value of 7.6 nM. Other peptide and nonpeptide hormones tested gave somewhat lower maximal ECAR values, but also reached about 130% of the basal value at 1 μM.

The maximal responses were considerably higher with HIT-T15 than with RINm5F. The ECAR values with RINm5F were found frequently at about 110% of the basal range. The results with HIT-T15 cells indicate that the pancreatic β cells respond dose-dependently to increasing concentrations of many hormones and that these cells express a large number of different functional receptors. It appears that HIT-T15 cells can be employed

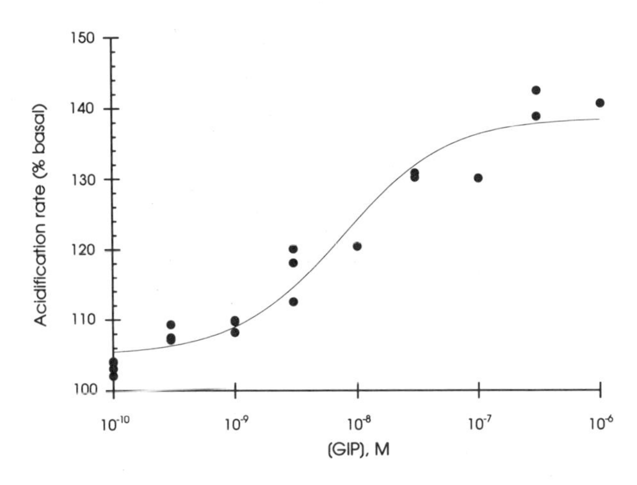

Figure 4. Dose-response dependence of extracellular acidification in HIT-T15 cells to increasing concentrations of GIP.

for comparisons of potencies of different ligands. In combination with the Cytosensor the cells can possibly be used as a model system for cellular activation. Importantly, both peptide hormones that cause insulin secretion from the islets and those that, inhibit insulin secretion from the pancreas, such as galanin, stimulate extracellular acidification and cause ECAR in a similar fashion. This makes it possible to use the instrument and this cell line for both types of effectors.

Extracts from pig intestinal tissue were fractionated with alcoholic solutions, further fractionated by size exclusion chromatography on Sephadex G25f, and assayed with HIT-T15 cells (Figure 5). ATP and carbachol were used as standard substances to monitor the metabolic state of the cells during the experiment. It is evident that 1 nM ATP gives only a small ECAR value reaching up to about 5–10% over the basal level. Carbachol at 10 times lower concentration (0.1 nM) gave much stronger signals reaching about 120% of basal. Comparing the data with ATP responses in CHO cells, a dramatic difference is evident. While the responses in CHO cells are within 3–4 pump cycles, i.e. 6–8 min, in some cases even longer, the corresponding curves for HIT-T15 contain only a single point, i.e. the response lasts only for 2 min.

It is also seen that the response to carbachol is biphasic. After the initial positive 2-min response, a long-lasting negative response period follows. Such curve shapes were typically observed for HIT-T15 and RINm5F cells, but not for CHO. It is evident that chromatographic fractions can give different responses in this cell line. The first two fractions did not have any effect, fraction 3 had a small effect, but the duration of the response was much longer than that for 1 nM ATP or 0.1 nM carbachol, while fractions 4 and 5 had a marked effect on extracellular acidification rate, reaching about 140% of the basal rate. The ECAR to these fractions is multiphasic, with an initial rapid burst and several subsequent plateau periods. Such behavior is particularly well expressed for fraction 4. A test with 1 nM ATP after fraction 4 indicates that the maximal response for the standard substance is still exactly at the same level as before. However, a memory effect can also be seen, when a single peak in the activity curves is replaced by a biphasic response. After the first burst response a second peak appears, although, with lower intensity, and with much longer duration. Such a response suggests that proteins involved in intracellular signalling pathways have been modified, possibly by phosphorylation, by the influence of fraction 4.

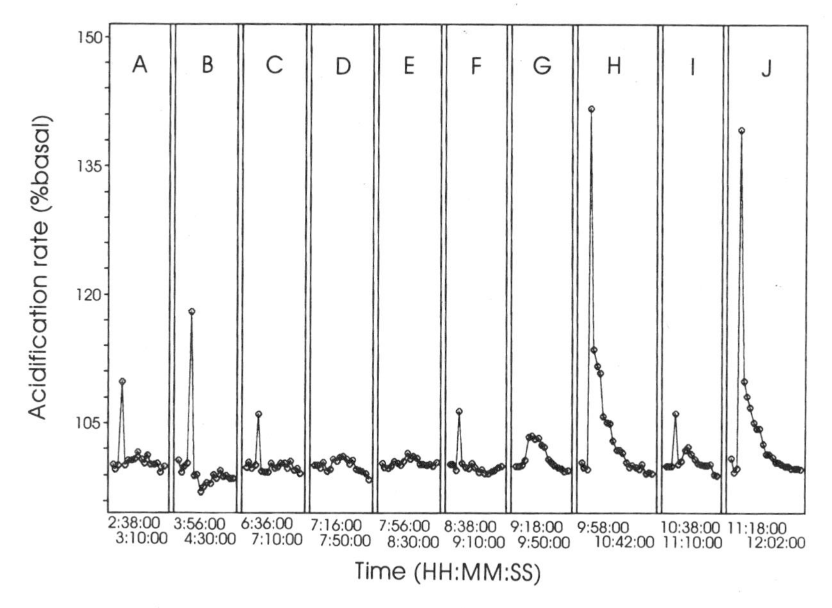

Figure 5. Acidification rate responses in HIT-T15 cells to 1 nM ATP (A, C, F and I), to chromatographic fractions (1 - D, 2 - E, 3 - G, 4 - H, 5 - J) from porcine intestinal extract at 0.05 mg/ml, and 0.1 nM carbamoyl choline (B).

The results from this section demonstrate that chromatographic fractions containing bioactive peptides can be successfully assayed with the pancreatic beta cells by the Cytosensor. In tissue extracts we do not know exactly which peptides in the chromatographic fractions cause the acidification in HIT-T15 cells, but we know that the active fractions contain VIP, secretin and possibly other, yet undiscovered peptide hormones. Hence, the technique of interaction analysis may be of general interest also in metabolic regulation such as that during ethanol metabolism.

3.3. Conclusions

Several nonpeptide hormones or their analogs caused a significant response in CHO cells showing the presence of receptors for all the tested nonpeptide compounds. However, several peptide ligands did not elicit responses in these cells suggesting that CHO cells may be used for design of transfected cells with suitable receptors for screening experiments. Among the pancreatic β cell clones, HIT-T15 seems to function best. A large number of hormones are readily detectable dose-dependently. Tissue extracts result in distinct responses and in combination with standard compounds can be employed in screening experiments with the Cytosensor. Acidification responses can be monitored during a long time interval, exceeding 10–15 h. This method is now available for further use in metabolic controls and other responses.

ACKNOWLEDGMENTS

This work was supported by the Swedish Medical Research Council, Swedish Society for Medical Research, the Swedish Institute, Novo Nordisk Foundation, Clas Groschinskys Foundation, Karolinska Institute, and the Commission of the European Union (B104-CT97-2123). Supply of peptides by Prof. V. Mutt, and VIP receptor cDNA by Dr. S. Nagata are gratefully acknowledged.

REFERENCES

Blache, P., Gros, L., Salazar, G., and Bataille, D. (1998) Cloning and tissue distribution of a new rat olfactory receptor-like (OL2), *Biochem. Biophys. Res. Commun.* **242**, 669–672.

Bonetto, V., Jörnvall, H., Mutt, V., and Sillard, R. (1995) Two alternative processing pathways for a preprohormone: a bioactive form of secretin, *Proc. Natl. Acad. Sci. USA* **92**, 11985–11989.

Fehmann, H. C., Jiang, J., Schweinfurth, J., Dorsch, K., Wheeler, M. B., Boyd, A. E 3rd, and Goke, B. (1994) Ligand-specificity of the rat GLP-1 receptor recombinantly expressed in Chinese hamster ovary (CHO-) cells, *Z. Gastroenterol.* **32**, 203–207.

Lagny-Pourmir, I., Amiranoff, B., Lorinet, A. M., Tatemoto, K., and Laburthe, M. (1989) Characterization of galanin receptors in the insulin-secreting cell line Rinm5F: evidence for coupling with a pertussis toxin-sensitive guanosine triphosphate regulatory protein, *Endocrinology* **124**, 2635- 2641.

Inagaki, N., Kuromi, H., and Seino, S. (1996) PACAP/VIP receptors in pancreatic beta-cells: their role in insulin secretion, *Ann. N. Y. Acad. Sci.* **805**, 44–53.

McConnel, H. M., Owicki, J. C., Parce, J. W., Miller, D. L., Baxter, G. T., Wada, H. G., and Pitchford S. (1992) The Cytosensor microphysiometer: biological application of silicon technology, *Science* **257**, 1906–1912.

Moens, K., Heimberg, H., Flamez, D., Huypens, P., Quartier, E., Ling, Z., Pipeleers, D., Gremlich, S., Thorens, B., and Schuit, F. (1996) Expression and functional activity of glucagon, glucagon-like peptide I, and glucose-dependent insulinotropic peptide receptors in rat pancreatic islet cells, *Diabetes* **45**, 257–261.

Pedersen, U. B., Norby, B., Jensen, A. A., Schiodt, M., Hansen, A., Suhr-Jessen, P., Scheideler, M., Thastrup, O., and Andersen, P. H. (1994) Characteristics of stably expressed human dopamine D1a and D1b receptors: atypical behaviour of the dopamine D1b receptor, *Eur. J. Pharmacol.* **267**, 85–93.

Richards, M. H. and van Giersbergen, P. L. (1995) Human muscarinic receptors expressed in A9L and CHO cells: activation by fully and partial agonists, *Brit. J. Pharmacol.* **114**, 1241–1249.

Sillard, R., Rökaeus, Å., Xu, Y., Carlquist, M., Bergman, T., Jörnvall, H., and Mutt, V. (1992) Variant forms of galanin isolated from porcine brain, *Peptides* **31**, 1055–1060.

Thermos, K., Meglasson, M. D., Nelson, J., Lounsbury, K. M., and Reisine, T. (1990) Pancreatic beta-cell somatostatin receptors, *Am. J. Physiol.* **259**, 216–224.

Vilardaga, J. P., Di Paolo, E., Bialek, C., De Neef, P., Waelbroeck, M., Bollen, A., and Robberecht, P. (1997) Mutational analysis of extracellular cysteine residues of rat secretin receptor shows that disulfide bridges are essential for receptor function, *Eur. J. Biochem.* **246**, 173–180.

Wick, M. J., Mihic, S. J., Ueno, S., Mascia, M. P., Trudell, J. R., Brozowski, S. J., Ye, Q. , Harrison, N. L., and Harris, R. A. (1998) Mutation of γ-aminobutyric acid and glycine receptors change alcohol cutoff: evidence for an alcohol receptor? *Proc. Natl. Acad. Sci. USA* **98**, 6504–6509.

44

MULTIPLICITY AND COMPLEXITY OF SDR AND MDR ENZYMES

Hans Jörnvall

Department of Medical Biochemistry and Biophysics
Karolinska Institutet
S-171 77 Stockholm, Sweden

1. INTRODUCTION

Many alcohol dehydrogenases in nature are members of either the MDR (medium-chain dehydrogenases/reductases) or SDR (short-chain dehydrogenases/reductases) protein families. In fact, both families were recognized in the early 80ies when an ADH was found to be related to another enzyme and thus gave the first enzyme pair known in each family. For SDR the first pair was *Drosophila* ADH and ribitol dehydrogenase, which showed the conservation of the Tyr-X-X-X-Lys active site (Jörnvall *et al.*, 1981). For MDR the initial pair was mammalian ADH class I and sorbitol dehydrogenase, which showed the conservation of the Zinc at the active site and its Cys~46/His~67 ligands (Jeffery *et al.*, 1984; Jörnvall *et al.*, 1981). Since then, known members in both families have grown enormously. From the start, interpretations for MDR were much ahead of those for SDR because of a known three-dimensional structure of MDR already at that time (Eklund *et al.*, 1976).

Subsequent leaps in the depth of knowledge on MDR were the incorporation of the isozyme (Smith *et al.*, 1973) and class (Vallee & Bazzone, 1983) concepts into the family relationships, recognition that class III ADH really represented another enzyme (GSH-dependent formaldehyde dehydrogenase; Koivusalo *et al.*, 1989), and recognition that zinc is not obligatory at the active site (as shown by ζ-crystallin being an MDR member; Borrás *et al.*, 1989). For SDR, two important leaps were the determination of a three-dimensional structure (Ghosh *et al.*, 1991) and the characterization of 15-hydroxyprostaglandin dehydrogenase (Krook *et al.*, 1990). The three-dimensional structure made functional assignments meaningful for SDR, like they already from the start had been for MDR, and the knowledge of the prostaglandin dehydrogenase made the SDR family of general interest by extending it into mammalian and vertebrate enzymes from only the insect and prokaryotic forms known before. By the previous Carbonyl Metabolism meeting, both families

Enzymology and Molecular Biology of Carbonyl Metabolism 7
edited by Weiner *et al.* Kluwer Academic / Plenum Publishers, New York, 1999.

had grown to large numbers of known forms, over 100 in SDR, and with all species variants almost similar numbers in MDR. Now they have passed 1000 for SDR and 500 for MDR (cf. Persson *et al.*, 1998).

In parallel with this increase in multiplicity, a corresponding increase in complexity has been revealed. Three aspects of that have been further clarified recently, *i.e.* MDR multiplicity also in prokaryotes, induction of MDR ADH, and the genetic representation. As summarized below, the recent data on these aspects give a new, balanced knowledge on especially the MDR family and its ADH forms and functions. The prokaryotic MDR multiplicity gives insight into evolution/origins of ADHs, and their functional roles, the induction offers a possibility to study MDR regulatory functions, and the genetic representation gives insight into the relationship patterns of MDR ADHs.

2. MATERIALS AND METHODS

Novel ADH forms were structurally analyzed after purification by multi-step ion-exchange and affinity chromatographies as described (Shafqat *et al.*, 1998). In this manner, ADH forms have been analyzed for primary structure and enzymology. Amino acid sequences thus obtained, as well as relevant further forms from databanks, were aligned with those of other MDR ADHs, evaluated in relation to structural and functional properties, modeled, and incorporated into the phylogenetic tree relationships already known. Major conclusions on structural relationships, enzymology, modeling and functional properties have been separately reported elsewhere (Shafqat *et al.*, 1998; Norin *et al.*, 1997) whereas the complexity and multiplicity are now further evaluated.

3. RESULTS AND DISCUSSION

3.1. MDR Multiplicity Also in Prokaryotes

Two aspects of recent structural analyses emphasize novel insight into the complexity of MDR ADHs. One is the purifications and analyses of prokaryotic ADHs, the other the availability in databanks of complete genomes for entire organisms.

Regarding the new purifications and analyses, an observation during expression of recombinant ADH forms in *E. coli* demonstrated that MDR ADHs are not only present in prokaryotes, but also in multiple form and with considerable ethanol dehydrogenase activity (Shafqat *et al.*, 1998), much like the well-known mammalian class I enzymes. Although ethanol dehydrogenase activity has been found before in *E. coli* (Still, 1940; Goodlove *et al.*, 1989), its origin and molecular nature have not been established, and the only ADH molecule directly analyzed in *E. coli* has been the class III form (Gutheil *et al.*, 1992). Similarly, the recent genome analysis of *E. coli* revealed the presence of 16 MDR gene homologs in the *E. coli* genome, one of which corresponds to the new ethanol dehydrogenase just purified (Shafqat *et al.*, 1998).

Notably, this multiplicity in prokaryotes appears to be of the same order of magnitude or even more extensive than that presently known in humans or vertebrates (with something like up to 9 ADH genes of 7 classes). Similarly, the finding of quite active ethanol dehydrogenase in *E. coli* and of that activity clearly related to an MDR enzyme was not directly expected from previous knowledge on the evolution of ADH at large. Thus, previous studies have shown that the ethanol-active animal forms appear to have evolved during vertebrate radiation through a number of gene duplications, with class I apparently formed from class III at early vertebrate times (Danielsson and Jörnvall, 1992). Similarly, with that pattern at hand, class III has resembled an ancestral form and it has then appeared quite natural that class III is the only one detectable in marine animals (Fernández *et al.*, 1993) and, also, that it is present in a more or less identical form in plants (Martínez *et al.*, 1996; Shafqat *et al.*, 1996) and prokaryotes (Gutheil *et al.*, 1992). How, then, with the interpretations of class III as very early, and of animal and plant ethanol-active enzymes as later formed, if ethanol-active forms and much multiplicity also exist in prokaryotes? Were those earlier conclusions on ancestral properties and origins of ethanol-activity incorrect? No, the recent finding of ADH multiplicity and MDR ethanol activity in prokaryotes do not contradict previous conclusions on ADH origins and activities in animals, plants, and prokaryotes. Thus, the MDR multiplicity in prokaryotes is not the same as in plants and animals, and the previous conclusions on timings and ancestral relationships in the vertebrate and whole animal lines still appear correct.

Instead, the new data show that the complexity is great, even greater than previously realized. In fact, the complexity with functional multiplicity of several enzymes and classes like in animals and plants also extends to prokaryotes. They, too, have been subject to a number of MDR gene duplications and they, too, have evolved ethanol-active ADH forms. Although the ethanol-active forms (and other activities) and a multiplicity are present in prokaryotes, as in the higher forms, the patterns are different, the classes not distinguished in the same relationships, and the structures not as closely related as they are for the formaldehyde-active forms between prokaryotes and eukaryotes. Combined with previous data, these patterns now reveal three extensions or refinements to previous conclusions:

a. Class III is still an early enzyme, present in most life forms analyzed. This reinforces its fundamental importance in formaldehyde elimination and is fully consistent with the constant evolutionary properties and basic conservation as a "house-hold" enzyme. These renewed observations agree with interpretations from the enzyme isolation (Uotila and Koivusalo, 1989) and from the pattern of changes of essentially only non-functional segments (Danielsson *et al.*, 1994). Apparently, formaldehyde elimination is important, and is the major function of the class III form. Although that reaction can be carried out also with other enzymes (cytochrome P450 in particular), the dehydrogenase reaction proceeds without radical formation and, in one form or another, appears essential to life.
b. Class multiplicity and enzymogenesis of ethanol-activity is present in many life forms. Independent evolution of ethanol-active forms in yeasts, plants and animals has been emphasized previously (Jörnvall, 1994) but the multi-enzyme sub-families in prokaryotes constitute a recent knowledge that must be explained. Apparently, the ability to handle a variety of alcohols other than the hydroxymethylglutathione from the GSH-coupled formaldehyde, is important. It appears significant that nearly all life forms turn out to have a family of multienzyme-related ADH forms. Frequently, they have been formed comparatively recent, like the closely related yeast forms (Young and Pilgrim, 1985). The repeated formation of ethanol-active forms in yeasts, plants and animals explains the non-identical patterns of these enzymes in different life forms. In spite of the non-identity, the mere existence of these enzymes and their repetitive formation suggests important fundamental properties also for these forms. Those properties are likely not to be fundamental to life itself (because of the different and late origins), but to important sub-specializations. This pattern is not only

compatible with the regulatory roles suggested for the class IV forms in vertebrates (Duester, 1996), but with a possible role in alcohol/aldehyde elimination (for class I) which in turn could be compatible with not finding additional forms in marine invertebrates. Small members of the latter can eliminate aldehydes/alcohols directly to the environment (Fernández *et al.*, 1993).

c. In spite of the early origin of class III (a) and repeated divergence with further enzymogenesis (b) there is nothing to say that class III is **the** origin of the ADH MDR line. In contrast, recent data may well question that view. Thus, some prokaryotes, like Gram-positive bacteria do not have class III ADH MDR, but instead have a mycothiol-dependent ADH which, however, is functionally of both the class III type and also an MDR form (Norin *et al.*, 1997). Similarly, the many forms of MDR enzymes now obvious in *E. coli* show that further analysis is necessary to find which of all forms is the ancestor at that early level. Obviously, much species variability of each of the novel forms discovered in the *E. coli* and other genomes need be studied in detail, like the multiplicity has previously been studied in the vertebrate line, to reveal the evolutionary pattern and ancestral relationships at the earlier levels.

Combined, all these points show that more ADH studies are needed to clarify the earliest events, and that studies of prokaryotes will then be necessary. The recent findings in prokaryotes of multiplicity reconfirm the importance of class III activity type in general and of ADH multiplicity in particular.

3.2. Induction of MDR ADH

During analysis of expressions in *E. coli*, we found and purified a new ethanol-active *E. coli* ADH, as mentioned above. However, it was not only ethanol-active, but also ethanol-inducible, at least in certain strains and under certain conditions (Shafqat *et al.*, 1998). In the strain studied, maximal induction was at 17 mM ethanol. This property opens a new possibility for study. Obviously, *E. coli* must have a sensing system for the presence of ethanol and possibly this could include some "receptor" mechanism. Naturally, it could also work via indirect modes, but even so, it opens a new mode for study of the regulation of expression of ADH activity. In this context, it may be of interest to note that novel modes of measuring receptor activity via pH alterations upon ligand binding exist, using a cytosensor outfit (Bonetto *et al.*, 1998). Much still remains to be studied regarding both the ethanol induction and its possible mechanism, but independent of mode, this is another novel concept recently opened to further studies by the observations on the *E. coli* enzyme and its different expression properties.

3.3. Genetic Representation of MDR and SDR Enzyme Genes in Completed Genomes Recently Available

During the last few years, several complete genomes of intact organisms have been reported, including the *E. coli* genome. As stated above, they reveal unexpected multiplicity in MDR genes, yielding new information on the ADH complexity. However, they also reveal another unexpected fact: that both the MDR and SDR families are richly represented in the genomes completed, even richer than many other well-known genes widely accepted as important and of wide spread. Therefore, both MDR and SDR are early families well represented in prokaryotes and elsewhere, and obviously of similar abundance and hence importance as more well-known regulatory systems in living cells. This further suggests important functions for ADH-like proteins in basic life forms. A second aspect of this rich genetic representation is that many MDR and SDR structures now are known, allowing further conclusions on the evolutionary patterns. With the increasing number of known MDR forms, it is then obvious that many of these MDR forms are as unrelated to each other as the SDR forms are to each other within their family. This is in contrast to previous data when fewer MDR forms were known and the ones known to a large part represented species, class and isozyme variants of ADH. At that stage, it appeared as if MDR relationships were generally closer than those for SDR, *i.e.* of more recent origin than the corresponding SDR relationships. The phylogenetic trees then gave a pattern of successive changes in MDR, while the SDR forms appeared more distantly related and giving another phylogenetic tree with a more star-like pattern (Jörnvall *et al.*, 1996). However, in the MDR tree, some relationships still are successive, reflecting stepwise changes as originally shown. Thus, overall the differences are now smaller between the evolutionary patterns of SDR and MDR but they still exist (Persson *et al.*, 1998). This has functional consequences, suggesting that also the MDR enzymes, like the SDR forms, already early represented multiple activities and had important functions common to many life forms.

In conclusion, all three sets of novel data recently obtained highlight the importance of both SDR and MDR as fundamental protein families of great multiplicity and complexity in living cells in general. ADH activity is already only a minor part of the SDR family, and soon ADH and its zinc dependency may constitute only a minor part of the MDR family, too. Instead, the basic folds and enzymatic mechanisms, with coenzyme binding in common but separate active site relationships, appear to constitute the architecturally central pattern of both families.

ACKNOWLEDGMENTS

Studies in the author's laboratory have been supported by the Swedish Medical Research Council (project 3532) and the European Commission (BIO4-CT97-2123) and have been carried through in collaborations with many scientists and coauthors, as evident from the separate publications. I am grateful to all individual contributors.

REFERENCES

Bonetto, V. Eriste, E., Metsis, M. and Sillard, R. (1998) Extracellular acidification studies in CHO and pancreatic beta cells: a novel detection system for bioactive peptides and hormones. In *Enzymology and Molecular Biology of Carbonyl Metabolism 7* (Weiner, H., Lindahl, R. and Crabb, D.W., eds) Plenum, New York, in press.

Borrás, T., Persson, B. and Jörnvall, H. (1989) Eye lens ζ-crystallin relationships to the family of "long-chain" alcohol/polyol dehydrogenases. Protein trimming and conservation of stable parts. *Biochemistry* 28:6133–6139.

Danielsson, O. and Jörnvall, H. (1992) "Enzymogenesis": Classical liver alcohol dehydrogenase origin from the glutathione-dependent formaldehyde dehydrogenase line. *Proc. Natl. Acad. Sci. USA* 89:9247–9251.

Danielsson, O., Atrian, S., Luque, T., Hjelmqvist, L., Gonzàlez-Duarte, R. and Jörnvall, H. (1994) Fundamental molecular differences between alcohol dehydrogenase classes. *Proc. Natl. Acad. Sci. USA* 91:4980–4984.

Duester, G. (1996) Involvement of alcohol dehydrogenase, short-chain dehydrogenase/reductase, aldehyde dehydrogenase, and cytochrome P450 in the control of retinoid signalling by activation of retinoic acid synthesis. *Biochemistry* 35:12221–12227.

Eklund, H., Nordström, B., Zeppezauer, E., Söderlund, G., Ohlsson, I., Boiwe, T., Söderberg, B.-O., Tapia, O., Brändén, C.-I. and Åkeson, Å. (1976) Three-dimensional structure of horse liver alcohol dehydrogenase at 2.4 Å resolution. *J. Mol. Biol.* 102:27–59.

Fernández, M.R., Jörnvall, H., Moreno, A., Kaiser, R. and Parés, X. (1993) Cephalopod alcohol dehydrogenase: purification and enzymatic characterization. *FEBS Lett.* 328:235–238.

Ghosh, D., Weeks, C.M., Groculski, P., Duax, W., Erman, M., Rimsay, R.L. and Orr, J.C. (1991) Three-dimensional structure of holo 3α,20β-hydroxysteroid dehydrogenase: A member of a short-chain dehydrogenase family. *Proc. Natl. Acad. Sci. USA* 88:10064–10068.

Goodlove, P.E., Cunningham, P.R., Parker, J. and Clark, D.P. (1989) Cloning and sequence analysis of the fermentative alcohol dehydrogenase-encoding gene of *Escherichia coli*. *Gene* 85:209–214.

Gutheil, W.G., Holmquist, B. and Vallee, B.L. (1992) Purification, characterization, and partial sequence of the glutathione-dependent formaldeyde dehydrogenase from *Escherichia coli*: a class III alcohol dehydrogenase. *Biochemistry* 31:475–481.

Jeffery, J., Cederlund, E. and Jörnvall, H. (1984) Sorbitol dehydrogenase. The primary structure of the sheep liver enzyme. *Eur. J. Biochem.* 140:7–16.

Jörnvall, H. (1994) The alcohol dehydrogenase system. In *Toward a Molecular Basis of Alcohol Use and Abuse* (Jansson, B., Jörnvall, H., Rydberg, L. and Terenius L, and Vallee, B.L., eds) Birkhäuser, Basel, pp. 221–229.

Jörnvall, H., Persson, M. and Jeffery, J. (1981) Alcohol and polyol dehydrogenases are both divided into two protein types, and structural properties cross-relate the different enzyme activities within each type. *Proc. Natl. Acad. Sci. USA* 78:4226–4230.

Jörnvall, H., Danielsson, O., Hjelmqvist, L., Persson, B. and Shafqat, J. (1996) Alcohol dehydrogenases: isozymes, divergence, convergence, and molecular building units. In *Gene families: structure, function, genetics, and evolution* (Holmes, R.S., Lim, H.A., eds) World Scientific, Singapore, pp. 35–41.

Koivusalo, M., Baumann, M. and Uotila, L. (1989) Evidence for the identity of glutathione-dependent formaldehyde dehydrogenase and class III alcohol dehydroenase. *FEBS Lett.* 257:105–109.

Krook, M., Marekov, L. and Jörnvall, H. (1990) Purification and structural characterization of placental NAD^+-linked 15-hydroxyprostaglandin dehydrogenase. The primary structure reveals the enzyme to belong to the short-chain alcohol dehydrogenase family. *Biochemistry* 29:738–743.

Martínez, M.C., Achkor, H., Persson, B., Fernández, M.R., Shafqat, J., Farrés, J., Jörnvall, H. and Parés, X. (1996) *Arabidopsis* formaldehyde dehydrogenase. Molecular properties of plant class III alcohol dehydrogenase provide further insights into the origins, structure and function of plant class P and liver class I alcohol dehydrogenases. *Eur. J. Biochem.* 241:849–857.

Norin, A., van Ophem, P.W., Piersma, S.R., Persson, B., Duine, J.A. and Jörnvall, H. (1997) Mycothiol-dependent formaldehyde dehydrogenase, a prokaryotic medium-chain dehydrogenase/reductase, phylogenetically links different eukaryotic alcohol dehydrogenases. Primary structure, conformational modelling and functional correlations. *Eur. J. Biochem.* 248:282–289.

Persson, B., Nordling, E., Kallberg, Y., Lundh, D., Oppermann, U.C.T., Marschall, H.-U. and Jörnvall, H. (1998) Bioinformatics in studies of SDR and MDR enzymes. In *Enzymology and Molecular Biology of Carbonyl Metabolism 7* (Weiner, H., Lindahl, R. and Crabb, D.W., eds) Plenum, New York, in press.

Shafqat, J., El-Ahmad, M., Danielsson, O., Martínez, M.C., Persson, B., Parés, X. and Jörnvall, H. (1996) Pea formaldehyde-active class III alcohol dehydrogenase: Common derivation of the plant and animal forms but not of the corresponding ethanol-active forms (classes I and P). *Proc. Natl. Acad. Sci. USA* 93:5595–5599.

Shafqat, J., Höög, J.-O., Hjelmqvist, L., Oppermann, U.C.T., Ibañez, C. and Jörnvall, H. (1998) Studies on variants of alcohol dehydrogenases and its domains. In *Enzymology and Molecular Biology of Carbonyl Metabolism 7* (Weiner, H., Lindahl, R. and Crabb, D.W., eds) Plenum, New York, in press.

Smith, M., Hopkins, D.A. and Harris, H. (1973) Studies on the subunit structure and molecular size of the human alcohol dehydrogenase isozymes determined by the different loci, ADH1, ADH2, and ADH3. *Ann. Hum. Genet.* 36:401–414.

Still, J.L. (1940) Alcohol enzyme of *Bact. coli*. *Biochem. J.* 34:1177–1182.

Uotila, L. and Koivusalo, M. (1989) Glutathione-dependent oxidoreductases: Formaldehyde dehydrogenase. In *Coenzymes and Cofactors* (Dolphin, D., Poulson, R. and Avramovic O., eds) vol. 3, John Wiley & Sons, New York, pp. 517–551.

Vallee, B.L. and Bazzone, T.J. (1983) Isozymes of human liver alcohol dehydrogenase, in *Isozymes: Current Topics in Biological and Medical Research*, Vol. 8 (Rattazzi, M., Scandalios, J.G. and Whitt, G.S., eds), Alan R. Liss, New York, pp. 219–244.

Young, E.T. and Pilgrim, D. (1985) Isolation and DNA sequence of ADH3, a nuclear gene encoding the mitochondrial isozyme of alcohol dehydrogenase in *Saccharomyces cerevisiae*. *Mol. Cell. Biol.* 5:3024–3034.

45

REGULATORY FACTORS AND MOTIFS IN SDR ENZYMES

Udo Oppermann,[1] Samina Salim,[1] Malin Hult,[1] Guenther Eissner,[2] and Hans Jörnvall[1]

[1]Department of Medical Biochemistry and Biophysics
Karolinska Institutet
S 171 77 Stockholm, Sweden
[2]GSF Institute for Clinical Molecular Biology
81377 Muenchen-Grosshadern, Germany

1. INTRODUCTION

At present, well over 700 different enzymes are grouped into the superfamily of short-chain dehydrogenases/reductases (SDR), comprising isomerases, lyases and oxido-reductases, thereby constituting one of the largest protein superfamilies known to date (Jörnvall et al., 1995; Persson et al., this volume). Sequence comparisons between the different SDR members reveal typically ~ 25 % residue identity, with some highly conserved sequence motifs in common. The general architecture of this protein family has been greatly furthered by crystallographic analysis of some 10 distinct enzymes, revealing a close to identical pattern of α/β folding, a common nucleotide binding site and a Tyr-dependent acid-base catalytic mechanism. In many cases, these crystallographic data allow the computer-aided molecular modelling of three-dimensional structures of still further members, thereby opening the possibility of predicting or exploring substrate specificities of proteins with unknown function.

However, beside these features in common, individual characteristics, like subcellular distribution, membrane attachment, protein modification, and interactions with other cellular components, make the prediction of the individual enzyme properties difficult at present. We aim at an understanding of such factors and motifs, important for subcellular targetting, translational modifications, membrane attachment, regulation of enzymatic activities and expression.

SDR enzymes under study in our group are the endoplasmic reticulum (ER) bound human 11β-hydroxysteroid dehydrogenase type 1 (11β-HSD-1) (Monder and White, 1993; Oppermann et al., 1997a, b) and the ERAB protein, an intracellular target for amyloid β-

Enzymology and Molecular Biology of Carbonyl Metabolism 7
edited by Weiner *et al.* Kluwer Academic / Plenum Publishers, New York, 1999.

peptide (Yin et al., 1997). Interactions between amyloid β-peptide and ERAB were shown to mediate neurotoxicity and apoptosis in neural cell lines (Yin et al.,1997), thus being implicated in the pathogenesis of neurodegenerative disorders like Alzheimer's disease (AD). Sequence analysis revealed ERAB to be a member of the SDR family, but no specific enzymatic function could be assigned, leaving its physiological role undeciphered. Therefore our attempts have focussed on finding an enzymatic role for ERAB in addition to determining the interactions between this protein and the amyloid β-peptide. Sequence comparisons suggested it to be related to bacterial hydroxysteroid dehydrogenases and eukaryotic hydroxyacyl-CoA dehydrogenases. During the course of our studies we found that recombinant ERAB expresses no hydroxysteroid dehydrogenase activities but hydroxyacyl-CoA dehydrogenase activity which is specifically inhibited by binding of amyloid β-peptide (Oppermann et al., submitted).

In this study we analyzed the N-terminal regions of the 11β-HSD-1 and ERAB proteins (cf. Figures 1 and 2) for their importance in targetting to subcellular compartments.

2. EXPERIMENTAL

2.1. Molecular Cloning and Plasmid Constructs

Experimental procedures were as described (Hult et al., 1998; Oppermann et al., submitted). In brief, constructs for recombinant expression and *in vitro* translation were produced by PCR cloning using human liver cDNA as template source. After amplifica-

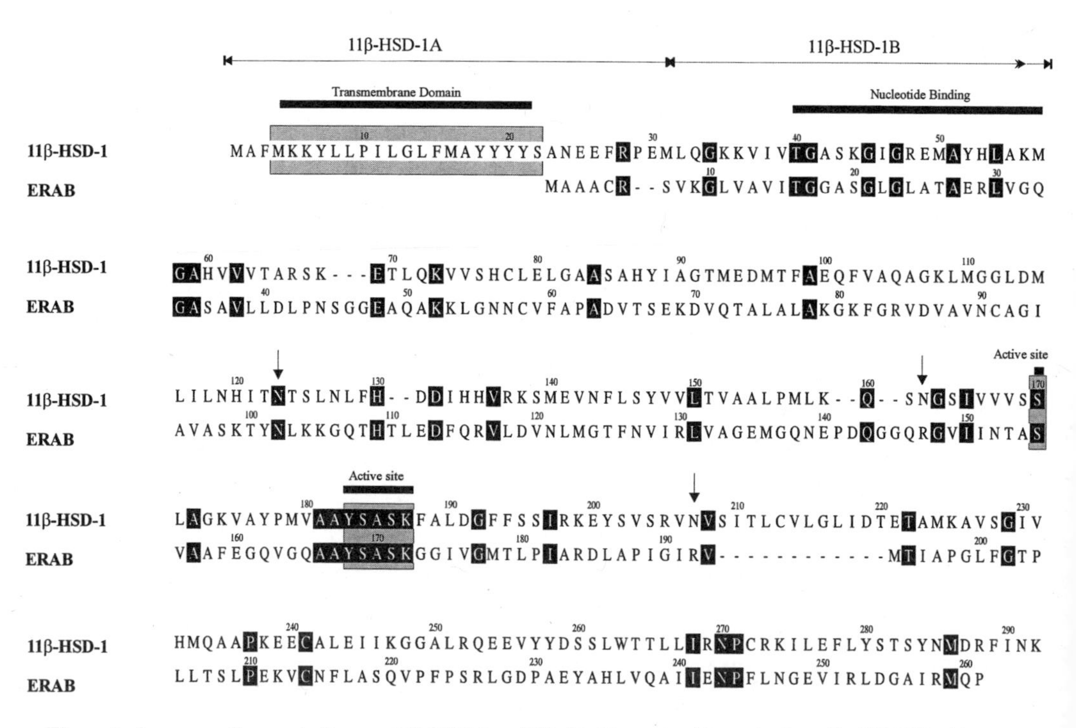

Figure 1. Sequence alignment of human 11β-HSD-1 and ERAB, illustrating N-terminal motifs. 11β-HSD-1A contains the N-terminal transmembrane domain, whereas the 1B form represents an SDR domain as indicated. Structural motifs are boxed and indicated, Asn linked glycosylation sites are marked by arrows.

tion, ligation into bacterial or yeast vectors was carried out. Plasmids used were pGEM4z (*in vitro* translation), pET15b (ERAB expression in *E. coli*) and pPIC9 and pPIC3.5 (11β-HSD1 expression in *Pichia pastoris*).

2.2. Heterologous Expression

Bacterial expression of recombinant proteins was performed in LB medium after induction with IPTG. Purification of His-tagged proteins was achieved by metal chelate chromatography. Protein expression in yeast was carried out in complex glycerol medium containing 0.5% (v/v) methanol for induction. Cells were harvested after 4 days of cultivation and mechanically disrupted using glass beads in the presence of 1 mM PMSF. After centrifugation and removal of beads and cell debris microsomes were prepared from the supernatant fraction by PEG 4000 precipitation.

2.3 *In Vitro* Translation

This was carried out after *in vitro* transcription using Sp6 or T7 RNA polymerases with plasmids carrying the ERAB and 11β-HSD-1 cDNAs under control of the respective viral promoters. *In vitro* translated proteins were labelled with ^{35}S-methionine and separated by SDS/PAGE, followed by autoradiography.

2.4. Enzymatic Assays

For 11β-hydroxysteroid dehydrogenase and 11-oxo reductase activity measurements a reverse phase HPLC method was used, consisting of a C18 column and an isocratic mobile phase of 30% acetonitrile in 0.1% ammoniumacetate, pH 7.0. ERAB activities were determined spectrophotometrically by recording the change of NAD+ or NADH absorbance at 340 nm. Substrates used were 3-hydroxyacyl-CoA, ketoacyl-CoA, testosterone, androsterone and isoursodeoxycholic acid. Enzymatic constants were derived using the EnzPack for Windows software package (Biosoft, Cambridge, UK).

3. RESULTS AND DISCUSSION

3.1. ER Targeting of 11β-HSD-1

The role of the 30 residue N-terminal hydrophobic domain of 11β-HSD-1 was analyzed by *in vitro* translation, protease protection and glycosylation experiments using full-length and truncated forms (11β-HSD-1A and 11β-HSD-1B, respectively, cf. Figure 1). The results obtained by these experiments are summarized in Table 1.

In vitro translation of both 11β-HSD-1A and 11β-HSD-1B yielded correctly translated products, but only the 1A construct resulted in a glycosylated integral membrane protein in the presence of microsomes. The 1B form was not incorporated into microsomes. These results clearly show that the N-terminal 30 residues in 11β-HSD-1A function as an ER targetting signal, which subsequently results in Asn-linked glycosylation of the mature protein. Glycosylation appears to be important for the enzymatic functions carried out by 11β-HSD-1A. Deglycosylation by endoglycosidase F treatment of microsomes derived from human liver or from yeast (Hult et al., 1998) containing recombinant human 11β-HSD-1 resulted in up to 70% loss of reductase activity whereas the de-

Table 1. Properties of full-length (11β-HSD-1A) and truncated (11β-HSD-1B) 11β-hydroxysteroid dehydrogenase constructs expressed in *Pichia pastoris* and analyzed by *in vitro* translation and glycosylation experiments

Structural properties	11β-HSD-1A	11β-HSD-1B
N-terminal transmembrane domain	yes	no
In vitro translation	product of expected size	product of expected size
Glycosylation	yes	no
Expression in yeast system	++++	(+)
Enzymatic activity	yes	no

hydrogenase activity was reduced by 10–15 % compared to control reactions (Oppermann et al., unpublished results). These data indicate different stabilities for the oxidative and reductive components of the enzyme. Glycosylation appears to be important in this respect, supposedly by constituting a recognition signal for lumenal ER chaperones like Bip or the calnexin/calreticulin system, which interact with unfolded but glycosylated peptides and assist in their correct folding (Helenius et al., 1997). A hypothetical synthesis pathway for 11β-HSD-1A is schematically given in Figure 2 summarizing the results obtained with truncated and full-length 11β-HSD-1 forms.

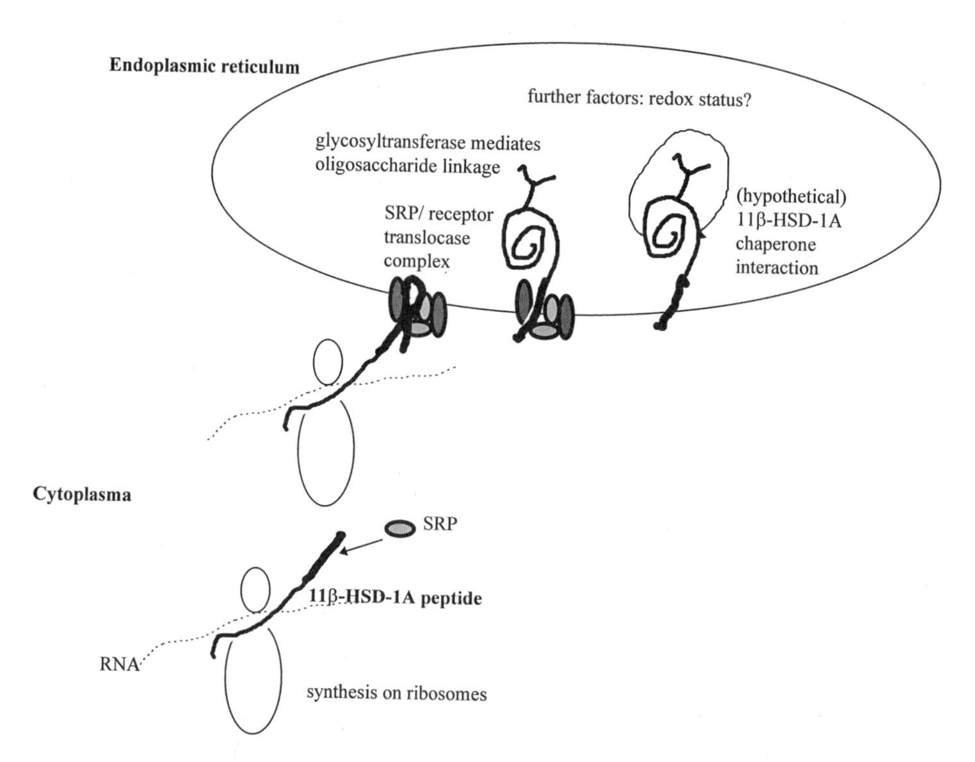

Figure 2. Intracellular 11β-HSD-1 synthesis and ER targeting. The N-terminal sequence functions as signal peptide and interacts with the signal recognition particle (SRP) and the ER translocase complex. It further serves as transmembrane domain. After ER insertion, 11β-HSD-1 oligosaccharide transfer takes place by glycosyltransferase. This modification might be a necessary factor for ER chaperone interaction, necessary for folding and assembly. The role of lumenal ER redox status might also be important for correct folding as indicated.

3.2. Subcellular Localization of ERAB

The intracellular amyloid β-peptide binding protein ERAB was identified by a yeast two-hybrid assay as a SDR protein with unknown enzymatic function and found to be localized mainly to the endoplasmic reticulum and to a lesser extent to mitochondria (Yin et al., 1997). As a consequence, amyloid β-peptide binding leads to cell death. Furthermore, upon binding, ERAB translocates to the plasma membrane. The topological signals or motifs for the ER targetting events are unknown, since no transmembrane domains or retention signals are detectable. However, comparison of the N-terminal ERAB sequence with that of other mitochondrial proteins, like type I hydroxyacyl-CoA dehydrogenase (HADHI) (cf. Figure 3), reveals a stretch of hydrophobic and basic residues typical for mitochondrial import signals (Furuta et al., 1997). Surprisingly, the 3β-HSD from *Comamonas testosteroni* displays high similarities to ERAB proteins in the overall structure and in the N-terminal regions, with a comparable pattern of basic and hydrophobic residues (cf. Figure 3). The significance of this finding is unknown, as is the possible mitochondrial import of 3β-HSD *in vitro*. Whether cleavage in the N-terminal ERAB sequence occurs upon mitochondrial import is also unknown at present. Studies are underway to determine the subcellular targetting and translocation of ERAB.

In this and other studies it was found that ERAB acts as a hydroxyacyl-CoA dehydrogenase (He et al., 1998; Oppermann et al., submitted). The mitochondrial localization of ERAB possibly provides a link between amyloid β-peptide mediated apoptosis and disruption of mitochondrial homoeostasis. Mitochondria are now considered to play a central role in the event of programmed cell death by generation of reactive oxygen species (ROS), and/or release of cytochrome c or other apoptosis initiating factors upon disturbance of mitochondrial homoeostasis (Beal, 1996; Simonian and Coyle, 1996; Mignotte and Vayssiere, 1998). Thus, we conclude that interaction of ERAB with amyloid β-peptide, inhibition of ERAB hydroxyacyl-CoA dehydrogenase activity (Oppermann et al., submitted) and disruption of the electron transport might lead to severe mitochondrial disturbance, triggering the apoptotic program. A hypothetical model for ERAB targetting is depicted in Figure 4, summarizing the present knowledge on ERAB distribution and its implication in triggering and execution of programmed cell death.

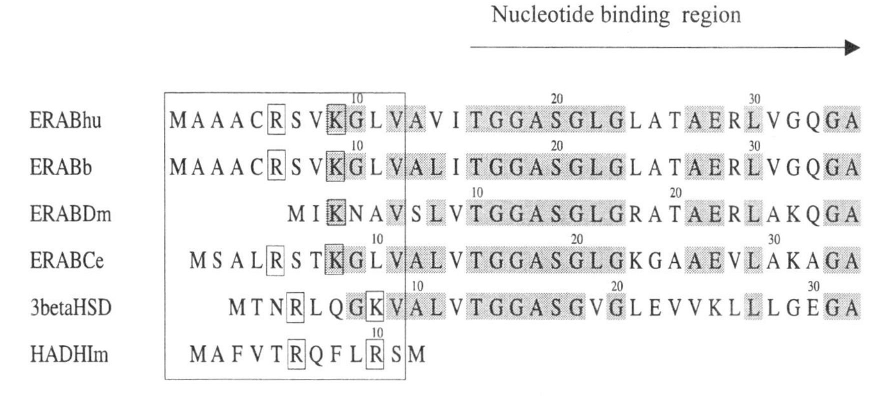

Figure 3. Alignment of N-terminal ERAB sequences from different species (from top to bottom: human, bovine, *Drosophila melanogaster* and *C. elegans*, followed by bacterial 3β-HSD and mouse type I hydroxy-acyl-CoA dehydrogenase (HADHIm). The SDR coenzyme binding region starting at Thr13 in 3β-HSD is indicated by an arrow. The proximal region with putative mitochondrial targeting motifs is indicated by boxing, highlighting basic residues possibly involved. Boxing in gray shows the sequence similarities in SDR structures.

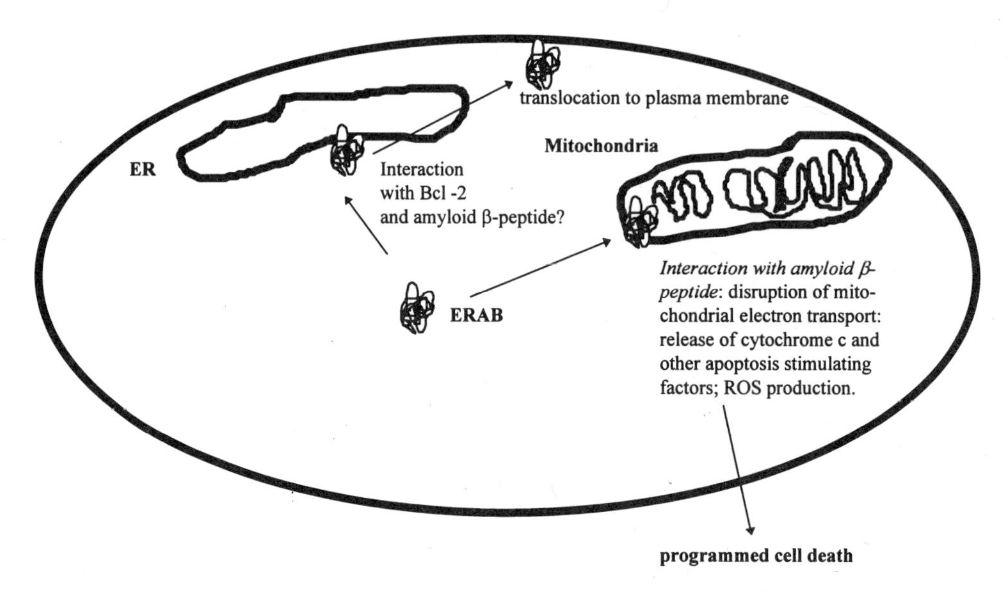

Figure 4. Model of ERAB mediated apoptosis. ERAB is found in two different cellular organelles, the endoplasmic reticulum (ER) and mitochondria. ERAB ER localization might be necessary for Bcl-2 interaction. Binding of amyloid β-peptide leads to disruption and plasma membrane translocation. Mitochondrial localization of ERAB suggests participation in electron transport, which might be interrupted by amyloid β-peptide binding. This could lead to production ROS (reactive oxygen species) and release of cytochrome c and other apoptosis mediating factors, which trigger the execution of the cell death program.

The selectivity of amyloid plaque formation in defined brain regions in Alzheimer`s disease might be related to the high energy and oxygen consumption of these tissues with the need for fatty acids as energy sources. However, the mechanism(s) of tissue specificity of amyloid fibril formation in relation to ERAB and amyloid β-peptide expression is at present unknown and needs to be explored.

3.3. Motifs in SDR Enzymes

The SDR protein superfamily is one of the largest known at present with more than 700 non-redundant database entries (Persson et al., this volume). The average amino acid chain-length was shown to be about 250–300 residues (Jörnvall et al., 1995). However, considerable chain-length variations occur making it necessary to group the particular members in distinct SDR classes (Persson et al., this volume), e.g. as shown for the membrane forms of hydroxysteroid dehydrogenases, isomerases and lyases. X-ray analysis of soluble SDR proteins reveals a one-domain structure with a Rossman fold, nucleotide-binding structure and a common active site consisting of a Ser, Tyr and Lys triad, with a strict conservation of Tyr as a general acid-base catalyst (Jörnvall et al., 1995; Oppermann et al., 1997a). In this study we have discussed further motifs and factors such as transmembrane, signal or insertion peptides, which extend the common SDR domain and determine subcellular organization and function of SDR enzymes.

ACKNOWLEDGMENTS

This study was supported by an EC biotechnology grant (EC 97-2123), the Swedish Medical Research Council (13X-3532), the Swedish Cancer Society (project 1806), Karolinska Institutet and NOVARTIS, Sweden.

REFERENCES

Beal, M. F. (1996) Mitochondria, free radicals and neurodegeneration. *Curr. Opinion Neurobiol.*, 6, 661–666.

Furuta, S., Kobayashi, A., Miyazawa, S., Hashimoto, T. (1997) Cloning and expression of cDNA for a newly identified isozyme of bovine liver 3-hydroxyacyl-CoA dehydrogenase and its import into mitochondria. *Biochem. Biophys. Acta*, 1350, 317–324.

He, X.Y., Schulz, H., Yang, S.Y. (1998) A human brain L-3-Hydroxyacyl-coenzyme A dehydrogenase is identical to an amyloid β-peptide-binding protein involved in Alzheimer's disease. *J. Biol. Chem.* 273, 10741–10746.

Helenius, A., Trombetta, E.S., Hebert, D.N. and Simons, J.F. (1997) Calnexin, calreticulin and the folding of glycoproteins. *Trends Cell Biol.* 7, 193–200.

Hult, M., Jörnvall, H. and Oppermann, U.C.T. (1998) Characterization of human 11β-hydroxysteroid dehydrogenase type 1 overexpressed in the yeast *Pichia pastoris. Endocrinology,* submitted.

Jörnvall, H., Persson, B., Krook, M., Atrian, S., Gonzalez-Duarte, R., Jeffery, J. and Ghosh, D. (1995) Short-chain dehydrogenase/reductases (SDR). *Biochemistry 34,* 6003 – 6013.

Mignotte, B. and Vayssiere, J.L. (1998) Mitochondria and apoptosis. *Eur. J. Biochem.* 252, 1–15.

Monder, C. and White, P.C. (1993) 11β-hydroxysteroid dehydrogenase. *Vitam. Horm.*, 47, 187–271.

Oppermann, U.C.T., Filling, C., Berndt, K.D., Persson, B., Benach, J., Ladenstein, R. and Jörnvall, H. (1997 a) Active-site directed mutagenesis of 3β/17β-hydroxysteroid dehydrogenase establishes differential effects on SDR reactions. *Biochemistry 36,* 34–40.

Oppermann, U.C.T., Persson, B. and Jörnvall, H. (1997 b) Function, gene organization and protein structures of 11β-hydroxysteroid dehydrogenase isoforms. *Eur. J. Biochem.* 249, 355–360.

Oppermann, U.C.T., Salim, S., Tjaernberg, L., Terenius, L. and Jörnvall, H. Inhibition of mitochondrial ERAB enzyme activity by amyloid β peptide: novel interaction and its consequence for neuronal apoptosis. *Proc. Natl. Acad. Sci. USA, submitted.*

Persson, B., Kallberg, Y., Nordling, E., Blunt, D., Oppermann, U.C.T. and Jörnvall, H. (1999) Bioinformatics in the study of MDR and SDR enzymes. *Adv. Exp. Med. Biol.* (in press).

Simonian, N.A., Coyle, J.T. (1996) Oxidative stress in neurodegenerative diseases. *Ann. Rev. Pharmacol. Toxicol.*, 36, 83–106.

Yin, S.D., Fu,J., Soto, C., Chen, X., Zhu, H., Al-Mohanna, F., Collison, K., Zhu, A., Stern, E., Saido, T., Tohyama, M., Ogawa, S., Roher, A., Stern, D.(1997) An intracellular protein that binds amyloid β peptide and mediates neurotoxicity in Alzheimer's disease. *Nature*, 389, 689–695.

46

BIOINFORMATICS IN STUDIES OF SDR AND MDR ENZYMES

Bengt Persson,[1] Erik Nordling,[1] Yvonne Kallberg,[1] Dan Lundh,[2]
Udo C. T. Oppermann,[1] Hanns-Ulrich Marschall,[1] and Hans Jörnvall[1]

[1]Department of Medical Biochemistry and Biophysics
Karolinska Institutet
S-171 77 Stockholm, Sweden
[2]Department of Computer Science
University of Skövde
S-541 28 Skövde, Sweden

1. INTRODUCTION

Bioinformatics utilises information in databases to understand biological processes and interpret experimental data. Methods include sequence comparisons, structural and functional predictions, and molecular modelling. Applied to the families of short-chain and medium-chain dehydrogenases/reductases (SDR and MDR), much information is obtained.

The SDR enzymes have subunits of 250-odd amino acid residues, an N-terminal coenzyme-binding pattern of GxxxGxG, and an active-site pattern of YxxxK. The SDR family is highly divergent with a typical pairwise residue identity of 15–30%. The enzymes cover a wide range of substrate specificity, including steroids, alcohols and aromatic compounds. Despite the low residue identity level of 15–30% between different SDR members, the three-dimensional structures thus far analyzed reveal a highly similar architecture with a one-domain α/β folding pattern. There also exist distantly related SDR forms with 350-odd residue subunits exhibiting dehydrogenase, dehydratase, epimerase or isomerase activity (Jörnvall *et al.*, 1995).

The MDR superfamily consists of members with subunits of 350-odd amino acid residues and 0, 1 or 2 zinc atoms. The subunits consist of one catalytic domain and one coenzyme-binding domain. The substrates of the MDR enzymes are often alcohols and aldehydes but the family also contains other members, including e.g. ζ-crystallin of the eye lens.

Enzymology and Molecular Biology of Carbonyl Metabolism 7
edited by Weiner *et al.* Kluwer Academic / Plenum Publishers, New York, 1999.

2. RESULTS AND DISCUSSION

2.1. Molecular Modelling of the γ Subunit of Class I Human Alcohol Dehydrogenase

A model of human γγ alcohol dehydrogenase was constructed with the program ICM (version 2.7, Molsoft LLC, Metuchen, NJ, USA) using the ββ alcohol dehydrogenase structure (Hurley *et al.*, 1994) as a template. The protein sequences of the β and γ subunits have 95 % positional identity without insertions/deletions. The subunits are therefore expected to have backbones without large differences. A modified version of the ICM homology script was used. The zinc atoms and the coenzyme were manually positioned according to the crystal structure. The γγ alcohol dehydrogenase model obtained showed good energy values and no steric hinderance. This model was subsequently used in docking calculations (cf. below).

2.2. Docking of on *iso*-Ursodeoxycholic Acid into the Model of the γ Subunit

Human γγ alcohol dehydrogenase is known to have catalytic activity on *iso*-ursodeoxycholic acid (iso-UDCA) (Marschall *et al.*, 1998). The molecular model of γγ alcohol dehydrogenase was used for docking calculations with *iso*-UDCA in order to evaluate the substrate-binding residues. During the first part of the docking calculation, a distance constraint of 2–2.4 Å was imposed between the catalytic zinc and OH on C-3 of the substrate. During the Monte-Carlo calculations, flexibility was only allowed in angles of substrate-binding residues, all flexible bonds of the substrate, and the position of the substrate. It was found that the bile acid fits tightly in the active site (Figure 1). Most of the residues within 4 Å from the steroid are hydrophobic (Table 1). We also see that the exchange of Ser/Thr48 between the γ and β isozyme subunits seems to disturb the binding of *iso*-UDCA.

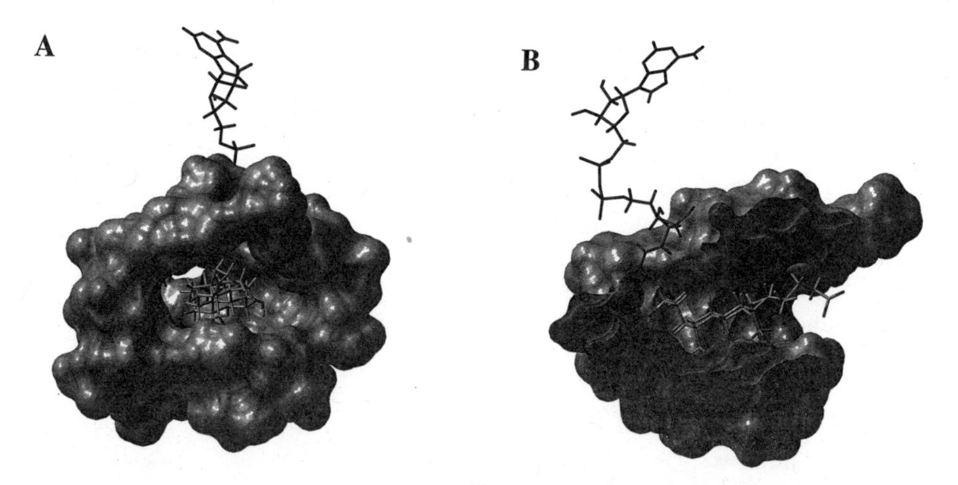

Figure 1. *iso*-Ursodeoxycholic acid docked into the active site of the class I γγ alcohol dehydrogenase model. The views in A and B are orthogonal to each other. The *iso*-ursodeoxycholic acid is shown as light sticks while the van der Waals surfaces of the residues within 4 Å are shown in gray. The catalytic zinc atom is dark and the coenzyme NAD^+ is shown as dark sticks.

Table 1. Residues within 4 Å from the steroid after the docking in the active-site of human class I $\gamma\gamma$ alcohol dehydrogenase model

Cys	46
Ser	48
Leu	57
His	67
Phe	93
Leu	116
Gly	117
Phe	140
Val	141
Cys	174
Val	294
Pro	295
Asp	297
Ile	318
Met	306
Leu	309
Thr	310

2.3. Molecular Modelling of 11β-Hydroxysteroid Dehydrogenase

The 11β-hydroxysteroid dehydrogenase type 1 is an NADPH-dependent SDR enzyme with 282-residue subunits. It is attached to membranes of the endoplasmatic reticulum and converts cortisone to cortisol. A molecular model of the subunit has been constructed using ICM, based on the known folds of other SDR members. Structurally and functionally critical residues are conserved, supporting the model. Further support is derived from the fact that the predicted glycosylation sites (Oppermann *et al.*, 1997) are superficially located in the model.

The model was also used for docking calculations with cortisol utilising the non-rigid ICM docking procedure and a distance restraint between O_η of Tyr183 and O-11 of cortisol. Free movements were allowed for the substrate and the χ angles of the enzyme. Cortisol fits at the active site with a distance of 2.4 Å between O-11 and O_η of Tyr 183 (Figure 2).

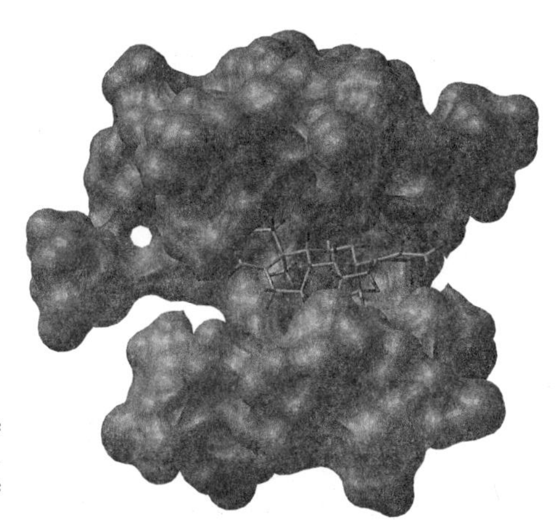

Figure 2. Cortisol docked into the active site of the subunit of the 11β-hydroxysteroid dehydrogenase model. Cortisol is shown in sticks and residues within 9 Å are shown as gray van der Waals surfaces.

2.4. Evolutionary Trees of MDR and SDR Super-Families

We have carried out extensive database searches in order to define all members of the MDR and SDR super-families using the "nr" database of NCBI (National Center for Biotechnology Information, Bethesda, MD, USA) with 309,224 sequences. The search result was checked using novel programs for exclusion of redundant sequences (Kallberg *et al.*, 1998). Both the MDR and SDR superfamilies have grown considerably and at present 537 MDR forms and over 1000 SDR forms exist. Of the SDR forms, 757 belong to the "classical" type with subunits of 250-odd amino acid residues while 299 are of the "extended" type with 350-odd residues typical of dehydratases, epimerases and isomerases.

Multiple sequence alignments and evolutionary trees were calculated with ClustalW (Thompson *et al.*, 1994). The evolutionary tree of SDR and MDR are shown in Figure 3, with subgroups encircled. In the SDR tree, the many species variants of *Drosophila* ADH are visible as a special subgroup. In the MDR tree, large subgroups are formed by plant and animal ADHs, reflecting much interest in those forms and hence many such forms determined. Also, the sorbitol dehydrogenase-related forms (SDH) and quinone reductase-related forms (QOR) have increased in number. The identification of new and distantly related MDR forms gives a complex pattern to this evolutionary tree.

3. CONCLUSIONS

With the exponential increase in sequence data, bioinformatics provide techniques to compare large numbers of sequences in order to define patterns and sub-families. In-

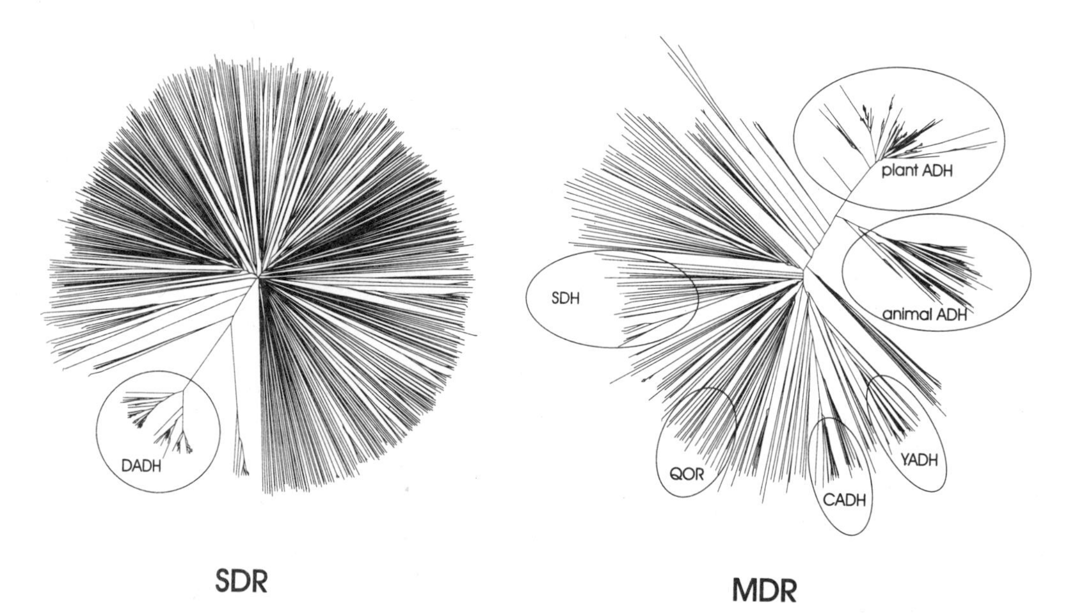

Figure 3. Evolutionary trees of the SDR and MDR superfamilies. The evolutionary trees are calculated with ClustalW (Thompson *et al.*, 1994) and drawn with NJPlot/Unrooted (Galtier *et al.*, 1996). Branch lengths are proportional to the sequence dissimilarity. The discussed subgroups are encircled. Abbreviations: CADH, cinnamyl alcohol dehydrogenase; DADH, *Drosophila* alcohol dehydrogenase; QOR, quinone reductase/ζ-crystallin; SDH, sorbitol dehydrogenase; YADH, yeast alcohol dehydrogenase.

creased knowledge of three-dimensional structures and improved molecular modelling techniques are valuable to evaluate active-site residues and correlations between structure and substrate specificity.

ACKNOWLEDGMENTS

This study was supported by grants from the Swedish Medical Research Council (projects 12564 and 3532), the European Union (BIO4-CT97-2123), the Alcohol Research Council of the Swedish Alcohol Retailing Monopoly, and Magn. Bergvalls Foundation.

REFERENCES

Galtier, N., Gouy, M. and Gautier, C. (1996) SEAVIEW and PHYLO_WIN: two graphic tools for sequence alignment and molecular phylogeny. *Comput. Appl. Biosci.* 12: 543–548.

Hurley, T.D., Bosron, B.F., Stone, C.L. and Amzel, L.M. (1994) Structures of three human beta alcohol dehydrogenase variants. Correlations with their functional differences. *J. Mol. Biol.* 235:983–1002.

Jörnvall, H., Persson, B., Krook, M., Atrian, S., Gonzàlez-Duarte, R., Jeffery, J. and Ghosh, D. (1995) Short-chain dehydrogenases/reductases (SDR). *Biochemistry* 34: 6003–6013.

Kallberg, Y, Jörnvall, H. and Persson, B. (1998) Implementing the Boyer-Moore algorithm. Manuscript in preparation.

Marschall, H.-U., Oppermann, U. C. T., Svensson, S., Nordling, E., Persson, B., Höög, J.-O. and Jörnvall, H. (1998) Human liver class I alcohol dehydrogenase $\gamma\gamma$ isozyme: the sole cytosolic 3β-hydroxysteroid dehydrogenase of *iso*-bile acids. *Biochemistry*, submitted.

Oppermann, U. C. T., Persson, B. and Jörnvall, H. (1997) Function, gene organization and protein structures of 11β-hydroxysteroid dehydrogenase isoforms. *Eur. J. Biochem.* 249: 355–360.

Thompson, J.D., Higgins, D.G. and Gibson, T.J. (1994) CLUSTAL W: improving the sensitivity of progressive multiple sequence alignment through sequence weighting, position-specific gap penalties and weight matrix choice. *Nucleic Acids Res.* 22: 4673–4680.

THE PROTECTIVE ROLE OF 11ß-HYDROXYSTEROID DEHYDROGENASE/CARBONYL REDUCTASE AGAINST TOBACCO-SMOKE RELATED LUNG CANCER

Edmund Maser

Department of Pharmacology and Toxicology
Philipps-University
D-35033 Marburg, Germany

1. INTRODUCTION

Lung cancer is one of the most common cancers worldwide and has the highest mortality rate among malignant tumors in the United States (Boring et al., 1994). More than 85 % of lung cancers may be attributed directly to cigarette smoking (Wynder and Hoffmann, 1994). One of the most potent carcinogens in tobacco products is the nitrosamine 4-(methylnitrosamino)-1-(3-pyridyl)-1-butanone (NNK) (Rivenson et al., 1988) which is formed by nitrosation of nicotine during the processing of tobacco (Burton et al., 1989) (Figure 1).

The carcinogenic potency of NNK in laboratory animals and its specificity for the lung (Hecht and Hoffmann, 1988) has strongly implicated NNK as an important etiological factor in the causation of lung cancer in human smokers (Hecht and Hoffmann, 1989). NNK requires metabolic activation by cytochrome P450 enzymes to express its mutagenic and carcinogenic properties (Crespi et al., 1991). The critical phase of this activation is an hydroxylation of the two carbons bound to the N-nitroso group and the formation of reactive electrophilic species that methylate and pyridyloxobutylate purine bases of DNA (Murphy et al., 1990).

However, the competing pathway for NNK activation is NNK detoxification, which proceeds via carbonyl reduction of NNK to 4-(methylnitrosamino)-1-(3-pyridyl)-1-butanol (NNAL), followed by glucuronidation of NNAL (Hecht, 1994). Although N-oxidation has also been considered as a NNK inactivating reaction, this pathway plays only a minor role in humans (Carmella et al., 1997).

Carbonyl reduction to NNAL is a very efficient pathway of NNK metabolism in animal and human tissues. In laboratory animals, the main metabolites in blood and urine after

Enzymology and Molecular Biology of Carbonyl Metabolism 7
edited by Weiner *et al.* Kluwer Academic / Plenum Publishers, New York, 1999.

Figure 1. Formation of 4-(methylnitrosamino)-1-(3-pyridyl)-1-butanone (NNK) by nitrosation of nicotine. NNK concentrations in tobacco products may amount up to 1–100 ppm.

i.v. administration of NNK resulted from carbonyl reduction (followed by glucuronidation or N-oxidation), indicating that carbonyl reduction to NNAL greatly exceeds metabolism by NNK α-hydroxylation (Adams et al., 1985; Adams et al., 1985; Hecht et al., 1993).

The predominance of carbonyl reduction has been confirmed by a characterization and quantification of NNK metabolites in the urine of smokers and smokeless tobacco users. The results clearly demonstrate that detoxification of NNK *in vivo* occurs mainly by carbonyl reduction and glucuronidation and, to a lesser extent, by pyridine N-oxidation in humans (reviewed in Maser, 1997). Interestingly, the fact that NNAL has been found in the urine of nonsmokers who are exposed to sidestream cigarette smoke provides evidence for the link between exposure to environmental tobacco smoke and the risk of lung cancer (Hecht et al., 1993).

In vitro studies revealed that NNAL was the major metabolite in hepatocytes and liver microsomal preparations indicating that the liver plays the predominant role in metabolism of NNK. In rodent lung tissue, carbonyl reduction was in the same range as α-hydroxylation and N-oxidation, whereas in rat nasal mucosa no NNAL was detectable (reviewed in Maser, 1998). The absence of NNAL formation in the nasal mucosa might explain why nasal mucosa is particularly susceptible to NNK-induced carcinogenesis. It has therefore been postulated that, depending on the extent of the competing pathways α-hydroxylation versus carbonyl reduction, the susceptibility of a tissue to tumor formation may be influenced.

In numerous studies there has been a great deal of interest in investigations on several cytochrome P450 isozymes mediating NNK activation via α-hydroxylation. However, although constituting the major pathway of NNK metabolism in several tissues, the enzyme that catalyzes the carbonyl reduction of NNK to NNAL in microsomal fractions has awaited its characterization since 1980 (Hecht et al., 1980). In recent investigations with a homogenously purified enzyme preparation from mouse liver we could finally demonstrate that the microsomal NNK carbonyl reductase is identical to 11ß-hydroxysteroid dehydrogenase type 1 (11ß-HSD 1) (EC 1.1.1.146) (Maser et al., 1996), the physiological function of which is the oxidoreduction of glucocorticoids.

Evidence is provided in the present investigation that 11ß-HSD 1, which belongs to the protein superfamily of the short-chain dehydrogenases/reductases, also functions as NNK carbonyl reductase in lung microsomes. Hence, 11ß-HSD 1 can be considered to be the central initiating enzyme of the final detoxification of NNK because, by ketone reduction of NNK, it provides the butanoyl-hydroxy function necessary for glucuronidation and excretion.

2. EXPERIMENTAL PROCEDURES

The purification of 11ß-HSD 1 from mouse liver for analytical studies, immunization and preparation of affinity purified antibodies was performed as described previously (Maser and Bannenberg, 1994a).

Activity assays of 11ß-HSD 1 were performed according to (Maser et al., 1996). Time and enzyme protein concentrations were chosen so that reaction velocities were time linear, which was found to be true at least for one hr. The apparent kinetic parameters were determined by using the GraphPad Inplot kinetic computer software.

For immunoinhibition studies 5–25 μl of affinity purified antibodies (corresponding to final concentrations of about 1–5 μg of protein, respectively) were preincubated with 0.1 mg of lung microsomal protein for 10 min, before starting the reaction by simultaneously adding the substrate NNK (at two final concentrations: 0.5 mM and 1 mM) and the NADPH-regenerating system.

3. RESULTS

3.1. Purification of 11ß-HSD 1

11ß-HSD 1 was isolated from mouse liver microsomes, as described previously, by following a new purification method which affords a gentle membrane protein solubilization as well as providing a favourable detergent surrounding during the subsequent chromatographic steps (Maser and Bannenberg, 1994a). The purity of 11ß-HSD 1 enzyme preparation was checked by SDS-PAGE, yielding a single band in the 34 kDa molecular mass region (Figure 2), and by N-terminal amino acid sequence analysis, a technique which requires homogenous protein preparations (data not shown). Antibodies directed against the purified 11ß-HSD 1 enzyme identified the respective protein in the same molecular mass region (Figure 2).

3.2. Kinetic Properties of 11ß-HSD 1

The kinetic properties of 11ß-HSD 1 together with the estimated intrinsic clearance values of the respective substrates are presented in Table 1. These data show that both glucocorticoid 11-oxidoreduction as well as xenobiotic carbonyl reduction can be catalyzed by 11ß-HSD 1. Whereas glucocorticoid 11ß-dehydrogenation is favoured over glucocorticoid 11-oxoreduction, the latter has approximately equal intrinsic clearance values for NNK carbonyl reduction. Our results demonstrate that purified 11ß-HSD catalyzes the carbonyl reduction of the tobacco-specific nitrosamine NNK to its alcohol metabolite NNAL and thus initiates NNK detoxification.

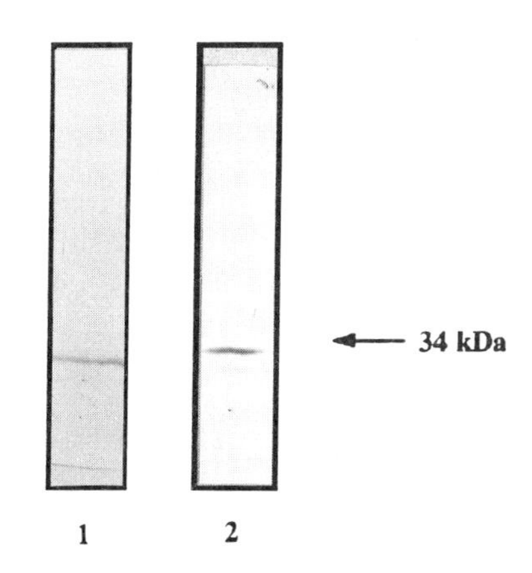

Figure 2. SDS-PAGE and immunoblot of 11ß-hydroxysteroid dehydrogenase 1 (11ß-HSD 1) from mouse liver. Lane 1 = SDS-PAGE and Coomassie blue staining of purified 11ß-HSD 1 (15 μg applied). Lane 2 = Western blot with affinity purified antibodies against 11ß-HSD 1 (15 μg of microsomal protein applied).

Table 1. Kinetic properties of glucocorticoid oxidoreduction and NNK carbonyl reduction

Substrate	Cosubstrate	V_{max}	K_m	V_{max}/K_m
Purified liver 11ß-HSD				
Corticosterone	$NADP^+$	329.33	0.66	498.48
Dehydrocorticosterone	NADPH-R.S.	5.33	0.22	24.23
NNK	NADPH-R.S.	135.13	1.75	77.22
Lung microsomes				
Corticosterone	$NADP^+$	0.135	0.007	19.29
Dehydrocorticosterone	NADPH-R.S.	0.048	0.036	1.33
NNK	NADPH-R.S.	0.655	0.629	1.04

Enzyme activities were assayed in the standard reaction mixture containing purified mouse liver 11ß-HSD 1 or mouse lung microsomes, the indicated cosubstrate (NADPH-R.S. = NADPH-regenerating system) and varying substrate concentrations (glucocorticoids: 12.5 to 200 µM; NNK: 10 µM to 1 mM). Kinetic parameter estimations were calculated from three experiments using the GraphPad Inplot kinetic computer software. V_{max} = nmol/min x mg; K_m = mM.

3.3. Expression of 11ß-HSD 1 Protein in Mouse and Rat Lung

Affinity purified antibodies directed against 11ß-HSD 1 from mouse liver were used to investigate the expression of 11ß-HSD 1 on the protein level in lung. Figure 3 shows the immunoblot analysis, indicating the presence of 11ß-HSD 1 in mouse and rat lung. Mouse liver microsomes were used as positive control. It turned out that the extent of immunoreactivity differed between mouse liver and lung. Based on the amount of protein subjected to gel electrophoresis and based on the densitometric signal of 11ß-HSD 1 (given in arbitrary units), 11ß-HSD 1 is expressed in mouse liver at considerably higher levels (by a factor of around 20-fold) compared to that in mouse lung.

3.4. Inhibition of NNK Carbonyl Reduction by Glucocorticoids and Glycyrrhetinic Acid

This result provides the first indication that 11ß-HSD 1 is responsible for NNK carbonyl reduction in mouse lung microsomes, because the glucocorticoids corticosterone

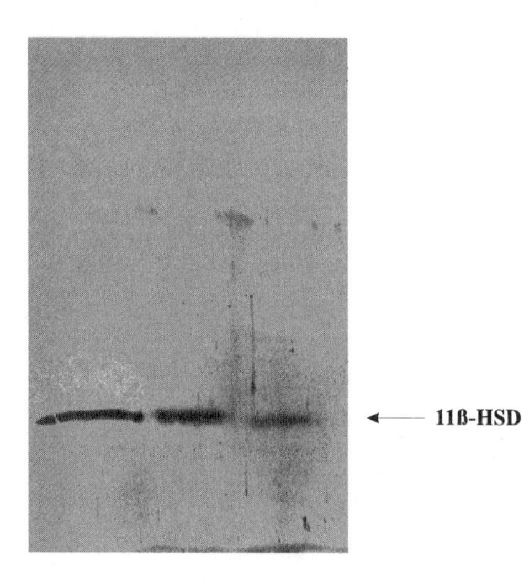

Figure 3. Immunoblot analysis of 11ß-HSD 1 in mouse and rat lung. Microsomes from mouse and rat lung were examined for 11ß-HSD 1 protein. NMRI mice liver microsomes served as positive control. All bands appear in the 34 kDa molecular mass region (as indicated by the arrow) (Maser, 1998).

and dehydrocorticosterone, as well as the selective inhibitor of 11ß-HSD, glycyrrhetinic acid, turned out to be potent inhibitors of NNK carbonyl reduction. Whereas the Ki values of NNK inhibition for glucocorticoids (corticosterone 37.78 μM, dehydrocorticosterone 21.33 μM) are significantly higher than those for glycyrrhetinic acid (10.91 μM), the low Ki value of glycyrrhetinic acid corresponds well to those of this steroidal 11ß-HSD inhibitor obtained in other studies. The competitive nature of inhibition of NNK carbonyl reduction by glucocorticoids and glycyrrhetinic acid suggests that both steroids and NNK bind to the catalytically active site of 11ß-HSD 1, and that NNK indeed is a substrate of this enzyme.

3.5. Inhibition of NNK Carbonyl Reduction by Anti-11ß-HSD 1 Antibodies

In order to conclusively substantiate the involvement of 11ß-HSD 1 in NNK carbonyl reduction in mouse lung microsomes, affinity purified antibodies against mouse liver 11ß-HSD 1 were employed to inhibit enzyme activity (Figure 4). By using two different substrate concentrations of NNK (0.5 mM and 1.0 mM) which were roughly around the Km for NNK carbonyl reduction in these fractions, the rate of NNAL formation by 11ß-HSD 1 was decreased by increasing amounts of the antibodies. For example, when mixing about 25 μl of antibody solution (corresponding to 5 μg of antibody protein) with about 0.1 mg of lung microsomal protein, the enzyme activity was decreased to a NNAL formation capacity of only 38 pmol/min/mg protein (at 0.5 mM NNK) or 120 pmol/min/mg protein (at 1.0 mM) which means a residual capacity of NNK carbonyl reduction of around 7.9 and 17.7 per cent, respectively, compared to the uninhibited control values.

4. DISCUSSION

The present study is the first to provide evidence for the involvement of 11ß-HSD 1 in the carbonyl reduction of NNK in lung.

Physiologically, 11ß-HSD catalyzes the reversible interconversion of cortisol to cortisone (in humans) and corticosterone to dehydrocorticosterone (in rodents) (reviewed in Seckl, 1993). Two distinct isozymes of 11ß-HSD have been described. 11ß-HSD 1 is a low affinity NADP/NADPH-dependent dehydrogenases/oxo-reductase with Km values in the μM range. The predominant role of this isozyme *in vivo* has been shown to be 11-oxo-re-

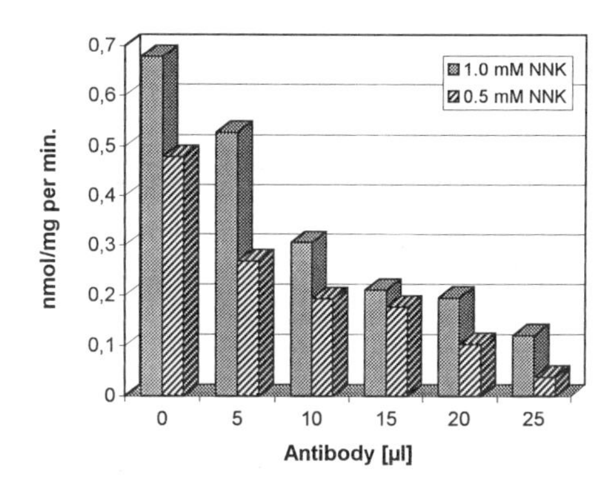

Figure 4. Immunoinhibition of NNK carbonyl reduction by affinity purified antibodies against 11ß-HSD 1. Antibodies (0, 5, 10, 15, 20 and 25 μl of antibody solution corresponding to 0, 1, 2, 3, 4 and 5 μg of protein, respectively) were mixed with 0.1 mg of lung microsomal protein for 10 min before starting the reactions. Two different substrate concentrations, 1.0 and 0.5 mM NNK, were used. Values are means of two different preparations and are expressed as nmol NNAL formed per min per mg protein (Maser, 1998).

duction, i.e. the generation of active glucocorticoid. In contrast, 11ß-HSD 2 is a high affinity, unidirectional, NAD-dependent dehydrogenase with Km values in the nM range. It is this isozyme that is found principally in mineralocorticoid target tissues such as the kidney and colon, where it protects the mineralocorticoid receptor from cortisol excess.

The distribution of 11ß-HSD 1 within tissues has been extensively studied, and its expression shown in glucocorticoid target tissues, such as liver, lung, gonad, cerebellum, and pituitary. Interestingly, evidence is emerging to suggest a fundamental role for 11ß-HSD 1 in the detoxification of nonsteroidal xenobiotic carbonyl compounds in mammals (Maser et al., 1996; Maser, 1996; Maser and Bannenberg, 1994b; Rekka et al., 1996; Maser, 1995; Oppermann et al., 1995; Maser et al., 1994). According to the fact that 11ß-HSD 1 acts as carbonyl reductase of the tobacco-specific nitrosamine NNK, it is hypothesized that any impact on 11ß-HSD 1 expression and/or activity results in a shift of the NNK/NNAL equilibrium toward NNK, which then undergoes activation via cytochrome P450 mediated α-hydroxylation.

It is clear that in response to stress, elevated levels of endogenous glucocorticoids occupy the 11ß-HSD enzyme. This would result in a shift of the NNK/NNAL equilibrium towards NNK, which, upon metabolic activation via α-carbon hydroxylation exerts its cancerogenic effect. High levels of cholic acids in cholestatic states, or estrogens and progesterone during pregnancy, would also affect NNK detoxification, as those steroids have been shown to inhibit 11ß-HSD. The same consequences would be expected upon administration of exogenous glucocorticoids during glucocorticoid therapy, or other pharmacologic drugs such as the diuretics furosemide and ethacrynic acid or the anti-ulcer drug

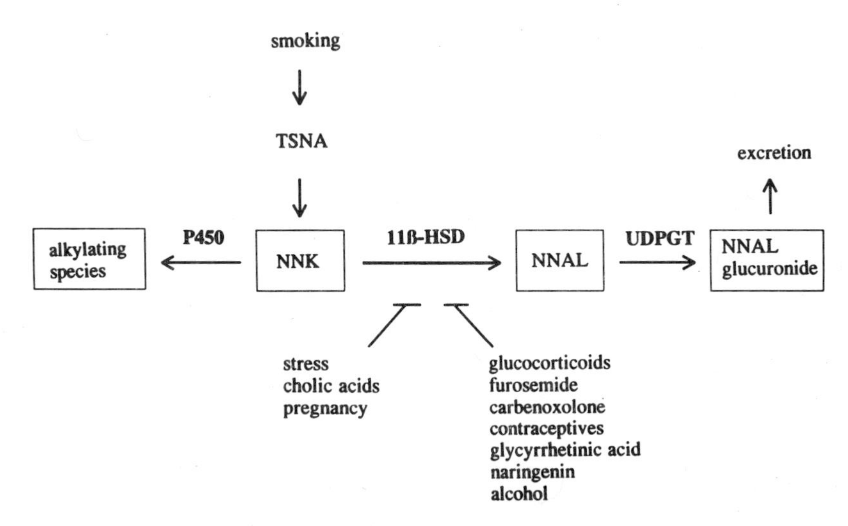

Figure 5. The metabolic activation and inactivation of 4-(methylnitrosamino)-1-(3-pyridyl)-1-butanone (NNK). NNK, which enters the body upon smoking, is either activated by cytochrome P450 enzymes (P450) to alkylating species, or inactivated by 11ß-HSD mediated carbonyl reduction to 4-(methylnitrosamino)-1-(3-pyridyl)-1-butanol (NNAL), followed by glucuronidation and final excretion. This situation has three consequences: First, the expression of 11ß-HSD 1, respectively the coexpression with glucuronosyltransferase (UDPGT) in certain tissues, plays an important role in the organospecificity of NNK induced cancerogenesis. Second, interindividual differences, i.e. genetic polymorphisms, in the expression of 11ß-HSD 1 may be of significance, if a smoker has a higher risk for the development of lung cancer. Third, endogenous (left-hand column) or exogenous (right-hand column) compounds, which are substrates or inhibitors of 11ß-HSD, may shift the equilibrium between NNK and NNAL towards NNK, a fact which then would favour NNK activation. (TSNA, tobacco-specific nitrosamines) (Maser, 1997).

carbenoxolone. A similar situation might also result upon usage of oral contraceptives, anabolic steroids, or the exposure to food constitutents like glycyrrhetinic acid (the principal constitutent of licorice) and naringenin (a flavonoid with high concentrations in grapefruit juice). For most of these compounds, especially for the pharmacologic drugs, it could be shown that they already in therapeutic doses impair the activity of 11ß-HSD *in vivo* (reviewed in Maser, 1997).

Interestingly, the fact that chronic alcohol consumption greatly enhances the carcinogenic response to NNK might now be explained on the base that ethanol has been shown to be a potent inhibitor of 11ß-HSD (Riddle and McDaniel, 1993; Valentino et al., 1995).

In addition, 11ß-HSD 1 is under multifactorial regulation. In primary cultures of rat hepatocytes, it has been shown that dexamethasone stimulates 11ß-HSD 1 activity, and growth hormone and insulin inhibit 11ß-HSD 1 activity. Sex steroids have been found to regulate 11ß-HSD 1 activities and mRNA levels in the liver, kidney and hippocampus. It may therefore be anticipated that these regulatory factors affect the potency of 11ß-HSD 1 to catalyze NNK carbonyl reduction, thus concomitantly promoting NNK α-hydroxylation (reviewed in Maser and Oppermann, 1997).

The results of the present study have important implications for understanding the organospecificity and/or interindividual susceptibility of NNK-induced carcinogenesis. It is hypothesized that the extent of the competing pathways, α-hydroxylation versus carbonyl reduction followed by glucuronidation, positively correlates with NNK induced tumorigenesis in several tissues. For example, the absence or low degree of NNAL formation in the nasal mucosa, oral tissue, esophagus and pancreas, in contrast to high levels of activating P450 isozymes, may be a causative factor of the organospecific incidence of tumors in these tissues. In liver itself NNK has been found to be predominantly metabolized via carbonyl reduction and glucuronidation, thus possibly explaining the lower tumor incidence of the liver after NNK exposure. Moreover, of all tissues studied the liver has been shown to express highest levels of 11ß-HSD protein and activity, and hepatic clearance has been expected to reduce the exposure of pulmonary tissue to NNK.

On the other hand, it has been demonstrated that NNK has a high degree of selectivity for induction of lung tumors, independent of the route of administration. The lung is composed of a variety of different cell types, and it is postulated that, compared to all others, tumors arise preferably from Clara cells, which are known to have high capacities of α-hydroxylation and low levels of NNK carbonyl reduction. In contrast, in alveolar macrophages and type II cells carbonyl reduction is considerably higher than α-hydroxylation, thus possibly explaining the lower tumor incidence in the latter two cell types. Therefore, tissue-specific differences in the expression of 11ß-HSD 1 may play a role in the organospecificity of NNK induced cancerogenesis.

Moreover, several studies on the profile of NNK metabolites in smokers' urine have revealed high interindividual variations with respect to NNAL, NNAL-glucuronide, and NNAL N-oxide (Carmella et al., 1995; Meger et al., 1996). It is these variations that are of greatest interest, because they may be linked not only to lung cancer susceptibility but also to a genetic polymorphism in terms of 11ß-HSD 1 expression. Here, genetic polymorphism profiles of 11ß-HSD 1 expression would help to define the role of NNK detoxification by 11ß-HSD 1 and its role in lung cancer prevention upon smoking.

In conclusion, the present findings may have potentially important implications for active and passive smokers who express low levels of 11ß-HSD 1/NNK carbonyl reductase and/or are concurrently exposed to 11ß-HSD 1 modulators.

ACKNOWLEDGMENTS

The skillful technical assistance of J. Friebertshäuser, M. Losekam and T. W. Tan is gratefully acknowledged. This study was supported by grants from the Deutsche Forschungsgemeinschaft (DFG) (Ma 1704/3-1), the Stiftung Verhalten und Umwelt (Verum) (München, Germany), and the European Commission (BIO4-97-2123).

REFERENCES

Adams, J.D., LaVoie, E.J., and Hoffmann, D., 1985, On the pharmacokinetics of tobacco-specific *N*-nitrosamines in Fisher rats, *Carcinogenesis*, 6: 509–511.

Adams, J.D., LaVoie, E.J., O'Mara-Adams, K.J., Hoffmann, D., Carey, K.D., and Marshall, M.V., 1985, Pharmacokinetics of *N*-nitrosonornicotine and 4-methylnitrosamino-1-(3-pyridyl)-1-butanone in laboratory animals, *Cancer Lett.*, 28: 195–201.

Boring, C.C., Squires, T.S., Tong, T., and Montgomery, S., 1994, Cancer Statistics, 1994, *CA Cancer J. Clin.*, 44: 7–26.

Burton, H.R., Childs, G.H., Anderson, R.A., and Fleming, P.D., 1989, Changes in chemical composition of Burley tobacco during senescence and curing. 3. Tobacco specific nitrosamines, *J. Agric. Food Chem.*, 37: 426–430.

Carmella, S.G., Akerkar, S.A., Richie, J.P., and Hecht, S.S., 1995, Intraindividual and interindividual differences in metabolites of the tobacco-specific lung carcinogen 4-(methylnitrosamino)-1-(3-pyridyl)-1-butanone (NNK) in smoker's urine, *Cancer Epidemiol. Biomarkers & Prev.*, 4: 635–652.

Carmella, S.G., Borukhova, A., Akerkar, S.A., and Hecht, S.S., 1997, Analysis of human urine for pyridine-N-oxide metabolites of 4-(methylnitrosamino)-1-(3-pyridyl)-1-butanone, a tobacco-specific lung carcinogen, *Cancer Epidemiol. Biomarkers Prev.*, 6: 113–120.

Crespi, C.L., Penman, B.W., Gelboin, H.V., and Gonzalez, F.J., 1991, A tobacco smoke-derived nitrosamine, 4-(methylnitrosamino)-1-(3-pyridyl)-1-butanone, is activated by multiple human cytochrome P450s including the polymorphic human cytochrome P4502D6, *Carcinogenesis*, 12: 1197–1201.

Hecht, S.S., 1994, Metabolic activation and detoxification of tobacco-specific nitrosamines - a model for cancer prevention strategies, *Drug Metab. Rev.*, 26: 373–390.

Hecht, S.S., Carmella, S.G., Murphy, S.E., Akerkar, S., Brunnemann, K.D., and Hoffmann, D., 1993, A tobacco-specific lung carcinogen in the urine of men exposed to cigarette smoke, *N. Engl. J. Med.*, 329: 1543–1546.

Hecht, S.S. and Hoffmann, D., 1988, Tobacco-specific nitrosamines, an important group of carcinogens in tobacco and tobacco smoke, *Carcinogenesis*, 9: 875–884.

Hecht, S.S. and Hoffmann, D., 1989, The relevance of tobacco-specific nitrosamines to human cancer, *Cancer Surv.*, 8: 273–294.

Hecht, S.S., Trushin, N., Reid Quinn, C.A., Burak, E.S., Jones, A.B., Southers, J.L., Gombar, C.T., Carmella, S.G., Anderson, L.M., and Rice, J.M., 1993, Metabolism of the tobacco-specific nitrosamine 4-(methylnitrosamino)-1-(3-pyridyl)-1-butanone in the patas monkey: pharmacokinetics and characterization of glucuronide metabolites, *Carcinogenesis*, 14: 229–236.

Hecht, S.S., Young, R., and Chen, C.B., 1980, Metabolism in the F344 rat of 4-(N-methyl-N-nitrosamino)-1-(3-pyridyl)-1-butanone, a tobacco-specific carcinogen, *Cancer Res.*, 40: 4144–4150.

Maser, E., 1995, Xenobiotic carbonyl reduction and physiological steroid oxidoreduction - the pluripotency of several hydroxysteroid dehydrogenases, *Biochem. Pharmacol.*, 49: 421–440.

Maser, E., 1996, 11ß-Hydroxysteroid dehydrogenase acts as carbonyl reductase in microsomal phase I drug metabolism, *Exp. Toxic. Pathol.*, 48: 266–273.

Maser, E., 1997, Stress, hormonal changes, alcohol, food constituents and drugs: factors that advance the incidence of tobacco smoke-related cancer?, *Trends Pharmacol. Sci.*, 18: 270–275.

Maser, E., 1998, 11ß-Hydroxysteroid dehydrogenase responsible for carbonyl reduction of the tobacco-specific nitrosamine 4-(methylnitrosamino)-1-(3-pyridyl)-1-butanone in mouse lung microsomes, *Cancer Res.*, (in press)

Maser, E. and Bannenberg, G., 1994a, The purification of 11ß-hydroxysteroid dehydrogenase from mouse liver microsomes, *J. Steroid Biochem. Molec. Biol.*, 48: 257–263.

Maser, E. and Bannenberg, G., 1994b, 11ß-Hydroxysteroid dehydrogenase mediates reductive metabolism of xenobiotic carbonyl compounds, *Biochem. Pharmacol.*, 47: 1805–1812.

Maser, E., Friebertshäuser, J., and Mangoura, S.A., 1994, Ontogenic pattern of carbonyl reductase activity of 11ß-hydroxysteroid dehydrogenase in mouse liver and kidney, *Xenobiotica*, 24: 109–117.

Maser, E. and Oppermann, U.C.T., 1997, Role of type-1 11ß-hydroxysteroid dehydrogenase in detoxification processes, *Eur. J. Biochem.*, 249: 356–369.

Maser, E., Richter, E., and Friebertshauser, J., 1996, The identification of 11 beta-hydroxysteroid dehydrogenase as carbonyl reductase of the tobacco-specific nitrosamine 4-(methylnitrosamino)-1-(3-pyridyl)-1-butanone, *Eur. J. Biochem.*, 238: 484–489.

Meger, M., Meger-Kossien, I., Dietrich, M., Tricker, A.R., Scherer, G., and Adlkofer, F., 1996, Metabolites of 4-(*N*-methylnitrosamino)-1-(3-pyridyl)-1-butanone in urine of smokers, *Eur. J. Cancer Prev.*, 5 (Suppl.1): 121–124.

Murphy, S.E., Palomino, A., Hecht, S.S., and Hoffmann, D., 1990, Dose-response study of DNA and hemoglobin adduct formation by 4-(methylnitrosamino)-1-(3-pyridyl)-1-butanone in F344 rats, *Cancer Res.*, 50: 5446–5452.

Oppermann, U.C.T., Netter, K.J., and Maser, E., 1995, Cloning and primary structure of murine 11ß-hydroxysteroid dehydrogenase/microsomal carbonyl reductase, *Eur. J. Biochem.*, 227: 202–208.

Rekka, E.A., Soldan, M., Belai, I., Netter, K.J., and Maser, E., 1996, Biotransformation and detoxification of insecticidal metyrapone analogues by carbonyl reduction in the human liver, *Xenobiotica*, 26: 1221–1229.

Riddle, M.C. and McDaniel, P.A., 1993, Acute reduction of renal 11 beta-hydroxysteroid dehydrogenase activity by several antinatriuretic stimuli, *Metabolism*, 42: 1370–1374.

Rivenson, A., Hoffmann, D., Prokopczyk, B., Amin, S., and Hecht, S.S., 1988, Induction of lung and exocrine pancreas tumors in F344 rats by tobacco-specific and Areca-derived N-nitrosamines, *Cancer Res.*, 48: 6912–6917.

Seckl, J.R., 1993, 11ß-Hydroxysteroid dehydrogenase isoforms and their implications for blood pressure regulation, *Eur. J. Clin. Invest.*, 23: 589–601.

Valentino, R., Tommaselli, A.P., Savastano, S., Stewart, P.M., Ghiggi, M.R., Galletti, F., Mariniello, P., Lombardi, G., and Edwards, C.R., 1995, Alcohol inhibits 11-beta-hydroxysteroid dehydrogenase activity in rat kidney and liver, *Horm. Res.*, 43: 176–180.

Wynder, E.L. and Hoffmann, D., 1994, Smoking and lung cancer: scientific challenges and opportunities, *Cancer Res.*, 54: 5284–5295.

STRUCTURE-FUNCTION RELATIONSHIPS OF 3β-HYDROXYSTEROID DEHYDROGENASES INVOLVED IN BILE ACID METABOLISM

Charlotta Filling, Hanns-Ulrich Marschall, Tim Prozorovski, Erik Nordling, Bengt Persson, Hans Jörnvall, and Udo C. T. Oppermann

Department of Medical Biochemistry and Biophysics
Karolinska Institutet
S-171 77 Stockholm, Sweden

1. INTRODUCTION

In vertebrates, 3β-hydroxysteroid dehydrogenases (3β-HSDs) fulfill several physiological and metabolic functions. They are primarily involved in the synthesis of all classes of steroid hormones by catalyzing the 3β-OH dehydrogenation/Δ4–5 isomerization of steroid hormone precursor molecules (Simard et al., 1996). In mammals they furthermore mediate C3-hydroxyl group epimerization of bile acids and other steroids during enterohepatic circulation (Figure 1). The enzymes involved in this reaction comprise 3α-hydroxysteroid dehydrogenases (3α-HSD), 3-keto reductases and 3β-HSDs. Epimerization of steroids secreted in the bile occurs in the intestine, catalyzed by microbial hydroxysteroid dehydrogenases and in the liver. The importance of these epimerization reactions is unknown, but it is anticipated that liver 3β-HSDs are involved in intracellular binding and transport of bile acids from the sinusoidal to the canalicular side of the hepatocyte and thus might play a role in cholestatic processes (Marschall et al., 1998).

Thus far, different 3β-HSDs have been characterized. The bacterial and mammalian enzymes described belong to the short-chain dehydrogenase/reductase (SDR) superfamily (Jörnvall et al., 1995; Oppermann et al., 1997a) with the exception of two class I ADH isozymes, the human γ and equine S isoforms (Park and Plapp, 1991), which are members of the medium-chain dehydrogenases/reductase (MDR) family.

Our research concentrates on structure-function relationships of 3β-HSDs of bacterial and mammalian origin. In this paper we describe molecular aspects of 3β/17β-HSD from *Comamonas testosteroni* and human liver 3β-HSD (class I γ ADH), two enzymes active with 3β-hydroxy ("iso") bile acids as substrates.

Enzymology and Molecular Biology of Carbonyl Metabolism 7
edited by Weiner *et al.* Kluwer Academic / Plenum Publishers, New York, 1999.

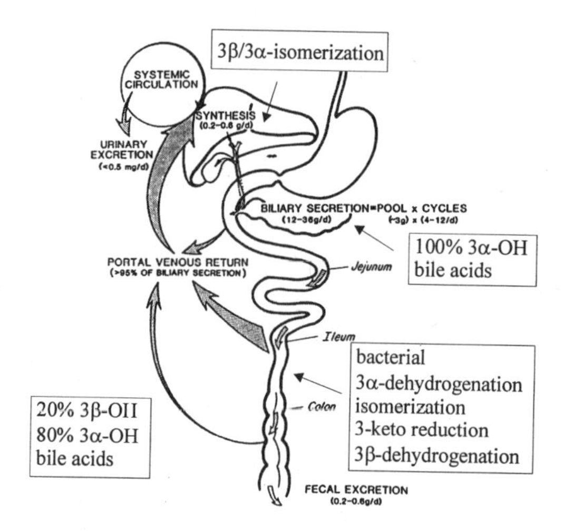

Figure 1. Enterohepatic circulation of bile salts showing typical kinetic values for healthy humans.

2. EXPERIMENTAL

2.1. Cloning, Expression and Site-Directed Mutagenesis of Bacterial 3β/17β-HSD from *Comamonas testosteroni*

The 3β/17β-HSD gene from *Comamonas testosteroni* was cloned into the pET 15b vector. Overexpression, purification and site-directed mutagenesis was carried out as described in a former paper (Oppermann et al., 1997b). Integrity and purity was assessed by SDS/PAGE and CD spectroscopy. Enzyme activities were assayed as NAD^+/NADH-dependent 3β-HSD and 3-keto reductase activities at 340 nm. Steroid products were confirmed by TLC. Reactions were performed at 25°C with 20 mM Tris-HCl and 250 μM NAD^+ or NADH. Substrates were isoursodeoxycholic acid (isoUDCA) for 3β-HSD activity at pH 8.5 and 5α-dihydro-testosterone for 3-keto reductase activity at pH 7.0.

2.2. ADH γγ: The Sole 3β-HSD from Human Liver Cytosol

Human class I ADH $\gamma_2\gamma_2$ isozyme was cloned, overexpressed in *E. coli* and purified by DEAE and 5'-AMP-Sepharose chromatography (Marschall et al.,1998). Human liver samples were obtained from organ donors. Cytosolic fractions were obtained after homogenisation and differential centrifugation. NAD^+ dependent oxidation of iso bile acids at 0.5–100 μM was measured using GC/MS. All reactions were performed in 0.1 M glycine-NaOH buffer, pH 10, and 2.5 mM NAD^+. Human liver cytosol and recombinant human ADH γγ isozyme were subjected to native PAGE and isoelectric focusing. Gels were cut into slices and incubated at 37º C in 15 ml glycine/NaOH, pH 10, containing 1.0 mM NAD^+ or $NADP^+$, and steroids at 50 μM or ethanol at 50 mM concentrations. Nitro-blue tetrazolium and 0.5 mg phenazine methosulphate were added and incubated until formazane bands became visible.

2.3 Molecular Modelling of Class I ADH γ

A three-dimensional model of the class I ADH γ subunit was obtained by adopting its amino acid sequence into the known fold of the human class I ADH β subunit using the program ICM (Marschall et al.,1998; Hurley et al., 1994). Docking of isoUDCA to the enzyme model was performed using a non-rigid procedure, allowing free movement of the substrate, the rotable bonds of the substrate, and the χ angles of the residues close to the active site.

3. RESULTS AND DISCUSSION

3.1. Site-Directed Mutagenesis of the SDR Type 3β/17β-HSD from *Comamonas testosteroni*

In order to study the basic architecture of SDR proteins site-directed mutagenesis in conserved regions has been carried out using 3β/17β-HSD as a model system (cf. Figure 2).

The residues mutated and kinetic data obtained (K_m and relative V_{max} values) of wild-type and mutant forms are presented in Table 1. The conserved residues are located in four

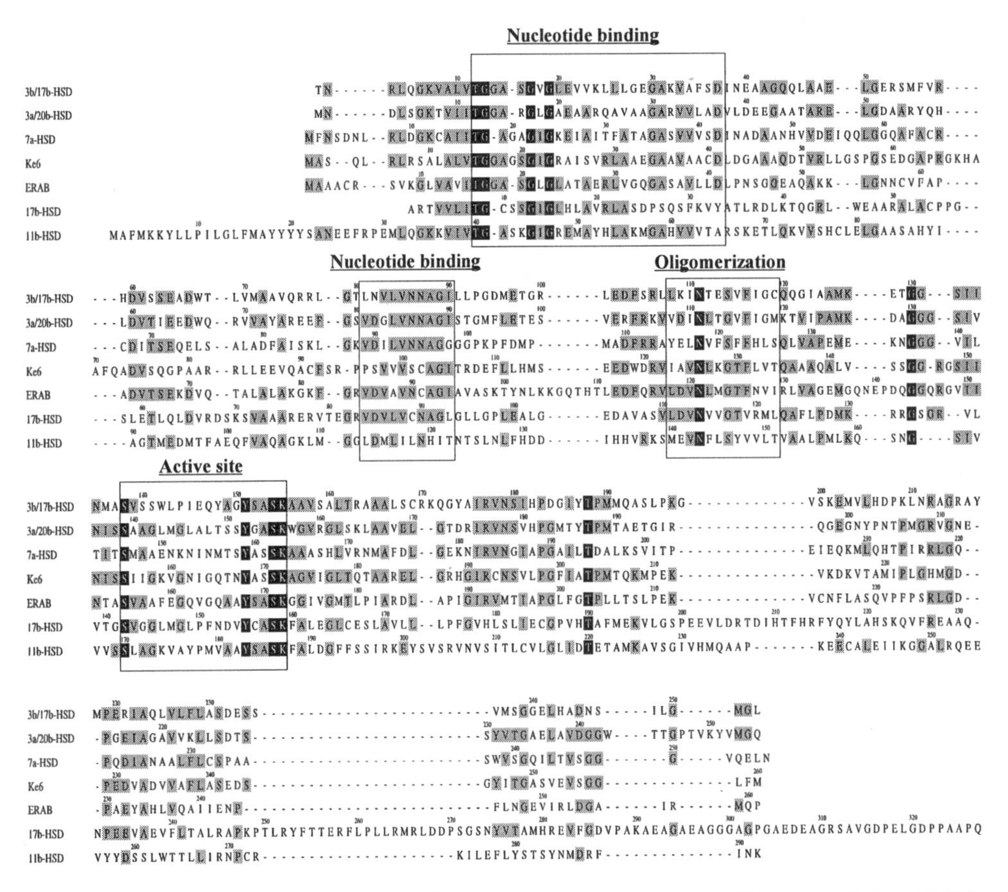

Figure 2. Alignment of SDR enzymes illustrating conserved sequence segments. From top to bottom: 3b/17b-HSD (3β-HSD from *Comamonas testosteroni*), 3a/20b-HSD (3α/20β-HSD from Streptomyces), 7a-HSD (7α-HSD from Eubacterium), Ke6 from mouse, ERAB from human, 17b-HSD (17β-HSD) and 11b-HSD (11β-HSD) from human. Conserved regions are highlighted by gray/black shading and boxing.

Table 1. Mutagenetic replacements and determination of 3β-HSD activity using isoUDCA as substrate with 3β/17β-HSD from *Comamonas testosteroni*[a]

Mutation	Motif/site	Km (μM)	Relative activity (%)	Suggested function
Wild type	–	22.3	100	–
Thr12Ala	coenzyme binding	–	na	H-bond to backbone amide of Asn87
Thr12Ser	coenzyme binding	28.3	97	H-bond to backbone amide of Asn87
Ser16Ala	coenzyme binding	27.6	88	Interaction with 2'phosphate of NAD^+
Asp60Ala	coenzyme binding	29.2	76	?
Asn87Ala	coenzyme binding	28.8	16	H-binding to coenzyme?
Asn111Leu	subunit interaction	–	na	oligomerization?
Ser138Ala	active site	–	na	H-bonds to substrate and NAD-ribose
Ser138Thr	active site	18.8	126	H-bonds to substrate and NAD-ribose
Tyr148Phe	active site	28.9	68	possibly no participation in catalysis
Tyr151Phe	active site	–	na	catalytic base
Asn179Ala	substrate binding	–	na	?
His182Leu	substrate binding	110	61	loop closure mechanism? participation in catalysis?

[a]Abbreviations: na: no activity; nd: not determined. Functions suggested are derived from interpretation of enzymatic data and the crystal structure of 3α/20β-HSD (Ghosh et al., 1994).

different regions of the enzyme: coenzyme binding site, subunit interaction area, catalytic site and the C-terminal region which defines substrate specificity. Mutations have been performed in all four regions and the results can be summarized as in Table 1. The location of these conserved residues in a model of 3α/20β-HSD, the first crystallized SDR protein, (Ghosh et al., 1994) is presented in Figure 3.

3.2. Identification of Class I γ ADH as the Human Liver Cytosol 3β-HSD and Molecular Modelling

Activity staining showed enzymatic activity only in the presence of NAD^+, with a single band migrating in the basic region of isoelectric focusing gels (pI ~ 9). With the *in situ* enzyme activity screening, a comparison of cytosolic proteins and recombinant ADH

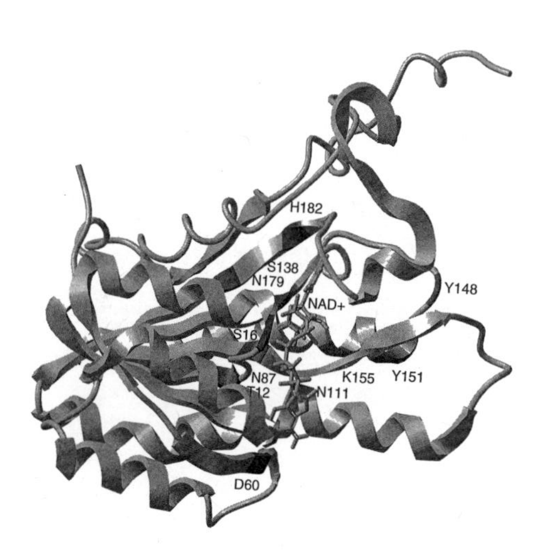

Figure 3. Ribbon model of a monomer of 3α/20β-HSD from *Streptomyces hydrogenans* complexed with NAD^+ (gray). Conserved residues mutated in this study are highlighted by indicating their position in black.

Table 2. Kinetic data of recombinant human ADH γγ isozyme

Substrate	K_m (μM)	V_{max} (nmol • min^{-1} • mg^{-1})	K_{cat} (min^{-1})	K_{cat}/K_m (μM •min^{-1})
iso-LCA	18.3	22.3	18	0.9
iso-DCA	28.3	69.1	55	1.9
iso-CDCA	15.3	31.4	25	1.6
iso-UDCA	23.8	75.1	60	2.5

3β-hydroxysteroid dehydrogenase activities of ADH γγ with various iso bile acids were determined. Velocity rates were measured by GLC and data obtained were fitted to the equation $V = V_{max} \bullet X/(K_m + X)$ with X being substrate concentration, V_{max}: maximal velocity and K_m Michaelis constant. Abbreviations: iso-LCA: iso lithocholic acid; iso-DCA: iso deoxycholic acid; iso-CDCA: iso chenodexycholic acid; iso-UDCA: iso ursodexycholic acid.

isozymes revealed that only the ADH γγ isozyme displays 3β-HSD activity towards etiocholanolone and iso bile acids, with NAD^+ as cofactor (Marschall et al., 1998). GC/MS was used to identify and quantify the 3-oxo-bile acids formed, suitable for kinetic analysis. The kinetic data obtained are given in Table 2. Molecular modelling of human class I γ ADH subunit and docking calculations demonstrated that isoUDCA fits into the active site (Figure 4) tightly surrounded by mainly hydrophobic residues.

3.3. Conclusions

The enzymes involved in epimerization reactions of bile acids during enterohepatic circulation are grouped into three distinct protein superfamilies, the aldo-keto reductases (AKR) (comprising 3α-HSDs), the short-chain dehydrogenases/reductases (SDR) and medium-chain dehydrogenases/reductases (MDR). Using isoUDCA as substrate we studied structure-function relationships of 3β.HSDs involved in the epimerization of iso-bile acids. Studies of 3β/17β-HSD from *Comamonas* reveal the importance of conserved residues for basic SDR architecture and protein function.

Hepatic cytosolic bile acid transport has been shown to be mediated by some mammalian 3α-HSDs (Stolz et al., 1993). As shown in this and another study (Marschall et al., 1998) ADH γγ isozyme acts as the sole cytosolic iso bile acid 3β-HSD and may also be involved in cytosolic bile acid transport. Computer modelling of this class I ADH shows that bile acids are substrates and bind to the active site consisting of a hydrophobic pocket.

3β-Hydroxy bile acids are taken up at the sinusoidal site of the hepatocyte but are excreted as 3α-hydroxy epimers at the canalicular site. Thus, in cytosol, 3β-hydroxy epi-

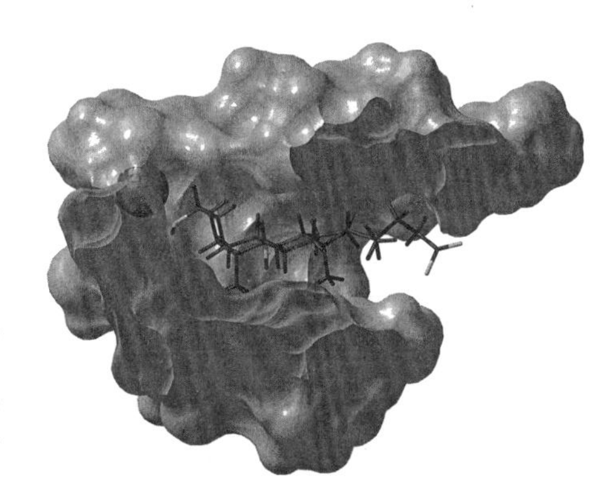

Figure 4. iso-UDCA docked into the active site of the class I ADH γ model. The steroid is shown in black with the van der Waals surfaces of the active site residues within 4 Å in gray. The catalytic Zn atom is shown as dark sphere close to the C3 hydroxyl of the iso-UDCA.

mers will first get in contact with 3β-HSD and than as ketones with 3α-HSD, and both contacts will be part of a directed movement through the cell. Further studies on intracellular bile acid transport mediated by HSDs and implications in hepatic liver diseases and cholestatic processes are on the way.

ACKNOWLEDGMENTS

This study was supported by an EC biotechnology grant (EC 97-2123), the Swedish Medical Research Council (13X-3532), the Swedish Cancer Society (project 1806), Karolinska Institutet and NOVARTIS, Sweden.

REFERENCES

Ghosh D. Wawrzak Z. Weeks CM. Duax WL. Erman M. (1994) The refined three-dimensional structure of 3 alpha, 20 beta-hydroxysteroid dehydrogenase and possible roles of the residues conserved in short-chain dehydrogenases. *Structure* 2, 629–40

Hurley TD. Bosron WF. Stone CL. Amzel LM. (1994) Structures of three human beta alcohol dehydrogenase variants. Correlations with their functional differences. *J. Mol. Biol.* 239, 415–29.

Jörnvall, H., Persson, B., Krook, M., Atrian, S., Gonzalez-Duarte, R., Jeffery, J. and Ghosh, D. (1995) Short-chain dehydrogenase/reductases (SDR). *Biochemistry 34,* 6003 – 6013.

Marschall, H.U., Oppermann, U.C.T., Svensson, S., Nordling, E., Persson, B., Höög, J.O. and Jörnvall, H. (1998) Human liver class I alcohol dehydrogenase γγ isozyme: the sole cytosolic 3β-hydroxysteroid dehydrogenase of iso bile acids. *Biochemistry*, submitted.

Oppermann, U.C.T., Persson, B., Filling, C. and Jörnvall, H. (1997a) Structure-function relationships of SDR hydroxysteroid dehydrogenases. *Adv. Exp. Med. Biol.* 414, 403–415.

Oppermann, U.C.T., Filling, C., Berndt, K.D., Persson, B., Benach, J., Ladenstein, R. and Jörnvall, H. (1997 b) Active-site directed mutagenesis of 3β/17β-hydroxysteroid dehydrogenase establishes differential effects on SDR reactions. *Biochemistry 36,* 34–40.

Simard J., Durocher F., Mebarki F., Turgeon C., Sanchez R., Labrie Y., Couet J., Trudel C., Rheaume E., Morel Y., Luu-The V. and Labrie F. (1996) Molecular biology and genetics of the 3-beta-hydroxysteroid dehydrogenase/delta5-delta4 isomerase gene family. *J. Endocrinol.*, 150, 189–207.

Stolz A., Hammond L., Lou H., Takikawa H., Ronk M. and Shively J.E. (1993) cDNA cloning and expression of the human hepatic bile acid-binding protein. A member of the monomeric reductase gene family. *J. Biol. Chem.* 268, 10448–10457.

CLONING AND SEQUENCING OF A NEW *COMAMONAS TESTOSTERONI* GENE ENCODING 3α-HYDROXYSTEROID DEHYDROGENASE/CARBONYL REDUCTASE

Eric Möbus and Edmund Maser

Department of Pharmacology and Toxicology
Philipps University
D-35033 Marburg, Germany

1. INTRODUCTION

Comamonas testosteroni is a Gram-negative bacterium capable of growing on C_{19} to C_{27} steroids in the absence of other suitable carbon and energy sources (Marcus and Talalay, 1956). Several steroid metabolizing enzymes, as well as steroid binding and transporting activities have been described and characterized (Pousette et al., 1986; Pousette and Carlström, 1984; Thomas et al., 1989; Watanabe and Watanabe, 1974; Mc Donald Francis et al., 1985). In addition to these steroid catabolic features several other non-steroid metabolizing activities such as polychlorinated biphenyl or aromatic hydrocarbon degradation were reported (Sondossi et al., 1992; Ahmad et al., 1990; Hollender et al., 1994). Both, transport system and degradative enzymes for either steroids or xenobiotic compounds are induced by the presence of these substances in the growth medium (Möbus et al., 1997; Oppermann and Maser, 1996). At present, there is no information available on the genetic organization of the genes encoding steroid degradative enzymes.

In previous investigations steroid inducible 3α-hydroxysteroid dehydrogenase (3α-HSD) has been purified from *C. testosteroni* and shown to mediate the oxidoreduction at position 3 of the steroid nucleus of a great variety of C_{19} to C_{27} steroids (Oppermann and Maser, 1996). This reaction is of significance in the initiation of the complete degradation of these substrates. Additionally, this enzyme was also capable of catalyzing the carbonyl reduction of non-steroidal xenobiotic aldehydes and ketones (Oppermann et al., 1996). Based on this pluripotent substrate specificity, the enzyme was named 3α-hydroxysteroid dehydrogenase / carbonyl reductase (3α-HSD/CR).

Enzymology and Molecular Biology of Carbonyl Metabolism 7
edited by Weiner *et al.* Kluwer Academic / Plenum Publishers, New York, 1999.

In the present work, we have identified at the molecular level the steroid-inducible gene encoding the 3α-HSD/CR from *C. testosteroni* ATCC 11996. The gene is located 156 bp upstream from the 5'-end of the delta-5–3-ketosteroid isomerase gene (Choi and Benisek, 1988), and was found on a 5.2 kb *Eco*RI fragment. For kinetic analyses 3α-HSD/CR was overexpressed in the *E. coli* pET15b vector system.

2. MATERIALS AND METHODS

2.1. Bacterial Strains, Plasmids, and Culture Conditions

The wild type *Comamonas testosteroni* (ATCC 11996) was from the Deutsche Sammlung für Mikroorganismen. It was grown at 30°C in LB medium under the conditions described by Boyer et al., 1965. *Escherichia coli* strains XL1-blue MRF', INFαF' and BL21(DE3)pLysS were grown in LB medium at 37°C. When necessary, antibiotics were added to the media: ampicillin 50 μg/ml, chloramphenicol 34 μg/ml, kanamycin 50 μg/ml, tetracycline 12.5 μg/ml and carbenicillin 50 μg/ml.

2.2. DNA Manipulations and Sequence Analysis

Plasmid DNA was isolated using the alkaline lysis method (Birnboim and Doly, 1979). The complete nucleotide sequence was obtained by the dideoxy chain termination method with Sequenase Kit version 2.0 (Amersham). The BLAST program was used to screen protein and DNA data bases for proteins that share sequence similarity. Alignment and open reading frame regions were determined by using the MegAlign program (DNASTAR package).

2.3. Construction of the 3α-HSD/CR-Probe

PCR reactions were run against a preparation of wild type *C. testosteroni* genomic DNA with primers derived from the NH_2-terminal sequence of 3α-HSD/CR (Möbus et al., 1997). The resulting 75 bp product was cloned into pCR2.1 vector (Invitrogen), amplified in INFαF' cells and sequenced. Part of the fragment containing the 5'-sequence of 3α-HSD/CR was digoxygenin labeled and used as homologous probe for Southern blot and gene bank screening.

2.4. Southern Blot Analysis and Screening of a *C. testosteroni* *Eco*RI Gene Bank

The genomic DNA of *Comamonas testosteroni* was prepared as described in Sambrook et al., 1989. Genomic DNA was then digested with various restriction enzymes and resolved by electrophoresis in a 1% agarose gel and transferred onto a nylon membrane (Hybond N^+, Amersham). The membrane was prehybridized at 60°C with 5 × NaCl/Cit, 0.02 % SDS, 0.1% N-Laurylsarcosine and 1% Blocking reagent (Boehringer Mannheim) and hybridized overnight in the same buffer with the digoxygenin labeled 3α-HSD/CR-probe. The membrane was washed twice at 68°C in a solution containing 2 × NaCl/Cit, 0.1% SDS and with 1 × NaCl/Cit, 0.1% SDS, respectively. For detection of the hybridizing fragments the standard protocol of Boehringer (Mannheim) was followed, finally resulting in the identification of a hybridizing 5.2-kilobase *Eco*RI fragment.

An *Eco*RI-gene bank was constructed using the Lambda ZAP Express system (Stratagene) and amplified in XL1-blue MRF' cells. About 12,000 plaque forming units (pfu) of recombinant phages from the *Eco*RI-library were screened with the homologous, digoxygenin labeled 3α-HSD/CR probe under the conditions described for Southern analysis. Positive clones were further purified by two cycles of screening at low plaque density until all phages produced positive signals. The positive clones were amplified and converted to pBK-CMV phagemids carrying the *Eco*RI-fragment containing the 3α-HSD/CR gene (pBK-E52). The nucleotide sequence of the insert was then determined.

2.5. Gene Expression, Activity Assay, and Protein Determination

The 3α-HSD/CR gene was subcloned from the pBK-CMV phagemid into the NdeI/BamHI restriction sites of the pET15b vector (Novagen). The IPTG induced overexpression was performed using BL21(DE3)pLysS cells. After 5 h of induction the cells were harvested, lysed and the debris removed. The histidine-tagged enzyme was purified out of the lysate using a Ni^{2+}-Sepharose (His-Trap Kit, Pharmacia) column. Pure protein was eluted by applying a linear gradient of imidazole (0–500 mM), pooled and the buffer exchanged to 0.5 M Tris/HCl buffer, pH 7.4. The enzyme activities were tested as described previously (Oppermann and Maser, 1996).

Protein concentration was measured by the method of Bradford (Bradford, 1976) with Roti-Quant-solution (Roth) using bovine serum albumine as the reference standard.

3. RESULTS AND DISCUSSION

3.1. Cloning of the 3α-HSD/CR Gene

The sequence of the first 28 amino acids in the amino (NH_2) terminus from 3α-HSD/CR of *C. testosteroni* has previously been determined by Edman degradation (Möbus et al., 1997). Two degenerated oligonucleotides corresponding to the respective 5'- and 3'-nucleotide sequences were synthesized as primers for polymerase chain reaction against a preparation of genomic DNA of *C. testosteroni*. The resulting 75 bp fragment was ligated into pCR2.1 vector, amplified in INVαF' cells and sequenced. 36 bp of the fragment were choosen to serve as homologous probe for Southern blot analysis and gene bank screening. The sequence of the probe was 5'-ACCGGCATTGGTGCCGCTACGG-CCAAGGTGCTGCTGGAG-3'. Southern blot analysis revealed that the probe hybridized to a 5.2 kb *Eco*RI fragment. To clone the gene of 3α-HSD/CR, we constructed an *Eco*RI gene bank of *C. testosteroni*. Screening of this library with the above described 3α-HSD/CR probe under the same conditions as in Southern blot, yielded the same hybridizing 5.2 kb fragment. The positive clones were amplified and converted to pBK-CMV phagemids carrying the *Eco*RI-fragment containing the 3α-HSD/CR gene (pBK-E52).

3.2. Nucleotide Sequence

The nucleotide sequence of the total 5.2 kb fragment was determined from both strands. Figure 1 presents the restriction map of the fragment. The gene encoding 3α-HSD/CR is located from base pair position 2615 to 3388. 156 bp downstream of the stop codon from the 3α-HSD/CR gene, the start codon of the delta-5–3-ketosteroid isomerase gene is situated (Choi and Benisek, 1988). One of the primary metabolic steps in the deg-

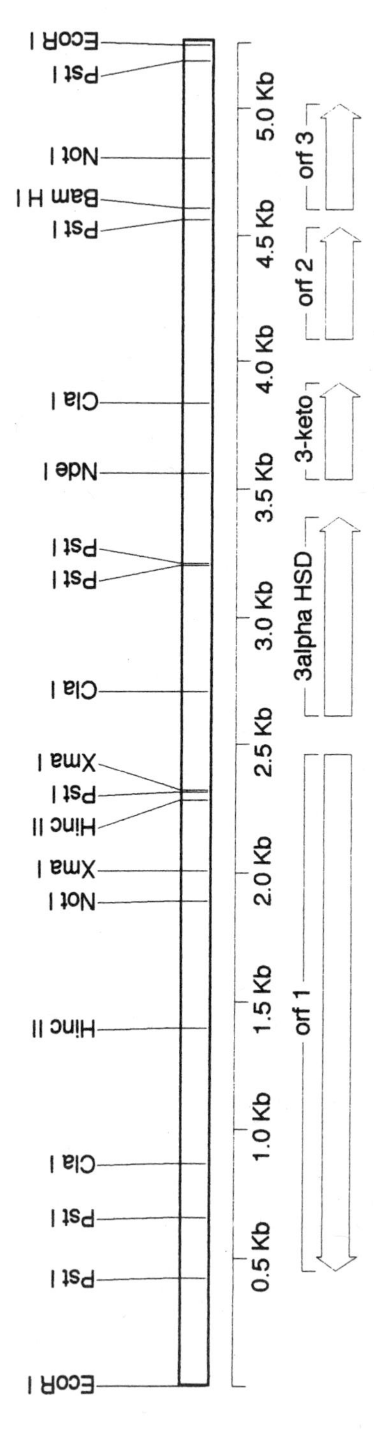

Figure 1. Restriction map of the 5.2 kb *Eco*RI fragment. Position of the gene for 3α-HSD/CR, delta-5-3-ketosteroid isomerase and open reading frames (orfs) are indicated by arrows.

radation of the steroid skeleton is the oxidation of the 3-hydroxy group to the 3-ketone, which is completed by 3α-HSD/CR. For complete degradation of the steroid nucleus, the A-ring has to be aromatized prior to cleavage by dioxygenases. This requires unsaturation of the A-ring, which is achieved amongst others by delta-5-3-ketosteroid isomerase. Furthermore, the 5.2 kb fragment contains three open reading frames (orfs). Orf 1 is situated 155 bp upstream the 3α-HSD/CR gene, on the reverse strand. It compromises 2007 bp and shows interesting relationship to 2,4-dienoyl-CoA reductase from *Escherichia coli* (He et al., 1997). Orf 2 (438 bp) is located downstream the delta-5–3-ketosteroid isomerase gene, and is followed further downstream by orf3 (417 bp). Neither orf 2, nor orf 3 reveal homologies to hitherto known proteins, suggesting that these are novel proteins.

3.3. Analysis of the 3α-HSD/CR Primary Structure

Figure 2 shows the sequence of the 3α-HSD/CR gene. It compromises 774 bp coding for a protein of 257 amino acids with a calculated molecular weight of 26.4 kDa. A ribosome binding site is situated seven bases upstream of the start codon (-7 to -12). The primary structure reveals that the 3α-HSD/CR belongs to the short chain dehydrogenase/reductase superfamily (SDR). The SDR proteins are a growing superfamily of NAD(P)(H)-dependent non-metallo oxidoreductases that are about 250–350 amino acids in length and which bind the cofactor with a Rossman-fold-motif (Persson et al., 1991;

Figure 2. The 3α-HSD/CR encoding gene from *C. testosteroni*. Numbers to the left correspond to the nucleotide sequence, numbers to the right refer to the amino acid sequence. Residues corresponding to the NH_2-terminal amino acid sequence obtained from the isolated protein by Edman degradation (Möbus et al., 1997) are underlined. The ribosome binding site AGGAGA seven bases upstream of the start codon is indicated by bold underlining. Stop codon is indicated by an asterisk.

Table 1. Kinetic constants for *3α*-HSD/CR from *C. testosteroni* with steroids, and non-steroidal aldehydes and ketones

Substrate	K_m µmol/L	V_{max} µmol/min × mg	CLi
3α-Oxo reduction			
Androstanedione	42.2	85.7	2.0
5α-DHT	22.3	64.5	2.9
3α-OH oxidation			
Androsterone	31.1	119.0	3.8
Cholic acid	18.4	27.4	1.5
Fusidic acid	6.1	2.5	0.4
Carbonyl reduction of xenobiotics			
Metyrapone	1882	3983	2.1
p-nitrobenzaldehyde	3286	1623	0.5

Abbrevations are: 5α-DHT, 5α-dihydrotestosterone; CLi, clearance.

Jörnvall et al., 1995). Amino acid residues that are highly conserved in the SDR family are present in the 3α-HSD/CR structure. The NH_2-terminal Gly-X-X-X-Gly-X-Gly motif (Gly8, Gly12 and Gly14, numbering according to the 3α-HSD/CR sequence) is related to the binding of the cofactor. The Tyr-X-X-X-Lys segment (residues 155 to 159 in the 3α-HSD/CR sequence) forms the catalytic center of the SDR proteins.

3.4. Kinetic Analysis of 3α-HSD/CR

The recombinant 3α-HSD/CR was active when tested with a variety of steroid substrates. Enzyme kinetic constants, *Km* and *Vmax*, for steroid 3-oxo reduction and 3α-hydroxysteroid dehydrogenation were determined with the recombinant protein (as shown in Table 1) and compared to values obtained with native 3α-HSD/CR *from C. testosteroni* in previous studies (Oppermann and Maser, 1996). The recombinant protein efficiently catalyzes 3-oxo reduction and 3α-hydroxysteroid dehydrogenation of steroids of the androstane and bile acid series with *Km* values in the low µM range. On the other hand, the recombinant enzyme catalyzes the carbonyl reduction of non-steroidal aldehyde and ketone compounds. Although *Km* values for the model substrates of carbonyl reducing enzymes metyrapone and p-nitrobenzaldehyde are relatively high, their CLi values are comparable to those of the steroid substrates.

3.5. Relationship of 3α-HSD/CR to Other Proteins

Blast search with the protein sequence of 3α-HSD/CR from *C. testosteroni* revealed that it shares 50.2 % similarity to the amino acid sequence of 3α-HSD from *Pseudomonas* sp., strain B-0831 (Suzuki et al., 1993), as shown in in Figure 3. 3α-HSD from *Pseudomonas* spec. contains 255 amino acid residues and, according to SDR consensus sequences, does also belong to the SDR superfamily.

ACKNOWLEDGMENT

The excellent technical assistance of Jutta Friebertshäuser and Eva Braun is gratefully acknowledged. This work was supported by the European Community, Programme Biotech 2, Contract BIO4-97-2132 and the Deutsche Forschungsgemeinschaft (SFB 395).

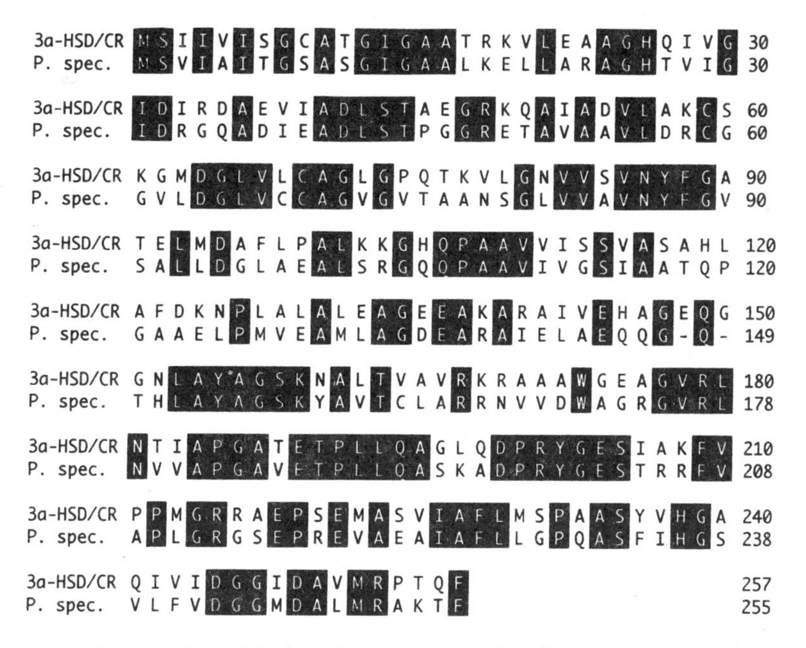

Figure 3. Sequence alignment of 3α-HSDs from *C. testosteroni* and *Pseudomonas* spec., respectively. Protein sequences were aligned with Clustal method of the DNASTAR software. The black boxes show the residues that are identical to one another in the sequences. Numbers refer to amino acid residues. Abbreviations are: 3α-HSD/CR, 3α-hydroxysteroid dehydrogenase/carbonyl reductase from *C. testosteroni*; P. spec., 3α-hydroxysteroid dehydrogenase from *Pseudomonas* spec. The sequences share a similarity of 50.2 %.

REFERENCES

Ahmad, D., Masse, R. and Sylvestre, M. (1990) Cloning and expression of genes involved in 4-chlorobiphenyl transformation by *Pseudomonas testosteroni*, homology to polychlorobiphenyl-degrading genes in other bacteria, *Gene* 86, 53–61.

Birnboim, H.C. and Doly, J. (1979) A rapid alkaline extraction procedure for screening recombinant plasmid DNA, *Nucleic Acids Res.* 7, 1513–1523.

Boyer, J., Baron, D.N. and Talalay, P. (1965) Purification and properties of a 3α- hydroxysteroid dehydrogenase from *Pseudomonas testosteroni*, *Biochemistry* 4, 1825–1833.

Bradford, M.M. (1976) A rapid and sensitive method for the quantitation of microgram quantities of proteins using the principle of dye-binding, *Anal. Biochem.* 72, 248–254.

Choi, K.Y. and Benisek, W.F. (1988) Nucleotide sequence of the gene for the delta 5- 3-ketosteroid isomerase of *Pseudomonas testosteroni*, *Gene* 69, 121–129.

He, X.Y., Yang, S.Y. and Schulz, H. (1997) Cloning and expression of the fadH gene and characterization of the gene product 2,2-dienoyl coenzyme A reductase from *Escherichia coli*, *Eur. J. Biochem.* 248, 516–520.

Hollender, J., Dott, W. and Hopp, J. (1994) Regulation of chloro- and methylphenol degradation in *Comamonas testosteroni* JH5, *Appl. Environ. Microbiol.* 60, 2330–2338.

Marcus, P.I. and Talalay, P. (1956) Induction and purification of α and ß- hydroxysteroid dehydrogenases, *J. Biol. Chem.* 218, 661–674.

Mc Donald Francis, M., Kowalsky, N. and Watanabe, M. (1985) Extraction of a steroid transport system from *Pseudomonas testosteroni* membranes and incorporation into synthetic liposomes, *J.Steroid Biochem.* 23, 523–528.

Möbus, E., Jahn, M., Schmid, R., Jahn, D. and Maser, E. (1997) Testosterone- regulated expression of enzymes involved in steroid and aromatic hydrocarbon catabolism in *Comamonas testosteroni*, *J. Bacteriol.* 179, 5951–5955.

Oppermann, U.C.T. and Maser, E. (1996) Characterization of a 3α-hydroxysteroid dehydrogenase/carbonyl reductase from the gram-negative bacterium *Comamonas testosteroni, Eur. J. Biochem.* 241, 744–749.

Oppermann, U.C.T., Belai, I. and Maser, E. (1996) Antibiotic resistance and enhanced insecticide catabolism as consequences of steroid induction in the Gram-negative bacterium *Comamonas testosteroni, J. Steroid Biochem. Molec. Biol.* 58, 217–223.

Pousette, A. and Carlström, K. (1984) Partial characterization of an androgen- binding protein in *Pseudomonas testosteroni, Acta Chimica Scand.* B38, 433–438.

Pousette, A. and Carlström, K. (1986) High affinity protein binding and enzyme- inducing activity of methyltrienolone in *Pseudomonas testosteroni, Acta Chimica Scand.* B40, 515–521.

Sondossi, M., Sylvestre, M. and Ahmad, D. (1992) Effects of chlorobenzoate transformation on the *Pseudomonas testosteroni* biphenyl and chlorobiphenyl degradation pathway, *Appl. Environ. Microbiol.* 58, 485–495.

Suzuki, K., Ueda, S., Sugiyama, M. and Imamura, S. (1993) Cloning and expression of a *Pseudomonas* 3 alpha-hydroxysteroid dehydrogenase-encoding gene in *Escherichia coli, Gene* 130, 137–140.

Thomas, J.E., Carrol, R., Sy, L.P. and Watanabe, M. (1989) Isolation and characterization of a 50 kDa testosterone-binding protein from *Pseudomonas testosteroni, J.Steroid Biochem.* 32, 27–34.

Watanabe, M. and Watanabe, H. (1974) Periplasmic steroid-binding proteins and steroid transforming enzymes of *Pseudomonas testosteroni, J.Steroid Biochem.* 5, 439–446.

50

COMPARATIVE PROPERTIES OF THREE PTERIDINE REDUCTASES

Chi-Feng Chang, Tom Bray, Kottayil I. Varughese, and John M. Whiteley

Department of Molecular and Experimental Medicine
The Scripps Research Institute
10550 North Torrey Pines Road
La Jolla, California 92037

1. DISCUSSION

Naturally occurring pteridines in eukaryotic systems usually contain 2-amino, 4-hydroxy and 6-alkyl substituents composed of either a methylene-*p*-aminobenzoylglutamate (or polyglutamate) or a dihydroxypropyl group. The former class are known collectively as the folates and occur widely as reduced and 5-alkylated derivatives, in which form they participate in important metabolic one-carbon transfers (Blakley, 1984). The latter, known as biopterin, also occurs in reduced forms and is an important cofactor in aromatic amino acid hydroxylations en route to the catecholamines (Shiman, 1985; Kaufman and Kaufman, 1985; Kuhn and Lovenberg, 1985) and in the nitrite synthase pathway (Marletta, 1993). Dihydrofolate reductase (DHFR), dihydropteridine reductase (DHPR) and pteridine reductase (PTR1), best characterised from *Leishmania*, are three enzymes that initiate the reduction of a pteridine in association with a reduced dinucleotide cofactor. Their comparative reaction pathways are illustrated in Figure 1.

DHFR is one of the most thoroughly investigated enzymes in the literature with both its structural and mechanistic features being fully characterised (Blakley, 1984). Motivation for this effort probably stemmed from it being metabolically en route to the synthesis of the DNA structural entity thymidylate, which caused it to become a major target in chemotherapy via its substrate analog inhibitor methotrexate (MTX); the first truly effective agent to control systemic human tumors (Farber *et al.,* 1948). It is included here for comparative purposes to enhance and illustrate the differing approaches to biological pteridine reduction. DHFR and PTR1 both take the fully oxidised pteridine via the 7,8-dihydro intermediate to the 5,6,7,8 tetrahydro product. The former in general favors a folate substrate, whereas the latter prefers biopterin, however, DHFR shows far greater enzymatic activity with the 7,8-dihydro derivative as substrate in contrast to PTR1, which favors the

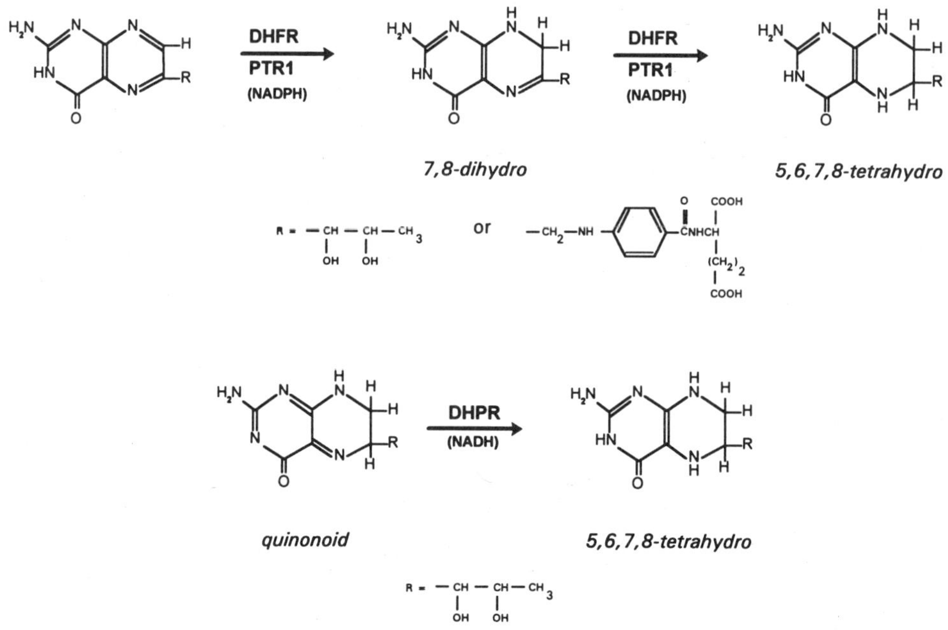

DHFR PTR1 (NADPH)

DHFR PTR1 (NADPH)

7,8-dihydro

5,6,7,8-tetrahydro

DHPR (NADH)

quinonoid

5,6,7,8-tetrahydro

Figure 1. The reductive pathways of DHFR, PTR1, and DHPR with their preferred reduced dinucleotide cofactors and substrates.

fully oxidised form. DHPR differs from both the other reductases insofar as it employs quinonoid dihydrobiopterin (an isomer of the 7,8-dihydropteridine) as its substrate, a molecule generated from the participation of tetrahydrobiopterin in the previously noted hydroxylations of the aromatic amino acids. The comparative properties of the three reductases are illustrated in Tables 1 and 2.

It is interesting to note that both DHFR and PTR1 prefer NADPH whereas DHPR prefers NADH, yet from sequence comparison it is clear that PTR1 and DHPR exhibit greater similarity and are members of the short-chain dehydrogenase/reductase (SDR) family (Figure 2). DHFR has markedly different sequence characteristics (Blakley, 1984). Both DHFR and PTR1 have similar affinities for NADPH (K_d = 1.4 and 5.3 μM respectively) whereas DHPR has a much higher affinity for the reduced dinucleotide (K_d = 0.025 μM). DHPR also exhibits a much higher turnover rate (k_{cat} = 22 sec^{-1}) probably reflecting the physiological requirements for rapid tetrahydropteridine synthesis associated with neurological responses (Rosenkranz-Weiss *et al.,* 1994; Marletta, 1993). PTR1 has a clear preference for a lower pH to achieve optimal activity.

Table 1. Comparative properties of DHFR, PTR1, and DHPR

Enzyme	Favored substrate	Reduced dinucleotide	K_d Nucl. (μM)	K_m (μM) Subst.	K_m (μM) Nucl.	K_{cat} (sec^{-1})	pH optimum
DHFR[(a)]	7,8 dihydrofolate	NADPH	1.4	3.2	6.8	1.1	~7
PTR1[(b)]	biopterin	NADPH	5.3	3.5	17	2.3	~6
DHPR[(c)]	*q*-dihydrobiopterin	NADH	0.025	0.32	13	22	~7

(a) Baccanari, et al., 1975; (b) Wang, et al., 1997; (c) Whiteley, et al., 1993.

Table 2. Comparative properties of DHFR, PTR1, and DHPR (*continued*)

Enzyme	Active form	Monomer M_r (kDa)	$K_{i\,(mtx)}$ (μM)	Type	Hydride transfer
DHFR	monomer	18.1	< 0.001	competitive	pro – R [(a)]
PTR1	tetramer	29.2	0.003	competitive	pro – S [(b)]
DHPR	dimer	25.4	~ 25	mixed	pro – S [(c)]

(a) Filman, et al., 1982; (b) Luba, et al., 1998; (c) Varughese, et al., 1992.

Structurally and mechanistically the three reductases show distinctive characteristics. DHFR usually functions as a monomer and dependent on its source it can have molecular weights from 15 to 22 kDa (Blakley, 1984), whereas PTR1 from either *L. tarentolae* and *L. major* is present in solution as a tetramer of subunit $M_r \sim 29$ kDa (Wang *et al.,* 1997) and mammalian DHPR occurs as a dimer with a subunit $M_r \sim 25$ kDa (Varughese *et al.,* 1992). Each of the reductases can be inhibited by 2,4-diamino analogs of their substrates but their responses are different. The folate analog, methotrexate (MTX), is a profound virtually irreversible competitive inhibitor of DHFR whereas it is a poor somewhat non-specific inhibitor of DHPR. With PTR1 it is also an inhibitor but not at the level observed with DHFR and elevation of the PTR1 intracellular molecular concentration that occurs when *Leishmania* are exposed to MTX is a novel mode of resistance generated to this agent by these parasites (Nare *et al.,* 1997). A further distinct difference between DHFR and the other two pteridine reductases lies in the fact that NADPH bound to DHFR donates its pro-R hydrogen in the reductive process. The reduced dinucleotide cofactors of PTR1 and DHPR transfer the pro-S hydrogen (Luba *et al.,* 1998; Varughese, *et al.,* 1992) as do other members of the SDR family.

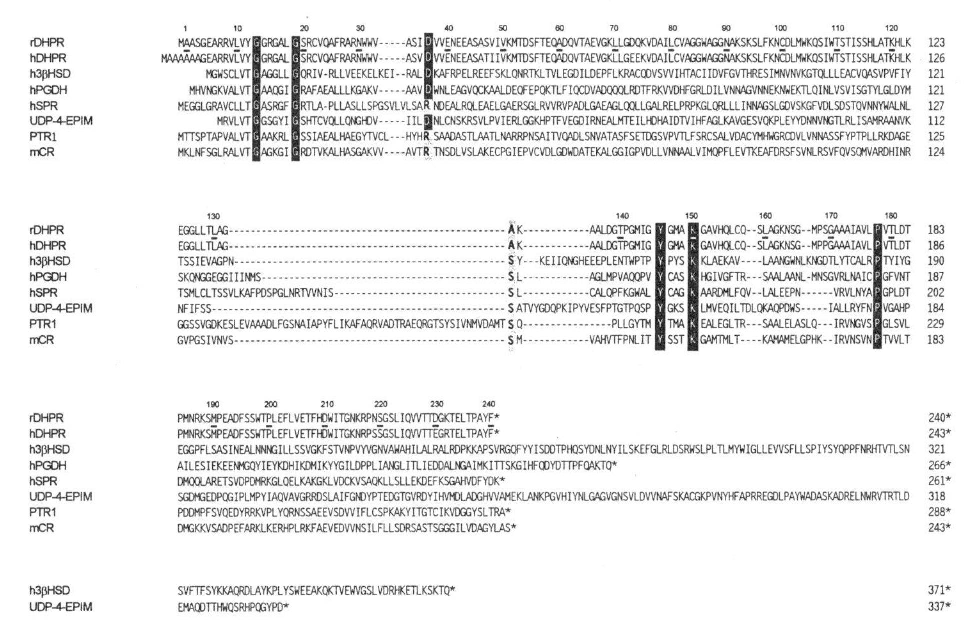

Figure 2. Aligned sequences of rat (r) and human (h) DHPRs, human 3β-hydroxysteroid dehydrogenase Δ^{5-4} isomerase (3βHSD), 15-hydroxyprostaglandin dehydrogenase (PGDH), sepiapterin reductase (SPR), *Escherichia coli* UDP-4-epimerase (UDP-4-EPIM), *L. tarentolae* PTR1 and mouse carbonyl reductase (CR).

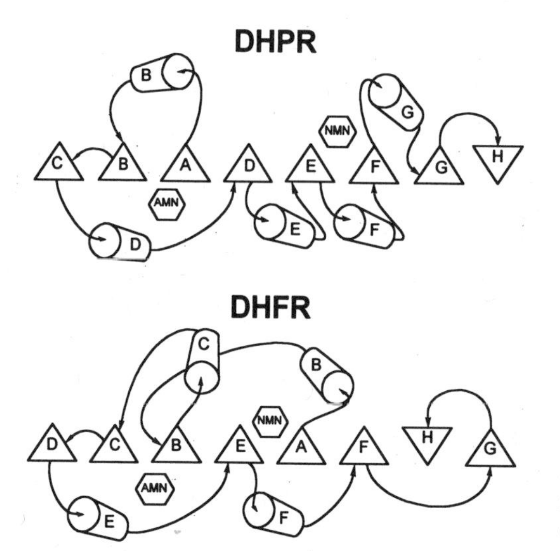

Figure 3. Comparison of the backbone folding of DHPR with DHFR. The α-helices are represented as cylinders and the β-sheets as triangles. The approximate positions of dinucleotide cofactors are indicated.

DHPR and PTR1 exhibit a typical dinucleotide cofactor binding sequence in the N-terminal region (Wierenga *et al.,* 1986), a feature illustrated in Figure 2 by comparison with additional members of the SDR family. In contrast, as shown in Figure 3, the overall protein folding of DHFR and DHPR differs markedly. Both DHPR and PTF1 contain the $Y(Xaa)_3K$ motif that has now become an identifying feature of the SDR family of proteins (Jörnvall *et al.,* 1995). This is not present in DHFR. The crystal structures of DHFR (Bystroff *et al.,* 1990) and DHPR (Varughese *et al.,* 1992) reinforce the differing structural properties of the two reductases. In particular the orientation of the reduced dinucleotide cofactor differs with the adenine component being close to the N-terminal in DHPR, whereas, in contrast, the nicotinamide component resides closer to the N-terminal in DHFR (Figure 3). The previously reported structures of DHPR also indicate the importance of the conserved $Y(Xaa)_3K$ motif to the reaction mechanism of this enzyme and by implication to other members of the SDR family (Varughese *et al.,* 1994).

The complete 3-dimensional structural features of the PTR1 protein are not yet available and therefore its properties have been deduced by more conventional approaches by this laboratory and others (Luba *et al.,* 1998; Nare *et al.,* 1997; Wang *et al.,* 1997). To provide ample supplies of wild-type and mutant PTR1 samples, selected constructed PTR1 genes have been subcloned into a pET16b vector followed by overexpression of the complementary protein induced in the usual way by IPTG. SDS gel analysis has shown the PTR1 samples to be the major cellular soluble protein component after induction (> 50%) and large scale purification that includes ammonium sulfate precipitation, DEAE-Trisacryl chromatography and Sephacryl S-100 filtration regularly affords homogeneous material (~ 100 mg from a 4 L culture) (Figure 4). As noted earlier the enzyme always favors NADPH and biopterin as substrates with an optimal pH = 6. A specific activity ~ 3.2 units/mg is regularly obtained at ~ 21° under these conditions. Double reciprocal plots for NADPH and biopterin show the usual intersecting patterns. This observation coupled with product inhibition by $NADP^+$ and substrate competitive inhibition with MTX suggests the reaction follows a sequential bi-bi pattern. K_d values for the wild-type and several pertinent mutant derivatives have been measured fluorimetrically. The results are illustrated in Table 3.

As can be seen the wild type protein shows a strong affinity for NADPH (5.3 μM), less for NADH (31 μM) and an even lower affinity for the cofactor reaction product

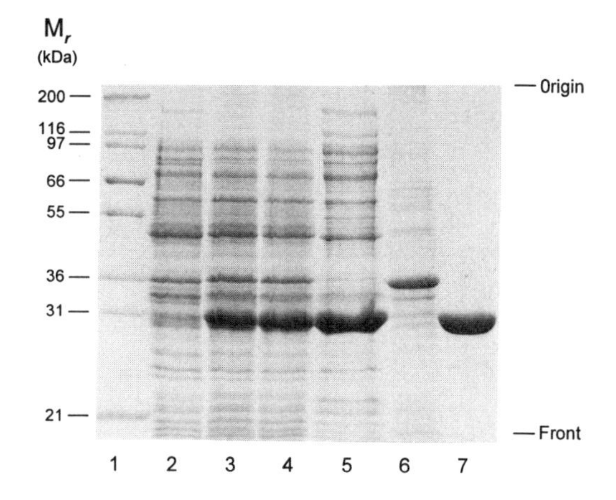

Figure 4. SDS PAGE gel illustrating the composition of differing samples of the *L. tarentolae* PTR1 protein during expression and purification (1) M_r markers, (2) total protein extract from the cells before induction, (3) 1 h after induction, (4) 4 h after induction, (5) 32% v/v ammonium sulphate precipitate, (6) DEAE-Trisacryl washing and (7) the peak fraction from DEAE-Trisacryl chromatography.

$NADP^+$ (81 μM). Mutation at K16 or R39 leads to a lesser affinity for NADPH (~ 2 fold) that is also present in the double mutant K16A, R39A. Associated losses of specific activity also occur (from 100 to 27%). It is interesting to note that protein affinity for NADH increases with these mutants. If the additional mutation Y37D is included, affinity for NADH is more pronounced and for NADPH is markedly reduced. These observations support PTR1 being a typical SDR protein with the typical N-terminal binding site for a reduced dinucleotide. PTR1 has two $Y(Xaa)_3K$ motifs (Figure 2) but only the one later in the sequence appears to be associated with activity. This is borne out by the Y152F mutation which lessens affinity for NADPH a little but has no effect on enzyme activity. In contrast, Y194F, Y194H and K198Q all diminish activity significantly. In fact Y194F activity drops to zero. It is interesting to note that K198Q also alters the dinucleotide affinity, which correlates with our earlier observation with DHPR that the lysine contained in the motif in SDR proteins assists in orienting the ribose adjacent to the nicotinamide in the dinucleotide cofactor. The participation of this $Y(Xaa)_3K$ motif in the enzymatic activity is further contributory evidence that PTR1 is a true SDR protein.

Since the ε-amino group of lysine 150 in DHPR, a part of the active-site $Y(Xaa)_3K$ motif in this enzyme, was found to form a strong hydrogen-bond with the cofactor at the

Table 3. Specific activities and dissociation constants for wild-type and mutant PTR1 proteins

	K_d (μM)[(a)]			Specific activity[(b)]	
Sample	NADPH	$NADP^+$	NADH	(μM/min/mg)	% of wild-type activity
Wild-type	5.3	81	31	3.2	100
K16A	12	~	28	1.46	46
R39A	9	~	15	1.13	35
K16A, R39A	10	~	16	0.86	27
Y37D, R39A	36	~	13	0.42	13
Y152F	10	~	~	2.93	92
Y194F	16	~	32	0	0
Y194H	21	~	~	0.85	27
K198Q	26	~	~	0.11	3

(a) The K_d values were measured in 100 mM Bis-Tris, pH 6.0 buffer solutions at 10° C.
(b) The specific activities were measured in 100 mM Bis-Tris, pH 6.0 containing 200 μM NADPH at 21° C.

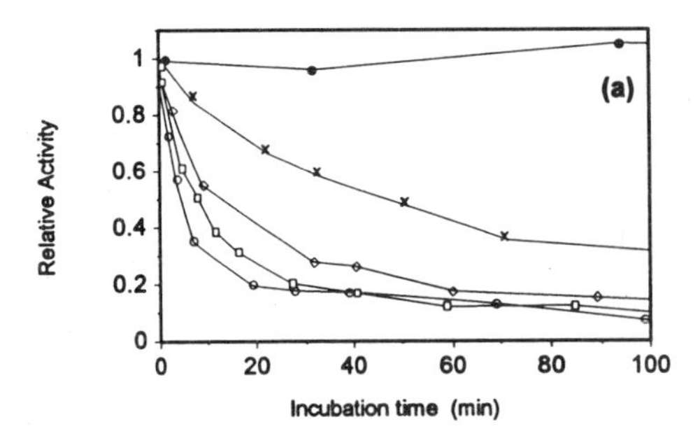

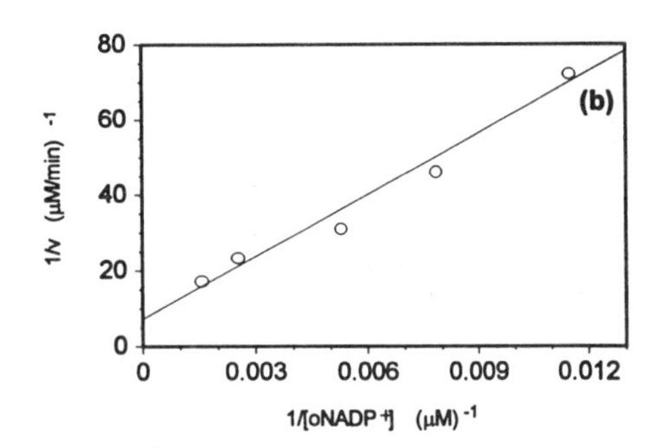

Figure 5. Saturation kinetics of the inactivation of PTR1 by incubation with increasing quantities of $oNADP^+$: 0 (•); 0.1 (x); 0.3 (◊); 0.5 (□); and 0.8 (o) mM. (a) Residual activity versus time. (b) Double reciprocal plot of the rate and $oNADP^+$ concentration.

nicotinamide ribose 2'- and 3'-hydroxy groups (Varughese *et al.*, 1992), it was considered possible that the 2',3'-dialdehyde derivative of $NADP^+$ (Chang *et al.*, 1989) might form an imino bond with the somewhat analogous lysine 198 residue of PTR1. The following experimental observations indicated that $oNADP^+$ was indeed a good affinity label for PTR1 and, in addition, suggested a residue essential for activity might be derivatised: (1) saturation kinetics of the PTR1 and $oNADP^+$ interaction indicated the formation of a [PTR1/$oNADP^+$] binary complex before inactivation took place (Figure 5 a and b); (2) the inactivation of $oNADP^+$ could be prevented by $NADP^+$ indicating the modification was at the same binding site; (3) totally inactivated PTR1 incorporated only one mole of $oNADP^+$ per enzyme monomer (Figure 6); (4) the mutant K198Q showed no inactivation of $oNADP^+$ indicating that lysine 198 was essential for the modification; and (5) the results of the mass spectral analysis of peptides isolated from the derivatised and *Staphyloccocal aureus* V8 proteolytically cleaved enzyme supported the hypothesis that modification had taken place in the C-terminal region.

Despite the differing properties of the three pteridine reductases their common function is to reduce the polarised C = N bond. Considerable evidence has accrued to suggest that in the case of DHFR the compensatory proton to complement hydride donation, despite a somewhat tenuous path, comes from a carboxylate moiety in the vicinity of the active site. In fact D27 has been the suggested source in the case of the *E. coli* enzyme (Bystroff *et al.,* 1990), although complete consensus on this issue appears to be still to

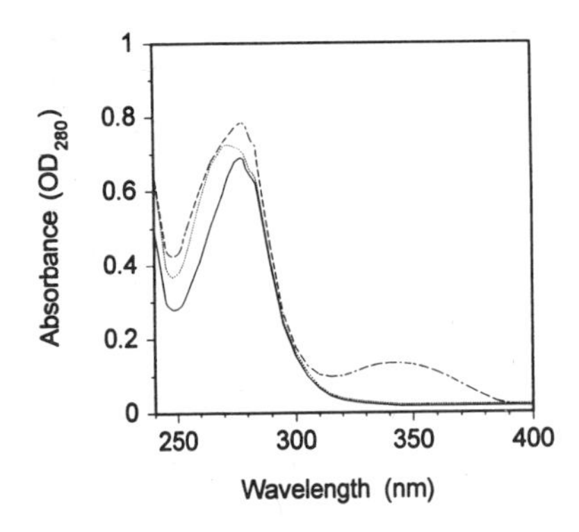

Figure 6. Absorbance spectral changes observed with the PTR1/$oNADP^+$ interaction: PTR1 (—); PTR1/$oNADP^+$ complex (····); sodium borohydride reduced PTR1/$oNADP^+$ complex (-··-).

come. Nevertheless, two steps of reduction occur to afford the 5,6,7,8-tetrahydrofolate product via the 7,8 dihydro-intermediate and the hydride transfers are made to the non-exchangeable 6- and 7-positions of the folate pteridine ring. With DHPR an unstable quinonoid dihydropteridine is the substrate and the reductive step is mediated via the $Y(Xaa)_3K$ motif oriented towards the active site (Varughese *et al.,* 1994). In this instance tyrosine provides the compensatory proton to complement hydride donation from NADH and hydride transfer occurs to the solvent exchangeable N-5 position of the pteridine (Armarego, 1979). PTR1 appears to have the qualities of both DHFR and DHPR insofar as reduction of the oxidised pteridine substrate occurs sequentially to the 7,8 and 5,6 positions, yet the mutant studies outlined above suggest the $Y(Xaa)_3K$ motif participates in the reduction. A recent publication (Luba *et al.,* 1998) confirms that when folate is a substrate for PTR1 the stereochemistry of reduction also follows that of DHFR, allowing the product of the reaction to participate in the further conversion of uridylate to thymidylate in the methylenetetrahydrofolate mediated reaction of thymidylate synthase. Clearly the precise orientation of the $Y(Xaa)_3K$ motif in PTR1 relative to that present in DHPR presents a fascinating structural relationship that remains to be resolved.

ACKNOWLEDGMENTS

This investigation was supported in part by Grants GM52699 and CA11778 from the National Institutes of Health and the Sam and Rose Stein Charitable Trust. The authors are indebted to Dr. Marc Ouellette, Université Laval, Québec, Canada, for supplying the PTR1 gene from *Leishmania tarentolae*. This is publication number 11708-MEM from the Department of Molecular and Experimental Medicine at The Scripps Research Institute.

REFERENCES

Armarego, W.L.F. (1979) Hydrogen transfer from 4-R and 4-S ($4-^3H$) NADH in the reduction of *d,l-cis*-6,7-dimethyl-6,7 (8H)-dihydropterin with dihydropteridine reductase from human liver and sheep liver. *Biochem. Biophys. Res. Commun.* **89**, 246–249.

Baccanari, D., Phillips, A., Smith, S., Sinski, D., and Burchall, J. (1975) Purification and properties of *Escherichia coli* dihydrofolate reductase. *Biochemistry* **14**, 5267–5273.

Blakley, R.L. (1984) Dihydrofolate reductase. In: *Folates and Pterins, Vol. 1: Chemistry and Biochemistry of Folates*, 191–253. Edited by Blakley, R.L. and Benkovic, S.J. New York, John Wiley & Sons.

Bystroff, C., Oatley, S.J., and Kraut, J. (1990) Crystal structure of *Escherichia coli* dihydrofolate reductase: the $NADP^+$ holoenzyme and the folate-$NADP^+$ ternary complex. Substrate binding and a model for the transition state. *Biochemistry* **29**, 3263–3277.

Chang, G., Shiao, M., Liaw, J., and Lee, H. (1989) Periodate-oxidized 3-aminopyridine adenine dinucleotide phosphate as a fluorescent affinity lable for pigeon liver malic enzyme. *J. Biol. Chem.* **264**, 280–287.

Farber, S., Diamond, L.K., Mercer, R.D., Sylvester, R.F., Jr., and Wolff, J.A. (1948) Temporary remissions in acute leukemia in children. *N.E.J. Med.* **238**, 787–793.

Filman, D.J., Bolin, J.T., Matthews, D.A., and Kraut, J. (1982) Crystal structures of *Escherichia coli* and *Lactobacillus casei* dihydrofolate reductase refined at 1.7 A resolution. II. Environment of bound NADPH and implications for catalysis. *J Biol Chem* **257**, 13663–13672.

Jörnvall, H., Persson, B., Krook, M., Atrian, S., Gonzàlez-Duarte, R., Jeffery, J., and Ghosh, D. (1995) Short-chain dehydrogenases/reductases (SDR). *Biochemistry* **34**, 6003–6013.

Kaufman, S. and Kaufman, E.E. (1985) Tyrosine Hydroxylase. In: *Folates and Pterins, Vol. 2*, 251–352. Edited by Blakley, R.L. and Benkovic, S.J. New York, Wiley Interscience.

Kuhn, D.M. and Lovenberg, W. (1985) Tryptophan Hydroxylase. In: *Folates and Pterins, Vol. 2*, 353–382. Edited by Blakley, R.L. and Benkovic, S.J. New York, Wiley Interscience.

Luba, J., Nare, B., Liang, P.-H., Anderson, K.S., Beverley, S.M., and Hardy, L. (1998) *Leishmania major* pteridine reductase 1 belongs to the short chain dehydrogenase family: Stereochemical and kinetic evidence. *Biochemistry* **37**, 4093–4104.

Marletta, M.A. (1993) Nitric oxide synthase structure and mechanism. *J. Biol. Chem.* **268**, 12231–12234.

Nare, B., Hardy, L.W., and Beverley, S.M. (1997) The roles of pteridine reductase 1 and dihydrofolate reductase-thymidylate synthase in pteridine metabolism in the protozoan parasite *Leishmania major*. *J. Biol. Chem.* **272**, 13883–13891.

Rosenkranz-Weiss, P., Sessa, W.C., Milstien, S., Kaufman, S., Watson, C.A., and Prober, J.S. (1994) Regulation of nitric oxide synthesis by proinflammatory cytokines in human umbilical vein endothelial cells. Elevations in tetrahydrobiopterin levels enhance endothelial nitric oxide synthase specific activity. *J. Clin. Invest.* **93**, 2236–2243.

Shiman, R. (1985) Phenylalanine Hydroxylase and Dihydropteridine Reductase. In: *Folates and Pterins, Vol. 2*, 179–249. Edited by Blakley, R.L. and Benkovic, S.J. New York, Wiley Interscience.

Varughese, K.I., Skinner, M.M., Whiteley, J.M., Matthews, D.A., and Xuong, N.H. (1992) Crystal structure of rat liver dihydropteridine reductase. *Proc. Natl. Acad. Sci., USA* **89**, 6080–6084.

Varughese, K.I., Xuong, N.H., Kiefer, P.M., Matthews, D.A., and Whiteley, J.M. (1994) Structural and mechanistic characteristics of dihydropteridine reductase: A member of the Tyr-$(Xaa)_3$-Lys-containing family of reductases and dehydrogenases. *Proc. Natl. Acad. Sci. ,USA* **91**, 5582–5586.

Wang, J.Y., Leblanc, E., Chang, C.F., Papadopoulou, B., Bray, T., Whiteley, J.M., Lin, S.X., and Ouellette, M. (1997) Pterin and folate reduction by the *Leishmania tarentolae* H locus short-chain dehydrogenase/reductase PTR1. *Arch. Biochem. Biophys.* **342**, 197–202.

Whiteley, J.M., Varughese, K.I., Xuong, N.H., Matthews, D.A., and Grimshaw, C.E. (1993) Dihydropteridine reductase. *Pteridines* **4**, 159–173.

Wierenga, R.K., Terpstra, P., and Hol, W.G.J. (1986) Prediction of the occurrence of the ADP-binding βαβ-fold in proteins, using an amino acid sequence fingerprint. *J. Mol. Biol.* **187**, 101–107.

51

EFFECT OF ANDROGEN STRUCTURES ON THE INHIBITION OF METYRAPONEINREDUCTASE IN RAT, MOUSE, AND HUMAN LIVER

Sachiko Nagamine, Chisako Yamagami, Yuka Hirai, and Seigo Iwakawa

Kobe Pharmaceutical University
4-19-1, Motoyama-kitamachi
Higashinada-ku, Kobe, 658-8558, Japan

1. INTRODUCTION

Numerous carbonyl containing drugs undergo carbonyl reduction as a major biotransformation step. Carbonyl reductases are widely distributed in each tissue of mammalian species (Felsted et al., 1980). These carbonyl reductases play an important role in steroids metabolism and detoxification processes.

Metyrapone (2-methyl-1,2-bis(3-pyridyl)-1-propanone;MO) is a potent inhibitor of adrenal steroid 11β-hydroxylase in man and animals (Liddle et al., 1958). The reduction of a carbonyl group of MO forms an asymmetric carbon in its metabolite, metyrapol (2-methyl-1-[3-(6-oxopyridyl)]-2-(3-pyridyl)-1-propanone; MOH) (Figure 1). In rat, the keto-reduction of MO showed a product stereoselectivity (Nagamine et al., 1997). MO was showed to inhibit 5-keto reduction of testosterone in a competitive manner. In order to clarify the characteristics of MO metabolism in rat, mouse and human liver, we have studied the product-stereoselectivity and subcellular location in the reductive metabolism of MO. We have compared inhibitory effects of several androgens on MO reductase. We have investigated the conformational similarities between MO and androgens using a 3D modeling analysis software (Anchore II, Fujitsu, Japan).

2. MATERIALS AND METHODS

2.1. Chemicals

Mo was obtained from Sigma Chemical Co.(St. Louis, MO). Racemic MOH was prepared by reducing MO with NaBH4 in methanol and its mp. was 99–101°C (from ethyl acetate) and [α]D= ± 0°(C=1.0, ethanol). The enantiomers of MOH were purified by a

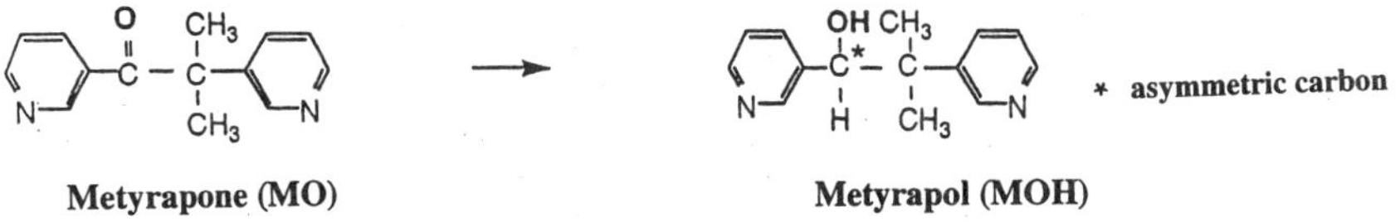

Metyrapone (MO) Metyrapol (MOH) * asymmetric carbon

Figure 1. Chemical structure of MO and MOH. * asymmetric carbon.

preparative liquid chromatography (PLC) with a Chiralpak AD column (250 × 10 mm i.d., Daicel Chemical Ind., Tokyo) from racemic MOH. The optical rotation of (-)-MOH was $[\alpha]_D = -29.0°$ (c=0.70, ethanol) and that of (+)-MOH was $[\alpha]_D = +27.1°$ (c=0.69, ethanol). The optical purity of enantiomers was over 98.0%. All other chemicals were of analytical reagent grade.

2.2. Chromatographic Conditions

The HPLC-system of a M600 multi-solvent delivery system, and a 990J photodiode array detector(Waters, Tokyo) served for chromatographic analysis. (-)-MOH and (+)-MOH were analyzed on a Chiralpak AD column (250 × 4.6 mm i.d., Daicel Chemical Ind., Tokyo). The mobile phase was n-hexane-ethanol (7:3, v/v) at a flow rate of 1.0 mL/min, and the absorbance was detected at a wavelength of 265nm. The recoveries of these compounds extracted from plasma were over 98.0%. PLC proceeded under the conditions with the mobil phase of n-hexane-ethanol (1:1, v/v) at a flow of 2.5 mL/min. The retention time of (-)-MOH and (+)-MOH was 8.5 and 12.5 min, respectively.

2.3. Enzyme Assays

The inhibitory potencies on the androgens of MO were compared using rat, mouse and human liver. Incubation mixture consisted of phosphate buffer (0.1 mol/L, pH 7.4), an NADPH regenerating system, MgC12 (5 μ mol), liver microsomes or liver cytosols (finally about 0.1 mg protein/mL), the inhibitor (final concentration 0.4 mmol/L) and 0.4 mmol/L MO to give a final volume of 1.0 mL. After 10 min preincubation of liver subcellular fraction with NADPH regenerating system and inhibitor, the reaction was started by adding the substrate and the mixture was incubated at 37°C for 2 hours in liver microsome or cytosol. The reaction was stopped by adding 1mL of acetonitrile to the mixture, and the product was extracted by adding 6 mL of chloroform containing internal standard (phenacetin, 0.54 μg/mL) and determined by HPLC as described.

Liver microsome and cytosol fractions were obtained by centrifuging 9000 × g supernatant fractions of liver. The protein concentration was determined by the Bio-Rad protein assay using bovine serum as the standard.

3. RESULTS AND DISCUSSION

3.1. Stereoselective Reductive Metabolism of MO

The reducing activity of MO are mainly located in the cytosolic fraction in liver (Felsted et al., 1980). In rat and mouse, the reducing activity were also founded in both the microsome and cytosol fraction (Oppermann et al., 1991). There are few reported studies on the product stereoselectivity in MO keto-reduction. We examined the stereoselective reductive metabolism of MO by using rat , mouse and human liver microsomes and cytosols. In rat, the activity was mainly localized in the microsomal fraction and was undetectable in cytosolic fraction. In mice, the activity was found in both fractions. The activity in human liver was mainly found in the cytosolic fraction. The proportion of (-)-MOH to (+)-MOH formed from MO by the reduction was about 4:1 in rat and mouse liver microsomes, 7:1 in mouse liver cytosol and 2:1 in human liver cytosol. The enantiomeric ratio showed some difference between the three species or the subcelluler fractions (Figure 2). The enantiomeric ratio[(-)-MOH/(+)-MOH] of reduced MO were about the same in rat and mouse liver microsome, but showed some difference between the mouse and human liver cytosol. The (-)-MOH enantiomer were always predominant. These results suggested the existence of the stereoselective reductive metabolism of MO. The difference of the stereoselectivity in MO keto-reduction was observed for species and subcellular fractions in liver. The stereoselectivity in MO keto-reduction was different among liver subcellular fractions.

3.2. Inhibitory Effects of Barbital, Quercitrin, and Pyrazol on Reductive Metabolism of MO

Barbital (an inhibitor of aldehyde reductases), quercitrin (an inhibitor of ketone reductases) or pyrazol (an inhibitor of alchohol dehydrogenase) was added to incubation mixtures as inhibitors. Their inhibitory effects on MO reductase activity were compared by measuring MOH formation from MO. Barbital decreased MOH formation in rat and mouse liver microsomes, but Quercitrin did not show a significant inhibitory effect on the formation of MOH. In mouse and human liver cytosols, MOH formation from MO were inhibited by quercitrin, but were not inhibited by barbital. Pyrazol did not inhibited the reductive metabolism of MO in rat, mouse and human liver (Figure 3).

According to the classic inhibitor subclassification of carbonyl reductase in rat, mouse and human liver, MO were reduced by aldehyde reductase in liver microsome and by ketone reductase in liver cytosol.

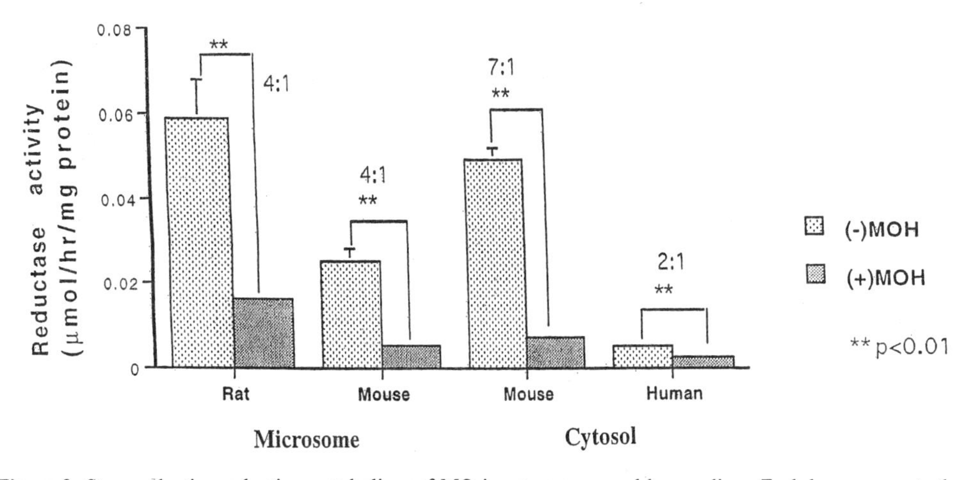

Figure 2. Stereoselective reductive metabolism of MO in rat, mouse, and human liver. Each bar represents the mean ± SD of 3–5 experiments.

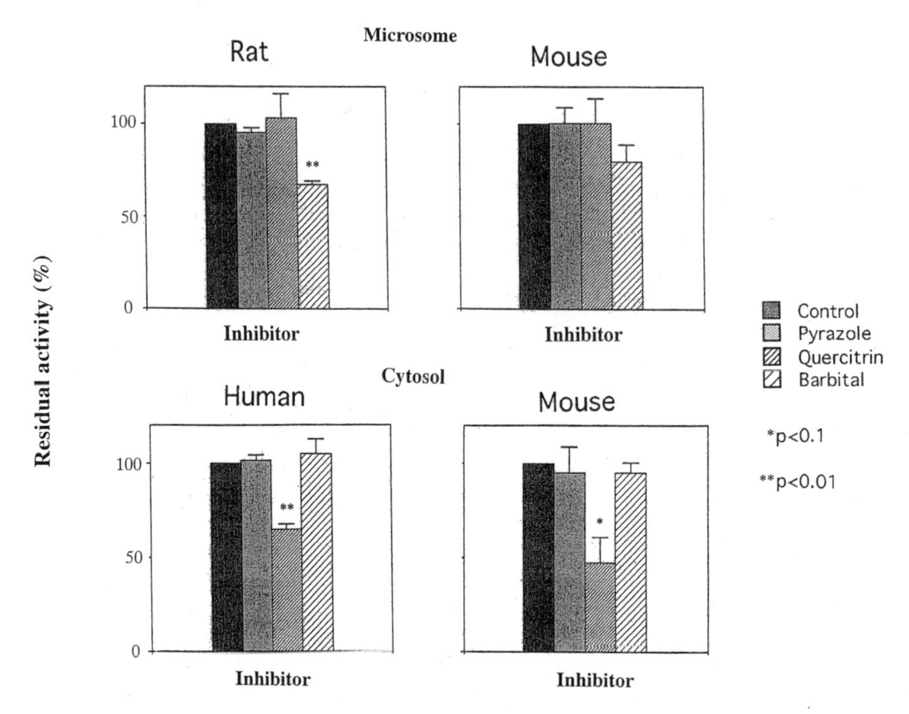

Figure 3. Effects of various inhibition on reductive metabolism of MO. Each bar represents the mean ± SD of 3–5 experiments.

3.3. Inhibitory Effects of Androgens on Reductive Metabolism of MO

Pietruszko *et al.* reported that peak II enzyme of the aldehyde reductases in rat liver catalyzed reversible reduction of 3-ketosteroids of A/B *cis* configuration, and the product of reduction appears to be 3α-alcohol, rather than 3β-alcohol (Pietruszko et al., 1976). According to Imamura *et al.*, there were sex-related differences in rat hepatic metabolism of MO and androgen had a role in sex-related difference (Imamura et al., 1993). The inhibitory effect of various androgens for the reductive metabolism of MO were investigated. Androgen structures are shown in Figure 4.

Inhibitory effects of 3-ketosteroids, such as testosterone, 5 α-dihydrotestosterone (5 α-DHT) and 5 β-DHT on MO metabolism were examined in rat , mouse and human liver. In rat liver microsomes, testosterone decreased (-)-MOH formation to 58%, 5 α-DHT to 47% and 5 β-DHT to 38%. In mouse liver microsomes, these 3-ketosteroids decreased (-)-MOH formation to about the same extend (about 40%) (Figure 5a). In mouse liver cytosol, testosterone, 5 α-DHT and 5 β-DHT decreased (-)-MOH formation to 80, 66 and 60%. In human liver cytosol, testosterone, 5-D-DHT and 5-D-DHT decreased (-)-MOH formation to 60, 80 and 52% (Figure 5b). However, these 3-ketosteroids showed insignificant effects on the formation of (+)-MOH. The inhibition on the reductive metabolism of MO by these 3-ketosteroids showed a competitive inhibition.

In 3-ketosteroids, 5 β-DHT with A/B *cis* configuration showed stronger inhibitory effect on MO reductive metabolism than 5 α-DHT with A/B *trans* configuration (Figure5a, 5b). These results suggest that the stereochemical configuration of 3-ketosteroids is profoundly concerned in the inhibition of MO metabolism.

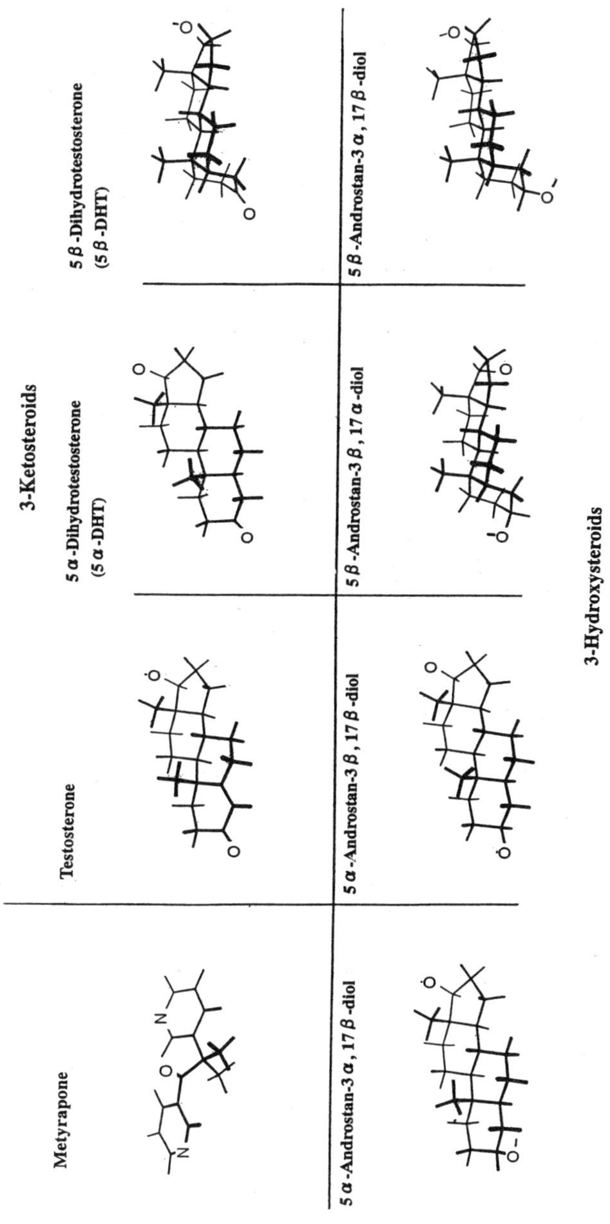

3-Ketosteroids

Metyrapone

Testosterone

5α-Dihydrotestosterone (5α-DHT)

5β-Dihydrotestosterone (5β-DHT)

5α-Androstan-3α, 17β-diol

5α-Androstan-3β, 17β-diol

5β-Androstan-3β, 17α-diol

5β-Androstan-3α, 17β-diol

3-Hydroxysteroids

Figure 4. The structures of MO and androgens.

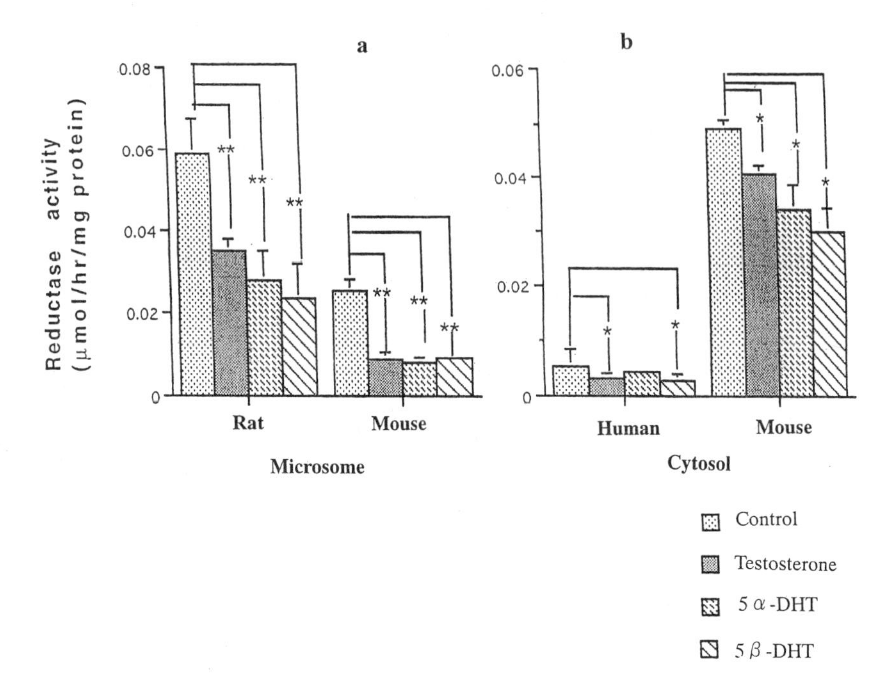

Figure 5(a and b). Effects of 3-ketosteroids on reductive metabolism of MO. Each bar represents the mean ± SD of 3–5 experiments.

Inhibitory effects of 3-hydroxysteroids were examined in rat and mouse liver microsomes. These 3-hydroxysteroids, such as 5 α-androstan-3-α, 17 β-diol and 5 β-androstan-3-β, 17 β-diol, significantly inhibited (-)-MOH formation on the reductive metabolism of MO, but 3-hydroxysteroids such as as 5α-androstan-3-β, 17 β-diol and 5β-androstan-3-α, 17 β-diol did not significantly inhibit. In 3-hydroxysteroids, 5 α-androstan-3-α, 17 β-diol and 5 β-androstan-3-β, 17 β-diol with a 5α, 3α and 5β, 3β configuration inhibited MO reductive metabolism, but compounds with 5α, 3β and 5β, 3α configuration did not. These results indicate that the configuration at 3- and 5-position of 3-hydroxysteroids is profoundly concerned with the interaction between the MO reductase and these steroids. These 3-hydroxysteroids showed insignificant inhibitory effect on the formation of (+)-MOH (Figure 6a). In mouse and human liver cytosol, these 3-hydroxysteroids did showed insignificant inhibitory effects in the reductive metabolism of MO (Figure 6b).

3.4. 3-D Modeling of 3-Ketosteroids and MO

The substrate and inhibitors specificity was rationalized by substantial conformational similarities between them (Penning et al., 1983). The 3D structures comparison of 3-ketosteroids with MO showed their conformational similarity. The superimposition of the substrate, MO and 3-ketosteroids, competitive inhibitors on the reductive metabolism of MO, were optimized by AM1 methods of the position four pairs of atoms. MO well matched with A/B ring of 3-ketosteroids (Figure 7). On the superimposed MO and each 3-ketosteroid, the distance between carbonyl group of both compounds was calculated. The inhibitory activity of 3-ketosteroids for MO metabolism was correlated with the closeness of the distance between their carbonyl groups (Table 1).

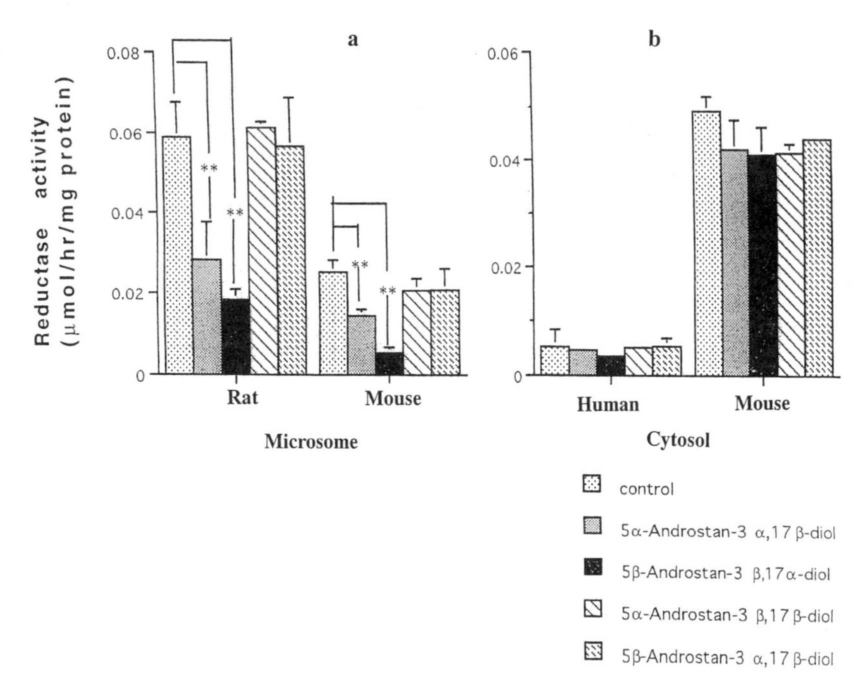

Figure 6 (a and b). Effects of 3-hydroxysteroids on reductive metabolism of MO. Each bar represents the mean ± SD of 3–5 experiments.

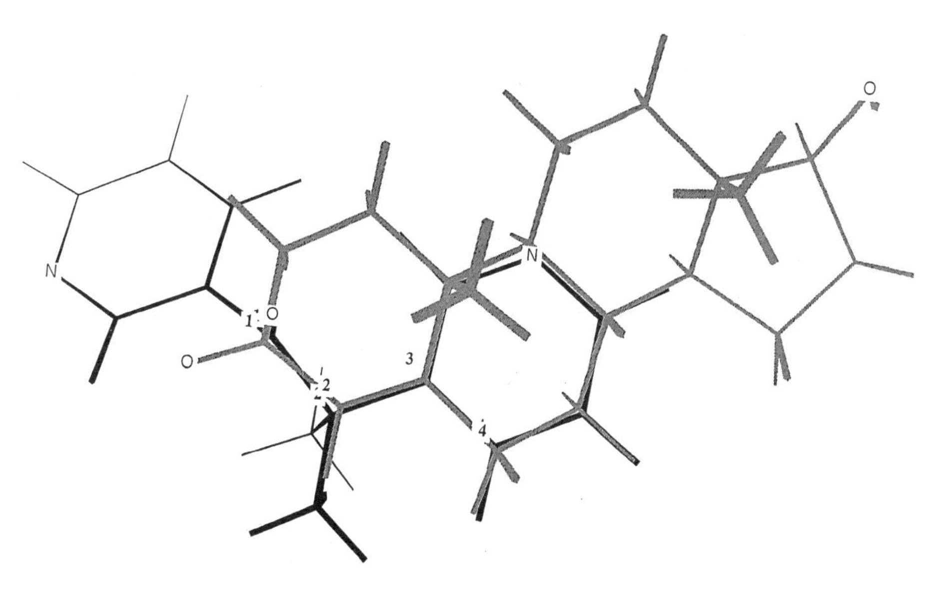

Figure 7. Superimposition of the structure of testosterone with MO. Testosterone is shown by the gray line and MO is shown by the black line. The picture created using the program Anchor II.

Table 1. The relationship between the inhibitory rate for MO reductive metabolism and the distance between 3-ketosterois- and MO-carbonyl groups on the superimposed their both compounds

3-ketosteroids	Distance (Å) between 3-ketosteroid- and MO-CO	Inhibitory rate(%)			
		Microsome		Cytosol	
		Mouse	Rat	Human	Mouse
Testosterone	2.260	39.6	12.8	42.9	40.6
5 α-Dihydrotestosterone	2.363	17.0	29.7	47.6	52.6
5 β-Dihydrotestosterone	1.479	47.2	36.2	47.6	60.3

4. CONCLUSION

The product stereoselectivity in the reductive metabolism of MO in rat, mouse and human was observed. The stereospecificity of MO reductase resided in rather subcellular fractions than species. According to inhibitory effects of androgens structures on the reductive metabolism of MO, it was suggested that A/B ring stereochemical configuration of 3-ketosteroids affected the binding sites of MO reductase and substrate. Comparing 3D structures of MO with androgens provide topographical understanding of the catalytic site of MO reductase, and may give useful information about the prediction of drug interaction in the reductive metabolic process.

REFERENCES

Felsted, R.L., Bachur, N.R., 1980, Mammalian carbonyl reductases, *Drug Metab. Rev.*, 11, 1–60.

Imamura, Y., Murata, H., Otagiri, M., 1993, Postnatal development and sex-related difference metyrapone reductase activities in liver microsomes and cytosol of the rat, *Res. Commun. Chem.Pathol. Pharmacol.*, 82, 101–110.

Liddle, G.W., Island, D., Ester, H., Tomkins, G.M., 1958, Modefication of adrenal steroid patterns in man by means of an inhibitor of 11-β-hydroxylase, *J. Clin. Invest.*, 37, 912.

Nagamine, S., Horisaka, E., Fukuyama, Y., Maetani, K., Matsuzawa, R., Iwakawa, S., Asada, S., 1997, Stereoselective reductive metabolism of metyrapone and inhibitory activity of metyrapone metabolites, metyrapol enantiomers, on steroid 11-β-hydroxylase in rat, *Biol. Pharm. Bull.*, 20, 188–192.

Oppermann, U.D.T., Maser, E., Mangoura, S.A., Netter, K.J., 1991, Heterogeneity of carbonyl reduction in subcellular fractions and different organs in rodents, *Biochem. Pharmacol.*, 42, s189-s195.

Penning, T.M., Talalay, P., 1983, Inhibition of a major NAD(P)-linked oxidoreductase from rat liver cytosol by steroidal and nonsteroidal anti-inflammatory agents and by prostaglandins, *Proc. Natl. Acad. Sci. USA.*, 80, 4504–4508.

Pietruszko, R., Chen, F.F., 1976, Aldehyde reductase from rat liver is a 3 α-hydroxysteroid Dehydrogenase, *Biochem. Pharmacol.*, 25, 2721–2725.

52

PHYSIOLOGICAL RELEVANCE OF ALDEHYDE REDUCTASE AND ALDOSE REDUCTASE GENE EXPRESSION

Junichi Fujii, Motoko Takahashi, Rieko Hamaoka, Yoshimi Kawasaki, Nobuko Miyazawa, and Naoyuki Taniguchi

Department of Biochemistry
Osaka University Medical School
2-2 Yamadaoka, Suita, Osaka 565-0871, Japan

1. INTRODUCTION

Carbonyl compounds which are produced as intermediate molecules during ordinary metabolism or are present in food or drugs are known to be toxic to living organisms because of their high degree of reactivity. It has also been suggested that elevation in protein carbonyl groups is also a likely cause of aging (Stadtman, 1992). Cells contain defense systems against these compounds (Flynn, 1982) in the form of aldo-keto reductases, which include aldehyde and aldose reductases, which catalyze the reduction of a variety of aldehydes to alcohols in an NADPH-dependent manner (Jez et al., 1997). Cytotoxic compounds which contain an aldehyde moiety, such as tripeptidyl aldehyde (Inoue et al., 1993), trioses (Vander Jagt et al., 1992) and methotrexate (Callahan and Beverley, 1992) are detoxified by enzymes in this gene family. The glycation reaction represents another source of carbonyl compounds. This reaction occurs during normal aging and at accelerated rates in diabetes, and is involved in the pathogenesis of diabetic complications (Fujii et al., 1998). Glycation alters the activity of some enzymes such as Cu,Zn-SOD (Arai et al., 1987; Ookawara et al., 1992), carbonic anhydrase (Kondo et al., 1987), sorbitol dehydrogenase (Hoshi et al., 1996), and aldehyde reductase (Takahashi et al., 1995b), and is also involved in production of dicarbonyl compounds such as 3-deoxyglucosone and methylglyoxal (Figure 1). The cross-linking of long-lived proteins such as collagen and lens crystallins are induced by these compounds and correlates with aging and diabetes. These dicarbonyl compounds are highly toxic and induce apoptosis in susceptible cells (Okado et al., 1996). The production of aldehydes is enhanced during pathological conditions, including diabetes and cancer. Concomitantly, aldehyde-reducing activity is also increased in hepatoma cell lines (Canuto et al., 1994). This is mainly due to an elevated expression of the aldose reductase gene (*AKR1B*) as has been demonstrated in chemically induced hepatoma tissues (Zeindl-Eberhart et al., 1994; Takahashi et al., 1995a).

Enzymology and Molecular Biology of Carbonyl Metabolism 7
edited by Weiner *et al.* Kluwer Academic / Plenum Publishers, New York, 1999.

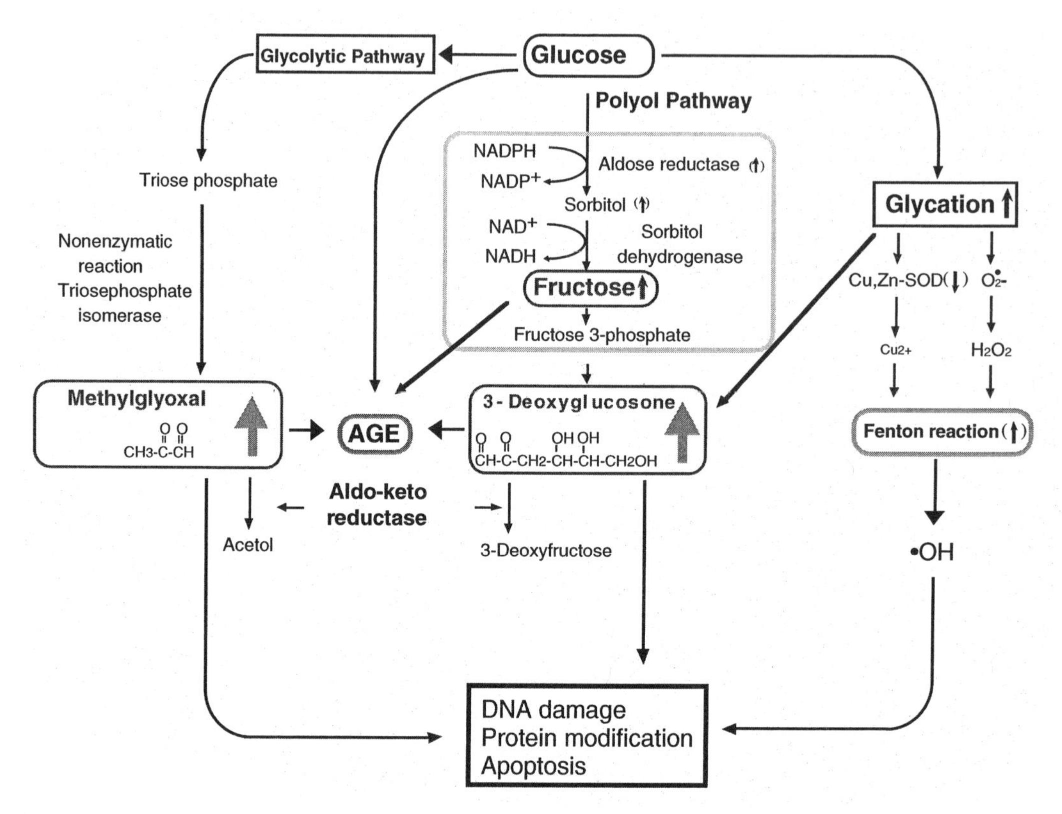

Figure 1. Metabolic map of carbohydrates and the glycation reaction.

Aldose reductase plays a key role in diabetic complications by converting glucose to sorbitol in the polyol metabolizing pathway. Since the accumulation of sorbitol is thought to play a role in the pathogenesis of diabetic complications in some organs, such as kidney, peripheral nerve, and the eye, extensive studies on aldose reductase, including the examination of inhibitors for the enzyme have been carried out, in order to elucidate relationships between sorbitol accumulation and diabetic complications. Aldehyde reductase (Tomlinson et al., 1994), on the other hand, is constitutively expressed in many organs and is involved in the reduction of a wide variety of aldehyde compounds. The present study describes our attempts to clone and characterize the gene.

2. ISOLATION OF THE HUMAN ALDEHYDE REDUCTASE GENE (*AKR1A1*)

To isolate the human aldehyde reductase gene, about 10^6 λphages of a human genomic DNA library constructed in EMBL3 were screened with a rat aldehyde reductase cDNA fragment (Takahashi et al., 1993) as a probe. The probe DNA, which contained the entire coding sequences, was labeled with ^{32}P by the random hexamer method. Using this screen, a clone was isolated and characterized by restriction endonuclease digestion analysis with EcoRI, HindIII, and BamHI, followed by agarose gel electrophoresis. After denaturation with 0.5 M NaOH and neutralization in 0.5 M Tris-Cl, pH7.4, in the presence of

1.5 M NaCl, the DNA fragments were transferred onto nylon membranes. These Southern blots were probed under high stringency conditions with a series of cDNA fragments. All DNA fragments were subcloned into the Bluescript plasmid vector for further restriction endonuclease analysis and sequencing. The DNA subclones were purified by CsCl ultracentrifugation and were used for the sequencing reactions. After sequence determination, the potential exon/intron boundaries were identified by computer alignment of the genomic sequence with the reported human cDNA sequence (Bohren et al., 1989) using the DNASIS program.

3. COMPARISON OF THE ALDEHYDE REDUCTASE GENE WITH THE ALDOSE REDUCTASE GENE

Since both aldehyde reductase and aldose reductase genes are members of aldo-keto reductase gene superfamily, they have significant amino acid homology (60% identities) (Nishimura et al., 1990; Bohren et al., 1989). The gene which encodes the human aldose reductase spans about 17 kb and consists of 10 exons (Graham et al., 1991a) (Figure 2).

It had been expected that the structure of these genes would also be similar. However, a comparison of the exon/intron boundaries between these genes shows matching for only 2 of the 7 boundaries (Figure 3). The aldehyde reductase gene is smaller than the aldose reductase gene and appears to have no pseudogene. In addition, regulation of gene expression is quite different. While aldose reductase is induced under several pathological conditions such as high osmotic pressure, hyper glycemia, and malignancy, the expression of the aldehyde reductase gene is constant. The putative regulatory elements responsive to osmotic pressure have been identified in the aldose reductase gene (Iwata et al., 1997; Ko et al., 1997). Barski et al. (1998) showed that the aldehyde reductase gene has two additional exons which encode 5' nontranslated sequences and thus is comprised of 11 exons.

To localize the aldehyde reductase gene, FISH was performed on human R-banded metaphase chromosomes, using a 10 kb-aldehyde reductase genomic DNA subclone (EcoRI to BamHI) as a probe. This localized the human aldehyde reductase gene unambiguously on human chromosome 1p32-p33 (Table I). By using a cDNA clone which encodes for aldose reductase, the human aldose reductase gene was mapped to human chromosomes 1q32-q42, 3p12, 7q31-q35, 9q22, 11p14-p15, and 13q14-q21 (Bateman et al., 1993). Graham *et al.* (1991b) mapped the bona fides gene to chromosomal region 7q35 using PCR in somatic cells as well as in situ hybridization using a genomic clone as a probe. Although a putative pseudogene has been mapped to chromosome 3 (Brown et al., 1992), it is not clear what the other hybridizable genes on the chromosomes may code. The present study shows that the aldehyde reductase gene can be unambiguously mapped to 1p32-p33. This clearly shows that the gene, which was previously mapped to 1q32-q42, is not identical to the aldehyde reductase gene. Thus, the structure and chromosomal localization of two genes are quite different and suggest that they are separated at a fairly early period in the evolutionary process.

4. SIGNIFICANCE OF ALDO-KETO REDUCTASES UNDER PATHOLOGICAL CONDITIONS

Aldo-keto reductases function as a self defense system against carbonyl compounds produced in living organisms. We have purified and cloned a rat liver enzyme that cata-

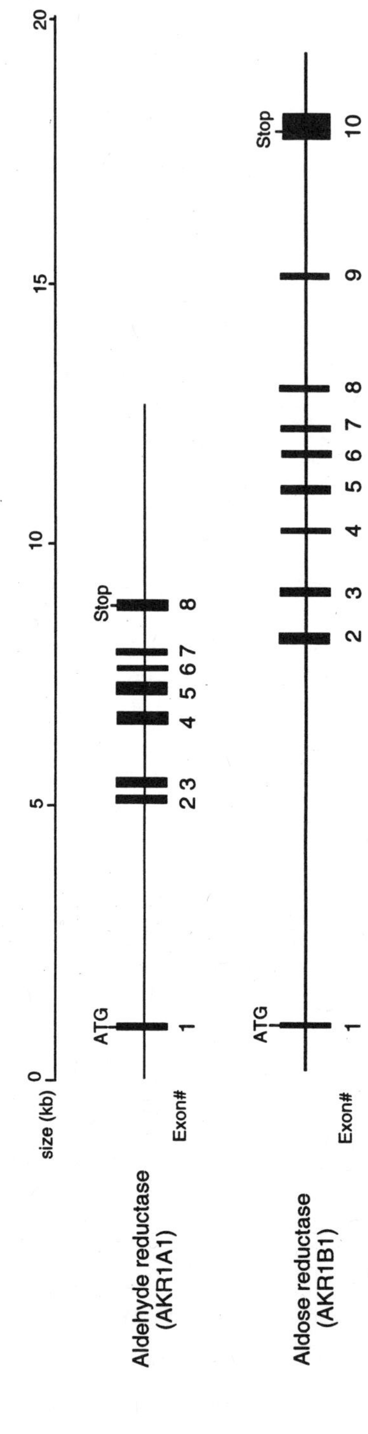

Figure 2. Comparison of gene structure between the human aldehyde reductase gene and the aldose reductase gene. The positions of exon/intron boundaries in the human aldehyde reductase gene (AKR1A1) and the corresponding position for the human aldose reductase gene (AKR1B1) are indicated.

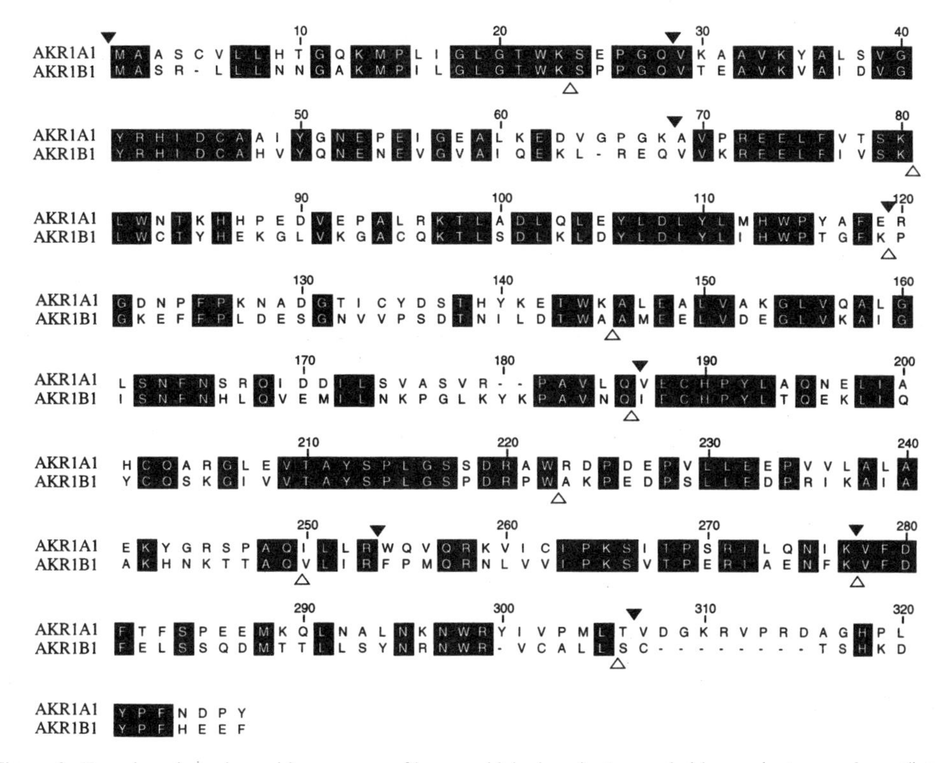

Figure 3. Homology in amino acid sequences of human aldehyde reductase and aldose reductase, and exon/intron boundaries.

lyzes the NADPH-dependent reduction of 3-deoxyglucosone and methylglyoxal, major intermediates in the Maillard reaction and potent cross-linkers responsible for the polymerization of proteins (Takahashi et al., 1993). COS-1 cells transfected with the rat aldehyde reductase cDNA exhibited NADPH-dependent 3-deoxyglucosone-reducing activity and the extracted proteins were cross-reacted with antiserum raised against the purified rat 3-deoxyglucosone-reducing enzyme. Thus the 3-deoxyglucosone-reducing enzyme is identical with aldehyde reductase, isolated from rat liver. In other tissues including kidney, 3-deoxyglucosone appears to be detoxified by both aldehyde reductase and aldose reductase (Sato et al., 1993; Matsuura et al., 1995). Northern blot analysis of total RNA from a variety of rat tissues showed fairly high levels of expression of aldehyde reductase mRNA. This suggests that the aldehyde reductase present in the body detoxifies 3-deoxyglucosone at locations where it is formed, via the glycation reaction.

Table 1. Comparison of chromosomal loci of aldose reductase and aldehyde reductase genes

Gene	Chromosomal loci	Reference
AKR1A1	1p32-p33	this work
AKR1B1	1q32-q42, 3p12, 7q31-q35, 9q22, 11p14-p15, 13q14-q21	Bateman et al., 1993
	7q35	Graham et al.,1991b
	chromosome 3	Brown et al., 1992

The aldehyde reductase purified from normal rat liver was found to be partially glycated as judged by binding to a boronate column and reactivity to anti-hexitol lysine IgG (Takahashi et al., 1995b). Sites of *in vivo* glycation of rat liver aldehyde reductase were identified by sequencing lysylendopeptidase-digested peptides reduced with $NaBH_4$ and by mass spectrometry. The major glycated sites were found to be lysines 67, 84, and 140. The glycated enzyme had low catalytic efficiency as compared to the nonglycated form. In streptozotocin-induced diabetic rats, the glycated form was significantly increased in kidney. Since the enzyme plays a role in detoxifying 3-deoxyglucosone formed in the *in vivo* glycation reaction, the glycation of aldehyde reductase and the reduction in its activity result in an enhancement of cytotoxicity of carbonyls under diabetic conditions, by virtue of the diminished activity of the enzyme.

Aldose reductase also can reduce 3-deoxyglucosone by the same mechanism as aldehyde reductase. While the level of aldehyde reductase remains unchanged, aldose reductase gene induction was observed in hepatoma tissues and cultured hepatoma cell lines (Canuto et al., 1994; Takahashi et al., 1995a; 1996; Zeindl-Eberhart et al., 1994). Thus this enzyme functions in the detoxification of naturally occurred carbonyl compounds and may have a role in immortality of malignant cells.

We also found that a pancreatic β–cell line, HIT, transfected with aldose reductase cDNA became vulnerable to high glucose conditions and underwent apoptotic cell death (Hamaoka et al., manuscript in preparation). Since overproduced aldose reductase would alter the redox state of the cells during the conversion of glucose to sorbitol by consuming NADPH, this redox imbalance may be related to apoptosis. The augmented expression of aldose reductase also causes an elevation in fructose levels, which is a more potent glycating sugar than glucose and results in the induction of apoptosis in pancreatic β cells (Kaneto et al., 1994). We recently established a polyclonal antibody raised against fructated protein and, using it, actually found that fructose-induced glycation reaction was enhanced under diabetic conditions and aging (Miyazawa et al., submitted for publication). The fructation reaction was more prominent in diabetic rat lens because the half life of such proteins are extremely long and sorbitol and fructose accumulate at the millimolar range. Since the major source of fructose production is the polyol pathway, aldose reductase could be responsible for the enhanced fructation. An inhibitor for aldose reductase actually suppressed the accumulation of sorbitol and fructated protein (Kawasaki et al., manuscript in preparation). Thus enzymes in aldo-keto reductase gene superfamily have two opposing functions. One function is to protect cells through detoxifying harmful carbonyl compounds and the other is to produce polyol and dysfunction redox balance, resulting in the development of diabetic complications.

ACKNOWLEDGMENT

This work was supported, in part, by Grants-in-Aid for Scientific Research (B) and (C) from the Ministry of Education, Science, Sports, and Culture, Japan and Research Grant (9A-1) for Nervous and Mental Disorders from the Ministry of Health and Welfare, Japan.

REFERENCES

Arai, K., Maguchi, S., Fujii, S., Ishibashi, H., Oikawa, K., and Taniguchi, N. (1987) Glycation and inactivation of human Cu-Zn-superoxide dismutase. Identification of the *in vitro* glycation sites. *J. Biol. Chem.* **262**, 16969–16972.

Barski, O. A., Gabbay, K. H., and Bohren, K. M. (1998) Characterization of the human aldehyde reductase gene promoter. (Abstract) 9th International Symposium. Enzymology and Molecular Biology of Carbonyl Metabolism. Italy.

Bateman, J.B., Kojis, T., Heinmann, C., Klisak, I., Diep, A., Carper, D., Nishimura, C., Mohandas, T., and Sparkes, R.S. (1993) Mapping of aldose reductase gene sequences to human chromosomes 1, 3, 7, 9, 11, and 13. *Genomics* **17,** 560–565.

Bohren, K.M., Bullock, B., Wermuth, B., and Gabbay, K.H. (1989) The aldo-keto reductase superfamily. *J. Biol. Chem.* **264**, 9547–9551.

Brown, L., Hedge, P.J., Markham, A.F., and Graham, A. (1992) A human aldehyde dehydrogenase (aldose reductase) pseudogene: Nucleotide sequence analysis and assignment to chromosome 3. *Genomics* **13,** 465–468.

Callahan, H.L., and Beverley, S.M. (1992) A member of the aldoketo reductase family confers methotrexate resistance in *Leishmania*. *J. Biol. Chem.* **267**, 24165–24168.

Canuto, R.A., Ferro, M., Muzio, G., Bassi, A.M., Leonarduzzi, G., Maggiora, M., Adamo, D., Poli, G., and Lindahl, R. (1994) Role of aldehyde metabolizing enzymes in mediating effects of aldehyde products of lipid peroxidation in liver cells. *Carcinogenesis* **15**, 1359–1364.

Flynn, T.G. (1982) Aldehyde reductase: Monomeric NADPH-dependent oxidoreductases with multifunctional potential. *Biochem. Pharmacol.* **31**, 2705–2712.

Fujii, J., Asahi, M., Okado, A., and Taniguchi, N. (1998) Dysfunction of redox regulation: a common mechanism for apoptotic cell death by nitric oxide and the glycation reaction. in *Pathophysiology of lipid peroxides and related free radicals.* (Yagi, K. ed.) p125–134, Japan Sci. Soc. Press, Tokyo.

Graham, A., Brown, L., Hedge, P.J., Gammack, A.J., and Markham, A. F. (1991a) Structure of the human aldose reductase gene. *J. Biol. Chem.* **266,** 6872–6877.

Graham, A., Heath, P., Morten, J.E.N., and Markham, A.F. (1991b) The human aldose reductase gene maps to chromosome region 7q35. *Hum. Genet.* **86,** 509–514.

Hoshi, A., Takahashi, M., Fujii, J., Myint, T., Kaneto, H., Suzuki, K., Yamasaki, Y., Kamada, T., and Taniguchi, N. (1996) Glycation and inactivation of sorbitol dehydrogenasae in normal and diabetic rats. *Biochem. J.* **318,** 119–123.

Inoue, S., Sharma, R.C., Schimkc, R.T., and Simoni, R.D. (1993) Cellular detoxification of tripeptidyl aldehydes by an aldo-keto reductase. *J. Biol. Chem.* **268**, 5894–5898.

Iwata, T., Minucci, S., McGowan, M., and Carper, D. (1997) Identification of a novel cis-element required for the constitutive activity and osmotic response of the rat aldose reductase promoter. *J. Biol. Chem.* **272,** 32500–32506.

Jez, J.M., Bennett, M.J., Schlegel, B.P., Lewis, M., and Penning, T.M. (1997) Comparative anatomy of the aldo-keto reductase superfamily. *Biochem. J.* **326**, 625–636.

Kaneto, H., Fujii, J., Myint, T., Miyazawa, N., Islám, K. N., Kawasaki, Y., Suzuki, K., Nakamura, M., Tatsumi, H., Yamasaki, Y., andTaniguchi, N. (1996) Reducing sugars trigger oxidative modification and apoptosis in pancreatic β-cells by provoking oxidative stress through the glycation reaction. *Biochem. J.* **320**, 855–863.

Kondo, T., Murakami, K., Ohtuk, Y., Tsuji, M., Gasa, S., Taniguchi, N., and Kawakami, Y. (1987) Estimation and characterization of glycosylated carbonic anhydrase I in erythrocytes from patients with diabetes mellitus. *Clin. Chim. Acta* **166**, 227–236.

Nishimura, C., Matsuura, Y., Kokai, Y., Akera, T., Carper, D., Morjana, N., Lyons, C., and Flynn, T. G. (1990) Cloning and expression of human aldose reductase. *J. Biol. Chem.* **265,** 9788–979.

Okado, A., Kawasaki, Y., Hasuike, Y., Takahashi, M., Teshima, T., Fujii, J., and Taniguchi, N. (1996) Induction of apoptotic cell death by methylglyoxal and 3-deoxyglucosone in macrophage- derived cell lines. *Biochem. Biophys. Res. Commun.* **225**, 219–224.

Ookawara, T., Kawamura, N., Kitagawa, Y., and Taniguchi, N. (1992) Site-specific and random fragmentation of Cu,Zn-superoxide dismutase by glycation reaction. Implication of reactive oxygen species. *J. Biol. Chem.* **267**, 18505–18510.

Sato, K., Inazu, A., Yamaguchi, S., Nakayama, T., Deyashiki, Y., Sawada, H., and Hara, A. (1993) Monkey 3-deoxyglucosone reductase: tissue distribution and purification of three multiple forms of the kidney enzyme that are identical with dihydrodiol dehydrogenase, aldehyde reductase, and aldose reductase. *Arch. Biochem. Biophys.* **307**, 286–294.

Stadtman, E.R. (1992) Protein oxidation and aging. *Science* **257**, 1220–1224.

Takahashi, M., Fujii, J., Miyoshi, E., Hoshi, A., and Taniguchi, N. (1995a) Elevation of aldose reductase gene expression in rat primary hepatoma and hepatoma cell lines. Implication of cytotoxic aldehydes. *Int. J. Cancer.* **62,** 749–754.

Takahashi, M., Fujii, J., Teshima, T., Suzuki, K., Shiba, T., and Taniguchi, N. (1993) Identity of a major 3-deoxyglucosone-reducing enzyme with aldehyde reductase in rat liver established by amino acid sequencing and cDNA expression. *Gene* **127**, 249–253.

Takahashi, M., Hoshi, A., Fujii, J., Miyoshi, E., Kasahara, T., Suzuki, K., Aozasa, K., and Taniguchi, N. (1996) Induction of aldose reductase gene expression in LEC rats during the development of the hereditary hepatitis and hepatoma. *Jpn. J. Cancer Res.* **87**, 337–341.

Takahashi, M., Lu, Y., Myint, T., Fujii, J., Wada, Y., and Taniguchi, N. (1995b) *In vivo* glycation of aldehyde reductase, a major 3-deoxyglucosone reducing enzyme: Identification of glycation sites. *Biochemistry* **34**, 1433–1438.

Tomlinson, D.R., Stevens, E.J., and Diemel, C.T. (1994) Aldose reductase inhibitors and their potential for the treatment of diabetic complications. *Trends Pharmacol. Sci.* **15**, 293–297.

Vander Jagt, D.L.V., Robinson, B., Taylor, K.K., and Hunsaker, L.A. (1992) Reduction of trioses by NADPH-dependent aldo-keto reductases: Aldose reductase, methylglyoxal, and diabetic complications. *J. Biol. Chem.* **267**, 4364–4369.

Zeindl-Eberhart, E., Jungblut, P. R., Otto, A., and Rabes, H. M. (1994) Identification of tumor- associated protein variants during rat hepatocarcinogenesis: Aldose reductase. *J. Biol. Chem.* **269**, 14589–14594.

53

THE ALDO-KETO REDUCTASES AND THEIR ROLE IN CANCER

David Hyndman and T. Geoffrey Flynn

Department of Biochemistry
Queen's University
Kingston, Ontario, Canada K7L 3N6

1. INTRODUCTION

A definitive role for many members of the aldo-keto reductase (AKR) superfamily has been elusive despite evidence of the involvement of these enzymes in the metabolism of steroids, biogenic amines and even the products of lipid peroxidation. It is generally believed that the primary role of AKR's is the detoxication of aldehydes but there is also evidence for the involvement of one of them, aldose reductase (ADR), in pathological processes i.e., the development of diabetic complications.

An emerging area of research where a link is being established between AKR's and carcinogenesis indicates yet another possible role for these enzymes in pathological events (Takahashi et al., 1995; Zeindl-Eberhart et al., 1994; Ciaccio et al., 1994; Ellis et al., 1993). We have previously isolated an aldose reductase-like enzyme (CHO reductase) from Chinese hamster ovary cells that was inducible by treatment with an aldehyde containing protease inhibitor (Hyndman et al., 1997). Since this overexpressing cell line showed resistance to the anticancer drug daunorubicin it was of interest to determine if there was a corresponding human enzyme and to determine if it was overexpressed in cancer cell lines as has been observed with aldose reductase (Takahashi et al., 1995; Takahashi et al., 1996). Significant overexpression would indicate a potential role in the detoxification of aldehyde and ketone containing anticancer drugs or involvement in the neoplastic changes of these cells. We have now cloned a human orthologue of this enzyme from human small intestine total RNA, which we have named HSI reductase. This protein shows high sequence identity to CHO reductase, mouse fibroblast growth factor-regulated-1 protein (FR-1) and mouse vas deferens protein (MVDP). It is strongly expressed in the adrenal gland, as is aldose reductase, suggesting a potential role in steroid metabolism but it shows a different expression pattern in cancer cell lines when compared to aldose reductase.

Enzymology and Molecular Biology of Carbonyl Metabolism 7
edited by Weiner *et al.* Kluwer Academic / Plenum Publishers, New York, 1999.

2. MATERIALS AND METHODS

2.1. Cell Proliferation Assay

Parental and CHO reductase expressing CHO-K1 cells were grown as described (Hyndman et al., 1997). Cells were seeded into a 96 well microtitre plate at 2.5×10^4 cells/well and allowed to attach for one day. Media was replaced with fresh media containing daunorubicin at the indicated concentrations in quadruplicate and the cells were allowed to grow for 4 days. Cell viability was assessed using the MTT assay as described (Campling et al., 1991).

2.2. Cloning of HSI Reductase

Initial screening of human total RNA (small intestine, fetal liver and testis; Clontech) for the presence of an ADR-like sequence was based on the tissue expression patterns previously reported for FR-1 and MVDP (Lau et al., 1995) and CHO reductase (Hyndman et al., 1997). A PCR product from a human small intestine cDNA pool was obtained using primers designed for CHO reductase. The resulting sequence was for a new member of the aldo-keto reductase family of monomeric oxidoreductases based on a Genbank search. Overlapping sequence fragments were obtained using 3'- and 5'-RACE procedures (Frohman, 1990). The 5'-RACE procedure was modified to include titration of the primer (Templeton et al., 1993) and a modified primer annealing step [denaturation of the RNA with the primer at 80°C for 3 min followed by annealing 5°C below the primer T_M for 1 hr in 0.3M NaCl, 10 mM Tris-HCl, pH 7.5 and 0.1 mM EDTA] (Chen, 1996). The complete sequence was reamplified from the small intestine cDNA pool and resequenced on both strands.

2.3. cRNA Probe Generation

The HSI reductase cDNA was used to generate a template for cRNA synthesis by linker oligonucleotide mediated addition (Horton et al., 1989) of a T7 initiation sequence to a coding region PCR fragment. This template generated a 201 nt riboprobe. A clone of human ADR in pGEM4z (Nishimura et al., 1990) was digested by *Bam*H1 and religated to remove the entire 3'-untranslated region prior to PCR amplification from the T7 promoter sequence contained in the vector to an insert specific primer to generate a 340 nt riboprobe. A pTRI-cyclophilin-human template (Ambion) was used to generate a 165 nt cRNA internal standard probe.

2.4. Human mRNA Dot-Blot Analysis

A Human RNA Master Blot (Clontech) was prehybridized and hybridized using ExpressHyb (Clontech) at 68°C. The appropriate riboprobe was generated with the Strip-EZ RNA probe generation kit (Ambion), which allowed for subsequent probe removal, and was used at 1×10^6 cpm/ml in the hybridization. Quantitation of the dot-blot was performed using an InstantImager (Packard Instrument Co.) and the supplied software.

2.5. Ribonuclease Protection Assay

The HSI reductase and human ADR riboprobes were prepared at high specific activity ($>10^8$ cpm/μg) and the cyclophilin riboprobe was prepared at 10-fold lower specific activity. RNA from lung cancer cell lines (Dr. Barbara Campling, Cancer Research Laboratories, Queen's University) was prepared using TriPure reagent (Boehringer Mannhein). Total RNA from all other cell lines was provided by the National Cancer Institute in the United States. The polyacrylamide gel-purified riboprobes (5×10^4 cpm) and cell line RNA (20 μg) were denatured at 85°C for 10 minutes and annealed overnight at 45°C in hybridization buffer. RNA digestion was performed at 30°C for 30 min using 0.08 kunitz units of RNase A and 24 units of RNase T1. RNase digestion and RNA extraction were performed as described by Bordonaro et al. (1994). The resulting protected fragments were heat denatured and separated on a 6% sequencing gel. The gel was transferred to paper, dried and exposed to a Molecular Dynamics Storm phosphorimager screen overnight. Band intensity was determined using the supplied software.

3. RESULTS AND DISCUSSION

The parental CHO-K1 and CHO reductase expressing CHO-K1 cell lines provided a matched pair for a preliminary determination of whether AKR expression would confer resistance to carbonyl containing anticancer drugs. As a preliminary test we employed the MTT chemosensitivity assay with the drug daunorubicin and observed almost a ten-fold difference in sensitivity to this drug (Figure 1). The CHO reductase expressing line was grown in the presence of verapamil and did not express elevated levels of P-glycoprotein (Inoue et al., 1993). Other forms of resistance cannot be ruled out but these promising results indicated that identifying the human orthologue of CHO reductase would be important in determining if it is overexpressed in cancer cells or could also confer resistance to carbonyl containing cancer drugs as determined by transfection experiments.

A human orthologue of CHO reductase was obtained from a small intestine cDNA pool. The complete HSI reductase sequence contained an open reading frame of 951 bp which encoded a 316-amino acid protein which is the same length as many other members of the AKR family. Multiple sequence alignment using CLUSTAL W (Horton et al., 1989) indicated that the deduced protein sequence was part of the AKR1B subfamily using the

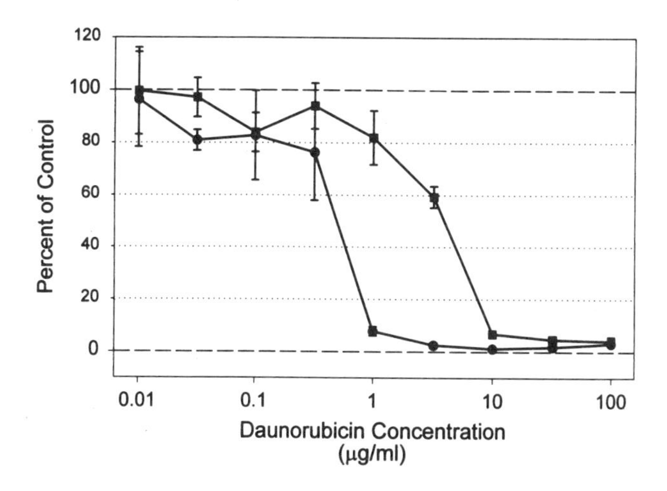

Figure 1. Effect of daunorubicin on the parental CHO-K1 (●——●) and CHO reductase overexpressing CHO-K1 (■——■) cell lines after a 4-day exposure. Points are mean ± SD of four determinations.

recently proposed nomenclature of Jez et al. (1997b). The new protein showed 82, 81, 79 and 70 percent sequence identity to CHO reductase (AKR1B9), FR-1 (AKR1B8), MVDP (AKR1B7) and human aldose reductase (AKR1B1) respectively (Figure 2). The translated protein sequence contained all the essential residues proposed for catalytic activity including Asp-44, Tyr-49, Lys-78 and His-111 (Jez et al., 1997a). Position 976 always sequenced as an A (major peak) and a G (minor peak) resulting in Asn or Asp at position 313 of the translated sequence. A search of the EST database resulted in a fragment match (c75075) in this region except for a G at position 976, suggesting allelic variation. Aspartic acid is present in this position in CHO reductase and FR-1 while histidine is present in MVDP and the ADR members. Also of note is the considerable variation of the sequences in the region 116–135 and near the C-terminal end of the protein (residues 299–308). These regions form the external folds for the substrate binding pocket (Wilson et al., 1993) and the variability in this region suggests that this enzyme will exhibit a unique substrate specificity profile even compared to its closest AKR1B homologues. After completion of this work two separate research groups published a complete and a partial sequence of this protein respectively (Cao et al., 1998; Scuric et al., 1998), confirming the results presented here but indicating the presence of aspartic acid at position 313.

```
                                                          *    *                     70
HSI red    MATFVELSTKAKMPIVGLGTWKSPLGKVKEAVKVAIDAGYRHIDCAYVYQNEHEVGEAIQEKIQEKAVKR
CHO red    -S-------------------Q--P-Q--------------------A-Y-------------K----R-
FR-1       ------------------------PNQ------A-------------A-C--N----------K----Q-
MVDP       --------------L--------SP-Q------A---------------H--N----------K-N----
hADR       --SRLL-NNG-----L--------P-Q-T--------V--------H-----N---V-----LR-QV---

                  *                                *                             140
HSI red    EDLFIVSKLWPTFFERPLVRKAFEKTLKDLKLSYLDVYLIHWPQGFKSGDDLFPKDDKGNAIGGKATFLD
CHO red    ------------C---K-LKE--Q---T----D---L--------LQP-KE------Q--VLTS-I----
FR-1       ------------C--KK-LKE--Q---T----D---L--------LQP-KE------Q-RILTS-T---E
MVDP       ----------A----KS--K---DN--S----D---L--V------QA-NA-L---N--KVLLS-S----
hADR       -E--------C-YH-KG--KG-CQ---S----D---L------T---P-KEF--L-ES--VVPSDTNI--

                                                                                 210
HSI red    AWEAMEELVDEGLVKALGVSNFSHFQIEKLLNKPGLKYKPVTNQVECHPYLTQEKLIQYCHSKGITVTAY
CHO red    ---V------------------N-----RI-------H-------------------E------------
FR-1       ---G------Q-----------N-----R--------H---------------------------S----
MVDP       ----------Q-------I---N-----R--------H------I-S-------------Q----A----
hADR       T-A-------------I-I---N-L-V-MI----------AV--I---------------Q----V----

                                                                                 280
HSI red    SPLGSPDRPWAKPEDPSLLEDPKIKEIAAKHKKTAAQVLIRFHIQRNVIVIPKSVTPARIVENIQVFDFK
CHO red    ------N---------------------------S-------------V-----------H--F-----Q
FR-1       ---------S---------------------E--S-------------V--------S--Q--------Q
MVDP       ---------Y-----VVM-I--------------V--------V----V--------S--Q--L-----Q
hADR       ----------------------R--A-----N--T-------PM---LV--------E--A--FK----E

                                          316
HSI red    LSDEEMATILSFNRNWRACNVLQSSHLEDYPFNAEY
CHO red    ---Q------G--------LLPETVNM-E--YD---
FR-1       -------------------LLPETVNM-E--YD---
MVDP       --E-D--A-----------DL-DARTE-----HE--
hADR       --SQD-T-L--Y-----V-AL-SCTSHK----HE-F
```

Figure 2. Comparison of the amino acid sequence of human small intestine reductase (HSI red) with other AKR members from the AKR1B subfamily: Chinese hamster ovary reductase (CHO red) (Hyndman et al., 1997) fibroblast growth factor 1-regulated protein (FR-1) (Donohue et al., 1994) mouse vas deferens protein (MVDP) (Pailhoux et al., 1990) and human aldose reductase (hADR) (Nishimura et al., 1990). Indicated by an asterisk are invariant residues involved in the catalytic mechanism. Indicated in bold at position 313 is the Asn residue that would be translated as Asp if the minor nucleotide sequence is used. The nucleotide sequence data for this protein translation has been deposited in Genbank with accession number AF052577.

For HSI reductase the strongest hybridization to the human RNA Master Blot was to adrenal gland with stomach, placenta, small intestine and pancreas at moderate levels and lower hybridization to all other tissues (Figure 3). Comparison to the other members of the subgroup indicated that the expression pattern was closest to that of MVDP (Lau et al., 1995) where high levels were observed in adrenal gland and moderate levels in the intestine. Cao et al. (1998) showed strong expression in colon and small intestine by Northern blot but did not test adrenal gland. It is of interest that we could detect specific products by RT-PCR in testis and fetal liver, but not in adult liver while Cao et al. (1998) also saw expression in testis and in liver but not in fetal liver.

For comparative purposes the dot-blot was reprobed with the human ADR riboprobe. Hybridization with this probe was even stronger to adrenal with most other tissues showing lower signals than with HSI reductase. It has been shown (Matsuura et al., 1996) that in the adrenal gland ADR is the major enzyme involved in the oxidoreduction of isocaproaldehyde formed from the side-chain cleavage of cholesterol in the first step of steroidogenesis. Most steroid hormone biosynthesis is carried out by the adrenals where in the human it is estimated that between 50 and 60 mg are produced per day (White et al., 1995). The present results suggest that HSI reductase may also play an important role, in addition to ADR, in the reductive metabolism of isocaproaldehyde.

Screening of a wide range of cancer cell lines was performed by ribonuclease protection assay in order to determine if either HSI reductase or ADR were overexpressed in these tumours. Overexpression could be part of the metastatic process in response to increased lipid peroxidation products or to the increased metabolic activity of neoplastic cells. In addition, knowledge of the relative expression of these enzymes in specific neoplasms could be of value if carbonyl containing anticancer drugs, such as daunorubicin or cyclophosphamide, are routinely used. In general, ADR expression was higher that HSI reductase in all the cell lines tested (Figure 4). Most notable was the higher expression of ADR in all the renal lines tested. This likely reflects the known physiological role of ADR in kidney osmolyte control (Bagnasco et al., 1987; Ferraris et al., 1996). HSI reductase expression was strong in one NSCL line (A549) and in one of the renal lines (A498). In a similar screening for aldehyde dehydrogenase activity Sreerama and Sladek (1997) showed that the NSCL line A549 had the highest of AlDH1 and AlDH3 levels and that the cytosolic fraction from this cell line had the highest ability to catalyse the oxidation of aldophosphamide.

The other sequences published (Scuric et al., 1998; Cao et al., 1998) were both obtained from hepatocellular carcinoma (HCC) samples. Scuric et al. (1998) found overexpression in 5 of 5 HCC samples screened by differential display and reverse Northern while Cao et al. (1998) showed overexpression of HSI reductase in 13 of 24 HCC samples tested by Northern analysis. ADR was overexpressed in 7 of 24 samples (Cao et al., 1998). These data in combination with the results presented here suggest that overexpression of HSI reductase, and ADR, is tissue specific with significant incidence of overexpression seen in liver carcinomas with other tissues showing elevated levels less frequently.

ACKNOWLEDGMENTS

The authors would like to thank Lilly Li for her expert technical assistance. This research was supported by grants from the Medical Research Council of Canada and the National Cancer Institute of Canada. Figures 2 and 3 used with permission of Elsevier Science B.V.

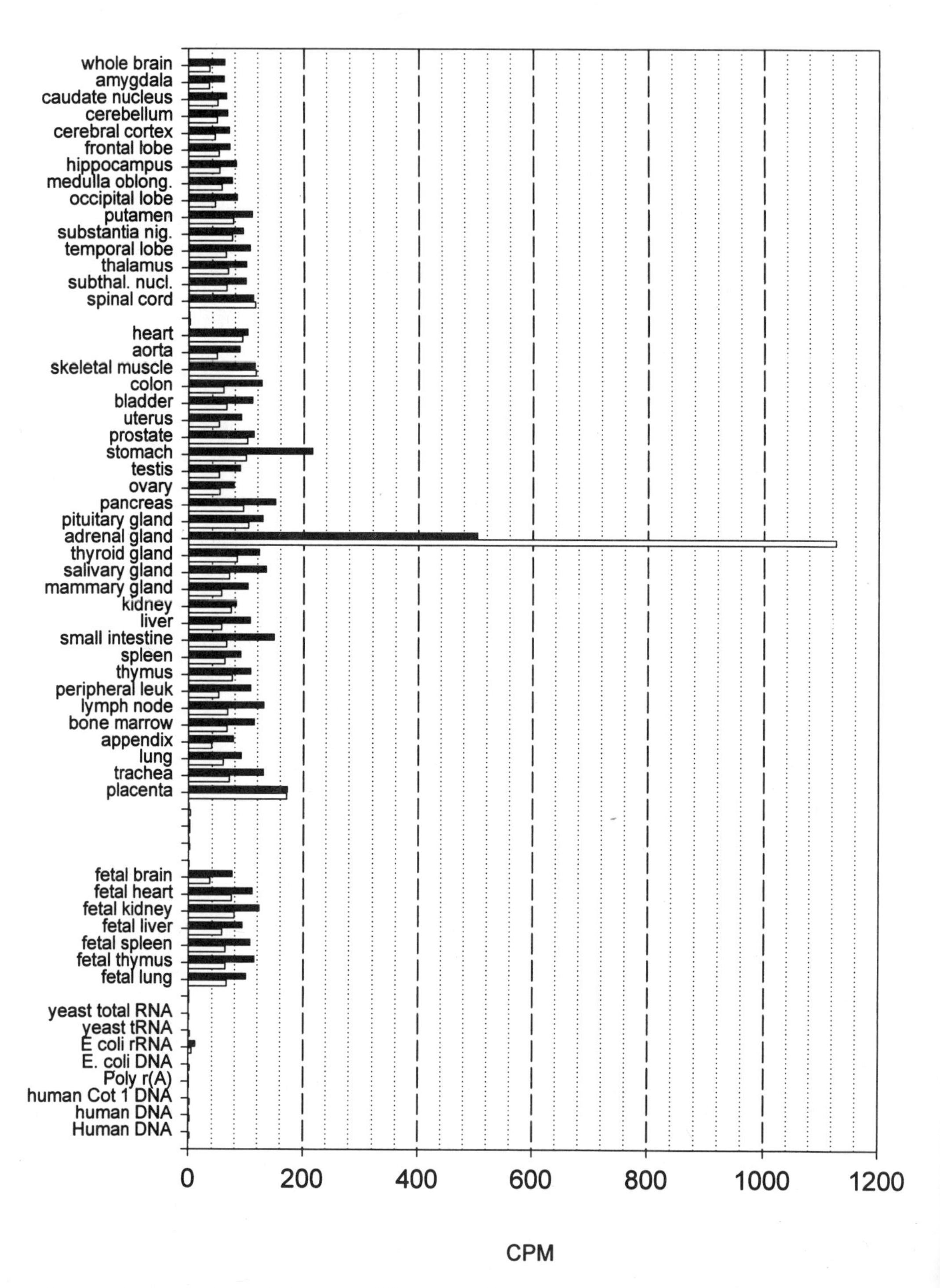

Figure 3. Histogram of the hybridization of HSI reductase (■) and human ADR (□) riboprobes to a Human RNA Master Blot (Clontech). Hybridization was 68 °C with a final stringency wash in 0.2 × SSC, 0.5% SDS for 30 min at 68 °C.

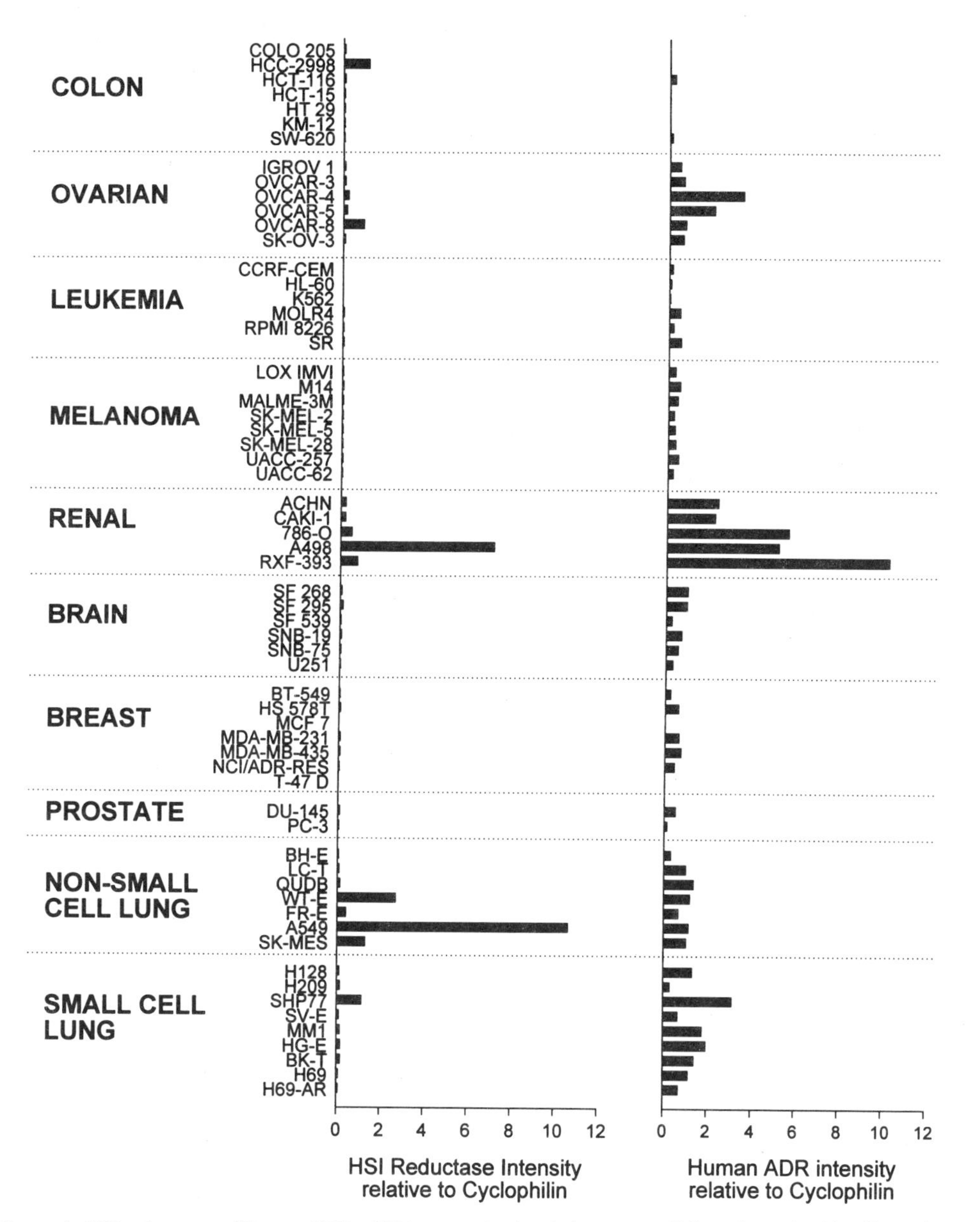

Figure 4. HSI reductase and human ADR mRNA expression levels in cancer cell lines determined by ribonuclease protection assay. Band intensity from the phosphorimager for HSI reductase or ADR was normalized for sample loading against cyclophilin.

REFERENCES

Bagnasco, S.M., Uchida, S., Balaban, R.S., Kador, P.F. and Burg, M.B. 1987, Induction of aldose reductase and sorbitol in renal inner medullary cells by elevated extracellular NaCl, *Proc. Natl. Acad. Sci. USA,* **84**, 1718–1720.

Bordonaro, M., Saccomanno, C.F. and Nordstrom, J.L. 1994, An improved T1/A ribonuclease protection assay, *BioTechniques,* **16**, 428–430.

Campling, B.G., Pym, J., Baker, H.M., Cole, S.P. and Lam, Y.M. 1991, Chemosensitivity testing of small cell lung cancer using the MTT assay, *Brit. J. Cancer,* **63**, 75–83.

Cao, D., Fan, S.T. and Chung, S.S. 1998, Identification and characterization of a novel human aldose reductase-like gene, *J. Biol. Chem.* **273**, 11429–11435.

Chen, Z. 1996, Simple modifications to increase specificity of the 5' RACE procedure, *Trends Gen.* **12**, 87–88.

Ciaccio, P.J., Jaiswal, A.K. and Tew, K.D. 1994, Regulation of human dihydrodiol dehydrogenase by Michael acceptor xenobiotics, *J. Biol. Chem.* **269**, 15558–15562.

Donohue, P.J., Alberts, G.F., Hampton, B.S. and Winkles, J.A. 1994, A delayed-early gene activated by fibroblast growth factor-1 encodes a protein related to aldose reductase. *J. Biol. Chem.* **269**, 8604–8609.

Ellis, E.M., Judah, D.J., Neal, G.E. and Hayes, J.D. 1993, An ethoxyquin-inducible aldehyde reductase from rat liver that metabolizes aflatoxin B1 defines a subfamily of aldo-keto reductases, *Proc. Natl. Acad. Sci. USA,* **90**, 10350–10354.

Ferraris, J.D., Williams, C.K., Jung, K.Y., Bedford, J.J., Burg, M.B. and Garcia-Perez, A. 1996, ORE, a eukaryotic minimal essential osmotic response element. The aldose reductase gene in hyperosmotic stress, *J. Biol. Chem.* **271**, 18318–18321.

Frohman, M.A. 1990, RACE: Rapid amplification of cDNA ends, in PCR protocols: A guide to methods and applications, (Innis, M.A., Gelfand, D.H., Sninsky, J.J. & White, T.J. eds.), pp. 28–38, Academic Press, Inc. San Diego.

Horton, R.M., Hunt, H.D., Ho, S.N., Pullen, J.K. and Pease, L.R. 1989, Engineering hybrid genes without the use of restriction enzymes: gene splicing by overlap extension, *Gene,* **77**, 61–68.

Hyndman, D.J., Takenoshita, R., Vera, N.L., Pang, S.C. and Flynn, T.G. 1997, Cloning, sequencing, and enzymatic activity of an inducible aldo-keto reductase from Chinese hamster ovary cells, *J. Biol. Chem.* **272**, 13286–13291.

Inoue, S., Sharma, R.C., Schimke, R.T. and Simoni, R.D. 1993, Cellular detoxification of tripeptidyl aldehydes by an aldo-keto reductase, *J. Biol. Chem.* **268**, 5894–5898.

Jez, J.M., Bennett, M.J., Schlegel, B.P., Lewis, M. and Penning, T.M. 1997a, Comparative anatomy of the aldo-keto reductase superfamily, *Biochem. J.* **326**, 625–636.

Jez, J.M., Flynn, T.G. and Penning, T.M. 1997b, A new nomenclature for the aldo-keto reductase superfamily, *Biochem. Pharmacol.* **54**, 639–647.

Lau, E.T., Cao, D., Lin, C., Chung, S.K. and Chung, S.S. 1995, Tissue-specific expression of two aldose reductase-like genes in mice: abundant expression of mouse vas deferens protein and fibroblast growth factor-regulated protein in the adrenal gland, *Biochem. J.* **312**, 609–615.

Matsuura, K., Deyashiki, Y., Bunai, Y., Ohya, I. and Hara, A. 1996, Aldose reductase is a major reductase for isocaproaldehyde, a product of side-chain cleavage of cholesterol, in human and animal adrenal glands, *Arch. Biochem. Biophys.* **328**, 265–271.

Nishimura, C., Matsuura, Y., Kokai, Y., Akera, T., Carper, D., Morjana, N., Lyons, C. and Flynn, T.G. 1990, Cloning and expression of human aldose reductase, *J. Biol. Chem.* **265**, 9788–9792.

Pailhoux, E.A., Martinez, A., Veyssiere, G.M. and Jean, C.G. 1990, Androgen-dependent protein from mouse vas deferens. cDNA cloning and protein homology with the aldo-keto reductase superfamily, *J. Biol. Chem.* **265**, 19932–19936.

Scuric, Z., Stain, S.C., Anderson, W.F. and Hwang, J.J. 1998, New member of aldose reductase family proteins overexpressed in human hepatocellular carcinoma, *Hepatology,* **27**, 943–950.

Sreerama, L. and Sladek, N.E. 1997, Class 1 and class 3 aldehyde dehydrogenase levels in the human tumor cell lines currently used by the National Cancer Institute to screen for potentially useful antitumor agents, *Adv. Exp. Med. Biol.* **414**, 81–94.

Takahashi, M., Fujii, J., Miyoshi, E., Hoshi, A. and Taniguchi, N. 1995, Elevation of aldose reductase gene expression in rat primary hepatoma and hepatoma cell lines: implication in detoxification of cytotoxic aldehydes, *Intl. J. Cancer,* **62**, 749–754.

Takahashi, M., Hoshi, A., Fujii, J., Miyoshi, E., Kasahara, T., Suzuki, K., Aozasa, K. and Taniguchi, N. 1996, Induction of aldose reductase gene expression in LEC rats during the development of the hereditary hepatitis and hepatoma, *Jap. J. Cancer Res.* **87**, 337–341.

Templeton, N.S., Urcelay, E. and Safer, B. 1993, Reducing artifact and increasing the yield of specific DNA target fragments during PCR-RACE or anchor PCR, *BioTechniques,* **15**, 48–50.

White, P.C., Pescovitz, O.H. and Cutler Jr., G.B. 1995, Synthesis and metabolism of corticosteroids, in Principles and practice of endocrinology and metabolism, (Becker, K.L. ed.), pp. 647–661, J.B. Lippincott, Philadelphia.

Wilson, D.K., Tarle, I., Petrash, J.M. and Quiocho, F.A. 1993, Refined 1.8 A structure of human aldose reductase complexed with the potent inhibitor zopolrestat. *Proc. Natl. Acad. Sci. USA,* **90**, 9847–9851.

Zeindl-Eberhart, E., Jungblut, P.R., Otto, A. and Rabes, H.M. 1994, Identification of tumor-associated protein variants during rat hepatocarcinogenesis. Aldose reductase. *J. Biol. Chem.* **269**, 14589–14594.

54

STRUCTURE-FUNCTION STUDIES OF FR-1

A Growth Factor-Inducible Aldo-Keto Reductase

J. Mark Petrash,[1] Theresa M. Harter,[1] Sanjay Srivastava,[2] Animesh Chandra,[2] Aruni Bhatnagar,[3] and Satish K. Srivastava[2]

[1]Department of Ophthalmology and Visual Sciences
Washington University School of Medicine
St. Louis, Missouri 63110
[2]Department of Human Biological Chemistry and Genetics
[3]Department of Physiology and Biophysics
University of Texas Medical Branch
Galveston, Texas 77550

1. INTRODUCTION

The aldo-keto reductase (AKR) gene superfamily represents a collection of proteins expressed in a wide variety of plants, animals, yeast, and procaryotic organisms. Most AKRs were originally identified as enzymes capable of catalyzing the NADPH-dependent reduction of carbonyl groups contained in a broad range of substrates (Bachur, 1976). However, recent genetic studies mediated by genome and expression sequencing approaches have identified several new members of the AKR superfamily. Many of these new proteins are characterized by high sequence homology to AKR enzymes although little or no information is available about their potential catalytic activities. One such new protein, designated FR-1* was identified as the product of a gene upregulated in serum-starved mouse fibroblasts following treatment with fibroblast growth factor I (FGF-I) (Donohue *et al.*, 1994). High amino acid sequence identity (~70%) was observed between FR-1 and aldose reductase as well as other AKRs. Many amino acid residues known to contribute to the catalytic mechanism in other AKR enzymes including aldose reductase (AKR1B1), aldehyde reductase (AKR1A1) and 3α-hydroxysteroid dehydrogenase (AKR1C9) are conserved in FR-1. These residues include Tyr-48, His-110, Lys-77 and Asp-43 (numbering is that of aldose reductase) (Barski *et al.*, 1995; Pawlowski & Penning, 1994; Schlegel *et al.*, 1998;

* According to a new system of AKR nomenclature, FR-1 is designated AKR1B8. Additional information about the AKR superfamily is maintained at a web site found at http://pharme26.med.upenn.edu

Tarle *et al.*, 1993). The present study was undertaken to evaluate whether FR-1 is a catalyst of carbonyl reduction and to measure the affinity of FR-1 for various ligands such as nucleotide cofactors, carbonyl substrates and aldose reductase inhibitors. Our studies show that FR-1 catalyzes the NADPH-dependent reduction of substrates representative of diverse structural classes of aliphatic and aromatic aldehydes. Both saturated and unsaturated aldehydes were excellent substrates. Unlike aldose reductase and aldehyde reductase, FR-1 catalyzed the reduction of simple ketones such as acetone and butanone; however virtually no catalytic activity could be detected using steroid and aldose substrates. FR-1 was inhibited by various aldose reductase inhibitors in a manner similar to human aldose reductase. Besides being an excellent substrate, 4-hydroxy-2-nonenal (HNE) inactivated the enzyme through a mechanism involving Michael addition to Cys-298.

2. MATERIALS AND METHODS

Recombinant FR-1 was over-expressed in bacterial cultures and purified essentially as described previously (Wilson *et al.*, 1995). Mutagenesis was carried out by PCR using commercially available kits (Clontech). Routine enzyme assays were carried out in 50 mM potassium phosphate, 100 mM potassium chloride, 0.1 mM EDTA, 0.15 mM NADPH. Where indicated, enzymes were treated with 100 mM DTT for 1 h, then rapidly desalted by chromatography.

Steady state kinetic parameters were obtained by fitting rate data to a general Michaelis-Menten equation. For K_m determinations, aldehyde substrates were varied over a concentration range covering from 0.2 to 7–10 times K_m. Initial velocity was measured at 6–8 different concentrations of substrate.

3. RESULTS AND DISCUSSION

3.1. Expression and Purification

FR-1 was obtained over-expression of the cDNA in *Escherichia coli* host cultures. The expression plasmid was constructed by ligating the cDNA (kindly provided by Jeffrey A. Winkles; Donohue *et al.*, 1994) into pMON 20,400, a plasmid vector used previously to over-express human aldose reductase (Tarle *et al.*, 1993). Expression was carried out in *E. coli* strain JM101 host cells contained in shaker flasks grown overnight following IPTG-induction. Host cells were lysed by treatment with DNAse and lysozyme as described previously (Merck *et al.*, 1992) and recombinant FR-1 was recovered from the supernatant following centrifugation. Cation exchange chromatography (MacroQ resin, BioRad) and affinity chromatography (AffiGel Red, Sigma) were used sequentially to obtain the enzyme in essentially homogeneous form. In some cases, purification by chromatofocusing was used instead of affinity chromatography. Quantities in excess of 10 mg were typically isolated from each liter of expression culture. The purified enzyme was usually stored at –80 °C with no apparent loss of activity after freezing.

3.2. Structure

Initial studies revealed that FR-1 catalyzed the NADPH-dependent reduction of simple aldehydes such as DL-glyceraldehyde (K_m of 0.9 ± 0.1 mM; Wilson *et al.*, 1995). Fluorescence quenching measurements at pH 7.0 showed that it bound nucleotide cofactors with high affinity (K_d values of 0.45± 0.03 μM and 0.23± 0.02 μM for NADPH and $NADP^+$ respectively) (Wilson *et al.*, 1995). Like aldose reductase, FR-1 was susceptible to inhibition by Zopolrestat, a new drug presently in clinical trials for prevention of diabetic complications (Mylari *et al.*, 1991). Zopolrestat bound tightly to FR-1 (K_d of 30 ± 10 nM in the presence of NADPH). These observations, together with the relatively high degree of sequence identity with aldose reductase, suggested that the structural organization of FR-1 was likely to be similar to that of aldose reductase. Indeed, a 1.7 Å resolution structure of the ternary complex containing FR-1, NADP(H) and Zopolrestat revealed the striking similarity of FR-1 with human aldose reductase (Wilson *et al.*, 1995). FR-1 adopts a $(\beta/\alpha)_8$ barrel structural fold with the active site located in a hydrophobic cavity at the C-terminal end of the β-barrel. Many interactions originally observed between the FR-1, the nucleotide cofactor and the inhibitor are similar to those previously observed with the similar ternary complex involving human aldose reductase (Wilson *et al.*, 1993). A major deviation between the FR-1 and aldose reductase structures was localized to a loop near the C-terminal end of the proteins (residues 295–315) which results in a somewhat more open active site cavity. Given the striking overall similarity in structure between aldose reductase and FR-1 and the lack of information about AKR enzymes in murine tissues, we next sought to evaluate the kinetic properties of FR-1.

3.3. Initial Steady State Kinetics

We examined a broad range of aldoses, aliphatic and aromatic aldehydes as potential substrates for FR-1. Table 1 reports the results for some aldoses, ketones and simple aromatic aldehydes in terms of K_m and catalytic efficiency (k_{cat}/K_m) and provides comparative data for human aldose reductase. Some major trends are evident from this data set. First, it is clear that FR-1 is not functionally orthologous to aldose reductase studied from other species. Activity of FR-1 with aldose sugars was markedly reduced in comparison with aldose reductase. Indeed, reactivity of FR-1 with glucose and galactose was virtually indistinguishable from background whereas kinetic constants for these substrates were readily determined with aldose reductase. In comparison with aldose reductase, FR-1 was found to be a poor catalyst of steroid reduction. The catalytic efficiency with 17α-hydroxyprogesterone as substrate was over 5 orders of magnitude better for aldose reductase than for FR-1. Conversely, activity of aldose reductase with short chain ketones such as acetone and butanone was not detectable whereas FR-1 activity with these compounds was readily measured.

Table 1. Kinetic constants for FR-1 and aldose reductase

	FR1		Aldose reductase	
	K_m (mM)	K_{cat}/K_m (mM^{-1} min^{-1})	K_m (mM)	K_{cat}/K_m (mM^{-1} min^{-1})
DL-glyceraldehyde	0.965	22	0.057	1434
p-nitrobenzaldehyde	0.018	1387	0.022	636
Acetone	527	0.001	NDA	
Butanone	149	0.009	NDA	
Galactose	786	0.008	91	0.6
Glucose	Trace	–	267	0.1
Xylose	248	0.022	16	4
17α OH progesterone	236	0.006	0.078	102

Table 2. Kinetic constants for aliphatic aldehyde reduction by FR-1

	K_m (mM)	K_{cat}/K_m (mM^{-1} min^{-1})
Propanal	2.9	6
Butanal	0.50	47
Pentanal	0.05	365
Hexanal	0.006	2067
Nonanal	0.009	3682
Trans-2-nonenal	0.010	2062
4-Hydroxy-*trans*-2-nonenal	0.009	2494
Trans,trans 2,4-nonadienal	0.010	4021

The substrate preference for FR-1 is directed toward hydrophobic compounds (Table 2). Both aromatic and aliphatic aldehydes were excellent substrates with K_m values generally in the low micromolar concentration range. As the chain length increased from 3 carbons (propanal) to 9 carbons (nonanal), the catalytic efficiency rose over 600-fold from 6 $mM^{-1}min^{-1}$ to over 3600 $mM^{-1}min^{-1}$. Kinetic constants for the C9 alkanal (nonanal) were modestly affected with structural elaboration of the hydrocarbon chain. For example, introduction of an α,β unsaturated bond as in *trans*-2-nonenal had no effect on K_m and reduced the catalytic efficiency slightly. Further elaboration by addition of a hydroxyl group at the 4th carbon as in 4-hydroxy-*trans*-2-nonenal made no further change in either K_m or k_{cat}/K_m. Introduction of a second double bond as in *trans, trans* 2,4-nonadienal also had no effect on K_m but returned the catalytic efficiency back to levels observed with the saturated C9 alkanal (nonanal).

3.4. Inactivation of FR-1 by HNE

Although 4-hydroxy-*trans*-2-nonenal (HNE) was an excellent substrate when combined with FR-1 in the presence of NADPH, it caused inactivation of the apoenzyme in a time- and concentration-dependent manner. Approximately 70% of initial activity was lost after 15 min exposure of the enzyme to 100 μM HNE (Figure 1). A biphasic inactivation pattern was observed at all HNE concentrations examined. Complete inactivation was observed following prolonged incubation with HNE. FR-1 was inactivated by other unsaturated aldehydes as well. Complete inactivation was observed when the apoenzyme was incubated for 60 min in the presence of 4-hydroxyoctenal and *trans*-2-nonenal. However, activity levels were not significantly altered following incubation with saturated aldehydes such as nonanal (100 μM, 60 min).

Substantial protection against HNE inactivation was provided by NADP (Figure 2). In the absence of NADP, approximately 50% initial activity is lost within 20 min when FR-1 was treated with 40 μM HNE. However, addition of 50 μM NADP completely blocked this activity loss.

In an effort to establish the stoichiometry of HNE binding to FR-1, we carried out electrospray ionization-mass spectrometry (ESI-MS) of the enzyme before and after inactivation with HNE. The spectrum of untreated FR-1 showed a well-defined peak corresponding to a molecular mass of 35,989.5 Da, a value in good agreement with the predicted mass (35,989.4 Da) from the primary sequence. ESI-MS analysis of FR-1 after treatment with HNE revealed two major species corresponding to a molecular mass of 35,984.9 Da and 36,143.0 Da. The mass difference between these species (158.1 Da)

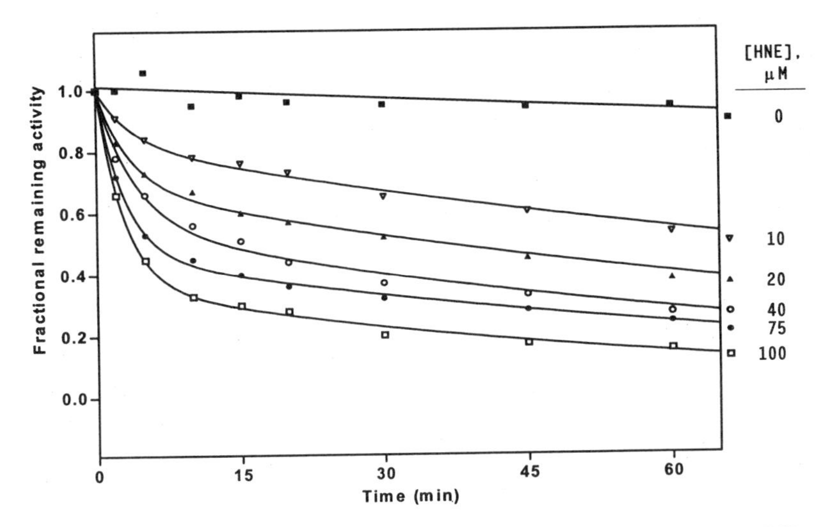

Figure 1. Inactivation of FR-1 by 4-hydroxy-*trans*-2-nonenal. FR-1 (1 μM) was treated with HNE at the indicated concentration in nitrogen-purged 0.1 M potassium phosphate, pH 7.0. Aliquots were removed at the indicated time intervals and assayed for enzyme activity using 60 mM DL-glyceraldehyde as substrate.

agrees well with the theoretical prediction for addition of one mole HNE (156 Da) per mole FR-1. Therefore, we conclude that HNE-inactivation involves binding of one mole HNE per mole enzyme.

3.5. Role of Cys-298 in HNE-Inactivation

To identify the HNE binding site on FR-1, we treated the enzyme with 4-[^{3}H]-HNE and subjected the modified protein to peptide mapping. Fragments resulting from exhaustive trypsin digestion of the ^{3}H-HNE modified protein were resolved by C18-HPLC. Scintillation counting revealed a single radioactive peak in the column eluate. Sequential

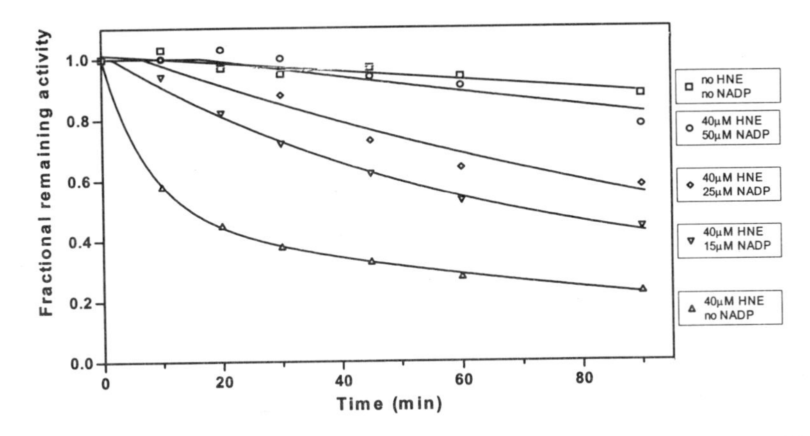

Figure 2. NADP protection of FR-1 against HNE-induced inactivation. FR-1 (1 μM) was incubated in the presence of HNE and/or NADP at the indicated concentrations. Aliquots were removed at the indicated time intervals and assayed for enzyme activity using 60 mM DL-glyceraldehyde as substrate.

Edman degradation of this material yielded the following sequence: Ala-Unidentified-Leu-Leu-Pro-Glu-Thr. Alignment of this sequence with the FR-1 primary structure showed that the peptide mapped to the C-terminal region of FR-1. All residues identified by Edman degradation perfectly matched the corresponding FR-1 sequence. The unidentified amino acid derivative in the second cycle of Edman degradation corresponds to Cys-298 in the FR-1 sequence.

Previous studies have shown that α,β unsaturated aldehydes such as HNE modify proteins through electrophilic reaction with Cys residues (Witz, 1989). FR-1 has six Cys residues, including Cys-298 which is located near the cofactor binding site in the active site cavity (Wilson *et al.*, 1995). Cys-298 in human aldose reductase plays an important role in regulating the catalytic rate and binding affinity to some aldose reductase inhibitors (Grimshaw *et al.*, 1995b; Petrash *et al.*, 1992). Due to its location in a solvent exposed opening near the active site, Cys-298 seemed a likely candidate as a binding target for HNE. To test this hypothesis, we created a Cys ºSer mutation at this site (C298S) and measured the susceptibility of the mutant protein to HNE inactivation. Essentially no loss in activity was observed following incubation of the C298S protein with HNE (100 μM) in the presence or absence of NADP. Furthermore, ESI-MS showed that HNE-treatment of the mutant enzyme had essentially no effect on the mass of the protein.

These results provide strong evidence that the mechanism of HNE-inactivation of FR-1 involves covalent modification of Cys-298. Since HNE-treatment of the C298S mutant did not result in a meaningful change in mass as judged by ESI-MS, it seems unlikely that an alternate site exists for HNE binding. It is unlikely that the Cys ºSer mutation resulted in structural reorganization of an allosteric HNE binding site. Steady state kinetic constants measured with the C298S mutant with a variety of substrates were not substantially different as compared to wild type. Crystallography of the ternary FR-1complex with cofactor and inhibitor showed that the Cys298 Sγ is located relatively close to the nicotinamide ring of NADP (Wilson *et al.*, 1995). Although it has not been functionally demonstrated as yet, the crystal structure revealed that cofactor binding and release requires movement of a large surface loop. It seems likely that on formation of the binary complex of FR-1 with NADP, the solvent accessibility of the Cys-298 Sγ becomes reduced relative to its accessibility in the apoenzyme. This would be expected to result in protection from modification by HNE. This effect is not unique to FR-1, as Del Corso *et al* as well as Srivastava *et al* recently reported that NADP protects against HNE-inactivation of bovine aldose reductase and that the inactivation probably occurs through modification of Cys-298 (Del Corso *et al.*, 1998; Srivastava *et al.*, 1998).

As shown in Figure 1, loss of FR-1 activity on incubation with HNE followed a biphasic pattern. Indeed, the rate of inactivation was found to fit very well to a biexponential function. The observed rate constants and intercept values indicate that HNE causes a rapid initial phase of inactivation followed by a subsequent slower inactivation phase. A hyperbolic dependence on HNE was observed on a replot of the observed rates of inactivation. We interpret this to be consistent with the formation of a dissociable enzyme-HNE complex before irreversible inactivation. We propose that rapid binding of HNE to the hydrophobic active site represents the initial rapid phase of inactivation. A slow conformational change, which leads to complete inactivation, represents the second phase of the inactivation process. Conformational changes induced by ligand binding have been well described in the AKR literature. As mentioned above, a large conformational change involving movement of a large loop occurs on binding and release of nucleotide cofactors (Borhani *et al.*, 1992; Grimshaw *et al.*, 1995a; Grimshaw *et al.*, 1995b; Wilson *et al.*, 1992). Association of Zopolrestat with the binary aldose reductase?NADP(H) complex induces a slight conformational change to accommodate the inhibitor (Wilson *et al.*, 1993). Therefore, it would not be surprising if a slow conformational change were required to bring the HNE and Cys-298 side chain into proper orientation to effect covalent modification of the Cys-298 Sγ.

3.6. Potential Physiological Role of FR-1 and Other AKRs

Long chain hydrophobic and α,β unsaturated aldehydes are produced during lipid peroxidation and oxidative stress (Esterbauer *et al.*, 1991) and as byproducts of food and drug metabolism (Witz, 1989). Although direct measurement of cytosolic HNE levels is difficult to achieve, it is estimated that concentrations well above the K_m values observed in our study are encountered *in vivo* (Esterbauer *et al.*, 1991). Thus, FR-1 appears to be well suited to function as a reductase in the detoxification of dietary and stress-related aldehydes. If necessary for dietary aldehyde detoxification, it should follow that gastrointestinal tissues would be a likely source of FR-1 expression. Examination of the tissue distribution of FR-1 gene expression in the adult mouse is ambiguous on this point. Winkles and coworkers report very high levels of FR-1 in adult mouse intestine (Donohue *et al.*, 1994) whereas FR-1 was virtually undetectable in the intestine in a study reported by Chung and colleagues (Lau *et al.*, 1995). Further study will certainly be needed to clarify this important discrepancy. Complementary studies from these laboratories revealed high transcript levels in adult mouse testis, ovary, adrenal gland, heart and intestine (Donohue *et al.*, 1994; Lau *et al.*, 1995). This pattern of expression is similar but not identical to that observed for MVDP, the mouse vas deferens protein (AKR 1B7) characterized by high sequence homology to FR-1 and aldose reductase (Lau *et al.*, 1995). The potential catalytic function and physiological role of MVDP has not been established to date. Why either of these proteins would be expressed at high levels in the adrenal gland, which also is one of the richest sources of aldose reductase (Grimshaw & Mathur, 1989), is not known.

FR-1 was initially discovered as a protein upregulated in quiescent mouse fibroblasts following stimulation with fibroblast growth factor I (Donohue *et al.*, 1994). While it may have been the first known growth factor-inducible AKR, it no longer holds this distinction. Recent studies have shown that both FR-1 and aldose reductase respond differently to various growth-promoting agents *in vitro* and that aldose reductase appears unique among AKRs in responding to osmoregulatory signals (Hsu *et al.*, 1997). New members of the AKR superfamily have been identified on the basis of their induction in tissue culture cell lines. Thus, the CHO reductase induced in a chinese hamster ovary cell line characterized by its insensitivity to the toxic effects of an aldehyde selection agent (Inoue *et al.*, 1993) (Hyndman *et al.*, 1997) is a functional aldo-keto reductase. Overexpression of one or more aldo-keto reductases has also been observed in human cancer cell lines (Cao *et al.*, 1998; Scuric *et al.*, 1998). Whether AKRs, exemplified by FR-1, both respond to and participate in aspects of cell signaling or cell cycle progression promises to be the topic of future study.

ACKNOWLEDGMENTS

Support for this work was provided by NIH grants EY05856, P30EY02687, HL59378, DK20579 and DK36118. Additional support to the Department of Ophthalmology and Visual Sciences from Research to Prevent Blindness, Inc. is gratefully acknowledged. The authors wish to thank Terry Griest and Jen Chyong-Wang for technical assistance.

REFERENCES

Bachur, N.R. (1976). Cytoplasmic Aldo-Keto Reductases: A Class of Drug Metabolizing Enzymes. *Science* **193**, 595–597.

Barski, O.A., Gabbay, K.H., Grimshaw, C.E. & Bohren, K.M. (1995). Mechanism of Human Aldehyde Reductase-Characterization of the Active Site Pocket. *Biochemistry* **34**, 11264–11275.

Borhani, D.W., Harter, T.M. & Petrash, J.M. (1992). The crystal structure of the aldose reductase.NADPH binary complex. *J.Biol.Chem.* **267**, 24841–24847.

Cao, D., Fan, S.T. & Chung, S.S.M. (1998). Identification and Characterization of a Novel Human Aldose Reductase-like Gene. *J.Biol.Chem.* **273**, 11429–11435.

Del Corso, A., Dal Monte, M., Vilardo, P.G., Cecconi, I., Moschini, R., Banditelli, S., Cappiello, M., Tsai, L. & Mura, U. (1998). Site-specific inactivation of aldose reductase by 4-hydroxynonenal. *Arch.Biochem.Biophys.* **350**, 245–248.

Donohue, P.J., Alberts, G.F., Hampton, B.S. & Winkles, J.A. (1994). A delayed-early gene activated by fibroblast growth factor-1 encodes a protein related to aldose reductase. *J.Biol.Chem.* **269**, 8604–8609.

Esterbauer, H., Schaur, R.J. & Zollner, H. (1991). Chemistry and biochemistry of 4-hydroxynonenal, malonaldehyde and related aldehydes. [Review] [400 refs]. *Free Radical Biol.Med.* **11**, 81–128.

Grimshaw, C.E., Bohren, K.M., Lai, C.J. & Gabbay, K.H. (1995a). Human aldose reductase - rate constants for a mechanism including interconversion of ternary complexes by recombinant wild-type enzyme. *Biochemistry* **34**, 14356–14365.

Grimshaw, C.E., Bohren, K.M., Lai, C.J. & Gabbay, K.H. (1995b). Human aldose reductase - subtle effects revealed by rapid kinetic studies of the c298a mutant enzyme. *Biochemistry* **34**, 14366–14373.

Grimshaw, C.E. & Mathur, E.J. (1989). Immunoquantitation of aldose reductase in human tissues. *Anal.Biochem.* **176**, 66–71.

Hsu, D.K., Guo, Y., Peifley, K.A. & Winkles, J.A. (1997). Differential control of murine aldose reductase and fibroblast growth factor (FGF)-regulated-1 gene expression in NIH 3T3 cells by FGF-1 treatment and hyperosmotic stress. *Biochem.J.* **328**, 593–598.

Hyndman, D.J., Takenoshita, R., Vera, N.L., Pang, S.C. & Flynn, T.G. (1997). Cloning, sequencing, and enzymatic activity of an inducible aldo-keto reductase from Chinese hamster ovary cells. *J.Biol.Chem.* **272**, 13286–13291.

Inoue, S., Sharma, R.C., Schimke, R.T. & Simoni, R.D. (1993). Cellular Detoxification of Tripeptidyl Aldehydes by an Aldo-keto Reductase. *J.Biol.Chem.* **268**, 5894–5898.

Lau, E.T., Cao, D., Lin, C., Chung, S.K. & Chung, S.S. (1995). Tissue-specific expression of two aldose reductase-like genes in mice: abundant expression of mouse vas deferens protein and fibroblast growth factor-regulated protein in the adrenal gland. *Biochem J* **312**, 609–615.

Merck, K.B., De Haard Hoekman, W.A., Oude Essink, B.B., Bloemendal, H. & de Jong, W.W. (1992). Expression and aggregation of recombinant alpha A-crystallin and its two domains. *Biochim.Biophys.Acta* **1130**, 267–276.

Mylari, B.L., Larson, E.R., Beyer, T.A., Zembrowski, W.J., Aldinger, C.E., Dee, M.F., Siegel, T.W. & Singleton, D.H. (1991). Novel, potent aldose reductase inhibitors: 3,4-dihydro- 4-oxo-3- [[5-(trifluoromethyl)-2-benzothiazolyl] methyl]- 1- phthalazineacetic acid (zopolrestat) and congeners. *J.Med.Chem.* **34**, 108–122.

Pawlowski, J.E. & Penning, T.M. (1994). Overexpression and mutagenesis of the cDNA for rat liver 3 alpha- hydroxysteroid/dihydrodiol dehydrogenase. Role of cysteines and tyrosines in catalysis. *J.Biol.Chem.* **269**, 13502–13510.

Petrash, J.M., Harter, T.M., Devine, C.S., Olins, P.O., Bhatnagar, A., Liu, S.Q. & Srivastava, S.K. (1992). Involvement of cysteine residues in catalysis and inhibition of human aldose reductase - site-directed mutagenesis of cys-80, cys-298, and cys-303. *J.Biol.Chem.* **267**, 24833–24840.

Schlegel, B.P., Jez, J.M. & Penning, T.M. (1998). Mutagenesis of 3 alpha-hydroxysteroid dehydrogenase reveals a "push-pull" mechanism for proton transfer in aldo-keto reductases. *Biochemistry* **37**, 3538–3548.

Scuric, Z., Stain, S.C., Anderson, W.F. & Hwang, J.J. (1998). New member of aldose reductase family proteins overexpressed in human hepatocellular carcinoma. *Hepatology* **27**, 943–950.

Srivastava, S., Chandra, A., Ansari, N.H., Srivastava, S.K. & Bhatnagar, A. (1998). Identification of cardiac oxidoreductase(s) involved in the metabolism of the lipid peroxidation-derived aldehyde-4-hydroxynonenal. *Biochem.J.* **329**, 469–475.

Tarle, I., Borhani, D.W., Wilson, D.K., Quiocho, F.A. & Petrash, J.M. (1993). Probing the active site of human aldose reductase. Site-directed mutagenesis of Asp-43, Tyr-48, Lys-77, and His-110. *J.Biol.Chem.* **268**, 25687–25693.

Wilson, D.K., Bohren, K.M., Gabbay, K.H. & Quiocho, F.A. (1992). An unlikely sugar substrate site in the 1.65 A structure of the human aldose reductase holoenzyme implicated in diabetic complications. *Science* **257**, 81–84.

Wilson, D.K., Nakano, T., Petrash, J.M. & Quiocho, F.A. (1995). 1.7 A structure of FR-1, a fibroblast growth factor-induced member of the aldo-keto reductase family, complexed with coenzyme and inhibitor. *Biochemistry* **34**, 14323–14330.

Wilson, D.K., Tarle, I., Petrash, J.M. & Quiocho, F.A. (1993). Refined 1.8 A structure of human aldose reductase complexed with the potent inhibitor zopolrestat. *Proc Natl Acad Sci U S A* **90**, 9847–9851.

Witz, G. (1989). Biological interactions of alpha,beta-unsaturated aldehydes. [Review] [114 refs]. *Free Radical Biol.Med.* **7**, 333–349.

4-HYDROXYNONENAL METABOLISM BY ALDO/KETO REDUCTASE IN HEPATOMA CELLS

Giuliana Muzio,[1] Raffaella A. Salvo,[1] Naoyuki Taniguchi,[2]
Marina Maggiora,[1] and Rosa A. Canuto[1]

[1]Dip. Scienze Cliniche e Biologiche
Università di Torino
Ospedale S. Luigi
Orbassano-Torino, Italy
[2]Department of Biochemistry
Osaka University Medical School, Japan

1. INTRODUCTION

Aldehydic products derived from lipid peroxidation of ω-6 polyunsaturated fatty acids (PUFA) present in biomembranes exhibit cytotoxic or stimulatory effects on cellular functions in relation to the concentration reached in the cells (Esterbauer, 1985; Spitz et al., 1990). In particular, 4-hydroxynonenal (HNE), one of the most reactive metabolites of PUFA oxidative breakdown, has been shown to cause non specific cytotoxic effects at concentrations above 100 μM (supraphysiological) by inhibiting DNA and protein synthesis (Poot et al., 1988), and mitochondrial respiration (Canuto et al., 1985). At concentrations between 1 and 20 μM the effect of HNE seems to be more specific: it inhibits c-myc expression (Barrera et al., 1991) and stimulates chemotactic activity towards rat neutrophils (Curzio et al., 1987; Schauer et al., 1994). The stimulatory effect has also been reported at concentrations below 1 μM, i.e. it acts on the generation of second cellular messengers by stimulating adenylate cyclase (Paradisi et al., 1985) and phospholipases (Rossi et al., 1988; Rossi et al., 1990). HNE is also important in controlling cell proliferation in both normal and neoplastic cells (Canuto et al., 1995; Cambiaggi et al., 1997); it has recently been shown that its effect is also related to culture conditions, in particular to the presence of serum growth factors (Kreuzer et al., 1998).

Tumor cells, and in particular hepatoma cells, are characterized by a low susceptibility to lipid peroxidation (Canuto et al., 1991; Canuto et al., 1995), which means that very low concentrations of potentially cytostatic and cytotoxic aldehydes are reached.

Enzymology and Molecular Biology of Carbonyl Metabolism 7
edited by Weiner *et al.* Kluwer Academic / Plenum Publishers, New York, 1999.

In parallel, hepatoma cells also show an increase in some enzymes involved in aldehyde metabolism, such as aldehyde dehydrogenase (Lindahl, 1992; Canuto et al. 1989), aldose reductase (Takahashi et al., 1995) and glutathione transferase (Tsuchida and Sato, 1992; Pitot, 1996), potentially a further opportunity for the cells to avoid the antiproliferative action of aldehydes gererated endogenously during lipid peroxidation.

Aldo/keto reductases (AKR) are a family of isoenzymes, including aldehyde reductase and aldose reductase, metabolizing a wide range of substrates including aliphatic aldehydes, monosaccharides, steroids and prostaglandins (Jez et al., 1997). It has been shown that HNE is a good substrate for purified aldose reductase of bovine lens (Srivastava et al., 1995; Vander Jagt et al., 1995) and that it is able to induce transcription and expression of this enzyme (Spycher et al., 1996). On the contrary, it has recently been reported that AR is inactivated by HNE in a time- and concentration-dependent manner, and that the inhibition can be prevented by thiol-reducing treatment (Del Corso et al., 1998). Previous reports had shown that, in the AKR family, aldose reductase is overexpressed in rat hepatoma, whereas aldehyde reductase is lower than in normal tissues (Takahashi et al., 1995). A novel human protein with a high homology with AR has recently been found to be overexpressed in a high percentage of liver cancers (Cao et al., 1998).

In the research reported here, we compared the capability of different hepatoma cells, cultured or derived from an experimental hepatocarcinogenesis protocol, to metabolize HNE by AKR, both in basal conditions and during culture. We also analyzed the effect of endogenously restored lipid peroxidation on the activity and the expression of aldo/keto reductase. Increased levels of lipid peroxidation were obtained by enriching hepatoma cells with arachidonic acid and by exposing them to a prooxidant system.

2. MATERIALS AND METHODS

2.1. Basal Culture Conditions

Rat hepatoma cells were grown routinely under an atmosphere of 5% $C0_2$ and 95% air, as monolayer at 37°C in DMEM/F12 medium supplemented with 10% newborn calf serum, 1% antibiotic/antimycotic solution and 2 mM glutamine.

2.2. Enrichment of Hepatoma Cells with Arachidonic Acid

7777 and JM2 hepatoma cell lines were seeded (day 0) and maintained for 24h in medium A [DMEM/F12 plus 2 mM glutamine and 1% antibiotic/antimycotic solution] plus 10% newborn calf serum; 24h later (day 1) they were put into medium B [medium A plus 0.4% albumin, 1% ITS (insulin, transferrin, sodium selenite), 1% non essential amino acid, 1% vitamin solution] supplemented or not with arachidonic acid (50 μM); 24h later again (day 2), the cells were put in unsupplemented medium B and given 4 doses of prooxidant (500 μM ascorbate/100 μM iron sulphate) at 12h intervals, then cultured for a further two days (day 3 and 4). The cells were harvested at the following times: 24h after the enrichment with arachidonic acid, 12h after the 2nd dose of ascorbate/iron sulphate, 4h after the 3rd dose of ascorbate/iron sulphate for 7777 hepatoma cells and 12h after the 4th dose of ascorbate/iron sulphate for JM2 hepatoma cells.

2.3. Hepatocarcinogenesis Experimental Protocol

Nodules and hepatoma were obtained in male F-344 rats following Roomi's technique (Roomi et al., 1985).

2.4. Preparation of Cytosolic Fractions

Cytosolic fractions from cultured hepatoma cells were obtained as previously decribed (Canuto et al., 1994), those from nodules and hepatoma were obtained as previously decribed (Canuto et al., 1993)

2.5. Enzyme Activity Determination

Aldo/keto reductase activity was determined as reported in Canuto et al. (1993).

2.6. Western Blot Analysis

The cells were homogenized in a lysing buffer and used for Western blot analysis, as decribed in Canuto et al. (1996); antibodies against aldose reductase and aldehyde reductase were used.

2.7. Statistical Analysis

All data were expressed as mean ± S.D.. The significance of differences between group means was assessed by variance analysis, followed by the Neuman-Keuls test.

3. RESULTS AND DISCUSSION

Figure 1 reports the capability of different hepatoma cells to metabolize HNE by AKR. This capability is directly correlated with the degree of deviation: AKR activity is higher in the more deviated HTC and JM2 cells than in the less deviated 7777 and MH_1C_1 cells, being similar to that of normal hepatocytes in the latter. The correlation between degree of deviation and increased AKR activity is confirmed by the data obtained during experimental hepatocarcinogenesis (Figure 2). The capability to metabolize HNE increases 3 times in preneoplastic nodules and 7 times in hepatoma in comparison with normal liver.

A correlation between AKR activity and cell proliferation was also evidenced by the trend of enzyme activity during the culture-time in medium supplemented with 10% bovine serum (Figure 3). In 7777 cells, the enzyme activity increased earlier and declined when cells reached confluence (day 4). In JM2 cells, the enzyme activity continued to increase and reached the highest value at day 4, probably because JM2 cells have greater growth capability than 7777 cells. The activity of JM2 cells was higher than in 7777 cells at all times.

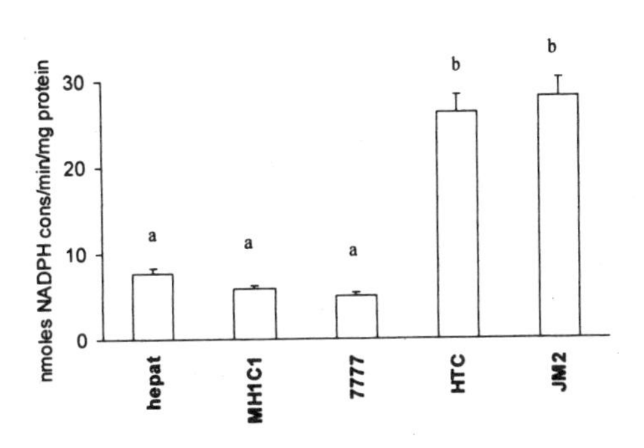

Figure 1. Aldo/keto reductase activity in cytosol from normal hepatocytes and hepatoma cell lines. Data are expressed as nmoles of NADPH consumed/min/mg of protein and represent mean ± S.D. of 5 experiments. 0.1 mM 4-hydroxynonenal was used as substrate. Means with different letters are statistically different ($p<0.001$) from one another as determined by variance analysis followed by the Neuman-Keuls test.

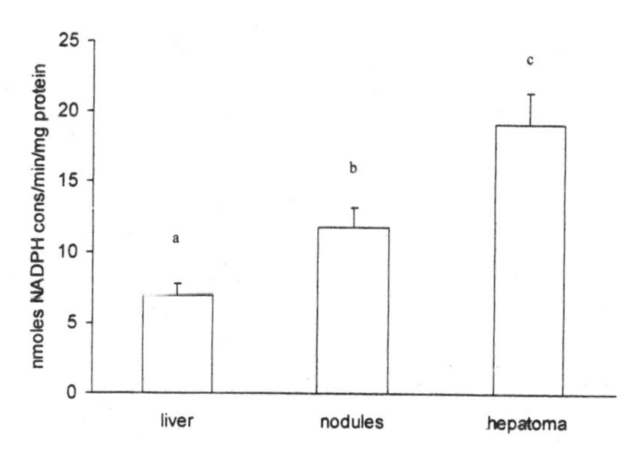

Figure 2. Aldo/keto reductase activity during chemically-induced hepatocarcinogenesis. Data are expressed as nmoles of NADPH consumed/min/mg of protein and represent mean ± S.D. of 5 experiments. 0.1 mM 4-hydroxynonenal was used as substrate. Means with different letters are statistically different ($p<0.001$) from one another as determined by variance analysis followed by the Neuman-Keuls test.

It has been reported that a change in AKR isoenzyme expression takes place during tumor development; in less deviated hepatoma cells, aldehyde reductase is more expressed than aldose reductase, whereas in more deviated hepatoma cells aldose reductase predominates (Takahashi et al., 1995). In the light of this finding, we also determined the content of aldose reductase and aldehyde reductase during the growth of 7777 and JM2 cells maintained in 10% serum supplemented medium by Western Blot analysis (Figure 4). We confirmed that in the more deviated cells, there was a shift from aldehyde reductase to aldose reductase expression, whereas no change in protein content was evident during the growth of 7777 and JM2 cells.

Since we had previously demonstrated that the products of restored lipid peroxidation in hepatoma cells inhibit the activity of class 3 aldehyde dehydrogenase, we also wanted to examine the effect of these products on AKR activity. For this purpose we enriched 7777 and JM2 cells with arachidonic acid and exposed them to an ascorbate/iron sulphate prooxidant system.

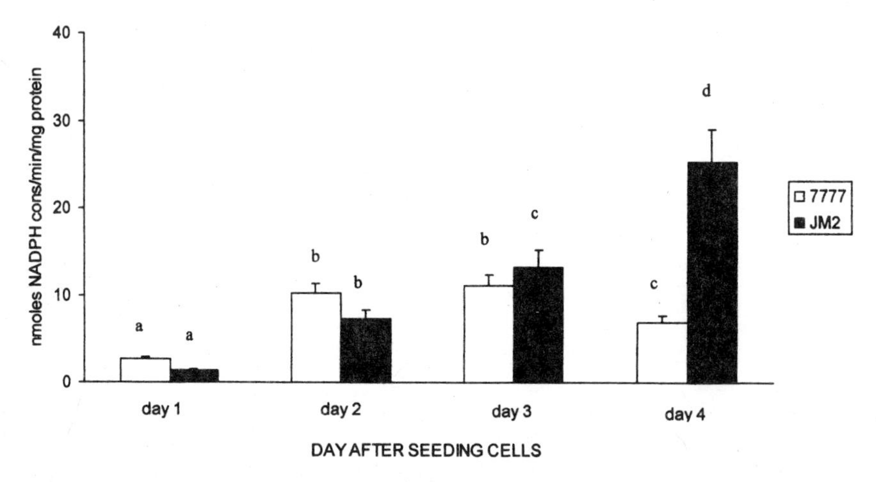

Figure 3. Aldo/keto reductase activity during culture-time of 7777 and JM2 hepatoma cell lines. Hepatoma cell lines were grown in 10% serum supplemented medium (see Materials and Methods section). Data are expressed as nmoles of NADPH consumed/min/mg of protein and represent mean ± S.D. of 5 experiments. 0.1 mM 4-hydroxynonenal was used as substrate. Means with different letters are statistically different ($p<0.001$) from one another as determined by variance analysis followed by the Neuman-Keuls test.

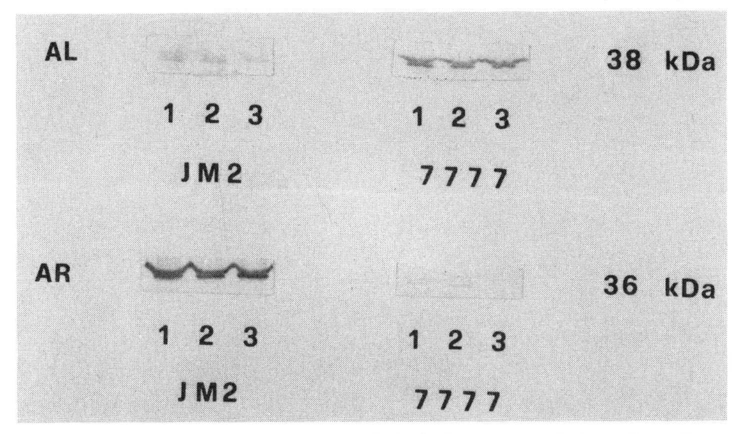

Figure 4. Western blot analysis of aldose reductase and aldehyde reductase during culture-time of 7777 and JM2 hepatoma cell lines. Lane 1, cells harvested 1 day after seeding; lane 2, cells harvested 3 days after seeding; lane 3, cells harvested 4 days after seeding. Abbreviation: AL, aldehyde reductase; AR, aldose reductase.

Figures 5 and 6 show that no significant change occurred during our experimental protocol treatment in either cell type.

The lack of significant change in AKR activity in 7777 and JM2 cells was confirmed by the Western Blot protein determination (Figures 7 and 8).

It is thus confirmed that, in hepatoma cells, AKR activity increases in direct correlation with the degree of deviation, in cultured cells and during the carcinogenesis process. The determination of activity during the growth of cultured hepatoma cells seems to confirm the correlation between AKR activity and cell proliferation.

We may thus conclude that highly deviated hepatoma cells are characterized by an increased capability to inactivate the cytostatic and cytotoxic properties of lipid peroxidation products, and that neither activity nor protein content of AKR is affected by these products.

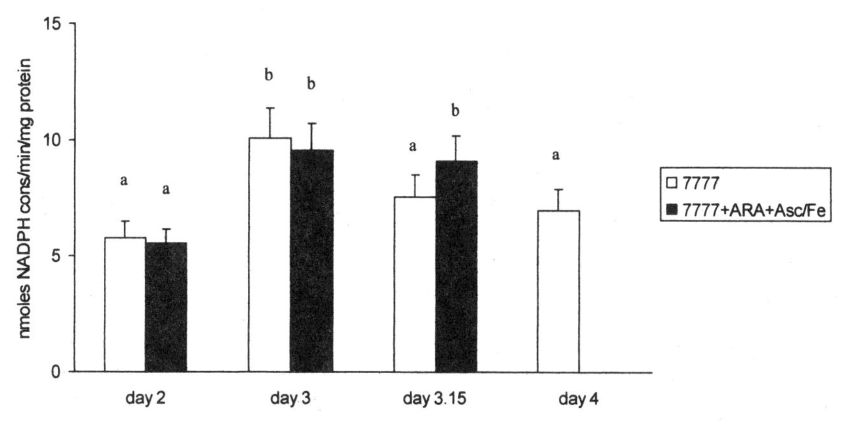

Figure 5. Aldo/keto reductase activity in 7777 hepatoma cells enriched with arachidonic acid and exposed to ascorbate/iron sulphate. Data are expressed as nmoles of NADPH consumed/min/mg of protein and represent mean ± S.D. of 5 experiments. 0.1 mM 4-hydroxynonenal was used as substrate. Means with different letters are statistically different ($p<0.001$) from one another as determined by variance analysis followed by the Neuman-Keuls test. Day 2, cells harvested 24h after enriching with arachidonic acid; day 3, cells harvested 12h after the 2nd doses of ascorbate/iron sulphate; day 3.15, cells harvested 4h after the 3rd dose of ascorbate/iron sulphate; day 4, cells harvested 12h after the 4th dose of ascorbate/iron sulphate.

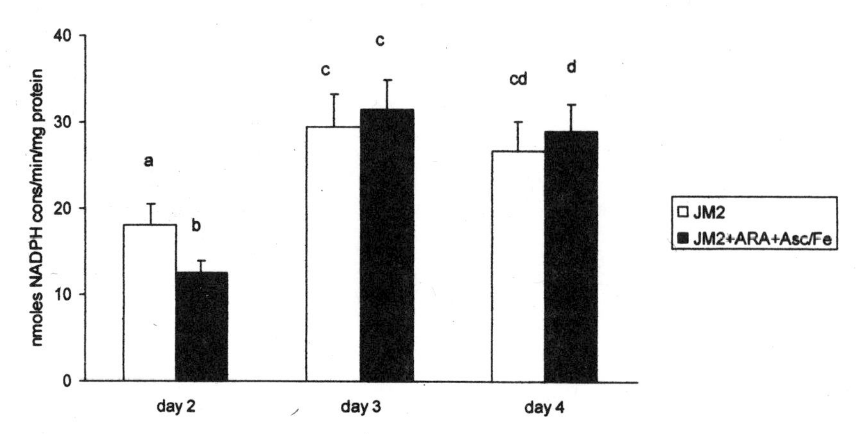

Figure 6. Aldo/keto reductase activity in JM2 hepatoma cells enriched with arachidonic acid and exposed to ascorbate/iron sulphate. Data are expressed as nmoles of NADPH consumed/min/mg of protein and represent mean ± S.D. of 5 experiments. 0.1 mM 4-hydroxynonenal was used as substrate. Means with different letters are statistically different ($p<0.001$) from one another as determined by variance analysis followed by the Neuman-Keuls test. Day 2, cells harvested 24h after enriching with arachidonic acid; day 3, cells harvested 12h after the 2nd doses of ascorbate/iron sulphate; day 4, cells harvested 12h after the 4th dose of ascorbate/iron sulphate.

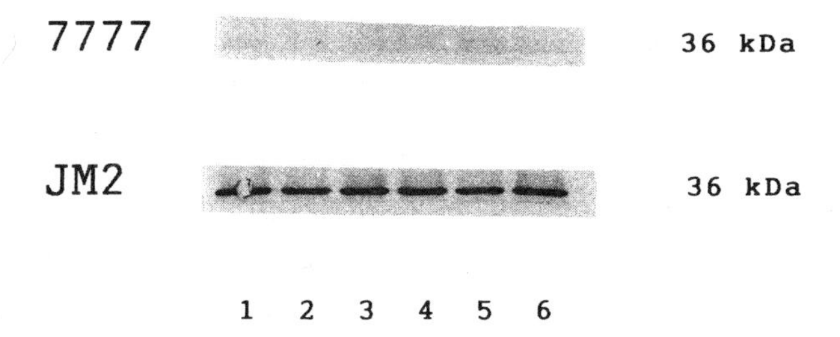

Figure 7. Western blot analysis of aldose reductase in 7777 and JM2 hepatoma cells enriched with arachidonic acid and exposed to ascorbate/iron sulphate. 1: control cells for lane 2; 2: cells harvested 24h after enrichment with arachidonic acid; 3: control cells for lane 4; 4: cells harvested 12h after the 2nd dose of ascorbate/iron sulphate; 5: control cells for lane 6; 6: cells harvested 4h after the 3rd dose of ascorbate/iron sulphate for 7777 cells and 12h after 4th dose of ascorbate/iron sulphate for JM2 cells.

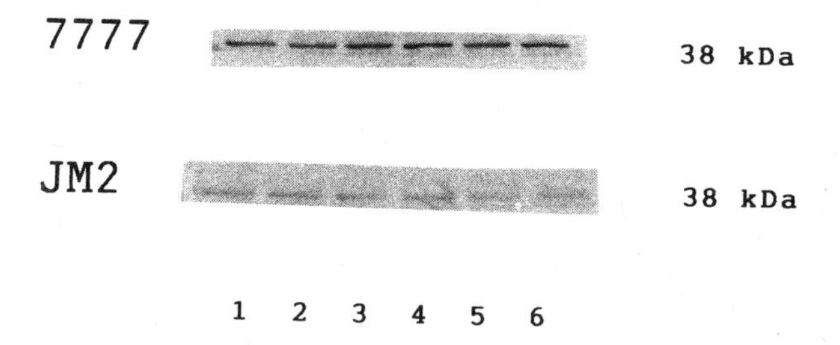

Figure 8. Western blot analysis of aldehyde reductase in 7777 and JM2 hepatoma cells enriched with arachidonic acid and exposed to ascorbate/iron sulphate. 1: control cells for lane 2; 2: cells harvested 24h after enrichment with arachidonic acid; 3: control cells for lane 4; 4: cells harvested 12h after the 2nd dose of ascorbate/iron sulphate; 5: control cells for lane 6; 6: cells harvested 4h after the 3rd dose of ascorbate/iron sulphate for 7777 cells and 12h after 4th dose of ascorbate/iron sulphate for JM2 cells.

ACKNOWLEDGMENTS

This study was supported by grants from the Italian Ministry for University, Scientific and Technological Research, and from the Italian Association for Cancer Research. R.A. Salvo was recipient of a fellowship from the Cavalieri Ottolenghi Foundation, Turin, Italy. We wish to thank Dr. Dennis R. Petersen for kindly giving us some 4-hydroxynonenal.

REFERENCES

Barrera, G., Di Mauro, C., Muraca, R., Ferrero, D., Cavalli, G., Fazio, V.M., Paradisi, L., and Dianzani, M.U. (1991) Induction of differentiation in human HL-60 cells by 4-hydroxynonenal, a product of lipid peroxidation, Exp. Cell Res. **197,** 148–152.

Cambiaggi, C., Dominici, S., Comporti., M., Pompella, A. (1997) Modulation of human T lymphocyte proliferation by 4-hydroxynonenal, the bioactive product of neutrophil-dependent lipid peroxidation, Life Sci. **61**, 777–785.

Canuto, R.A., Biocca, M.E., Muzio, G., Garcea, R., and Dianzani, M.U. (1985) The effect of various aldehydes on the respiration of rat liver and hepatoma AH-130 cells, Cell Biochem. Funct. **3**, 3–8.

Canuto, R.A., Muzio, G., Biocca, M.E., and Dianzani, M.U. (1989) Oxidative metabolism of 4-hydroxy-2,3-nonenal during diethyl-nitrosamine-induced carcinogenesis in rat liver, Cancer Lett. **46**, 7–13.

Canuto, R.A., Muzio, G., Biocca, M.E., and Dianzani, M.U. (1991) Lipid peroxidation in rat AH-130 hepatoma cells enriched *in vitro* with arachidonic acid, Cancer Res. **51**, 4603–4608.

Canuto, R.A., Muzio, G., Maggiora, M., Biocca, M.E., and Dianzani, M.U. (1993) Glutathione-S-transferase, alcohol dehydrogenase and aldehyde reductase activities during diethylnitrosamine-carcinogenesis in rat liver, Cancer Lett. **68**, 177–183.

Canuto, R.A., Ferro, M., Muzio, G., Bassi, A.M., Leonarduzzi, G., Maggiora, M., Adamo, D., Poli, G., and Lindahl, R. (1994) Role of aldehyde metabolizing enzymes in mediating effects of aldehyde products of lipid proxidation in liver cells, Carcinogenesis **15**, 1359–1364.

Canuto, R.A., Muzio, G., Bassi, A.M., Maggiora, M., Leonarduzzi, G., Lindahl, R., Dianzani, M.U., and Ferro, M. (1995) Enrichment of arachidonic acid increases the sensitivity of hepatoma cells to the cytotoxic effects of oxidative stress, Free Rad. Biol. Med. **18**, 287–293.

Canuto, R.A., Ferro, M., Maggiora, M., Federa, R., Brossa, O., Bassi, A.M. Lindahl, R., and Muzio, G. (1996) In hepatoma cell lines restored lipid peroxidation affects cell viability inversely to aldehyde metabolizing enzyme activity, Adv. Exp. Med. Biol. **414**, 113–122.

Cao, D., Fan, S.T., Chung, S.S. (1998) Identification and characterization of a novel human aldose reductase-like gene, J. Biol. Chem. **273**, 11429–11435.

Curzio, M., Esterbauer, H., Poli, G., Biasi, F., Cecchini, G., Di Mauro, C., Cappello, N., and Dianzani, M.U. (1987) Possible role of aldehydic lipid peroxidation products as chemoactractans, Int. J. Tissue React. **9**, 295–306.

Del Corso, A., Dal Monte, M., Vilardo, P.G, Cecconi, I., Moschini, R., Banditelli, S., Cappiello, M., Tsai, L., Mura, U. (1998) Site-specific inactivation of aldose reductase by 4-hydroxynonenal, Arch. Biochem. Biophys. **350**, 245–248.

Esterbauer, H. (1985) Lipid peroxidation products formation, chemical properties and biological activities. In: G. Poli, K.H. Cheeseman, M.U. Dianzani and T.F. Slater, editors, Free Radicals in Liver Injury. IRC Press, Oxford, pp. 29–47.

Jez, J.M., Bennet, M.J., Schlegel, B.P., Lewis, M., Penning, T.M. (1997) Comparative anatomy of the aldo/keto reductase superfamily, Biochem. J. **326**, 625–636.

Kreuzer, T., Grube, R., Wutte, A., Zarkovic, N., Schaur, R.J. (1998) 4-hydroxynonenal modifies the effects of serum growth factors on the expression of the c-fos proto-oncogene and the proliferation of HeLa carcinoma cells, Free Rad. Biol. Med. **25**, 42–49.

Lindahl, R. (1992) Aldehyde dehydrogenases and their role in carcinogenesis, CRC Crit. Rev. Biochem. Mol. Biol. **27**, 283–335.

Paradisi, L., Panagini, C., Parola, M., Barrera, G., and Dianzani, M.U. (1985) Effects of 4-hydroxynonenal on adenylate cyclase and 5' nucleotidase activities in rat liver plasma membranes, Chem. Biol. Interact. **53**, 209–217.

Pitot, H.C. (1996) Stage-specific gene expression during hepatocarcinogenesis in the rat, J. Cancer Res. Clin. Oncol. **122**, 257–265.

Poot, M., Verkerk, A., Koster, J.F., Esterbauer, H., Jongkind, J.F. (1988) Reversible inhibition of DNA and protein synthesis by cumene hydroperoxide and 4-hydroxy-nonenal, Mech. Ageing Dev. **43**, 1–9.

Roomi, M.,W., Ho, R.K., Sarma, D.S.R., and Farber, E. (1985) A common biochemical pattern in preneoplastic hepatocyte nodules generated in four different models in rat, Cancer Res. **45**, 564–571.

Rossi., M.A., Garramone, A., and Dianzani, M.U. (1988) Stimulation of phospholipase C activity by 4-hydroxynonenal: influence of GTP and calcium concentration, Int. J. Tiss. React. **10**, 321–325.

Rossi., M.A., Fidale, F., Garramone, A., Esterbauer, H., and Dianzani, M.U. (1990) Effect of 4-hydroxyalkenals on hepatic phosphatidylinositol-4,5 bis phosphate phospholipase C, Biochem. Pharmacol. **39**, 1715–1719.

Schaur, R.J., Dussing, G., Kink, E., Schauenstein, E., Posch, W., Kukovetz, E., and Egger, G. (1994) The lipid peroxidation product 4-hydroxynonenal is formed by, and is able to actract, rat neutrophils *in vivo*, Free Rad. Res. **20**, 365–373.

Spitz, D.R., Malcolm, R.R., and Roberts, R.J. (1990) Cytotoxicity and metabolism of 4-hydroxy-2-nonenal and 2-nonenal in H_20_2-resistant cell lines, Biochem. J. **267**, 453–459.

Spycher, S., Tabataba-Vakili, S., O'Donnell, V.B., Palomba, L., and Azzi, A. (1996) 4-hydroxy-2,3-trans-nonenal induces transcription and expression of aldose reductase, Biochem. Biophys. Res. Commun. **226**, 512–516.

Srivastava, S., Chandra, A., Bhatnagar, A., Srivastava, S.K., Ansari, N.H. (1995) Lipid peroxidation product, 4-hydroxynonenal and its conjugated with GSH are excellent substrates of bovine lens aldose reductase, Biochem. Biophys, Res., Commun. **217**, 741–746.

Takahashi, M., Fujii, J., Miyoshi, E., Hoshi, A., and Taniguchi, N. (1995) Elevation of aldose reductase gene expression in rat primary hepatoma and hepatoma cell lines: implication in detoxification of cytotoxic aldehydes, Int. J. Cancer **62**, 749–754.

Tsuchida, S., and Sato, K. (1992) Glutathione transferases and cancer, CRC Crit. Rev. Biochem. Mol. Biol. **27**, 337–394.

Vander Jagt, D.L., Kolb, N.S., Vander Jagt, T.J., Chino, J., Martinez, F.J., Hunsaker, L.A., Royer, R.E. (1995) Substrate specificity of human aldose reductase: identification of 4-hydroxynonenal as an endogenous substrate, Biochim. Biophys. Acta. **1249**, 117–126.

INTERCONVERSION PATHWAYS OF ALDOSE REDUCTASE INDUCED BY THIOL COMPOUNDS

Antonella Del Corso, Pier Giuseppe Vilardo, Catia Barsotti, Mario Cappiello, Ilaria Cecconi, Massimo Dal Monte, Isabella Marini, Stefania Banditelli, and Umberto Mura

Dipartimento di Fisiologia e Biochimica
Università di Pisa
via S. Maria, 55
56100 Pisa, Italy

1. INTRODUCTION

The occurrence of protein S-thiolation as a consequence of oxidative stress is widely recognized (Thomas et al., 1990; Lou et al., 1990; Schuppe et al., 1992; Giblin, et al., 1995). However, the role and the relevance of this process are still a matter of debate.

The most relevant question is whether protein thiolation occurs as a side reaction of the thiol/disulfide red/ox cycle of physiological thiol compounds, or as a programmed process toward preferential targets.

A proposed role for protein thiolation, related to the buffering ability of proteins in the maintenance of proper GSSG/GSH ratios, is the protection of proteins cysteine residues against irreversible modification (Coan et al., 1992; Del Corso et al., 1998). Furthermore, more specific functions are evoked when the protein targets of S-thiolation are enzymes. In those cases a potential modulation of the enzyme activity can be postulated and a concerted metabolic function envisaged (Brigelius, 1985; Del Corso et al., 1994).

The lens is especially susceptible to oxidative insult and thiols, namely glutathione, which is very highly represented in this organ, have been selected as the most relevant cellular antioxidant defence system. It is therefore not surprising that in the lens there is a massive occurrence of thiolation phenomena, and protein mixed disulfides with both GSH and Cys have been reported (Lou et al., 1990; Lou and Dickerson, 1992).

Aldose reductase (alditol: NADP$^+$ oxidoreductase, EC 1.1.1.21) (ALR2), which catalyzes the NADPH-dependent reduction of a variety of aldoses and alyphatic and aromatic aldehydes, is an enzyme which is especially susceptible to S-thiolation and represents a good protein model to verify the potential functions of S-thiolation. ALR2 has

Enzymology and Molecular Biology of Carbonyl Metabolism 7
edited by Weiner *et al.* Kluwer Academic / Plenum Publishers, New York, 1999.

been shown to be modified at a level of Cys298, both by thiols in oxidative conditions (Giannessi et al., 1993; Bohren, and Gabbay, 1993) and disulfides (Bhatnagar et al., 1992; Cappiello et al., 1994; Cappiello et al., 1996). These thiol induced modifications, which consist in the formation of a mixed disulfide bond between Cys298 and the thiol compound, have different effects on the structural and kinetic properties of the enzyme (Del Corso et al., 1989; Del Corso et al., 1990; Del Corso et al., 1993; Cappiello et al., 1994). While 2-mercaptoethanol is able, in the presence of Fe^{2+}/EDTA, to induce an increase in thc catalytic activity of bovine lens ALR2 up to 2.5 fold, oxidized glutathione leads to an inactivation of the enzyme. Since the different effect exerted on the enzyme activity by different thiol compounds, it appears that S-thiolation can be relevant in modulating the ALR2 activity.

Not only glutathione but also cysteine has been reported to affect ALR2 activity. In fact, both cysteine in oxidative conditions and cystine lead to, as observed for glutathione, an inactivation of the enzyme (Giannessi et al., 1993; Cappiello et al., 1996). In the current work we report that the cysteine-induced modification of bovine lens ALR2 results in the formation of two different modified enzyme forms.

2. EXPERIMENTAL PROCEDURES

2.1. Materials

NADPH, D,L-glyceraldehyde, and GSSG were purchased from Sigma Chemical Co. L-Cysteine and L-cystin were from Carlo Erba. Matrex Orange A was from Amicon. All other chemicals were of reagent grade from BDH. Sorbinil ((S)-6-fluorospiro[chroman-4,4'-imidazolidine]-2',5'-dione) was a gift from Dr. G. Caccia, Laboratori Baldacci SpA, Pisa. Calf lenses were obtained from freshly slaughtered animals and kept at –20°C until used.

2.2. Measurement of Enzyme Activity

The ALR2 activity was measured at 37°C as previously described (Del Corso et al., 1989), using 4.7 mM D,L-glyceraldehyde as substrates. One unit of enzyme activity is the amount of enzyme which catalyzes the oxidation of 1 μmol of NADPH per min. The sensitivity of ALR2 to inhibition by Sorbinil was tested in the above assay conditions in the presence of 10 μM of the inhibitor.

2.3. Purification of Bovine Lens ALR2

In order to purify bovine lens ALR2, frozen lenses, after incision of the capsule, were suspended in 10 mM sodium phosphate buffer pH 7.0, containing 5 mM DTT (1 gr tissue/3.5 ml) and stirred in an ice-cold bath for 1 h. The suspension was then centrifuged at 22,000 × g at 4°C for 40 min and the supernatant was processed as previously described (Del Corso et al., 1990).

2.4. Synthesis of Cysteinylglutathione

Cysteinylglutathione (CySSG) was synthesized, starting from the cystine disulfoxide, according to Eriksson and Eriksson (Eriksson, B., and Eriksson, S.A., 1967)

2.5. Other Methods

Protein concentration was determined according to Bradford's method (Bradford, 1976) using bovine serum albumin as the standard.

3. RESULTS AND DISCUSSION

Similarly to oxidized glutathione, which is able to modify ALR2, leading to an enzyme form (GS-ALR2) with a reduced catalytic activity with respect to the native enzyme, cystine, which has been shown to be able to inactivate the bovine lens enzyme, and cysteinylglutathione appear to induce a modification of the enzyme. The time curves of inactivation at 37°C of native purified ALR2 (130 mU/ml) in the presence of 0.4 mM of GSSG, cystine and CySSG are reported in Figure 1.

The loss of enzyme activity proceeds essentially at the same rate for all the tested disulfides. From the data of the initial part of each time curve, it was possible to measure a rate of inactivation of 0.35, 0.46 and 0.44 mU × ml^{-1} × min^{-1} for GSSG, cysteine and cysteinylglutathione, respectively. Moreover, the same maximal extent of enzyme inactivation (approximately 30% of the initial activity) was observed with all disulfides. The inactivation, as observed also for the GSH and the 2-mercaptoethanol-induced modifications (Giannessi et al., 1993; Cappiello et al., 1994), is accompanied by a complete loss of sensitivity to Sorbinil. The removal of disulfides by extensive dialysis is not sufficient to restore the initial enzyme activity. On the contrary, as shown in Figure 1, a further incubation at 37°C in the presence of 5 mM DTT allowed a complete recovery of both the enzyme activity and the sensitivity to Sorbinil. These results clearly indicate the occurrence of a covalent modification of the enzyme, which can be easily reversed by thiol reducing agents, which are able to regenerate the native enzyme form.

An affinity chromatography step on Matrex Orange A, a method previously used to separate the GS-ALR2 from the native enzyme (Cappiello et al., 1995), turned out to be an useful approach for analyzing the cystine as well as the CySSG-inactivated ALR2.

This chromatographic analysis highlighted that a significant fraction of the enzyme preparation after incubation with CySSG did not bind to the column and was unaffected by Sorbinil (Figure 2). These features, together with the susceptibility to DTT reduction, which allows the recovery of both enzyme activity and sensitivity to Sorbinil, are consis-

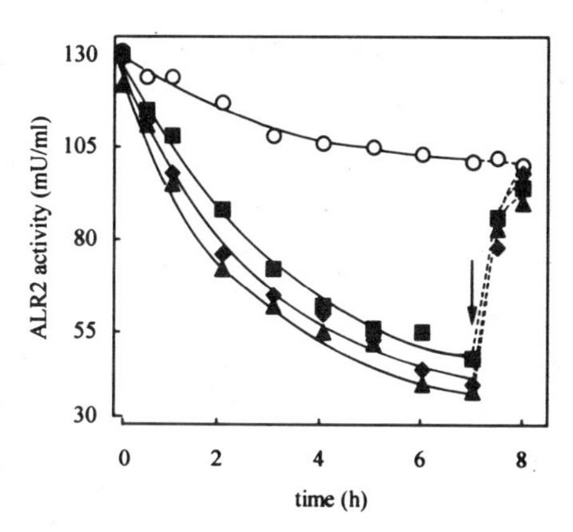

Figure 1. Inactivation of ALR2 by disulfides. Purified bovine lens ALR2 (130mU/ml) was incubated at 37°C in 100mM sodium phosphate buffer pH 6.8 in the presence of 0.4 mM GSSG (■), cystine (▲) and cysteinylglutathione (◆), and the enzyme activity measured. At the time indicated by the arrow, each mixture was supplemented of 5 mM DTT and incubated again at 37°C.

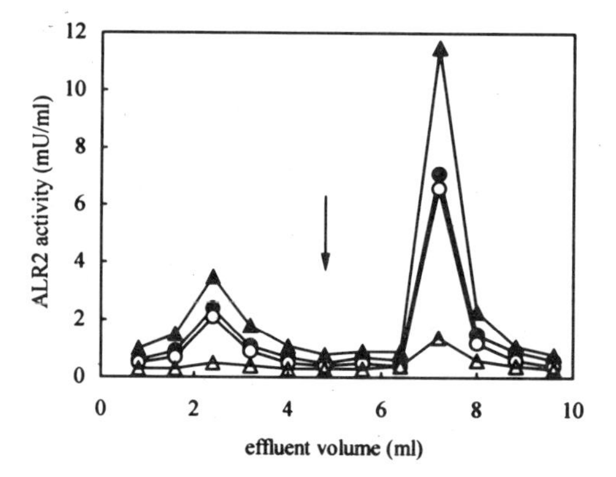

Figure 2. Affinity chromatographic analysis of the CySSG-modified ALR2. An aliquot of CySSG-modified ALR2 (11 mU) was applied to a Matrex Orange A column (1 × 2.3 cm) and the column was eluted with 10 mM sodium phosphate buffer pH 7.0. The arrow indicates the application to the column of 0.1 mM NADPH in phosphate buffer, as elution buffer. The enzyme activity was measured before (●, ○) and after (▲, Δ) 1 h incubation at room temperature in the presence of 1 mM DTT, both in the absence (closed symbols) and in the presence (open symbols) of 10 μM Sorbinil.

tent with the generation, as a consequence of the CySSG treatment, of GS-ALR2 thus accounting for a glutathionylation of 30–40 % of the enzyme (Figure 2). Figure 2 also shows the occurrence of an enzyme fraction which interacts with the chromatographic support, in the same way as the native enzyme (Cappiello et al., 1995). However, it appeared to be unaffected by Sorbinil. In this case too, although the DTT treatment allowed the rescue of some enzyme activity, it restored the sensitivity to Sorbinil. When the cystine-inactivated ALR2 was analyzed, as above, by affinity chromatography, all the enzyme bound to the column (Figure 3).

However, on the basis of the different susceptibilities to Sorbinil shown by the cystine-modified and the native ALR2, it was possible to rule out the presence of residual unmodified ALR2 in the protein sample treated with cystine. Moreover, by considering both the sensitivity to inhibition by Sorbinil and the susceptibility to DTT-dependent reactivation of the eluted fractions, the analysis by affinity chromatography of the enzyme preparation subjected to incubation at 37°C with cystine, revealed the presence of at least two

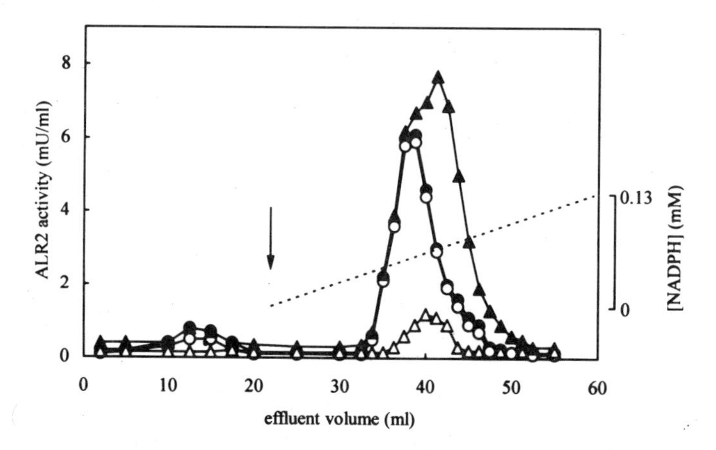

Figure 3. Affinity chromatographic analysis of the cysteine-modified ALR2. An aliquot of the cystine-modified ALR2 (70 mU) was applied to a Matrex Orange A column (1.4 × 10 cm) and the elution was performed with 10 mM sodium phosphate buffer pH 7.0. The arrows indicates the application of a linear gradient from 0 to 0.13 mM of NADPH. The enzyme activity was measured before (●, ○) and after (▲, Δ) 1 h incubation at room temperature in the presence of 1 mM DTT, both in the absence (closed symbols) and in the presence (open symbols) of 10 μM Sorbinil.

modified enzyme forms (Figure 3). These two enzyme forms were both able to bind to the Matrex Orange A column but with different properties with respect to the native enzyme. It was possible to partially resolve them using a linear gradient of NADPH as an eluting system. The eluted fractions, before and after treatment with DTT, were assayed for enzyme activity both in the absence and in the presence of 10 μM Sorbinil. Although the front of the elution peak was unaffected by DTT, a significant recovery of enzyme activity was observed at higher elution volumes. Moreover, a recovery of sensitivity to inhibition was observed after treatment with DTT in the entire range of active fractions. These results indicate the presence of one enzyme form that was still active, but no longer sensitive to the ARI, and one form, characterized by a low ALR2 activity, but which is still susceptible to reactivation in thiol reducing conditions. Thus, it appears that the reactivity of bovine lens ALR2 with respect to thiol-induced modification cannot be simply confined to S-thiolation of Cys298. We are currently studying whether the cystine-induced modification involves the S-thiolation of more than one Cys residue of the enzyme, or whether the thiolated protein, which is likely modified at level of Cys298, can undergo intra- or inter-molecular thiol/disulfide rearrangment. In any event it appears that the modifications induced on ALR2 by physiological disulfides result in the inactivation of the enzyme. Thus, when associated to oxidative conditions, S-thiolation of ALR2 may represent a way for the cell to slow down the flux of NADPH, which would be channelled in those processes (i.e. GSSG reductase or thioredoxine reductase-catalyzed reactions) devoted to rescue reducing power.

ACKNOWLEDGMENTS

This work was supported in part by the "Target Project on Cellular Oxidative Stress" of the Italian National Research Council (C.N.R.) and in part by Italian Board of Education (MURST) and by Pisa University.

REFERENCES

Bhatnagar, A., Liu, S-Q., Petrash, J.M., and Srivastava, S.K. (1992) Mechanism of inhibition of aldose reductase by menadione (Vitamin K3), *Molec. Pharmacol.*, 42, 917–921.

Bohren, K.M., and Gabbay, K.H., (1993) Cys298 is responsible for reversible thiol-induced variation in aldose reductase activity, in "Enzymology and Molecular Biology of Carbonyl Metabolism" (Weiner, H., Ed.), vol. 4, pp. 267–277, Plenum Press, New York.

Bradford, M.M., (1976) A rapid and sensitive method for the quantitation of microgram quantities of protein utilizing the principle of protein-dye binding, *Anal. Biochem.*, 72, 248–254.

Brigelius, R., (1985) Mixed disulfides: biological functions and increase in oxidative stress in "Oxidative Stress" (Sies H., Ed) pp. 243–271, Academic Press, London.

Cappiello, M., Voltarelli, M., Cecconi, I., Vilardo, P.G., Dal Monte, M., Marini, I., Del Corso, A., Wilson, D.K., Quiocho, F.A., Petrash, J.M., and Mura, U. (1996) Specifically targeted modification of human aldose reductase by physiological disulfides, *J. Biol. Chem.*, 271, 33539–33544.

Cappiello, M., Voltarelli, M., Giannessi, M., Cecconi, I., Camici, G., Manao, G., Del Corso, A., and Mura, U., (1994) Glutathione dependent modification of bovine lens aldose reductase, *Exp. Eye Res.*, 58, 491–501.

Cappiello,M., Vilardo, P.G., Cecconi, I., Leverenz, V., Giblin, F.J., Del Corso, A., and Mura, U., (1995) Occurrence of glutathione-modified aldose reductase in oxidatively stressed bovine lenses, *Biochem. Biophys. Res. Commun.* 207, 775–782.

Coan, C., Ji, J-H., Hideg, K., and Mehlhorn, R.J., (1992) Protein sulfhydryls are protected from irreversible oxidation by conversion to mixed disulfides, *Arch. Biochem. Biophys.*, 295, 369–378.

Del Corso, A., Barsacchi, D., Camici, M., Garland, D., and Mura, U. (1989) Bovine lens aldose reductase: identification of two enzyme forms, *Arch. Biochem. Biophys.*,270, 604–610.

Del Corso, A., Barsacchi, D., Giannessi, M., Tozzi, M.G., Camici, M., Houben, J.L., Zandomeneghi, M., and Mura, U., (1990) Bovine lens aldose reductase: tight binding of the pyridine coenzyme, *Arch. Biochem. Biophys*, 283, 512–518.

Del Corso, A., Cappiello, M., and Mura, U., (1994) Thiol dependent oxidation of enzymes: the last chance against oxidative stress, *Int. J. Biochem*., 26, 745–750.

Del Corso, A., Dal Monte, M., Vilardo, P.G., Cecconi, I., Moschini, R., Banditelli, S., Cappiello, M., Tsai, L., and Mura, U., (1998) Site-specific inactivation of aldose reductase by 4-hydroxynonenal, *Arch. Biochem. Biophys*., 350, 245–248.

Del Corso, Voltarelli, M., Giannessi, M., Cappiello, M., Barsacchi, D., Zandomeneghi, M., Camici, M., and Mura,U., (1993) Thiol-dependent metal-catalyzed oxidation of bovine lens aldose reductase. II. Proteolytic susceptibility of the modified enzyme form, *Arch. Biochem. Biophys*. 300, 430–433.

Eriksson, B., and Eriksson, S.A., (1967) Synthesis and characterization of the L-cysteine-glutathione mixed disulfide, *Acta Chem. Scand*., 21, 1304–1312.

Giannessi, M., Del Corso, A., Cappiello, M., Voltarelli, M., Marini, I., Barsacchi, D., Garland, D., Camici, M., and Mura, U., Thiol-dependent metal-catalyzed oxidation of bovine lens aldose reductase. I. Studies on the modification process, *Arch. Biochem. Biophys*., 300, 423–429.

Giblin, F.J., Padgaonkar, V.A., Leverenz, V.R., Lin, L-R., Lou, M.F., Unakar, N.J., Dang, L., Dickerson, J.E.Jr., and Reddy, V.N., (1995) Nuclear light scattering, disulfide formation and membrane damage in lenses of older guinea pigs treated with hyperbaric oxygen, *Exp. Eye Res*. 60, 219–235.

Lou, M.F., and Dickerson, J.E.Jr., (1992) Protein-thiol mixed disulfides in human lens, *Exp. Eye Res* 55, 889–896.

Lou, M.F., Dickerson, J.E.Jr., and Garadi, R., (1990) The role of protein thiol mixed disulfide in cataractogenesis, *Exp. Eye Res*. 50, 819–826.

Schuppe, I., Moldeus, P., and Cotgreave, I.A., (1992) Protein-specific S-thiolation in human endothelial cells during oxidative stress, *Biochem. Pharmacol*. 44, 1757–1764.

Thomas, J.A., Park, E.M., Chai, Y.C., Brooks, R., Rokutan, K., and Johnston, R.B.Jr., (1990) S-Thiolation of protein sulfhydryls, in "Biological Reactive Intermediates IV: Molecular and Cellular Effects, and Human Impact" (Witmer, C.M., Snyder, R., and Sipes, G., Eds.) pp. 95–101, Plenum Press, New York.

57

ALDO-KETO REDUCTASES IN NOREPINEPHRINE METABOLISM

Sanai Sato,[1] Minoru Kawamura,[2,3] Graeme Eisenhofer,[2] Irwin J. Kopin,[2] Shigeki Fujisawa,[1] and Peter F. Kador[1]

[1]Laboratory of Ocular Therapeutics
National Eye Institute
[2]Clinical Neuroscience Branch
National Institute of Neurological Disorders and Stroke
National Institutes of Health
Bethesda, Maryland 20892
[3]Institute of Bio-Active Science
Nippon Zoki Pharmaceutical Co., Ltd.
Hyogo 673-14, Japan

1. INTRODUCTION

Monoamine oxidase (MAO, EC 1.4.3.4) and catechol-O-methyltransferease (COMT, EC 2.1.1.6) are the two primary enzymes responsible for the first step of the metabolism of the catecholamines epinephrine, norepinephrine and dopamine. Catecholamines and their methylated metabolites, produced by COMT, are metabolized to the corresponding aldehyde intermediates by MAO. In the subsequent second step these aldehydes are removed through either their oxidation to acids by aldehyde dehydrogenase (EC 1.2.1.3) or their reduction to alcohols by aldehyde reductase (Kopin, 1985). Here, we focus on the metabolism of norepinephrine and summarize evidences that aldose reductase is the key enzyme responsible for the metabolism of norepinephrine.

Norepinephrine is a main sympathetic neurotransmitter originating primarily in sympathetic nerve cells. In the first step of metabolism, norepinephrine can be either methylated by COMT or deaminated by MAO. However, MAO is the only enzyme that is active in intraneural locations while COMT only works as an extraneural enzyme. As a result, norepinephrine is primarily metabolized to 3,4-dihydroxyphenylglycolaldehyde by MAO.

2. RESULTS

2.1. The Reduction to 3,4-Dihydroxyphenylglycol Is the Major Pathway of the Aldehyde Intermediate Formed from Norepinephrine

3,4-Dihyroxyphenylglycolaldehyde formed by MAO is either oxidized to 3,4-dihydroxymandelic acid (DHMA) by aldehyde dehydrogenase or reduced to 3,4-dihydroxyphenylglycol (DHPG) (Figure 1). The first question to be addressed is whether the major pathway for this aldehyde intermediate is oxidation or reduction. To answer this question, the levels of norepinephrine metabolites in rats were assayed by HPLC using a C18 reversed-phase column (250 × 4.6 mm I.D.) as previously described by Eisenhofer, et al. (1986). With this assay system, DHPG can easily be detected in normal rat plasma (4.94 ± 0.67 pmol/ml) compared to the levels of DHMA that are extremely low (0.15 ± 0.06 pmol/ml). When rats are treated with the aldose reductase inhibitor AL 1576, the balance of DHPG and DHMA levels quickly changes (Figure 2). Within an hour, DHPG levels drastically decrease while the levels of DHMA significantly increase. This clearly demonstrates that, although the deaminated product from norepinephrine 3,4-dihydroxyphenylglycolaldehyde can be either oxidized to DHMA or reduced to DHPG, the reduction to DHPG is the major pathway in normal conditions. However, when the reduction pathway is blocked, aldehyde dehydrogenase becomes the major pathway and the aldehyde intermediate is oxidized to DHMA.

2.2. The Metabolite Formed from Norepinephrine by Aldose Reductase Is 3,4-Dihydroxyphenylglycol (DHPG)

Because the reduction of the aldehyde intermediate from norepinephrine to DHPG is dominant in normal conditions and the aldose reductase inhibitor AL 1576 inhibits DHPG formation, it has been speculated that aldose reductase (EC 1.1.1.21) is the enzyme responsible for the formation of DHPG. To confirm that aldose reductase generates DHPG

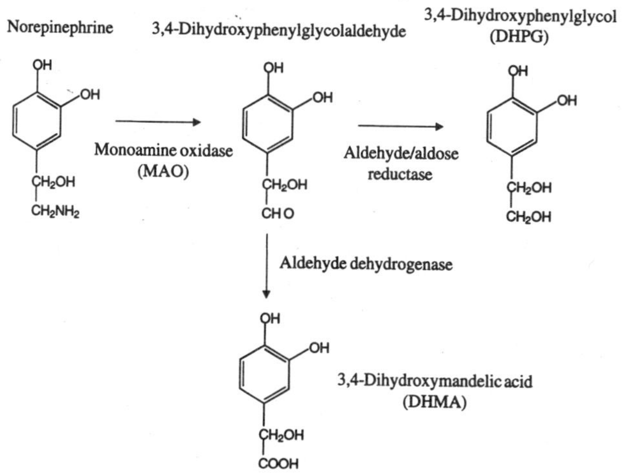

Norepinephrine
3,4-Dihydroxyphenylglycolaldehyde
3,4-Dihydroxyphenylglycol (DHPG)
Monoamine oxidase (MAO)
Aldehyde/aldose reductase
Aldehyde dehydrogenase
3,4-Dihydroxymandelic acid (DHMA)

Figure 1. Intraneural metabolism of norepinephrine in sympathetic nerves.

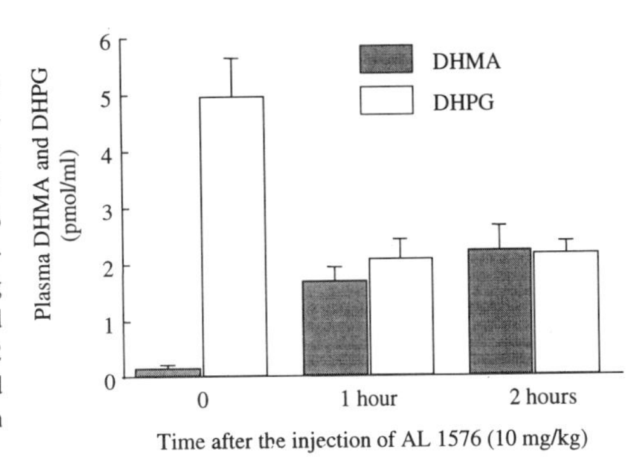

Figure 2. Plasma levels of 3,4-dihydroxymandelic acid (DHMA, gray bars) and 3,4-dihydroxyphenylglycol (DHPG, open bars) of rats. Under the anesthesia by intraperitoneal injection of sodium pentobarbital (65 mg/kg), 10 mg/kg of AL 1576 dissolved with 33% polyethylene glycol was given to male Sprague-Dawley rats weighing 340–420 g by directly injecting into femoral vein. The blood samples were obtained from femoral artery of rats before and after 1 and 2 hours of the injection. DHPG and DHMA were assayed as previously described (Eisenhofer, et al. 1986). Mean ± S.D. (n=5).

from norepinephrine, purified rat lens aldose reductase was incubated *in vitro* with norepinephrine and MAO in the presence of NADPH (Figure 3). When norepinephrine is incubated with MAO, the norepinephrine peak almost completely disappears, but DHPG can not be detected. The formation of DHPG clearly be detected when a mixture containing both MAO and aldose reductase is used. This confirms that aldose reductase generates DHPG. The data also demonstrate that norepinephrine is first metabolized by MAO and aldose reductase utilizes only the aldehyde intermediate formed by MAO from norepinephrine.

2.3. The Activity of Aldose Reductase to Form 3–4-Dihydroxyphenylglycol (DHPG) Is Much Higher than that of Aldehyde Reductase

Like aldose reductase, aldehyde reductase (EC 1.1.1.2) also generates DHPG. To compare the activities of aldose reductase versus aldehyde reductase, the formation of DHPG by similar amount of enzyme activity determined by DL-glyceraldehyde as substrate was examined (Figure 4). The DHPG concentration produced by 0.01 mU of aldose reductase was still higher than that by 0.1 mU of aldehyde reductase, indicating that al-

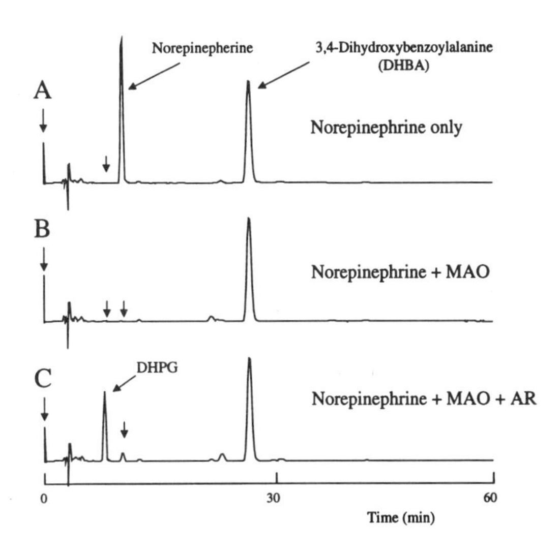

Figure 3. HPLC recordings of the reaction mixture (100 µl) containing 100 µM norepinephrine (NE), 2 mM ascorbic acid, 25 mU recombinant rat lens aldose reductase (Old, et al. 1990), and rat liver mitochondria (Kobayashi, et al. 1994) as MAO. The incubation was performed at 37°C for 3 hours and the reaction was terminated by the addition of 50 µl of 0.4 M perchloric acid containing 0.5 mM EDTA. A illustrates the chromatogram obtained from the reaction mixture containing NE but not aldose reductase and MAO. B indicates the mixture containing norepinephrine and MAO but not aldose reductase while C is the complete mixture containing norepinephrine, monoamine oxidase and aldose reductase. DHBA: 3,4-dihydroxybenzylamine.

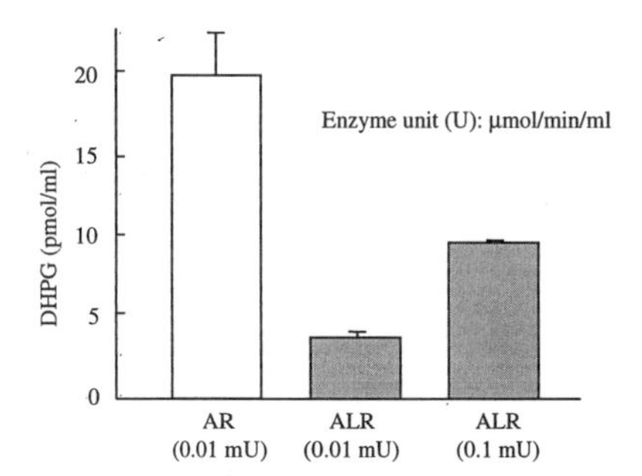

Figure 4. 3,4-Dihydroxyphenylglycol (DHPG) formation by rat lens aldose reductase (AR) and rat liver aldehyde reductase (ALR). The incubations of aldose reductase and aldehyde reductase with norepinephrine were conducted as described in Figure 3. The formed DHPG was assayed by HPLC. Enzyme unit (U) is expressed as µmol/min with DL-glyceraldehyde as substrate. Mean ± S.D. (n=4).

dose reductase is at least 10-fold more active in producing DHPG than aldehyde reductase. This is consistent with kinetic studies that the Km of aldose reductase for this intermediate aldehyde is better than that of aldehyde reductase (Turner, et al., 1974; Wermuth, 1985).

2.4. Aldose Reductase, Not Aldehyde Reductase, Is the Major Enzyme in Rat Sympathetic Nerve Cells

The major source for norepinephrine is the sympathetic nerves. Since the formation of the aldehyde intermediate by MAO occurs in sympathetic nerve cells, and aldose reductase forms DHPG more readily than aldehyde reductase, is aldose reductase also present in sympathetic nerve cells? To answer this question, studies using Western blots of rat superior cervical ganglia were conducted (Figure 5). Superior cervical ganglia are the source of sympathetic nerve fibers to the tissues of the head (e.g. innervation to salivary glands) and some parts of the upper bodies. Crude extracts of rat superior cervical ganglia demonstrate positive bands with antibody against rat lens aldose reductase corresponding exactly to purified aldose reductase. In contrast, no recognizable immunoreactive bands were found when the same crude extract of rat superior cervical ganglia were reacted with antibody against rat kidney aldehyde reductase despite a positive staining with purified rat liver aldehyde reductase. This confirms that aldose reductase is present in rat sympathetic nerve ganglia, and that aldose reductase is dominant over aldehyde reductase in these ganglia cells.

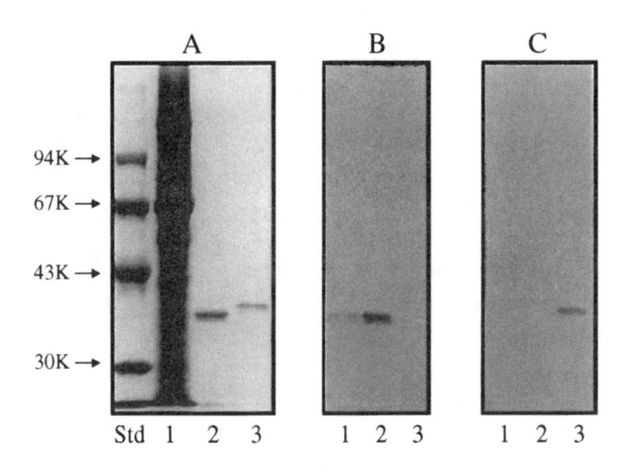

Figure 5. Western blots of the crude extract of rat superior cervical ganglia (lane 1), rat lens aldose reductase (lane 2) and rat liver aldehyde reductase (lane 3). Electrophoresis was conducted on a 14% acrylamide gel by the method of Laemmli (1970). A, B and C represent coomassie blue stain, immunoblot with antibody against rat lens aldose reductase (Shiono, et al. 1986) and immunoblot with antibody against rat kidney aldehyde reductase (Sato, 1992). Std: molecular weight standards.

3. CONCLUSION

The primary metabolite from norepinephrine, the main neurotransmitter of the sympathetic nerve system, is 3,4-dihydroxyphenylglycolaldehyde. Although this aldehyde intermediate can be metabolized to either DHPG or DHMA, DHPG is the major metabolite detected in both plasma and brain tissues of rats under physiological conditions (Kawamura, et al. 1997).

Both aldose reductase and aldehyde reductase generate DHPG when incubated with norepinephrine in the presence of MAO and NADPH. However, the activity of aldose reductase is much higher than that of aldehyde reductase in the formation of DHPG. Aldose reductase is also the major enzyme present in sympathetic nerve cells that are the major source of norepinephrine. These findings indicate that aldose reductase, not aldehyde reductase, is responsible for the formation of DHPG in rat sympathetic nerve cells.

In general, many biogenic aldehydes are toxic and they are quickly metabolized. It is well known that aldehyde dehydrogenase, aldehyde oxidase and aldehyde/aldose reductase are three major aldehyde-scavenging enzymes. The aldehyde formed from norepinephrine is not the exception. The major pathway for this particular aldehyde is the reduction to DHPG by aldose reductase in physiological conditions. However, once the reduction pathway to DHPG is blocked, the flux of norepinephrine quickly changes into the oxidation to DHMA. This occurs within an hour.

REFERENCES

Eisenhofer, G., Goldstein, D.S., Stull, R., Keiser, H.R., Sunderland, T., Murphy, D.L., and Kopin, I.J. (1986) Simultaneous liquid-chromatographic determination of 3,4-dihydroxyphenylglycol, catecholamines, and dihydroxyphenylalanine in plasma and their responses to inhibition of monoamine oxidase, *Clin. Chem.*, **32**, 2030–2033.

Kawamura, M., Kopin, I.J., Kador, P.K., Sato, S., Tjurmina, O., and Eisenhofer, G. (1997) Effects of aldehyde/aldose reductase inhibition on neuronal metabolism of norepinephrine, *J. Auton. Nerv. Syst.* **66**, 145–148.

Kobayashi, A. and Fujisawa, S. (1994) Effect of l-carnitine on mitochondrial acyl CoA esters in the ischemic dog heart, *J. Mol. Cell Cardiol.* **26**, 499–508.

Kopin, I.J. (1985) Catecholamine metabolism: basic aspects and clinical significance, *Pharmacol. Rev.* **37**, 333–364.

Laemmli, U.K. (1970) Cleavage of structural proteins during assembly of the head of bacteriophage T4, *Nature* **15**, 680–685.

Old, S.E., Sato, S., Kador, P.F., and Carper, D.A. (1990) *In vitro* expression of rat lens aldose reductase in *Escherichia coli*, *Proc. Natl. Acad. Sci. USA* **87**, 4942–4945.

Sato, S. (1992) Rat kidney aldose reductase and aldehyde reductase and polyol production in rat kidney, *Am. J. Physiol.* **236** (Renal fluid Electrolyte Physiol. 32), F799-F805.

Shiono, T., Sato, S., Reddy, V.N., Kador, P.F., and Kinoshita, J.H. (1987) Rapid purification and properties of rat lens aldose reductase, *Prog. Clin. Biol. Res.* **232**, 317–324.

Turner, A.J., Illingworth, J.A., and Tipton, K.A. (1974) Simulation of biogenic amine metabolism in the brain, *Biochem. J.* **144**, 353–360.

Turner, A.J. and Flynn, T.G. (1982) The nomenclature of aldehyde reductases, *Prog. Clin. Biol. Res.* **114**, 401–402.

Wermuth, B. (1985) Aldo-keto reductases, *Prog. Clin. Biol. Res.* **174,** 209–30.

ROTAMERS OF TOLRESTAT AND THEIR BINDING MODE TO ALDOSE REDUCTASE

Yong S. Lee,* Katsumi Sugiyama, and Peter F. Kador

National Eye Institute
National Institutes of Health
Bethesda, Maryland 20892

1. INTRODUCTION

Tolrestat is known to bind tightly ($IC_{50} < 10^{-7}$ M) to aldose reductase (E.C. 1.1.1.21; ALR2), an enzyme which has been linked to diabetic complications (Kinoshita, 1986; Dvornik, 1987; Kador, 1988). Recently, it has been reported that tolrestat co-crystallizes at the active site of pig lens ALR2 (Urzhumtsev et al., 1997) and aldehyde reductase (El-Kabbani et al., 1997). The reported crystal structure of ALR2 complexed with $NADP^+$ and tolrestat reveals that the carboxylate of tolrestat is hydrogen bonded to Tyr48, His110 and Trp111 while the aromatic portion is surrounded by the hydrophobic residues such as Trp20, Trp219, Trp111, Phe122, and Leu300. The observed hydrogen bonding interaction between tolrestat and ALR2 is essentially identical to that reported in the crystal structure of ALR2 complexed with zopolrestat (Wilson et al. 1993; Wilson et al. 1995). This similarity in hydrogen bonding interaction indicates that hydrogen bonding or charge interaction between ALR2 and the ionized portion of inhibitors plays an important role in enhancing the binding affinity of inhibitors (Lee et al., 1998).

Tolrestat (**1**) has a number of rotatable bonds suggesting that it can have several conformations in solution. A thorough understanding of rotamers of tolrestat and their binding modes to ALR2 can help designing aldose reductase inhibitors with tighter binding and greater selectivity which would reduce potential side effects. Molecular dynamics simulations on ALR2 complexed with tolrestat have shown a number of distinct binding modes of tolrestat (Rastelli & Costantino, 1998). Experimentally, tolrestat is known to exist as two rotamers, **1** and **2**, about the C–N bond of the thioamide group in solution with an en-

* Address correspondence to: Yong S. Lee, Ph.D., Building 10, Room 10B11, National Eye Institute, National Institutes of Health, 10 Center Dr. MSC 1850, Bethesda, Maryland 20892-1850. Phone: (301) 496-8548; Fax: (301) 402-2399; E-mail: yongslee@helix.nih.gov.

ergy barrier of 25–26 kcal/mol (Lee & Querijero, 1985). The larger rotational energy barrier about ϕ_1 has been attributed to a partial double bond character of the C–N bond and a restricted rotation of the thioamide group about the bond joining it to the naphthalene ring (Lee & Querijero, 1985). As the rotation about ϕ_1 is hindered, the rotation of the thioamide group about the C-C bond (ϕ_2) is also likely restrained due to steric repulsion between the thioamide group and the hydrogens at position 2 and 8 of the naphthalene ring. This indicates that more than two rotamers of tolrestat exists in solution, and suggests that the binding mode of rotamer **1** to ALR2 observed in the crystal structure differs from that of other rotamers.

In the present study, the geometries and energetics of the rotamers of tolrestat were investigated by varying the torsional angle ϕ_1 and ϕ_2 of tolrestat with the use of the PM3 quantum mechanical method (Stewart, 1989). Each of the PM3 optimized rotamers of tolrestat was then docked to the active site of human aldose reductase and energy minimized by CHARMM (Brooks et al., 1983) in order to determine their extent of binding to aldose reductase complexed with $NADP^+$.

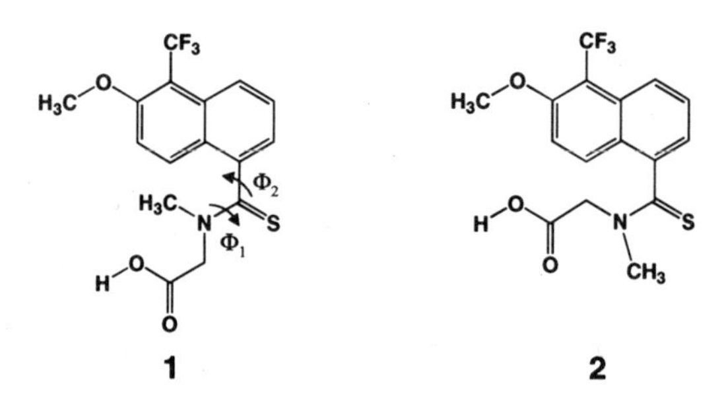

Formulas 1 and 2.

2. METHODS

2.1. Geometry and Energetics of the Rotamers of Tolrestat by the PM3 Hamiltonian

A potential energy surface was constructed using an 18×18 grid generated by varying ϕ_1 (90°) and ϕ_2 (-10°) of the negatively charged tolrestat at the center of the grid in increments of -20° and 20° toward both axes. At each grid point a full geometry optimization was performed with ϕ_1 and ϕ_2 fixed at the respective grid value with the PM3 Hamiltonian along with key words EF and GNORM = 0.1 as implemented in AMPAC 6.02 (Semichem, 1998). To compare the calculated energetics in the gas phase to those in water, single point energy calculations were performed on the grid points having the geometry of minimum or maximum energy with the PM3-SM3 solvation model.

2.2. Binding Modes of the Rotamers of Tolrestat by CHARMM

The coordinates of human aldose reductase complexed with NADP(H) and crystal water (1ads) were obtained from the Brookhaven Protein Data Bank. Hydrogens were added to the amino acid residues of human aldose reductase using the HBUILD routine of CHARMM (Brooks et al., 1983). PM3 optimized rotamers of tolrestat were manually docked into the active site of aldose reductase complexed with $NADP^+$ so that the aromatic portion of tolrestat became positioned between Leu300 and Phe122 as reported in the crystal structure (Urzhumtsev et al., 1997). When possible, the carboxyl anion of the rotamers was positioned within the distance of the hydrogen bonding interactions with Tyr48 and His110. With the exception of His110, all His residues were treated as neutral with hydrogen assigned to the Nδ1 of neutral histidine. His110 was treated as positively charged based on the crystallographic evidence that the carboxylate of zopolrestat is salt-linked to the Nε2 hydrogen of His110 (Wilson et al. 1993; Wilson et al. 1995). All Asp and Glu residues were assumed to be negatively charged while Arg and Lys were positively charged. Charge assignment on $NADP^+$, tolrestat, the hydration of the complex of ALR2–$NADP^+$– tolrestat, and the subsequent energy minimization by CHARMM with an all-atom parameter set (Molecular Simulations, 1992) were carried out according to the procedures published elsewhere (Lee et al., 1998). The interaction energy was obtained by taking out all hydrated water molecules from the energy optimized complex and then subtracting the energy of the complex of ALR2–$NADP^+$ and tolrestat from the total energy. The interaction energy is the sum of the van der Waals and electrostatic interactions.

3. RESULTS

A conformational energy surface of tolrestat was obtained by plotting the heat of formation (ΔH_f) of tolrestat at each grid point relative to that at the center of the grid. Figure 1 is the contour plot of the potential energy surface illustrating the existence of four stable rotamers of tolrestat with a sizable energy barrier ($\geq$19 kcal/mol) among them. Figure 2 depicts the geometries of the four rotamers at the bottom of each potential well. These rotamers were labeled as E and Z with respect to the rotation of the N-methyl group about the C–N bond. E and Z rotamers were then further configured as *R* and *S* to indicate the spatial orientation of the N-methylglycine. Table 1 lists the relative energetics of the

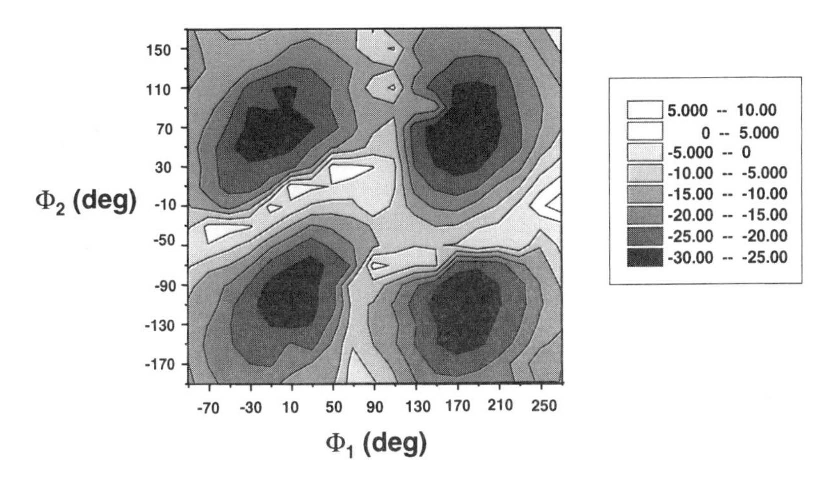

Figure 1. Energy contour plot of a potential energy surface of tolrestat.

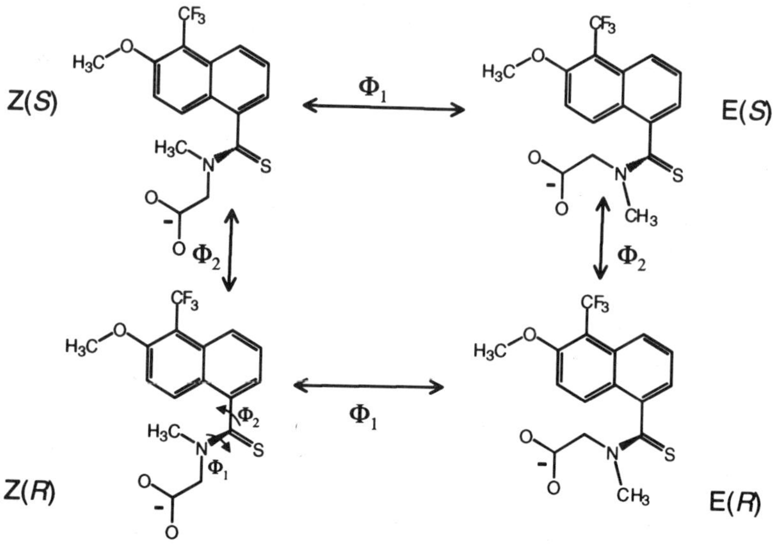

Z(*S*) E(*S*) Z(*R*) E(*R*)

Figure 2. Schematic diagram depicting the geometries of the four rotamers at the bottom of each potential well in Figure 1.

four rotamers in both gaseous and aqueous phase. Table 2 lists the rotational energy barriers of the Z(*R*) rotamer to both E(*R*) and Z(*S*).

The binding modes of each of the rotamers modeled (Figure 2) at the active site of ALR2 are shown in Figures 3 and 4. Figure 3 illustrates that the carboxylate of the Z(*R*) and Z(*S*) rotamers forms hydrogen bonds with Tyr48, His110 and Trp111 while the aromatic portion of tolrestat is positioned to form van der Waals and aromatic–aromatic interactions with Trp20, Trp111 and Phe122. The trifluoromethyl group of both rotamers also hydrogen bonds to Ser302. The major difference between the two rotamers is that Cys298 hydrogen-bonds to the thioamide group of Z(*S*). The aromatic portion of both E(*R*) and

Table 1. Relative energetics of the rotamers of tolrestat (kcal/mol)

	PM3	PM3-SM3*
E(*R*)	0	0
E(*S*)	0.7	0
Z(*R*)	2.0	0.8
Z(*S*)	1.7	0.8

*Single point energy calculation in water

Table 2. Energy barrier between the rotamers of tolrestat (kcal/mol)

	PM3	PM3-SM3	Exp.
Z(*R*) to E(*R*)	19.3	18.5	25-26[c]
Z(*R*) to Z(*S*)	29.1	26.0	

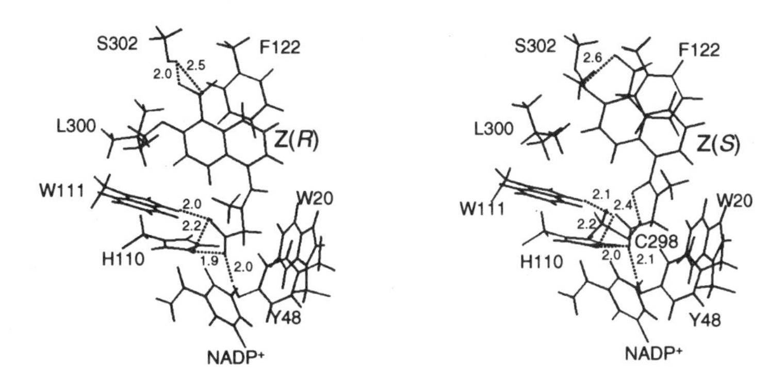

Figure 3. CHARMM energy minimized structure of aldose reductase complexed with the Z(*R*) and Z(*S*) rotamers. Dashed lines indicate the hydrogen bonding interactions (3.0 Å).

E(*S*) rotamers occupies a similar position at the active of ALR2 (Figure 4). While the carboxylate of the E(*R*) rotamer forms limited hydrogen bonds with His110 and Trp111, the carboxylate of E(*S*) is not oriented to form hydrogen bonds with ALR2 except Cys298. Table 3 lists the interaction energy between the individual rotamers of tolrestat and aldose reductase.

4. DISCUSSION

The variation of the rotational angles ϕ_1 and ϕ_2 result in four distinct rotamers of tolrestat as depicted in Figure 2. Energetics calculations in the gas phase indicate that the E rotamers are more stable than the Z rotamers by 1 to 2 kcal/mol (Table 1). The energy difference was reduced to 0.8 kcal/mol when the solvent effect was included. These energetics differ from the report that the Z rotamer is more stable by *ca.* 0.6 kcal/mol than the E rotamer (Lee & Querijero, 1985). The contour plot (Fig.1) illustrates that rotation about ϕ_2 is more hindered than about ϕ_1. Both calculations in the gas phase and in water (Table 2) indicate that the energy barrier between the Z(*R*) and Z(*S*) rotamers is *ca.* 8 to 10

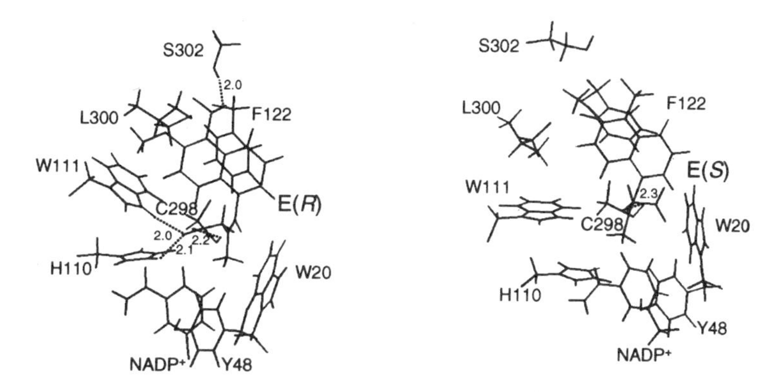

Figure 4. CHARMM energy minimized structure of aldose reductase complexed with the E(*R*) and E(*S*) rotamers. This figure and Figure 3 were produced using Molden.

Table 3. CHARMM interaction energy between the rotamers of tolrestat and ALR2 (kcal/mol)

	Total	ES	VDW
E(*R*)	−125.5	−92.4	−33.1
E(*S*)	−99.9	−63.8	−36.1
Z(*R*)	−144.7	−111.7	−32.9
Z(*S*)	−144.5	−111.7	−32.8

kcal/mol larger than that between the Z(*R*) and E(*R*). The larger energy barrier about ϕ_2 may arise from a stronger steric repulsion between the N-methyl and the sulfur atom of the thioamide group and the hydrogens at position 2 and 8 of the naphthalene ring. The calculated rotational energy barrier of the Z(*R*) rotamer to the E(*R*) rotamer is *ca.* 6 to 7 kcal/mol lower than the observed experimental value. This can be attributed to the fact that semi-empirical quantum chemical methods such as PM3 generally underestimate the rotational energy barrier.

As illustrated in Figure 2, rotation about ϕ_1 and ϕ_2 converts the Z(*R*) rotamer to the E(*R*) diastereomer and Z(*S*) enantiomer, respectively. The Z(*R*) rotamer becomes the E(*S*) diastereomer upon concerted rotation about ϕ_1 and ϕ_2. However, the sizable energy barrier (> 19 kcal/mol) among these rotamers should reduce their rate of interconversion. In addition to the sizable energy barrier, the comparable energetics of the four rotamers suggest that they should exist in solution. The existence of four rotamers of tolrestat is supported by the observation that N,N-disubstituted-2-methyl-naphthalenecarboxamide (**3**) was separated into four rotamers by the use of a stationary chiral phase (Pirkle et al., 1993). The topology of compound **3**, which possess a naphthalene ring and an amide group with two rotatable bonds, resembles tolrestat.

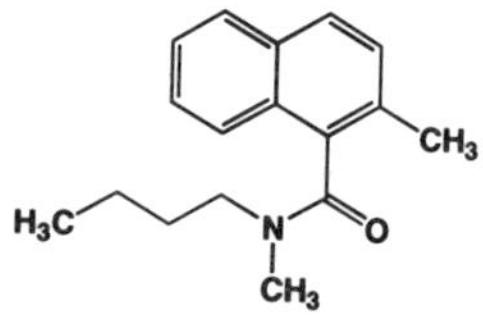

3

Formulae 3.

The hydrogen bonding pattern and the spatial orientation of the aromatic portion of the Z(*R*) rotamer at the active site of ALR2 (Figure 3) are in good agreement with those reported in the crystal structure (Urzhumtsev et al., 1997). Figure 3 also illustrates that the binding mode of the Z(*S*) rotamer to ALR2 is also comparable to that of the Z(*R*) rotamer although the two rotamers differ in the orientation of the N-methylglycine. This binding mode similarity indicates that the binding affinity of the Z(*R*) and Z(*S*) rotamers to ALR2 is comparable. This premise is supported by the similar CHARMM interaction energy of the two rotamers with ALR2 (Table 3). Tolrestat crystallized with either aldose reductase or aldehyde reductase corresponds to the Z(*R*) rotamer. Most likely, the Z(*R*) rotamer is slightly favored to crystallize with aldose reductase or aldehyde reductase over the Z(*S*) rotamer. Interestingly, the interaction energy of the Z(*R*) rotamer is 0.2 kcal/mol stronger than that of the Z(*S*) rotamer with aldose reductase.

Figure 4 shows that the carboxylate of the E(*R*) rotamer forms hydrogen bonds with His110 and Trp111 while the carboxylate of the E(*S*) rotamer is not positioned to form hydrogen bonds with either Tyr48, His110 or Trp111 at the active site. This loss of the hydrogen bonding interaction between the E(*S*) rotamer and ALR2 is reflected in a weaker electrostatic interaction energy (-63.8 kcal/mol) compared to that of -92.4 kcal/mol between the E(*R*) rotamer and ALR2. The electrostatic interaction between the Z rotamers and ALR2 is *ca.* 20 to 48 kcal/mol stronger than that of the E rotamers and ALR2. This indicates that the Z(*R*) and Z(*S*) rotamers selectively bind to ALR2 compared to the E (*R*) and E(*S*) rotamers because of the stronger hydrogen bonding interactions. This premise is supported by binding studies (Sugiyama et al., manuscript in preparation) which have shown only the Z rotamers bind to the complex of ALR2–NADP(H).

5. CONCLUSION

Four distinct rotamers of tolrestat can exist in solution due to a sizable energy barrier about the rotational angles ϕ_1 and ϕ_2, and comparable energetics among the rotamers. Comparisons of the binding modes and interaction energies of the four rotamers with aldose reductase indicate that the Z(*R*) and Z(*S*) rotamers selectively bind to aldose reductase by forming strong hydrogen bonding interactions with aldose reductase.

ACKNOWLEDGMENTS

Useful discussions with Drs. Jen-Yue Tsai, Andrea Sinz, and Milan Hodoscek are gratefully acknowledged.

REFERENCES

Brooks, B.R., Bruccoleri, R.E., Olafson, B.D., States, D.J., Swaminathan, S. and Karplus, M. CHARMM: a program for macromolecular energy, minimizations, and molecular dynamics calculations. J. Comput. Chem. 4 (1983) 187–217.

Dvornik, D. In Aldose Reductase Inhibition An Approach to the Prevention of Diabetic Complications. Biomedical Information Corporation; Porte, D., Ed.; McGraw-Hill: New York, (1987), Chapter 4.

El-Kabbani, O., Carper, D., McGowan, M.H., Devedjiev, Y., Rees-Milton, K.J. and Flynn, T.G. Studies on the Inhibitor -binding site of porcine aldehyde reductase. crystal structure of the holoenzyme-inhibitor ternary complex. Proteins: Struct., Func. Gen. 29 (1997) 186–192.

Kador, P.F. The role of aldose reductase in the development of diabetic complications. Med. Res. Rev. 8 (1988) 325–352.

Kinoshita, J.H. Aldose reductase in the diabetic eye. Am. J. Ophthalmol. 102(1986) 685–692.

Lee, H.K and Querijero G. Kinetics and mechanism of thioamide rotational isomerism: *N*-thianaphthoyl-*N*-methyl glycine derivative. J. Pharm. Sci. 74 (1985) 273–276.

Lee, Y.S., Chen Z. and Kador P.F. Molecular modeling studies of the binding modes of aldose reductase inhibitors at the active site of human aldose reductase. in press.

Molecular Simulation Inc., Parameter file for CHARMm version 22 (1992) Waltham, MA.

Pirkle, W.H., Welch, C. J. and Zych, A. Chromatographic investigation of the slowly interconverting atropisomers of hindered naphthamides. J. Chromatogr. 648 (1993) 101–109.

Rastelli, G. and Constantino, L. Molecular dynamics simulations of the structure of aldose reductase complexed with the inhibitor tolrestat Bioorg. Med. Chem. Lett. 8 (1998) 641–646.

Semichem Inc., AMPAC 6.02, (1998), Shawnee, KS.

Stewart, J.J.P Optimization of parameters for semiempirical methods. J. Comp. Chem. 10 (1989) 209–220.

Urzhumtsev, A., Tete-Favier, F., Mitschler, A., Barbanton, J., Barth, P., Urzhumtseva, L., Bielmann, J.F., Podjarny, A.D. and Moras, D. A specificity pocket inferred from the crystal structures of the complexes of aldose reductase with the pharmaceutically important inhibitors of tolrestat and sorbinil. Structure 5 (1997) 601–612.

Wilson, D. K., Tarle I., Petrash, J.M. and Quiocho, F.A. Refined 1.8 Å structure of human aldose reductase complexed with the potent inhibitor zopolrestat. Proc. Natl. Acad. Sci. U.S.A. 90 (1993) 9847–9851.

Wilson, D.K., Nakano, T., Petrash, J.M. and Quiocho, F.A. 1.7 Å structure of FR-1, a fibroblast growth factor induced member of the aldo-keto reductase family complexed with coenzyme and inhibitor. Biochemistry 34 (1995) 14323–14330.

STEREOSELECTIVE HIGH-AFFINITY REDUCTION OF KETONIC NORTRIPTYLINE METABOLITES AND OF KETOTIFEN BY ALDO-KETO REDUCTASES FROM HUMAN LIVER

Ursula Breyer-Pfaff and Karl Nill

Department of Toxicology
University of Tuebingen
D-72074 Tuebingen, Germany

1. INTRODUCTION

Oxidative biotransformation of drugs and other xenobiotics by hydroxylation at an aliphatic carbon atom may produce secondary alcohols that can undergo reversible oxidation by an aldo-keto reductase. An example are the major oxidative metabolites of the antidepressant nortriptyline. These consist of two pairs of enantiomeric alcohols, (-)- and (+)-E- and -Z-10-hydroxynortriptyline, which on administration to human volunteers were only partially excreted in urine in unchanged form or as their glucuronides (Nusser et al., 1988; Dahl-Puustinen et al., 1989). On incubation with human liver fractions, the (+)-enantiomer of E-10-hydroxynortriptyline was preferentially oxidized in the presence of NAD^+ to the corresponding ketone E-10-oxonortriptyline (E-10-Oxo-NT) which occurs as a minor metabolite in urine of patients (Prox and Breyer-Pfaff, 1987).

The reverse reductive reaction is carried out by human liver cytosol in the presence of NADPH and converts E- and Z-10-Oxo-NT nearly exclusively to the (+)-enantiomers of E- and Z-10-hydroxynortriptyline. This reduction is also catalysed by liver cytosol from rabbit, but not from rat or guinea pig. Preferential incorporation of the pro-4R hydrogen of tritium-labelled NADPH pointed to the involvement of an aldo-keto reductase (Breyer-Pfaff and Nill, 1995).

Alternatively, ketonic groups can be present in drug substrates and can by their reduction give rise to alcohols as major metabolites as in the cases of the neuroleptic haloperidol (Forsman and Larsson, 1978) and the antiasthmatic ketotifen (Le Bigot et al., 1987).

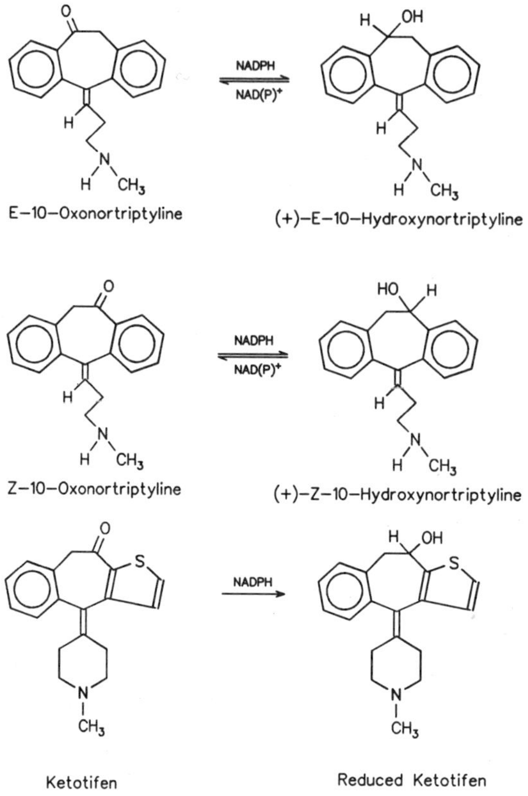

Figure 1. Structural formulas of ketonic substrates and their reduction products. Ketotifen forms two conformational enantiomers and their reduction products consist of isomeric alcohols that are interconvertible.

In view of the great structural similarity between E- and Z-10-Oxo-NT on the one hand and ketotifen on the other (Figure 1), the reduction of these compounds by individual carbonyl-reducing enzymes from human liver was to be studied. Since ketotifen consists of the stable conformational R-(+)- and S-(-)-enantiomers (Polivka et al., 1989), these were investigated separately. The procedure for the isolation of aldo-keto reductases from human liver tissue (Hara et al., 1990) was modified and the isolation of carbonyl reductase was included.

2. METHODS

2.1. Materials

Human liver samples were kindly supplied by Prof. Dr. Lauchart, Department of Surgery, University of Tübingen. These were either samples from livers excluded from transplantation for medical reasons or excess normal tissue obtained on partial hepatectomy for tumor metastases. They were cut into pieces of 5–10 g and stored at -80°C.

R-(+)- and S-(-)-ketotifen and reduced ketotifen were generously donated by Novartis Pharma AG (Basel, Switzerland) and E- and Z-10-oxonortriptyline (E- and Z-10-Oxo-NT) by A. Jorgensen (Kobenhavn, Denmark). Sephadex G-100, Q Sepharose, CM Sepharose, DEAE Sepharose, polybuffer exchanger PBE 94, and the polybuffers 74 and 96 were purchased from Pharmacia (Freiburg, Germany), hydroxyapatite Bio-Gel HTP from Bio-Rad (München, Germany), Reactive Red 120-Agarose and S-(+)- and R-(-)-indanol from Sigma-Aldrich (Deisenhofen, Germany), a 50-ml ultrafiltration chamber and filters YM10 from Amicon (Witten, Germany).

2.2. Enzyme Isolation

Liver homogenates were prepared with four volumes of Tris-buffered sucrose and cytosol was obtained by fractionated centrifugation, in the last step at 85000 *g* for 1 h. Samples were concentrated to half their volume by ultrafiltration and stored at -80°C. Quantities corresponding to 5–7 g of liver were applied to a 2.5 × 100 cm column with Sephadex G-100 in 25 mM Tris-HCl buffer pH 7.4 and eluted in fractions of 5 ml. Those exhibiting Z-10-Oxo-NT reductase were concentrated to about 0.5 ml by ultrafiltration. As in all further fractionations, the medium for storage of the concentrates was changed to buffer A (see below) and samples were stored at -80°C.

Separation of gel filtrates was achieved on Q Sepharose with NaCl gradients in buffer A (5 mM Tris-HCl pH 8 containing 0.5 mM EDTA, 5 mM 2-mercaptoethanol and 20% glycerol) according to Hara et al. (1990) and three fractions were obtained:

Q1 eluted without NaCl containing Z-10-Oxo-NT reductase and S-indanol oxidase activities;

Q2 eluted at 0.05–0.1 M NaCl with Z-10-Oxo-NT and 4-benzoylpyridine reductase activities;

Q3 eluted at 0.5 M NaCl with S-indanol oxidase and 4-benzoylpyridine reductase activities.

Further purification by ion-exchange chromatography or on Reactive Red 120-Agarose was performed in the presence of EDTA and 2-mercaptoethanol and chromatofocussing on PBE 94 in the presence of 2-mercaptoethanol. A gradient of NaCl in sodium phosphate pH 6.5 eluted from CM Sepharose single active peaks on purification of Q1 (Q1-CM with 0.1 M NaCl) and Q2 (Q2-CM with 0.05–0.09 M NaCl) and two peaks from Q3 (Q3-CM1 not retained on the column and Q3-CM2 eluted with 0.05–0.09 M NaCl that was combined with Q2-CM).

Pure AKR1C1 (DD1 of Hara et al., 1990) was obtained from Q1-CM on PBE 94 equilibrated with 25 mM imidazole-HCl pH 7.8. With Polybuffer 96 (diluted 1+12) pH 6 as eluent, AKR1C1 appeared following a protein peak without enzymatic activity.

AKR1C2 (DD2 of Hara et al., 1990) and carbonyl reductase (EC 1.1.1.184) contained in Q2-CM and Q3-CM2 were separated on Bio-Gel HTP by a sodium phosphate gradient in Tris-HCl pH 7.5 containing 5 mM 2-mercaptoethanol. 4-Benzoylpyridine reductase activity (corresponding to carbonyl reductase) was eluted with 0.06–0.1 M phosphate and Z-10-Oxo-NT reductase activity (AKR1C2) with 0.11–0.16 M phosphate. The two fractions were purified separately on PBE 94 equilibrated with 25 mM imidazole pH 7.4 and eluted with PB 74 (diluted 1+7) pH 5. Carbonyl reductase appeared as two peaks at pH 7.6 and 7.4–7.25 that were collected separately. AKR1C2 was eluted at pH 7.45–7.3.

Q3-CM1 in 10 mM sodium phosphate pH 6.5 containing 20% glycerol was applied to a 3-ml column of Reactive Red 120-Agarose and left at 4°C for 17 h. The unadsorbed fraction Q3-CM1-R1 containing 4-benzoylpyridine reductase activity was eluted with 10 mM Tris-HCl pH 7.5 in 20 % glycerol and purified on PBE 94 under the same conditions as carbonyl reductase from Q2-CM-HA1. Two clearly separated peaks appeared around pH 7.25 and 6.9. Buffer with 2 M NaCl eluted from the Reactive Red column a smaller quantity of 4-benzoylpyridine reductase activity and an indanol oxidase (Q3-CM1-R2). For chromatofocussing of Q3-CM1-R2, PB 74 was adjusted to 4 and contained 20 % glycerol, while 2-mercaptoethanol was omitted. Following peaks of 4-benzoylpyridine reductase activity around pH 7.1 and 6.9, indanol oxidase activity was eluted in 2–3 peaks between pH 6.3 and 5.6, the last one containing large quantities of contaminating proteins. From the first peak, AKR1C4 (DD4 of Hara et al., 1990) could be purified on DEAE Sepharose with an NaCl gradient in 10 mM Tris-HCl pH 7.9, the enzyme being eluted at 0.08 M NaCl.

Protein was measured according to Bradford (1976) with bovine serum albumin as standard. SDS-PAGE was standardized by a 10 kDa protein ladder.

2.3. Measurement of Enzyme Activities

All tests were carried out at 37°C in 100 mM Tris-HCl pH 7.4 in the presence of 8 mM $MgCl_2$ and 25 mM KCl. NADPH was generated from 0.2 mM $NADP^+$, 2 mM glucose 6-phosphate and 1 U/ml glucose 6-phosphate dehydrogenase when reductase activities were determined by HPLC measurement of products. For reductase determination by the optical test at 339 nm, NADPH was added at 0.1 mM, while oxidase activity was measured with 0.25 mM $NADP^+$. Reduction of E- and Z-10-Oxo-NT and ketotifen: Incubations in a volume of 0.36 ml were, after a pre-incubation of 5 min, started by the addition of substrate. The reaction was stopped after 10 min (4–20 min in kinetic measurements) by mixing with 0.04 ml 2 N perchloric acid and cooling. Following centrifugation, the supernatant was injected for HPLC in the previously described system (Breyer-Pfaff and Nill, 1995). Reduced ketotifen was eluted as two peaks at 8.5 and 11.7 min connected by a low plateau. The sum of the heights of the two peaks was used for quantitation. The isomers contained in them are interconvertible as shown by isolation from the HPLC eluate and re-injection.

Product quantities increased linearly with time; the increase with protein concentration was linear except when substrate inhibition was present.

2.4. Calculations

Kinetic parameters were calculated according to the Michaelis-Menten equation by non-linear least-squares regression analysis. Intrinsic clearance (Cl_{int}) was calculated as V_{max}/K_m. In cases of substrate inhibition, a formula for non-competitive inhibition was used. Substrate concentrations were corrected for substrate consumption during the incubation. Using 6–11 different substrate concentrations, r^2 values of 0.99 or higher were obtained in all cases.

3. RESULTS

The fractionation scheme and the enzyme quantities recovered are presented in Table 1. Z-10-Oxo-NT reductase activities measured in gel filtrates exceeded those found in cytosol by a factor of 2–3, apparently due to the removal of inhibitors. Recoveries of reductase activity in the following steps usually were between 50 and 90 % and in the cases of AKR1C1 and 1C2 were the same when measured with S-(+)-indanol as a substrate.

3.1. Aldo-Keto Reductases

AKR1C1 and 1C2 (DD1 and DD2) were obtained in electrophoretically pure form with estimated molecular weights of 36 kDa. The preparation could be simplified in com-

Table 1. Fractionation of cytosol from HL 25 (35 g) and specific activities measured with Z-10-Oxo-NT 10 μM, S-(+)-indanol 0.5 mM, or 4-benzoylpyridine 1 mM

Purification by	Resulting fraction or enzyme	Protein (mg)	Specific activity (mU/mg) with		
			Z-10-Oxo-NT	S-Indanol	4-Benzoyl-pyridine
	cytosol	810	0.6	16	
Sephadex G-100	gel filtrate	113	13	128	
Q Sepharose	Q1	6.0	53	630	
CM Sepharose	Q1-CM	2.6	69	810	
PBE 94	AKR1C1	0.6	180	2540	
Q Sepharose	Q2	31	26	210	
CM Sepharose	Q2-CM	9.2	62	326	
Bio-Gel HTP	Q2-CM-HA1	1.3			3200
PBE 94	CR	0.11			14000
		0.15			7000
Bio gel HTP	Q2-CM-HA2	3.7	123		
PBE 94	AKR1C2	1.7	145	710	
Q Sepharose	Q3	45		37	
CM Sepharose	Q3-CM1	26	53		303
Reactive red	Q3-CM1-R1	14			165
PBE 94	CR	0.05			8600
		0.18			4100*
Reactive red	Q3-CM1-R2	3.5		240	1020
PBE 94	CR	0.14			9600
PBE 94	Q3-CM1-R2-PB	0.27		660	
		0.86		150	
DEAE Sepharose	AKR1C4	0.15		2000*	

*Specific activities of final products exhibiting contaminations on electrophoresis.
Gel filtration was performed in three fractions. Intermediate products are specified as in "Methods", final products as AKR1C1, 1C2 or 1C4 or as CR (carbonyl reductase).

parison to that of Hara et al. (1990) by omission of the initial ammonium sulfate precipitation and of one purification step for AKR1C1. In contrast to findings of Hara et al. (1990), this protein was eluted as a single peak from CM Sepharose. AKR1C4 exhibited on electrophoresis a small contamination of 39 kDa besides the main band with 36 kDa. Further criteria for the identity of the proteins were:

The amino acid sequence of a tryptic fragment of AKR1C2 matched amino acids 305–318 of DD2 according to Hara et al. (1996).

K_m values for the oxidation of S-(+)-indanol by AKR1C1, 1C2 and 1C4 were in accordance with those of Hara et al. (1990) as was the stereoselectivity when the oxidation rates of R-(-)-indanol were compared with those of the S-enantiomer.

The kinetics of alcohol production from ketonic substrates differed between the closely related AKR1C1 and 1C2 and between isomeric substrates. Each reaction was investigated with purified enzymes from three different livers and the data were in good accordance (Table 2). The highest affinities towards both enzymes (K_m around 3 μM) were observed with Z-10-Oxo-NT closely followed by that of (-)-ketotifen towards AKR1C2, whereas the K_m value for its reduction by AKR1C1 was 15fold higher. V_{max} values varied 20fold among the four ketotifen-enzyme combinations and 12fold among those of 10-Oxo-NT isomers and enzymes. AKR1C1 was inhibited by high concentrations of Z- and E-10-Oxo-NT, and a small inhibitory effect was exerted by E-10-Oxo-NT on AKR1C2 (Table 2). An example of kinetics with substrate inhibition is depicted in Figure 2.

Table 2. Kinetic parameters for the reduction of (+) and (-)-ketotifen and of Z- and E-10-Oxo-NT by AKR1C1 isolated from cytosol of HL 15, 17 and 23, and by AKR1C2 isolated from HL 15, 17, and 24. Enzyme quantities varied betweeen 0.6 and 12 μg per 0.36 ml incubate

Substrate	Enzyme	K_m (μM)	V_{max} (mU/mg)	K_i (μM)	Cl_{int} (ml/min per mg)
(+)-ketotifen	AKR1C1	10.9 (2.0)	5 (1)		0.44
	AKR1C2	7.5 (0.1)	58 (8)		7.7
(-)-ketotifen	AKR1C1	53 (2.6)	92 (10)		1.7
	AKR1C2	3.6 (0.2)	100 (5)		28
Z-10-Oxo-NT	AKR1C1	2.6 (0.5)	313 (46)	32 (2.5)	120
	AKR1C2	3.1 (0.3)	173 (3)		57
E-10-Oxo-NT	AKR1C1	6.4 (1.3)	140 (10)	30 (5)	22
	AKR1C2	10.2 (1.4)	26 (0.7)	250 (126)	2.6

Values are means (SD), N = 3.

3.2. Carbonyl Reductase

Enzymes with apparent molecular weights of 33–35 kDa in gel electrophoresis and with high specific activities in reducing 4-benzoylpyridine or menadione were recovered from various fractions. They were inhibited by 80–90 % by 10 μM rutin, a potent inhibitor of carbonyl reductase (Wermuth 1981), whereas 1 mM barbital, an aldehyde reductase inhibitor, reduced the activity by 20–30% only.

The enzymes were devoid of Z-10-Oxo-NT reductase or S-indanol oxidase activity. Since there was a complete separation of 4-benzoylpyridine reductases on ion-exchange chromatography on CM Sepharose, enzymes isolated from the Q2-CM-HA1 and Q3-CM1-R1 fractions were subjected to partial amino acid sequence analysis. In both cases, the sequences matched amino acids 113–119 of carbonyl reductase (EC 1.1.1.184) (Wermuth et al. 1988). Since chromatofocussing resulted in further separation into 2–3 peaks, carbonyl reductase occurred in at least four different forms. The two exhibiting identical amino acid sequences were also compared with regard to their substrate spectrum. The relative rates at which they reduced 1 mM 4-benzoylpyridine, 0.25 mM menadione and 1 mM 4-nitrobenzaldehyde were 1 : 1.8 : 0.56 and 1 : 1.6 : 0.54, respectively.

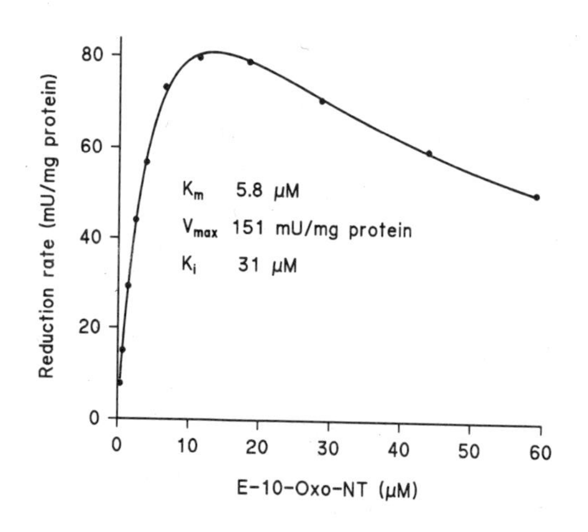

Figure 2. Kinetics of E-10-Oxo-NT reduction by AKR1C1 from HL 23. The protein concentration was 4 μg/ml and the incubation time 10 min.AKR1C4 produced traces of alcoholic metabolites from Z-10-Oxo-NT and (-)-ketotifen, but none from (+)-ketotifen.

4. DISCUSSION

To our knowledge, this is the first report describing a procedure for the parallel isolation of aldo-keto reductases and carbonyl reductase from human liver. Wermuth (1981) described separation by DEAE Sepharose chromatography of purified carbonyl reductase from human brain into three fractions with isoelectric points of about 7, 8 and 8.5. Less purified preparations contained an additional protein of pI 5.2 with very similar enzymatic properties. Carbonyl reductase from human testis was separated by chromatofocussing into fractions eluting at pH 7.74, 7.63 and 7.28 (Inazu et al., 1992). Modification of the native most basic enzyme form by binding of 2-oxomono- or -dicarboxylic acids to a lysine residue and autocatalytic reduction of the condensation product is probably responsible for the occurrence of multiple forms (Wermuth et al. 1993). The present investigation revealed the occurrence of at least one additional isoform. The relative velocities in the reduction of 4-benzoylpyridine, menadione and 4-nitrobenzaldehyde were in good accordance with those found with carbonyl reductase from human brain (Inazu et al., 1992).

Though the aldo-keto reductases AKR1C1 and 1C2 differ by seven amino acids only (Hara et al., 1996; Matsuura et al., 1997), their substrate specificities for steroids and inhibitor sensitivities are clearly different. These differences could largely be abolished by replacing Leu-54 in AKR1C1 by Val, the corresponding amino acid in AKR1C2 (Matsuura et al., 1997). Whereas AKR1C2 reduces some ketonic steroids with K_m values in the low or sub-micromolar range (Deyashiki et al., 1992; Hara et al., 1996), an investigation with drug substrates from various classes resulted in K_m values between 190 and 3200 μM for AKR1C1 and 1C2 (Ohara et al., 1995). Thus, the structurally similar substrates Z- and E-10-Oxo-NT and (+)- and (-)-ketotifen (Figure 1) are distinguished by their low K_m values of 2.6–53 μM (Table 2). These may be the reason why in human urine the corresponding alcohols are predominantly detected. This applies to ketotifen (Le Bigot et al., 1987) which is administered as the racemic ketone as well as to the isomers of 10-Oxo-NT which are produced from (+)-Z- and (+)-E-10-hydroxynortriptyline in human liver cytosol under physiological conditions in the presence of NAD^+ (Breyer-Pfaff and Nill, 1995). The stereoselectivity of the high-affinity reduction of these substrates may make them appropriate tools for the elucidation of the active site structure of the enzymes. AKR1C1 apparently possesses an additional binding site with relatively high affinity the occupation of which results in an inhibition of the catalytic activity.

REFERENCES

Bradford, M.M. (1976) Rapid and sensitive method for the quantification of microgram quantities of protein utilising the principle of protein-dye binding, *Anal. Biochem.* 72, 248–254.

Breyer-Pfaff, U. and Nill, K. (1995) Stereoselective reversible ketone formation from 10-hydroxylated nortriptyline metabolites in human liver, *Xenobiotica* 25, 1311–1325.

Dahl-Puustinen, M.-L.,Perry, Jr., T.L., Dumont, E., von Bahr, C., Nordin, C. and Bertilsson, L. (1989) Stereoselective disposition of racemic E-10-hydroxynortriptyline in human beings, *Clin. Pharmacol. Ther.* 45, 650–656.

Deyashiki, Y., Taniguchi, H., Amano, T., Nakayama, T., Hara, A. and Sawada, H. (1992) Structural and functional comparison of two human liver dihydrodiol dehydrogenases associated with 3α-hydroxysteroid dehydrogenase activity, *Biochem. J.* 282, 741–746.

Forsman, A. and Larsson, M. (1978) Metabolism of haloperidol, *Curr. Ther. Res.* 24, 567–568.

Hara, A., Matsuura, K., Tamada, Y., Sato, K., Miyabe, Y., Deyashiki, Y. and Ishida, N. (1996) Relationship of human liver dihydrodiol dehydrogenases to hepatic bile-acid-binding protein and an oxidoreductase of human colon cells, *Biochem. J.* 313, 373–376.

Hara, A., Taniguchi, H., Nakayama, T. and Sawada, H. (1990) Purification and properties of multiple forms of dihydrodiol dehydrogenase from human liver, *J. Biochem.* 108, 250–254.

Inazu, N., Ruepp, B., Wirth, H. and Wermuth, B. (1992) Carbonyl reductase from human testis: purification and comparison with carbonyl reductase from human brain and rat testis, *Biochim. Biophys. Acta* 1116, 50–56.

Le Bigot, J.F., Begue, J.M., Kiechel, J.R. and Guillouzo, A. (1987) Species differences in metabolism of ketotifen in rat, rabbit and man: demonstration of similar pathways in vivo and in cultured hepatocytes, *Life Sci.* 40, 883–890.

Matsuura, K., Deyashiki, Y., Sato, K., Ishida, N., Miwa, G. and Hara, A. (1997) Identification of amino acid residues responsible for differences in substrate specificity and inhibitor sensitivity between two human liver dihydrodiol dehydrogenase isoenzymes by site-directed mutagenesis, *Biochem. J.* 323, 61–64.

Nusser, E., Nill, K. and Breyer-Pfaff, U. (1988) Enantioselective formation and disposition of (E)- and (Z)-10-hydroxynortriptyline, *Drug Metab. Dispos.* 16, 509–511.

Ohara, H., Miyabe, Y., Deyashiki, Y., Matsuura, K. and Hara, A. (1995) Reduction of drug ketones by dihydrodiol dehydrogenases, carbonyl reductase and aldehyde reductase of human liver, *Biochem. Pharmacol.* 50, 221–227.

Polívka, Z., Budesínsky, M., Holubek, J., Schneider, B., Sedivy, Z., Svátek, E., Matousová, O., Metys, J., Valchar, M., Soucek, R. and Protiva, M. (1989) 4H-Benzo[4,5]cyclohepta[1,2-b]thiophenes and 9,10-dihydro derivatives - sulfonium analogues of pizotifen and ketotifen; chirality of ketotifen; synthesis of the 2-bromo derivative of ketotifen, *Collect. Czech. Chem. Commun.* 54, 2443–2469.

Prox, A. and Breyer-Pfaff, U. (1987) Amitriptyline metabolites in human urine. Identification of phenols, dihydrodiols, glycols, and ketones, *Drug Metab. Dispos.* 15, 890–896.

Wermuth, B. (1981) Purification and properties of an NADPH-dependent carbonyl reductase from human brain. Relationship to prostaglandin 9-ketoreductase and xenobiotic ketone reductase, *J. Biol. Chem.* 256,, 1206–1213.

Wermuth, B., Bohren, K.M. and Ernst, E. (1993) Autocatalytic modification of human carbonyl reductase by 2-oxocarboxylic acids, *FEBS Lett.* 335, 151–154.

Wermuth, B., Bohren, K.M., Heinemann, G.,von Wartburg, J.-P. and Gabbay, K.H. (1988) Human carbonyl reductase. Nucleotide sequence analysis of a cDNA and amino acid sequence of the encoded protein. *J. Biol. Chem.* 263, 16185–16188.

FORMATION OF LENS ALDOSE REDUCTASE MIXED DISULFIDES WITH GSH BY UV IRRADIATION AND ITS PROTEOLYSIS BY LENS CALPAIN

Tadashi Mizoguchi,[1] Isamu Maeda,[1] Kiyohito Yagi,[1] and Peter F. Kador[2]

[1]Graduate School of Pharmaceutical Sciences
Osaka University
Suita, Osaka 565-0871, Japan
[2]National Eye Institute
National Institutes of Health
Bethesda, Maryland 20892

1. INTRODUCTION

UV irradiation can damage ocular lens proteins and inactivate lens enzymes (Li et al., 1990; Zigman et al., 1991; Rafferty et al., 1993; Hightower, 1995). We have previously reported that bovine lens aldose reductase activity is enhanced by UV irradiation under relatively high D-fructose conditions (Mizoguchi et al., 1996). In our study on the photoreactivity of lens proteins, we have observed that significant amounts of lens protein mixed disulfides with GSH are formed when crude pig lens extract is UV irradiated for 3 min. To determine whether the protein mixed disulfides formed with GSH are sensitive or resistant to proteolysis and to gain insight into the physiological role of lens protein mixed disulfides, we have monitored calpain proteolysis of aldose reductase mixed disulfides with GSH.

2. MATERIALS AND METHODS

2.1. Material

NADPH and bovine serum albumin were obtained from Sigma Chemical Co. D,L-Glyceraldehyde, casein and 5,5'-dithiobis (2-nitrobenzoic acid) were purchased from Wako Pure Chemical Co. Dithiothreitol was from Nacalai Tesque, Pharmalyte from Pharmacia Fine Chemicals, and Dyematrex gel Red A from Amicon, Inc. All other chemicals were of reagent grade. Fresh pig eyes were obtained from a local slaughterhouse and the lenses were removed and kept frozen until needed. Aldose reductase from pig lens was purified routinely by the series of chromatographic procedures described previously (Tanimoto et al., 1990), except that Dyematrex gel Red A and Pharmalyte were used instead of Matrex gel Orange A and Mono P, respectively. Aldose reductase activity was measured as previously described, using D,L-glyceraldehyde as a substrate (Tanimoto et al., 1990). Lactate dehydrogenase from pig lens was purified by gel filtration using Sephadex G-75 followed by affinity chromatography using Dyematrex gel Red A. Pig lens crystallins α and β were separated by gel filtration using Sephadex G-75 followed by Q-Sepharose chromatography. Pig lens crystallin γ was purified by Q-Sepharose chromatography followed by Sephacryl S-100 chromatography. Proteins were determined by the Coomassie blue binding assay (Bradford, 1976), using bovine serum albumin as a standard.

2.2. Purification of Pig Lens Calpain and Protein Digestion by Calpain

Calpain from pig lens was purified routinely by a series of chromatographic procedures described previously (Yoshimura et al., 1983), except that Q-Sepharose, Sephacryl S-100 and CM-Sephadex were used instead of DE52, TSK-gel G-3000 SWG, blue Sepharose CL-6B and DEAE-Bio-Gel A. Calpain activity was measured with casein as a substrate as previously described (Yoshimura et al., 1983). One unit of calpain activity was defined as the quantity of enzyme increasing the absorbance at 280 nm by 0.001 after 30 min of incubation at 30 °C. The final calpain preparation had a specific activity of 1640. The isoelectric point of pig lens calpain obtained with Pharmalyte was 3.96. The incubation of protein mixed disulfides with calpain was carried out as follows: the reaction mixture having a final volume of 1.2 ml contained 50 mM Tris-HCl (pH 7.8), 5.0 mM $CaCl_2$, substrate proteins at the specified concentration, and pig lens calpain (100 units). After gentle shaking for 30 min at 30 °C, the digestion was stopped by the addition of 0.9 ml 10% trichloroacetic acid followed by centrifugation at 14000 rpm for 1 min. The absorbance of the supernatant was measured at 280 nm. Control experiments were carried out using the same mixtures without incubation.

2.3. UV Irradiation of Aldose Reductase in the Presence of GSH

After 30 min reduction by 10 mM dithiothreitol followed by Sephadex G-25 gel filtration using 100 mM phosphate (pH 7.3), purified pig lens aldose reductase at the specified concentration was irradiated with UV light (253.7 nm, 0.7 W/m^2) from a UV lamp (National GL-15, Japan) in the presence of GSH (about 2 mM). The irradiated mixtures were employed for enzyme assay without GSH removal. The concentration of about 2 mM GSH did not affect enzyme activity in the dark. After 20-min UV irradiation, enzyme proteins were immediately gel filtered with 50 mM Tris-HCl (pH 7.8), lyophilized, and stored at -20 °C. Lyophilized proteins were dissolved in water prior to use.

2.4. $NaBH_4$ Reduction of Protein Mixed Disulfides and Estimation of Released GSH

Four hundred μl of the protein mixed disulfides solution was reduced at concentration of about 3.0 mg/ml with one drop of octyl alcohol to prevent foaming. For reduction, about 100 molar excess of solid $NaBH_4$ was added to the protein solution at room temperature. After 1 h, the mixture was acidified by the addition of 9.6% $HClO_4$ under ice-cooling. After removal of the denatured proteins by centrifugation, the resulting supernatant was neutralized by the addition of 1N NaOH and subjected to GSH estimation according to the enzymatic cycling assay of Owens and Belcher (Owens & Belcher 1965) with slight modification. Total free GSH and GSSG in pig lens extract were routinely determined by the enzymatic cycling assay as described above.

2.5. Preparation of Pig Lens Aldose Reductase Mixed Disulfides with GSH by Iodometry

The mixture of aldose reductase (1.3 mg/ml) and GSH (5 mM) in 100mM phosphate (pH 7.3) was titrated by adding 0.05 N iodine to slight excess, and the end point was monitored by the appearance of a pale yellow color. The titration was finished within 5–6 seconds and the mixture was immediately gel-filtered using Sephadex G-25 with 50 mM Tris-HCl (pH 7.8). The protein fraction was lyophilized and stored at -20 °C. The resulting mixed disulfides of aldose reductase were estimated to be about 12.1 mol GSH/mg protein by the enzymatic cycling assay.

3. RESULTS AND DISCUSSION

Several investigations have been conducted on GSH protein-mixed disulfides in rat blood (Simplicio et al., 1998), human red blood cells (Lii and Hung, 1997), and mouse blood (Simplicio et al., 1997). In the course of oxidative stress studies, we have also examined GSH mixed disulfides of glutathione S-transferase (Nishihara et al., 1991), rat erythrocytes (Terada et al., 1993) and rabbit liver fructose 1,6-bisphosphatase (Terada et al., 1994). Lens protein mixed disulfides with GSH, however, have to date not been well-characterized, except for bovine lens aldose reductase (Cappiello et al., 1995). The enzymatic cycling assay for GSH with pig lens proteins after $NaBH_4$ reduction revealed that the amount of mixed disulfides was 1.14 nmol GSH/mg protein in whole proteins, which was about 3.9% that of free GSH and GSSG (Table 1).

The report that the amount of bound GSH in fresh pig lens was ~4% of free GSH (Wang et al. 1997) is in good agreement with our data. Pig lens aldose reductase and lactate

Table 1. The content of mixed disulfides with GSH in several proteins

	Mixed disulfide (nmol GSH/mg protein)
Aldose reductase	1.41
Lactate dehydrogenase	0.83
α-Crystallin	0.46
β-Crystallin	2.30
γ-Crystallin	1.56
Whole proteins	1.14
Free GSH and GSSG	29.20

[a]Pig lens proteins were isolated and the content of mixed disulfides with GSH in the proteins or free GSH including GSSG were determined by enzymatic cycling assay as described in Materials and Methods. Values are the means of three experiments.

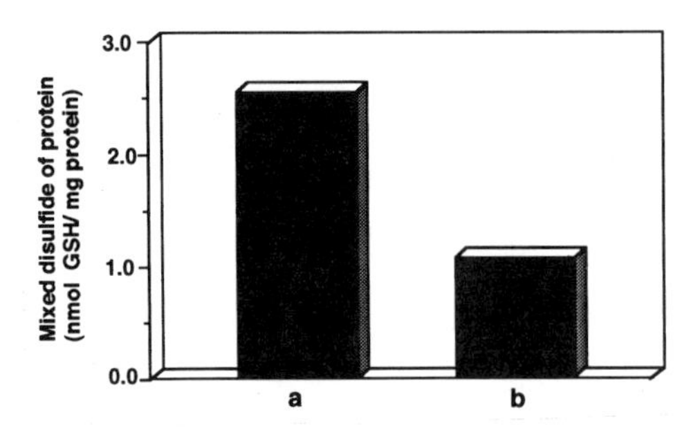

Figure 1. Formation of mixed disulfides of pig lens whole proteins during UV irradiation in the presence of GSH. Pig lens protein (7.6 mg/ml) was irradiated for 3 min with UV light (253.7 nm, 0.7 W/m^2) in the presence of 5 mM GSH (a) or in the absence of GSH (b), in 100 mM phosphate, (pH 7.3), at 25 °C. After dialysis of the irradiated mixtures, the GSH content in the proteins was determined by the enzymatic cycling assay as described in Materials and Methods. Values are the means of three experiments.

dehydrogenase contained GSH mixed disulfides of 1.41 and 0.83 nmol GSH/ mg protein, respectively. This finding suggests that pig lens aldose reductase in lens is modulated by its mixed disulfides with GSH. In fact, as mentioned later, pig lens aldose reductase exhibited transient activity upon UV irradiation in the presence of GSH. The reported occurrence of bovine lens aldose reductase mixed disulfides with 2-mercaptoethanol is worth noting (Giannessi et al., 1993). In the presence of 5 mM GSH, UV irradiation for 3 min was carried out on pig lens whole protein after sufficient dialysis against 100 mM phosphate buffer (pH 7.3). In the presence of GSH, the amount of the mixed disulfides of pig lens increased by about 2-fold as shown in Figure 1. This indicates that UV irradiation in the presence of GSH induced the rapid formation of mixed disulfides of pig lens whole proteins.

As shown in Figure 2, UV irradiation of pig lens aldose reductase in the presence of GSH resulted in an initial increase in enzyme activity followed by a rapid decrease. The net maximal increase in enzyme activity was observed at about 0.2 mM GSH, which is approximately 6-fold molar excess of the enzyme (data not shown). Although a slight increase in bovine lens aldose activity mediated by UV irradiation was previously observed (Mizoguchi et al., 1996), this enzyme activity was not increased by UV irradiation in the absence of GSH.

Figure 3 shows the formation of pig lens aldose reductase mixed disulfides with GSH during UV irradiation in the presence or absence of GSH. The amounts of mixed disulfides of pig lens aldose reductase, intact and UV-irradiated in the absence of GSH, were similar, being about 1.5 nmol GSH per mg protein. After 20-min UV irradiation, the amount of mixed disulfides of pig lens aldose reductase increased to about 4.5 nmol GSH per mg protein, which was about 3-fold that of the control. The time-dependent formation

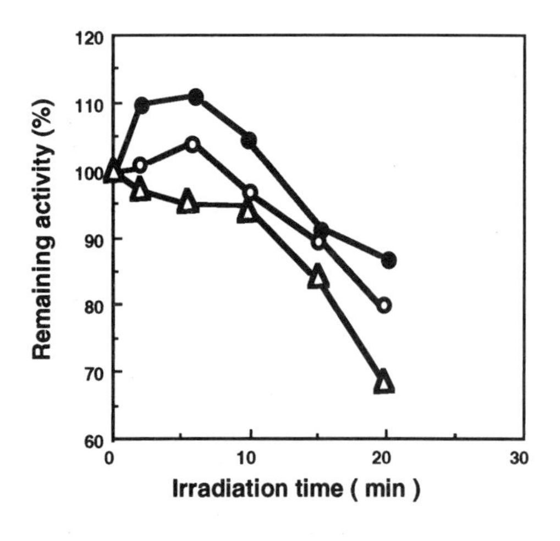

Figure 2. Time course of pig lens aldose reductase activity during UV irradiation in the presence or absence of GSH. Pig lens aldose reductase (3.7 mg/ml) was irradiated with UV light (253.7 nm, 0.7 W/m^2) in the presence of 0.4 mM (●) or 2 mM (○) GSH or its absence (△), in 100 mM phosphate, (pH 7.3), at 25 °C. The irradiated mixtures were employed for enzyme assay without GSH removal. Data points are means of duplicate determinations.

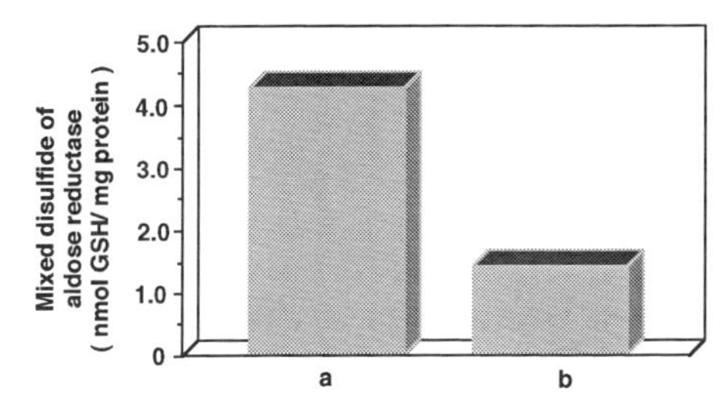

Figure 3. Formation of mixed disulfides of pig lens aldose reductase during UV irradiation in the presence or absence of GSH. Pig lens aldose reductase (0.28 mg/ml) was irradiated with UV light (253.7 nm, 0.7 W/m^2) in the presence (a) or absence (b) of GSH (0.28 mM), in 100 mM phosphate, (pH 7.3), at 25°C for 20 min. After gel filtration on Sephadex G-25 of the irradiated mixtures, the content of GSH in the proteins was determined by the enzymatic cycling assay as described in Materials and Methods. Values are the means of five experiments.

of mixed disulfides with the enzyme was observed during UV irradiation in the presence of GSH; however, there was no direct correlation between mixed disulfides formation and change in enzyme activity (data not shown). The transient activity of pig lens aldose reductase shown in Figure 2, however, is considered to be attributed to the binding of GSH to the enzyme via primary configuration changes in the enzyme structure.

When pig lens aldose reductase mixed disulfides obtained after 20-min UV irradiation was incubated with purified pig lens calpain, the digestion of the mixed disulfides was increased by about 3-fold, compared to that of the UV-irradiated enzyme in the absence of GSH, which was approximately similar to intact enzyme digestion. The mixed disulfides of the enzyme prepared by iodometry was about 12.1 nmol GSH per mg protein. The percentage digestion of the mixed disulfides prepared by iodometry was higher than that of the mixed disulfides formed by UV irradiation in the presence of GSH (Figure 4). The susceptibility to calpain digestion of aldose reductase mixed disulfides was considered to be dependent on the binding of GSH to the enzyme (data not shown). From exhaustive investigations on several substrates for calpain in the lens, such as crystallins, vimetin, actin, beaded filaments and MP26, it was found that calpain is mainly involved in cortical cataract (Andersson et al., 1996). Our finding indicates that proteolysis by lens calpain of lens proteins involved in cataracts is primarily mediated by protein mixed disulfides with GSH that was formed during oxidative stress. Experiments are underway to clarify the susceptibility to calpain of lens crystallin mixed disulfides with GSH during UV irradiation.

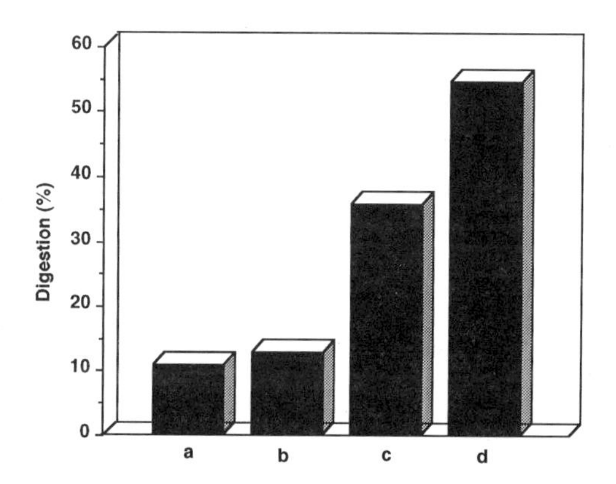

Figure 4. Digestion by pig lens calpain of pig lens aldose reductase mixed disulfides with GSH. The incubated amounts of native (a), and UV-irradiated (b) aldose reductase were 1.34, and 1.25mg/ml, respectively. Mixed disulfides of aldose reductase obtained by UV irradiation (c) or iodometry (d) was 1.39mg/ml. The digestion was carried out as described in Materials and Methods and expressed as percent of casein digestion. Values are the means of three determinations.

4. CONCLUSION

Judging from the release of GSH after $NaBH_4$ reduction of the protein fractions, UV irradiation of pig lens aldose reductase in the presence of GSH resulted in the formation of enzyme mixed disulfides with GSH. Iodometry of the mixture of the enzyme and GSH also gave the same product. On incubation with pig lens calpain, the proteolysis of enzyme mixed disulfides with GSH was significantly higher than that of intact enzyme.

REFERENCES

Andersson, M., Sjostrand, J. and Karlsson, J.O., 1996. Calpains in the human lens: relations to membranes and possible role in cataract formation. *Ophthalmic. Res.*, **28**: Suppl. 1: 51–54.

Bradford, M.M., 1976. A rapid and sensitive method for the quantitation of microgram quantities of protein utilizing the principle of protein-dye binding. *Anal. Biochem.* **72**: 248–254.

Cappiello, M., Vilardo, P.G., Cecconi, I., Leverenz, V., Giblin, F.J., Corso, A.D. and Mura, U. 1995. Occurrence of glutathione-modified aldose reductase in oxidatively stressed bovine lens. *Biochem. Biophys. Res. Commun* . **207**: 775–782.

Giannessi, M., Del, C.A., Cappiello, M., Voltarelli, M., Marini, I., Barsacchi, D., Garland, D., Camici, M. and Mura. U., 1993. Thiol-dependent metal-catalyzed oxidation of bovine lens aldose reductase. I. Studies on the modification process. *Arch. Biochem. Biophys.* **300**: 423–429.

Hightower, K. R., 1995. The role of the lens epithelium in development of UV cataract. *Curr. Eye. Res.* **14**: 71–78.

Li, D.Y., Borkman, R.F., Wang, R.H. and Dillon, J., 1990. Mechanisms of photochemically produced turbidity in lens protein solutions. *Exp. Eye Res.* **51**: 663–669.

Lii, C.K. and Hung, C.N., 1997. Protein thiol modifications of human red blood cells treated with t-butyl hydroperoxide. *Biochim. Biophys. Acta,* 1336: 147–156.

Mizoguchi,T., Ogura, T., Yagi, K. and Kador, P.F. 1996. D-Fructose-mediated stimulation of bovine lens aldose reductase activation by UV-irradiation. In *Enzymology and Molecular Biology of Carbonyl Metabolism* **6,** ed. by Weiner, H., Lindahl, R., Crabb, D. W. and Flynn, T. G. Plenum Press, New York, 529–535

Nishihara, T., Maeda, H., Okamoto, K., Oshida, T., Mizoguchi, T. and Terada, T., 1991. Inactivation of human placenta glutathione S-transferase by SH/SS exchange reaction with biological disulfides. *Biochem. Biophys. Res. Commun.* **174**: 580–585.

Owens, C.W.I. and Belcher, R.V. 1965. A colorimetric micro-method for the determination of glutathione. *Biochem. J.* **94:** 705–709.

Rafferty, N.S., Zigman, S., McDaniel, T. and Scholz, D.L., 1993. Near-UV radiation disrupts filamentous actin in lens epithelial cells. *Cell Motil. Cytoskel.* **26**: 40–48.

Simplicio, P.D., Giannerini, F., Giustarini, D., Lusini, L. and Rossi, R., 1998. The role of cysteine in the regulation of blood glutathione-protein mixed disulfides in rats treated with diamide. *Toxicol. Appl. Pharm.* **148**: 56–64.

Simplicio, P.D., Rossi, R., Falcinelli, S., Ceserani, R. and Formento, M.L., 1997. Antioxidant status in various tissues of the mouse after fasting and swimming stress. *Eur. J. Appl. Physiol.* **76**: 302–307.

Tanimoto, T., Sato, S. and Kador, P.F., 1990. Purification and properties of aldose reductase and aldehyde reductase from EHS tumor cells. *Biochem. Pharmacol.* **39**: 445–453.

Terada, T., Nishimura, M., Oshida, H., Oshida, T. and Mizoguchi, T., 1993. Effect of glucose on thioltransferase activity and protein mixed disulfides concentration in GSH-depleting reagents treated rat erythrocytes. *Biochem. Mol. Biol. Int.* **29**: 1009–1014.

Terada, T., Hara, T., Yazawa, H., and Mizoguchi, T., 1994. Effect of thioltransferase on the cystamine-activated fructose 1,6-bisphosphatase by its redox regulation. *Biochem. Mol. Biol. Int.* **32**: 239–244.

Wang, G.M., Raghavachari, N. and Lou, M. F., 1997. Relationship of protein-glutathione mixed disulfide and thioltransferase in H2O2-induced cataract in cultured pig lens. *Exp. Eye Res.* **64**:693–700.

Yoshimura, N., Kikuchi, T., Sasaki, T., Kitahara, A., Hatanaka, M. and Murachi, T., 1983. Two distinct Ca^{2+} proteases (calpain I and calpain II) purified concurrently by the same method from rat kidney. J. Biol. Chem. **258**: 8883–8889.

Zigman, S., Paxhia, T., McDaniel, T., Lou, M.F. and Yu, N.T., 1991. Effect of chronic near-ultraviolet radiation on the gray squirrel lens *in vivo*. *Invest. Ophthalmol. Vis. Sci.* **32**: 1723–1732.

INHIBITION OF ALDOSE REDUCTASE BY GOSSYPOL AND GOSSYPOL-RELATED COMPOUNDS

Lorraine M. Deck,[1] Brian B. Chamblee,[1] Robert E. Royer,[2]
Lucy A. Hunsaker,[2] and David L. Vander Jagt[2]

[1]Department of Chemistry
University of New Mexico
[2]Department of Biochemistry and Molecular Biology
University of New Mexico School of Medicine
Albuquerque, New Mexico 87131

1. INTRODUCTION

The aldo-keto reductases are a widely distributed family of oxidoreductases (Wirth and Wermuth, 1985; Wermuth, 1985; Grimshaw and Mathur, 1989). Aldose reductase (EC 1.1.1.21; ALR2; or AKR1B1 (Jez et al., 1997)), which has a role in the etiology of diabetic complications (Gabbay, 1973; Kador and Kinoshita, 1985), has received most attention as a potential drug target. The emphasis on ALR2 results from the ability of ALR2 to catalyze the reduction of glucose to sorbitol, which is an important reaction under conditions of hyperglycemia. ALR2 continues to be a major target for development of drugs that may be useful in preventing diabetic complications, an effort aided by the crystal structures of ALR2-cofactor-inhibitor ternary complexes (Petrash et al., 1994; Wilson et al., 1993). The more recent crystallographic studies of ALR2-cofactor-inhibitor complexes indicate a complex pattern of inhibitor binding, including the ability of ALR2 to complex more than one equivalent of inhibitor (Harrison et al., 1997; Potier et al., 1997) and the ability of ALR2 to adopt different conformations in response to different inhibitors (Urzhumtsev et al., 1997). In all cases, these inhibitors are bound to the extended substrate binding site (Nakano and Petrash, 1996).

ALR2 is comprised of an α/β-barrel without a classic dinucleotide binding site (Rondeau et al., 1992; Wilson et al., 1992); ALR2 and other aldo-keto reductases utilize an ordered mechanism with NADPH binding preceding aldehyde binding, followed by hydride transfer, alcohol release and rate-determining isomerization of the ALR2-NADP binary complex prior to NADP release (Barski et al., 1995; Bohren et al., 1994; Grimshaw et al., 1990, 1992; Kubisecki et al., 1992). In principle, inhibitors may bind differently to ALR2-NADP and ALR2-NADPH binary complexes. In all reports to date, inhibitors are bound to binary ALR2-cofactor complexes rather than to the apoprotein.

Recently, we described examples of inhibition of dehydrogenases by the natural product gossypol and its derivatives (Gomez et al., 1997; Torres et al., 1998; Yu et al., 1998). This involves competitive inhibition of cofactor binding. Gossypol and derivatives also are ALR2 inhibitors (Deck et al., 1991). This raised the question whether these compounds bind to the same site as the inhibitors that have been studied crystallographically or whether gossypol and derivatives represent a new type of ALR2 inhibitor directed at the NADP(H) binding site.

2. MATERIALS AND METHODS

2.1. Enzyme Purification, Assays, and Kinetics

Human ALR2 was purified from skeletal muscle (Vander Jagt et al., 1990a,b; 1992). Reductase activity was extracted from the 100,000xg supernatant fraction with Red SepharoseCL-6B, followed by chromatofocusing on PBE 94 (Pharmacia LKB). Final purification was accomplished by chromatography on a Bio-Gel HPHT HPLC column (Bio-Rad). Kinetic analysis of ALR2 was carried out at pH7, 25°C, in 0.1 M sodium phosphate buffer. K_m and k_{cat} values were determined by nonlinear regression analysis with the Enzfitter program (Vander Jagt et al., 1995).

2.2. Chemicals

Gossypol was obtained from the Southern Regional Research Center, US Department of Agriculture. Gossylic lactone and dideoxygossylic acid were prepared as described previously (Deck et al., 1991; Royer et al., 1995). The synthesis of

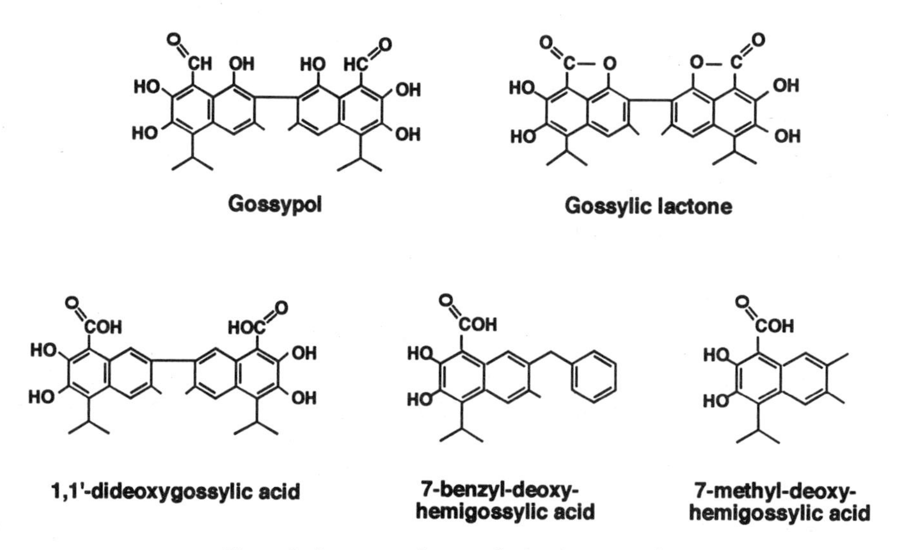

Figure 1. Structures of gossypol-related compounds.

7-benzyl-8-deoxyhemigossylic acid and 7-methyl-8-deoxyhemigossylic acid will be described elsewhere (Deck et al., in preparation). Structures of gossypol and related compounds are shown in Figure 1.

3. RESULTS AND DISCUSSION

3.1. Inhibition of ALR2 by Gossypol and Gossylic Lactone: Kinetic Study

The inhibition of ALR2 by gossypol with D,L-glyceraldehyde as substrate is shown in Figure 2. Inhibition is noncompetitive with respect to NADPH, $K_i = 0.5\ \mu M$; the pattern of inhibition, analyzed by Dixon plots, is a classical pattern with no evidence of deviation from linearity. Thus, the kinetic pattern supports the conclusion that gossypol does not bind at the cofactor site of ALR2, unlike the situation with numerous dehydrogenases where gossypol binds at the dinucleotide binding site.

Gossypol contains two aldehyde functional groups that can complicate inhibition and binding studies due to covalent interactions between gossypol and nucleophilic groups in proteins. Therefore, the inhibition of ALR2 by a less reactive gossypol derivative was investigated. Gossylic lactone is a stable derivative of gossypol in which the aldehyde groups are at the carboxylic acid oxidation state, cyclized into a lactone. The inhibition of ALR2 by gossylic lactone with D,L-glyceraldehyde as substrate is shown in Figure 2. Inhibition again is noncompetitive with NADPH, $K_i = 0.2\ \mu M$.

3.2. Fluorescence Quenching Study of the Binding of Gossylic Lactone to ALR2

The binding of gossylic lactone to ALR2 was also analyzed by fluorescence quenching to address the question whether the inhibitor binds to apo-ALR2 as well as to the ALR2-cofactor binary complexes. Gossylic lactone quenched the intrinsic protein fluorescence of ALR2 completely. Addition of NADPH to ALR2 resulted in partial quenching of the intrinsic protein fluorescence. Addition of gossylic lactone to the ALR2-NADPH binary complex resulted in quenching of the remaining fluorescence, with a binding isotherm very similar to that of the binding of gossylic lactone to ALR2. Thus it appears that gossylic lactone binds equally well to ALR2 and to the ALR2-NADPH complex, consistent with the conclusion from the kinetic studies that these inhibitors do not compete with cofactor binding.

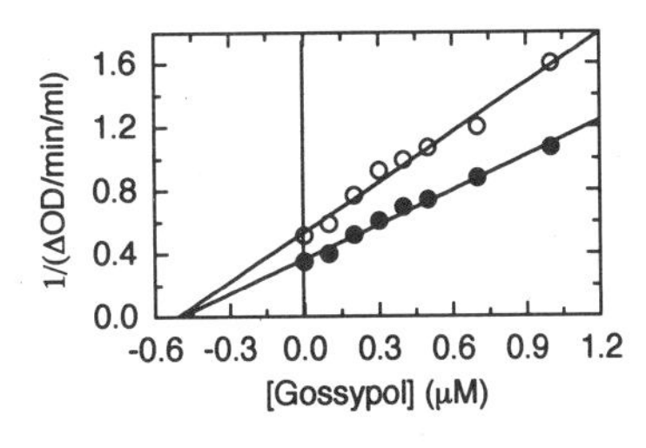

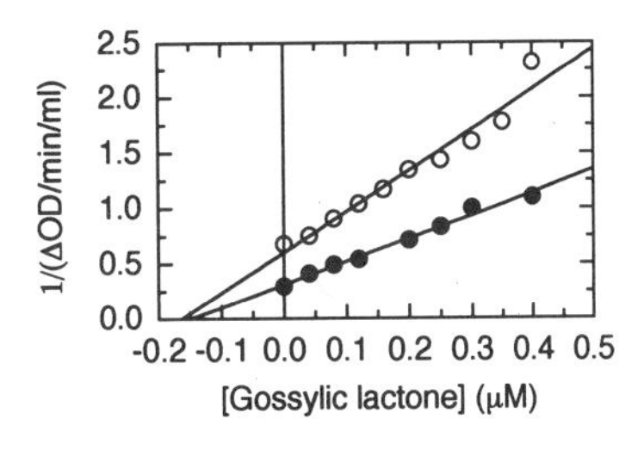

Figure 2. Dixon plots of the inhibition of ALR2 by gossypol (left) and by gossylic lactone (right) indicate that inhibition is noncompetitive with the binding of NADPH; data obtained at 10uM (open circles) and 40uM (closed circles) NADPH, at constant 10mM glyceraldehyde.

3.3. Inhibition of ALR2 by Analogs of Gossypol

Gossypol and gossylic lactone exist as optical isomers due to restricted rotation about the binaphthyl bond (atropisomerism). Removal of groups from the 1,1'-positions or 3,3'-positions lowers the rotational barrier about the binaphthyl bond. 1,1'-Dideoxygossylic acid, which is the product of a de novo synthetic scheme (Royer et al., 1995), is similar to gossypol except that the reactive aldehydes present in gossypol are replaced by carboxylic acid functional groups and the phenolic groups at the 1,1'-positions are replaced by hydrogen which removes the rotational barrier. The binding of dideoxygossylic acid to the ALR2-NADPH binary complex was analyzed by fluorescence quenching; $K_D = 1.3\mu M$, similar to gossypol and gossylic lactone. Therefore, the rotational freedom about the binaphthyl bond does not appear to have a major effect, raising the question whether only one half of the gossypol backbone is involved in binding to ALR2.

Kinetic analysis of the inhibition of ALR2 by dideoxygossylic acid revealed a more complex pattern than was observed by fluorescence quenching, with deviations from linearity in Dixon plots. This is similar to the pattern reported earlier for inhibition of ALR2 by sorbinil (Vander Jagt et al., 1990a). To examine this behavior further, 7-methyl-8-deoxyhemigossylic acid and 7-benzyl-8-deoxyhemigossylic acid were tested. These inhibitors are comprised of one half of the dideoxygossylic acid structure with nonpolar methyl or benzyl groups at the 7-position replacing the other half of the dideoxygossylic acid structure. The results were similar to those for dideoxygossylic acid with deviations from linearity. A representative Dixon plot is shown in Figure 3 with 7-benzyl-8-deoxyhemigossylic acid. For comparison, a Dixon plot of the nonlinear inhibition pattern of sorbinil is shown. With sorbinil, deviation from linearity is observed after about 80% of the activity is inhibited. With 7-benzyl-8-deoxyhemigossylic acid, sharp deviation from linearity is observed after about 20% inhibition of activity.

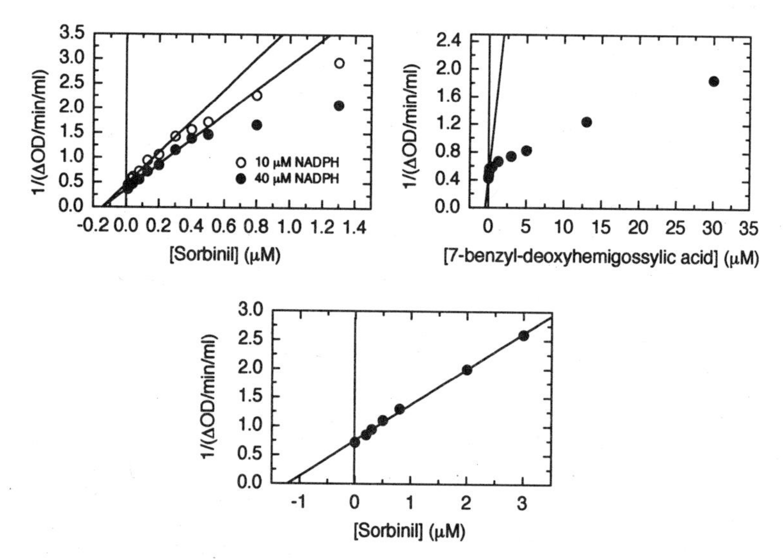

Figure 3. Dixon plots of the inhibition of ALR2 by sorbinil and by 7-benzyl-8-deoxyhemi- gossylic acid reveal deviations from linearity. These deviations from linearity are especially prominent in the case of 7-benzyl-8-deoxyhemigossylic acid (upper right) where the Dixon plot is shown at only one concentration of NADPH. Partial inhibition (20%) of ALR2 by 7-benzyl-8-deoxyhemigossylic acid produces a complex that exhibits a linear inhibition pattern by sorbinil (bottom), shown as a Dixon plot at one concentration of NADPH.

The pattern of inhibition with 7-benzyl-8-deoxyhemigossylic suggests that this inhibitor binds tightly (Ki = 30nM) but inhibits only 20% of the activity. If ALR2 is first treated with 7-benzyl-8-deoxyhemigossylic acid to inhibit this apparent high affinity site, the resulting complex can be inhibited by sorbinil and now exhibits a linear pattern of inhibiton (Figure 3).

These results suggest that ALR2 can bind both sorbinil and 7-benzyl-8-deoxyhemigossylic acid at different sites, with the two sites most likely residing in the extended substrate specificity site.

In summary, the results of this study support the conclusion that gossypol-related compounds inhibit ALR2 by binding to the substrate site rather than to the cofactor site. In this respect, the inhibition of ALR2 by gossypol-related compounds differs markedly from the pattern observed for the inhibition of dehydrogenases that contain a Rossmann fold where inhibition by gossypol-related compounds involves the cofactor site.

ACKNOWLEDGMENTS

This work was supported by NIH grant GM52576 from the Minority Biomedical Support Program, National Institutes of Health.

REFERENCES

Barski, O.A., Gabbay, K.H., Grimshaw, C.E. and Bohren, K.M. (1995) Mechanism of human aldehyde reductase: Characterization of the active site pocket, *Biochemistry* **34**, 11264–11275.

Bohren, K.M., Grimshaw, C.E., Lai, C-J., Harrison, D.H., Ringe, D., Petsko, G.A. and Gabbay, K.H. (1994) Tyrosine-48 is the proton donor and histidine-110 directs substrate stereochemical selectivity in the reduction reaction of human aldose reductase: Enzyme kinetics and crystal structure of the Y48H mutant enzyme, *Biochemistry* **33**, 2021–2032.

Deck, L.M., Vander Jagt, D.L. and Royer, R.E. (1991) Gossypol and derivatives: a new class of aldose reductase inhibitors, *J. Med. Chem.* **34**, 3301–3305.

Gabbay, K.H. (1973) The sorbitol pathwayand the complications of diabetes. *N. Engl. J. Med.* **288**, 831–836.

Gomez, M.S., Piper, R.C., Hunsaker, L.A., Royer, R.E., Deck, L.M., Makler, M.T. and Vander Jagt, D.L. (1997) Substrate and cofactor specificity and selective inhibition of lactate dehydrogenase from the malarial parasite P. falciparum, *Mol. Biochem. Parsitol.* **90**, 235–246.

Grimshaw, C.E. (1992) Aldose reductase: Model for a new paradigm of enzymic perfection In detoxification catalysis, *Biochemistry* **31**, 10139–10145.

Grimshaw, C.E. and Mathur, E.J. (1989) Immunoquantitation of aldose reductase in human tissues, *Anal. Biochem.* **176**, 66–71.

Grimshaw, C.E., Shahbaz, M. and Putney, C.G. (1990) Mechanistic basis for non-linear kinetics of aldehyde reduction catalyzed by aldose reductase, *Biochemistry* **29**, 9947–9955.

Harrison, D.H.T., Bohren, K.M., Petsko, G.A., Ringe, D. and Gabbay, K.H. (1997) The alrestatin double-decker: binding of two inhibitor molecules to human aldose reductase reveals a new specificity determinant, *Biochemistry* **36**, 16134–16140.

Jez, J.M., Flynn, T.G. and Penning, T.M. (1997) A new nomenclature for the aldo-keto reductase superfamily. *Biochem. Pharmacol.* **54**, 639–647.

Kador, P.F. and Kinoshita, J.H. (1985) Role of aldose reductase in the development of diabetes-associated complications, *Am. J. Med.* **79**, 8–12.

Kubisecki, T.J., Hyndman, D.J., Morjana, N.A. and Flynn, T.G. (1992) Studies of pig muscle aldose reductase . Kinetic mechanism and evidence for a slow conformational change upon coenzyme binding, *J. Biol. Chem.* **267**, 6510–6517.

Nakano, T. and Petrash, J.M. (1996) Kinetic and spectroscopic evidence for active site inhibition of human aldose reductase, *Biochemistry* **35**, 11196–11202.

Petrash, J.M., Tarle, I., Wilson, D.K. and Quiocho, F.A. (1994) Aldose reductase catalysis and crystallography. Insights from recent advances in enzyme structure and function, *Diabetes* **43**, 955–959.

Potier, N., Barth, P., Tritsch, D., Biellmann, J-F. and Van Dorsselaer, A. (1997) Study of non-covalent enzyme-inhibitor complexes of aldose reductase by electrospray mass spectrometry. *Eur. J. Biochem.* **243**, 274–282.

Rondeau, J-M., Tete-Favier, F., Podjarny, A., Reymann, J-M., Barth, P., Biellmann, J-F. and Moras, D. (1992) Novel NADPH-binding domain revealed by the crystal structure of aldose reductase, *Nature* **355**, 469–472.

Royer, R.E., Deck, L.M., Vander Jagt, T.J., Martinez, F.M., Mills, R.G., Young, S.A. and VanderJagt, D.L. (1995) Synthesis and anti-HIV activity of 1,1'-dideoxygossypol and related compounds. *J. Med. Chem.* **38**, 2427–2432.

Torres, J.E., Deck, J, Caprihan, K., Hunsaker, L.A., Deck, L.M. and Vander Jagt, D.L. (1998) Dihydroxynaphthoic acids: a new class of active site dehydrogenase inhibitors, *FASEB J.***12**, 271.

Urzhumtsev, A,. Tete-Favier, F., Mitschler, A., Barbanton, J., Barth, P., Urzhumtseva, L., Biellmann, J-F., Podjarny, A.D. and Moras, D. (1997) A 'specificity' pocket inferred from the crystal structures of the complexes of aldose reductase with the pharmaceutically important inhibitors tolrestat and sorbinil, *Structure* **5**, 601–612.

Vander Jagt, D.L., Hunsaker, L.A., Robinson, B., Stangebye, L.A. and Deck, L.M. (1990a) Aldehyde and aldose reductases from human placenta, *J. Biol. Chem.* **265**, 10912–10918.

Vander Jagt, D.L., Robinson, B., Taylor, K.K. and Hunsaker, L.A. (1990b) Aldose reductase from human skeletal and heart muscle, *J. Biol. Chem.* **265**, 20982–20987.

Vander Jagt, D.L., Robinson, B., Taylor, K.K. and Hunsaker, L.A.(1992) Reduction of trioses by NADPH-dependent aldo-keto reductases. Aldose reductase, methylglyoxal, and diabetic complications, *J. Biol. Chem.* **267**, 4364–4369.

Vander Jagt, D.L., Kolb, N.S., Vander Jagt, T.J., Chino, J., Martinez, F.J., Hunsaker, L.A and Royer, R.E. (1995) Substrate specificity of human aldose reductase: identification of 4- hydroxynonenal as an endogenous substrate, *Biochim. Biophys. Acta* **1249**, 117–126.

Wermuth, B., Aldo-keto reductases, in: Enzymology of Carbonyl Metabolism 2: Aldehyde Dehydrogenase, Aldo-Keto Reductases, and Alcohol Dehydrogenase, T.G. Flynn and H.Weiner, eds., 1985, A.R. Liss, New York, pp. 209–230.

Wilson,D.K., Bohren, K.M., Gabbay, K.H. and Quiocho, F.A. (1992) An unlikely sugar substrate site in the 1.65Å structure of the human aldose reductase holoenzyme implicated in diabetic complications, *Science* **257**, 81–84.

Wilson, D.K., Tarle, I., Petrash, J.M. and Quiocho, F.A. (1993) Refined 1.8Å structure of human aldose reductase complexed with the potent inhibitor zopolrestat, *Proc. Natl. Acad. Sci. USA* **90**, 9847–9851.

Wirth, H-P. and Wermuth, B. (1985) Immunochemical characterization of aldo-keto reductases from human tissues, *FEBS Lett.* **187**, 280–282.

Yu, Y., Torres, J.E., Hunsaker, L.A., Deck, L.M., Goldberg, E. and Vander Jagt, D.L. (1998) Selective active site inhibitors of the three isozymes of human lactate dehydrogenase, *FASEB J.* **12**, 836.

INHIBITION OF HUMAN ALDOSE AND ALDEHYDE REDUCTASES BY NON-STEROIDAL ANTI-INFLAMMATORY DRUGS

D. Michelle Ratliff, Francella J. Martinez, Timothy J. Vander Jagt, Christina M. Schimandle, Brian Robinson, Lucy A. Hunsaker, and David L. Vander Jagt

Department of Biochemistry and Molecular Biology
University of New Mexico
School of Medicine
Albuquerque, New Mexico 87131

1. INTRODUCTION

Aldose reductase (EC 1.1.1.21; alditol:NAD(P)$^+$ oxidoreductase; ALR2), a member of the multigene family of NADPH-dependent aldo-keto reductases, is one of the major enzymes implicated in the intracellular events leading to the development of diabetic complications. ALR2 catalyzes the reduction of glucose to sorbitol in the first reaction of the polyol pathway which may be important under conditions of hyperglycemia (Kador and Kinoshita, 1985). There have been a number of reports that some non-steroidal anti-inflammatory drugs (NSAIDs) can delay the development of diabetic complications. These reports include in vitro studies with isolated lens or nerve tissue (Chaudhry et al., 1983; Sharma and Cotlier, 1982; Crabbe et al., 1985; Jacobson et al., 1983), animal studies (Gupta and Joshi, 1991a; Blakytny and Harding, 1992; Gupta and Joshi, 1991b; Mansour et al., 1990; Sharma et al., 1989a), and clinical studies (Sharma et al., 1989b; Cunha-Vaz et al., 1985; Cohen and Harris, 1987). Several NSAIDs have been shown to be inhibitors of ALR2 (Chaudhry et al., 1983; Sharma and Cotlier, 1982; Crabbe et al., 1985; Gupta and Joshi, 1991a; Gupta and Joshi, 1991b). Most studies of inhibition of ALR2 by NSAIDs have used either animal ALR2 or crude homogenates of human tissues (Sharma and Cotlier, 1982; Crabbe et al., 1985; Gupta and Joshi, 1991a; Gupta and Joshi, 1991b). There are marked differences in the inhibitor-binding properties of ALR2 from various sources (Kador et al., 1980). In addition, we have shown that human ALR2 can be oxidized easily, resulting in the formation of a form of ALR2 with markedly altered kinetic properties (Vander Jagt et al., 1990a). There is also the problem that crude homogenates generally contain a mixture of ALR2 and the immunochemically distinct, but kinetically related, aldehyde reductase, ALR1 (Vander Jagt et al.,1990b).

Enzymology and Molecular Biology of Carbonyl Metabolism 7
edited by Weiner *et al.* Kluwer Academic / Plenum Publishers, New York, 1999.

In the present study we have evaluated the structural features of a wide range of NSAIDs that may contribute to inhibition of human ALR2, using purified ALR2 that has been shown to retain its native kinetic properties. In addition, we have compared these NSAIDs as inhibitors of human ALR1. Our goal was to determine if there are structural features of NSAIDs that may be useful in the design of new ALR2 inhibitors (ARI).

2. MATERIALS AND METHODS

2.1. Chemicals

Samples of human skeletal muscle were obtained through the Office of the Medical Investigator, State of New Mexico. Human kidneys were obtained from the National Disease Research Interchange, Philadelphia, PA. Tissue samples were generally frozen within 12 hours post mortem. Sulindac was obtained from Merck Sharp & Dohme Research Lab. Sorbinil was obtained from Pfizer Chemical Company, tolrestat from Ayerst, and statil from Imperial Chemical Industries. NADPH, D,L-glyceraldehyde, indomethacin, zomepirac, tolmetin, flurbiprofen, diflunisal, phenylbutazone, oxyphenbutazone, mefenamic acid, meclofenamic acid, diclofenac, piroxicam, ketoprofen, fenoprofen, ibuprofen, and naproxen were purchased from Sigma.

2.2. Enzyme Purification and Assay

Aldehyde reductase and aldose reductase were purified as reported previously (Vander Jagt et al., 1990a; 1990b; 1992). Routine assays of ALR1 and ALR2 during their purification were carried out in 1 ml volumes of 0.1 M sodium phosphate buffer pH 7, containing 10 mM D,L-glyceraldehyde and 0.1 mM NADPH, by measuring the rate of the enzyme-dependent decrease of NADPH absorption at 340 nm with a Perkin-Elmer Lamda 6 UV/VIS spectrophotometer at 25 °C. For the determination of dissociation constants of inhibitors, the same assay buffer containing 0.04 mM NADPH, 10 mM D,L-glyceraldehyde and various concentrations of an inhibitor was used. More detailed analysis of inhibition of ALR2 by sulindac was carried out using glucose and methylglyoxal as substrates. Dissociation constants were determined by linear regression analysis of Dixon plots of the initial rate data using the Enzfitter program (Elsevier-Biosoft).

2.3. Fluorescence Quenching Studies

Addition of ligands to ALR2 was monitored by following changes in the intrinsic protein fluorescence of ALR2, excitation wavelength 280 nm, emission wavelength 330 nm, using a Perkin Elmer Luminescence Spectrometer. Corrections for inner filter effects were made using established procedures (Kirby, 1971).

3. RESULTS AND DISCUSSION

3.1. Inhibition of Aldose Reductase by NSAIDs

Sixteen structurally diverse NSAIDs were examined for their ability to inhibit the native, reduced form of human muscle ALR2. The structures of these NSAIDs are shown in Figure 1. The dissociation constants of these sixteen compounds as inhibitors of ALR2

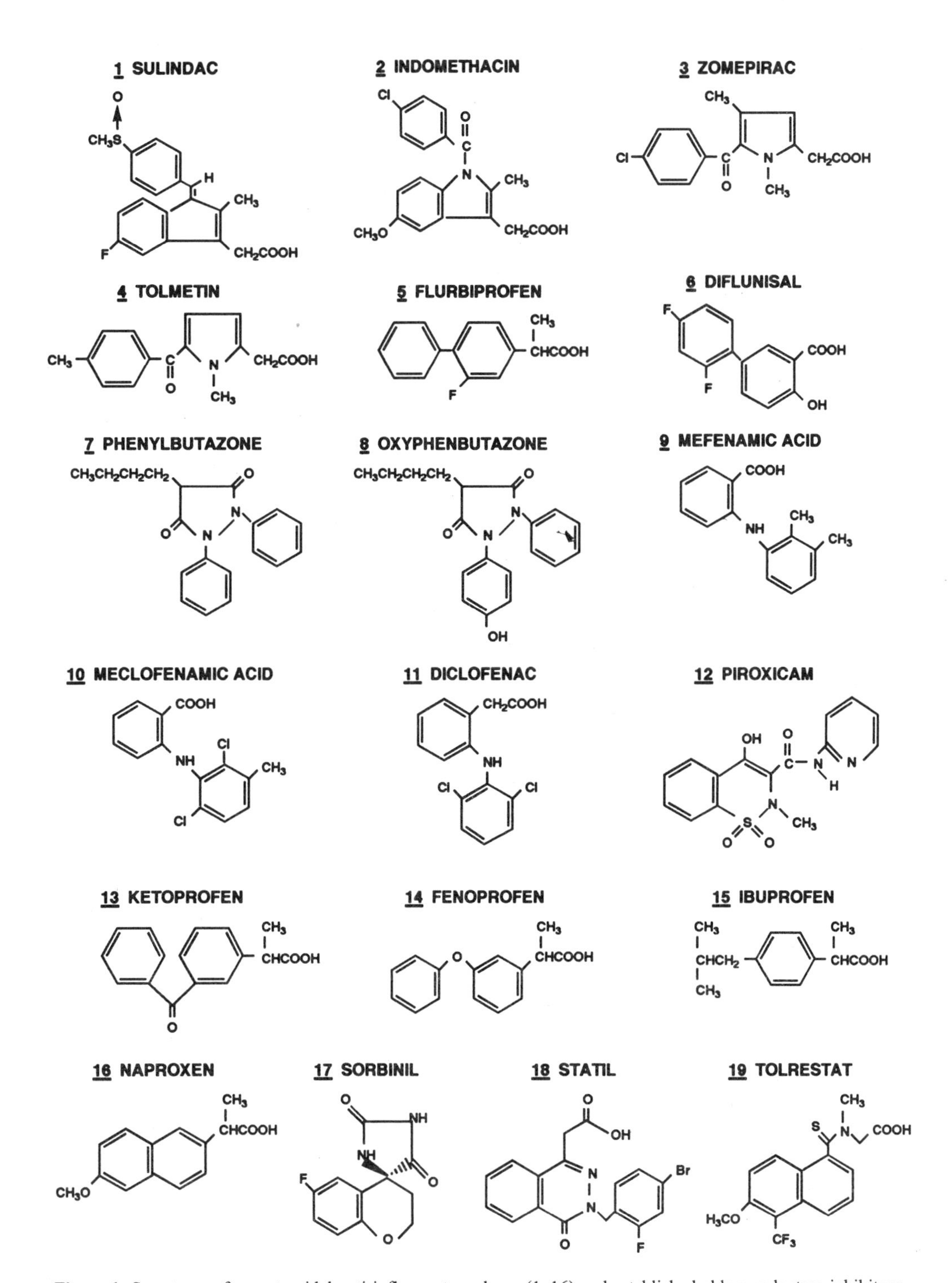

Figure 1. Structures of non-steroidal anti-inflammatory drugs (1–16) and established aldose reductase inhibitors (17–19).

Table 1. Dissociation constants for inhibition of human aldose reductase by non-steroidal anti-inflammatory drugs compared to established aldose reductase inhibitors

Inhibitor	$K_i(\mu M)$
1 Sulindac	0.3
2 Indomethacin	4
3 Zomepirac	12
4 Tolmetin	7
5 Flurbiprofen	35
6 Diflunisal	73
7 Phenylbutazone	>1000
8 Oxyphenbutazone	>1000
9 Mefenamic acid	>1000
10 Meclofenamic acid	>1000
11 Diclofenac	300
12 Piroxicam	74
13 Ketoprofen	51
14 Fenoprofen	70
15 Ibuprofen	110
16 Naproxen	800
17 Sorbinil[a]	0.2
18 Statil[a]	0.056
19 Tolrestat[a]	0.018

[a]From Vander Jagt *et al.* 1990b.

ranged from 0.3 μM to > 1mM (Table 1). Of the compounds examined, sulindac was the most potent inhibitor (K_i=0.3 μM), followed by indomethacin (K_i=4 μM), tolmetin (K_i=7 μM), and zomepirac (K_i=12 μM), using DL-glyceraldehyde as the substrate. The dissociation constant for sulindac compares favorably with that of sorbinil (K_i=0.2 μM). Sulindac showed similar inhibition of ALR2 regardless of substrate; K_i=0.3 μM with DL-glyceraldehyde, K_i=0.2 μM with glucose, and K_i=0.6 μM with methylglyoxal.

3.2. Inhibition of Aldehyde Reductase by NSAIDs

The four best NSAID inhibitors of ALR2 and the established ARIs sorbinil, statil and tolrestat were compared as inhibitors of ALR1 using DL-glyceraldehyde as a substrate. The results are shown in Table 2. Sulindac again is the best inhibitor among the NSAIDs that were examined (K_i=14 μM). Sulindac exhibits an ALR2/ALR1 selectivity ratio of 70 which compares favorably with that of statil and tolrestat.

3.3. Kinetic and Fluorescence Quenching Study of the Binding of Sulindac to ALR2

Inhibition of ALR2 by sulindac was measured at two different concentrations of NADPH (10 μM and 40 μM) with saturating levels of DL-glyceraldehyde as substrate. Dixon plots of the data are shown in Figure 2. Inhibition of ALR2 by sulindac shows classical noncompetitive inhibition with respect to NADPH.

The question of whether there is a binding site(s) for sulindac in the absence of cofactor binding was addressed by determining the ability of sulindac to quench the intrinsic fluorescence of ALR2. Sulindac exhibits significant absorptivity at 330 nm (molar extinc-

Table 2. Dissociation constants for inhibition of human aldehyde reductase by selected non-steroidal anti-inflammatory drugs and by aldose reductase inhibitors

Inhibitors	K_i (μM) ALR1	ALR2/ALR1 selectivity ratio[b]
1 Sulindac	14	70
2 Indomethacin	32	8
3 Zomepirac	118	10
4 Tolmetin	171	24
17 Sorbinil	2.5[a]	13
18 Statil	4[a]	71
19 Tolrestat	2[a]	111

[a]From Vander Jagt *et al.* 1990b.
[b]The selectivity ratio is defined as K_i(ALR2/K_i(ALR1)

tion coefficient 11,870 $M^{-1}cm^{-1}$) where ALR2 exhibits maximum fluorescence. The binding of sulindac results in quenching of the intrinsic fluorescence (Figure 3). Plots of reciprocal fluorescence corrected for inner filter effects vs reciprocal free ligand concentrations suggest that there are 2 binding sites for sulindac, K_D = 0.44 and 8.8 μM. The ALR2-NADP and ALR2-NADPH binary complexes retain significant intrinsic fluorescence at 330 nm. Therefore, the binding of sulindac to these binary complexes was evaluated by monitoring fluorescence quenching upon addition of sulidac (Figure 4). In both cases, single binding sites were observed with K_D values 0.30 and 0.58 μM for the binding of sulindac to the ALR2-NADP and ALR2-NADPH binary complexes, respectively. These are in good agreement with the kinetically determined K_i values for inhibition of ALR2 by sulin-

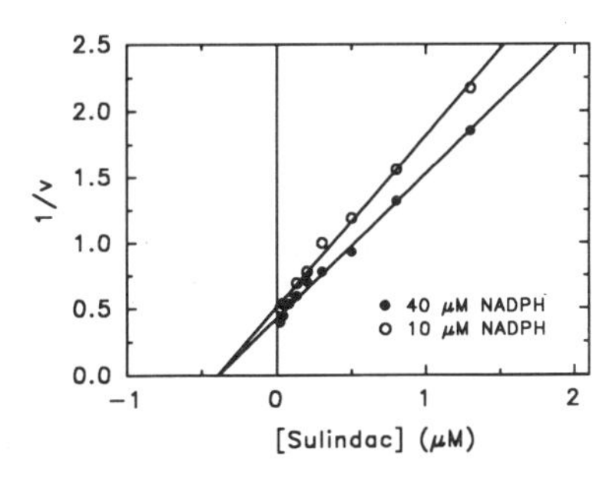

Figure 2. Dixon plot of inhibition of aldose reductase by sulindac.

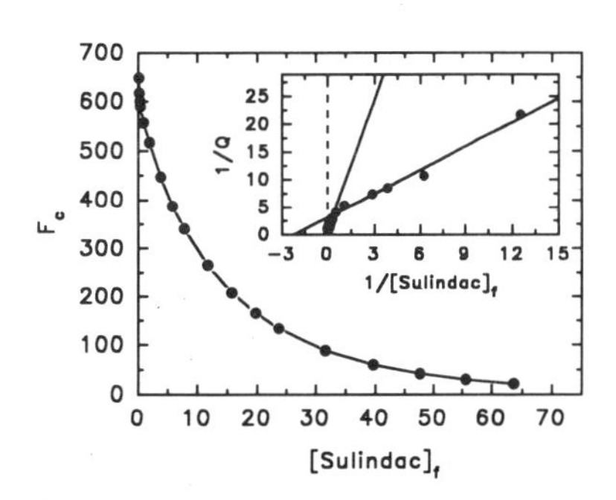

Figure 3. Fluorescence quenching of apo-aldose reductase by sulindac. Sulindac was added to 0.35 μM aldose reductase, pH 7, 25° C; Fc is the intrinsic protein fluorescence corrected for inner filter effects; Q is the fractional quenching; [sulindac]$_f$ is the concentration of unbound sulindac. Two binding sites are apparent, K_D = 0.44 and 8.8 μM.

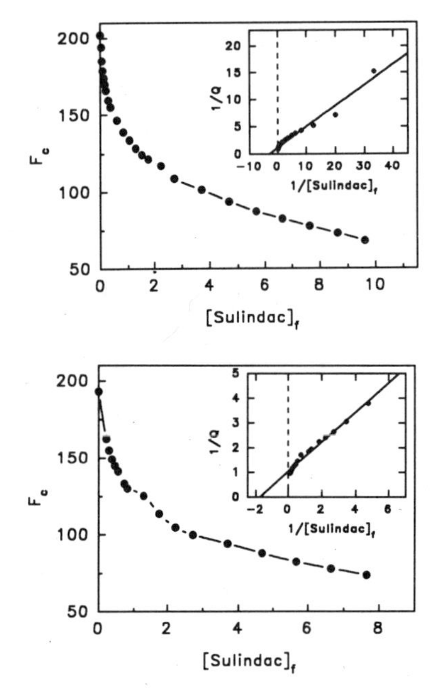

Figure 4. Fluorescence quenching of ALR2-NADP (left) and ALR2-NADPH (right) binary complexes by sulindac reflect single binding sites, K_D = 0.30 and 0.58 μM, respectively.

dac, and suggest that the binding site for sulindac is not strongly dependent on the presence of cofactor. The biphasic pattern observed for the binding of sulindac to apo-ALR2 suggests that there are two binding sites in the absence of cofactor.

In summary, sulindac is a classical noncompetitive inhibitor of ALR2 with respect to the binding of cofactor. The dissociation constants of the ALR2-sulindac binary complex and the ALR2-cofactor-sulindac ternary complexes are less than 1 micromolar, comparable to sorbinil. Sulindac appears to bind to ALR2 equally well regardless of the presence of NADP or NADPH. Sulindac may provide structural features that can be exploited in the design of new ALR2 inhibitors.

ACKNOWLEDGMENTS

This work was supported by NIH grant GM52576 from the MBRS Program, NIH.

REFERENCES

Blakytny, R., and Harding, J. (1992) Prevention of cataract in diabetic rats by aspirin, paracetamol and ibuprofen, *Exp. Eye Res.* **54**, 509–518.

Chaudhry, P.S., Cabrera, J., Juliani, H.R., and Varma, S.D. (1983) Inhibition of human lens aldose reductase by flavinoids, sulindac and indomethacin, *Biochem. Pharmacol.* **32**, 1995–1998.

Cohen, K.L., and Harris, S. (1987) Efficacy and safety of nonsteroidal anti-inflammatory drugs in the therapy of diabetic neuropathy, *Arch. Intern. Med.* **147**, 1442–1444.

Crabbe, M.J.C., Freeman, G., Halder, A.B., and Bron, A. (1985) The inhibition of bovine lens aldose reductase by clinoril, its absorption into the human red cell and its effect on human red cell aldose reductase activity, *Ophthalmic Res.* **17**, 85–89.

Cunha-Vaz, J.G., Mota, C.C., Leite, E.C., Abreu, J.R., and Ruas, M.A. (1985) Effect of sulindac on the permeability of the blood-retinal barrier in early diabetic retinopathy, *Arch. Ophthalmol.* **103**, 1307–1311.

Gupta, S.K., and Joshi, S. (1991a) Naproxen: an aldose reductase inhibitor and potential anti-cataract agent, *Dev. Ophthalmol.* **21**, 170–178.

Gupta, S.K., and Joshi, S. (1991b) Relationship between aldose reductase inhibiting activity and anti-cataract action of various non-steroidal anti-inflammatory drugs, *Dev. Ophthalmol.* **21**, 151–156.

Jacobson, M., Sharma, Y.R., Cotlier, E., and Hollander, J.D. (1983) Diabetic complications in lens and nerve and their prevention by sulindac or sorbinil: two novel aldose reductase inhibitors, *Invest. Ophthalmol. Vis. Sci.* **24**, 1426–1429.

Kador, P.F., and Kinoshita, J.H. (1985) Role of aldose reductase in development of diabetes-associated complications, *Amer. J. Med.* **79** (suppl. 5A), 8–12.

Kador, P.F., Kinoshita, J.H., Tung, W.H., and Chylack, L.T. (1980) Differences in the susceptibility of various aldose reductases to inhibition, *Invest. Ophthalmol. Vis. Sci.* **19**, 980–982.

Kirby,E.P., in: Excited States of Proteins and Nucleic Acids, Steiner, R.F. and Weinryb, I., eds., Plenum, New York, 1971, pp 31–56.

Mansour, S.Z., Hatchell, D.L., Chandler,D., Saloupis, P., and Hatchell, M.C. (1990) Reduction of basement membrane thickening in diabetic cat retina by sulindac, *Invest. Ophthalmol. Vis. Sci.* **31**, 457–463.

Sharma, Y.R., and Cotlier, E. (1982) Inhibition of lens and cataract aldose reductase by protein-bound anti-rheumatic drugs: salicylate, indomethacin, oxyphen-butazone, sulindac. *Exp. Eye Res.* **35**, 21–27.

Sharma, Y.R., Bhatnagar, R., Vajpayee, R.B., Mohan, M., Gupta, S.K., and Mukesh, K. (1989a) In vivo effectiveness of sulindac in sugar cataracts-the implication, *Dev. Ophthalmol.* **17**, 173–175.

Sharma, Y.R., Vajpayee, R.B., Bhatnagar, R., Mohan, M., Azad, R.V., Mukeshkumar and Ramnath (1989b) Topical sulindac therapy in senile cataracts: Cataract-IV. *Ophthalmol.* **37**, 127–133.

Vander Jagt, D.L., Robinson, B., Taylor, K.K., and Hunsaker, L.A. (1990a) Aldose reductase from human skeletal and heart muscle, *J. Biol. Chem.* **265**, 20982–20987.

Vander Jagt, D.L., Hunsaker, L.A., Robinson, B., Stangebye, L.A., and Deck, L.M. (1990b) Aldeyde and aldose reductases from human placenta, *J. Biol. Chem.* **265**, 10912–10918.

Vander Jagt, D.L., Robinson, B., Taylor, K.K., and Hunsaker, L.A. (1992) Reduction of trioses by NADPH-dependent aldo-keto reductases, *J. Biol. Chem.* **267**, 4364–4369.

REGULATION OF ALDOSE REDUCTASE BY ALDEHYDES AND NITRIC OXIDE

Satish K. Srivastava,[1] Animesh Chandra,[1] Sanjay Srivastava,[1]
J. Mark Petrash,[3] and Aruni Bhatnagar[1,2]

[1]Department of Human Biological Chemistry and Genetics
[2]Department of Physiology and Biophysics
University of Texas Medical Branch
Galveston, Texas 77555-0647
[3]Department of Ophthalmology and Visual Sciences
Washington University
St. Louis, Missouri 63110

1. INTRODUCTION

Aldose reductase (AR) is a member of the aldo-keto reductase superfamily. It was initially discovered in the seminal vesicles, where it is believed to catalyze the NADPH-mediated reduction of glucose to sorbitol—which in turn is metabolized to fructose required for spermatogenesis (*for review* see Bhatnagar and Srivastava, 1992). The conversion of glucose to fructose via sorbitol, or the so called polyol pathway was later identified in several other tissues and AR was found to be universally expressed in most tissues examined and in species ranging from yeast to man. Based on the observations that high concentration of polyols accumulate during sugar cataractogenesis, it was suggested that during hyperglycemia, the flux of glucose via the polyol pathway is increased and that due to the low permeability of the membrane to polyols, sugar alcohols accumulate in diabetic tissues leading to marked hydration, cell swelling and lysis (Kinoshita and Nishimura, 1988).

In agreement with this view, pharmacological inhibition of AR has been found to prevent, delay and in some cases even reverse tissue injury associated with several diabetic complications such as cataractogenesis, neuropathy, and nephropathy (Bhatnagar and Srivastava, 1992). However, not all trials with inhibitors of AR have proved to be beneficial. Although, the reasons for the variable efficacy of these drugs remain obscure, the inhibitor data, as well as the observed increases in polyols during diabetes, have been taken as conclusive evidence that the primary cellular function of AR is the reduction of glucose—hence the enzyme is called (and accepted as an) aldose reductase. However, several lines of evidence suggest that AR may be involved in other important cellular functions.

Enzymology and Molecular Biology of Carbonyl Metabolism 7
edited by Weiner *et al.* Kluwer Academic / Plenum Publishers, New York, 1999.

The enzyme is highly expressed in several tissues such as skeletal muscle and heart (Cao *et al.*, 1998; van der Jagt *et al.*, 1990; Srivastava *et al.*, 1998) in which (due to their dependence on insulin) the intracellular concentrations of glucose do not attain high enough levels to cause significant flux via the polyol pathway. The K_m of AR for glucose is very high (50 to 100 mM), and therefore, at least under euglycemic conditions, AR may have little role in glucose metabolism. On the other hand, the enzyme displays high affinity for various aromatic and aliphatic aldehydes. Furthermore, kinetic studies with the pure enzyme suggested that the active site of the enzyme is highly hydrophobic, and this was further confirmed by X-ray analysis of the enzyme crystals (Wilson *et al.*, 1992). These studies revealed that the enzyme contains an atypical carbohydrate binding site. Unlike other carbohydrate binding enzymes, the active site of AR is highly hydrophobic and lacks binding groups capable of providing efficient anchorage to hydrophilic carbohydrate via multiple hydrogen bonds. Based on these results, it appears that the "ideal" AR substrate is unlikely to be a carbohydrate, but rather long chain hydrophobic aldehyde(s).

While several aldehydes are generated in the environment and during metabolism, the most abundant (and in terms of pathology the most significant) endogenous source of hydrophobic aldehydes is lipid peroxidation. Mono and polyusaturated fatty acids (PUFA) are particularly vulnerable to oxidation, and in the presence of adventitious metals and unquenched reactive oxygen species (ROS) the PUFAs under go a characteristic chain reaction, which involves several short lived radical intermediates. The peroxyl radicals generated are converted to hydroperoxides, which in most cells are either reduced catalytically by glutathione peroxidase or are converted back to radical species by metal ions. The other reactive intermediates generated during lipid peroxidation, are the alkoxyl radicals which spontaneously dismutate to generate saturated and unsaturated aldehydes (Grosch, 1987). Several previous studies have shown that aldehydes are the major end products of lipid peroxidation and may be "toxic second messengers" that mediate the cellular effects of their short lived precursors (Esterbauer *et al.*, 1991). Of the aldehydes generated during lipid peroxidation, the most reactive are the α, β unsaturated aldehydes which are generated during the oxidation of ω-6 fatty acids. These aldehydes react avidly with cellular nucleophiles (such as glutathione) and form adducts with cellular proteins. The conjugation of these aldehydes with glutathione is catalyzed by glutathione reductases (Danielson *et al.*, 1987).

During the past several years the major interest of our research group has been to elucidate the physiological and pathological roles of aldose reductase. Given the high hydrophobicity of the active site of the enzyme, we speculated that the long chain α, β unsaturated aldehydes generated during lipid peroxidation (of which, 4-hydroxy *trans*-2-nonenal; HNE is the most abundant and one of the most toxic) may be the natural and endogenous substrates of AR. Since the active site of AR contains a highly reactive thiol (Cys-298; Petrash *et al.*, 1992), we reasoned that its reaction with unsaturated aldehydes may serve as a regulatory mechanism for controlling AR enzyme activity. In this communication we report some of our recent observations on the role of AR in the catalysis of unsaturated aldehydes, and how this catalysis is regulated not only by the unsaturated aldehydes but also other endogenous thiol reactive groups such as nitric oxide, nitrosothiols and peroxynitrile.

2. MATERIALS AND METHODS

Human recombinant AR and its cysteine mutants were prepared as described before. Before each experiment, stored AR was reduced by DTT and filered through Sephadex G-

25 column (PD-10) as described previously (Petrash *et al.*, 1992). The activity of the enzyme was determined at 25°C in a 1 ml system containing 0.1M potassium phosphate, pH 7.0, 10 mM D,L-glyceraldehyde and 0.1 mM NADPH. The aldehyde substrates were varied over a concentration range covering from 0.2 to 5–7 times the K_m of each aldehyde. Initial velocity was measured at 6 to 8 different concentrations of each substrate. Individual saturation curves used to obtain the steady state kinetic parameters were fitted to a general Michaelis-Menton equation. In all cases, the best fit to the data was chosen on the basis of the standard error of the fitted parameter and the lowest value of O, which is defined as the sum of squares of the residuals divided by the degrees of freedom.

For modification studies, reduced AR (0.5 mg/ml) was incubated with 100 μM freshly prepared aldehyde or NO donor in 0.1 M potassium phosphate, pH 7.0, at room temperature and aliquots of the reaction mixture were withdrawn at various time intervals to measure the enzyme activity as described above. The electrospray ionization mass spectra were obtained on a Finnigan-TSQ70 (upgraded to TSQ700) triple quadropole instrument with a Vestec electrospray ionization source with tapered fused silica capillary needle (50 μM i.d.) as described previously (Chandra *et al.*, 1997b). In all cases, the standard errors in mass determinations ranged between 1.5 to 2.5 mass units, which were similar to the errors calculated with the protein standard (apomyoglobin and bovine serum albumin) used to calibrate the instrument.

3. RESULTS AND DISCUSSION

Our studies show that AR efficiently catalyzes the NADPH-mediated reduction of α, β unsaturated aldehydes derived from lipid peroxidation, as well as the gluathione conjugates of these aldehydes. The K_m for HNE was found to be 6 to 8 μM and that of GS-HNE is 25 to 30 μM (Table I). Note that K_m acrolein is approximately 800 μM, whereas conjugation of acrolein with GSH led to a 100 fold decrease in the K_m. Since the K_m of the enzyme for these compounds is 3 orders of magnitude lower than that for glucose, we suggest that one of the physiological functions of AR may be the metabolism of lipid-derived unsaturated aldehydes and their glutathione conjugates (Figure 1).

Table I. Kinetic parameters for the reduction of lipid-derived aldehydes and their glutathione conjugates catalyzed by aldose reductase from several tissues

	K_m (mM)	k_{cat} (min^{-1})	k_{cat}/K_m
Bovine lens			
HNE	0.009 ± 0.0009	19.2	2133
GS-HNE	0.034 ± 0.004	20.6	606[a]
Bovine heart			
HNE	0.007 ± 0.002	24.7	3526[b]
GS-HNE	0.019 ± 0.002	27.9	1468
Human placenta			
Acrolein	0.8 ± 0.2	37.6	47
GS-acrolein	0.007 ± 0.001	29.8	4257
HNE	0.022 ± 0.007	34.3	1559
GS-HNE	0.016 ± 0.005	28.8	1800

The kinetic parameters were determined in 0.1 M potassium phosphate, pH 7.0 with 0.1 mM NADPH. Note: a: from Srivastava et al., 1995; b: from Srivastava et al., 1998.

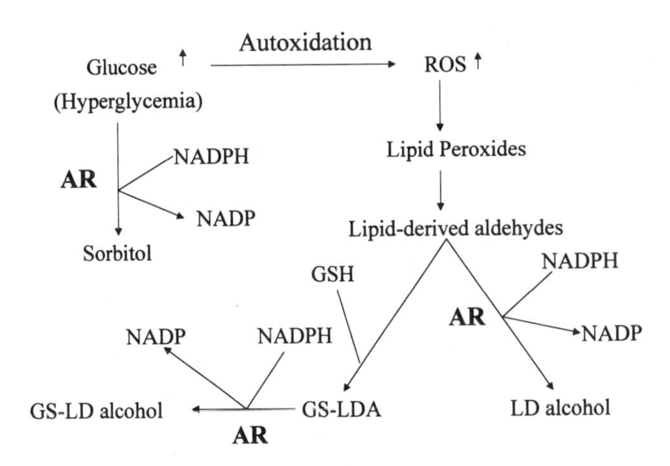

Figure 1. Reaction scheme for the generation and metabolism of lipid derived aldehydes during diabetes. Hyperglycemia leads to the generation of reactive oxygen species (ROS), which induce lipid peroxidation resulting in the formation of lipid derived aldehydes (LDA). Aldose reductase (AR) catalyzes the NADPH-dependent reduction of the LDAs and/or their glutathione conjugates. In addition, AR also catalyzes the reduction of glucose to sorbitol.

The high catalytic efficiency with which the enzyme reduces unsaturated aldehydes and their glutathione conjugates also suggests that the AR may represent a protective mechanism for the cellular detoxification of the noxious products of lipid peroxidation. In support of this is our observation that the HNE-induced decrease in the light transmittance by the ocular lens is exacerbated by AR inhibitors (Ansari *et al.*, 1996), suggesting that inhibition of AR enhances the toxicity of HNE by diminishing its cellular detoxification.

In addition to general detoxification, reduction of HNE and GS-HNE by AR may also represent an important component of intracellular signaling. Recent evidence suggests that ROS play an important and essential role in cell signaling related to growth factor induced mitogenesis, inflammation, cell differentiation and apoptosis (Lander, 1997). While the specific signaling molecules utilized by ROS have not been identified, it appears likely that some of the effects of ROS may be mediated by products of lipid peroxidation. In contrast to their radical precursors, HNE and related aldehydes are long-lived and can diffuse to sites far away from their site of origin. These aldehydes can elicit specific alterations in signaling molecules by their ability to form covalent adducts with cellular protein cysteine, histidine and lysine residues. Aldose reductase, by regulating the cellular concentrations of these aldehydes and their glutathione adducts, could, therefore, regulate the cellular redox state and affect multiple signaling events. In this regard it is interesting to point out that AR and its murine homolog FR-1 are upregulated by mitogenic stimuli due to FGF as well as components of cell signaling such as activation of protein kinase C (Donohue *et al.*, 1994; Jacquin-Becker and Labourdette, 1997).

3.1. Regulation of AR Activity

We had demonstrated it earlier that Cys-298, present at the active site of AR, while not a direct participant in substrate binding or catalysis, plays an important role in regulating the catalytic cycle as well as the binding of some of the active site inhibitors such as sorbinil, statil and alrestatin (Petrash *et al.*, 1992; Liu *et al.*, 1993; Bhatnagar *et al.*, 1989, 1994). Carboxymethylation of Cys-298 or its replacement with serine increases the K_m of the enzyme for aldehydes along with an increase in the k_{cat}. In addition, we have shown that the enzyme is inactivated by oxidized glutathione (GSSG) specifically due to the formation of a mixed disulfide with Cys-298.

Our recent results indicate that besides being substrates, the aldehydes could also be regulators of AR activity. Modification of the enzyme with short chain (< C5) aldehydes such as acrolein and crotonaldehyde in the absence of NADP or NADPH enhances the catalytic activity of the enzyme (similar to the effect observed with site-directed mutagenesis). On the other hand, covalent modification of the enzyme with long chain (> C6) unsaturated aldehydes such as HNE causes enzyme inactivation. Furthermore, we find that in both cases (the long and short chain aldehydes) the alteration in the enzyme activity is specifically due to the covalent modification of the enzyme at Cys-298, since sequence analysis of the modified enzyme reveals modification at Cys-298 (Srivastava *et al.*, 1998) and the AR:C298S mutant is completely insensitive to the modifying effects of both acrolein and HNE. Thus, besides being substrates of AR, the α, β- aldehydes could also be potential regulators of AR catalysis. *In vivo* modification of the enzyme by these aldehydes is, however, likely only if a significant fraction of the enzyme exists without the nucleotides bound to it, since NADP protects against modification due to both acrolein and HNE.

Another physiological regulator of AR activity could be thiol reactive signaling molecules such as nitric oxide (NO), nitrosothiols and peroxynitrile. These species react aggressively with sulfhydryls. Whereas NO displays little or no direct activity with thiols, its thiol modifying reactions are greatly facilitated by the presence of oxygen (Kharitonov *et al.*, 1995). Nitrosothiols such as S-nitroso glutathione, GSNO, react avidly with protein thiols via trans-nitrosylation and peroxynitrile (generated by the reaction of NO with superoxide) reacts rapidly and non-specifically with several cellular nucleophiles such as thiols, and tyrosine. Due to the well known sensitivity of AR to thiol modifying reagents including, as above, the α, β unsaturated aldehydes, we investigated the sensitivity of the enzyme to NO donors.

When reduced AR was incubated with the NO donors, SNP, SNAP, and SIN-1, a time and concentration-dependent loss of enzyme activity was observed (Chandra *et al.*, 1997a). To assess whether this was due to modification of the enzyme at Cys-298, we studied the effects of the NO donors on the carboxymethylated enzyme. In our previous investigations we had found that in the presence of 1.0 mM iodoacetate, the enzyme is selectively carboxymethylated at Cys-298, which leads to enzyme activation and loss of sensitivity to sorbinil (Liu *et. al.*, 1993; Bhatnagar *et al.*, 1994). When treated with SNP, however, the activity of the carboxymethylated enzyme was largely unaffected, suggesting that the enzyme with a carboxymethylated (protected) Cys-298 is insensitive to NO donors. It is also significant to point out that exposure to SIN-1 (which generates peroxynitrile) led to a rapid inactivation of the enzyme, indicating that high intracellular concentrations of peroxynitrile may be particularly damaging to AR.

While NO is involved in several physiological functions and serves by itself as a neurotransmitter, many of the effects of NO are mediated by nitrosothiols (Jia *et al.*, 1996). It has been suggested that nitrosothiols may mediate some of the vascular and antimitogenic effects of NO, and nitrosation of cellular thiols may serve signaling functions analogous to phosphorylation and dephosphorylation. Therefore, we tested whether GSNO, which is likely to be the most abundant nitrosothiol *in vivo* alters AR activity. We found that GSNO causes a profound decrease in enzyme activity. The modification of AR by GSNO was found to be due to specific modification of Cys-298, since AR:C298S was found to be insensitive to the inactivating effects of GSNO. In contrast, other cysteine mutants of AR (C303S and C80S) were as sensitive to GSNO as the wild type enzyme (Chandra *et al.*, 1997b).

The selectivity of GSNO was further confirmed by electrospray ionization-mass spectrometry (ESI-MS). When the enzyme exhaustively modified by GSNO was subjected to ESI-MS, two distinct enzyme species were observed (Figure 2). One of the protein forms (B), which accounted for 70 % of the signal corresponded to a molecular mass of

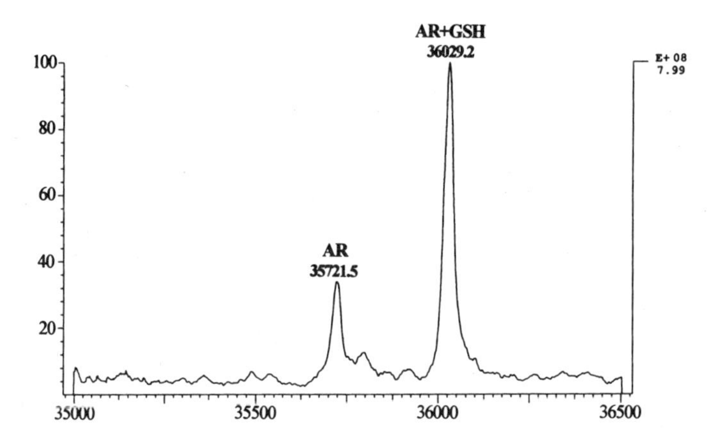

Figure 2. ESI-M spectrum of aldose reductase treated with GSNO. Human recombinant AR was incubated with 50 μM GSNO for 1 hr, after which unreacted GSNO was removed by rapid gel filtration. Two distinct species were observed upon ESI-MS. A minor peak corresponding to a molecular mass of 35721.5 was assigned to unmodified AR, whereas the peak with a molecular mass of 36029.2 was identified to be due the formation of a mixed disulfide between AR and glutathione. The increase in the molecular mass of AR from 35721 to 36029.2 is consistent with the addition of a single glutathione molecule (molecular mass 307) to AR.

36028.8, whereas the other species (A) displayed a molecular mass of 35721.0. Based on the amino acid sequence of the enzyme, the expected molecular mass of AR is 35,723.0; thus species A was assigned to unmodified AR. Species B, which displayed a molecular mass 306 Da in excess of the molecular mass of the enzyme was assigned to a mixed disulfide between the enzyme and glutathione (expected molecular mass 36028.3). On the basis of these results we conclude that exposure of AR to GSNO leads to the formation of a mixed disulfide with glutathione.

The generation of a mixed disulfide with GSNO was somewhat surprising since we expected GSNO to act like an NO donor and cause enzyme nitrosation. It is well known that GSNO does not liberate much free NO is solution, and modifies proteins by transnitrosation (Singh *et al.*, 1996). Apparently, in case of AR, the transnitrosation reaction does not go to completion, and NO, rather than GS- is released from the complex. However, the mechanism of AR-SSG formation does not seem to involve a direct interaction between AR and GS- generated in solution due to the dissociation of GSNO, since incubation of AR with 50 μM GSSG, which is twice the amount of GSSG that could be formed from 50 μM GSNO (the maximum amount we used to obtain 60–70% inhibition of AR activity), did not form AR-SSG in 2 hrs at 37°C. Since the *in vivo* concentration of GSSG does not usually reach more than 50 μM even under most of the pathological conditions of oxidative stress, it seems that the formation of mixed disulfides between AR (and other proteins for that matter) may be mediated by nitrosothiols. Our results, therefore, suggest the interesting possibility that protein S-thiolation may be a novel and important component of NO mediated signaling.

In conclusion, our present results suggest that the ubiquitously distributed AR is a tightly regulated enzyme. Under physiological conditions AR may play a pivotal role in the detoxification of xenobiotics, environmental pollutants and metabolically generated aldehydes. Thus, in view of this beneficial role of AR, inhibition of this enzyme, as a therapeutic modality for the long term clinical management of hyperglycemia, should be carefully evaluated.

ACKNOWLEDGMENTS

Support for this work was provided by NIH grants DK36118, HL55477, and EY05856.

REFERENCES

Ansari, N.H. Wang, L-F., and Srivastava, S.K. (1996). Role of lipid aldehydes in cataractogenesis: 4-hydroxynonenal-induced cataract. *Biochem. Mol. Med.* **58**, 25–30.

Bhatnagar, A., Liu, S-Q., Das, B. and Srivastava, S.K. (1989) Involvement of sulfhydryl residues in aldose reductase-inhibitor interaction. *Mol. Pharmacol.* **36**, 825–830.

Bhatnagar, A. and Srivastava, S.K. (1992) Aldose reductase: Congenial and injurious profiles of an enigmatic enzyme. *Biochem. Med. Metabol. Biol.* **48**, 91–121.

Bhatnagar, A., Liu, S-Q., Ueno, N., Chakrabarti, B. and Srivastava, S.K. (1994) Human placental aldose reductase: Role of Cys-298 in substrate and inhibitor binding. *Biochim. Biophys. Acta* **1205**, 207–214.

Cao, D., Fan, S.T. and Chung, S.S.M. (1998) Identification and Characterization of a Novel Human Aldose Reductase-like Gene. *J.Biol.Chem.* **273**, 11429–11435.

Chandra, A., Srivastava, S., Petrash, M., Bhatnagar, A., and Srivastava, S.K. (1997a) Active site modification of aldose reductase by nitric oxide. *Biochim. Biophys. Acta* **1341**, 217–222.

Chandra, A., Srivastava, S., Petrash, M., Bhatnagar, A., and Srivastava, S.K. (1997b) Modification of aldose reductase by nitrosoglutathione. *Biochemistry* **36**, 15801–15809.

Danielson, U.H., Esterbauer, H., and Mannervik, B. Structure-activity relationships of 4- hydroxyalkenals in the conjugation catalyzed by mammalian glutathione transferases. (1987) *Biochem J.* **247**, 707–713.

Donohue, P.J., Alberts, G.F., Hampton, B.S. and Winkles, J.A. (1994) A delayed-early gene activated by fibroblast growth factor-1 encodes a protein related to aldose reductase. *J.Biol.Chem.* **269**, 8604–8609.

Esterbauer, H., Schaur, R.J., and Zollner, H. (1991) Chemistry and biochemistry of 4-hydroxynonenal, malonaldehyde and related aldehydes. *Free Radic Biol Med.* **11**, 81–128.

Grosch, W. (1987) Reactions of hydroperoxides - products of low molecular weight. In Autoxidation of Unsaturated Lipids. (H.W.-S. Chan Ed.) Academic Press, London, pp95–139.

Jacquin-Becker C. and Labourdette, G. (1997) Regulation of aldose reductase expression in rat astrocytes in culture. *Glia* **20**, 135–144.

Jia L. Bonaventura C. Bonaventura J., and Stamler JS. (1996) S-nitrosohaemoglobin: a dynamic activity of blood involved in vascular control. *Nature* **380**, 221–226.

Kharitonov, V.G., Sundquist, A.R., and Sharma, V.S. (1995) Kinetics of nitrosation of thiols by nitric oxide in the presence of oxygen. *J. Biol. Chem.* **270**, 28158–28164.

Kinoshita, J.H. and Nishimura, C. (1988) The involvement of aldose reductase in diabetic complications. *Diab-Meta. Rev.* **4**, 323–37.

Lander, H.M. (1997) An essential role of free radicals and derived species in signal transduction. *FASEB J.* **11**, 118–124.

Liu, S-Q., Bhatnagar, A., Ansari, N.H., and Srivastava, S.K. (1993) Identification of the reactive cysteine residue in human placenta aldose reductase. *Biochim. Biophys. Acta.* **1164**, 268–272.

Petrash, J.M., Harter, T.M., Devine, C.S., Olins, P.O., Bhatnagar, A., Liu, S-Q. and Srivastava, S.K. (1992) Involvement of cysteine residues in catalysis and inhibition of human aldose reductase: Site-directed mutagenesis of Cys-80, -298 and -303. *J. Biol. Chem.* **267**, 24833–24840.

Singh, R.J. Hog, N., Joseph, J., and Kalyanaraman, B. (1996) Mechanism of nitric oxide release from S-nitrosothiols. *J. Biol. Chem.* **271**, 18596–18603.

Srivastava, S., Chandra A., Bhatnagar, A., Srivastava, S.K., and Ansari, N.H. (1995) Lipid peroxidation product, 4-hydroxynonenal and its conjugate with GSH are excellent substrates of bovine lens aldose reductase. *Biochem. Biophys. Res. Comm.* **217**, 741–746.

Srivastava, S., Chandra, A., Ansari, N.H., Srivastava, S.K. and Bhatnagar, A. (1998) Identification of cardiac oxidoreductase(s) involved in the metabolism of the lipid peroxidation derived aldehyde - 4-hydoxynonenal. *Biochem. J.* **329**, 469–475.

Vander Jagt, D.L. Robinson, B., Taylor, K.K. and Hunsaker, L.A. (1990) Aldose reductase from human skeletal and heart muscle. Interconvertible forms related by thiol-disulfide exchange. *J. Biol. Chem.* **265**, 20982–20987.

Wilson, D.K., Bohren, K.M., Gabbay, K.H. and Quiocho, F.A. (1992) An unlikely sugar substrate site in the 1.65 A structure of the human aldose reductase holoenzyme implicated in diabetic complications. *Science* **257**, 81–84.

THE CYTOTOXICITY OF METHYLGLYOXAL AND 3-DEOXYGLUCOSONE IS DECREASED IN THE ALDEHYDE REDUCTASE GENE-TRANSFECTED CELLS

Keiichiro Suzuki, Young Ho Koh, and Naoyuki Taniguchi

Department of Biochemistry
Osaka University Medical School
2-2, Yamadaoka, Suita, 565-0871, Japan

1. INTRODUCTION

Glycation (the initial portion of the Maillard reaction) plays a major role in diabetic complications since some of the reaction intermediates are responsible for the modification and cross-linking of long-lived proteins, which, in turn, result in a deterioration of normal cell function. Included among these reaction intermediates are methylglyoxal(MG) and 3-deoxyglucosone(3-DG) both of which are cytotoxic dicarbonyl compounds. MG, a reactive α,β-dicarbonyl metabolite and physiological substrate for the glyoxalase system (Thornalley, 1993), is formed via the non-enzymatic and enzymatic elimination of phosphate from dihydroxyacetone phosphate, glyceraldehyde-3-phosphate (Phillips and Thornalley, 1993; Pompliano et al., 1990), as well as by the oxidation of hydroxyacetone and aminoacetone (Reichard et al., 1986; Koop et al., 1985; Lyles et al., 1992). The serum concentration of MG increased 5–6 fold in patients with insulin-dependent diabetes mellitus and 2–3 fold in patients with non-insulin-dependent diabetic mellitus (McLellan et al., 1994). 3-DG, another glucose-derived dicarbonyl compound, is a major and highly reactive intermediate in the Maillard reaction and a potent cross-linking agent responsible for the polymerization of proteins and the formation of advanced glycation end products (Niwa et al., 1993). Plasma 3-DG levels are also increased under diabetic conditions (Yamada et al., 1994).

Aldehyde reductase (ALR) catalyzes the reduction of these both compounds. ALR is expressed in many organs and serves to reduce aldehyde carbonyl groups using NADPH as a hydrogen donor (Bohren et al., 1989; Bohren et al., 1991). Recently, we identified aldehyde reductase as a major enzyme in rat liver which detoxifies 3-deoxyglucosone (3-

Enzymology and Molecular Biology of Carbonyl Metabolism 7
edited by Weiner *et al.* Kluwer Academic / Plenum Publishers, New York, 1999.

DG) (Takahashi et al., 1993) and reported the inactivation of this enzyme as the result of a glycation reaction (Takahashi et al., 1995). In addition, we demonstrated that 3-DG and methylglyoxal (MG) induce apoptotic cell death in macrophage-derived cell lines (Okado et al., 1996). Very recently we have also reported that 3-DG and methylglyoxal (MG) selectively induce heparin-binding epidermal growth factor-like growth factor (HB-EGF) in rat aortic smooth muscle cells and that this may be implicated in diabetic macroangiopathy (Che et al., 1997).

A considerable body of biochemical and clinical evidence suggests that the increased levels of MG or 3-DG in diabetes mellitus may be linked to the development of diabetic complications and that ALR plays a protective role in this process. However the exact nature of its protective role remains unclear. The present study presents evidence to show that the overexpression of ALR exerts a protective effect against the cytotoxicity of MG or 3-DG in PC12 cells.

2. MATERIALS AND METHODS

2.1. Materials

MG was purchased from Sigma and further purified by distillation and the solution concentration was determined as described previously (Che et al., 1997). 3-DG was chemically synthesized and its structure and purity were confirmed using ^{1}H NMR (Takahashi et al., 1993). 2'7'-Dichlorofluorescin diacetate (DCFH-DA) was purchased from Molecular Probes, Inc. Other chemicals were of the highest grade available.

2.2. Cell Culture

Rat pheochromocytoma PC12 cells were obtained from the Japanese Cancer Research Resources Bank (Tokyo). PC12 cells and the ALR gene-transfected cells were cultured in RPMI1640 medium, supplemented with 10% fetal calf serum, 5% horse serum, and 0.1 mg/ml of kanamycin under a humidified atmosphere of 95% air and 5% CO_2.

2.3. Measurement of Intracellular ROS by Flow Cytometry

Levels of reactive oxygen species (ROS) were assessed using an oxidation-sensitive fluorescent probe DCFH-DA, as previously described (Kayanoki et al., 1994). In the presence of a variety of ROS, such as intracellular peroxides, DCFH is oxidized to 2'7'-dichlorofluorescein, a highly fluorescent compound. PC12 cells which had been treated with MG were incubated with 5 μM DCFA-DA and their cellular fluorescence intensities were determined using a FACScan (Becton-Dickinson).

2.4. Gene Transfection

Rat ALR cDNA (Takahashi et al., 1993) was inserted into the pRc/CMV (Invitrogen). The PC12 cells used for the transfection of cDNA were plated in a collagen-coated 10-cm plastic culture dish to a density of 1 × 10^6/ml cells. After 24 h the medium was removed and the cells washed twice with cold PBS, pH7.4, and the medium changed to serum free DMEM. The ALR-pRc/CMV vector (20μg) was mixed with Lipofectamine (Life Technologies, Inc.) and 100 μl of this solution was added to the PC12 cells. After a 5 h incubation, the medium was changed to the original as described above. Stable transfectants were screened with 0.5 mg/ml Neomycin. Out of 58 Neomycin-resistant clones, two clones with normal cell shapes and good cell growth were randomly selected and designated as PC12-ALR(H) and PC12-ALR(L).

2.5. Activity of ALR

The enzyme activity of ALR was determined by measurement of the rate of decrease in absorbance at 340 nm. The standard assay mixture contained 100mM Na-phosphate, pH 7.0/0.1 mM NADPH/10 mM MG.

2.6. Determination of 3-DG Levels in PC12 Cells

Determination of 3-DG was performed using the HPLC methodology reported by Yamada et.al., which was modified by us for use with crude tissues (Yamada et al., 1994). Briefly, after harvesting the PC12 cell, the cells were homogenized with phosphate-buffered saline (PBS), after which 6% perchloric acid was added to the solution. The supernatant was neutralized using saturated NaHCO3 after centrifugation. One ml of this solution was incubated over night with 100 μl of 0.1% 2,3-diaminonaphthalene at 4 °C in the presence of 50 μl of 0.005% 2,3-butanedione as an internal standard. The reaction mixture was extracted using 4 ml of ethyl acetate, followed by evaporation to dryness. The dried extract was redissolved in 200 μl of methanol and used for HPLC. HPLC was performed using a Shimadzu Chromatopac C-R7A plus system. Separation was accomplished on a TSK ODS-80 TM column using a solvent system of 50 mM phosphate, acetonitrile and methanol (80:17.5;2.5) at a flow rate of 1 ml/min. The effluent was monitored at 268 nm with a Shimadzu SPD-10A UV detector. Quantitation of 3-DG was performed by calculating the ratio of the peak height of the 3-DG derived peak to that of an internal standard.

2.7. Cell Viability Assay

The viability of cells in 96-well culture plates was determined 24 h after the addition of MG or 3-DG by the methylene blue method (Nakata et al., 1993). Twenty microliters of a 25% glutaraldehyde solution were added to each well, followed by incubation for 15 min to fix the surviving cells. After washing with PBS, the fixed cells were stained for 15 min with 100 ml of a 0.05% methylene blue solution. After washing 3 to 4 times with PBS, 200 ml of 0.33 M HCl was added to each well, in order to extract the methylene blue, the level of which was determined by the measurement of the absorbance at 665 nm.

3. RESULTS

3.1. Flow Cytometric Analysis of Intracellular ROS

MG and 3-DG show cytotoxicity including apoptosis toward PC12 cells (Suzuki, et al. 1998). To investigate the mechanism of cytotoxicity, changes in intracellular ROS levels in PC12 cells after treatment with MG was investigated using DCFH-DA (Figure 1). Treatment of PC12 cells with MG shifted the peak, thus indicating that the cells treated with MG had undergone more oxidative stress than the control cells.

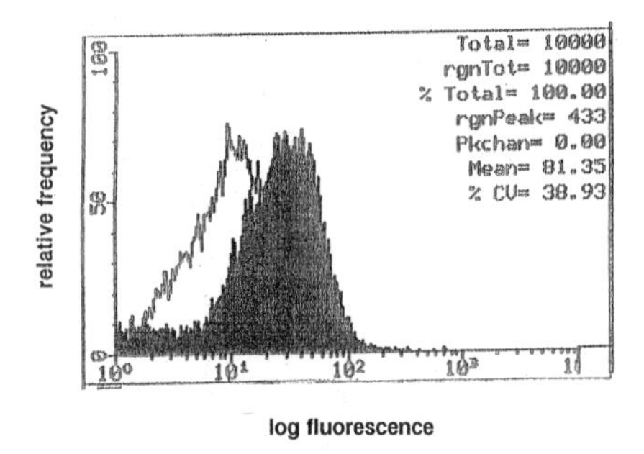

Figure 1. Flow cytometric analysis of intracellular peroxides. The production of peroxides by MG-treated PC12 cells was evaluated by flow cytometry. Cells treated with MG (0.5 mM) for 4 h showed significantly increased levels of ROS. White, control; Black, treated with MG.

3.2. ALR Activity of ALR Transfected Cells

ALR activities of control(wild), mock, PC12-ALR(H) and PC12-ALR(L) cells were 18.0, 15.4, 22.2 and 16.5 units/mg cellular protein, respectively. However ALR is a member of the aldo-keto reductase superfamily and MG or 3-DG could also serve as substrates for other reductases, thus making it difficult to distinguish the activity of an individual enzyme. Therefore these results were confirmed by western blotting using a specific antibody and northern blotting (data not shown). Western blotting analysis shows that PC12-ALR(H) cells contain higher levels of ALR protein than mock, control(wild), and PC12-ALR(L) cells. In addition the northern blotting of PC12-ALR(H) cells showed that higher levels of ALR mRNA were expressed, relative to control PC12 cells both in normal conditions and after treatment with MG. This result is in agreement with the results of the immunoblotting and activity

3.3. Determination of 3-DG Levels in PC12 Cells under Hyperglycemic Condition

PC12-ALR(H) cells and mock cells were cultured under hyperglycemic conditions (25 mM) for 3 days and 3-DG levels of these cells were determined. The 3-DG level of the PC12-ALR(H) cells was 215 ng/mg protein, significantly higher than that of mock cells (154 ng/mg protein).

3.4. Effects of ALR Overexpression on the Cytotoxicity by MG in PC12 Cells

The cytotoxicity of MG to PC12 cells was examined as a function of their content (Figure 2). The PC12-ALR(H) cells, which highly overexpressed ALR, were resistant to MG, compared to the control, mock or the PC12-ALR(L) cells. These findings suggest that overexpression of ALR has a protective effect relative to the cytotoxicity of MG.

3.5. Effects of ALR Overexpression on the Cytotoxicity by 3-DG in PC12 Cells

The cytotoxicity of 3-DG to PC12 cells which contained varying levels of ALR was also examined (Figure 3). While the difference is not as clear as in the case of MG, the PC12-ALR(H) cells, which highly overexpressed ALR, also show more resistance to 3-DG than the control, mock or the PC12-ALR(L) cells. These findings are consistent with a

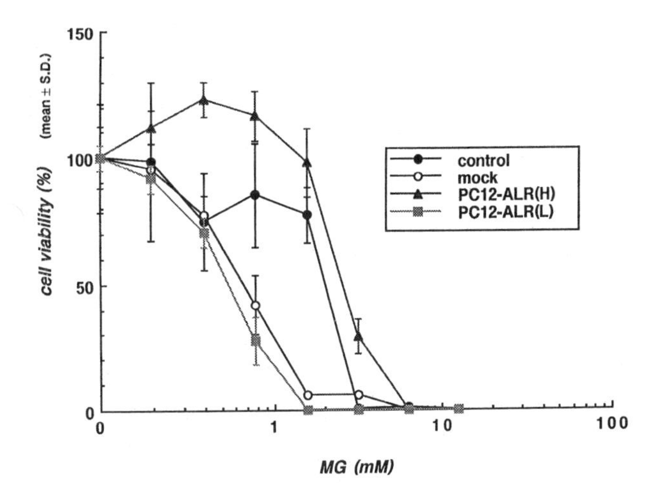

Figure 2. Effects of ALR overexpression on cytotoxicity by MG in PC12 cells. The effect of ALR overexpression on the viability of PC12 cells was examined after treatment with MG. ALR overexpression increased cell viability (PC12-ALR(H)).

relationship between the overexpression of ALR and its protective effect against the cytotoxicity of 3-DG.

4. DISCUSSION

In this study cellular levels of 3-DG were elevated under hyperglycemic culture conditions and thc cytotoxicity of MG and 3-DG was decreased in the ALR gene-tranfected cells. The data collected herein show that the protective effect by ALR in PC12 cells has its origin in rat pheochromocytoma in which relatively low levels of ALR were found. MG induced the accumulation of ROS and as a result, cell death including apoptosis occurred in PC12 cells, similar to the effect observed in macrophage-derived cell lines as reported previously by our group (Okado et al., 1996). A number of 2-oxoaldehyde compounds are elevated under diabetic conditions (McLellan et al., 1994; Yamada et al., 1994). These compounds are capable of inducing cellular damage, as well as accelerating the glycation process. These compounds can be detoxified by ALR, which is distributed over a variety

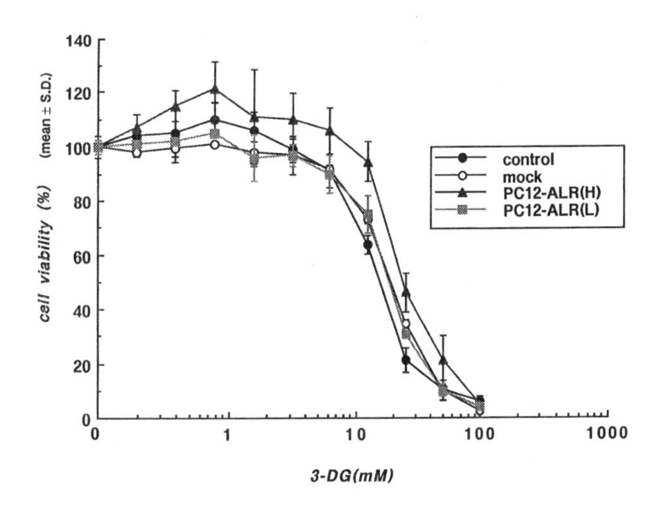

Figure 3. Effects of ALR overexpression on the cytotoxicity by 3-DG in PC12 cells. The effect of ALR overexpression on the viability of PC12 cells was observed after treatment with 3-DG. ALR overexpression increased cell viability (PC12-ALR(H)).

of organs. However, ALR levels in nervous tissues are relatively low (Takahashi et al., 1993) suggesting that nerve tissue might be susceptible to the cytotoxic effects of dicarbonyl compounds, and that this may represent a link to diabetic complications. While the concentrations of MG used in this study might appear to be high, we previously reported that the incorporation of MG into cells is only a fraction of the total MG in the culture medium. For example, for the case of rat smooth muscle cells (Che et al., 1997), only 1.8% of the total MG is incorporated. Therefore, the concentration of MG employed in this study results in physiologically relevant intracellular levels.

In conclusion, intracellular ALR protects neural cells from the cytotoxicity of 3-DG or MG, and neural cells, which normally express low levels of ALR, would be predicted to be susceptible to diabetic complications evoked by intermediate products in the Maillard reaction, such as 3-DG and MG. Since in diabetic conditions ALR is inactivated by the glycation reaction, as we previously reported (Takahashi et al., 1995), this may accelerate cell death and may be an important issue with respect to diabetic complications in neural tissues.

ACKNOWLEDGMENTS

We thank Dr. Motoko Takahashi for advice on the construction of the ALR-pRc/CMV vector and Dr. Paul J. Thonalley for advice on the purification of MG. We also thank Dr. Milton S. Feather for editing this manuscript.

REFERENCES

Bohren, K. M., Bullock, B., Wermuth, B. and Gabbay, K. H., 1989, The aldo-keto reductase superfamily. cDNAs and deduced amino acid sequences of human aldehyde and aldose reductases. *J. Biol. Chem.* 264: 9547–9551.

Bohren, K. M., Page, J. L., Shankar, R., Henry, S. P. and Gabbay, K. H., 1991, Expression of human aldose and aldehyde reductases. Site-directed mutagenesis of a critical lysine 262. *J. Biol. Chem.* 266: 24031–24037.

Che, W., Asahi, M., Takahashi, M., Kaneto, H., Okado, A., Higashiyama, S., and Taniguchi, N., 1997, Selective induction of heparin-binding epidermal growth factor-like growth factor by methylglyoxal and 3-deoxyglucosone in rat aortic smooth muscle cells. *J. Biol. Chem.* 272: 18453–18459

Kayanoki, Y., Fujii, J., Suzuki, K., Kawata, S., Matsuzawa, Y. and Taniguchi, N., 1994, Suppression of antioxidative enzyme expression by transforming growth factor-beta 1 in rat hepatocytes. *J. Biol. Chem.* 269: 15488–15492.

Koop, D. R., and Casazza, J.P., 1985, Identification of ethanol-inducible P-450 isozyme 3a as the acetone and acetol monooxygenase of rabbit microsomes. *J. Biol. Chem.* 260: 13607–03612.

Lyles, G. A. and Chalmers, J., 1992, The metabolism of aminoacetone to methylglyoxal by semicarbazide-sensitive amine oxidase in human umbilical artery. *Biochem. Pharmacol.* 43: 1409–1414.

McLellan, A. C., Thornalley, P. J., Benn, J. and Sonksen, P. H., 1994, Glyoxalase system in clinical diabetes mellitus and correlation with diabetic complications. *Clin. Sci. Colch.* 87: 21–29.

Nakata, T., Suzuki, K., Fujii, J., Ishikawa, M. and Taniguchi, N., 1993, Induction and release of manganese superoxide dismutase from mitochondria of human umbilical vein endothelial cells by tumor necrosis factor-alpha and interleukin-1 alpha. *Int. J. Cancer* 55: 646–650.

Niwa, T., Takeda, N., Yoshizumi, H., Tatematsu, A., Ohara, M., Tomiyama, S. and Niimura, K., 1993, Presence of 3-deoxyglucosone, a potent protein crosslinking intermediate of Maillard reaction, in diabetic serum. *Biochem. Biophys. Res. Commun.* 196: 837–43.

Okado, A., Kawasaki, Y., Hasuike, Y., Takahashi, M., Teshima, T., Fujii, J., and Taniguchi, N., 1996, Induction of apoptotic cell death by methylglyoxal and 3-deoxyglucosone in macrophage-derived cell lines. *Biochem. Biophys. Res. Commun.* 225: 219–224

Phillips, S. A. and Thornalley, P. J., 1993, The formation of methylglyoxal from triose phosphates. Investigation using a specific assay for methylglyoxal. *Eur. J. Biochem.* 212: 101–105.

Pompliano, D. L., Peyman, A. and Knowles, J. R., 1990, Stabilization of a reaction intermediate as a catalytic device: definition of the functional role of the flexible loop in triosephosphate isomerase. *Biochemistry* 29: 3186–3194.

Reichard, G. J., Skutches, C. L., Hoeldtke, R. D. and Owen, O. E., 1986, Acetone metabolism in humans during diabetic ketoacidosis. *Diabetes* 35: 668–74.

Suzuki, K., Koh, Y. H., Mizuno, H., Hamaoka R., and Taniguchi, N., 1998, Overexpression of aldehyde reductase protects PC12 cells from the cytotoxicity of methylglyoxal or 3-deoxyglucosone. *J. Biochem.* 123: 353–357.

Takahashi, M., Fujii, J., Teshima, T., Suzuki, K., Shiba, T., and Taniguchi, N., 1993 Identity of a major 3-deoxyglucosone-reducing enzyme with aldehyde reductase in rat liver established by amino acid sequencing and cDNA expression. *Gene* 127: 249–253

Takahashi, M., Lu, Y. B., Myint, T., Fujii, J., Wada, Y., and Taniguchi, N., 1995, In vivo glycation of aldehyde reductase, a major 3-deoxyglucosone reducing enzyme: identification of glycation sites. *Biochemistry* 34: 1433–1438

Thornalley, P. J., 1993, The glyoxalase system in health and disease. *Mol. Aspects Med.* 14: 287–371.

Yamada, H., Miyata, S., Igaki, N., Yatabe, H., Miyauchi, Y., Ohara, T., Sakai, M., Shoda, H., Oimomi, M. and Kasuga, M., 1994, Increase in 3-deoxyglucosone levels in diabetic rat plasma. Specific in vivo determination of intermediate in advanced Maillard reaction. *J. Biol. Chem.* 269: 20275–20280.

65

ENZYMES METABOLIZING ALDEHYDES IN HL-60 HUMAN LEUKEMIC CELLS

Giuseppina Barrera,[1] Stefania Pizzimenti,[1] Giuliana Muzio,[1]
Marina Maggiora,[2] Mario Umberto Dianzani,[1] and Rosa Angela Canuto[2]

[1]Dipartimento di Medicina e Oncologia Sperimentale
[2]Dipartimento di Scienze Cliniche e Biologiche
Università di Torino
Ospedale S.Luigi Gonzaga
Orbassano, Torino, Italy

1. INTRODUCTION

Lipid peroxidation produces several toxic carbonyl compounds including alpha-beta unsaturated aldehydes (Dianzani, 1982). Lipid peroxidation also occurs under physiological conditions, particularly in cells that are not rapidly proliferating (Esterbauer et al., 1991). By contrast, proliferating neoplastic cells, such as K562 and HL-60 do not show detectable levels of lipid peroxidation even if exposed to prooxidants, and their endogenous production of aldehydes is, therefore, very low (Dianzani, 1993).

Enzymes metabolizing aldehydes produced during lipid peroxidation (aldehyde dehydrogenase, aldo/keto reductase, alcohol dehydrogenase and glutathione-S-transferase) contribute to maintain the steady-state aldehyde concentration inside the cells (Canuto et al., 1994). Studies on the activities of aldehyde metabolizing enzymes have been performed in normal tissues like liver (Canuto et al., 1994; Hartley et.al., 1995; Lindahl, 1992) kidney (Lindahl, 1992; Ullrich et. al., 1994) and retina (Singhal et al., 1995; King and Holmes, 1996)) as well as in hepatoma cell lines (Canuto et al., 1994; Canuto et al., 1991; Grune et al., 1994) and in liver tissue during chemically-induced carcinogenesis (Canuto et al., 1989); little information is however available for myeloyd cells . Generally, in tumour tissues marked changes of both aldehyde dehydrogenase and aldehyde reductase activities have been reported (Canuto et al., 1991; Grune et al., 1994). Aldehyde dehydrogenase activity increases in direct correlation with the degree of deviation, whereas glutathione-S-transferase activity was found to be reduced in hepatoma cell lines (Canuto et al., 1991) but increased during liver carcinogenesis (Canuto et al., 1989).

Enzymology and Molecular Biology of Carbonyl Metabolism 7
edited by Weiner *et al.* Kluwer Academic / Plenum Publishers, New York, 1999.

Among the aldehydes produced from lipid peroxidation, 4-hydroxynonenal (HNE) is thought to be largely responsible for cytopathological effects observed during oxidative stress (Dianzani, 1982). Nevertheless, in recent years it has become evident that HNE is normally produced in many cells and tissues and acts on different biological parameters at concentrations similar to those found in a number of normal cells (Esterbauer et al., 1991). It has recently been shown that HNE, added to smooth muscle cells, increases the aldose reductase mRNA content (Spycher et al., 1996), and that lipid peroxidation products, induced in hepatoma cells by arachidonic acid and prooxidants, decreases the class-3 aldehyde dehydrogenase mRNA content (Canuto et al., 1996). Thus, HNE treatment itself may induce changes in such enzyme activities and expressions.

We have previously reported that HNE blocks cell growth and induces a granulocytic differentiation of HL-60 cells at concentrations similar to those detected in normal non proliferating cells (Barrera et al., 1991; Barrera et al., 1994).

Cell differentiation elicits a number of metabolic changes including modifications of different enzyme activities linked to the acquisition of a differentiated phenotype (Collins et al., 1978). HL-60 cells have become an archetype model for studying terminal differentiation of myeloid cells "in vitro" (Collins et al., 1977). They can be induced to differentiate either in monocytic- or in granulocytic-like cells by a variety of compounds such as TPA, retinoic acid and DMSO (Breitman et al., 1980). Some studies on HL-60 cell line demonstrated that differentiation leads to the acquisition of most of polymorphonuclear leukocytes functions: chemotaxis, phagocytosis, respiratory burst activity, and bacterial killing (Breitman et al., 1980). During induced and spontaneus HL-60 cell differentiation NADPH-dependend superoxide production increases as well as the maturation of O_2 generating system (Newburger et al., 1984.). On the other hand, glutathione peroxidase activity also increases indicating a rapid modulation of cellular genes involved in the protection against hydrogen peroxide and related toxic oxidants (Shen et al., 1994). These evidences suggested that the potential changes in the redox state might induce an alteration of the lipid peroxidation level and/or of the catabolism rate of its products.

In this research, we evaluated: i) the pattern of aldehyde metabolizing enzymes in HL-60 cells ii) the effect of HNE treatments iii) the effect of HNE- and DMSO-induced differentiation on aldehyde metabolizing enzymes by determination of enzyme activities when the differentiated phenotype was evident (5 days after the treatment).

2. MATERIALS AND METHODS

HL-60 cells (DSM, German collection of microorganisms and cell cultures, Braunschweig, Germany) were cultured at 37°C in a humidified atmosphere of 5% CO_2 in air in RPMI 1640 medium (Biochrom KG, Berlin) supplemented with 2 mM glutamine, antibiotics and 10% FCS (Biochrom KG, Berlin). Growth rate and cell viability were monitored daily by the trypan blue exclusion test.

Cells were seeded at 300,000/ml and cultured for 5 days. HNE, kindly provided by Prof. J. Schaur, (University of Graz, Austria) was prepared as previously described (Barrera et al., 1991). At the beginning of each experiment, 1 μM HNE was added 10 times to the cultures at 45 min intervals. HNE concentrations were monitored by HPLC after each addition, and never exceeded 1 μM (data not shown). HNE was no longer detectable in the culture medium 45 min after the last addition (7.5 hours from the beginning of experiment).

Cells induced with 1.25% DMSO (Sigma Chemical Co.) were exposed to the chemical for 5 days.

Phagocytosis was evaluated by counting the number of HL-60 cells that engulfed opsonized zymosan (Sigma Chemical Co.) as previously described (Barrera et al., 1991).

To obtain cytosolic fractions, cells (about 250 × 10^6) were harvested in exponentially growth phase (basal conditions), 7.5 hours after the beginning of HNE treatments, and 5 days after HNE and DMSO treatments. Cell homogenates were obtained by one freeze-thaw cycle followed by a 20 min incubation in a hypotonic solution (17.5 mM sucrose, 55 mM mannitol, 5 mM Tris-HCl pH 7.4, 0.5 mM EGTA and 0.025 % (w/v) bovine serum albumin (fraction V, fatty acid free); then they were diluted to 20% (w/v) with sucrose/mannitol to obtain an isotonic solution (70 mM sucrose, 220 mM mannitol, 20 mM Tris-HCl, pH 7.4, 2 mM EGTA and 0.1% bovine serum albumin) and mildly sonicated. Homogenates were centrifuged in a Beckman L8–55 centrifuge at 105.000 g for 60 min to obtain cytosolic fraction.

Aldehyde dehydrogenase (ALDH; EC 1.2.1.3), aldo/keto reductase (AKR; EC 1.1.1.2), alcohol dehydrogenase (ADH; EC 1.1.1.1.) and glutathione-S-transferase (GST; EC 2.5.1.18) were assayed in cytosolic fractions as previously described (Canuto et al., 1994) by using HNE as substrate, at the concentration of 0.05 mM for GST assay and 0.1 mM for the other enzymes.

Proteins were determined by biuret method.

All data were expressed as mean ± SD. Variance analysis followed by the Bonferroni test was carried out to evaluate the differences between group means.

3. RESULTS AND DISCUSSION

The metabolic pattern of HL-60 cells is shown in Figure 1 On the whole, the enzyme activities were lower than those reported for other cell types (Canuto et al., 1991; Canuto et al., 1989). In fact, in HL-60 cells ALDH activity is about one third of that present in hepatocytes, AKR activity is about 15 times lower and ADH is almost 100 times lower than in liver (Canuto et al., 1994). This may be consistent with the extremely low values of endogenous HNE production through lipid peroxidation in HL-60 cells (data not shown).

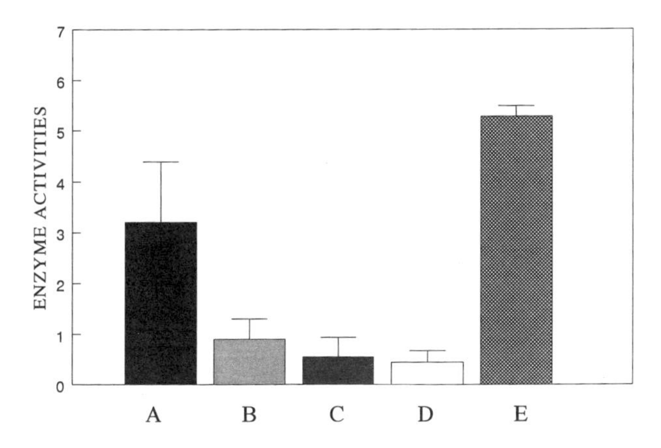

Figure 1. Pattern of aldehyde metabolizing enzymes in HL-60 cells. A: ALDH NAD; B: ALDH NADP; C: ADH; D: AKR; E: GST. Means ± S.D. of four experiments expressed as nmol of NAD(P) reduced/min/mg of protein for ALDH, as nmol of NAD(P)H consumed/min/mg of protein for ADH and AKR respectively and as nmol of HNE consumed/min/mg of protein for GST.

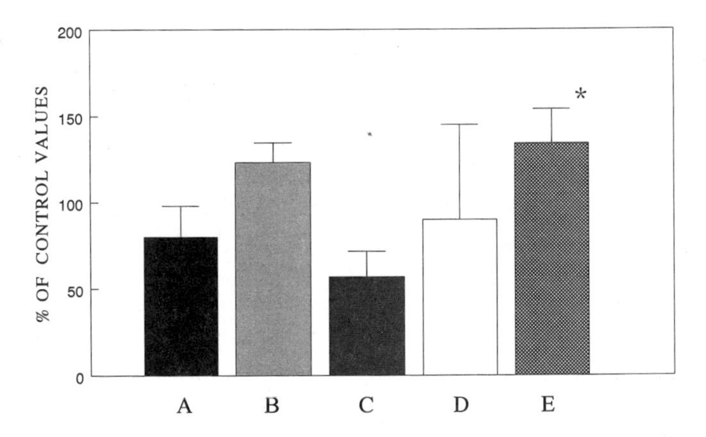

Figure 2. Early effects of HNE treatment. A: ALDH NAD; B: ALDH NADP; C: ADH; D: AKR; E: GST. The activities were evaluated at the end of 10 treatments with 1 μM HNE (7.5 hours from the beginning of experiment). Data represent the Mean ± S.D. of four experiments and are expressed as percentages of enzyme activities with respect to the control values.* $p < 0.05$ versus control values as determined by variance analysis followed by the Bonferroni test.

After HNE treatment (Figure 2), ALDH (NAD- or NADP-dependent), AKR and ADH activities remained unchanged, whereas GST activity increased (from 5.22 nmol of HNE consumed/min/mg of protein in control cells to 7.085). A similar increase was evident also in HNE- and DMSO- differentiated cells (Figure 3) The degree of differentiation obtained after either HNE or DMSO treatments was evaluated by measuring the induction of phagocytosis (Table 1). According with previous studies (Barrera et al., 1991), HNE and DMSO provoked a significant increase of the cells that aquire the phagocytosis capa-

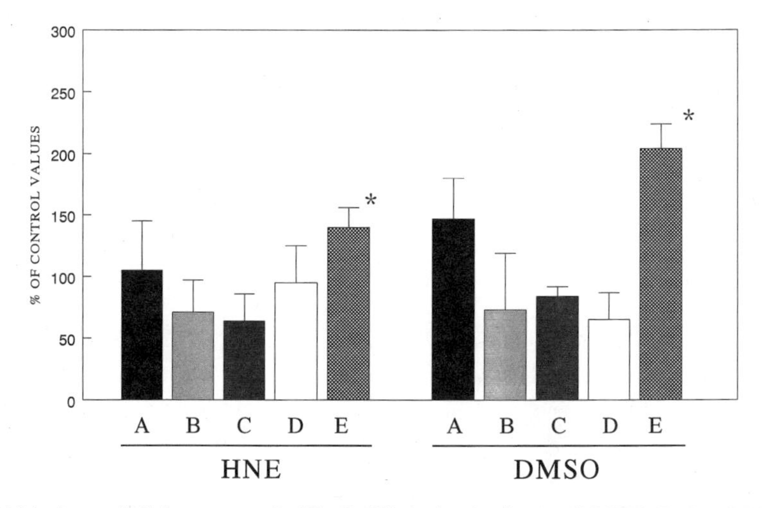

Figure 3. Aldehyde metabolizing enzymes in HL-60 differentiated cells. A: ALDH NAD; B: ALDH NADP; C: ADH; D: AKR; E: GST. The activities were evaluated after 5 days from the HNE and DMSO treatments. Data represent the means ± S.D. of four experiments and are expressed as percentages of enzyme activities with respect to the control values. * $p < 0.05$ versus control values as determined by variance analysis followed by the Bonferroni test.

Table 1. Assay of phagocytosis

	3 days	5 days
Control	5.1 ± 1.0	5.5 ± 1.5
HNE	16.2 ± 2.3*	28.4 ± 3.1*
DMSO	19.5 ± 1.5*	46.6 ± 6.2*

Results are expressed as percentages of phagocytosis (number of cells phagocytozing the opsonized zymosan/total number of cells × 100) and are the mean ± SD of four separate experiments. * $P < 0.01$ versus control values

bility, with respect to the control values, 3 days after the treatments. In the following days the percentage of phagocytosis gradually increased; 5 days after treatments it was 19.5% and 46.6% with HNE and DMSO, respectively.

ALDH (NAD- or NADP-dependent) AKR and ADH activities remained unchanged, whereas GST activity was increased (by 140% in HNE and by 204% in DMSO treated cells). The time course of GST activity was investigated at different intervals from the beginning of treatments. Enzyme activity increased since 45 minutes, peaked at 7.5 hours and remained almost constant in the following days (data not shown) .

Cytosolic GSTs are a family of important detoxification isoenzymes (Waxman, 1990) one of them (8:8) is known to be able to use HNE as a substrate (Poot et al., 1987). GST catalyzes reactions involving the direct coupling of glutathione to electrophilic drugs, carcinogens and endogenous compounds (Waxman, 1990). The increase in GST activity with HNE as a substrate, in HNE-treated cells, might be a consequence of a substrate-linked induction. However, the cells remain in the presence of added HNE only for 7.5 hours, after this period the aldehyde is not longer detectable in the culture medium (Barrera et al., 1991). GST activity remains however at high levels even 5 days after the treatment, in absence of aldehyde in the culture medium. This suggests that GST increase is more probably linked to the acquisition of the differentiated phenotype rather than to substrate depending induction. On the other hand, the fact that an even higher increase is observed in DMSO-treated cells is in favour of this interpretation.

The fact that GST increase appears shortly after HNE and DMSO induction, is in agreement with data reported by Newburger et al. (1984) indicating that the induction of differentiation along the monocytic or granulocytic pathway in HL-60 cells resulted in a rapid modulation of enzymes linked to the increase of the detoxification capability in differentiated cells.

ACKNOWLEDGMENTS

This work was supported by the Italian Association for Cancer Research (A.I.R.C.), and Italian Ministry of University and Technological Research (M.U.R.S.T.)

REFERENCES

Barrera, G., Di Mauro, C., Muraca, R., Ferrero, D., Cavalli, G., Fazio, V.M., Paradisi, L. and Dianzani, M.U. (1991): Induction of differentiation in human HL-60 cells by 4-hydroxy-nonenal, a product of lipid peroxidation. Exp. Cell Res., 197, 148–152.

Barrera, G., Muraca, R., Pizzimenti, S. Serra, A., Rosso, C., Saglio, G., Farace, M.G., Fazio, V. and Dianzani, M.U. (1994): Inhibition of c-myc expression induced by 4-hydroxyno nenal, a product of lipid peroxidation, in the HL-60 human leukemic cell line. Biochem. Biophys. Res. Commun., 203, 553–561.

Breitman, T.R., Selonick, E.S., Collins, S.J. (1980): Induction of differentiation of the human promyelocytic cell line (HL-60) by retinoic acid. Proc. Natl. Acad. Sci. U.S.A., 77, 2936–2940.

Canuto, R.A., Muzio, G., Biocca, M.E. and Dianzani M.U. (1989): Oxidative metabolism of 4-hydroxy-2, 3-nonenal during diethylnitrosamine-induced carcinogenesis in rat liver. Cancer Lett., 46, 7–13.

Canuto, R.A., Muzio, G., Bassi, A.M., Bocca, M.E., Poli, G., Esterbauer, H. and Ferro, M.: Metabolism of 4-hydroxynonenal in hepatoma cell lines. In: H.Weiner, B.Wermuth and D.W. Crabb (Eds), Enzymology and Molecular Biology of Carbonyl Metabolism, vol. 3, Plenum Press., New York, 1991. pp.75–84.

Canuto, R.A., Ferro, M., Muzio, G., Bassi, A.M., Leonarduzzi, G., Maggiora, M., Adamo, D., Poli, G. and Lindahl, R. (1994): Role of aldehyde metabolizing enzymes in mediating effects of aldehyde products of lipid peroxidation in liver cells. Carcinogenesis, 15, 1359–1364.

Canuto, R.A., Ferro, M., Maggiora, M., Federa, R., Brossa, O., Bassi, A.M., Lindahl, R. and Muzio, G. (1996): In hepatoma cell lines restored lipid peroxidation affects cell viability inversely to aldehyde metabolizing enzyme activity. Adv. Exp. Med. Biol., 414, 113–122.

Collins, S.J., Gallo, R.C. and Gallagher, R.E. (1977): Continuous growth and differentiation of human myeloid leukemia cells in suspension cultures. Nature, 270, 347–349.

Collins, S.J., Ruscetti, F.W., Gallaghen, R.E. and Gallo, R.C. (1978): Terminal differentiation of human promyelocytic leukemic cells induced by dimethyl sulfoxide and other polar compounds. Proc. Natl. Acad. Sci. U.S.A., 75, 2458–2462.

Dianzani, M.U.: Biochemical effect of saturated and unsaturated aldehydes. In: D.C.H. McBrien and T.F. Slater (Eds), Free radicals, lipid peroxidation and cancer. Acad. Press, London, 1982. pp. 129–158.

Dianzani, M.U. (1993): Lipid peroxidation and cancer. Critical Reviews in Oncology/Hematology, 15, 125–147.

Esterbauer H., Schaur, R.J., and Zollner, H. (1991): Chemistry and Biochemistry of 4-hydroxynonenal, malonaldehyde and related aldehydes. Free Rad. Biol. Med., 11, 81–128.

Grune, T., Siems, W.G., Zollner, H. and Esterbauer, H. (1994): Metabolism of 4-hydroxynonenal, a cytotoxic lipid peroxidation product, in Ehrlich Mouse Ascites cells at different proliferation stages. Cancer Res., 54, 5231–5235.

Hartley, D.P., Ruth, J.A. and Petersen, D.R. (1995): The hepatocellular metabolism of 4-hydroxynonenal by alcohol dehydrogenase, aldehyde dehydrogenase and glutathione-S-transferase. Arch. Bioch. Biophys., 316, 197–205.

King, G. and Holmes, R.: Human corneal and lens aldehyde dehydrogenases. Adv. Exp. Med. Biol., 414 (1996) 19–27.

Lindahl, R. (1992): Aldehyde dehydrogenases and their role in carcinogenesis. Crit. Rew. Biochem. Mol. Biol., 27, 283–335.

Newburger R.E., Speier C., Borregaard N., Walsh C.E., Whitin, J.C. and Simons E.R. (1984): Development of the superoxide generating system during differentiation of the HL-60 human promyelocytic leukemic cell line. J. Biol. Chem., 259, 3771–3776.

Poot, M., Verkerk A., Koster, J.F., Esterbauer, H. and Jorgkind, J.F. (1987): Influence of cumene hydroperoxide and 4-hydroxynonenal on the glutathione metabolism during "in vitro" aging of human skin fibroblasts Eur. J. Biochem, 162, 287–291.

Shen, Q., Chada, S., Withitney, C. and Newburger, P.E. (1984): Regulation of the human cellular glutathione peroxidase gene during in vitro myeloid and monocytic differentiation. Blood, 84, 3902–3908.

Singhal, S.S., Awasthi, S., Srivastava, S.K., Zimniak, P., Ansari, N.H., Awasthi, Y.C. (1995): Novel human glutathione S-transferase with high activity toward 4-hydroxynonenal. Invest. Ophthalmol. Vis. Sci., 36, 142–150.

Spycher, S., Tabataba-Vakili, S., O'Donnel, V.B., Palomba, L. and Azzi, A. (1996):4-hydrxy-2, 3-trans-nonenal induces transcription and expression of aldose reductase. Biochem. Biophys. Res. Commun., 13, 512–516.

Ullrich, D., Grune, T., Henke, W., Esterbauer, H. and Siems, W.G. (1994): Identification of metabolic pathways of the lipid peroxidation product 4-hydroxynonenal by mitochondria isolated from rat kidney cortex. FEBS Lett., 352, 84–86.

Waxman, D.J. (1990): Glutathione-S-transferases: role in alkylating agent resistence and possible target for modulation chemotherapy. Cancer Res., 50, 6449–6454.

RAT CARBONYL REDUCTASE

How Many Genes?

Micheline Schaller and Bendicht Wermuth

Department of Clinical Chemistry
University of Berne, Inselspital
CH-3010 Bern, Switzerland

1. INTRODUCTION

Carbonyl reductase (secondary alcohol:$NADP^+$ oxidoreductase, EC 1.1.1.184) catalyzes the NADPH-dependent reduction of a variety of endogenous and xenobiotic carbonyl compounds (Wermuth, 1981; Jarabak et al., 1983; Nakayama et al., 1985). Structurally it belongs to the short-chain dehydrogenase/reductase (SDR) family which includes steroid, prostaglandin, pterin and retinol metabolizing enzymes (Jörnvall et al. 1995); however the physiological substrate(s) of carbonyl reductase is (are) not known. In man, the enzyme is expressed in essentially all tissues with highest activity in liver, kidney and brain (Wirth and Wermuth, 1992). Evidence from sequence analysis of cDNAs (Wermuth et al., 1988; Forrest et al., 1990) as well as genomic DNA (Forrest et al., 1991) suggests the presence of a single gene which consists of three exons and two introns and spans about 3 kb on chromosome 21q22.1. The 5'-untranslated region contains a GC-rich island but no TATA or CAAT consensus sequences (Forrest et al., 1991) consistent with the constitutive expression of the enzyme in essentially all tissues.

In contrast to the ubiquitous distribution in human tissues, in the rat, carbonyl reductase predominates in the gonads, including accessory tissues (Iwata et al., 1989; 1990) and the adrenals (Inazu et al., 1994). Enzyme expression in the ovary is modulated by estrogens and corticosteroids (Inazu et al., 1990; 1992) and Ca^{2+} ions and testosterone appear to influence the expression in the testes (Inazu and Fujii, 1993). These findings strongly suggested a role of the rat enzyme in steroid-modulated pathways or steroid metabolism itself. To understand the structural and functional relationships between the human and rat carbonyl reductases, we previously cloned and sequenced the cDNA coding for rat testis carbonyl reductase and showed that the two enzymes exhibit 86% positional identity (Wermuth et al., 1995). More recently, a cDNA which differs from the rat testis cDNA by 5 nucleotides was obtained from rat ovaries (Aoki et al., 1997). The number of genes in the rat, their organization and the factors regulating their expression, however, are not known. Towards an answer to these questions we initiated a study to determine the sequence and organization of the rat carbonyl reductase (Cbr) gene(s).

2. RESULTS AND DISCUSSION

Overlapping fragments containing the complete coding and intron sequences were obtained by the PCR-mediated amplification of genomic DNA from male Sprague Dawley rats using cDNA-specific primers. In addition, an EMBL3 rat genomic library (Clontech) and pairs of cDNA- and vector-specific primers were used to obtain sequences containing 1200 bp upstream of the ATG start codon and a short stretch downstream of the polyadenylation site, respectively. The amplimers were sequenced in both directions by DNA walking and sequences of amplimers from the EMBL3 library were verified by comparison with the corresponding amplimers from genomic DNA.*

Alignment of the genomic sequences with the cDNA from rat testis indicated the presence of three exons which code for the same protein as the cDNA and two introns of 486 and 905 bp, respectively. Exon-intron boundaries occur at the same positions as in the homologous human gene (Forrest et al., 1991). Primer extension of total rat testis RNA, carried out to determine the transcription start site, yielded a prominent product extending 71 bp upstream of the translation start codon.

In order to identify possible regulatory elements of gene transcription, a computer analysis of the 5'-untranslated region was carried out. No characteristic RNA polymerase II promoters, e.g. a TATA box or a MED-1 element were detetctable. Screening for estrogen response elements (ERE) yielded 4 perfect 3'-halves (TGACC), 194, 349, 946 and 1038 bp, respectively, upstream of the translation start site, but no 5'-half of the palindrome. Similarly, we detected two 3'-halves (TGTCCT/TGTTCA) and one 5'-half (GTTACA) of the glucocorticoid response element (GRE) at positions -659, -1098 and -1195, respectively. In order to identify possible interactions of the ERE half palindromes with DNA binding proteins, we carried out band shift experiments. DNA fragments covering the various response elements were synthesized by the PCR in the presence of [^{35}S]dATPαS and incubated with decreasing amounts of total cellular extracts from rat ovary. The labeled DNA was partly retained at the top of the gel at the highest protein concentrations, but no discrete bands of retarded DNA were detectable at lower protein concentrations (Figure 1). Similar patterns with slightly better retention of the DNA at high protein concentrations were obtained with extracts from rat liver which does not express carbonyl reductase. The findings do not support the idea that the half palindromes of the estrogen response element are involved in steroid hormone-modulated expression of the gene.

During the amplification of rat genomic DNA using primers on both ends of the two exons, we consistently obtained fragments of the same size as the corresponding cDNA, in addition to the expected intron-spanning amplimers. Sequence analysis of these fragments revealed an open reading frame exhibiting 93% positional identity with the cDNA from rat testis and 86% identity between the deduced protein sequences. Most notably, the protein deduced from the intronless transcript lacked a glutamate residue (Glu-214) relative to rat testis carbonyl reductase. In order to investigate whether the intronless gene codes for an

* The Cbr gene sequence data have been submitted to the EMBL Nucleotide Sequence Database under the accession number X95986.

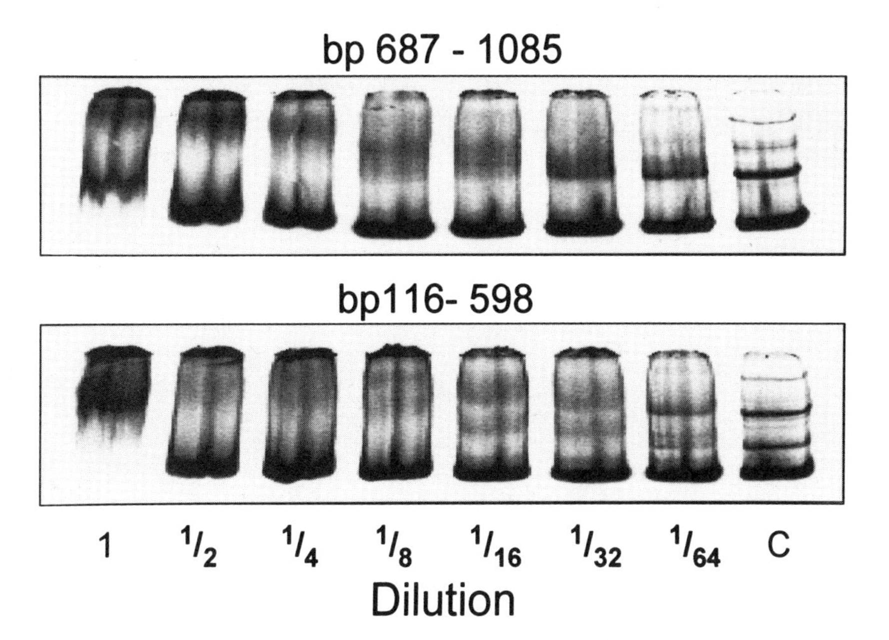

Figure 1. Autoradiographs of band shift analyses of ERE half palindromes from the 5'-UTR of the rat Cbr gene. Radiolabeled oligonucleotides, containing two half palindromes each, were incubated with decreasing concentrations of whole cell extract from rat ovary and run on a 5% polyacrylamide gel. Numbers above the panels indicate the position of the oligonucleotide relative to the translation start site. C, control without cellular extract.

enzymatically active protein, the DNA was cloned into the bacterial expression vector pET-11a (Novagen) and expressed in E. coli BL21(DE3). The PCR was used to introduce NdeI restriction sites for cloning into the vector, and the correct direction and sequence of the insert DNA were assessed by sequencing in both directions. Western blots of the bacterial extract using antibodies against human liver carbonyl reductase yielded a distinct band which migrated slightly slower on SDS-PAGE than recombinant rat testis carbonyl reductase (Figure 2). No increase of carbonyl reductase activity above the bacterial background was detectable using menadione and 4-benzoylpyridine, respectively, as substrates. The absence of introns in the genomic sequence and apparent lack of enzymatic activity of the expressed protein suggested that we had identified a pseudogene. However, Aoki et al. (1997) recently reported the presence of a noninducible cDNA in rat ovary, testis, lung, spleen and kidney which had the same sequence as the presumptive pseudogene. In all the tissues examined the cDNA was obtained by RT-PCR using RNA prepared by extraction with acid guanidinium thiocyanate and phenol-chloroform and primers which are specific for the highly homologous carbonyl reductase. Guanidinium thiocyanate extraction does not completely separate RNA from DNA, and it cannot be excluded, therefore, that Aoki et al. amplified genomic DNA rather than mRNA. Further work will be necessary to clarify this point.

In addition to the noninducible cDNA, Aoki et al. also cloned a cDNA from rat ovaries which was highly inducible by pregnant mare serum gonadotropin and differed by only 5 nucleotides from the cDNA previously isolated in our laboratory from rat testis (Wermuth et al., 1995). The same sequence was also amplified from total RNA of rat testis, but no transcript corresponding to our cDNA was obtained. From this the authors con-

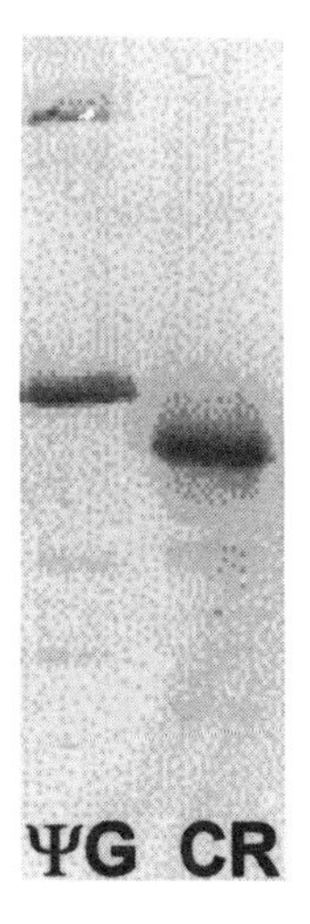

Figure 2. Western blot of rat carbonyl reductase (CR) and "pseudogene" protein (ψG). Extracts of E. coli expressing the respective protcin were subjected to SDS-PAGE and blotted to polyvinylidene membranes (Millipore). The membranes were incubated with antibodies against human liver carbonyl reductase, and bound antibodies were made visible using horseradish peroxidase coupled to protein A, H_2O_2 and diaminobenzidine (Wirth and Wermuth, 1992).

cluded that their inducible carbonyl reductase and our testicular enzyme are strain-specific forms of the same enzyme. In order to further investigate this heterogeneity, genomic DNA from a single rat was amplified using primers in which the 3'-terminal nucleotide matched either our cDNA from testis or Aoki's cDNA from ovary. Both pairs of primers yielded amplimers of the same mobility upon electrophoresis on agarose gels and equal intensity of the bands after staining with ethidium bromide. Sequence analysis showed that the amplimers correspond to the expected products and that, therefore, rat genomic DNA contains sequences corresponding to both the cDNA from testis isolated in our laboratory and the cDNA from ovary as described by Aoki et al.. Nothing is known about the structure of the gene encoding the ovarian enzyme. In order to obtain a rough estimate of the complexity of the carbonyl reductase genes, genomic DNA was digested with EcoRI and Hind III, respectively, and analyzed by Southern blotting. Neither enzyme cuts the Cbr gene or the "pseudogene" described above, and two bands are therefore expected from these genes. As shown in Figure 3 at least five bands are detectable after digestion

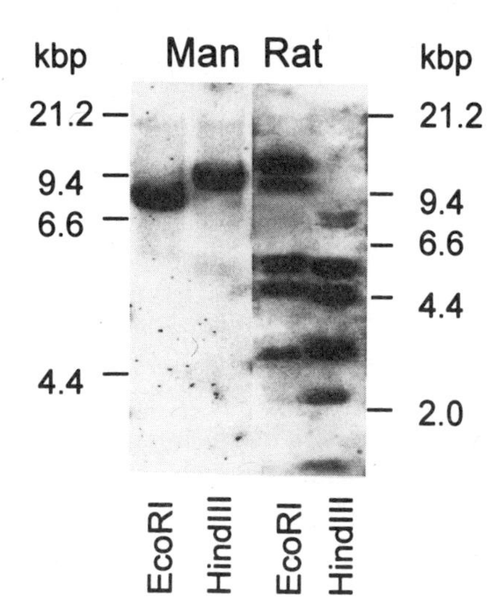

Figure 3. Southern blots of human and rat DNA. Genomic DNA was digested with EcoRI and HindIII, respectively, separated on a 0.8% agarose gel, blotted onto nylon membranes (Roche-Boehringer) and hybridized with digoxigenin-labeled carbonyl reductase cDNA. The hybridized probe was made visible using the DIG chemiluminescence detection system (Roche-Boehringer) using CDP-Star as substrate.

with either enzyme. In contrast, digestion of human genomic DNA with the same enzymes yielded a single band as expected from the published sequence of the human Cbr gene (Forrest et al., 1991). The results confirm that, in contrast to man, the rat genome contains multiple copies of the Cbr or Cbr-like genes. Further studies will be needed to determine their exact number and physiological significance.

ACKNOWLEDGMENTS

Supported by the Swiss National Science Foundation (project 31-36384.92) and the joint contributions of the EC Biotech 2 program (project PL 962123) and the Swiss Federal Office of Education and Science (project 97-0153).

REFERENCES

Aoki, H., Okada, T., Mizutani, T., Numata, Y., Minegishi, T. and Miyamoto K. (1997) Identification of two closely related genes, inducible and noninducible carbonyl reductase in the rat ovary, *Biochem. Biophys. Res. Commun.* **230**, 518–523.

Forrest, G.L., Akman, S., Krutzik, S., Paxton, R.J., Sparkes, R.S., Doroshow, J., Felsted, R.L., Glover, C.J., Mohandas, T. and Bachur, N.R. (1990) Induction of human carbonyl reductase gene located on chromosome 21, *Biochim. Biophys Acta* **1048**, 149–155.

Forrest, G.L., Akman, S., Doroshow, J., Rivera, H. and Kaplan, W.D. (1991) Genomic sequence and expression of cloned human carbonyl reductase gene with daunorubicin reductase activity, *Mol. Pharmacol.* **40**, 502–507.

Inazu, N. and Fujii, T. (1993) Effects of age and calcium ion on testis carbonyl reductase in rats, *Japan. J. Pharmacol.* **63,** 65–71.

Inazu, N. Iwata, N. and Satoh, T. (1990) Inhibitory effect of glucocorticoid and stimulatory effect of human chorionic gonadotropin on ovarian carbonyl reductase in rat, *Life Sciences* **46**, 841–848.

Inazu, N., Inaba, N, Satoh, T. and Fujii, T. (1992) Human chorionic gonadotropin causes an estrogen-mediated induction of rat ovarian carbonyl reductase, *Life Sciences,* **51**, 817–822.

Inazu, N., Nagashima, Y., Satoh, T. and Fujii, T. (1994) Purification and properties of six aldo-keto reductases from rat adrenal gland, *J. Biochem.* **115,** 991–999.

Iwata, N., Inazu, N. and Satoh, T. (1989) The purification and properties of NADPH-dependent carbonyl reductases from rat ovary, *J. Biochem.* **105**, 556–564.

Iwata, N., Inazu, N. Takeo, S. and Satoh, T. (1990) Carbonyl reductases from rat testis and vas deferens, *Eur. J. Biochem.* **193**, 75–81.

Jarabak, J., Luncsford, A. and Berkowitz, D. (1983) Substrate specificity of three prostaglandin dehydrogenases, *Prostaglandins* **26**, 849–868.

Jörnvall, H., Persson, B., Krook, M., Atrian, S., Gonzàlez-Duarte, R., Jeffery, J. and Ghosh, D. (1995) Short-chain dehydrogenases/reductases (SDR), *Biochemistry* **34**, 6004–6013.

Nakayama, T., Hara, A., Yashiro, K. and Sawada, H. (1985) Reductases for carbonyl compounds in human liver, *Biochem. Pharmacol.* **34**, 107–117.

Wermuth, B. (1981) Purification and properties of an NADPH-dependent carbonyl reductase from human brain: Relationship to prostaglandin 9-ketoreductase and xenobiotic ketone reductase, *J. Biol. Chem.* **256**, 1206–1213.

Wermuth, B., Bohren, K.M., Heinemann, G., von Wartburg, J.P., and Gabbay, K.H. (1988) Human carbonyl reductase, nucleotide sequence analysis of a cDNA and amino acid sequence of the encoded protein, *J. Biol. Chem.* **263**, 16185–16188.

Wermuth, B., Mäder Heinemann, G. and Ernst, E. (1995) Cloning and expression of carbonyl reductase from rat testis, *Eur. J. Biochem.* **228,** 473–479.

Wirth, H. & Wermuth, B. (1992) Immunohistochemical localization of carbonyl reductase in human tissues, *J. Histochem. Cytochem.* **40**, 1857–1863.

1996b). Interestingly, 13-hydroxy metabolites of anthracyclines, such as doxorubicinol and daunorubicinol (DRCOL), are significantly less potent than the parent drug, in terms of inhibiting tumor cell growth *in vitro* (Schott and Robert, 1989; Ozols et al., 1980; Dessypris et al., 1986; Olson et al., 1988; Beran et al., 1979; Yesair et al., 1980; Kuffel et al., 1992), suggesting that carbonyl reduction is an important biochemical mechanism in the detoxification of carbonyl group-bearing anthracyclines. Therefore, elevated levels of anthracycline carbonyl reducing enzymes constitute an additional mechanism in the development of non-classical MDR (Soldan et al., 1996a; Soldan et al., 1996b).

Three reductases capable of catalyzing the carbonyl reduction of the anthracycline daunorubicin (DRC) to its less toxic 13-hydroxy metabolite DRCOL have been described in human liver, namely aldehyde reductase (ALR1, AKR1A1, EC 1.1.1.2), carbonyl reductase (CR, EC 1.1.1.184), and an isoenzyme of dihydrodiol dehydrogenase (DD2, AKR1C2, EC 1.3.1.20) (Ohara et al., 1995). Moreover, overexpression of an unknown aldo-keto reductase in Chinese hamster ovary cells has recently been related to the inactivation of a cytotoxic synthetic tripeptide, by catalysing the carbonyl reduction of its active aldehyde group to the corresponding alcohol. This inducible reductase, which has been shown to mediate carbonyl reduction of DRCOL, exhibits about 70 % sequence identity to aldose reductase (ALR2, AKR1B1, EC 1.1.1.21) (Hyndman et al., 1997), the latter also being capable of reducing aldehyde and ketone substrates.

In the present study, DRC metabolism was investigated in different cancer cell lines with respect to toxicity, enzymatic DRC-reduction, and mRNA expression of the DRC-reductases ALR1, ALR2, and CR. In addition, we generated a DRC resistant stomach carcinoma cell line by continuous exposure to increasing concentrations of DRC and additionally in the presence of nontoxic verapamil concentrations. Verapamil is a P-gp and MRP blocking compound, which suppresses the expression and the activity of these drug efflux pumps. After a culture period of 7 month, resistance increased and therefore the resistance mechanisms responsible were evaluated. Transfection experiments with several DRC reductases supported the assumption that DRC carbonyl reduction plays an important role in the development of tumor drug resistance.

2. EXPERIMENTAL

2.1. Cells and Cell Culture

Cell lines used were: EPG85-257 (stomach carcinoma), EPG85-181P (pancreas carcinoma), EPG85-181DBR (DRC resistant pancreas carcinoma), MeWo (malignant melanoma), MCF 7 (breast cancer) and EFO 21 (ovarian carcinoma). The cells were free of mycoplasma as judged by staining with DAPI.

2.2. Generation of a DRC-Resistant Stomach Cancer Subline

DRC-resistant descendants were derived from DRC-sensitive cells over a period of 5 months by passage in increasing sublethal concentrations of DRC and permanent supplementation of 20 μM verapamil to the culture media. DRC concentrations were doubled every two weeks beginning with 0.1 ng/ml, finally resulting in DRC resistant sublines grown at 12.8 ng/ml DRC.

2.3. MTT Assay

The MTT [3-(4,5-dimethyl-2-thiazolyl)-2,5-diphenyl tetrazolium bromide] assay involves the conversion of tetrazolium salt to colored formazan by cells serving as an indirect measurement of cell proliferation and viability, and was performed according to (Hansen et al., 1989).

2.4. Preparation of Subcellular Fractions, DRC Carbonyl Reduction Assay, and Determination of DRCOL

These techniques were performed as described previously (Soldan et al., 1996a).

2.5. RT-PCR of P-glycoprotein, MRP, LRP, and DRC Reductases

All reverse transcriptase-polymerase chain reactions (RT-PCR) were carried out with a RT-PCR kit (Ready To Go, Pharmacia Biotech, Freiburg). Total cellular RNA was isolated by the RNeasy-Kit (Quiagen, Hilden). Aliquots (20 µl of RT-PCR products) were then subjected to electrophoresis in a 2% agarose gel and visualised by staining with ethidium bromide. Positive controls were performed using RNA pepared from sensitive pancreas carcinoma cells (EPG85-P181) known to express P-gp, MRP and LRP at high levels. Expression of all measured gene products (density units) were compared in sensitive and resistant cell lines after being normalized against β_2-microglobulin signals as the internal standard to account for RT-PCR and DNA loading variations.

2.6. Determination of Protein

Protein determination was carried out according to (Lowry et al., 1951).

2.7. Expression Vector

The cDNAs of the human liver reductases were amplified and modified by PCR to generate NheI and XbaI restriction sites on the 5' and 3' ends, respectively. Following the PCR amplification, the products were cloned into the pCl-neo mammalian expression vector (Promega, Madison, USA). After amplification of the modified reductase vectors in XL1-Blue, orientation and sequence of the respective cDNA was verified by DNA sequencing.

2.8. Transfection of Cancer Cells of with DRC Reductases

Transfections were performed in EPG85-181P cells using the transfection reagent DAC-30 (Eurogentec, Seraing, Belgium) according to the manufacturer's protocol. Cells were grown in 24 well plates (80000 cells per well) and transfected using 2 µg DNA and 4 µl transfection reagent. The DNA complex remained on the cells for 6 hours before the addition of an equal volume of culture media. After 72 hours, geneticin (G418 Sulphate, PAA, Linz, Austria) was added to a final concentration of 5 mg/ml. Cells were subcloned every 5 days in selection media.

3. RESULTS

3.1. DRC Metabolism in Different Cell Lines

The amount of cytostatic drug that actually reaches the tumor during chemotherapy *in vivo* could be very different. To consider this fact, we have simulated chemotherapy *in vitro* and have incubated tumor cells over a period of 72 hours at varying concentrations of DRC. For these experiments 4 cell lines were selected and screened for DRC carbonyl reducing activity, a pancreas carcinoma cell line (EPG85-181P), a breast carcinoma cell line (MCF-7), an ovarian carcinoma cell line (EFO-21) and a malignant melanoma cell line (MeWo).

DRCOL formation in these cell lines differed considerably and was strongly dependent on the pretreatment with DRC (Figure 1). In all cell lines tested, an increase of DRC carbonyl metabolism after contact with DRC occurred. Due to the lower toxicity of the formed metabolite DRCOL, this reaction may enable the tumor cells to survive in a toxic DRC environment.

3.2. Generation of DRC Resistant Stomach Carcinoma Cells

An induced metabolism of DRC is of significance only, if the cytostatic drug reaches the inside of the tumor cell. But this is often prevented by an overexpression of plasma membrane transporters like P-gp and MRP. These classical resistance mechanisms lower the intracellular drug concentration by active transport out of the cell. The consequence is that tumor cells tolerate higher drug concentrations. In this case an enhanced carbonyl metabolism seems to be of secondary importance. To determine the role of DRC metabolism in the absence of membrane transporters, we generated a resistant stomach cancer cell line in the presence of increasing concentrations of DRC, together with constant amounts of the drug efflux pump blocking compound verapamil. After eight passages (5 month) cells grown at 12.8 ng/ml DRC had IC_{50} values for DRC of 0.410 µg/ml, which means an 8-fold increase in DRC resistance compared to the IC_{50} values of the parental cells (0.055 µg/ml DRC) (not shown).

Parallel to the elevated resistance, an enhanced DRC carbonyl reducing activity of 6-fold could be observed in these cells. In contrast, P-gp and MRP showed no significant alterations in mRNA expression. The resistance factor LRP was not detectable, neither in the sensitive nor in resistant cells (Figure 2). Hence, the elevated detoxification of DRC by increased phase-I metabolism seems to be the major resistance mechanism in the generated resistant stomach carcinoma cell line.

Four known reductases are involved in DRC carbonyl reduction. A determination of their mRNA expression revealed that the sensitive as well as the resistant stomach carcinoma cells expressed all four. ALR1 showed no significant alteration in the resistant cells, whereas DD2 and ALR2 were expressed at higher levels in resistant compared to sensitive cells. The strongest induction could be observed with CR (Figure 3). This co-induction of three reductases, with CR in main, indicates that not only one DRC reductase alone accounts for the higher DRCOL formation.

3.3. Transfection of Cancer Cells with DRC Reductases

Transfections were carried out in a pancreas carcinoma cell line (known to constitutively express basal levels of the reductases) in order to simulate an overexpression. Single transfections were done with ALR1, ALR2 and CR. In addition, simultaneous transfection of all three enzymes was performed to simulate the overexpression of more

Figure 1. Effect of DRC pretreatment on DRC carbonyl reduction in cytosolic fractions of different cancer cell lines. Cells were cultured in the absence or presence of increasing concentrations of DRC. Specific activities are expressed as nmol / mg protein DRCOL formed in 30 min. Each bar represents the mean of 4–12 determinations.

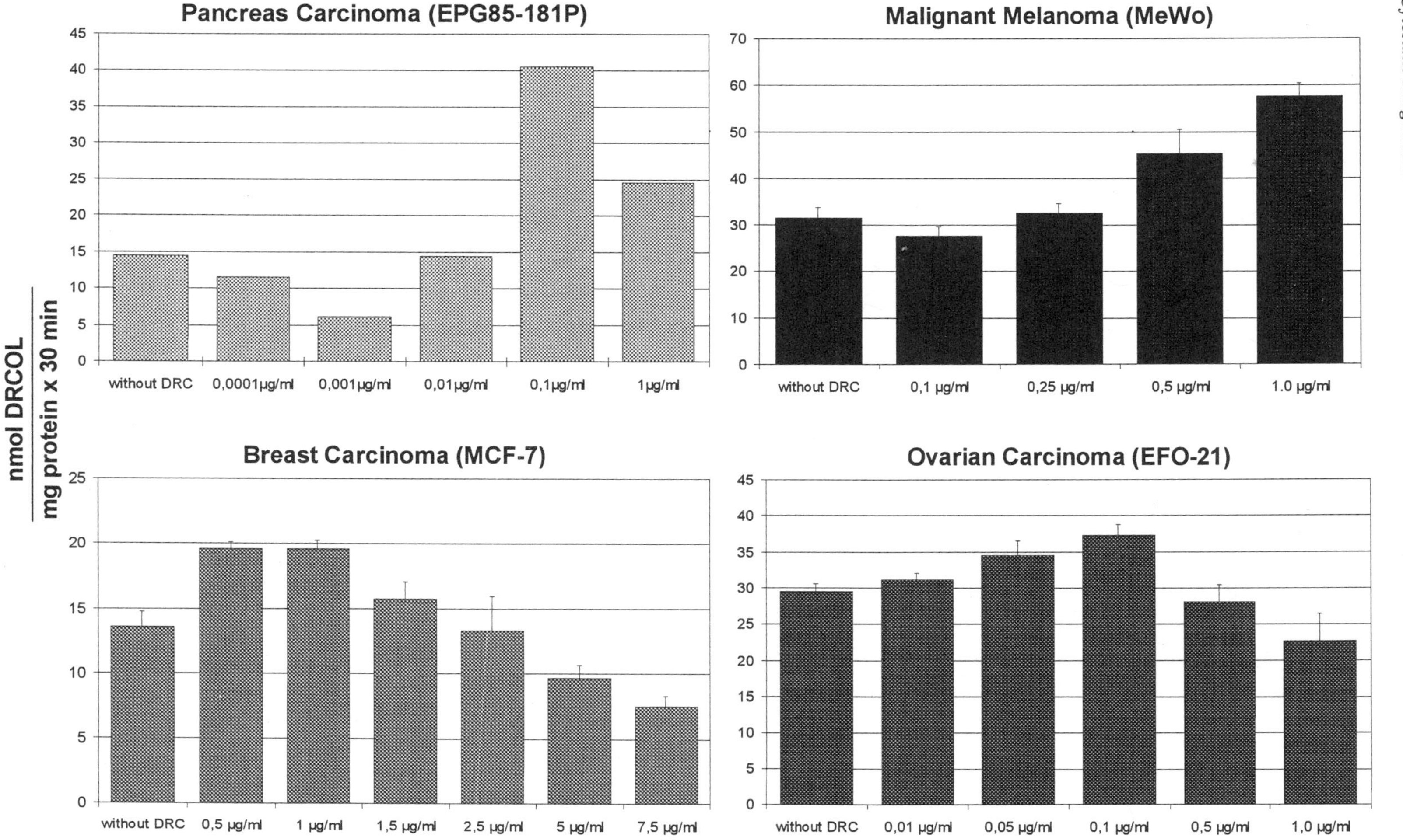
Pancreas Carcinoma (EPG85-181P)
45
40
35
30
25
20
15
10
5
0
without DRC
0,0001µg/ml
0,001µg/ml
0,01µg/ml
0,1µg/ml
1µg/ml
Malignant Melanoma (MeWo)
70
60
50
40
30
20
10
0
without DRC
0,1 µg/ml
0,25 µg/ml
0,5 µg/ml
1.0 µg/ml
nmol DRCOL
mg protein x 30 min
Breast Carcinoma (MCF-7)
25
20
15
10
5
0
without DRC
0,5 µg/ml
1 µg/ml
1,5 µg/ml
2,5 µg/ml
5 µg/ml
7,5 µg/ml
Ovarian Carcinoma (EFO-21)
45
40
35
30
25
20
15
10
5
0
without DRC
0,01 µg/ml
0,05 µg/ml
0,1 µg/ml
0,5 µg/ml
1,0 µg/ml

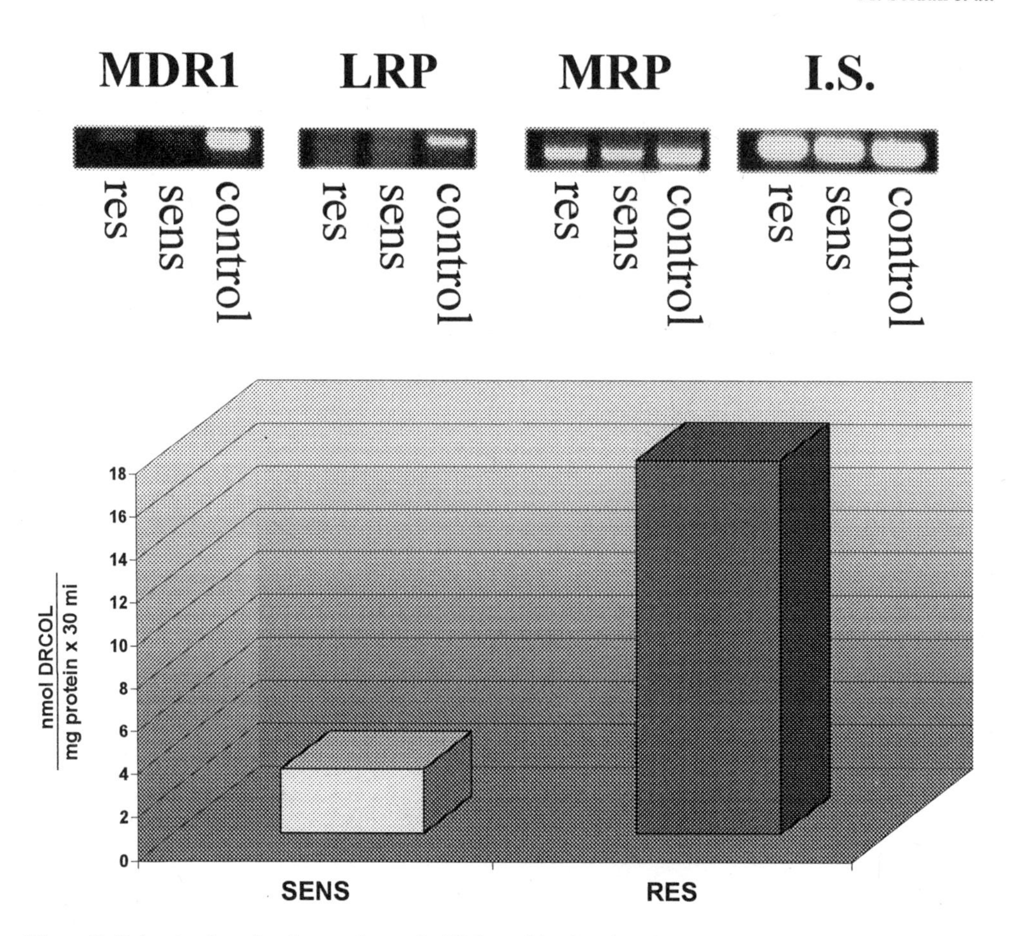

Figure 2. Determination of resistance factors in DRC sensitive (sens) and resistant (res) stomach carcinoma cells with control experiments of DRC-induced pancreas carcinoma cells. RT-PCR of MDR1 (P-gp), LRP, MRP and ß2-microglobulin as internal standard (I.S.) was run as described in the Experimental section. Lower diagram: specific activity of DRC carbonyl reduction in sensitive and resistant stomach carcinoma cells. Specific activity is expressed as nmol / mg protein DRCOL formed in 30 min.

than one reductase. A subsequent determination of DRC toxicity demonstrated the resulting increase of resistance. Each transfected DRC reductase was shown to protect the tumor cells from DRC toxicity, yet to a different extent (Table 1). CR exhibited the highest effectiveness, whereas ALR1 and ALR2 were somewhat weaker. Interestingly, simultaneous transfection of all three enzymes caused the strongest effect in protecting against DRC. These results support the suggestion that DRC reductases constitute mportant resistance mechanisms in the development of DRC resistance in tumor cells.

4. DISCUSSION

Our results show that DRC carbonyl reduction in pancreas carcinoma, breast cancer, malignant melanoma and ovarian carcinoma cells is inducible by the substrate itself and that elevated carbonyl reduction is a likely contributor of resistance against anthracyclines.

The generation of cell lines with acquired resistance to selected chemotherapeutics is one approach to study the important clinical problem of drug resistance. Therefore, we

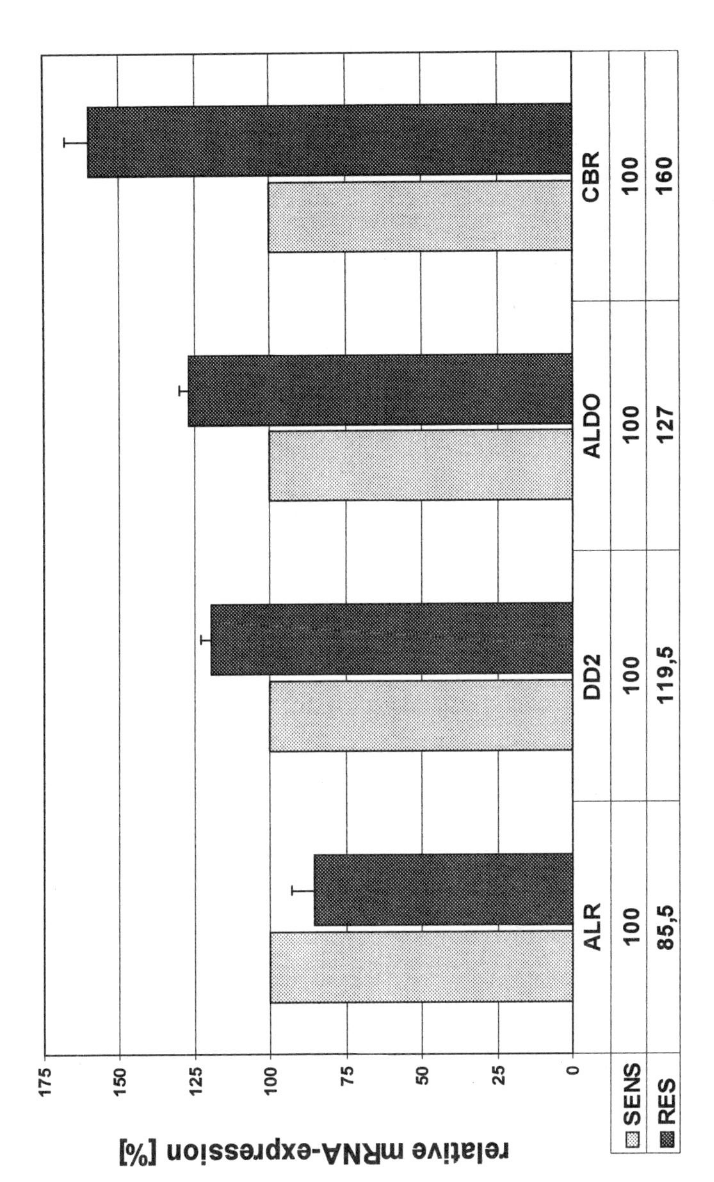

Figure 3. mRNA expression of DRC carbonyl reductases in sensitive (sens) and resistant (res) stomach carcinoma cells by RT-PCR. Values of sensitive cells were set as 100%.

Table 1. DRC toxicity in control (sensitive) and transfected pancreas carcinoma cells

	IC 20 µg/ml	IC 50 µg/ml	IC 80 µg/ml
Tumor sensitive	0.0036	0.176	8.128
ALR2 transf.	0.1380	1.686	19.055
ALR1 transf.	0.0871	1.475	20.417
CR transf.	0.1949	2.748	34.674
Mixed transf.	0.2512	3.484	44.668

Determination of the DRC concentration for 20, 50, and 80% inhibition of cell activity was performed by the MTT-test. Mixed transf. = co-transfection of sensitive cells with ALR2, ALR1 and CR.

have developed DRC-resistant stomach carcinoma cell lines by continuous passage in increasing sublethal concentrations of DRC. To circumvent the selection for the MDR phenotype, we have simultaneously added verapamil to all culture media. Verapamil at 5–20 µM reverses the MDR phenotype by inhibiting P-gp and MRP activity (Tsuruo et al., 1981) and by suppressing P-gp mRNA expression (Muller et al., 1994). In an overall culture period of 5 month, DRC concentrations supplemented to the media were doubled every two weeks, beginning with 0.1 ng/ml. Verapamil remained constantly at 20 µM in the media throughout the entire investigation. The results clearly indicate that inducible DRCOL formation is an important determinant in the acquired resistance to DRC in human stomach carcinoma cells. Compared to the sensitive cells, a 6-fold induction of DRC carbonyl reducing activities occurred in the DRC-resistant subline, thereby paralleling the 8-fold increase in resistance against DRC. No correlation between the expression of other known drug resistance factors (P-gp, MRP, LRP) and DRC-resistance was observed. The 8-fold increase in resistance against DRC is in discordance with the constant levels of MRP or the low levels of P-gp expression which have been blocked by verapamil. This indicates that the pattern of resistance does not conform to a classical MDR phenotype and that P-gp and MRP do not contribute to DRC resistance in these cells. Rather, the acquired resistance to DRC is consistent with the 6-fold increase in DRC-carbonyl reduction.

It is, at present, unclear whether this mechanism of non-classical MDR also accounts for other cancer cells expressing resistance against anthracyclines bearing a carbonyl group. Interestingly, non-MDR resistant Chinese hamster ovary cells resistant to calpain inhibitor (N-acetyl-leucyl-leucyl-norleucinal) have been described to overexpress a NADPH-dependent aldo-keto reductase which has been postulated to function in the detoxification of xenobiotic carbonyl compounds, including DRC (Inoue et al., 1993; Hyndman et al., 1997).

An additional question arising from our results concerns the nature of the DRC carbonyl reducing enzyme(s) undergoing induction by treatment with DRC. Carbonyl reduction of endo- and xenobiotics is generally mediated by members of two protein superfamilies, the aldo-keto reductases (Bohren et al., 1989) and the short chain dehydrogenases/reductases (Persson et al., 1991). In case of DRC, ALR1, ALR2 and DD2 (belonging to the aldo-keto reductases), as well as CR (a short chain dehydrogenase/reductase) have been shown to perform this reaction in human liver (Ohara et al., 1995). The elevated level of CR mRNA in the resistant stomach carcinoma cells is in accordance with the finding that transfection of CR in CR-deficient K562 leukemia cells protects them against DRC toxicity and could therefore contribute to DRC resistance (Gonzalez et al., 1995). But also members of the aldo-keto reductase superfamily are induced in our studies. The increased expression of ALR2 and DD2 mRNA are presumably associated with the higher DRC reduction in the resistant cells. Hence, increases in DRC-reducing activity in tumor cells might not be due to the induction of one single specific DRC reductase. Rather, the concerted action of several reductases with the ability to reduce DRC mediates the 6-fold higher activity. However, we have no indication about unknown reductases additionally involved in this reaction.

The transfection of a sensitive pancreas carcinoma cell line with CR, ALR1 and demonstrated that the overexpression of each single DRC reductase alone already confers DRC resistance to cancer cells. Among these, CR is the enzyme with the highest potency. This finding is in accordance with the results on the resistant stomach cancer cell line, where again CR showed the strongest increase in mRNA expression. Therefore, CR seems to be the dominant DRC reductase in tumor cells during generation of DRC resistance by induction of phase-I metabolism.

The clinical consequences of our findings remain to be established. Considering that the incidence of cardiotoxicity caused by DRC therapy is attributed to the occurrence of its alcohol metabolite DRCOL (Olson et al., 1988), inhibition of DRC carbonyl reduction to DRCOL could have two-fold beneficial consequences during DRC chemotherapy, which are the preservation of the antineoplastic potency of the parent drug and the prevention of cardiomyopathy caused by its reduced alcohol metabolite DRCOL.

Taken together, according to our results, acquired non-MDR resistance of the investigated stomach carcinoma cell line against DRC may be posited to result from an increased enzymatic phase I-detoxification of DRC via carbonyl reduction to the less toxic 13-hydroxy metabolite DRCOL.

ACKOWLEDGMENTS

The present study was supported by grants from the Alfred and Ursula Kulemann-Stiftung, Marburg, Germany, by the European Community, Programme Biotech 2, contract BIO4-97-2132, and by consumables from Wyeth-Lederle and Farmitalia Carlo Erba pharmaceutical companies.

REFERENCES

Batist, G., Tulpule, A., Sinha, B.K., Katki, A.G., Myers, G.E., and Cowans, K.H., 1986, Overexpression of a novel anionic glutathione transferase in multidrug-resistant human breast cancer cells, *J. Biol. Chem.*, 261: 15544–15549.

Beran, M., Andersson, B., Eksborg, S., and Ehrsson, H., 1979, Comparative studies on the *in vitro* killing of human normal and leukemic clonogenic cell (CFUC) by daunorubicin, daunorubicinol and daunorubicin-DNA complex, *Cancer Chemother. Pharmacol.*, 2: 19–24.

Bohren, K.M., Bullock, B., Wermuth, B., and Gabbay, K.H., 1989, The aldo-keto reductase superfamily. cDNAs and deduced amino acid sequences of human aldehyde and aldose reductases, *J. Biol. Chem.*, 264: 9547–9551.

Capranico, G., Riva, A., Tinelli, S., Dasdia, T., and Zunino, F., 1987, Markedly reduced levels of anthracycline-induced DNA strand breaks in resistant P388 leukemia cells and isolated nuclei, *Cancer Res.*, 47: 3752–3756.

Deffie, A.M., Alam, T., Seneviratne, C., Beenken, S.W., Batra, J.L., Shea, T.C., Henner, W.D., and Goldenberg, G.J., 1988, Multifactorial resistance to adriamycin: relationship of DNA repair, glutathione transferase activity, drug efflux, and P-glycoprotein in cloned cell lines of adriamycin-sensitive and -resistant P388 leukemia, *Cancer Res.*, 48: 3595–3602.

Dessypris, E.N., Brenner, D.E., and Hande, K.R., 1986, Toxicity of doxorubicin metabolites to human marrow erythroid and myeloid progenitors *in vitro*, *Cancer Treat. Rep.*, 70: 487–490.

Endicott, J.A. and Ling, V., 1989, The biochemistry of P-glycoprotein-mediated multidrug resistance, *Annu. Rev. Biochem.*, 58: 137–171.

Gessner, T., Vaughan, L.A., Beehler, B.C., Bartels, C.J., and Baker, R.M., 1990, Elevated pentose cycle and glucuronyltransferase in daunorubicin-resistant P388 cells, *Cancer Res.*, 50: 3921–3927.

Gonzalez, B., Akman, S., Doroshow, J., Rivera, H., Kaplan, W.D., and Forrest, G.L., 1995, Protection against daunorubicin cytotoxicity by expression of a cloned human carbonyl reductase cDNA in K562 leukemia cells, *Cancer Res.*, 55: 4646–4650.

Hansen, M.B., Nielsen, S., and Berg, K., 1989, Reexamination and further development of a precise and rapid dye method for measuring cell growth/cell kill, *J. Immunol. Methods*, 119: 203–210.

Hyndman, D.J., Takenoshita, R., Vera, N.L., Pang, S.C., and Flynn, T.G., 1997, Cloning and sequencing, and enzymatic activity of an inducible aldo-keto reductase from Chinese hamster ovary cells, *J. Biol. Chem.*, 272: 13286–13291.

Inoue, S., Sharma, R.C., Schimke, R.T., and Simoni, R.D., 1993, Cellular detoxification of tripeptidyl aldehydes by an aldo-keto reductase, *J. Biol. Chem.*, 268: 5894–5898.

Kramer, R.A., Zakher, J., and Kim, G., 1988, Role of glutathione redox cycle in acquired and *de novo* multidrug resistance, *Science*, 241: 694–697.

Kuffel, M.J., Reid, J.M., and Ames, M.M., 1992, Anthracyclines and their C-13 alcohol metabolites: growth inhibition and DNA damage following incubation with human tumor cells in culture, *Cancer Chemother. Pharmacol.*, 30: 51–57.

Lowry, O.H., Rosebrough, N.J., Farr, A.L., and Randall, R.J., 1951, Protein measurements with the Folin phenol reagent, *J. Biol. Chem.*, 193: 265–275.

McGrath, T. and Center, M.S., 1987, Adriamycin resistance in HL60 cells in absence of detectable P-glycoprotein., *Biochem. Biophys. Res. Com.*, 145: 1171–1176.

Muller, C., Bailly, J.D., Goubin, F., Laredo, J., Jaffrezou, J.P., Bordier, C., and Laurent, G., 1994, Verapamil decreases P-glycoprotein expression in multidrug resistant human leukemic cell lines, *Int. J. Cancer*, 56: 749–754.

Ohara, H., Miyabe, Y., Deyashiki, Y., Matsuura, K., and Hara, A., 1995, Reduction of drug ketones by dihydrodiol dehydrogenases, carbonyl reductase and aldehyde reductase of human liver., *Biochem. Pharmacol.*, 50: 221–227.

Olson, R.D., Mushlin, P.S., Brenner, D.E., Fleischer, S., Cusack, B.J., Chang, B.K., and Boucek Jr., R.J., 1988, Doxorubicin cardiotoxicity may be caused by its metabolite, doxorubicinol, *Proc. Natl. Acad. Sci. USA*, 85: 3585–3589.

Ozols, R.F., Willson, J.K.V., Weltz, M.D., Grotzinger, K.R., Myers, C.E., and Young, R.C., 1980, Inhibition of human ovarian cancer colony formation by adriamycin and its major metabolites, *Cancer Res.*, 40: 4109–4117.

Persson, B., Krook, M., and Jornvall, H., 1991, Characteristics of short-chain alcohol dehydrogenases and related enzymes, *Eur. J. Biochem.*, 200: 537–543.

Ramachandran, C., Yuan, Z.K., Huang, X.L., and Krishan, A., 1993, Doxorubicin resistance in human melanoma cells: *MDR-1* and glutathione S-transferase gene expression., *Biochem. Pharmacol.*, 45: 743–751.

Rekha, G.K., Sreerama, L., and Sladek, N.E., 1994, Intrinsic cellular resistance to oxazaphosphorines exhibited by a human colon carcinoma cell line expressing relatively large amounts of a class-3 aldehyde dehydrogenase, *Biochem. Pharmacol.*, 48: 1943–1952.

Scheffer, G.L., Wijngaard, P.L.J., Flens, M.J., Izquierdo, M.A., Slovak, M.L., Pinedo, H.M., Meijer, C.J.L.M., Clevers, H.C., and Scheper, R.J., 1995, The drug resistance-related protein LRP is the human major vault protein., *Nature Med.*, 1: 578–582.

Schott, B. and Robert, J., 1989, Comparative activity of anthracycline 13-hydroxymetabolites against rat glioblastoma cells in culture, *Biochem. Pharmacol.*, 38: 4069–4074.

Soldan, M., Netter, K.J., and Maser, E., 1996a, Induction of daunorubicin carbonyl reducing enzymes by daunorubicin in sensitive and resistant pancreas carcinoma cells, *Biochem. Pharmacol*, 51: 117–123.

Soldan, M., Netter, K.J., and Maser, E., 1996b, Enzymatic detoxification of daunorubicin as supplementary mechanism to multidrug resistance, *Exp. Toxic. Path.*, 48 (Suppl. II): 370–376.

Toffoli, G., Simone, F., Gigante, M., and Boiocchi, M., 1994, Comparison of mechanisms responsible for resistance to idarubicin and daunorubicin in multidrug resistant LoVo cell lines, *Biochem. Pharmacol.*, 48: 1871–1881.

Tsuruo, T., Iida, H., Tsukagoshi, S., and Sakurai, Y., 1981, Overcoming of vincristine resistance in P388 leukemia in vivo and in vitro through enhanced cytotoxicity of vincristine and vinblastine by verapamil, *Cancer Res.*, 41: 1967–1972.

Volm, M., Mattern, J., Efferth, T., and Pommerenke, E.W., 1992, Expression of several resistance mechanisms in untreated human kidney and lung carcinomas, *Anticancer Res.*, 12: 1063–1067.

Yesair, M., Thayer, P.S., McNitt, S., and Teague, K., 1980, Comparative uptake, metabolism and retention of anthracyclines by tumors growing *in* vitro, *Ger. J. Cancer*, 16: 901–907.

EXPRESSION OF mRNAs FOR DIHYDRODIOL DEHYDROGENASE ISOFORMS IN HUMAN TISSUES

Hiroaki Shiraishi,[1] Kazuya Matsuura,[1] Toshiyuki Kume,[2] and Akira Hara[1]

[1]Biochemistry Laboratory
Gifu Pharmaceutical University
Gifu 502-8585, Japan
[2]Discovery Research Laboratory
Tanabe Seiyaku Co.
Saitama 335-8505, Japan

1. INTRODUCTION

Dihydrodiol dehydrogenase (DD) [EC 1.3.1.20] catalyzes the $NADP^+$-linked oxidation of *trans*-dihydrodiols of polycyclic aromatic hydrocarbons to corresponding catechols, and controls the formation of both their carcinogenic dihydrodiol epoxides (Oesch, *et al.*, 1984) and cytotoxic *o*-quinones through autoxidation of the catechol metabolites (Flowers, *et al.*, 1996). In addition, DD in mammalian liver is implicated in the metabolism of xenobiotic carbonyl compounds, steroids and prostaglandins because of its broad substrate specificity (Penning, *et al.*, 1986; Hara, *et al.*, 1986; Ohara, *et al.*, 1994; 1995).

So far, four distinct DD isoforms have been described in human tissues and cultured cells. They belong to the aldo-keto reductase (AKR) superfamily, and have been named as AKR1C1–AKR1C4 (Jez, *et al.*, 1997). AKR1C1, AKR1C2 and AKR1C4 are identical to DD1, DD2 and DD4, respectively, purified from liver (Hara, *et al.*, 1990), but AKR1C3 protein has not been identified in human tissues despite the isolation of its cDNA (Qin, *et al.*, 1993) and gene (Khanna, *et al.*, 1995). The catalytic properties of the four isoforms are different from one another in spite of their 83–98% sequence identities. AKR1C1 shows 3(20)α-hydroxysteroid dehydrogenase (HSD) activity, AKR1C2 has low 3α-HSD activity and bile acid-binding ability, and AKR1C4 exhibits high 3α-HSD activity for various steroids including bile acids (Hara, *et al.*, 1990; 1996). AKR1C3 was originally called as 3α-HSD type 2 (Khanna, *et al.*, 1995), but has been shown to exhibit 3α(17β)-HSD (Lin, *et al.*, 1997) and prostaglandin D_2 11-keto reductase activities (Matsuura, *et al.*, 1998).

An analysis of multiple forms of DD in six liver specimens of Japanese has suggested inter-individual differences in the ratios of activities of AKR1C1, AKR1C2 and

Enzymology and Molecular Biology of Carbonyl Metabolism 7
edited by Weiner *et al.* Kluwer Academic / Plenum Publishers, New York, 1999.

Table 1. Multiple cDNAs for AKR1C1–AKR1C4 cloned from tissues or cultured cells

AKR type	Abbreviation of cDNA[a]	Tissue or cells	Encoded protein	AKR designation[b]
1C1	DD1 cDNA	Liver, colon cells	DD1	1C1
	HAKRc	Liver	(not studied)[c]	
1C2	DD2 cDNA	Liver	DD2/bile acid binding protein	1C2
	c81	Colon cells	DD2	1C2
	Type III	Prostate	DD2/3α-HSD type 3	1C2
	MCDR2	Liver	(not studied)[c]	
	HAKRd	Liver	(not studied)[c]	
1C3	HAKRb	Liver	3α-HSD type 2	1C3
	DBDH cDNA	Myeloid cells	PG $D_2$11-keto reductase	
	3α(17β)-HSD cDNA	Prostate	3α(17β)-HSD	
1C4	DD4 cDNA	Liver	DD4/3α-HSD type 1	1C4
	CCDR33	Liver	(not studied)[c]	
	DD4v	Liver	Variant form of DD4	

[a] The abbreviations of cDNAs were used throughout the text.
[b] The names of the proteins defined by the nomenclature of the AKR superfamily (Jez, *et al.*, 1997).
[c] The encoded protein has not been studied.

AKR1C4 (Miyabe, *et al.*, 1997). In addition, multiple cDNAs with sequences quite similar (more than 97% identities) to the respective cDNAs for the above four AKR isoforms have been isolated from human tissues and cells. As summarized in Table 1, the numbers of the cDNAs similar to AKR1C1, AKR1C2, AKR1C3 and AKR1C4 are two, five, three and three, respectively, which include a new DD2 cDNA (Shiraishi, *et al.*, 1998) and cDNA for variant DD4, DD4v (Iwasa, *et al.*, 1997). The accumulation of the reports on the isolation of the strongly similar cDNAs has suggested the existence of allelic variants of some or all of the genes for AKR1C1–1C4.

In this study, we have developed methods that distinguishably detect mRNAs corresponding to the multiple cDNAs by RT-PCR and diagnostic restriction with endonucleases, and analyzed the expression of the respective mRNA species in 25 Japanese liver specimens and 15 extra-hepatic tissue specimens of Japanese and non-Japanese.

2. HUMAN SAMPLES

Livers, kidneys, prostates, stomachs, adrenal gland and testes were obtained with informed consent from Japanese patients during biopsy or surgery of the tissues for pathological examination, and a placenta was from a healthy volunteer after parturition. Total RNAs of non-Japanese tissues (brain, lung, liver, heart, spleen and small intestine) were purchased from Sawady Technology (Tokyo, Japan). Total RNAs from the tissues were reverse transcribed using Moloney murine leukemia virus reverse transcriptase.

3. EXPRESSION OF mRNA SPECIES FOR AKR1C1–1C4 IN HUMAN TISSUES

3.1. mRNAs for AKR1C1-Type cDNAs

The AKR1C1-type cDNAs are DD1 cDNA (Stolz, *et al.*, 1993; Hara, *et al.*, 1996) and HAKRc (Qin, *et al.*, 1993), which show differences of 15 nucleotides and 8 amino acids. We designed the primer pairs, C1f and C1r, that specifically amplified the two AKR1C1-type cDNAs. C1f had a sequence corresponding to nucleotides 109–126 of the two cDNAs, and C1r was complimentary to the sequence from nucleotide 666 to 683. DD1 cDNA was distinguished from HAKRc by digesting the PCR products (575 bp) with *Pvu*II, because this restriction site exists only in DD1 cDNA at nucleotide 516. The results for representative liver and tissue specimens are shown in Figure 1A and Figure 2, respectively. The 575-bp PCR products (Figure 2) were amplified from all the human samples, and digested into two fragments with the expected sizes of 408 bp and 167 bp (Figure 1A). This indicates that DD1 cDNA represents the principal mRNA species for AKR1C1. In addition, we show for the first time that AKR1C1 mRNA is ubiquitously expressed in human tissues, of which testis and intestine showed relatively high levels of the message. AKR1C1 uniquely exhibits high 20α-HSD activity (Hara, *et al.,* 1990; 1996). Soluble $NADP^+$-linked 20α-HSD is present in mammalian testes including man (Tamaoki and Shikita, 1966), and the amino acid sequences deduced from cDNAs for the rabbit and rat enzymes (Lacy, *et al*., 1993; Miura, *et al*., 1994) show high sequence identity with AKR1C1. Thus, AKR1C1 may act as the predominant 20α-HSD in human tissues.

3.2. mRNAs for AKR1C2-Type cDNAs

Of the five cDNAs in this type, DD2 cDNA (Shiraishi, *et al.,* 1998), c81 (Ciaccio and Tew, 1994) and Type III (Dufort, *et al*., 1996) encode the same protein, but show 1 to 3 nucleotide differences. DD2 cDNA differs HAKRd (Qin, et al. 1993) and MCDR2 (Winters, *et al.* 1990) by 4 and 5 nucleotides respectively, which result in 1 and 3 amino acid differences. As shown in Figure 1B, the five similar cDNAs were discriminated by a series of reactions (Shiraishi, *et al.,* 1998), which consists of PCR with primer pairs specific for HAKRd (f), MCDR2 (e) or the other cDNAs (a), followed by *Pma*CI digestion (b) and nested PCR with primer pairs for DD2 cDNA (c) and Type III (d). Since the results of analyses for 40 specimens were the same as those for Liver 1 in Figure 1B, the principal mRNA species for AKR1C2 expressed in human tissues may have the sequence corresponding to that of DD2 cDNA. The ubiquitous tissue distribution of the mRNA (Figure 2) suggests the

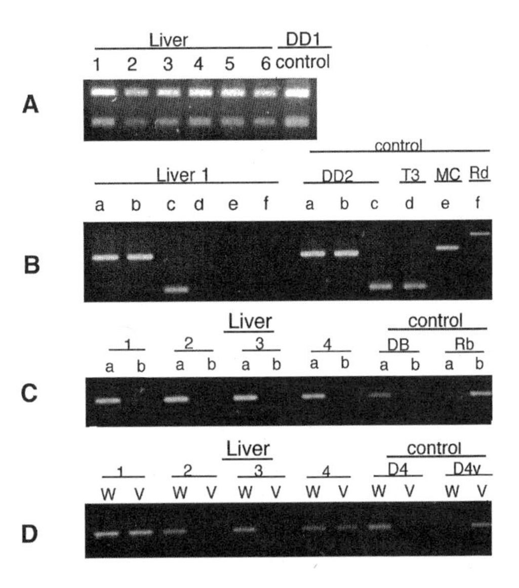

Figure 1. Identification of mRNA species for the respective types of AKR1C1–1C4 in representative liver specimens. (A) Analysis of the AKR1C1-type cDNAs. PCR products from DD1 cDNA (DD1) and hepatic cDNA samples (1 to 6) were digested with *Pvu*II. (B) Analysis of the AKR1C2-type cDNAs. Control cDNAs: DD2 cDNA (DD2); Type III (T3); MCDR2 (MC); and HAKRd (Rd). PCRs were performed with primers specific for HAKRd (f), and MCDR2 (e) and the other cDNAs (a). The products from PCR (a) was digested with *Pma*CI (b), or subjected to nested PCR with primers specific for DD2 cDNA (c) and Type III (d). (C) Analysis of the AKR1C3-type cDNAs. (a) and (b) are PCRs with primers specific for DBDH cDNA (DB) and HAKRb (Rb). (D) Analysis of the AKR1C4-type cDNAs. (w) and (v) are PCRs with primers specific for DD4 cDNA and DD4v.

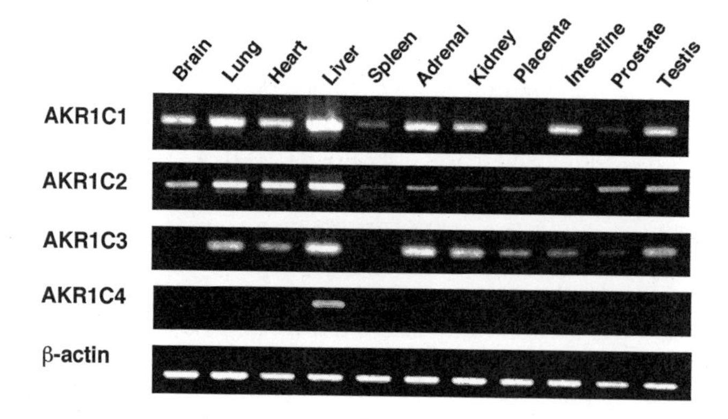

Figure 2. Expression of mRNAs for AKR1C1–1C4 in human tissues. The mRNAs for the enzymes, except AKR1C4, were detected in all the tissues, although very low signals in some tissues cannot be reproduced well in the photograph.

role of AKR1C2 in extra-hepatic metabolism of steroids because this enzyme shows higher catalytic efficiency for 3α-hydroxy/3-ketosteroids compared to AKR1C1 and AKR1C3. Despite sequence identity (98%) between AKR1C1 and AKR1C2, their mRNAs were co-expressed in all the samples, which indicates that the two isoforms are not allelic variants but derived from distinct genes.

3.3. mRNAs for AKR1C3-Type cDNAs

HAKRb (Qin, *et al.*, 1993), DBDH cDNA (Nagase, *et al.*, 1995) and 3α(17β)-HSD cDNA (Lin, *et al.*, 1997) are included in the AKR1C3-type cDNAs. The latter two cDNAs have the same sequence except a silent mutation, but show differences of 6 nucleotides and 2 amino acids from HAKRb. It was difficult to discriminate the cDNAs for DBDH and 3α(17β)-HSD by PCR, whereas the cDNAs were easily distinguished from HAKRb by PCR using specific primers (Figure 1C), which did not amplify the multiple cDNAs for the other types of AKR1C subgroup (data not shown). The analyses for liver and other tissue samples indicated the expression of only mRNA species for DBDH or 3α(17β)-HSD. The entire coding regions of the cDNAs cloned from the liver, kidney and lung of Japanese were identical to the sequence of DBDH cDNA, but not to that of 3α(17β)-HSD cDNA. Thus, DBDH cDNA may represent the principal AKR1C3 allele. The recombinant DBDH shows lower *K*m and higher catalytic efficiency for prostaglandin D_2 and 9α,11β-prostaglandin F_2 than AKR1C1 and AKR1C2, and the catalytic efficiency for the prostaglandins are also higher than those for 3α- and 17β-hydroxysteroids (Matsuura, *et al.*, 1998). AKR1C3 may play an important role in the metabolism of the prostaglandins in all the tissues.

3.4. mRNAs for AKR1C4-Type cDNAs

CCDR33 contains five replacements at positions 105, 261, 434, 878 and 931 in the sequence of DD4 cDNA (Winters, *et al.*, 1990), and DD4v has the two replacements at positions 434 and 931 (Iwasa, *et al.*, 1997). The discrimination of DD4 cDNA from the other two cDNAs was achieved by allele-specific PCR (Figure 1D). Since the replacement at position 261 on CCDR33 results in abolishing a restriction site of *Apa*LI, the digestion of the PCR products (nucleotides 5–449) with this endonuclease discriminates DD4 and DD4v from CCDR33. In addition to the previous analysis of 15 Japanese liver specimens (T. Kume, *et al.*, unpublished result) by this method, about 20% of totally 40 liver samples showed heterozygous expression of mRNAs for DD4v and DD4, as shown Liver 1 and 4 in Figure 1D. The expression of mRNA corresponding to DD4v, not CCDR33, was further confirmed by sequencing the PCR products. The mRNAs were extensively expressed in liver (Figure 2), but expression of small amount of the AKR1C4 mRNA was detected in some tissues and cultured prostatic cells, which was also confirmed by sequencing the PCR products. Although AKR1C4 has been recognized as liver-specific 3α-HSD (Khanna *et al.*, 1995), it would be localized in specific cells of some extra-hepatic tissues and regulate the steroid action.

4. CONCLUSION

The present analyses have identified the principal mRNAs for the four types of human dihydrodiol dehydrogenase isoforms, and suggested the polymorphism of AKR1C4 gene. However, it still remains unknown whether the undetectable cDNAs for the respective isoforms are derived from rare variants or sequencing artifacts. In order to elucidate this point and the physiological effect of the phenotype of DD4v, the analysis of genomic DNAs is now in progress.

ACKNOWLEDGMENTS

This work was supported by a Grant-in-Aid for Scientific Research from the Ministry of Education, Science, Sports, and Culture, and by a grant from the Suzuken Memorial Foundation.

REFERENCES

Ciaccio, P.J. and Tew, K.D., 1994, cDNA and deduced amino acid sequences of a human colon dihydrodiol dehydrogenase, *Biochim. Biophys. Acta*, 1186, 129–132.

Dufort, I., Soucy, P., Labrie, F., and Luu-The, V., 1996, Molecular cloning of human type 3 3α-hydroxysteroid dehydrogenase that differs from 20α-hydroxysteroid dehydrogenase by seven amino acids, *Biochem. Biophys. Res. Commun.*, 228, 474–479.

Flowers, L., Bleczinski, W.F., Burczynski, M.E., Harvey, R.G., and Penning, T.M., 1996, Disposition and biological activity of benzo[*a*]pyrene-7,8-dione. A genotoxic metabolite generated by dihydrodiol dehydrogenase, *Biochemistry*, 35, 13664–13672.

Hara, A., Hasebe, K., Hayashibara, M., Matsuura, K., Nakayama, T. and Sawada, H., 1986, Dihydrodiol dehydrogenases in guinea pig liver, *Biochem. Pharmacol.*, 35, 4005–4012.

Hara, A. Taniguchi, H., Nakayama, T., and Sawada, H., 1990, Purification and properties of multiple forms of dihydrodiol dehydrogenase from human liver, *J. Biochem. (Tokyo)*, 108, 250–254.

Hara, A., Matsuura, K., Tamada, Y., Sato, K., Miyabe, Y., Deyashiki, Y., and Ishida, N., 1996, Relationship of human liver dihydrodiol dehydrogenases to hepatic bile-acid-binding protein and an oxidoreductases of human colon cells, *Biochem. J.*, 313, 373–376.

Iwasa, H., Hara, A., Kume, T., Yokoi, T., Kamataki, T., Inoue, K., Terada, K., Takagi, T., and Otsuka, M., 1997, Isolation and properties of human liver dihydrodiol dehydrogenase variant. The 117th Annual Meeting of the Pharmaceutical Society of Japan, Tokyo, Abstract, Part 3, p. 39.

Jez, J.M., Flynn, T.G., and Penning, T.M., 1997, A new nomenclature for the aldo-keto reductase superfamily, *Biochem. Pharmacol.*, 54, 639–647.

Khanna, M., Qin, K.-N., Wang, R.W., and Cheng, K.-C., 1995, Substrate specificity, gene structure, and tissue distribution of multiple human 3α-hydroxysteroid dehydrogenases, *J. Biol. Chem.*, 270, 20162–20168.

Lacy, W.R., Washenik, R.G., and Dunbar, B.S., 1993, Molecular cloning and expression of an abundant rabbit ovarian protein with 20α-hydroxysteroid dehydrogenase activity, *Mol. Endcrinol.*, 7, 58–66.

Lin, H.-K., Jez, J.M., Schlegel, B.P., Peehl, D.M., Pachter, J.A., and Penning, T.M., 1997, Expression and characterization of recombinant type 2 3α-hydroxysteroid dehydrogenase (HSD) from human prostate: Demonstration of bifunctional 3α/17β-HSD activity and cellular distribution, *Mol. Endocrinol.*, 11, 1971–1984.

Matsuura, K., Shiraishi, H., Hara, A., Sato, K., Deyashiki, Y., Niomiya, M., and Sakai, S., 1998, Identification of a principal mRNA species for human 3α-hydroxysteroid dehydrogenase isoform (AKR1C3) that exhibits high prostaglandin D_2 11-ketoreductase activity. *J. Biochem. (Tokyo)*, in press.

Miura, R., Shiota, K., Noda, K., Yagi, S., Ogawa, T., and Takahashi, M., 1994, Molecular cloning of cDNA for rat ovarian 20α-hydroxysteroid dehydrogenase (HSD1), *Biochem. J.*, 299, 561–567.

Miyabe, Y., Matsuura, K., Nakayama, T., Ohya, I., and Hara, A., 1997, A sensitive fluorometric assay for dihydrodiol dehydrogenase, *Yakugaku Zasshi*, 117, 167–177.

Nagase, T., Miyajima, N., Tanaka, A., Sazuka, T., Seki, N., Sato, S., Tabata, S., Ishikawa, K., Kawarabayashi, Y., Kotani, H., and Nomura, N., 1995, Prediction of the coding sequences of unidentified genes. III. The coding sequences of 40 new genes (KIAA0081-KIAA0120) deduced by analysis of cDNA clones from human cell line KG-1, *DNA Res.*, 2, 37–43.

Oesch, F., Glatt, H.R., Vogel, K., Seidel, A., Petrovic, P., and Platt, K.L., 1984, Dihydrodiol dehydrogenase: A new level of control by both sequestration of proximate and inactivation of ultimate carcinogens. In Greim, H., Jung, R., Kramer, M., Marquardt, H., and Oesch, F. (Eds.), Biochemical Basis of Carcinogenesis, Raven Press, New York, pp. 23–31.

Ohara, H., Nakayama, T., Deyashiki, Y., Hara, A., Miyabe, Y., and Tukada, F., 1994, Reduction of prostaglandin D_2 to 9α,11β-prostaglandin F_2 by a human liver 3α-hydroxy-steroid/dihydrodiol dehydrogenase isozyme, *Biochim. Biophys. Acta*, 1215, 59–65.

Ohara, H., Miyabe, Y., Deyashiki, Y., Matsura, K., and Hara, A., 1995, Reduction of drug ketones by dihydrodiol dehydrogenases, carbonyl reductase and aldehyde reductase of human liver, *Biochem. Pharmacol.*, 50, 221–227.

Penning, T.M., Smithgall, T.E., Askonas, L.J., and Sharp, R.B., 1986, Rat liver 3α-hydroxysteroid dehydrogenase, *Steroids*, 47, 221–247.

Qin, K.-N., New, M.I., and Cheng, K.-C., 1993, Molecular cloning of Multiple cDNAs encoding human enzymes structurally related to 3α-hydroxysteroid dehydrogenase, *J. Steroid Biochem. Molec. Biol.*, 46, 673–679.

Shiraishi, H., Ishikura, S., Matsuura, K., Deyashiki, Y., Niomiya, M., Sakai, S., and Hara, A., 1998, Sequence of human dihydrodiol dehydrogenase isoform (AKR1C2) cDNA and tissue distribution of its mRNA. *Biochem. J.*, in press.

Stolz, A., Hammond., L., Lou, H., Takikawa, H., Ronk, M., and Shively, J.E., 1993, cDNA cloning and expression of the human hepatic bile acid-binding protein. A member of the monomeric reductase gene family, *J. Biol. Chem.*, 268, 10448–10457.

Tamaoki, B. and Shikita, M., 1966, Biosynthesis of steroids in testicular tissue in vitro. In Pincus, G., Nakao, T. and Tait, J.F. (Eds.), *Steroid Dynamics*, Academic Press, New York, pp. 493–530.

Winters, C.J., Molowa, D.T., and Guzelian, P.S.,1990, Isolation and characterization of cloned cDNAs encoding human liver chlordecone reductase, *Biochemistry*, 29, 1080–1087.

ALTERATIONS IN THE EXPRESSION OF DAUNORUBICIN PHASE-I METABOLISING ENZYMES IN DIFFERENT CARCINOMA CELL LINES

Lutz Koch, Edmund Maser, and Michael Soldan

Department of Pharmacology and Toxicology
University of Marburg
Karl-von-Frisch-Strasse 1
D-35033 Marburg, Germany

1. INTRODUCTION

Drug resistance is a major obstacle in the successful treatment of cancer by chemotherapy. The phenomenon of multidrug resistance is often related to the overexpression of two plasma membrane transporters, termed glycoprotein P-170 (MDR1) and multidrug resistance related protein (MRP). Both proteins are acting as ATP-driven efflux pumps, removing intracellular cytotoxic agents out of tumor cells (Endicott and Ling, 1989). Another protein mediating drug resistance is involved in the nucleo-cytoplasmic transport process. It is termed lung resistance related protein (LRP), because it was found first time in a lung cancer cell line (Scheffer et al., 1995). This resistance mechanism also occurs in several other cancer cells. But also the metabolism of cytostatic drugs could contribute to a lower toxicity and failure of chemotherapy. It is known that elevated levels of drug-inactivating enzymes support tumor cells in acquiring resistance (Deffie et al., 1988; Gessner et al., 1990; Rekha et al., 1994).

Anthracyclines, such as daunorubicin (DRC), are major components of combination chemotherapy regimes for a wide range of cancer diseases. However, the metabolism of 13-ketone anthracyclines via carbonyl reduction to the less toxic 13-hydroxy metabolites limits the potency of these drugs (Schott and Robert, 1989; Kuffel et al., 1992).

In the present study, we cultured three cancer cell lines (malignant melanoma, sensitive and resistant pancreas cancer) in the presence of increasing sublethal concentrations of DRC. Supplementation of DRC to the culture media led to concentration-dependent alterations of MDR1, MRP, LRP and DRC carbonyl reduction in all three examined cell

Enzymology and Molecular Biology of Carbonyl Metabolism 7
edited by Weiner *et al.* Kluwer Academic / Plenum Publishers, New York, 1999.

lines.To elucidate possible modifications in the spectrum of reductases after DRC pretreatment, the mRNA expression of four known DRC reductases, specifically carbonyl reductase (EC 1.1.1.184), aldehyde reductase (EC 1.1.1.2), aldose reductase (EC 1.1.1.21) and dihydrodiol dehydrogenase [DD2] (EC 1.3.1.20), were determined.

2. EXPERIMENTAL

2.1. Cell Lines

Two human pancreas carcinoma cell lines (EPP 85-181, DRC sensitive and EPP 85-181 RDB, DRC resistant) and a malignant melanoma cell line (MeWo) were used in this study.

2.2. Cell Culture

For induction of resistance proteins cells were grown for 72 h in the presence of increasing concentrations of DRC. Control values were obtained from cells cultured without DRC supplementation.

2.3. Reverse Transcription-Polymerase Chain Reaction (RT-PCR)

RNA was isolated from cells with the RNeasy Mini Kit (Quiagen). RT-PCR was performed using the Ready To Go RT-PCR beads (Pharmacia Biotech).

PCR products were analysed by agarose gel electrophoresis and visualised by ethidium bromide staining. The quantitative evaluation of the PCR products was carried out by scanning densitometry using the ImageMaster VDS (Pharmacia Biotech).

3. RESULTS

3.1. Pancreas Carcinoma Sensitive

In this cell line DRC pretreatment leads to a 2.9-fold induction of DRC carbonyl reduction. This increase in activity is paralleled by an enhanced mRNA expression of carbonyl reductase. The expression of the other reductases does not show significant alterations.

In addition, the mRNA of the two transmembrane transporters MDR1 and MRP is strongly enhanced at high DRC concentrations. Consequently, carbonyl reductase together with MDR1 and MRP are obviously the main resistance factors in sensitive pancreas carcinoma cells.

3.2. Pancreas Carcinoma Resistant

Compared to the sensitive cells, DRC carbonyl reduction is somewhat weaker induced (1.7 fold) in the resistant subline. As revealed from the pattern of mRNA expression, only carbonyl reductase seems responsible for this higher DRC detoxification.

In contrast to the sensitive cells, only MDR1 is co-induced. Interestingly, at lower DRC concentrations a transient induction of dihydrodiol dehydrogenase (DD2) and MRP

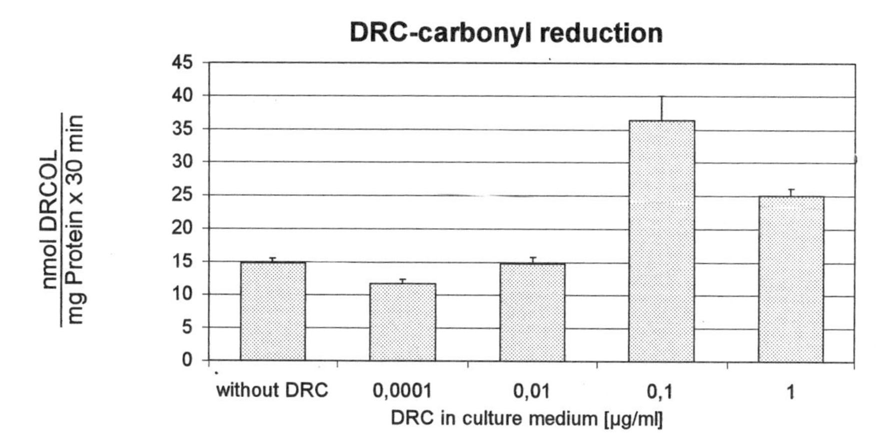

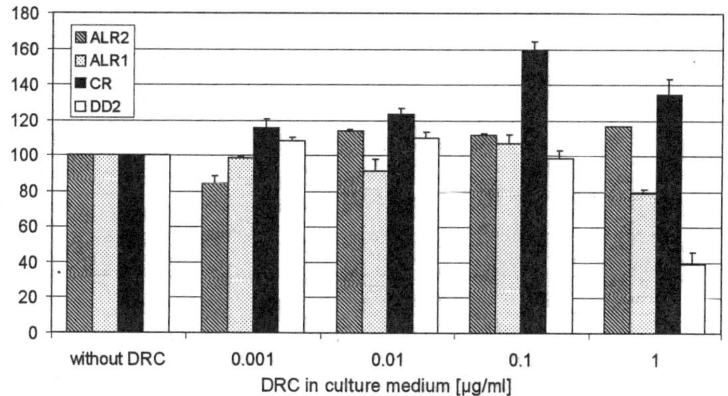

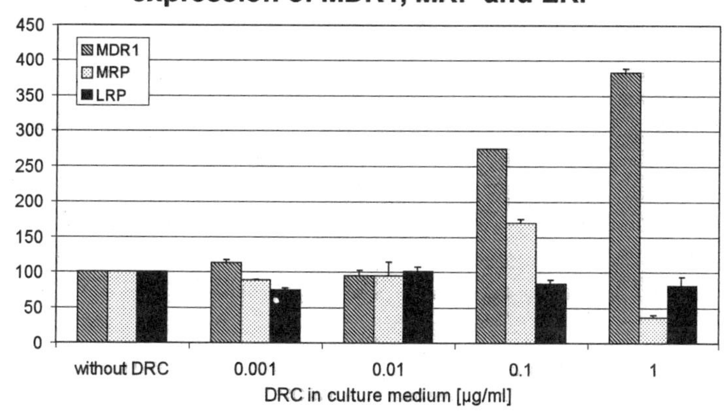

Figure 1. Effect of DRC pretreatment on DRC carbonyl reduction, mRNA expression of aldose reductase (ALR2), aldehyde reductase (ALR1), carbonyl reductase (CR), dihydrodiol dehydrogenase (DD2), and mRNA expression of MDR1, MRP and LRP in the sensitive pancreas carcinoma cell line.

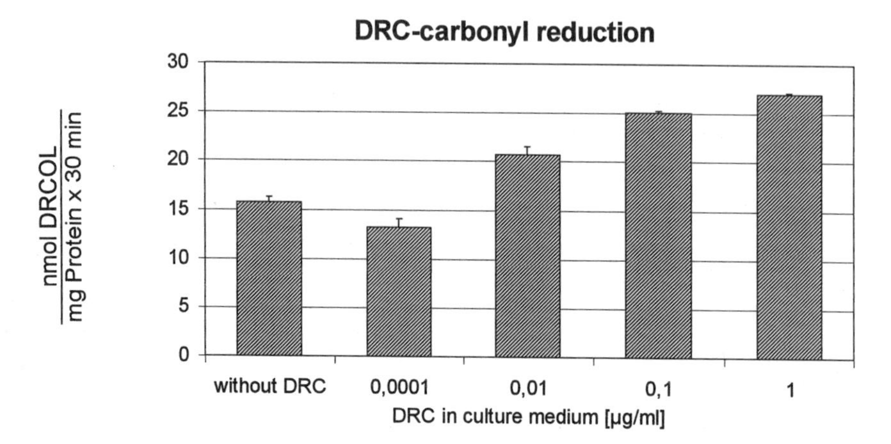

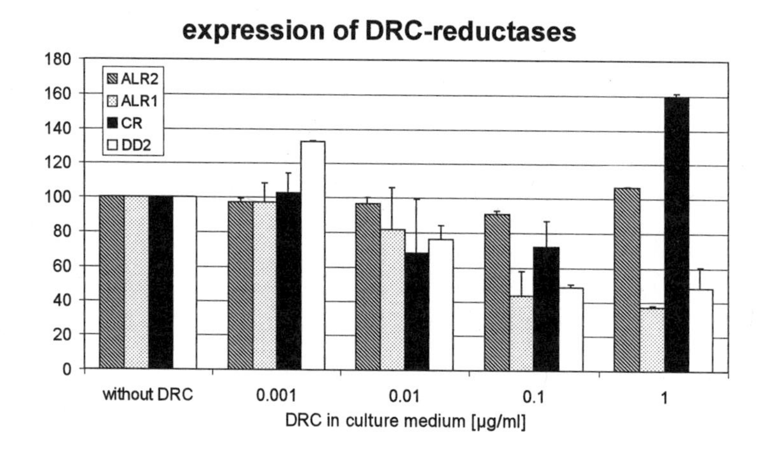

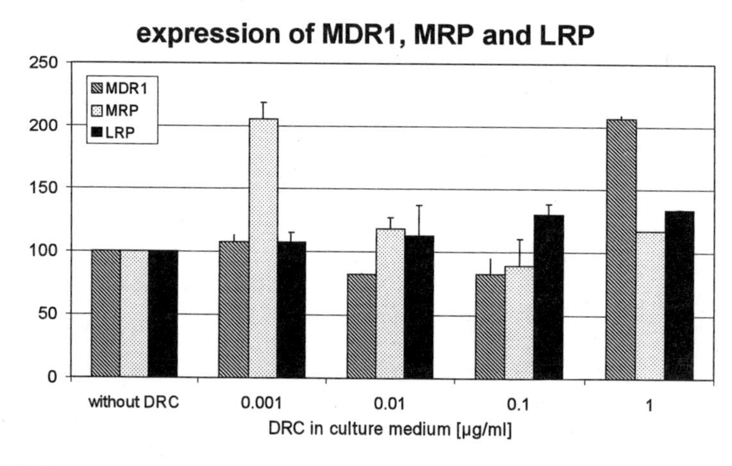

Figure 2. Effect of DRC pretreatment on DRC carbonyl reduction, mRNA expression of aldose reductase (ALR2), aldehyde reductase (ALR1), carbonyl reductase (CR), dihydrodiol dehydrogenase (DD2), and mRNA expression of MDR1, MRP and LRP in the resistant pancreas carcinoma cell line.

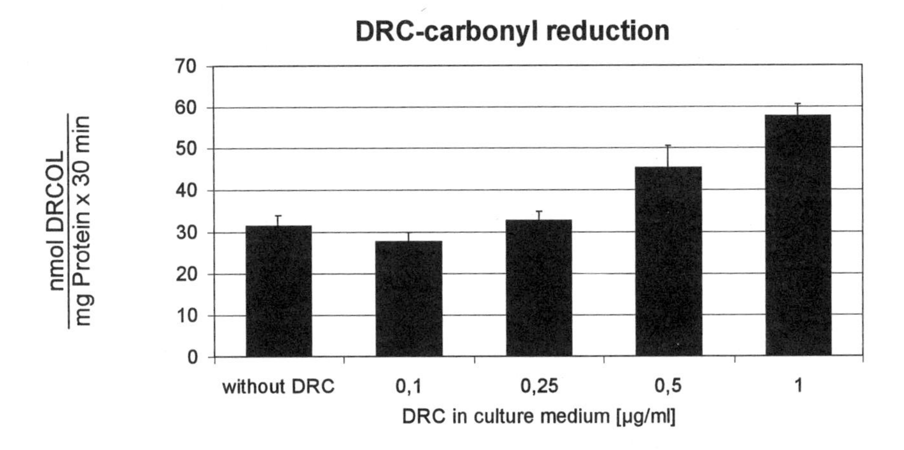

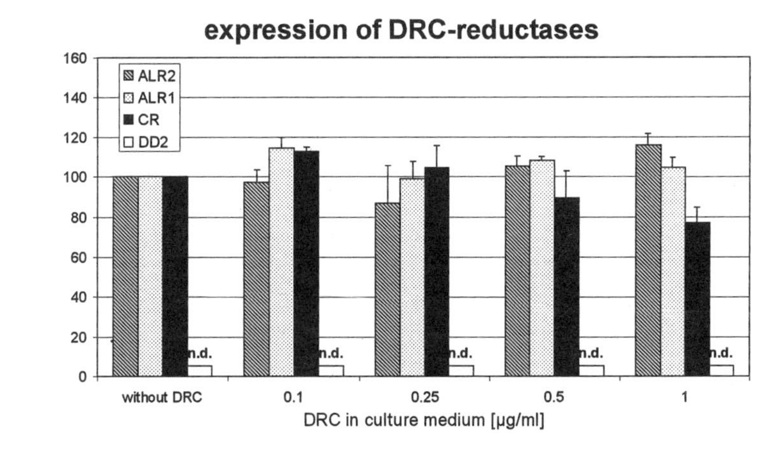

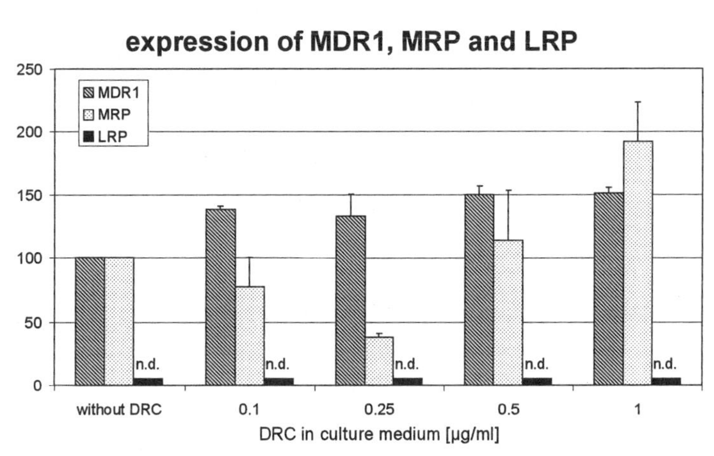

Figure 3. Effect of DRC pretreatment on DRC carbonyl reduction, mRNA expression of aldose reductase (ALR2), aldehyde reductase (ALR1), carbonyl reductase (CR), dihydrodiol dehydrogenase (DD2), and mRNA expression of MDR1, MRP and LRP in malignant melanoma cells. n.d. = not detectable.

occurs. LRP shows a moderate increase. In principle, resistant pancreas carcinoma cells reflect the situation of sensitive cells except that LRP is also involved.

3.3. Malignant Melanoma

In these cells a 1.9-fold increase in DRC reductase activity is in contrast to the lack of significant alterations of the tested DRC reductases. Moreover, dihydrodiol dehydrogenase (DD2) is completely absent. This supports the suggestion that a hitherto unknown DRC reductase is active.

Even low DRC amounts induce MDR1, whereas MRP increases at higher DRC concentrations. Like dihydrodiol dehydrogenase (DD2), LRP is not expressed.

4. DISCUSSION

In this study, we have incubated three cancer cell lines in different DRC concentrations and then determined the expression of several resistance factors.

Changes in mRNA expression and enzyme activities of all cell lines are strongly dependend on the DRC concentration used. The overexpression of the drug transporter MRP at low DRC (0.001 μg/ml) in the resistant pancreas carcinoma cell line disappeared completely at higher DRC, whereas MRP in the malignant melanoma cells decreased at low DRC and increased strongly at the highest DRC concentration used. One reason for this concentration dependend expression of resistance factors may be the different strategies of tumor cells to acquire resistance. The modifications in activity or expression of the resistance factors at highest DRC concentrations were of prior interest in this study, because the dramatic increase in DRC toxicity stimulates all available protection factors in tumor cells. At this concentration all three cell lines raised the expression of MDR1 but obviously to a different degree of significance. Whereas the sensitive pancreas carcinoma cells showed a 3.7-fold increase of MDR1, that of the malignant melanoma cells increased only 1.5-fold. This suggests that overexpression of MDR1 is an important resistance mechanism for the sensitive pancreas carcinoma cells.

In contrast, malignant melanoma cells tend to use other resistance mechanisms, e.g. MRP and the induction of DRC carbonyl reduction. DRC reduction in malignant melanoma cells is already without DRC induction by far the highest of all cell lines tested. This finding is in contrast with the expression of known carbonyl reductases. No mRNA signal of the DRC reductase DD2 could be detected, and pretreatment with DRC caused no relevant induction of carbonyl reductase, aldose reductase or aldehyde reductase. Consequently, the enhanced DRC metabolism in malignant melanoma seems to be caused by an overexpression of a hitherto unknown DRC reductase.

A detailed analysis of pancreas carcinoma cells revealed that differences in the expression of resistance factors against DRC between sensitive and resistant cells (the latter being generated from the sensitive cell line) are weaker than differences to cell lines derived from other tissues. However, an investigation of these differences between sensitive and resistant cancer cells is important to find biochemical modifications caused by a long term treatment with cytostatics.

Our results strongly support the concept that DRC carbonyl reductases contribute to the multifactorial process in the development of tumor resistance. However, there is no generalisation possible, since the extent of the different resistant factors varies between the cell lines.

ACKNOWLEDGMENTS

This work was supported by the European Commission (BIO4–97–2123) and a grant from the Alfred and Ursula Kulemann-Stiftung, Marburg.

REFERENCES

Deffie, A.M., Alam, T., Seneviratne, C., Beenken, S.W., Batra, J.L., Shea, T.C., Henner, W.D., and Goldenberg, G.J., 1988, Multifactorial resistance to adriamycin: relationship of DNA repair, glutathione transferase activity, drug efflux, and P-glycoprotein in cloned cell lines of adriamycin-sensitive and -resistant P388 leukemia, *Cancer Res.*, 48: 3595–3602.

Endicott, J.A. and Ling, V., 1989, The biochemistry of P-glycoprotein-mediated multidrug resistance, *Annu. Rev. Biochem.*,58: 137–171.

Gessner, T., Vaughan, L.A., Beehler, B.C., Bartels, C.J., and Baker, R.M., 1990, Elevated pentose cycle and glucuronosyltransferase in daunorubicin-resistant P388 cells, *Cancer Res.*, 50: 3921–3927.

Kuffel, M.J., Reid, J.M., and Ames, M.M., 1992, Anthracyclines and their C-13 alcohol metabolites: growth inhibition and DNA damage following incubation with human tumor cells in culture, *Cancer Chemother. Pharmacol.*, 30: 51–57.

Rekha, G.K., Sreerama, L., and Sladek, N.E., 1994, Intrinsic cellular resistance to oxazaphorines exhibited by a human colon carcinoma cell line expressing relatively large amounts of a class-3 aldehyde dehydrogenase, *Biochem. Pharmacol.*, 48: 1943–1952.

Scheffer, G.L., Wijngaard, P.L.J., Flens, M.J., Izquierdo, M.A., Slovak, M.L., Pinedo, H.M., Meijer, C.J.L.M., Clevers, H.C., and Scheper, R.J., 1995, The drug resistance related protein LRP is the human major vault protein., *Nature Med.*, 1: 578–582.

Schott, B. and Robert, J., 1989, Comparative activity of anthracycline 13-hydroxymetabolites against rat glioblastoma cells in culture, *Biochem. Pharmacol.*, 38: 4069–4074.

INDEX